Advances in Solid-State Lasers: Development and Applications

Edited by

Mikhail Grishin

Advances in Solid-State Lasers: Development and Applications

Edited by Mikhail Grishin

This is a reprint of the book published in 2010 by Intech

Republished by InTech

Janeza Trdine 9, 51000 Rijeka, Croatia

Technical Editor Teodora Smiljanic

Cover Designer Dino Smrekar

Additional hard copies can be obtained from orders@intechopen.com

Advances in Solid-State Lasers: Development and Applications, Edited by Mikhail Grishin
p. cm.
ISBN 978-953-7619-80-0

For EU product safety concerns:
IN TECH d.o.o., Prolaz Marije Krucifikse Kozulić 3, 51000 Rijeka, Croatia,
info@intechopen.com or visit our website at intechopen.com.

Preface

Invention of the solid-state laser has initiated the beginning of the laser era. Performance of solid-state lasers improved amazingly during five decades. Nowadays, solid-state lasers remain one of the most rapidly developing branches of laser science and become an increasingly important tool for modern technology.

This book represents a selection of chapters exhibiting various investigation directions in the field of solid-state lasers and the cutting edge of related applications. The materials are contributed by leading researchers and each chapter represents a comprehensive study reflecting advances in modern laser physics. Considered topics are intended to meet the needs of both specialists in laser system design and those who use laser techniques in fundamental science and applied research.

The book begins with the section devoted to new laser media and key components (ch. 1 - ch. 3), followed by theoretical and experimental studies the objective of which was to improve temporal and spatial performance of mid-infrared lasers (ch. 4, ch. 5). Novel schemes of side-pumped lasers are also considered in this part of the volume (ch. 6, ch. 7). Subsequent several chapters describe specific applications of solid-state lasers. In particular, remote sensing, absolute distance measurement, and ignition of automobile engines are described in chapters 8 - 10.

Nowadays, development of ultrafast laser systems, including amplification of ultrashort pulses to high energies, is based mainly on solid-state laser technology. A substantial part of the volume is devoted to ultrafast phenomena. Amplifying techniques, including regenerative amplification and optical parametric chirped-pulse amplification, are described in significant depth (ch. 11 - ch. 13). Modern methods of the femtosecond pulse characterization as well as the pulse shaping and spectrum control are also considered in detail (ch. 14 - ch. 18). Impressive achievements in femtosecond lasers intensify investigations in the scope of the high-field science. In turn, advances in this area allow development of coherent radiation sources which simultaneously produce extremely short wavelength (reaching soft X-rays particularly by means of high-order harmonic generation) and feature the attosecond pulse duration. Moreover, ultra-intense laser pulses provide new possibilities in the particle acceleration technique. Chapters 19 - 25 cover fundamentals and development of the laser-based high-field science.

This book is the result of efforts of experts from different countries. I would like to acknowledge the authors for their contribution to the book. I also wish to acknowledge Vedran Kordic for indispensable technical assistance in the book preparation and publishing.

Editor

Mikhail Grishin
Institute of Physics and EKSPLA uab
Vilnius, Lithuania

Contents

1

Concentration-Dependent Laser Performance of Yb:YAG Ceramics and Passively Q-switched Yb:YAG/Cr,Ca:YAG Lasers

Jun Dong[1], Ken-ichi Ueda[2], Hideki Yagi[3] and Alexander A Kaminskii[4]
[1]Department of Electronic Engineering, Xiamen University
[2]Institute for Laser Science, University of Electro-Communications
[3]Konoshima Chemical Co. Ltd.
[4]Institute of Crystallography, Russian Academy of Science
[1]P. R. China
[2,3]Japan
[4]Russia

1. Introduction

Ytterbium doped laser materials have been intensely investigated for developing high power laser-diode pumped solid-state lasers around 1 μm (Krupke 2000). Yb:YAG as crystals and polycrystalline ceramics are one of the dominant laser gain media used for solid-state lasers (Lacovara et al., 1991; Brauch et al., 1995; Bruesselbach et al., 1997; Taira et al., 1997; Dong et al., 2006) owing to the excellent optical, thermal, chemical and mechanical properties (Bogomolova et al., 1976). Owing to the small radius difference between yttrium ions and ytterbium ions (Dobrzycki et al., 2004), Yb:YAG single-crystal doped with different Yb concentrations can be grown by different crystal growth methods and efficient laser performance has been achieved (Brauch et al., 1995; Patel et al., 2001; Dong et al., 2006). Transparent laser ceramics (Lu et al., 2000; Lu et al., 2001; Lu et al., 2002; Takaichi et al., 2003; Dong et al., 2006) fabricated by the vacuum sintering technique and nanocrystalline technology (Yanagitani et al., 1998) have been proven to be potential replacements for counterpart single crystals because they have several remarkable advantages compared with single-crystal laser materials, such as high concentration and easy fabrication of large-size ceramics samples, multilayer and multifunctional ceramics laser materials (Yagi et al., 2006; Dong et al., 2007). Efficient and high power laser operation in Nd^{3+}- and Yb^{3+}-ions doped YAG ceramics has been demonstrated (Lu et al., 2002; Dong et al., 2006). Yb:YAG has been a promising candidate for high-power laser-diode pumped solid-state lasers with rod (Honea et al., 2000), slab (Rutherford et al., 2000), and thin disk (Giesen et al., 1994; Stewen et al., 2000) configurations. The quasi-three-level laser system of Yb:YAG requires high pumping intensity to overcome transparency threshold and achieve efficient laser operation at room temperature (Dong & Ueda 2005). The thin disk laser has been demonstrated to be a good way to generate high power with good beam quality owing to the efficiently cooling of gain medium and good overlap of the pump beam and laser beam (Giesen et al., 1994). However, in the thin disk case, the pump beam must be folded many times into thin laser gain medium disk with mirrors in order to absorb sufficient pump power, which makes the laser system extremely complicated. Some applications require that the lasers should be compact and economic; therefore, the cooling system is eliminated in compact and easily maintainable laser system. Therefore, laser-diode end-pumped microchip lasers are a better choice to achieve highly efficient laser operation under high pump power intensity. The thinner the gain medium, the better the cooling effect, therefore, heavy doped Yb:YAG gain media are the better choice for such lasers. The development of Yb:YAG ceramics doped with 1 at.% Yb^{3+} ions have been reported (Takaichi et al., 2003), but the efficiency of such Yb:YAG ceramic laser is low owing to the deficient activator concentration. In principle, there is no concentration quenching effect in Yb:YAG, however, the unwanted impurities (such as Er^{3+}, Tm^{3+}, Ho^{3+}, and so on) from raw materials will be deleterious to the laser performance owing to the high activator doping. Concentration dependent optical properties and laser performance of Yb:YAG crystals have been reported(Yin et al., 1998; Qiu et al., 2002; Yang et al., 2002; Dong et al., 2007). The concentration quenching of Yb:YAG crystals has been investigated and it was found that fluorescence lifetime decreases when the Yb concentration is greater than 15 at.% and lifetime decreases up to 15% when the Yb concentration reaches to 25 at.% (Sumida & Fan 1994; Yang et al., 2002). The fluorescence lifetime of Yb^{3+} doped materials is usually affected by the radiative trapping and concentration quenching effects (M. Ito et al., 2004). Radiative trapping and concentration quenching effects become stronger with Yb concentration and there is a concentration region (from 15 to 25 at.% for Yb:YAG crystal), two trends compete each other and consequently compensate each other, leading to a constant value of measured fluorescence lifetime. Therefore special technologies have been taken to eliminate the radiative trapping effect when the fluorescence lifetime is measured for Yb doped materials. Optical-thin samples or powder sandwiched between two undoped YAG crystals were used to measure the radiative lifetime of Yb:YAG crystals (Sumida & Fan 1994; Patel et al., 2001). The radiative lifetime of Yb:YAG crystal was found to decrease with Yb concentration. Optical spectra of Yb:YAG ceramics doped with different Yb^{3+}-lasant concentration (C_{Yb} = 9.8, 12, and 20 at.%) and efficient 9.8 at.% Yb:YAG ceramic microchip lasers (Dong et al., 2006) have been reported recently. The comparison of laser performance of Yb:YAG ceramic and single-crystal doped with 20 at.% Yb has been reported (Dong et al., 2007). However, there is no systematic comparison studies of microchip laser performance of Yb:YAG ceramics and single-crystal doped with different Yb concentrations.

Compact, high beam quality laser-diode pumped passively Q-switched solid-state lasers with high peak power are potentially used in optical communications, pollution monitoring, nonlinear optics, material processing and medical surgery, and so on(Zayhowski 2000). Passively Q-switched solid-state lasers are usually achieved by using neodymium or ytterbium doped crystals as gain media and Cr,Ca:YAG as saturable absorber(Zayhowski & Dill III 1994; Lagatsky et al., 2000; Takaichi et al., 2002; Dong et al., 2006) or semiconductor saturable absorber mirror (SESAM)(Spuhler et al., 2001) as saturable absorber. Compared with SESAM, Cr^{4+} doped bulk crystals as saturable absorber have several advantages, such as high damage threshold, low cost, and simplicity. The output pulse energy from passively Q-switched solid-state lasers is inversely proportional to the emission cross section of gain medium and reflectivity of the output coupler according to the passively Q-switched theory(Degnan 1995). Besides the broad absorption spectrum (Bruesselbach et al., 1997), longer fluorescence lifetime(Sumida & Fan 1994), high quantum efficiency (over 91% with pump wavelength of 941 nm and laser wavelength of 1030 nm) (Fan 1993) of Yb:YAG gain medium and easy growth of high quality and moderate concentration crystal without concentration quenching (Patel et al., 2001), smaller emission cross section of Yb:YAG (about one tenth of that for Nd:YAG) (Dong et al., 2003) is more suitable to obtain high pulse energy output than Nd:YAG in passively Q-switched solid-state lasers. Another interest in Yb:YAG lasers is that the frequency doubled wavelength of 515 nm matches the highest power line of Ar-ion lasers, thereby leading to the possibility of an all solid-state replacement (Fan & Ochoa 1995). Linearly polarized laser output was observed in these compact passively Q-switched lasers (Li et al., 1993; Yankov 1994; Kir'yanov et al., 1999; Yoshino & Kobyashi 1999; Dong et al., 2000; Bouwmans et al., 2001). The causes of the linearly polarized output in these passively Q-switched lasers were attributed to the influence of the pump polarization(Bouwmans et al., 2001), relative orientations of the switch and an intracavity polarizer(Kir'yanov et al., 1999), temperature change induced weak phase anisotropy(Yoshino & Kobyashi 1999), and the anisotropic nonlinear saturation absorption of Cr,Ca:YAG crystal under high laser intensity(Eilers et al., 1992). The anisotropic nonlinear absorption of Cr,Ca:YAG crystal induced linearly polarization in passively Q-switched lasers with Cr,Ca:YAG as saturable absorber held until appearing of transparent rare-earths doped YAG laser ceramics(Lu et al., 2002; Dong et al., 2006). Efficient laser operation in Nd^{3+}:YAG and Yb^{3+}:YAG ceramic lasers has been demonstrated(Lu et al., 2002; Dong et al., 2006; Nakamura et al., 2008). Chromium doped YAG ceramic has also been demonstrated to be a saturable absorber for passively Q-switched Nd:YAG and Yb:YAG ceramic lasers(Takaichi et al., 2002; Dong et al., 2006). Recently, laser-diode pumped passively Q-switched Yb:YAG/Cr:YAG all-ceramic microchip laser has been demonstrated(Dong et al., 2006), and pulse energy of 31 μJ and pulse width of 380 ps have been achieved with 89% initial transmission of the Cr,Ca:YAG ceramic as saturable absorber and 20% transmission of the output coupler. However, there is coating damage occurrence because of the high energy fluence with low transmission of the output coupler. There are two ways to solve the coating damage problem: one is to improve the coating quality on the gain medium which is costly; the other is to increase the transmission of the output coupler to decrease the intracavity pulse energy fluence. Therefore, 50% transmission of the output coupler was used to balance the output pulse energy and intracavity pulse energy, for this case, the initial transmission of Cr,Ca:YAG can be further decreased to obtain high energy output according to the passively Q-switched solid-state laser theory(Degnan 1995). The laser performance of passively Q-switched Yb:YAG/Cr,Ca:YAG all-ceramic microchip laser was further improved by using 20% initial transmission of the Cr,Ca:YAG ceramic as saturable absorber and 50% transmission of the output coupler, and no coating damage were observed with high pump power(Dong et al., 2007). Highly efficient, sub-nanosecond pulse width and high peak power laser operation has been observed in Yb:YAG/Cr^{4+}:YAG composite ceramics(Dong et al., 2007; Dong et al., 2007). Although linearly polarized states was reported in passively Q-switched Nd:YAG/Cr,Ca:YAG ceramic lasers(Feng et al., 2004), the extinction ratio was very small. The crystalline-orientation self-selected linearly polarized, continuous-wave operated microchip lasers were demonstrated by adopting [111]-cut Yb:YAG crystal(Dong et al., 2008) and [100]-cut Nd:YAG crystal(McKay et al., 2007) as gain medium.

Here, we report on the systematical comparison of the performance of miniature Yb:YAG (C_{Yb} = 9.8, 12, and 20 at.%) ceramic and Yb:YAG single-crystals (C_{Yb} = 10, 15, and 20 at.%)

lasers at 1030 nm with two-pass pumping scheme. The laser performance of Yb:YAG ceramics is nearly comparable to or better than their counterpart single crystals depending on the Yb doping concentration. The effect of Yb concentration on the optical-to-optical efficiency and laser emitting spectra was also addressed. The polarization states of passively Q-switched Yb:YAG/Cr,Ca:YAG microchip lasers with Yb:YAG crystal or ceramic as gain medium and Cr,Ca:YAG crystal or ceramic as saturable absorber were also presented. Based on our previous experiments and results of passively Q-switched Yb:YAG/Cr,Ca:YAG microchip lasers, 20% initial transmission of the saturable absorber and 50% transmission of the output coupler were used in the experiments to compare the polarization states and the effect of the polarization states on the laser performance of these passively Q-switched microchip laser. Linearly polarized states were observed in Yb:YAG/Cr,Ca:YAG combinations with at least one crystal. For Yb:YAG/Cr,Ca:YAG all-ceramics combination, the laser oscillates at random polarization state. The effect of polarized states of passively Q-switched Yb:YAG/Cr,Ca:YAG lasers on the laser performance was also investigated.

2. Experiments

To compare the laser performance of Yb:YAG ceramics and single-crystals, double-pass pumped miniature lasers were used in the experiments. To absorb sufficient pump power, high doping concentration was needed for thin gain medium. Therefore, high doping concentration Yb:YAG single-crystals and ceramics were used in the laser experiments. Three Yb:YAG ceramics samples (C_{Yb} = 9.8, 12, and 20 at.%) were used in the laser experiments. Comparable Yb:YAG single-crystals (C_{Yb} = 10, 15, and 20 at.%) were used to compare the laser characteristics with those of Yb:YAG ceramics. The size of the samples is 10 mm in diameter and 1 mm in thickness. The absorption spectra of Yb:YAG crystals and ceramics were measured by using a ANDO white light source and an optical spectral analyzer (ANDO AQ6137). Emission spectra were measured at 900 – 1150 nm with a fiber-coupled diode laser operating at 940 nm as the pump source. The pump light was focused into one of the ground surfaces of the sample close to the polished surface through which the emitted fluorescence was to be observed. As a result the fluorescent light detected was generated close to the surface from which it exited the sample, such that it experienced minimal radiation trapping. The fluorescence emission spectral signal was collected by using a focus lens and coupled into a multi-mode fiber which was connected to an optical spectral analyzer. The resolution of the optical spectral analyzer is 0.01 nm. The effective emission cross section of Yb:YAG crystals and ceramics was calculated by applying Fuechtbauer-Ladenburg formula. The absorption and emission spectra of Yb:YAG ceramics and single-crystals doped with different ytterbium concentrations were measured at room temperature. Fig. 1 shows the room temperature absorption and emission spectra of Yb:YAG ceramic and single-crystal containing 20 at.% of ytterbium activators. The absorption and emission spectra of Yb:YAG ceramics are nearly identical to those of Yb:YAG single crystals. However there were some differences between Yb:YAG crystals and ceramics which may have potential effects on the laser performance. Firstly, the absorption coefficient of Yb:YAG ceramics is higher than that of Yb:YAG single crystal for the same Yb doping concentration. The peak absorption coefficient at 940 nm increases linearly with Yb activator concentration for both Yb:YAG ceramics and single-crystals, as shown in the inset of Fig. 1(a), the peak absorption coefficients at 940 nm of low doping concentraiton Yb:YAG samples were taken from Ref. (Dong et al., 2003; Takaichi et al., 2003). However, the absorption coefficient of Yb:YAG single-crystals increases slowly with Yb concentration compared to Yb:YAG ceramics. And the absorption coefficient of Yb:YAG ceramics is about 10% more than that for the counterpart single-crystal with same doping levels at high doping levels. This was caused by the segregation in Yb:YAG single crystal during the crystal growth and the Yb^{3+} concentration is lower than that in melt. This effect becomes more obvious at high doping levels. However, Yb^{3+} ions are uniformly distributed in the mixed Yb:YAG nanocrystalline power. Secondly, there were some differences between the emission cross section of Yb:YAG ceramics and crystals. Fig. 1(b) shows the emission spectra of Yb:YAG ceramic and single-crystals doped with 20 at.% Yb^{3+}-ions. Two main emission peaks are centered at 1030 nm and 1049 nm. The effective peak emission cross section of Yb:YAG ceramics was estimated to be 2.2×10^{-20} cm^2 at 1030 nm, which was lower than that of Yb:YAG single-crystal (2.3×10^{-20} cm^2). The effective emission cross section at 1047.5 nm (0.37×10^{-20} cm^2) was about one sixth of that at 1030 nm for Yb:YAG ceramics. However, the effective emission cross section at 1046.9 nm (0.39×10^{-20} cm^2) of Yb:YAG crystal is 5% higher than that of Yb:YAG ceramics. The lower effective emission cross section of Yb:YAG ceramics limits the laser performance under the same laser conditions as that for Yb:YAG crystals. The emission cross section of Yb:YAG ceramics does not change with Yb concentration, which is in good agreement with the measured emission cross section of Yb:YAG single-crystals (Dong et al., 2003) although the emission intensity increases with increase of the Yb concentrations.

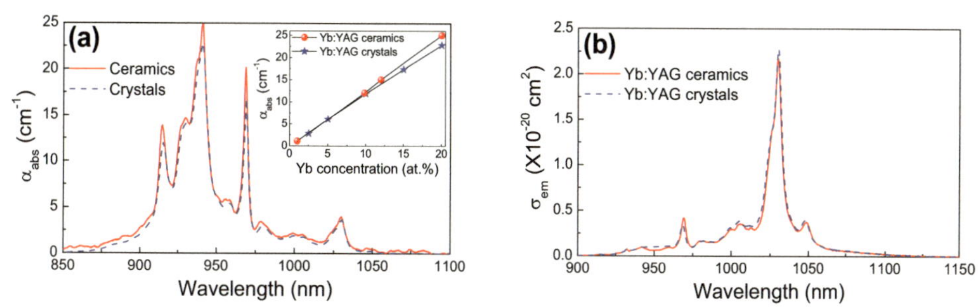

Fig. 1. The absorption and emission spectra of Yb:YAG ceramic and single-crystal doped with 20 at.% Yb at room temperature. Inset in Fig.1(a) shows the 940-nm peak absorption coefficient of Yb:YAG ceramics and single-crystals as a function of Yb concentrations.

Fig. 2 shows a schematic diagram of the experimental setup for laser-diode pumped Yb:YAG miniature laser. One surface of the sample was coated for antireflection both at 940 nm and 1.03 μm. The other surface was coated for total reflection at both 940 nm and 1.03 μm, acting as one cavity mirror and reflecting the pump power for increasing the absorption of the pump power. Plane-parallel fused silica output couplers with transmission (T_{oc}) of 5 and 10% were mechanically attached to the gain medium tightly. A 35-W high-power fiber-coupled 940 nm laser diode (Apollo, F35-940-1) with a core diameter of 100 μm and numerical aperture of 0.22 was used as the pump source. Optical coupling system with two lenses M1 (8-mm focal length) and M2 (15-mm focal length) was used to focus the pump beam on the ceramic rear surface and to produce a pump light footprint on the Yb:YAG of about 170 μm in diameter. The laser spectrum was analyzed by using an optical spectrum analyzer (ANDO AQ6137) with resolution of 0.01 nm. Output beam profile of these lasers was monitored by using a CCD camera, and beam quality factor, M^2, was determined by measuring the beam diameters at different positions along the laser propagation direction.

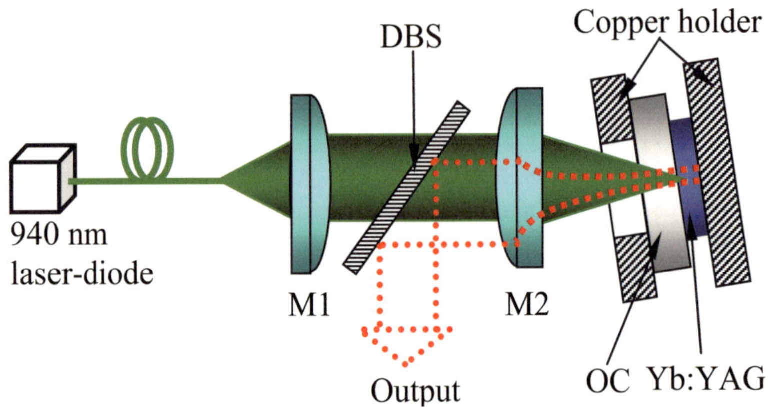

Fig. 2. Schematic diagram of laser-diode pumped Yb:YAG ceramics and single-crystals miniature lasers. DBS, dichroic beam splitter; OC, output coupler; M1, focus lenses with focal length of 8 mm; M2, focus lenses with focal length of 15 mm.

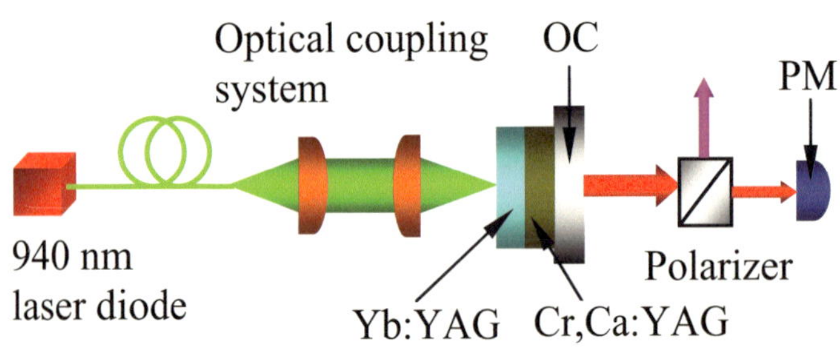

Fig. 3. Schematic diagram for passively Q-switched Yb:YAG microchip lasers with Cr,Ca:YAG as saturable absorber. OC, output coupler; PM, power meter.

Fig. 3 shows a schematic diagram of experimental setup for passively Q-switched Yb:YAG microchip laser with Cr,Ca:YAG as saturable absorber. Two Yb:YAG samples are used as gain media, one is Yb:YAG ceramic doped with 9.8 at.% Yb, the other is [111]-cut Yb:YAG crystal doped with 10 at.% Yb. The thickness of Yb:YAG samples is 1 mm, and the Yb:YAG samples are polished to plane-parallel. One surface of the gain medium was coated for anti-reflection at 940 nm and total reflection at 1.03 μm acting as one cavity mirror. The other surface was coated for high transmission at 1.03 μm. Two 1-mm-thick, uncoated Cr,Ca:YAG ceramic and [111]-cut Cr,Ca:YAG crystal with 80% initial transmission, acting as Q-switch, was sandwiched between Yb:YAG sample and a 1.5-mm-thick, plane-parallel fused silica output coupler with 50% transmission. Total cavity length was 2 mm. The initial charge concentration of $CaCO_3$ and Cr_2O_3 in growth of Cr,Ca:YAG crystal and fabrication of Cr,Ca:YAG ceramic are 0.2 at.% and 0.1 at.%, respectively. The absorption center of Cr,Ca:YAG sample centered at 1 μm is also strongly affected by the annealing process and the exact concentration of this absorption center is difficult to determine, the concentration center of this absorption is roughly about 4% of the initial Cr doping concentration

(Okhrimchuk & Shestakov 1994). Therefore, the initial transmission of the Cr,Ca:YAG saturable absorber is usually used in comparing the laser performance of passively Q-switched lasers. The initial transmission of Cr,Ca:YAG is governed by the doping concentration and the thickness of the sample, to fully compare laser performance with our previously passively Q-switched Yb:YAG/Cr,Ca:YAG all-ceramic microchip laser and the effect of polarization states on the passively Q-switched Yb:YAG/Cr,Ca:YAG microchip lasers, 1-mm-thick Cr,Ca:YAG crystal with 80% initial transmission was used in the experiment. It should be noted that the polarization behavior keeps the same if a different modulation depth of Cr,Ca:YAG saturable absorber is used. A high-power fiber-coupled 940 nm laser diode with a core diameter of 100 μm and numerical aperture of 0.22 was used as the pump source. Two lenses of 8-mm focal length were used to focus the pump beam on the Yb:YAG rear surface and to produce a pump light footprint on the Yb:YAG of about 100 μm in diameter. The laser was operated at room temperature. The Q-switched pulse profiles were recorded by using a fiber-coupled InGaAs photodiode with a bandwidth of 16 GHz, and a 7 GHz Tektronix TDS7704B digital phosphor oscilloscope. The laser spectrum was analyzed by using an optical spectrum analyzer. The laser output beam profile was monitored using a CCD camera both in the near-field and the far-field of the output coupler.

3. Results and discussion

3.1 Continuous-wave Yb:YAG miniature lasers

Fig. 4 shows the output power of miniature Yb:YAG ceramics and single-crystal lasers as a function of the absorbed pump power for different Yb concentrations and T_{oc}. The absorbed pump power for reaching laser thresholds of Yb:YAG ceramics (C_{Yb} = 9.8, 12, and 20 at.%) were 0.3, 0.33, and 0.64 W for T_{oc} = 5% and 0.33, 0.5, 0.68 W for T_{oc} = 10%. The pump power threshold increases with T_{oc} and Yb concentration for Yb:YAG ceramic lasers. This was caused by the increase of the losses introduced by the large T_{oc} and the increasing reabsorption of Yb^{3+} at lasing wavelength with Yb concentrations. For Yb:YAG ceramics doped with different Yb concentrations, the output power increases linearly with absorbed pump power for T_{oc} = 5 and 10%. The slope efficiencies respected to the absorbed pump power for Yb:YAG ceramics (C_{Yb} = 9.8, 12, and 20 at.%) were measured to be 50, 55, and 45% for T_{oc} = 5% and 52, 44, and 38% for T_{oc} = 10%. Slope efficiency increases with T_{oc} for 9.8 at.% Yb:YAG ceramic, however, the slope efficiencies decrease with T_{oc} for Yb:YAG ceramics doped with 12 and 20 at.% Yb^{3+} ions. Maximum output power of 2.54 W was measured for T_{oc} = 5% by using 12 at.% Yb:YAG ceramic as gain medium when the absorbed pump power was 5.3 W. The corresponding optical-to-optical efficiency was about 48%.

The absorbed pump power for reaching laser thresholds of Yb:YAG single-crystal (C_{Yb} = 10, 15, and 20 at.%) were 0.3, 0.51 and 0.76 W for T_{oc} = 5% and 0.35, 0.55, and 0.84 W for T_{oc} = 10%. The absorbed pump power threshold increases with the T_{oc} and Yb concentrations, the same tendency as that for Yb:YAG ceramics. However, the absorbed pump power thresholds of Yb:YAG single crystals were higher than those of Yb:YAG ceramics. This may be caused by the pump configuration used in the laser experiments and low pump power intensity with pump beam diameter of 170 μm. Because the incident pump beam from laser-diode is several degrees away from the normal direction of the laser beam, there is a mismatch between the pump beam and laser beam. From Fig. 4 (b), we can see at low pump power just above absorbed pump power threshold, the laser performance is lower than that

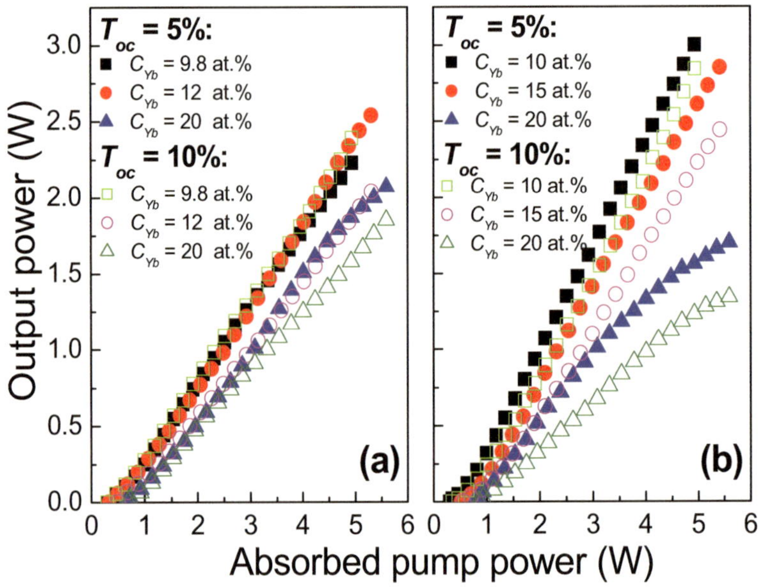

Fig. 4. Output power of (a) Yb:YAG ceramic and (b) Yb:YAG single crystal miniature lasers as a function of absorbed pump power for different Yb concentrations and transmissions of the output couplings.

high pump power levels, this is normal for the quasi-three-level system; efficient laser performance can be achieved at high pump power density (Dong & Ueda 2005). However, for Yb:YAG ceramics, owing to the random distribution of Yb:YAG crystalline particles, the absorbed pump power threshold can be achieved more easily. Output power increases linearly with absorbed pump power for Yb:YAG single-crystals doped with 10 and 15 at.% Yb. The slope efficiencies of miniature lasers based on Yb:YAG single-crystals doped with 10 and 15 at.% Yb were measured to be 69, 62% for T_{oc} = 5% and 67, 55% for T_{oc} = 10%. The slope efficiencies of Yb:YAG crystals doped with 10 and 15 at.% Yb^{3+} were higher than those for Yb:YAG ceramics although the pump power thresholds were higher than those of Yb;YAG ceramics. At higher pump power density, the inversion population excited by the pump beam is well-over the threshold, and the modes matched very well, therefore, the laser oscillates at high slope efficiency, especially for Yb:YAG crystal doped with 10 and 15 at.% Yb. The better laser performance of these crystals compared to their counterpart ceramic suggests that the intacavity loss for Yb:YAG crystal lower than that of Yb:YAG ceramics.

For Yb:YAG single-crystal doped with 20 at.% Yb, the output power increases with the absorbed pump power, and tends to increase slowly when the absorbed pump power is higher than a certain value (e.g. 3 W for T_{oc} = 5% and 2.3 W for T_{oc} = 10%), as shown in Fig. 4(b). However, besides the higher absorbed pump power threshold compared to its counterpart Yb:YAG ceramic, the slope efficiencies (45% for T_{oc} = 5% and 32 at.% for T_{oc} = 10%) of 20 at.% Yb:YAG single crystal were lower than those (47% for T_{oc} = 5% and 38 at.% for T_{oc} = 10%) for its counterpart Yb:YAG ceramic. The laser results show that heavy doped

Yb:YAG ceramic is better than its single crystal counterpart. The strong segregation of the impurities in Yb:YAG crystal with increase of the Yb concentration during crystal growth is the main reason for the worse laser performance. The other reasons for the less efficient laser operation may be the impurities increases with doping concentration (Yin et al., 1998), the impurities induced concentration quenching effect limit the laser performance of highly doped Yb:YAG crystals. The green emission was observed in the Yb:YAG crystals and ceramics when they were pumped with laser-diodes, and visible intensity increases with Yb concentration up to 15 at.% and then decreases with Yb concentration (Xu et al., 2005). Energy transfer from Yb^{3+} ions to Er^{3+} and Tm^{3+} impurities and cooperative energy transfer between Yb^{3+} ions are the causes of these visible luminescence. These are deleterious to the infrared laser operation. However, the distance between Yb^{3+} ions and impurities or other quenching centers is decreased with Yb concentration, the cooperative luminescence intensity decreases because the excited ions are more easily quenched by reaching a neighboring defect site. Therefore, the effect of cooperative energy transfer is not a main factor to limit the laser performance of highly doped Yb:YAG crystals.

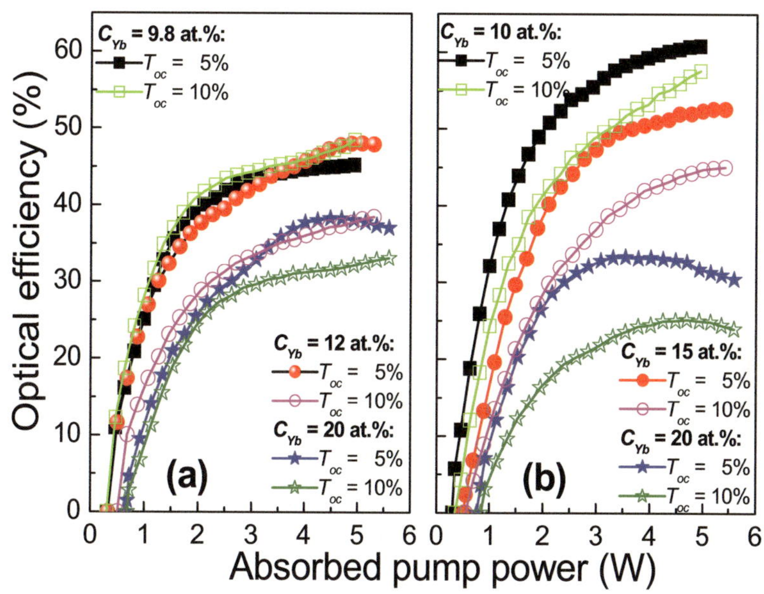

Fig. 5. Optical efficiencies of (a) Yb:YAG ceramic and (b) Yb:YAG single crystal miniature laser as a function of absorbed pump power for different Yb concentrations and transmissions of the output coupler.

Fig. 5 shows the optical-to-optical efficiencies of Yb:YAG lasers as a function of absorbed pump power. Under present laser experimental conditions, there is no saturation effect of Yb:YAG lasers with different output couplings for Yb concentration equal to or less than 15 at.% although the optical efficiency increases slowly with the absorbed pump power. However, for 20 at.% Yb:YAG, there is saturation effect for ceramic lasers with T_{oc} = 5% and for single-crystal lasers with different output couplings. Maximum optical efficiency of 48% was achieved for Yb:YAG ceramic doped with 12 at.% Yb at the absorbed pump power of

5.3 W. For single crystal doped with 20 at.% Yb lasants, there is a maximum optical efficiency for all output couplings [as shown in Fig. 5(b)]. The optical-to-optical efficiency decreases with further increase of the pump power. For Yb:YAG ceramics, except the comparable laser performance of 9.8 at.% Yb:YAG with T_{oc} = 5 and 10%, the optical-to-optical efficiency decreases with T_{oc} and Yb concentration. However, for Yb:YAG single-crystals, the optical-to-optical efficiency decreases with the T_{oc} and Yb concentration under different pump power levels. Optical-to-optical efficiency of Yb:YAG crystal doped with less than 15 at.% Yb is higher than that for Yb:YAG ceramics under certain pump power levels. For 20 at.% Yb:YAG, Yb:YAG ceramic has higher optical-to-optical efficiency than that of crystal under different pump power levels. The decrease of the optical-to-optical efficiency of Yb:YAG lasers with Yb concentration was attributed to the thick samples used for highly doped Yb:YAG samples. The better laser performance can be further improved through optimizing the thicknesses for Yb:YAG samples with different Yb concentrations. The highly efficient microchip lasers has been demonstrated by using the same crystals (Dong et al., 2007) as those here used.

Fig. 6 shows the maximum optical-to-optical efficiency under available pump power of Yb:YAG ceramics and single-crystals lasers as a function of Yb concentrations for different output couplings. For Yb:YAG single crystals, the maximum optical-to-optical efficiency decreases with Yb concentrations, there are 45% and 56% dropping for T_{oc} = 5 and 10% when Yb concentration increases from 10 at.% to 20 at.%. However, for Yb:YAG ceramics, the maximum optical-to-optical efficiency decreases with Yb concentration, the decrease is smaller for Yb:YAG ceramics compared to that for Yb:YAG single crystal. There are 15 and 32% dropping for T_{oc} = 5 and 10% when Yb concentration increases from 9.8 to 20 at.% for Yb:YAG ceramics. Because small different optical properties were observed in Yb:YAG ceramics and single-crystals doped with different Yb concentrations(Dong et al., 2003; Dong et al., 2006), the different laser performance of Yb:YAG ceramics and single-crystals may be caused by the Yb^{3+}-ions distribution in YAG host and optical quality of Yb:YAG samples. Although the distribution coefficient of Yb in Yb:YAG is close to unit, there is still concentration gradient observed in Yb:YAG single crystals along the growth axis and radius of the crystal boule (Xu et al., 2003). The Yb^{3+}-ion distribution inhomogeneity in Yb:YAG single-crystal becomes server with Yb concentration. The impurities such as Ho^{3+}, Er^{3+} increase with Yb concentration in Yb:YAG crystals because the strong segregation of rare-earth ions in YAG crystal was observed. This was observed in the reduced radiative lifetime in highly doped Yb:YAG crystals (Sumida & Fan 1994; Yin et al., 1998; Patel et al., 2001). This concentration quenching effect limits the efficient laser performance of highly doped Yb:YAG crystals. For ceramics, the distribution of Yb ions in the grains and grain boundary is a main factor to determine the optical properties. The gain boundary of YAG ceramics was measured to be less than 0.5 nm (Barabanenkov et al., 2004), and sintering temperature is about 200 °C lower than the melt point of Yb:YAG crystal, the segregation of Yb in grain boundary can only be achieved by diffusion or migration, therefore the distribution of Yb in gain and boundary should be close to homogeneous. When Yb ions were doped in YAG ceramics, the segregation of ytterbium ions in the grain boundary, accompanied by a reduction of the acoustic mismatch, leads to increased phonon transmission (Bisson et al., 2007). This will be further enhanced by introducing more ytterbium ions. This may be one of the main reasons for the better laser performance of heavy doped Yb:YAG ceramics compared to that of single-crystal with same doping levels.

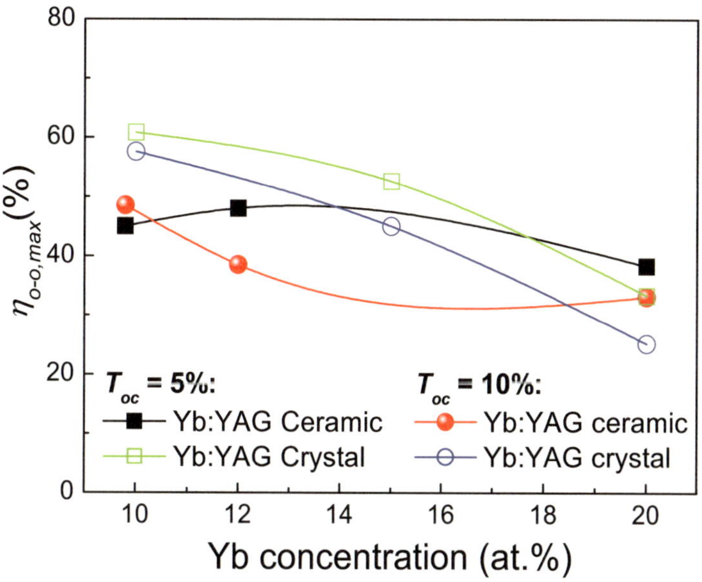

Fig. 6. Comparison of the maximum optical-to-optical efficiency of Yb:YAG miniature lasers as a function of Yb concentration. The solid lines were used for illustration.

Fig. 7 shows the comparison of the laser emitting spectra of 9.8 at.% Yb:YAG ceramic and 10 at.% Yb:YAG single-crystal miniature lasers under different absorbed pump power for T_{oc} = 5, and 10%. Lasers operated at multi-longitudinal modes under different pump levels. The number of longitudinal modes increases with the absorbed pump power because the inversion population provided with pump power can overcome the threshold for low gain away from the highest emission peak of Yb:YAG gain medium. The longitudinal mode oscillation for these miniature Yb:YAG lasers was mainly caused by the etalon effect of plane-parallel Yb:YAG thin plate. The separation of longitudinal modes was measured to be 0.29 nm, which is in good agreement with the free spectral range (0.292 nm) of 1-mm-long cavity filled with gain medium predicted by (Koechner 1999) $\Delta\lambda_c = \lambda^2/2L_c$, where L_c is the optical length of the resonator and λ is the laser wavelength. And the center wavelength of the lasers shifts to longer wavelength with the pump power which is caused by the temperature dependent emission spectra of Yb:YAG crystal (Dong et al., 2003). For T_{oc} = 5%, both Yb:YAG ceramic and crystal lasers are oscillating at longer wavelength comparing to those for T_{oc} = 10%. The cause of the wavelength shift to longer wavelength for T_{oc} = 5% is relating to the change of the intracavity laser intensity (Kong et al., 2004) because only the intracavity laser intensity is different for both cases. Intracavity laser intensity for T_{oc} = 5% is about two times higher than that for T_{oc} = 10%, therefore, more longitudinal modes will also be excited for T_{oc} = 5%. Because the better laser performance for 10 at.% Yb:YAG lasers compared to 9.8 at.% Yb:YAG ceramics lasers, the intracavity intensity is higher for crystal laser, therefore Yb:YAG crystal lasers oscillate at longer wavelength than those for Yb:YAG ceramics lasers, especially for T_{oc} = 5%. Strong mode competition and mode hopping in these Yb:YAG ceramic lasers were also observed. When the laser oscillates, the excited Yb^{3+}

ions jump back to the lower laser level, they always relax to other even-lower energy levels or ground level, this process is rapid compare to the lifetime of Yb^{3+} ion in YAG crystal or ceramics. The relaxation of Yb^{3+} ions to the lower energy or ground levels causes the lower-level population to increase with the lasing intensity, this increases the reabsorption. This enhanced reabsorption provides a negative feedback process for the lasing modes and effective gain profile of Yb:YAG medium. This negative feedback process accompanied with the effects of strong mode competition makes some stronger laser modes eventually faded or quenched. When the intracavity light intensity is high enough, the population distribution at lower energy levels is changed dramatically. At the same time, the effective gain curve of Yb:YAG under lasing condition was altered by the strong reabsorption and temperature rise induced by the absorption pump power. Some initially suppressed modes at longer wavelength governed by the emission spectra can oscillate under changed gain curve, therefore the laser wavelength shifts to longer wavelength and mode hopping was observed. Fig. 8 shows the laser emitting spectra of 20 at% Yb:YAG ceramic and single-crystal miniature lasers under different pump power levels and output couplings. The lasers oscillate at multi-longitudinal modes. The number of the longitudinal modes increases with the pump power. The laser oscillates at longer wavelength for 20 at.% Yb:YAG lasers compared to that for 10 at.% Yb:YAG lasers for both transmissions of the output couplers (as shown in Fig. 7 and Fig. 8). The red-shift of laser wavelength for T_{oc} = 10% with Yb concentration is smaller than that for T_{oc} = 5% because of the lower intracavity laser intensity Yb:YAG lasers with T_{oc} = 10%. The number of longitudinal modes is larger for T_{oc} = 5% than that for T_{oc} = 10%. This may be related to the gain curve change due to the strong reabsorption under strong intracavity intensity.

The output beam transverse intensity profiles were also monitored in all the pump power range. One example of the beam intensity profile at output power of 2.5 W for 12 at.%

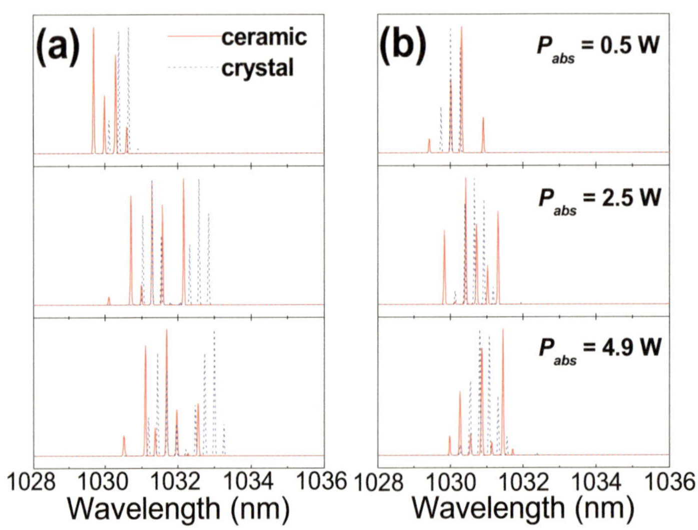

Fig. 7. Laser emitting spectra of 9.8 at.% Yb:YAG ceramic and 10 at.% Yb:YAG crystal miniature lasers under different pump power levels, (a) T_{oc} = 5%, (b) T_{oc} = 10%. The resolution of the optical spectral analyzer is 0.01 nm.

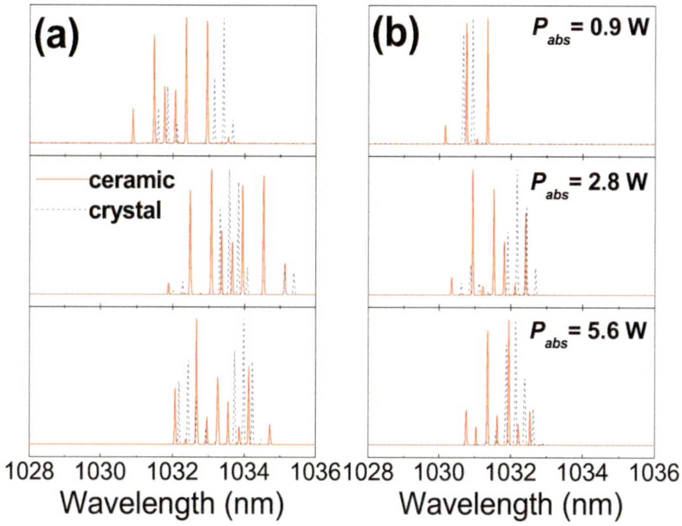

Fig. 8. Laser emitting spectra of 20 at.% Yb:YAG ceramic and crystal miniature lasers under different pump power levels, (a) T_{oc} = 5%, (b) T_{oc} = 10%. The resolution of the optical spectral analyzer is 0.01 nm.

Yb:YAG ceramics with T_{oc} = 5% was shown in Fig. 9, as well as a horizontal slice through the center. The output beam profile is close to TEM_{00} mode. The measured spatial profile can be fitted with Gaussian function very well, as shown in Fig. 9(b). Near-diffraction-limited beam quality with M^2 of less than 1.1 was achieved in these miniature lasers with Yb:YAG ceramics and single-crystals as gain media in the available pump power range.

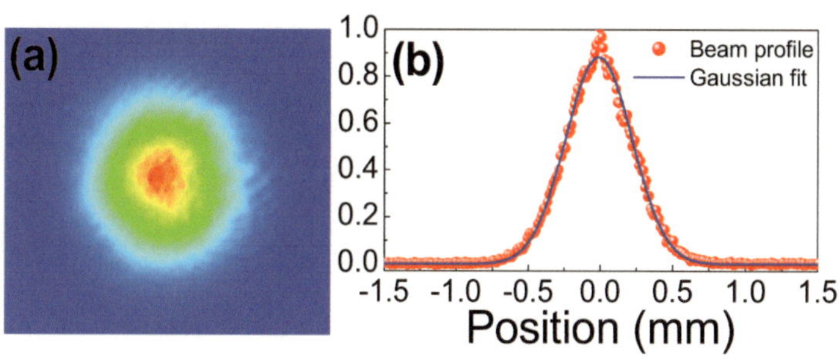

Fig. 9. (a) Output laser beam profile of Yb:YAG ceramic miniature lasers when output power is 2.5 W for 12 at.% Yb:YAG ceramics with T_{oc} = 5%. (b) Horizontal slice through center of beam profile with Gaussian fit.

3.2 Passively Q-switched Yb:YAG/Cr,Ca:YAG microchip lasers

Four combinations of Yb:YAG and Cr,Ca:YAG were used in the laser experiments to investigate the polarization states of passively Q-switched Yb:YAG/Cr,Ca:YAG microchip lasers: C1, Yb:YAG crystal + Cr,Ca:YAG crystal; C2, Yb:YAG ceramic + Cr,Ca:YAG ceramic; C3, Yb:YAG crystal + Cr,Ca:YAG ceramic; C4, Yb:YAG ceramic + Cr,Ca:YAG crystal. The polarization states of passively Q-switched Yb:YAG/Cr,Ca:YAG microchip lasers with different combinations were investigated by measuring the output power after polarizer. Table 1 summaries the polarization states observed in passively Q-switched Yb:YAG/Cr,Ca:YAG microchip lasers with different combinations of Yb:YAG, Cr,Ca:YAG crystals and ceramics. By rotating the combination of Yb:YAG/Cr,Ca:YAG, the polarization states of these lasers do not change, only the polarization directions are changed by arranging Yb:YAG or Cr,Ca:YAG. Rotating any one of sample does not affect the polarization states and no stronger influence on the polarization was observed.

No.	Combinations	Polarization
C1	Yb:YAG crystal + Cr,Ca:YAG crtstal	Linear
C2	Yb:YAG ceramic + Cr,Ca:YAG ceramic	Random
C3	Yb:YAG crystal + Cr,Ca:YAG ceramic	Linear
C4	Yb:YAG ceramic + Cr,Ca:YAG crystal	Linear

Table 1. Polarizations states of passively Q-switched Yb:YAG/Cr,Ca:YAG microchip lasers

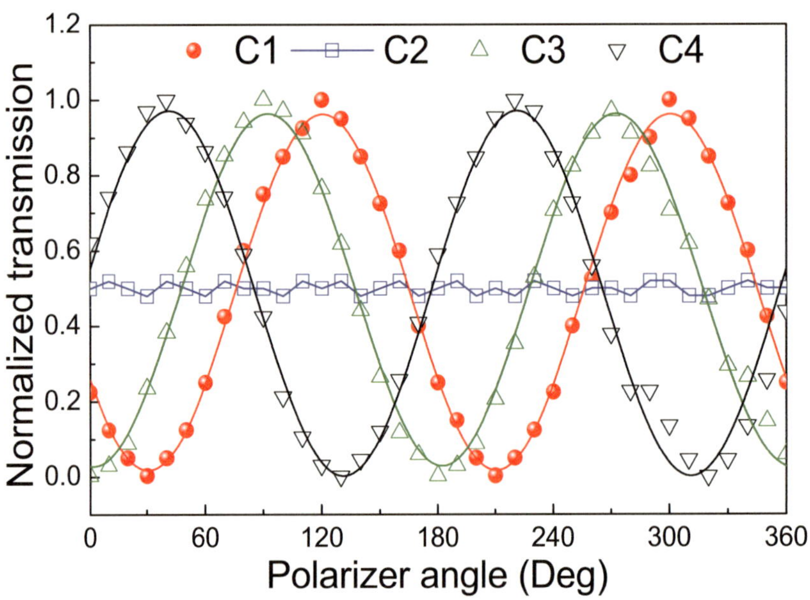

Fig. 10. Polarization states of passively Q-switched Yb:YAG/Cr,Ca:YAG microchip lasers. The solid lines show the sine function fitting of the experimental data.

Fig. 10 shows the typical polarization states of four combinations. Except for the random oscillation of Yb:YAG/Cr,Ca:YAG all-ceramics combination, other three combinations exhibit linearly polarization output. The extinction ratio of the linearly polarization is greater than 300:1. Some differences between the extinction ratios for different linearly polarization were observed. The extinction ratios of three different linearly polarized combinations are in the order of C1 > C4 > C3. The extinction ratios of three different linearly polarized combinations decrease a little with increase of the pump power, we did not observe significant decrease of the extinction ratio at the maximum pump power used here, this shows that the thermal effect under current available pump power is not strong enough to induce sufficient birefringence and depolarization for Yb:YAG crystals and ceramics. However, we did observe the thermal effect under high pump power level for cw Yb:YAG microchip lasers(Dong et al., 2008), therefore, the thermal effect induced birefringence and depolarization should be considered in high power pumped passively Q-switched Yb:YAG/Cr,Ca:YAG microchip lasers. The different polarization states between all-ceramics combination and three others are due to the random distribution of nanocrystalline particles in ceramics. To fully understand the nature of polarization states in passively Q-switched Yb:YAG/Cr,Ca:YAG microchip lasers, we measured the polarization states of Yb:YAG crystals and ceramics by removing Cr,Ca:YAG saturable absorber and found that Yb:YAG crystals oscillate at linearly polarization states selected by the crystalline-orientations in the (111) plane(Dong et al., 2008) and Yb:YAG ceramic oscillates at unpolarization states. Although there is saturation absorption in Cr,Ca:YAG ceramic, the same as that for Cr,Ca:YAG crystal, owing to the random distribution of Cr,Ca:YAG particles in ceramic, the saturation absorption does not exhibit crystalline-orientation dependent anisotropic properties when the sample is rotated, which is different from the anisotropic saturation absorption of Cr,Ca:YAG crystal when the laser propagate along [111] direction(Eilers et al., 1992). Therefore, the polarization states in passively Q-switched microchip lasers with Cr,Ca:YAG as saturable absorber are not only determined by the anisotropic saturation absorption of Cr,Ca:YAG saturable absorber, but also determined by the linearly polarized states of Yb:YAG crystals.

The continuous-wave operation of Yb:YAG crystal and ceramic has been investigated previously by using different transmissions of output coupler(Dong et al., 2006; Dong et al., 2007) and found that the laser performance 1-mm-thick Yb:YAG crystal doped with 10 at.% Yb is better than that of 1-mm-thick Yb:YAG ceramic doped with 9.8 at.% Yb. The absorbed pump power thresholds are 0.46 W and 0.54 W for 1-mm-thick Yb:YAG crystal and ceramic, respectively, the slope efficiencies were 49% and 44%, respectively by using 50% transmission of output coupler. The differences of cw laser performance between Yb:YAG crystal and ceramic suggest that the optical quality of ceramic used in the experiments is not as good as that of Yb:YAG crystal, and the slight different doping concentration may be another cause of the difference.

Here we show the effect of different polarization states of passively Q-switched Yb:YAG/Cr,Ca:YAG microchip lasers on the laser performance. Average output power as a function of absorbed pump power for these four combinations of Yb:YAG and Cr,Ca:YAG microchip lasers was shown in Fig. 11. The absorbed pump power thresholds were about 0.53, 0.66, 0.75, and 0.6 W for combinations C1, C2, C3 and C4. The higher pump power threshold of these passively Q-switched lasers was due to the low initial transmission of Cr,Ca:YAG and high transmission of the output coupler used in the experiments. Average output power increases linearly with absorbed pump power for the four combinations, the slope efficiencies with respect to the absorbed pump power were estimated to be about 39, 36, 36 and 29% for the four combinations of C1, C2, C3 and C4, respectively. The best laser performance (low threshold and high slope efficiency) of passively Q-switched Yb:YAG/Cr,Ca:YAG microchip lasers was obtained with C1 combination because of the enhancement of linearly polarized laser operation due to the combination of linearly oscillation of Cr:YAG crystal under high intracavity laser intensity(Eilers et al., 1992) and the crystalline-orientation selected linearly polarized states of Yb:YAG crystal(Dong et al., 2008). Maximum average output power of 310 mw was obtained with Yb:YAG/Cr,Ca:YAG all-crystal combination when the absorbed pump power was 1.34 W, corresponding to optical-to-optical efficiency of 23%. The optical-to-optical efficiency is 15% with respect to the incident pump power for C1. The optical-to-optical efficiencies with respect to the incident pump power were measured to be 12, 11 and 11% for C2, C3 and C4 respectively. There is no coating damage occurrence with further increase of the pump power owing to decrease of the intracavity energy fluence by using high transmission output coupler.

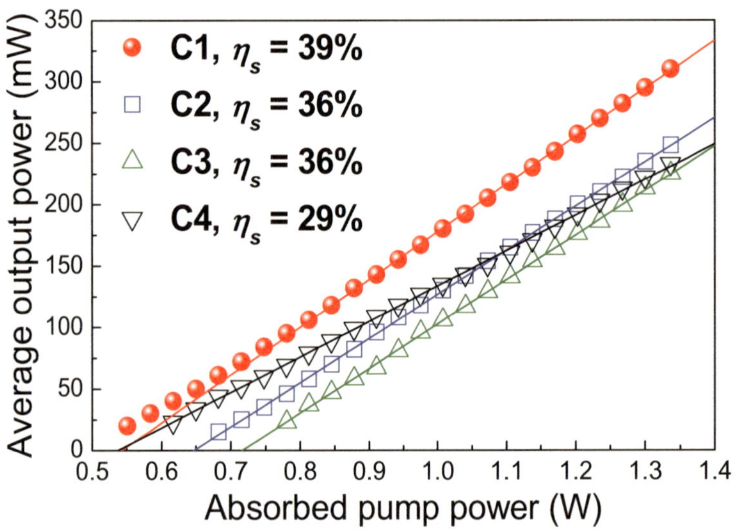

Fig. 11. Average output power as a function of the absorbed pump power for passively Q-switched Yb:YAG/Cr,Ca:YAG microchip lasers with different combinations of Yb:YAG and Cr,Ca:YAG. The solid lines show the linear fit of the experimental data.

Although linearly polarized laser operation was observed in Yb:YAG/Cr,Ca:YAG combinations with at least one crystal, the effect of linearly polarized states on the laser performance was different. The slope efficiency of C4 is lower than that of C3, however the laser threshold of C4 is lower than that of C3 and the average output power is higher than that of C3 for all the available pump power range, as shown in Fig. 11. The contribution of polarization states from Cr,Ca:YAG crystal and Cr,Ca:YAG ceramic is different, when Cr,Ca:YAG crystal is used as saturable absorber, even with Yb:YAG ceramic as gain medium, the laser threshold is low. For all-ceramics combination, C2, although the laser threshold is higher than those of C1 and C4, the slope efficiency is better those of C4 and C3. These results show that the polarized states have great effect on the laser performance. Even with random polarized states of all-ceramics combination, passively Q-switched Yb:YAG/Cr,Ca:YAG laser has nearly the same laser performance as that of all-crystals combination. The discrepancies between all-crystals (C1) and all-ceramics (C2) combinations were caused by the optical quality of Yb:YAG crystal and Yb:YAG ceramic, the laser performance of Yb:YAG crystal is better than its ceramic counterpart. The discrepancies between C3 and C4 were attributed to the linearly polarization, with Cr,Ca:YAG crystal as saturable absorber, the extinction ratio of the polarization is stronger than that of with Cr,Ca:YAG ceramic as saturable absorber, the laser prefers to oscillate more efficiently with orientation selected anisotropic saturbable absorption of Cr,Ca:YAG crystal along <111> direction under high intracavity intensity(Eilers et al., 1992).

The output beam profile is close to fundamental transverse electro-magnetic mode. Near diffraction-limited output beam quality with M_x^2 of 1.05 and M_y^2 of 1.04, respectively, was achieved in such compact passively-Q-switched Yb:YAG/Cr,Ca:YAG microchip lasers. The output beam diameter near the output mirror was measured to be 100 μm.

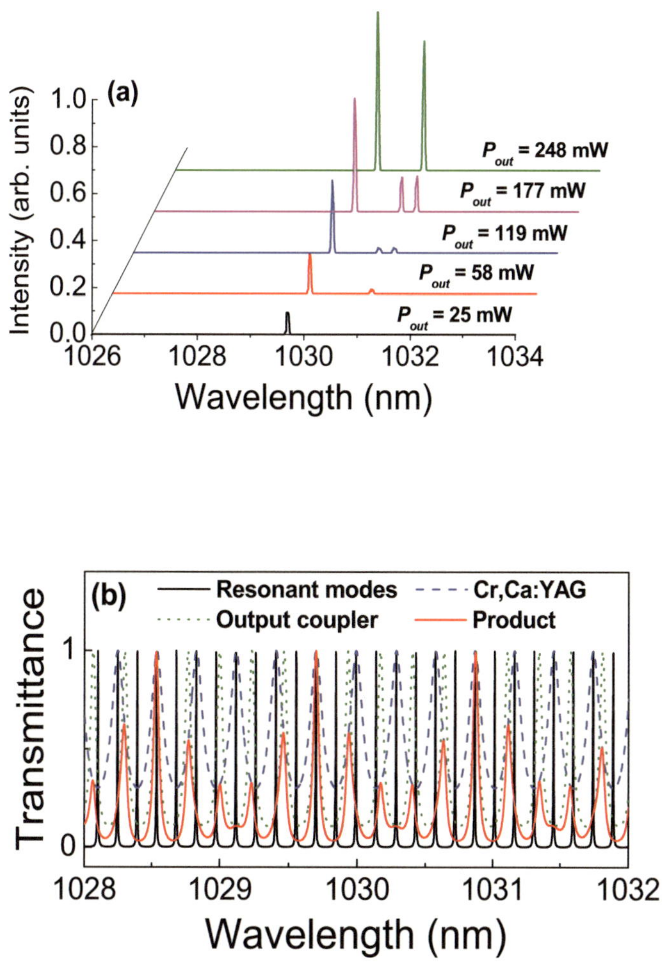

Fig. 12. (a) Laser emission spectra under different average output powers in passively Q-switched Yb:YAG/Cr,Ca:YAG all-ceramic microchip laser; (b) transmittance curves of 1-mm-thick Cr,Ca:YAG, 1.5-mm-thick fused silica output coupler, and their transmittance product. Resonant modes are also plotted for illustration.

Owing to the broad emission spectrum of the Yb:YAG materials around 1.03 μm (about 10 nm in FWHM), many longitudinal modes can be excited even for a 1-mm-thick Yb:YAG crystal. Microchip cw Yb:YAG lasers operate in a multi-longitudinal-mode over the whole pump power region(Dong et al., 2006). However, single-longitudinal-mode oscillation around 1029.7 nm was observed in passively Q-switched Yb:YAG/Cr,Ca:YAG microchip lasers when the average output power was kept below 50 mW for different Yb:YAG/Cr,Ca:YAG combinations, the same as that for all-ceramics combinations(Dong et al., 2007). Above this value, the laser exhibited two-mode oscillation and three-mode oscillation. A typical example of single-longitudinal-mode and multi-longitudinal-mode oscillations of passively Q-switched Yb:YAG/Cr,Ca:YAG all-ceramic microchip laser under different average output power levels is shown in Fig. 12(a). The separation between first and second modes was measured to be 1.16 nm, which is eight times wider than the free spectral range between the longitudinal modes (0.146 nm) in the laser cavity filled with gain medium predicted by(Koechner 1999) $\Delta\lambda_c = \lambda^2/2L_c$, where L_c is the optical length of the resonator and λ is the laser wavelength. The separation between second and third modes was measured to be 0.3 nm, which is twice of that determined by the laser cavity. The potential output longitudinal modes were selected by the combined etalon effect of the 1-mm-thick Cr,Ca:YAG as an intracavity etalon and 1.5-mm-thick fused silica output coupler as a resonant reflector(Koechner 1999). Fig. 12(b) shows the possible selected modes by the combining effect of 1-mm-thick Cr^{4+}:YAG and 1.5-mm-thick fused silica. The resonant modes, eight times of free spectral range (0.146 nm) away from the main mode centered at 1029.7 nm, will oscillate preferably because the wavelengths of these modes are very close to the high transmittance of the combined transmittance product. The resonant mode will oscillate at 1030.87 nm due to the asymmetric gain profile centered at 1029.7 nm of Yb:YAG. At high pump power levels, besides the oscillation of the main mode depleting the inversion population and suppressing the oscillation of the resonant modes close to it, the local temperature rise induced by the pump power will change the transmittance of the etalons. The relative gain and loss for different resonant modes will vary and determine the appearance of the third mode and elimination of the second mode. The linewidth of each mode was less than 0.02 nm, limited by the resolution of optical spectra analyzer. The central wavelength of 1029.7 nm shifts to longer wavelength with pump power, which is caused by the temperature dependent emission spectrum of Yb:YAG crystal(Dong et al., 2003). Therefore, stable single-longitudinal-mode oscillation can be maintained by increasing pump beam diameter incident on the laser medium at higher pump power.

The polarization states of passively Q-switched Yb:YAG/Cr,Ca:YAG microchip lasers have great effect on the characteristics of the output pulses. Fig. 13 shows the pulse characteristics (pulse repetition rate, pulse width, pulse energy and pulse peak power) of passively Q-switched Yb:YAG/Cr,Ca:YAG microchip lasers as a function of absorbed pump power. For all four combinations of Yb:YAG and Cr,Ca:YAG, the repetition rate of passively Q-switched laser increases linearly with the absorbed pump power. Pulse width (FWHM) of passively Q-switched Yb:YAG/Cr,Ca:YAG microchip lasers decreases with absorbed pump power at low pump power levels and tends to keep constant at high pump power levels. The shortest pulse width of 277 ps was achieved with C4 combination. Pulse widths for the four combinations are in the order of C4 < C1 < C2 < C3. Pulse energy increases with absorbed pump power and tends to keep constant at high pump power levels. The highest pulse energy was achieved with C1 combination. The pulse energy for the four combinations are in the order of C1 > C4 > C2 > C3. Peak power of passively Q-switched Yb:YAG/Cr,Ca:YAG microchip lasers exhibits the same tendency as those of pulse energy: C1 > C4 > C2 > C3 for the four combinations. Therefore, the overall best laser performance (highest peak power) in passively Q-switched Yb:YAG/Cr,Ca:YAG microchip lasers achieved by using C1 combination. Linearly polarization operation of passively Q-switched all-crystals lasers is more favorable for laser performance. The combination of Yb:YAG/Cr,Ca:YAG microchip lasers with Cr,Ca:YAG crystal as saturable absorber, C4, has better laser pulse characteristics than those of all-ceramics combination, C2 and combination C3. Although linearly polarized state was achieved in combination C3 with Yb:YAG crystal, the linearly polarized states was attributed to the linearly polarization of Yb:YAG crystal, not from the Cr,Ca:YAG ceramic. The contribution of the linearly polarized state from Yb:YAG crystal in C3 combination is less than that from the nonlinear anisotropic absorption of Cr,Ca:YAG crystal, therefore, therefore, the laser performance of combination C3 is less than those of combination C1 and C4. The effect of depolarization effect on the polarization states observed in Yb:YAG crystal (Dong et al., 2008) may be another cause to less efficient laser operation in C3 combination at high pump power levels.

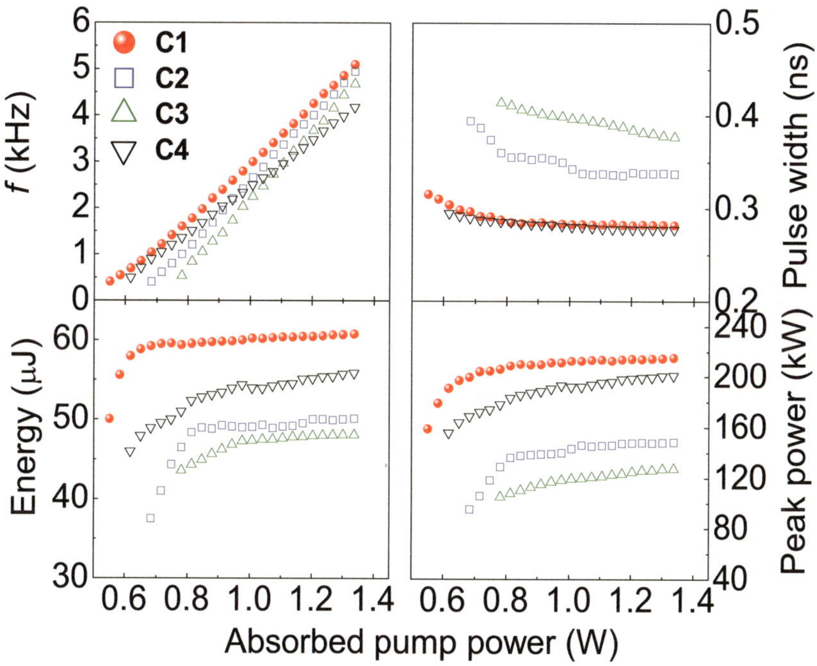

Fig. 13. Pulse characteristics (repetition rate, pulse width, pulse energy and peak power) of passively Q-switched Yb:YAG/Cr,Ca:YAG microchip lasers as a function of absorbed pump power for different combinations of Yb:YAG and Cr,Ca:YAG.

4. Conclusions

In conclusion, systematic comparison of laser performance was done for Yb:YAG ceramics and single-crystals doped with different concentrations. Although the pump power thresholds of Yb:YAG crystals were higher than their ceramics counterparts due to the pump configuration, the efficient laser operation was obtained by using both Yb;YAG ceramics and single-crystals. The laser performance of 1-mm-thick Yb:YAG ceramics and crystals becomes worse with Yb concentration under present miniature laser configuration. However, the laser performance of Yb:YAG crystals is more sensitive to the Yb concentrations, while the laser performance of Yb:YAG ceramics is less sensitive to the Yb concentrations. The laser performance of low doping Yb:YAG ceramics is worse than those obtaining from Yb:YAG singly crystals. The laser performance of 20 at.% Yb:YAG ceramics is better than its counterpart single crystal. Both Yb:YAG ceramics and crystals miniature lasers oscillate at multi-longitudinal modes, the number of longitudinal-mode increases with absorbed pump power. Strong mode competition and mode hopping were observed in these Yb:YAG lasers. The strong reabsorption and gain curve change under high intracavity laser intensity play important roles on the red-shift of the output laser wavelength. High beam quality lasers with M^2 less than 1.1 were achieved by adopting Yb:YAG ceramics and crystals as gain media. Heavy-doped Yb:YAG ceramic will be a potential candidate for microchip lasers by optimizing the thickness and Yb^{3+} concentration.

Random polarized oscillation was observed in passively Q-switched Yb:YAG/Cr,Ca:YAG all-ceramic microchip laser while linearly polarized oscillations were observed with at lease one crystal in the Yb:YAG/Cr,Ca:YAG combinations. The polarization states in passively Q-switched Yb:YAG/Cr,Ca:YAG microchip lasers show that the linearly polarization states in passively Q-switched laser are not only resulted from the anisotropic saturation absorption of Cr,Ca:YAG crystal, but also from linearly polarization states of Yb:YAG crystal. High peak power pulses with sub-nanosecond pulse-width and nearly diffraction-limited beam quality were obtained in these lasers. The best laser performance was achieved by using Yb:YAG crystal as gain medium and Cr,Ca:YAG crystal as saturable absorber because of the enhancement of linearly polarized state due to the crystalline-orientation selected polarized states of Yb:YAG crystal and linearly polarized oscillation of Cr,Ca:YAG crystal under high intracavity laser intensity. Other combinations of Yb:YAG and Cr,Ca:YAG have less efficient linearly polarized laser oscillation and also affect the laser performance of passively Q-switched Yb:YAG/Cr,Ca:YAG microchip lasers.

5. References

Barabanenkov, Y. N., Ivanov, S. N., Taranov, A. V., Khazanov, E. N., Yagi, H., Yanagitani, T., Takaichi, K., Lu, J., Bisson, J. F., Shirakawa, A., Ueda, K. & Kaminskii, A. A. (2004). Nonequilibrium acoustic phonons in $Y_3Al_5O_{12}$-based nanocrystalline ceramics. *JETP Lett.* Vol. 79, No. 7, (342 - 345).

Bisson, J., Yagi, H., Yanagitani, T., Kaminskii, A., Barabanenkov, Y. N. & Ueda, K. (2007). Influence of the grain boundaries on the heat transfer in laser ceramics. *Opt. Rev.* Vol. 14, No. 1, (1 - 13).

Bogomolova, G. A., Vylegzhanin, D. N. & Kaminskii, A. A. (1976). Spectral and lasing investigations of garnets with Yb^{3+} ions. *Sov. Phys. JETP* Vol. 42, No. 3, (440 - 446).

Bouwmans, G., Segard, B. & Glorieux, P. (2001). Polarisation dynamics of monomode Nd^{3+}:YAG lasers with Cr^{4+} saturable absorber: influence of the pump polarisation. *Opt. Commun.* Vol. 196, No. 1-6, (257 - 268).

Brauch, U., Giesen, A., Karszewski, M., Stewen, C. & Voss, A. (1995). Multiwatt diode-pumped Yb:YAG thin disk laser continuously tunable between 1018 and 1053 nm. *Opt. Lett.* Vol. 20, No. 7, (713 - 715).

Bruesselbach, H. W., Sumida, D. S., Reeder, R. A. & Byren, R. W. (1997). Low-heat high-power scaling using InGaAs-diode-pumped Yb:YAG lasers *IEEE J. Sel. Top. Quantum Electron.* Vol. 3, No. 1, (105 - 116).

Degnan, J. J. (1995). Optimization of passively Q-switched lasers. *IEEE J. Quantum Electron.* Vol. 31, No. 11, (1890 - 1901).

Dobrzycki, L., Bulska, E., Pawlak, D. A., Frukacz, Z. & Wozniak, K. (2004). Structure of YAG crystals doped/substituted with erbium and ytterbium. *Inorg. Chem.* Vol. 43, No. 24, (7656 - 7664).

Dong, J., Bass, M., Mao, Y., Deng, P. & Gan, F. (2003). Dependence of the Yb^{3+} emission cross section and lifetime on the temperature and concentration in ytterbium aluminum garnet. *J. Opt. Soc. Am. B* Vol. 20, No. 9, (1975 - 1979).

Dong, J., Deng, P., Lu, Y., Zhang, Y., Liu, Y., Xu, J. & Chen, W. (2000). LD pumped Cr^{4+},Nd^{3+}:YAG with self-Q-switched laser output of 1.4 W. *Opt. Lett.* Vol. 25, No. 15, (1101 - 1103).

Dong, J., Shirakawa, A., Takaichi, K., Ueda, K., Yagi, H., Yanagitani, T. & Kaminskii, A. A. (2006). All ceramic passively Q-switched Yb:YAG/Cr^{4+}:YAG microchip laser. *Electron. Lett.* Vol. 42, No. 20, (1154 - 1156).

Dong, J., Shirakawa, A. & Ueda, K. (2008). A crystalline-orientation self-selected linearly polarized Yb:$Y_3Al_5O_{12}$ microchip laser. *Appl. Phys. Lett.* Vol. 93, No. 10, (101105).

Dong, J., Shirakawa, A., Ueda, K. & Kaminskii, A. A. (2007). Effect of ytterbium concentration on CW Yb:YAG microchip laser performance at ambient temperature Part I: Experiments. *Appl. Phys. B* Vol. 89, No. 2-3, (359 - 365).

Dong, J., Shirakawa, A., Ueda, K., Xu, J. & Deng, P. (2006). Efficient laser oscillation of Yb:$Y_3Al_5O_{12}$ single crystal grown by temperature gradient technique. *Appl. Phys. Lett.* Vol. 88, No. 16, (161115).

Dong, J., Shirakawa, A., Ueda, K., Yagi, H., Yanagitani, T. & Kaminskii, A. A. (2006). Efficient Yb^{3+}:$Y_3Al_5O_{12}$ ceramic microchip lasers. *Appl. Phys. Lett.* Vol. 89, No. 9, (091114).

Dong, J., Shirakawa, A., Ueda, K., Yagi, H., Yanagitani, T. & Kaminskii, A. A. (2007). Laser-diode pumped heavy doped Yb:YAG ceramic lasers. *Opt. Lett.* Vol. 32, No. 13, (1890 -1892).

Dong, J., Shirakawa, A., Ueda, K., Yagi, H., Yanagitani, T. & Kaminskii, A. A. (2007). Near-diffraction-limited passively Q-switched Yb:$Y_3Al_5O_{12}$ ceramic lasers with peak power > 150 kW. *Appl. Phys. Lett.* Vol. 90, No. 13, (131105).

Dong, J., Shirakawa, A., Ueda, K., Yagi, H., Yanagitani, T. & Kaminskii, A. A. (2007). Ytterbium and chromium doped composite $Y_3Al_5O_{12}$ ceramics self-Q-switched laser. *Appl. Phys. Lett.* Vol. 90, No. 19, (191106).

Dong, J. & Ueda, K. (2005). Temperature-tuning Yb:YAG microchip lasers. *Laser Phys. Lett.* Vol. 2, No. 9, (429 - 436).

Dong, J., Ueda, K., Shirakawa, A., Yagi, H., Yanagitani, T. & Kaminskii, A. A. (2007). Composite Yb:YAG/Cr^{4+}:YAG ceramics picosecond microchip lasers. *Opt. Express* Vol. 15, No. 22, (14516 - 14523).

Eilers, H., Hoffman, K. R., Dennis, W. M., Jacobsen, S. M. & Yen, W. M. (1992). Saturation of 1.064 μm absorption in Cr,Ca:$Y_3Al_5O_{12}$ crystals. *Appl. Phys. Lett.* Vol. 61, No. 25, (2958 -2960).

Fan, T. Y. (1993). Heat generation in Nd:YAG and Yb:YAG. *IEEE J. Quantum Electron.* Vol. 29, No. 6, (1457 -1459).

Fan, T. Y. & Ochoa, J. (1995). Tunable single-frequency Yb:YAG laser with 1-W output power using twisted-mode technique. *IEEE Photon. Technol. Lett.* Vol. 7, No. 10, (1137 -1138).

Feng, Y., Lu, J., Takaichi, K., Ueda, K., Yagi, H., Yanagitani, T. & Kaminskii, A. A. (2004). Passively Q-switched ceramic Nd^{3+}:YAG/Cr^{4+}:YAG lasers. *Appl. Opt.* Vol. 43, No. 14, (2944 - 2947).

Giesen, A., Hugel, H., Voss, A., Wittig, K., Brauch, U. & Opower, H. (1994). Scalable concept for diode-pumped high-power solid-state lasers. *Appl. Phys. B* Vol. 58, No. 5, (365 - 372).

Honea, E. C., Beach, R. J., S. C. Mitchell, Sidmore, J. A., M. A. Emanuel, Sutton, S. B., Payne, S. A., Avizonis, P. V., Monroe, R. S. & Harris, D. (2000). High-power dual-rod Yb:YAG laser. *Opt. Lett.* Vol. 25, No. 11, (805 - 807).

Kir'yanov, A. V., Aboites, V. & Ii'ichev, N. N. (1999). A polarisation-bistable noedymium laser with a Cr^{4+}:YAG passive switch under the weak resonant signal control. *Opt. Commun.* Vol. 169, No. 1-6, (309 - 316).

Koechner, W. (1999). *Solid State Laser Engineering*. Berlin, Springer-Verlag.

Kong, J., Tang, D. Y., Lu, J. & Ueda, K. (2004). Random-wavelength solid-state laser. *Opt. Lett.* Vol. 29, No. 1, (65 - 67).

Krupke, W. F. (2000). Ytterbium solid-state lasers - The first decade. *IEEE J. Sel. Top. Quantum Electron.* Vol. 6, No. 6, (1287 - 1296).

Lacovara, P., Choi, H. K., Wang, C. A., Aggarwal, R. L. & Fan, T. Y. (1991). Room-temperature diode-pumped Yb:YAG laser. *Opt. Lett.* Vol. 16, No. 14, (1089 - 1091).

Lagatsky, A. A., Abdolvand, A. & Kuleshov, N. V. (2000). Passive Q-switching and self-frequency Raman conversion in a diode-pumped Yb:KGd$(WO_4)_2$ laser. *Opt. Lett.* Vol. 25, No. 9, (616 - 618).

Li, S., Zhou, S., Wang, P., Chen, Y. C. & Lee, K. K. (1993). Sefl-Q-switched diode-end-pumped Cr,Nd:YAG laser with polarized output. *Opt. Lett.* Vol. 18, No. 3, (203 - 204).

Lu, J., Prabhu, M., Song, J., Li, C., Xu, J., Ueda, K., Kaminskii, A. A., Yagi, H. & Yanagitani, T. (2000). Optical properties and highly efficient laser oscillation of Nd:YAG ceramics. *Appl. Phys. B* Vol. 71, No. 4, (469 - 473).

Lu, J., Prabhu, M., Song, J., Li, C., Xu, J., Ueda, K., Yagi, H., Yanagitani, T. & Kaminskii, A. A. (2001). Highly efficient Nd:$Y_3Al_5O_{12}$ ceramic laser. *Jpn. J. Appl. Phys.* Vol. 40, No. 6A, (L552 - L554).

Lu, J., Ueda, K., Yagi, H., Yanagitani, T., Akiyama, Y. & Kaminskii, A. A. (2002). Neodymium-doped yttrium aluminum garnet ($Y_3Al_5O_{12}$) nanocrystalline ceramics - a new generation of solid-state laser and optical materials. *J. Alloys Compd.* Vol. 341, No. 1-2, (220 - 225).

M. Ito, C. Goutaudier, Y. Guyot, K. Lebbou, T. Fukuda & Boulon, G. (2004). Crystal growth, Yb^{3+} spectroscopy, concentration quenching analysis and potentiality of laser emission in $Ca_{1-x}Yb_xF_{2+x}$. *J. Phys.: Condens. Matter* Vol. 16, No. 8, (1501 - 1521).

McKay, A., Dawes, J. M. & Park, J. (2007). Polarisation-mode coupling in (100)-cut Nd:YAG. *Opt. Express* Vol. 15, No. 25, (16342 - 16347).

Nakamura, S., Yoshioka, H., Matsubara, Y., Ogawa, T. & Wada, S. (2008). Efficient tunable Yb:YAG ceramic laser. *Opt. Commun.* Vol. 281, No. 17, (4411 - 4414).

Okhrimchuk, A. G. & Shestakov, A. V. (1994). Performance of YAG: Cr^{4+} laser crystal. *Opt. Mater.* Vol. 3, No. 1, (1 - 13).

Patel, F. D., Honea, E. C., Speth, J., Payne, S. A., Hutcheson, R. & Equall, R. (2001). Laser demonstration of $Yb_3Al_5O_{12}$ (YAG) and materials properties of highly doped Yb:YAG. *IEEE J. Quantum Electron.* Vol. 37, No. 1, (135 - 144).

Qiu, H., Yang, P., Dong, J., Deng, P., Xu, J. & Chen, W. (2002). The influence of Yb concentration on laser crystal Yb:YAG. *Mater. Lett.* Vol. 55, No. 1-2, (1 - 7).

Rutherford, T. S., Tulloch, W. M., Gustafson, E. K. & Byer, R. L. (2000). Edge-pumped quasi-three-level slab lasers: design and power scaling. *IEEE J. Quantum Electron.* Vol. 36, No. 2, (205 - 219).

Spuhler, G. J., Paschotta, R., Kullberg, M. P., Graf, M., Moser, M., Mix, E., Huber, G., Harder, C. & Keller, U. (2001). A passively Q-switched Yb:YAG microchip laser. *Appl. Phys. B: Lasers Opt.* Vol. 72, No. 3, (285 - 287).

Stewen, C., Contag, K., Larionov, M., Giessen, A. & Hugel, H. (2000). A 1-kW CW Thin Disc laser. *IEEE J. Sel. Top. Quantum Electron.* Vol. 6, No. 4, (650 -657).

Sumida, D. S. & Fan, T. Y. (1994). Effect of radiation trapping on fluorescence lifetime and emission cross section measurements in solid-state laser media *Opt. Lett.* Vol. 19, No. 17, (1343 - 1345).

Taira, T., Saikawa, J., Kobayashi, T. & Byer, R. L. (1997). Diode-pumped tunable Yb:YAG miniature lasers at room temperature: modeling and experiment. *IEEE J. Sel. Top. Quantum Electron.* Vol. 3, No. 1, (100 - 104).

Takaichi, K., Lu, J., Murai, T., Uematsu, T., Shirakawa, A., Ueda, K., Yagi, H., Yanagitani, T. & Kaminskii, A. A. (2002). Chromium-doped $Y_3Al_5O_{12}$ ceramics - a novel saturable absorber for passively self-Q-switched 1-μm solid-state lasers. *Jpn. J. Appl. Phys.* Vol. 41, No. 2A, (L96 - L98).

Takaichi, T., Yagi, H., Lu, J., Shirakawa, A., Ueda, K., Yanagitani, T. & Kaminskii, A. A. (2003). Yb^{3+}-doped $Y_3Al_5O_{12}$ ceramics - A new solid-state laser material. *Phys. Status Solidi (a)* Vol. 200, No. 1, (R5 - R7).

Xu, X., Zhao, Z., Song, P., Jiang, B., Zhou, G., Xu, J., Deng, P., Bourdet, G., Chanteloup, J. C., Zou, J. & Fulop, A. (2005). Upconversion luminescence in Yb^{3+} - doped yttrium aluminum garnets. *Physica B* Vol. 357, No. 3 -4, (365 - 369).

Xu, X., Zhao, Z., Zhao, G., Song, P., Xu, J. & Deng, P. (2003). Comparison of Yb:YAG crystals grown by CZ and TGT method. *J. Crystal Growth* Vol. 257, No. 3-4, (297 - 300).

Yagi, H., Yanagitani, T., Yoshida, K., Nakatsuka, M. & Ueda, K. (2006). Highly efficient flashlamp-pumped Cr^{3+} and Nd^{3+} codoped $Y_3Al_5O_{12}$ ceramic laser. *Jpn. J. Appl. Phys.* Vol. 45, No. 1A, (133 - 135).

Yanagitani, T., Yagi, H. & Hiro, Y. (1998). Production of yttrium aluminium garnet fine powders for transparent YAG ceramic. Japan Patent 10-101411.

Yang, P., Deng, P. & Yin, Z. (2002). Concentration quenching in Yb:YAG. *J. Lumin.* Vol. 97, No. 1, (51 - 54).

Yankov, P. (1994). Cr^{4+}:YAG Q-switching of Nd:host laser oscillators. *J. Phys. D* Vol. 27, No. 6, (1118 - 1120).

Yin, H., Deng, P. & Gan, F. (1998). Defects in YAG:Yb crystal. *J. Appl. Phys.* Vol. 83, No. 7, (3825 - 3828).

Yoshino, T. & Kobyashi, Y. (1999). Temperature characteristics and stabilization of orthogonal polarization two-frequency Nd^{3+}:YAG microchip lasers. *Appl. Opt.* Vol. 38, No. 15, (3266 - 3270).

Zayhowski, J. J. (2000). Passively Q-switched Nd:YAG microchip lasers and applications. *J. Alloys Comp.* Vol. 303-304, No. (393 - 400).

Zayhowski, J. J. & Dill III, C. (1994). Diode-pumped passively Q-switched picosecond microchip lasers. *Opt. Lett.* Vol. 19, No. 18, (1427 -1429).

2

New Infrared Luminescence from Bi-doped Glasses

Yasushi Fujimoto
Institute of Laser engineering, Osaka University
Japan

1. Introduction

Infrared luminescent materials are widely used as laser media, for example, Nd:YAG and Er-doped silica fibers. In the infrared region, luminescence offers many advantages as laser media due to the variety of optics and excitation sources, such as semiconductor lasers and flash lamps. A frequency-doubled laser-beam with a nonlinear crystal can also obtained in the visible region. Therefore, infrared light source materials are very useful for scientific research and industrial uses.

Fujimoto et al. discovered a new infrared luminescent from bismuth-doped silica glass (BiSG)(Fujimoto & Nakatsuka 2001). Wide luminescent exists between 1000 and 1600 nm, and absorption is in the visible region. The lifetime is 630 μs at room temperature. Since this silica glass-based material has luminescence around 1.3 μm of zero-dispersion wavelength, it can be used as a core material of an optical fiber amplifier for telecommunication in the 1.3-μm range. Fujimoto et al. also demonstrated an optical amplification at 1.3 μm with 0.8-μm excitation(Fujimoto & Nakatsuka 2003). Therefore, BiSG is expected to be a promising substitution for Pr^{3+}:ZBLAN fiber or Raman amplifiers in the 1.3-μm region.

The spectroscopic properties of BiSG are different from previously reported Bi luminescent material(Fujimoto & Nakatsuka 2001), because the previous Bi-doped glasses or crystals with low Bi_2O_3 concentration showed that luminescences exist in the visible region (400 ~ 600 nm) and that absorptions are in the ultraviolet region. The lifetimes (~4 μs) are much shorter than those of BiSG at room temperature. Such luminescences have been reported to originate from the Bi^{3+} ion in the media(M. J. Weber & Monchamp 1973; Parke & Webb 1973; van der Steen, van Hesteren et al. 1981). BiSG is also different from the Bi_2O_3-rich glass whose Bi_2O_3 concentration ranges from 18.7 to 42.5 mol%(Sugimoto, Kanbara et al. 1996) because the glass has no absorption in the visible region. Then the investigation of new glass compositions with Bi luminescence is also progressing vigorously in a past decade. Although the first discovered glass was a silica-based material with a SiO_2 of 97.5 mol%(Fujimoto & Nakatsuka 2001), many glass compositions were tested and shown to effectively generate Bi luminescence, such as silicate(Peng, Qiu et al. 2005; Ren, Yang et al. 2006; Suzuki & Ohishi 2006; Arai, Suzuki et al. 2007; Murata & Mouri 2007; Peng, Chen et al. 2007; Ren, Qiu et al. 2007; Zhou, Feng et al. 2007; Peng, Wu et al. 2008), germanate(Peng, Qiu et al. 2004; Meng, Qiu et al. 2005; Peng, Meng et al. 2005; Peng, Qiu et al. 2005; Peng, Wang et al. 2005; Wang & Xia 2006; Xia & Wang 2006; Murata & Mouri 2007; Ren, Wu et al. 2007;

Ren, Dong et al. 2007; Ren, Qiao et al. 2007; Ren, Qiu et al. 2007; Ren, Qiu et al. 2007; Peng, Wu et al. 2008; Qiu, Peng et al. 2008), borate(Meng, Qiu et al. 2005; Murata & Mouri 2007; Qiu, Peng et al. 2008), germanosilicate(Ren, Dong et al. 2007; Ren, Dong et al. 2008), and phosphate(Meng, Qiu et al. 2005; Qiu, Peng et al. 2008) glasses.

In this chapter, the author will introduce the basic properties of Bi doped silica glass (BiSG), such as a phase diagram and spectroscopic properties, and then mainly talk about the origine of luminescent center.

2. Basic properties of BiSG

2.1 Phase diagram

A phase diagram of BiSG was obtained for the Bi_2O_3-Al_2O_3-SiO_2 glass system. Silica powder (Aerosil 50; SiO_2, 99.8%), bismuth-oxide (Kojundo Chemical Lab. Co., Ltd.; α-Bi_2O_3, 99.99%), and aluminum oxide (Kojundo Chemical Lab. Co., Ltd.; α-Al_2O_3, 99.99%) were used as reagents. A properly mixed powder was inserted into a silica tube with an inner/outer diameter of 1/3 mmϕ, and then the tube was heated by a mixed gas burner of natural gas and oxygen. After heating, the powder was checked for a distinctive glassy wetting. If it had such wetting, the composition was determined to be the glassy phase in the Bi_2O_3-Al_2O_3-SiO_2 glass system.

The measured phase diagram of the Bi_2O_3-Al_2O_3-SiO_2 glass system is shown in Fig. 1. The circle, triangle, and cross points show the glassy wetting states (pink to reddish-brown), unclear glassy states (pink to reddish-brown), and no glassy state (reddish-brown), respectively. This phase diagram shows a tendency where the glass whose Al_2O_3 content is larger than Bi_2O_3 is well glassified above a 90% SiO_2 concentration. On the other hand, all the samples are glassified at a 80% SiO_2 concentration without dependence on the Bi_2O_3 or Al_2O_3 content. The infrared luminescent spectrum is observed until 10 mol% of Bi_2O_3 concentration.

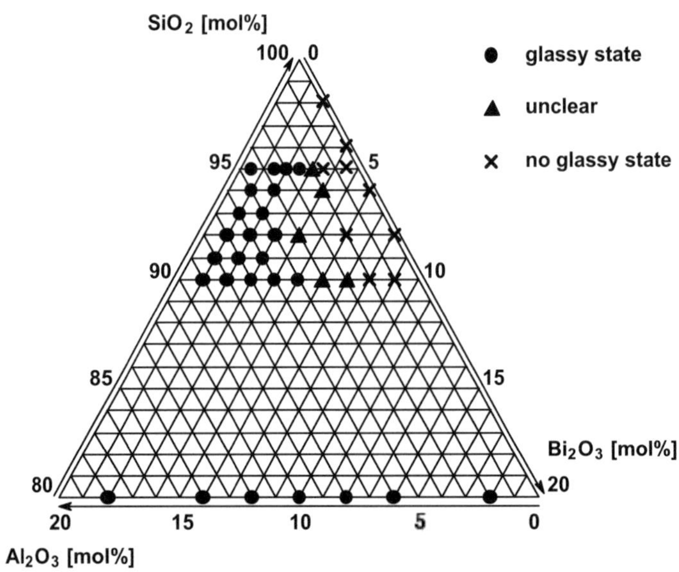

Fig. 1. Measured phase diagram of Bi_2O_3-Al_2O_3-SiO_2 system.

2.2 Influence of additives on luminescent intensitys

The Bi_2O_3-Al_2O_3-SiO_2 glass system is affected by alkaline metal additive. Figure 2 shows an effect of Li_2O additive in BiSG. The glass composition is Bi_2O_3(1.0 mol%)-Al_2O_3(7.0 mol%)-SiO_2(92-x mol%)-Li_2O(x mol%) glass system. The luminescent intensity decreases with increasing Li_2O additive drastically (Seo, Fujimoto et al. 2006). Therefore, in order to get luminescent center, we should carefully choose the host composition.

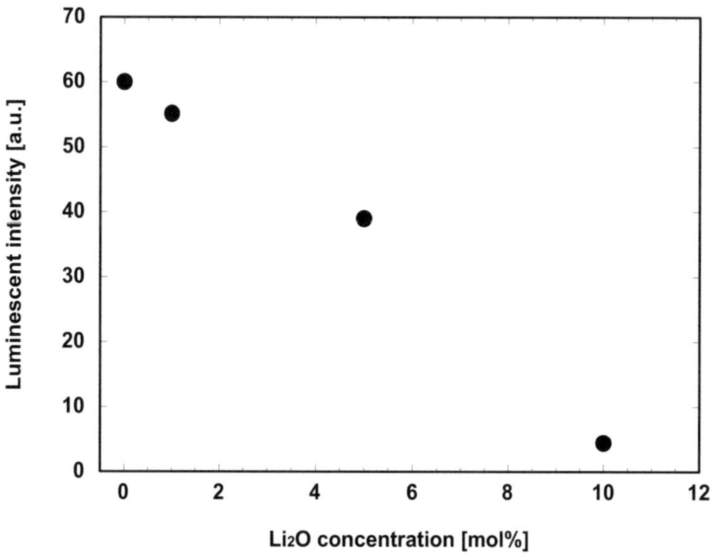

Fig. 2. Luminescent intensity dependence on Li_2O concentration for Bi_2O_3(1.0 mol%)-Al_2O_3(7.0 mol%)-SiO_2(92-x mol%)-Li_2O(x mol%) glass system.

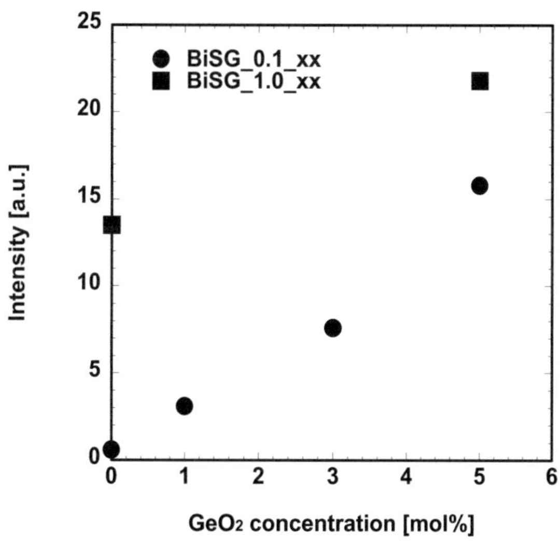

Fig. 3. Dependence of lumienscent intensity on GeO_2 additive. The filled circle denotes BiSG with 0.1 mol% Bi_2O_3 concentration, while the filled square represents that of 1.0 mol% Bi_2O_3 concentration. The nomenclature of the samples is as follows: for example, the sample name "BiSG_1.0_5.0" means that the composition is Bi_2O_3 = 1.0 mol%, GeO_2 = 5.0 mol%.

On the other hand, GeO_2 is the one of attractive additive to increase luminescent intensity of BiSG (Fujimoto, Hirata et al. 2007). Figure 3 shows the dependence of luminescent intensity on GeO_2 concentration. All the samples contain 7.0 mol% of Al2O3. The nomenclature of the samples is as follows: for example, the sample name "BiSG_1.0_5.0" means that the composition is Bi_2O_3 = 1.0 mol%, GeO_2 = 5.0 mol%. The luminescent intensity of BiSG grows with increasing GeO_2 concentration almost linearly for BiSG_0.1_xx samples, and the luminescent intensity of BiSG_0.1_5.0 is 26.3 times larger than that of BiSG_0.1_0.0 only because of the GeO_2 additive. The increment effect on BiSG with GeO_2 additive is also observed for a Bi_2O_3 concentration of 1.0 mol% (BiSG_1.0_5.0), but it is less remarkable than for the lower Bi_2O_3 concentration of 0.1 mol%. Furthermore, the increment is only about 1.5 times larger than for BiSG without GeO_2 (BiSG_1.0_0.0). On the other hand, the luminescent intensity of BiSG_0.1_5.0 is almost same as that of BiSG_1.0_0.0. According to these data, the GeO_2 additive effectively increases the luminescent intensity of BiSG, and this effect is especially remarkable at low Bi_2O_3 concentrations.

2.3 Spectroscopic properties

The typical spectroscopic properties of BiSG are shown in Fig. 4. Five main absorption bands exist between 250 and 2000 nm: 300 (A), 500 (B), 700 (C), 800 (D), and 1000 nm (E) (Fig. 5(a)). It is considered that absorption band A is mainly derived from O^{2-} in the base glass(Scholze 1991). The other four absorptions (B~E) are connected to the infrared emissions shown in Fig. 5(b). Each luminescent peak wavelength and full width at half maximum (FWHM) were recognized as 1182 and 300 nm at 500-nm excitation, 1144 and 192 nm at 700 nm, 1265 and 336 nm at 800 nm, and 1130 and 188 nm at 1000 nm. A similarity of the luminescent spectra can be seen between 700- and 1000-nm excitations. These spectroscopic properties are different from the Bi^{3+} ones(M. J. Weber & Monchamp 1973; Parke & Webb 1973; van der Steen, van Hesteren et al. 1981)(Table 1).

	Bi^{3+}:Glass*1 ($Na_2O \bullet P_2O_5$)	Bi^{3+}:Crystal*2 ($Bi_3Ge_4O_{12}$)	BiSG *3
Excitation wavelength [nm]	250	290	500~800
Luminescent wavelength [nm]	390	480	1000~1400
Lifetime [μs] (room temp.)	3.9	0.1	630

*1 (Parke & Webb 1973), *2 (M. J. Weber & Monchamp 1973), *3 (Fujimoto & Nakatsuka 2001)

Table 1. Differences of spectroscopic properties between BiSG and previously reported Bi^{3+} luminescent material.

3. The origin of luminescent center

After the discovery of a new infrared luminescent bismuth center, several research groups try to understand what is the origine of the luminescence. Although there have been many previous discussions on this distinctive luminescent center, it has remained unclear. Different authors tentatively assigned the different origins of this near infrared luminescence to the electronic transition derived from Bi^{5+}(Fujimoto & Nakatsuka 2001; Dvoyrin, Mashinsky et al. 2006; Fujimoto & Nakatsuka 2006; Wang & Xia 2006; Xia & Wang 2006; Ohkura, Fujimoto et al. 2007), Bi^{2+}(Ren, Qiu et al. 2007; Ren, Qiu et al. 2007; Ren, Qiu et al. 2007; Ren, Dong et al. 2008), and Bi^{+}(Meng, Qiu et al. 2005; Meng, Qiu et al. 2005;

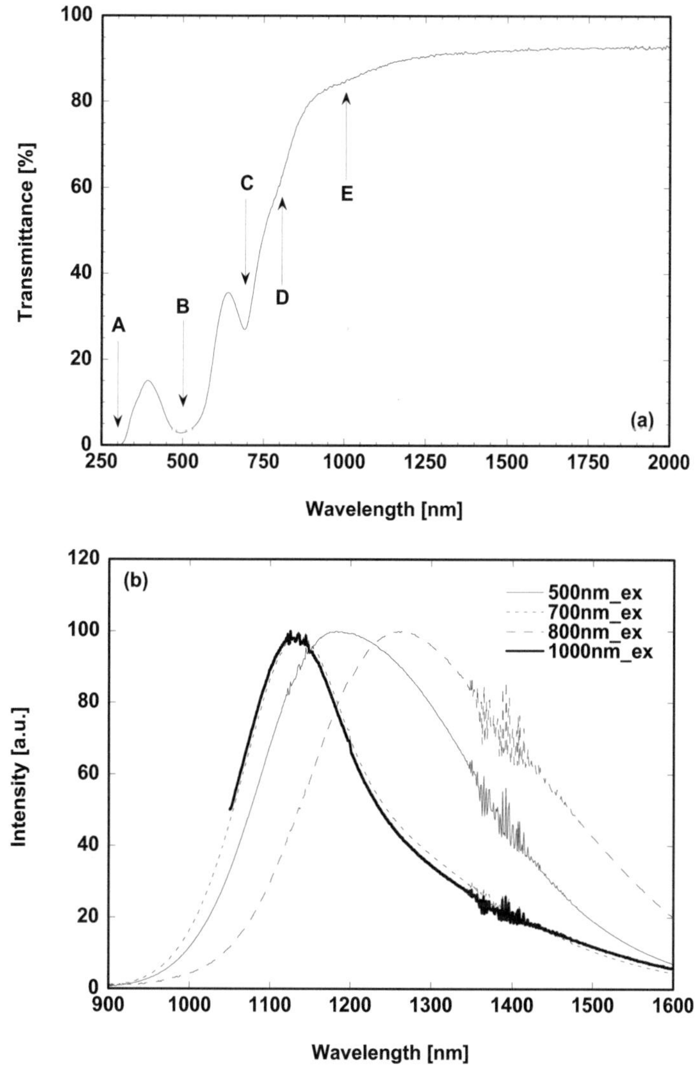

Fig. 4. Typical spectroscopic properties of BiSG (sample B1 in Table 2): (a) transmittance spectrum shows five main absorption bands between 250 and 2000 nm, including 300 nm (A), 500 nm (B), 700 nm (C), 800 nm (D), and 1000 nm (E), and (b) luminescent spectra excited by 500, 700, 800, and 1000 nm.

Dvoyrin, Mashinsky et al. 2006; Arai, Suzuki et al. 2007; Ren, Qiu et al. 2007; Ren, Qiu et al. 2007; Ren, Qiu et al. 2007; Yang, Chen et al. 2007; Zhou, Feng et al. 2007; Qiu, Peng et al. 2008; Ren, Dong et al. 2008; Truong, Bigot et al. 2008) or to that of BiO molecules(Ren, Yang et al. 2006; Murata & Mouri 2007; Peng, Chen et al. 2007; Peng, Wu et al. 2008) dispersed in the glass host. Since most of these proposals were however analogized from only spectroscopic mesurement analysis, uncertainty remains in the present discussions. In addition, even the same researchers offer several explanations, suggesting that spectroscpic analogy is insufficient to correctly understand the luminescent center or the luminescent mechanism.

Based on this viewpoint, our group discarded the spectroscopic analogical approach to understand the structure of the luminescent center or its mechanism and instead chose such instrumental analytical approaches as nuclear magnetic resonance (NMR), x-ray diffraction (XRD), electron spin resonance (ESR), x-ray photoelectron spectroscopy (XPS), and extended x-ray absorption fine structure (EXAFS). NMR data revealed the coordination structure of the specific ions in a material, ESR and XPS data provided the valence state of ions, and EXAFS offered the bonding distance between the specific and neighboring ions. Once these analytic instrumental data are integrated to build one possible figure of a luminescent center, then the luminescent mechanism will be discussed, because knowledge of the luminescent center is very useful for understanding the luminescent mechanism, which is obviously important for such optical applications as lasers or amplifiers.

In the following paragraphs, let's try to build up an image of the local structure of the Bi luminescent center based on the analytical data, such as, spectroscopic properties (SPCT), luminesent intensity (LMI), NMR, XRD, ESR, XPS, and the EXAFS data of BiSG in a simple Bi_2O_3-Al_2O_3-SiO_2 glass system without discrepancies. From this approach, we will reach one strongly possible local structure of a distinctive Bi luminescent center that consists of Bi and Al that exist closely to each other.

3.1 Sample preparation

In these analyses, composition of BiSG is very important, because Bi infrared luminescent center, as shown in section 2.2, is disappeared with increasing some additives in the host. This means that majority part of Bi ions in the host shifts to non-emissive site, thus even if we choose the multi-component glass system and analyze the Bi centers, we could not judge the results form an inherent luminescent center or the other non-emissive site, then we would go into the complicated maze. Therefore, the analyses should be carried out to concentrate on the simple three components system, Bi_2O_3-Al_2O_3-SiO_2 glass system.

The samples for the instrumental analyses were prepared as follows.

1. A-series: The ratio of Bi_2O_3 to Al_2O_3 was kept at 3/7, and the Bi_2O_3 concentration varied from 0.1, 0.3, 0.5, 1.0, and 3.0 mol%. These samples were called A1-A5, respectively. Their compositions are inside the glassy phase area in Fig. 1.
2. B-series: B1 was composed of Bi_2O_3 (1.0 mol%), Al_2O_3 (7.0 mol%), and SiO_2 (92.0 mol%). This was used for the measurement of the spectroscopic properties because Al_2O_3-rich BiSG includes few bubbles. B2 was composed of Bi_2O_3 (0.3 mol%), Al_2O_3 (2.3 mol%), and SiO_2 (97.4 mol%) and B3 of Bi_2O_3 (3.0 mol%), Al_2O_3 (7.0 mol%), and SiO_2 (90.0 mol%). They were used for the ESR measurements.
3. C-series: C1 was composed of Bi_2O_3 (1.0 mol%) and SiO_2 (99.0 mol%) without Al_2O_3, and C2 was composed of Al_2O_3 (2.3 mol%) and SiO_2 (97.7 mol%) without Bi_2O_3. C1 and C2 were dark brown and colorless, respectively. C1 had no glassy wetting, but C2 did.

The chemical composition of the investigated BiSG samples and the several standards are listed in Table 2. Silica powder (Aerosil 50; SiO_2, 99.8%), bismuth-oxide (Kojundo Chemical Lab. Co., Ltd.; α-Bi_2O_3, 99.99%), and aluminum oxide (Kojundo Chemical Lab. Co., Ltd.; α-Al_2O_3, 99.99%) were used as reagents to make the glass samples of the Bi_2O_3-Al_2O_3-SiO_2 glass system. The mixtures of these powders were melted in an aluminum crucible in an electric furnace at 1750°C in air, and the melts were annealed by natural cooling. All of these glasses were reddish-brown. $NaBiO_3$ (Kanto Chemical Co., Inc.; >80%), which was identified as $NaBiO_3 \bullet 2H_2O$ by XRD(PDF#30-1161 2000), α-Bi_2O_3, and Bi-metal (Kojundo Chemical Lab. Co., Ltd.; 99.9%) were used as standard materials for the analyses.

Sample	Bi_2O_3[mol%]	Al_2O_3[mol%]	SiO_2[mol%]	Measurements
<BiSG>				
A1	0.1	0.23	99.67	LMI, NMR, XRD
A2	0.3	0.7	99.0	LMI, NMR, XRD, EXAFS
A3	0.5	1.15	98.35	LMI, NMR, XRD, EXAFS
A4	1.0	2.3	96.7	LMI, NMR, XRD, EXAFS, XPS
A5, B3	3.0	7.0	90.0	LMI, NMR, ESR, XRD, EXAFS, XPS
B1	1.0	7.0	92.0	SPCT
B2	0.3	2.3	97.4	ESR
C1	1.0	0.0	99.0	LMI
C2	0.0	2.3	97.7	NMR, XRD
<Aluminium standard>				
α- Al_2O_3	0.0	100	0.0	NMR, XRD
<Bismuth standards>				
$NaBiO_3$	---	---	---	EXAFS, XPS
α- Bi_2O_3	100	0.0	0.0	EXAFS, XPS
Bi metal	---	---	---	EXAFS, XPS

LMI: Luminescent Intensity, NMR: ^{27}Al-NMR(Nuclear Magnetic Resonance), XRD: X-ray Diffraction, ESR: Electron Spin Resonance, XPS: X-ray Photoelectron Spectroscopy, EXAFS: Extended X-ray Absorption Fine Structure, SPCT: Spectroscopic Properties

Table 2. List of chemical composition and performed measurements for investigated Bi doped silica glasses and Al, Bi standards.

The prepared samples were evaluated by instrumental analyses, including spectroscopic measurements, NMR, XRD, ESR, XPS, and EXAFS. Luminescent intensity and spectra were measured by a spectrophotometer (JASCO Corporation; SS-25). Transmission spectra were obtained by a spectrophotometer (HITACHI; U-4100). The aluminum coordination state (ACS) in BiSG was determined by the ^{27}Al-NMR spectrum, and the ^{27}Al-NMR spectra were obtained on a nuclear magnetic resonance spectrometer (JEOL Ltd.; JNM-GSX-400). The XRD patterns were measured by an X-ray diffractometer (RIGAKU; RINT2500). The ESR signals were measured by an electron spin resonance spectrometer (JEOL Ltd.; JES-ME-2X) at both room and liquid nitrogen temperatures.

The XPS spectra were recorded on a SHIMADZU (KRATOS) AXIS 165 using Mg*K* radiation (1253.6 eV) with a beam size of 0.7 x 0.3 mm^2 and 50 eV analyzer pass energy. The operating vacuum pressure was maintained from 10^{-8} to 10^{-9} Torr. An electron beam was used to reduce the electrical charging during the measurements. Any binding energy shifts due to electrical charging were corrected by the C1*s* (284.6 eV) peak as an internal reference. 3-minute Ar-sputtering was used to remove the oxidized surface on the Bi-metal powder.

The Bismuth L_{III} XAFS spectra were measured in the transmission mode using a BL-7C beam line at the Photon Factory (PF). The ring energy was 2.5 GeV. The x-ray was monochromized by a Si(111) double crystal monochrometer. Two ionization chambers were used as detectors. A 170-mm length chamber was filled with Ar/N_2 (15%:85%) gas for incident x-ray energy (I_0), and the other was 310-mm long with Ar-100% gas for transmitted x-ray energy (I). The EXAFS data of the Bi L_{III} edge (13426.5 eV) were collected between 12926 and 14526 eV with 481 energy points. Data analysis was carried out on *UWXAFS*. The back-scattering amplitude and the phase shift were theoretically calculated using *FEFF 8.2* code. The Debye-Waller factor was estimated by the Debye code implemented in *FEFF 8.2* based on Raman spectroscopy results(Narang, Patel et al. 1994).

3.3 Luminescent intensity

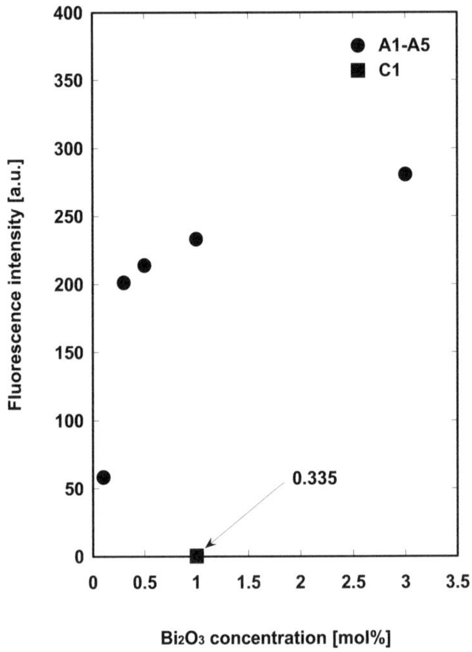

Fig. 5. Dependence of luminescent intensity (LMI) on Bi_2O_3 concentration detected at 1120-nm luminescence with 500-nm excitation.

The dependence of luminescent intensity (LMI) on Bi_2O_3 concentration is illustrated in Fig. 5. The measured samples were A-series (A1~A5) and C1. The excitation and detection wavelengths of the luminescence were at 500 and 1120 nm, respectively. The luminescent intensity nonlinearly increased with increased Bi_2O_3 concentration. At a 1.0 mol% of Bi_2O_3 concentration, the luminescent intensity from A4, which includes 2.3 mol% of Al_2O_3, is three orders of magnitude larger than that of C1 without Al_2O_3. Based on these results, we conclude the following:

1. Al_2O_3 additive can remarkably increase to generate a Bi luminescent center.
2. The generation of a Bi luminescent center has a nonlinear relation for Bi_2O_3 concentration.

3.4 ^{27}Al-NMR spectra

^{27}The Al-NMR spectra in BiSG are shown in Fig. 6(Fujimoto & Nakatsuka 2006). ^{27}Al chemical shifts were measured relative to $Al(H_2O)_6^{3+}$. The measured samples were A-series,

C2, and α-Al_2O_3. α-Al_2O_3 with a 6-fold coordinated state of corundum structure was used as a standard sample, and a peak exists at 15 ppm. The peaks at 70 and –40 ppm (marked by asterisks) were derived from spinning sidebands. The peaks of ^{27}Al-NMR from A1 to A3 only exist at 15 ppm and are the same as α-Al_2O_3. The A4 peak is still dominated by the 15 ppm peak, but a peak around 50 ppm begins to emerge, and then a peak of 56.4 ppm becomes dominant in A5 (Fig. 6(b)).

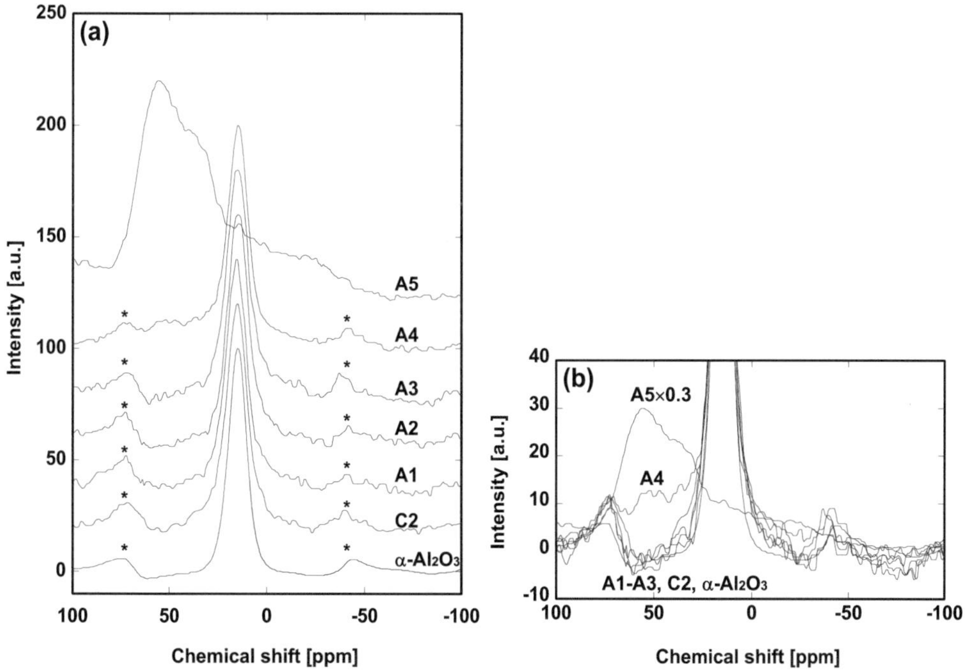

Fig. 6. ^{27}Al-NMR spectra of A1–A5, C2, and α-Al_2O_3: (a) whole view of spectra, and (b) expanded view of spectra in intensity scale

Sample A4, which includes Al_2O_3 of 2.3 mol%, has a weak 50 ppm peak in the ^{27}Al-NMR spectrum, while sample C2, which has the same amount of Al_2O_3 without Bi_2O_3, shows no signal around 50 ppm (Fig. 6(b)), suggesting that the Bi ion affects ACS over a 1.0 mol% of Bi_2O_3 concentration. On the other hand, since the C2 spectrum is dominated by a peak at 15 ppm, the aluminum ions in the silica glass naturally configure the 6-fold coordinated state of the corundum structure up to 2.3 mol% of Al_2O_3 without Bi_2O_3. This is also supported by the work of Mysen et al., who concluded the aluminum ions in silica glass work as a network modifier rather than a network former up to a 6.1 mol% Al_2O_3 concentration in the measurement of Raman spectra(Mysen, Virgo et al. 1980). The Al corundum structure dominates ACS at lower Bi concentration up to 0.5 mol%, so the Al corundum structure clearly has certain important roles for the generation of the Bi luminescent center in BiSG.

3.5 XRD measurements

The XRD data of the BiSG samples were measured to check for existing crystallizations, including undissolved alumina, mullite, or crystbalite because crystallization influences the ^{27}Al-NMR spectra. The measured samples are A-series. Fig. 7 shows the XRD data on samples A4, α-Al_2O_3 (alumina), and pure silica with a range between 10 and 80° in 2θ. Sample A4 is substituted for the other BiSG ones because these XRD patterns are almost the same. The peaks due to any kind of crystallization are not recognized in Fig. 7, especially undissolved alumina, where there is only a halo pattern. We previously confirmed an XRD pattern on a Nd_2O_3(3 wt%; 0.55 mol%)-SiO_2(97 wt%; 99.45 mol%) system that included undissolved 0.55 mol% of Nd_2O_3 at best(Fujimoto & Nakatsuka 1997). 7.0 mol% of alumina (at maximum in this experiment) is probably an adequate quantity for XRD detection if the quantity is changed to undissolved alumina or other crystals in the sample. Therefore, it is concluded that all our samples are in the amorphous phase.

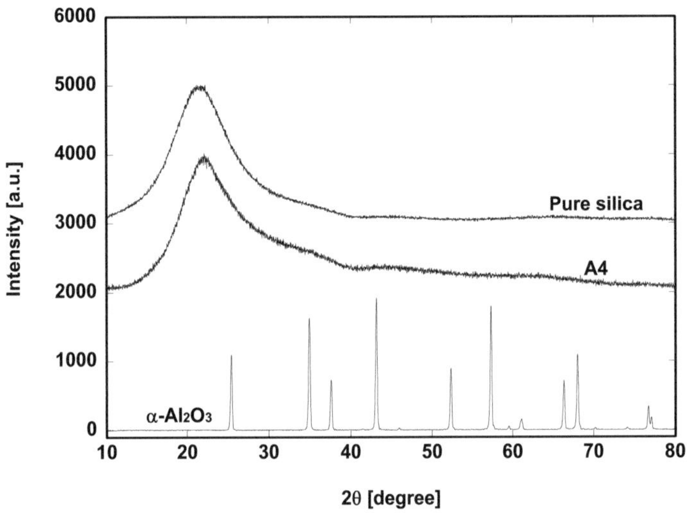

Fig. 7. XRD patterns on α-Al_2O_3 (alumina), sample A4 and pure silica with range between 10° and 80° in 2θ

3.6 ESR measurements

The presence of unpaired electrons in BiSG was verified by ESR signal. The measured sample was B2 and B3. There was no ESR signal due to the unpaired electrons for both B2 and B3, even at liquid N_2 temperature. The same phenomena without signals were also reported on Bi-doped multi-component glasses(Peng, Wang et al. 2005; Peng, Wu et al. 2008). According to Hund's rule, the valence states of bismuth ions without unpaired electrons should be Bi^{3+}(~$5d^{10}6s^2$) or Bi^{5+} (~$5d^{10}$)(Ohkura, Fujimoto et al. 2007).

3.7 XPS measurements

3.7.1 Analysis on chemical shift

The results of the XPS measurements are shown in Fig. 8. The measured samples were A4, A5, and three standards, $NaBiO_3$, Bi_2O_3, and Bi-metal. The main Bi($4f_{5/2}$, $4f_{7/2}$) peaks of Bi_2O_3 exist at 163.7 and 158.4 eV, respectively. Bi-metal was treated with 3-minute etching by Ar-beam in a vacuum chamber (1.0×10^{-7} Torr) to eliminate the oxidized Bi-metal surface before the measurement. Even after the treatment, weak residual peaks were found due to Bi_2O_3. The main Bi($4f_{5/2}$, $4f_{7/2}$) peaks of the Bi-metal exist at 162.4 and 157.1 eV, respectively. These Bi_2O_3 and Bi-metal peaks well agree with those previously reported(Wagner 1990; Saffarini

& Saiter 2000), and the chemical shifts of Bi_2O_3 and Bi-metal are very stable in the XPS measurement. $NaBiO_3$ is often used as a standard of the penta-valent state of Bi, but in our experiment, the main Bi($4f_{5/2}$, $4f_{7/2}$) peaks of $NaBiO_3$ were obtained at 163.8 and 158.5 eV corresponding to the Bi_2O_3 ones, and the second Bi($4f_{5/2}$, $4f_{7/2}$) peaks exist at higher bonding energy at 165.9 and 160.6 eV, respectively. The peaks of both BiSGs, that is, A4 and A5, are located at almost the same position at the second $NaBiO_3$ peaks. After arranging the chemical shifts for the measured samples, the order of the bonding energy is as follows:

[lower bonding energy] Bi metal (Bi^0) -> Bi_2O_3 (Bi^{3+}), first peaks of $NaBiO_3$ (Bi^{3+}) -> second peaks of $NaBiO_3$ (Bi^{5+}) = BiSG (A4, A5) [higher bonding energy]

In general, the valence state of the target ion becomes higher with increased bonding energy(Wagner 1990), and the same tendency is observed in my measurement. Therefore, Bi ions of the penta-valent state exist in BiSG.

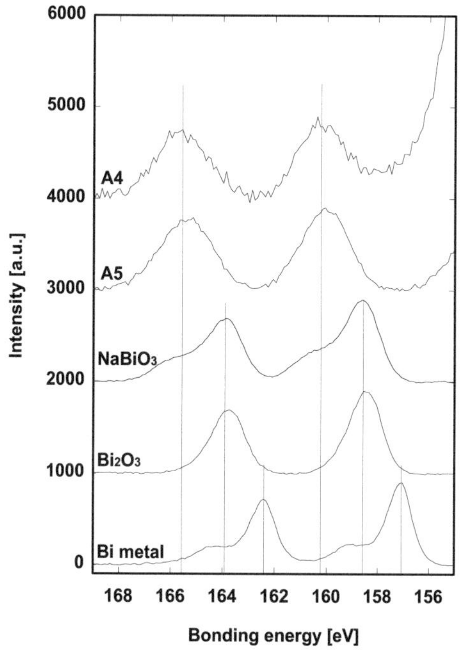

Fig. 8. XPS peaks of Bi($4f_{5/2}$, $4f_{7/2}$) on A4, A5, $NaBiO_3$, Bi_2O_3, and Bi-metal. In observation order of binding energy, [lower bonding energy] Bi metal (Bi^0) -> Bi_2O_3 (Bi^{3+}), first peaks of $NaBiO_3$ (Bi^{3+}) -> second peaks of $NaBiO_3$ (Bi^{5+}) = BiSG (A4, A5) [higher bonding energy].

The peak positions of $NaBiO_3$, however, seem unstable. The previously reported Bi($4f_{5/2}$, $4f_{7/2}$) peaks(Kulkarni, Vijayakrishnan et al. 1990) were 164.1 and 158.7 eV, respectively, and they showed single peaks with almost the same binding energy of the Bi_2O_3 ones. In fact, although we measured the XPS data on the $NaBiO_3$ several times, the ratio of the main peaks corresponding to Bi_2O_3 to the second peaks was unstable. Therefore, we put the most probable XPS data of $NaBiO_3$ in Fig. 8.

Kumada et al.(Kumada, Takahashi et al. 1996; Kumada, Kinomura et al. 1999) reported that $NaBiO_3$ and $LiBiO_3$ are synthesized at 120-200°C and that the Bi^{5+} state is changed to a Bi^{3+} state over 400°C; similar unstability may occur for the Bi^{5+} state in $NaBiO_3$. The standard material of $NaBiO_3$ was identified as $NaBiO_3 \bullet 2H_2O$ by XRD in Kumada's experiment. They also reported that Na ions in the A-site ($A^+B^{5+}O_3$) tend to be exchanged for Sr^{2+} or Ba^{2+} ions in $NaBiO_3 \bullet nH_2O$, and then the redistributed Bi ions in the A-site take the Bi^{3+} state(Kumada, Kinomura et al. 1999). Although they neglected to mention the redistribution of Bi ions in $NaBiO_3$ themselves, a similar phenomenon may occur for $NaBiO_3$ in their experiment.

3.7.2 Analysis on peak separation

By precisely observing the peak positions and the line widths, we recognized that the A4 peaks were slightly shifted to higher bonding energy than A5 and that the line widths of A4 and A5 were wider than the standards. Since the A4 and A5 peaks are composed of two or more peaks, we separated them with a Gaussian fitting curve to examine the origin of the peak shift. In this procedure, we make the following assumptions:

1. If such Bi ionic states as Bi^0 or Bi^{3+} are considered identical, the line widths of Bi($4f_{5/2}$, $4f_{7/2}$) are also identical.
2. The line widths of Bi $4f_{5/2}$ and Bi $4f_{7/2}$ are the same.
3. The ratio of Bi $4f_{5/2}$ to Bi $4f_{7/2}$ is constant for different ionic states in a sample. This ratio is theoretically calculated as $Bi^{4f_{5/2}}/Bi^{4f_{7/2}}=l/(l+1)=3/4$(Seah 1983).

The results are shown in Table 3. The peak separation results show five peak positions for all Bi $4f_{5/2}$ and Bi $4f_{7/2}$ peaks that are normalized at 100. No. 5 corresponds to Bi^0, No. 4 to Bi^{3+}, and No. 1 to Bi^{5+}. Nos. 2 and 3 are the intermediate states between numbers 1 and 4, and these states have intermediate coordination states rather than intermediate valence states such as Bi^{4+} due to the ESR measurements. These results show that all the Bi ions in BiSG are not the penta-valent state, and therefore the Bi valence states were mixed states of Bi^{3+} with Bi^{5+}. This mixed valence state of Bi^{3+} and Bi^{5+} is also supported by EXAFS analysis

Sample		Bi $4f_{5/2}$(Bi $4f_{7/2}$)				
		1	2	3	4	5
A4	Peak [eV]	166.0(160.6)	164.9(159.5)			
	FWHM [eV]	1.7	1.7			
	Height	100.0	64.6			
A5	Peak [eV]	166.2(160.8)	165.1(159.8)			
	FWHM [eV]	1.9	1.9			
	Height	50.6	100.0			
$NaBiO_3$	Peak [eV]	165.9(160.6)		164.6(159.3)	163.8(158.5)	
	FWHM [eV]	1.8		1.4	1.4	
	Height	38.6		35.0	100.0	
α-Bi_2O_3	Peak [eV]			164.7(159.4)	163.7(158.4)	
	FWHM [eV]			1.4	1.4	
	Height			19.9	100.0	
Bi metal	Peak [eV]			164.6(159.3)	163.4(158.1)	162.4(157.1)
	FWHM [eV]			1.4	1.4	1.1
	Height			22.4	22.4	100.0

Table 3. Peak separation results on Bi($4f_{5/2}$, $4f_{7/2}$) of A4, A5, $NaBiO_3$, α-Bi_2O_3, and Bi-metal. Peak position, FWHM, and normalized peak height are listed. Peak-heights are normalized at 100.

in the next section. The peak height ratio of Nos. 1 and 2 is counterchanged for A4 and A5 due to the existence ratio of Bi^{3+} and Bi^{5+}. Thus, the peaks of A4 are slightly shifted due to higher bonding energy than A5.

3.8 Bi-O distance from EXAFS

Figure 9 shows the radial structure functions (RSF) to which the EXAFS oscillations were Fourier-transformed. The measured samples were A-series (A2~A5) and the two standards of α-Bi_2O_3 and $NaBiO_3$. The peak shown in about 1.0 Å is derived from the XANES region because it is too short for any Bi-O distance. Therefore, it is a ghost peak, and we conclude that the largest peak around 1.6-1.7 Å (the first relevant peak) corresponds to the first neighboring Bi-O bond. α-Bi_2O_3 and $NaBiO_3$ have the second peak at 3.5 and 3.2 Å, respectively. Since the BiSG ones have no secondary peak, the local environment of the Bi ion does not have any periodical structure; that is, BiSG should be an amorphous phase. These results are supported by the XRD data in Section 3.5. The RSF shows that all the BiSG peaks are about 0.1 Å shorter than α-Bi_2O_3. The $NaBiO_3$ peak is also shifted to a shorter position, but the line width is wider than that of any of the BiSG ones. Since RSF $|F(r)|$ includes a phase shift, the radial distance in RSF shifted a shorter range than the actual Bi-O distance. To determine the length of the first neighboring Bi-O, the RSF of α-Bi_2O_3 and the BiSG samples were analyzed by the curve-fitting method in r-space with *FEFF 8.2*. In this curve-fitting calculation, we only took two coordination spheres due to the parameter number limitation in *FEFF 8.2*.

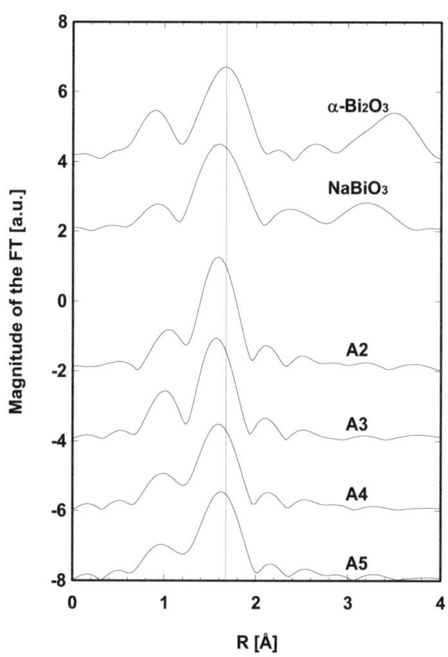

Fig. 9. Radial structure functions (RSF) of A-series (A2~A5), α-Bi_2O_3 and $NaBiO_3$

The fitting results of the BiSG samples, α-Bi_2O_3 and $NaBiO_3$, are listed in Table 4. We assumed amplitude reduction factor S_0^2 = 0.9(Manzini, Lottici et al. 1998) and absorption edge energy E_0 = 13426.5 eV. The fitting range was selected from 1.2 to 2.1 Å in RSF (Fig. 9). The Bi-O distances of the first and second coordination spheres for BiSG were calculated as about 2.1 and 2.3 Å, respectively; on the other hand, the Bi-O distances for α-Bi_2O_3 were 2.2 and 2.4 Å, respectively. The Bi-O distance of 2.1 Å in BiSG is in good agreement with the previously reported Bi^{5+}-O distance in LiBi(5+)O_3(Kumada, Takahashi et al. 1996) and Bi_2(3+,5+)O_4(Kumada, Kinomura et al. 1995). Therefore, the existence of the Bi^{5+} state is also indicated from the Bi-O distance in BiSG, and the first coordination sphere corresponds to the Bi^{5+}-O distance. The second coordination sphere of 2.3Å corresponds to the Bi^{3+}-O distance(Ohkura, Fujimoto et al. 2007). Therefore, the EXAFS curve-fitting results also show that the mixed valence state of the Bi ions exists in BiSG as Bi^{3+} and Bi^{5+}.

The Bi-O distance of the first coordination sphere in the A-series is slightly shifted to a longer range with increased Bi_2O_3 concentration in Table 4. The ratio of Bi^{3+} to Bi^{5+} increases with increased Bi_2O_3 concentration, and the change of the Bi^{3+} to Bi^{5+} ratio can also explain the non-linear increment of LMI. $NaBiO_3$ is a well-known material as a standard for penta-valent state Bi ions. The first coordination state distance of $NaBiO_3$ is longer than the expected value of the Bi^{5+}-O distance. This valence state of Bi ions in $NaBiO_3$ is also the mixed state of Bi^{3+} and Bi^{5+}. These phenomena are also supported by the peak separation data of XPS.

Samples	First coordination sphere			Second coordination sphere			R-factor(%)
	N_1	R_1	σ_1^2(Å^2)	N_2	R_2	σ_2^2(Å^2)	
α-Bi_2O_3	2.01	2.18	3.91E-03	0.74	2.40	4.07E-03	8.82
$NaBiO_3$	7.15	2.13	3.89E-03	3.12	2.38	4.06E-03	0.89
A2	2.19	2.08	3.89E-03	1.11	2.32	4.06E-03	1.65
A3	2.05	2.08	3.89E-03	1.33	2.31	4.06E-03	2.50
A4	1.81	2.11	3.89E-03	1.05	2.31	4.06E-03	2.84
A5	1.86	2.13	3.89E-03	0.99	2.33	4.06E-03	3.04

Table 4. FEFF fitting results providing two coordination spheres. Fitting results of A-series (A2~A5), α-Bi_2O_3, and $NaBiO_3$ are listed.

3.9 Discussion (local structure of luminescent center)

In the previous section, several physical phenomena were observed in BiSG, especially regarding the local structure of the distinctive luminescent center. Now we consider the structural configuration on the Bi luminescent center.

First, the roles of the Al_2O_3 additive can be understood by luminescent intensity measurement. Based on Fig. 5, the luminescent intensity of A4, which includes 2.3 mol% of Al_2O_3, is three orders of magnitude larger than that of C1 without Al_2O_3; clearly, the Al_2O_3 additive remarkably increases the generation of the Bi luminescent center. Second, Al_2O_3 assists the Bi ions to enter the silica glass network because C1 has no glassy wetting. This tendency is also supported by the phase diagram of the Bi_2O_3-Al_2O_3-SiO_2 glass system, because the glassy phase is likely achieved at the Al_2O_3-rich composition. Therefore, aluminum ions have two roles in BiSG:

1. They assist the configuration of the distinctive luminescent center of Bi ions with a coupling effect that denotes that an aluminum ion behaves like a "generator" of the luminescent center.
2. They increase compatibility with the silica network.

These aluminum ion roles imply that both the Bi and Al atoms should be close together in BiSG. Based on the above discussion, the image view between Bi and Al ions in BiSG is illustrated in Fig. 10(a). Peng et al.(Peng, Qiu et al. 2005) reported that Ta ions also work as a "generator." Although aluminum is not the only element that behaves as a generator, the aluminum ion accepts its important role in the Bi_2O_3-Al_2O_3-SiO_2 glass system.

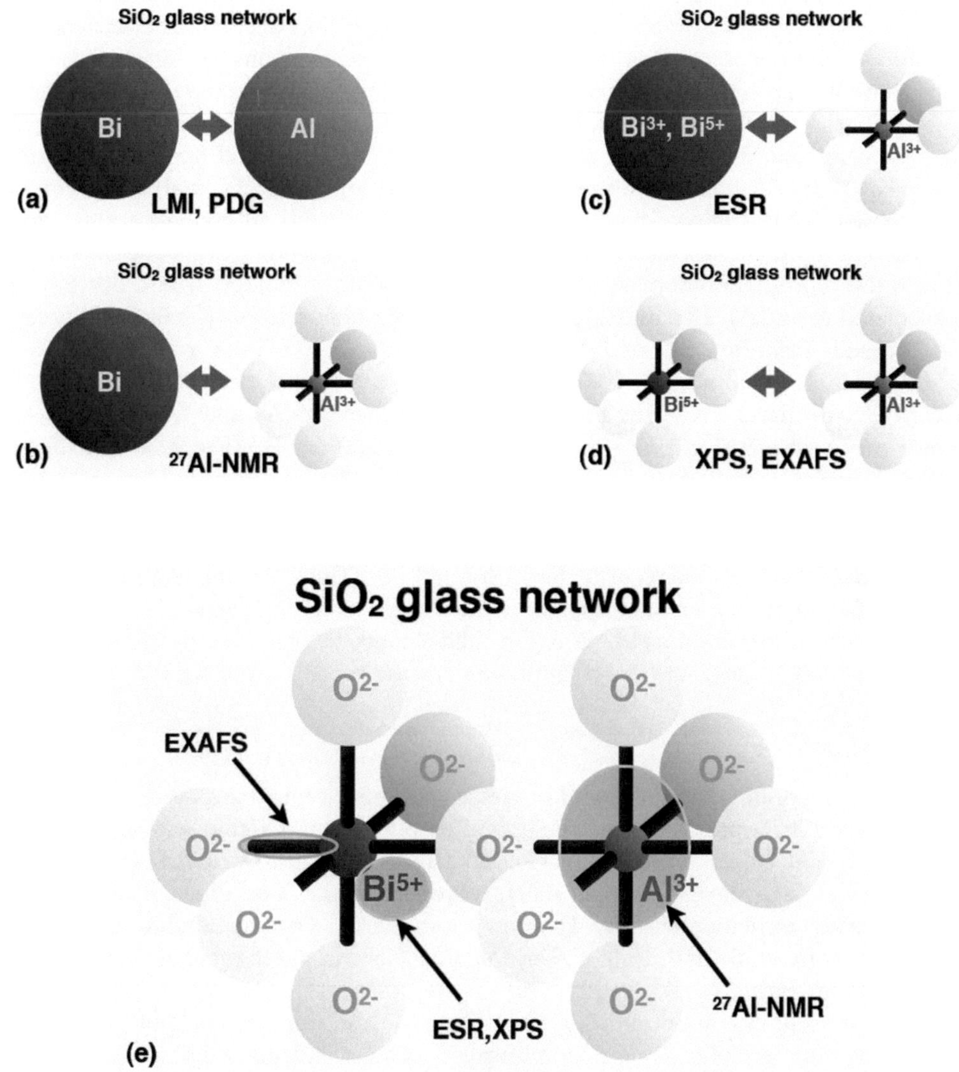

Fig. 10. Image view on local structure of infrared Bi luminescent center in BiSG: (a) image view determined by PDG (phase diagram) and LMI, (b) by ^{27}Al-NMR, (c) by ESR, (d) by XPS, EXAFS, (e) local structure of infrared Bi luminescent center.

Next, the aluminum cordination state (ACS) should be close to the Bi ion in BiSG, as seen from the ^{27}Al-NMR results. Since the relation between ACS and the chemical shift in the ^{27}Al-NMR measurement has been well studied(Laussac, Enjalbert et al. 1983), ACS is determined by a chemical shift in comparison between the standard materials and the target samples. In the case of BiSG, ACS is dominated by the α-Al_2O_3 corundum structure at lower Bi_2O_3 concentration up to 0.5 mol%. Al ions with corundum structure are crucial to generate a distinctive luminescent center, and the Al coordination state located near Bi should be a 6-fold corundum structure. Based on the above discussion, the image view between Bi and Al is illustrated in Fig. 10(b).

Next, information on the valence states of the Bi ions in BiSG is given by ESR measurements. Of course, it's not only for Bi ions, but since no signal exists for the unpaired electrons in the whole BiSG, the valence states of Bi ions without unpaired electrons are Bi^{3+} or Bi^{5+}. These results show Bi^{3+} or Bi^{5+} ions close to the 6-fold coordination state of the Al ions. Based on the above discussion, the image view between Bi and Al is illustrated in Fig. 10(c).

Three types of coordination states of Bi^{3+} exist, including 5-, 6-, and 8-fold; on the other hand, only 6-fold coordination exists for Bi^{5+}(Shannon 1976). Since BiSG is an oxide material, it is estimated that the neighboring ions of the Bi ion are oxygen. Although the ionic radius of O^{2-} has few differences with the coordination number, variation exists between 1.35 and 1.42 Å. If the coordination number of O^{2-} is 4, Bi^{3+}(5)-O^{2-}(4), Bi^{3+}(6)-O^{2-}(4), and Bi^{5+}(6)-O^{2-}(4) are calculated to be 2.34, 2.41, and 2.12 Å, respectively(Shannon 1976). The Bi-O distances in typical crystals including Bi^{3+} or Bi^{5+}, such as $LiBiO_3$(Kumada, Takahashi et al. 1996) or Bi_2O_4(Kumada, Kinomura et al. 1995), show that the Bi^{5+}-O distance is 2.1 Å. This value agrees well with the 2.1 Å of the first coordination sphere for A4 and A5. But the Bi^{3+}-O distance varies from 2.15 to 3.26 Å in α-Bi_2O_3(Harwig 1978) or Bi_2O_4(Kumada, Kinomura et al. 1995). Therefore, the EXAFS data show that the Bi^{5+} ionic state exists in BiSG, and this is also supported by the XPS data. The previously reported Bi^{3+} spectroscopic properties are quite different in luminescent and absorption spectra and lifetime. It is concluded that the Bi valence state of the Bi luminescent center is Bi^{5+}, not Bi^{3+}. Therefore, the luminescent center model of Bi^{5+} with 6-fold coordination is expected to be close to Al^{3+} with 6-fold coordination of the corundum structure (Fig. 10(d)). Since the neighboring atom is oxygen, the local structure of the distinctive bismuth luminescent center is expected (Fig. 10(e)).

4. Applications

After the discovery of a new infrared luminescent bismuth center, several research groups started to study its applications, such as optical amplification(Fujimoto & Nakatsuka 2003; Seo, Fujimoto et al. 2006; Seo, Fujimoto et al. 2006; Ren, Wu et al. 2007; Ren, Dong et al. 2007; Ren, Qiao et al. 2007; Seo, Lim et al. 2007), waveguide inscription(Psaila, Thomson et al. 2006), or laser oscillation(Dianov, Dvoyrin et al. 2005; Dianov, Shubin et al. 2007; Razdobreev, Bigot et al. 2007; Rulkov, Ferin et al. 2007; Truong, Bigot et al. 2008) using Bi luminescent materials.

With respects to device applications, optical fibers with Bi luminescent center in the core material are very curious. Optical amplification around 1.3 μm with Bi-doped multi-component glass fiber was achieved by Seo et al.(Seo, Fujimoto et al. 2006), and this is useful for metro area network optical amplifiers. Laser oscillation with Bi-doped optical fiber was firstly demonstrated by Dianov's group in 2005(Dianov, Dvoyrin et al. 2005), then the possibility of Bi-doped fiber is actively developed, now the oscillation power has achieved at 15 W at 1160 nm(Bufetov & Dianov 2009). It is known that the 570 – 590 nm band is very promising for ophthalmology and dermatology applications, thus the second harmonic of Bi fiber laser can be used in medical use. And the broad luminescence in near infrared region is also useful for a light source of optical coherence tomography.

5. Conclusions

In this chapter, we introduce the basic properties of BiSG and the analyzed local structure of Bi luminescent center. Several instrumental analyses, such as spectroscopic properties (SPCT), LMI, NMR, XRD, ESR, XPS, and EXAFS were advanced on a Bi_2O_3-Al_2O_3-SiO_2 glass system. The roles and the structure of the Al ions and the valence state of the luminescent Bi ions were examined.

The following are the roles and the structure of the Al ions: 1) to assist the configuration of the distinctive luminescent center of Bi ions with a coupling effect, which means that the aluminum ion behaves like a "generator" of the luminescent center; 2) to increase compatibility with the silica network; 3) to be a 6-fold corundum structure. The valence state exmination of Bi ions in BiSG reveals the following: 1) Bi^{3+} or Bi^{5+}; 2) a mixed state of Bi^{3+} and Bi^{5+}; 3) Bi^{5+} for the distinctive Bi luminescent center. Therefore, the distinctive bismuth luminescent center model was investigated with a 6-fold coordination state of Bi^{5+} that is combined with the 6-fold corundum structure of Al^{3+} through an oxygen ion (Fig. 10(e)). These results will bridge to verify the energy diagram and the mechanism of Bi luminescent center.

In the last place, this new infrared luminescent material, Bi-doped silica glass, which attains sensational progress in the past decade will continue to give us curious possibilities in the field of the optical science.

6. Acknowledgement

The EXAFS measurement in this work was performed under the approval of the Photon Factory Program Advisory Committee (Proposal No. 2006G123).

7. References

M. J. Weber and R. R. Monchamp (1973). Luminescence of Bi4Ge3O12 : Spectral and decay properties. *J. Appl. Phys.*, Vol.44, 5495-5499.

S. Parke and R. S. Webb (1973). Optical Properties of Thallium, Lead and Bismuth in Oxide Glasses. *Journal of Physics and Chemistry of Solids*, Vol.34, No.1, 85-95.

R. D. Shannon (1976). Revised Effective Ionic-Radii and Systematic Studies of Interatomic Distances in Halides and Chalcogenides. *Acta Crystallographica Section A*, Vol.32, No.Sep1, 751-767.

H. A. Harwig (1978). Structure of Bismuthsesquioxide - Alpha,Beta,Gamma and Delta-Phase. *Zeitschrift Fur Anorganische Und Allgemeine Chemie*, Vol.444, No.Sep, 151-166.

B. O. Mysen, D. Virgo, et al. (1980). Relations between the Anionic Structure and Viscosity of Silicate Melts - a Raman-Spectroscopic Study. *American Mineralogist*, Vol.65, No.7-8, 690-710.

A. C. van der Steen, J. J. A. van Hesteren, et al. (1981). Luminescence of the Bi-3+ Ion in Compounds LiLnO2 and NaLnO2 (Ln=Sc, Y, La, Gd, Lu). *Journal of the Electrochemical Society*, Vol.128, No.6, 1327-1333.

D. B. a. M. P. Seah (1983). Practical surface analysis : by auger and X-ray photo-electron spectroscopy. New York, Wiley.

J. P. Laussac, R. Enjalbert, et al. (1983). Tetracoordinated, Penta-Coordinated and Hexacoordinated Aluminum - Nmr and X-Ray-Diffraction Studies of the Complexes of Ethylenediamine with Aluminum Isopropoxide and Its Fluoro Analogs. *Journal of Coordination Chemistry*, Vol.12, No.3, 133-143.

C. D. Wagner (1990). Auger and x-ray photoelectron spectroscopy. New York, John Wiley.

G. U. Kulkarni, V. Vijayakrishnan, et al. (1990). State of Bismuth in BaBiO3 and BaBi1-xPbxO3 - Bi 4f Photoemission and Bi L3 Absorption Spectroscopic Studies. *Applied Physics Letters*, Vol.57, No.17, 1823-1824.

H. Scholze (1991). Glass : nature, structure, and properties. New York, Springer-Verlag.

S. N. Narang, N. D. Patel, et al. (1994). Infrared and Raman Spectral Studies and Normal-Modes of Alpha-Bi2O3. *Journal of Molecular Structure*, Vol.327, No.2-3, 221-235.

N. Kumada, N. Kinomura, et al. (1995). Crystal-Structure of Bi2O4 with Beta-Sb2O4-Type Structure. *Journal of Solid State Chemistry*, Vol.116, No.2, 281-285.

N. Sugimoto, H. Kanbara, et al. (1996). Ultrafast response of third-order optical nonlinearity in glasses containing Bi2O3. *Optics Letters*, Vol.21, No.20, 1637-1639.

N. Kumada, N. Takahashi, et al. (1996). Preparation and crystal structure of a new lithium bismuth oxide: LiBiO3. *Journal of Solid State Chemistry*, Vol.126, No.1, 121-126.

Y. Fujimoto and M. Nakatsuka (1997). A novel method for uniform dispersion of the rare earth ions in SiO2 glass using zeolite X. *Journal of Non-Crystalline Solids*, Vol.215, No.2-3, 182-191.

I. Manzini, P. P. Lottici, et al. (1998). EXAFS at the BiLIII edge in Bi4Ge3O12 and in xBi(2)O(3)-(100-x)GeO2 glasses. *Journal of Non-Crystalline Solids*, Vol.224, No.1, 23-30.

N. Kumada, N. Kinomura, et al. (1999). Ion-exchange reaction of Na+ in NaBiO3 center dot nH(2)O with Sr2+ and Ba2+. *Solid State Ionics*, Vol.122, No.1-4, 183-189.

PDF#30-1161 (2000). Powder diffraction file : alphabetical indexes for experimental patterns : inorganic phases : sets 1-50. Newtown Square, Pa. , ICDD, International Center for Diffraction Data.

G. Saffarini and J. M. Saiter (2000). X-ray photoelectron spectroscopic measurements on glassy Ge20S80-xBix (x=0,16). *Materials Letters*, Vol.46, No.6, 327-331.

Y. Fujimoto and M. Nakatsuka (2001). Infrared luminescence from bismuth-doped silica glass. *Japanese Journal of Applied Physics Part 2-Letters*, Vol.40, No.3B, L279-L281.

Y. Fujimoto and M. Nakatsuka (2003). Optical amplification in bismuth-doped silica glass. *Applied Physics Letters*, Vol.82, No.19, 3325-3326.

M. Y. Peng, J. R. Qiu, et al. (2004). Bismuth- and aluminum-codoped germanium oxide glasses for super-broadband optical amplification. *Optics Letters*, Vol.29, No.17, 1998-2000.

M. Y. Peng, X. G. Meng, et al. (2005). GeO2 : Bi, M (M = Ga, B) Glasses with Super-Wide Infrared Luminescence. *Chemical Physics Letters*, Vol.403, No.4-6, 410-414.

M. Y. Peng, J. R. Qiu, et al. (2005). Broadband infrared luminescence from Li2O-Al2O3-ZnO-SiO2 glasses doped with Bi2O3. *Optics Express*, Vol.13, No.18, 6892-6898.

M. Y. Peng, J. R. Qiu, et al. (2005). Superbroadband 1310 nm emission from bismuth and tantalum codoped germanium oxide glasses. *Optics Letters*, Vol.30, No.18, 2433-2435.

E. M. Dianov, V. V. Dvoyrin, et al. (2005). CW bismuth fibre laser. *Quantum Electronics*, Vol.35, No.12, 1083-1084.

X. G. Meng, J. R. Qiu, et al. (2005). Infrared broadband emission of bismuth-doped barium-aluminum-borate glasses. *Optics Express*, Vol.13, No.5, 1635-1642.

X. G. Meng, J. R. Qiu, et al. (2005). Near infrared broadband emission of bismuth-doped aluminophosphate glass. *Optics Express*, Vol.13, No.5, 1628-1634.

M. Y. Peng, C. Wang, et al. (2005). Investigations on bismuth and aluminum co-doped germanium oxide glasses for ultra-broadband optical amplification. *Journal of Non-Crystalline Solids*, Vol.351, No.30-32, 2388-2393.

Y. Fujimoto and M. Nakatsuka (2006). Al-27 NMR structural study on aluminum coordination state in bismuth doped silica glass. *Journal of Non-Crystalline Solids*, Vol.352, No.21-22, 2254-2258.

Y. S. Seo, Y. Fujimoto, et al. (2006). Optical amplification in a bismuth-doped silica glass at 1300 nm telecommunications window. *Optics Communications*, Vol.266, No.1, 169-171.

Y. S. Seo, Y. Fujimoto, et al. (2006). Simultaneous amplification at two wavelengths near 1300 nm in a 6.5-cm-long bismuth-doped silica glass. *Ieee Photonics Technology Letters*, Vol.18, No.17-20, 1901-1903.

X. J. Wang and H. P. Xia (2006). Infrared superbroadband emission of Bi ion doped germanium-aluminum-sodium glass. *Optics Communications*, Vol.268, No.1, 75-78.

T. Suzuki and Y. Ohishi (2006). Ultrabroadband near-infrared emission from Bi-doped Li2O-Al2O3-SiO2 glass. *Applied Physics Letters*, Vol.88, No.19, 191912.

H. P. Xia and X. J. Wang (2006). Near infrared broadband emission from Bi5+-doped Al2O3-GeO2-X (X=Na2O,BaO,Y2O3) glasses. *Applied Physics Letters*, Vol.89, No.5, 051917.

N. D. Psaila, R. R. Thomson, et al. (2006). Femtosecond laser inscription of optical waveguides in bismuth ion doped glass. *Optics Express*, Vol.14, No.22, 10452-10459.

V. V. Dvoyrin, V. M. Mashinsky, et al. (2006). Bismuth-doped-glass optical fibers - a new active medium for lasers and amplifiers. *Optics Letters*, Vol.31, No.20, 2966-2968.

T J. J. Ren, L. Y. Yang, et al. (2006). Effect of various alkaline-earth metal oxides on the broadband infrared luminescence from bismuth-doped silicate glasses. *Solid State Communications*, Vol.140, No.1, 38-41.

T. Ohkura, Y. Fujimoto, et al. (2007). Local structures of bismuth ion in bismuth-doped silica glasses analyzed using bi L-III x-ray absorption fine structure. *Journal of the American Ceramic Society*, Vol.90, No.11, 3596-3600.

Y. Fujimoto, Y. Hirata, et al. (2007). Effect of GeO2 additive on fluorescence intensity enhancement in bismuth-doped silica glass. *Journal of Materials Research*, Vol.22, No.3, 565-568.

E. M. Dianov, A. V. Shubin, et al. (2007). High-power cw bismuth-fiber lasers. *Journal of the Optical Society of America B-Optical Physics*, Vol.24, No.8, 1749-1755.

T. Murata and T. Mouri (2007). Matrix effect on absorption and infrared fluorescence properties of Bi ions in oxide glasses. *Journal of Non-Crystalline Solids*, Vol.353, No.24-25, 2403-2407.

M. Y. Peng, D. P. Chen, et al. (2007). Bismuth-doped zinc aluminosilicate glasses and glass-ceramics with ultra-broadband infrared luminescence. *Optical Materials*, Vol.29, No.5, 556-561.

I. Razdobreev, L. Bigot, et al. (2007). Efficient all-fiber bismuth-doped laser. *Applied Physics Letters*, Vol.90, No.3, -.

J. Ren, B. Wu, et al. (2007). Broadband optical amplification near 1300 nm in bismuth-doped germanate glass. *Applied Physics B-Lasers and Optics*, Vol.88, No.3, 363-366.

J. J. Ren, H. F. Dong, et al. (2007). Ultrabroadband infrared luminescence and optical amplification in bismuth-doped germanosilicate glass. *Ieee Photonics Technology Letters*, Vol.19, No.17-20, 1395-1397.

J. J. Ren, Y. B. Qiao, et al. (2007). Optical amplification near 1300 nm in bismuth-doped strontium germanate glass. *Journal of the Optical Society of America B-Optical Physics*, Vol.24, No.10, 2597-2600.

J. J. Ren, J. R. Qiu, et al. (2007). Infrared luminescence properties of bismuth-doped barium silicate glasses. *Journal of Materials Research*, Vol.22, No.7, 1954-1958.

J. J. Ren, J. R. Qiu, et al. (2007). Ultrabroad infrared luminescence from Bi-doped aluminogermanate glasses. *Solid State Communications*, Vol.141, No.10, 559-562.

J. J. Ren, J. R. Qiu, et al. (2007). Ultrabroad infrared luminescences from Bi-doped alkaline earth metal germanate glasses. *Journal of Materials Research*, Vol.22, No.6, 1574-1578.

A. B. Rulkov, A. A. Ferin, et al. (2007). Narrow-line, 1178nm CW bismuth-doped fiber laser with 6.4W output for direct frequency doubling. *Optics Express*, Vol.15, No.9, 5473-5476.

Y. S. Seo, C. Lim, et al. (2007). Bismuth-doped silica glass as a new laser material. *Journal of the Korean Physical Society*, Vol.51, No.1, 364-367.

G. Yang, D. P. Chen, et al. (2007). Effects of melting temperature on the broadband infrared luminescence of bi-doped and Bi/Dy co-doped chalcohalide glasses. *Journal of the American Ceramic Society*, Vol.90, No.11, 3670-3672.

S. F. Zhou, G. F. Feng, et al. (2007). Broadband near-infrared emission from Bi-doped aluminosilicate glasses. *Journal of Materials Research*, Vol.22, No.6, 1435-1438.

Y. Arai, T. Suzuki, et al. (2007). Ultrabroadband near-infrared emission from a colorless bismuth-doped glass. *Applied Physics Letters*, Vol.90, No.26, 261110.

V. G. Truong, L. Bigot, et al. (2008). Study of thermal stability and luminescence quenching properties of bismuth-doped silicate glasses for fiber laser applications. *Applied Physics Letters*, Vol.92, No.4, 041908.

J. J. Ren, G. P. Dong, et al. (2008). Inhomogeneous broadening, luminescence origin and optical amplification in bismuth-doped glass. *Journal of Physical Chemistry A*, Vol.112, No.14, 3036-3039.

J. R. Qiu, M. Y. Peng, et al. (2008). Novel Bi-doped glasses for broadband optical amplification. *Journal of Non-Crystalline Solids*, Vol.354, No.12-13, 1235-1239.

M. Y. Peng, B. T. Wu, et al. (2008). Bismuth-activated luminescent materials for broadband optical amplifier in WDM system. *Journal of Non-Crystalline Solids*, Vol.354, No.12-13, 1221-1225.

I. A. Bufetov and E. M. Dianov (2009). Bi-doped fiber lasers. *Laser Physics Letters*, Vol.6, No.7, 487-504.

3

Faraday Isolators for High Average Power Lasers

Efim Khazanov
Institute of Applied Physics of the Russian Academy of Science, N. Novgorod Russia

1. Introduction

The average power of solid-state and fiber lasers has considerably increased during the last ten years. The 10 kW power is not record-breaking any longer, and a topical problem nowadays is to create lasers with a power of 100 kW. Therefore, the study of thermal effects caused by absorption of laser radiation in the bulk of optical elements becomes ever more important. The Faraday isolator (FI) strongly depends on these effects because its magneto-optical elements (MOEs) are relatively long and its absorption α_0 is $10^{-3}...10^{-2}$ cm^{-1}, see Table 1. As a result, heat release power is at least tenths of percent of transmitted laser power P_0.

	V 1.06 µm	$\frac{1}{V}\frac{dV}{dT}$	κ	α_0	ξ	α_T	$\lvert Q \rvert$	dn/dT	P
	rad/T/m	10^{-3}/K	W/K/m	10^{-3}/cm		10^{-7}/K	10^{-7}/K	10^{-6}/K	10^{-6}/K
TGG	39 [1;2] 35 [3-5] 36 [6] 40 [7]	3.5 [8]	4.4±0.1 [9] 4.5±0.5 [10] 5.3±0.5 [11] 7.4 [5;12-15]	2 [15] 1-6 [14] 4.8 [13] 1.4-4.2 [16] 2.5 [17] 1.6 [18]	2.2 [16] 2.25 [13]	94 [15] 67-72 [9] 40[5]	17 [13 *)]	20 [15] 19 [13] 18-21 [9]	17 [13 *)]
MOC101	8.7 [1;2;19]	4 [19]		1 [1]; 2.3 [19]	1				
MOC105	17 [1]; 18 [2;19]	5 [19]	0.51 [20]	2.3 [19]	1	82 [21]	6 [20]	0.6 [21]	
MOC04	21 [1;2;19]		0.74 [20]	1 [1]; 2.3 [19]	1	49 [21]	9 [20]	8.7 [21]	
MOC10	28 [2;19]; 26 [1]		0.68 [20]	2 [1]; 4.6 [19]	1	56 [21]	8.5 [20]	8.5 [21]	
FR-5	21 [1;4]	3.4 [22]	0.84 [15]	3 [1]; 10 [15]	1	47 [15]	9 [20]	7.5 [15]	

Table 1. Property of magneto-optical materials. MOC 10 is analog of M-24 (Kigre, USA).

*) assuming κ=5W/Km;

[1](Zarubina & Petrovsky, 1992), [2](Zarubina et al., 1997), [3](Chen et al., 1998), [4](Jiang et al., 1992), [5](Kaminskii et al., 2005), [6](Yasuhara et al., 2007), [7](Raja et al., 1995), [8](Barnes & Petway, 1992), [9](Ivanov et al., 2009), [10](Slack & Oliver, 1971), [11](Chen et al., 1999), [12](Wynands et al., 1992), [13](Khazanov et al., 2004), [14] (Mueller et al., 2002), [15](Mansell et al., 2001), [16](Khazanov et al., 2002a), [17](Mukhin et al., 2009), [18](VIRGO-Collaboration, 2008), [19](Malshakov et al., 1997), [20](Andreev et al., 2000a), [21](Zarubina, 2000), [22](Davis & Bunch, 1984).

At P_0=100 W (and higher) this gives rise to polarization distortions deteriorating the isolation degree, and phase distortions - aberrations. Many applications require a combination of high average power, high isolation degree, and small aberrations. Below we shall demonstrate that although the methods well known for laser amplifiers can be used for analyzing thermal effects in FI, yet one has to take into account specific features imposed by the magnetic field (the Faraday effect). We shall overview theoretical and experimental results of investigations of thermal effects in FIs and methods for their compensation and suppression. Note that all the results reported below are valid not only for cw lasers but for pulse lasers with high repetition rate as well.

Unlike FI, a Faraday mirror proposed in (Giuliani & Ristori, 1980) is used not for optical isolation, but for compensation of birefringence in laser amplifiers (Carr & Hanna, 1985), oscillators (Giuliani & Ristori, 1980), regenerative amplifiers (Denman & Libby, 1999) and fiber optics as well (Gelikonov et al., 1987). Despite the great similarity between the Faraday mirror and FI, there are two primary differences between them. First, the isolation in FI is governed only by the depolarization in the second pass, whereas in the Faraday mirror the polarization distortions are accumulated during both the passes. Second, the radiation that is incident on the MOE in FI is linearly polarized, whereas the radiation that is incident on the Faraday mirror has already been depolarized. We shall consider only FI; a Faraday mirror for high power lasers is studied in (Khazanov, 2001; Khazanov et al., 2002b; Khazanov, 2004).

In the absence of thermal effects in the MOE after the first pass (from left to right), a beam retains its horizontal polarization (Fig. 1, 2) and passes through polarizer 4, while during the return pass (from right to left), the polarization is altered to vertical and the beam is reflected by polarizer 1.

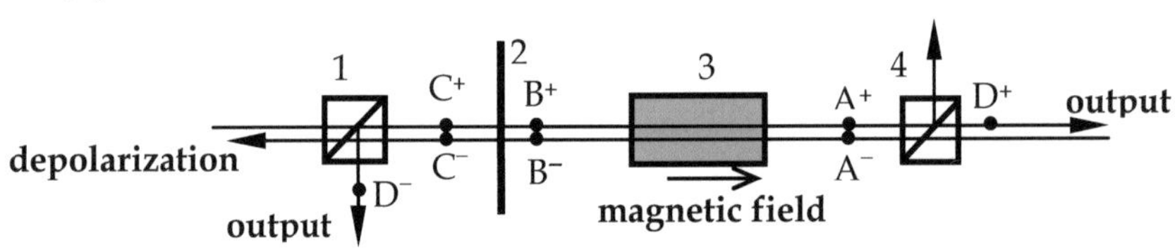

Fig. 1. Traditional design of a Faraday isolator. 1,4 – polarizers; 2 – λ/2 plate; 3 – MOE.

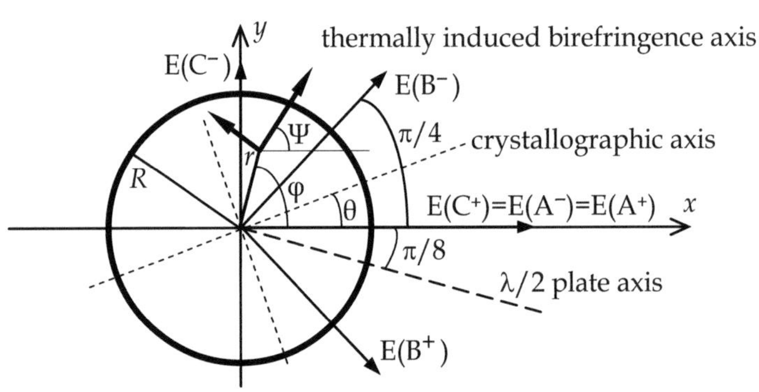

Fig. 2. Cross-section of magneto-optical crystal: r, φ are polar coordinates; θ is angle of inclination of the crystallographic axis; Ψ is angle of inclination of eigen polarization of thermally induced birefringence.

The light absorption in MOE generates a temperature distribution that is nonuniform over a transverse cross section. This leads to three physical mechanisms affecting the laser radiation: i) wave front distortions (thermal lens) caused by the temperature dependence of the refraction index; ii) nonuniform distribution of the angle of polarization rotation because of the temperature dependence of the Verdet constant and thermal expansion of the MOE; and iii) simultaneous appearance not only of the circular birefringence (Faraday effect), but also of the linear birefringence caused by mechanical strains due to the temperature gradient (photoelastic effect).

The first mechanism (Zarubina et al., 1997) does not induce any polarization changes in laser radiation and hence does not affect the isolation degree. The latter two mechanisms do alter the polarization state of radiation. The temperature dependence of the Verdet constant and thermal expansion lead to changes of the phase shift between eigen polarizations which remain circular (Wynands et al., 1992). The photoelastic effect not only changes the phase shift between eigen polarizations, but also alters the eigen polarizations themselves, which become elliptical (Khazanov, 1999; Khazanov et al., 1999). In section 2 we discuss the influence of all thermal effects on FI parameters and determine the figure of merit of magneto-optical materials for high average power lasers.

Thermal effects in FI may be compensated by some additional optical elements or suppressed (reduced) by choosing optimal FI parameters or geometries. Section 3 is devoted to compensation of thermal lens (by means of an ordinary negative lens or a negative thermal lens) and compensation of depolarization (by means of crystalline quartz placed inside a telescope or by means of replacing one 45^0 MOE by two 22.5^0 MOEs and a $\lambda/2$ plate or a 67.5^0 polarization rotator between them).

In section 4 we discuss the methods of thermal effects suppression: cooling FI to liquid nitrogen temperature, shortening MOE using a strong magnetic field, employing several thin discs cooled through optical surfaces, and using slabs and rectangular beams.

2. Thermal effects in Faraday isolators

2.1 Jones matrix of thermally loaded magneto-optical element

A non-uniformly heated MOE is a polarization phase plate that has simultaneously two types of birefringence: circular due to the Faraday effect, and linear due to the photoelastic effect. The circular birefringence is completely described by a phase shift between circular eigen polarizations δ_c; the polarization rotation angle is $\delta_c/2=VBL$, where B is magnetic field, V and L are Verdet constant and length of MOE. Linear birefringence is described by a phase shift between linear eigen polarizations δ_l and an inclination angle Ψ of eigen polarization relative to the x axis (Fig. 2). Such a polarization phase plate is described by the Jones matrix (Tabor & Chen, 1969)

$$F(\delta_c\,,\delta_l\,,\Psi)=\exp(ikLn_0)\exp(ikL[T(r)-T(0)]P)\sin\frac{\delta}{2}\begin{pmatrix}\cot\frac{\delta}{2}-i\frac{\delta_l}{\delta}\cos 2\Psi & -\frac{\delta_c}{\delta}-i\frac{\delta_l}{\delta}\sin 2\Psi\\ \frac{\delta_c}{\delta}-i\frac{\delta_l}{\delta}\sin 2\Psi & \cot\frac{\delta}{2}+i\frac{\delta_l}{\delta}\cos 2\Psi\end{pmatrix},\tag{1}$$

where

$$P=\frac{dn}{dT}-\alpha_T\frac{n_0^3}{4}\frac{1+\nu}{1-\nu}\cdot(p_{11}+p_{12})\tag{2}$$

is a thermo-optical constant of MOE, $\delta^2 = \delta_l^2 + \delta_c^2$, and n_0, ν, α_T, $p_{i,j}$ are "cold" refractive index, Poisson's ratio, thermal expansion coefficient, and photoelastic coefficients, respectively, $k=2\pi/\lambda$, λ is wavelength in vacuum. Here and further we assume that the temperature T is uniform along the direction of beam propagation z. The second exponential factor in (1) has no influence upon polarization distortions and is an isotropic thermal lens. A contribution to this lens is made by the temperature dependence of the refraction index and "isotropic" part of the photoelastic effect (see two corresponding terms in (2)). We also assume that the contribution of thermal expansion is negligibly small in comparison with the temperature dependence of the refractive index; and magnetic field B (and hence δ_c) does not depend on the longitudinal coordinate z. The case when B depends on z was considered in (Khazanov et al., 1999).

For rod geometry δ_l and Ψ are defined by the formulas (Soms & Tarasov, 1979):

$$\delta_l = 4\pi \frac{L}{\lambda} Q q(\varphi) \left(\frac{1}{r^2} \int_0^r r^2 \frac{dT}{dr} dr \right) \tag{3}$$

$$\tan(2\Psi - 2\theta) = \xi \tan(2\varphi - 2\theta), \tag{4}$$

where

$$q(\varphi) = \begin{cases} \sqrt{\cos^2(2\varphi - 2\theta) + \xi^2 \sin^2(2\varphi - 2\theta)} & for\ [001] \\ (1+2\xi)/3 & for\ [111] \end{cases} \tag{5}$$

$$Q = \left(\frac{1}{L}\frac{dL}{dT} \right) \frac{n_0^3}{4} \frac{1+\nu}{1-\nu} \cdot (p_{11} - p_{12}) \tag{6}$$

$$\xi = \frac{2p_{44}}{p_{11} - p_{12}}. \tag{7}$$

Parameter of optical anisotropy ξ shows the difference of the cubic crystal from glass (for all glasses ξ=1). It can be seen from (3-7) that expressions for δ_l and Ψ for the [111] crystal orientation can be obtained from the corresponding expressions for the [001] orientation by making a formal substitution:

$$\xi \to 1, \quad Q \to Q(1+2\xi)/3 \quad \text{(for the transition [001]} \to \text{[111])}. \tag{8}$$

Further we shall give all results only for the [001] orientation, having in mind that the corresponding formulas for the [111] orientation can be obtained by substituting (8). Arbitrary crystal orientation is analyzed in (Khazanov et al., 2002a).

For the Gaussian beam with radius r_0 and power P_0 one may substitute the solution of the heat conduction equation

$$\frac{dT}{dr} = -\frac{\alpha P_0}{2\pi\kappa} \cdot \frac{1 - \exp(-r^2 / r_0^2)}{r} \tag{9}$$

into (3):

$$\delta_l(u,\varphi) = p\frac{u+\exp(-u)-1}{u}\sqrt{\cos^2(2\varphi-2\theta)+\xi^2\sin^2(2\varphi-2\theta)}\,, \tag{10}$$

where

$$p = \frac{L}{\lambda}\frac{\alpha_0 Q}{\kappa}P_0\,, \tag{11}$$

$u=r^2/r_0^2$, α_o and κ are absorption and thermal conductivity. Dimensionless parameter p physically means normalized laser power. Assuming for a TGG crystal $L/\lambda=20000$, $\alpha_0=1.5\cdot10^{-3}\text{cm}^{-1}$, $Q=17\cdot10^{-7}\text{K}^{-1}$, and $\kappa=5\text{W/Km}$ we obtain $p=1$ when $P_0=1\text{kW}$.

Formula for δ_c follows from the Faraday effect, taking into account the temperature dependence of the Verdet constant and thermal expansion:

$$\delta_c(r) = \delta_{co}\left[1+\left(\frac{1}{V}\frac{dV}{dT}+\alpha_T\right)\left(-T(r)-T\left(r^*\right)\right)\right], \tag{12}$$

where δ_{co} is a doubled angle of polarization rotation at $r=r^*$; and r^* can be chosen such as to minimize depolarization, see below. Thus, Jones matrix of MOE is determined by (1) with (4, 10, 12).

2.2 Polarization distortions (depolarization)

Let us calculate the depolarization ratio of the beam after the second pass through the FI (Fig. 1). In the absence of thermal effects, the beam at a point C^- is vertically polarized and is reflected by polarizer 1. Because of the thermal effects there occurs depolarized radiation, which, being horizontally polarized at a point C^-, passes through polarizer 1. The local depolarization ratio $\Gamma(r,\varphi)$ is

$$\Gamma(r,\varphi) = \left|\mathbf{E_C x}_0\right|^2 / \left|E_C\right|^2\,, \tag{13}$$

where E_C is the complex amplitude of the field at point C^-. Of major interest is the integral depolarization γ (the isolation degree of the FI is $1/\gamma$) that is a fraction of horizontally polarized radiation power at point C^-:

$$\gamma = \int_0^{2\pi} d\varphi \int_0^{\infty} \left|\mathbf{E}_C\mathbf{x}_0\right|^2 rdr \Bigg/ \int_0^{2\pi} d\varphi \int_0^{\infty} \left|E_C\right|^2 rdr = \frac{1}{\pi r_0^2}\int_0^{2\pi} d\varphi \int_0^{\infty} \Gamma \exp\left(\frac{-r^2}{r_0^2}\right) rdr\,. \tag{14}$$

Here we assume that the FI aperture is such that aperture losses can be neglected, i.e. the integration in (14) over a polar radius r can be extended to infinity; and the beam at a point A^- has Gaussian shape and horizontal polarization:

$$\mathbf{E}(A^-) = const\;\mathbf{x}_0 \exp\left(-r^2/2r_0^2\right). \tag{15}$$

Knowing Jones matrices of all elements, the field at points C^- can be easily found:

$$\mathbf{E}(C^-) = \mathbf{L}_2(3\pi/8)\mathbf{F}(\delta_c=\pi/2, \delta_l)\mathbf{E}(A^-), \tag{16}$$

where $\mathbf{L_2}(\beta_L)$ is the matrix of a $\lambda/2$ plate with an angle of inclination of the optical axis β_L:

$$\mathbf{L}_2(\beta_L) = \begin{pmatrix} \cos 2\beta_L & \sin 2\beta_L \\ \sin 2\beta_L & -\cos 2\beta_L \end{pmatrix}. \tag{17}$$

Substituting (1, 15, 17) into (16), and the result into (13, 14) yields Γ and γ. Let us consider the case when the linear birefringence is small

$$\delta_l << 1 \tag{18}$$

and changes of the polarization rotation angle are small too, i.e. $(\delta_c(r)-\delta_{co})<<\delta_{co}$. In this case from (13) accurate to within terms of order δ_l^4 and $\delta_l^2(\delta_c-\delta_{c0})$ we obtain

$$\Gamma = \frac{2\delta_l^2}{\pi^2}\sin^2\left(2\Psi - \frac{\pi}{4}\right) + \left(\frac{\delta_c}{2} - \frac{\pi}{4}\right)^2. \tag{19}$$

The substitution of (4, 10, 12) into (19), and the subsequent substitution of the result into (14) yield

$$\gamma = p^2\frac{A_1}{\pi^2}\left(1+(\xi^2-1)\cos^2\left(\frac{\pi}{4}-2\theta\right)\right) + \left(\frac{\pi^2}{16}\right)\left(\frac{1}{V}\frac{dV}{dT}+\alpha_T\right)^2 \cdot \int_0^\infty \exp(-u)(T(r)-T(r^*))^2 du, \tag{20}$$

where A_i are given in Table 2. By rotating the MOE around z axis, i.e. by varying angle θ, one can minimize the first term in (20). By differentiating (20) over r^* and equating the derivative to zero, we obtain for the optimal value $r_{opt} \approx 0.918 r_0$. In practice, when choosing the value of the magnetic field or length of the MOE, one should secure rotation of polarization by an angle $\pi/4$ at point $r=0.918r_0$, see (12). As a result of these two optimizations we obtain

$$\gamma_{min} = \frac{A_1}{\pi^2}p^2 + A_3\left(\frac{\alpha_0 P_0}{16\kappa}\right)^2\left(\frac{1}{V}\frac{dV}{dT}+\alpha_T\right)^2. \tag{21}$$

Thus, depolarization (19, 20, 21) is an arithmetic sum of contributions of two effects: the photoelastic effect (the first term) and temperature dependence of Verdet constant (the second term). Note that both terms in (20, 21) are independent of the beam radius r_0 and are proportional to the square of laser power P_0. Expression (21) allows us to compare the impacts of these effects. Assuming $L/\lambda \cong 20000$ and taking into account data in Table 1 one can show that the photoelastic effect is dominating. This fact found numerous experimental evidences. The most illustrative one is the transverse distribution of Γ(r, φ). If temperature dependence of the Verdet constant is neglected, Γ(r,φ) according to (4, 19) has the form of a cross, and the axes of this cross (directions where Γ=0) are rotated relative to the x, y axes by an angle $\pi/8$. This completely conforms to the experimental data, see Fig. 3.

m	1	2	8	∞
$A_0(m) = \sigma_1 / \sigma_0$	1	0.56	0.48	1/2
$A_1(m) = \frac{1}{\sigma_0^3} \cdot \int_0^\infty \frac{h_m^2(u)du}{\exp(u^m)}$	0.137	0.111	0.087	1/12
$A_2(m) = \frac{1}{\sigma_0^5} \cdot \int_0^\infty \frac{h_m^4(u)du}{\exp(u^m)}$	0.042	0.0265	0.0145	1/80
$A_3(m) = \frac{1}{\sigma_0^3} \cdot \int_0^\infty \frac{f_m^2(u)}{\exp(u^m)} du - \frac{1}{\sigma_0^4} \left[\int_0^\infty \frac{f_m(u)}{\exp(u^m)} du \right]^2$	0.268	0.158	0.092	1/12
$A_4(m) = A_3(m) - \frac{\sigma_1^2}{\sigma_0^4} \frac{1}{\sigma_2 \sigma_0 - \sigma_1^2} \left[\int_0^\infty \frac{f_m(u)}{\exp(u^m)} (1 - \frac{u \sigma_0}{\sigma_1}) du \right]^2$	0.0177	0.0021	10^{-5}	0
$A_5(m) = A_1(m) - \frac{1}{\sigma_2 \sigma_0^3} \left(\int_0^\infty \frac{u h_m(u)}{\exp(u^m)} du \right)^2$	0.012	0.0017	10^{-5}	0
$A_6(m) = \frac{m^2}{\sigma_0} \int_0^\infty \frac{w_m^2(u)}{u^2 \exp(u^m)} du$	0.046	0.054	0.028	0
$A_7(m) = \frac{m^4}{\sigma_0} \int_0^\infty \frac{w_m^4(u)}{u^4 \exp(u^m)} du$	0.0031	0.0076	0.0082	0
$h_m = \frac{1}{u} \int_0^u dz \int_0^z \frac{dy}{\exp(y^m)}$ $\quad f_m = \int_0^u \left(\int_0^z \frac{dy}{\exp(y^m)} \right) \frac{dz}{z}$ $\quad w_m = \int_0^u \frac{z^m dz}{\exp(z^m)}$ $\quad \sigma_k(m) = \int_0^\infty \frac{y^k dy}{\exp(y^m)}$				

Table 2. Values $A_{0\text{-}7}$ for different m. $A_i \equiv A_i(m=1)$.

Fig. 3. Theoretical (a) and experimental (b) (Khazanov et al., 2000) intensity distributions of depolarized beam.

In addition, experiments on depolarization compensation (see sections 3.2, 3.3) also confirmed domination of the photoelastic effect. Further we shall assume that γ is given by

$$\gamma = A_1 p^2 / \pi^2 . \tag{22}$$

Thermal effects influence not only depolarization γ, but also power losses during the forward pass γ_1, i.e. losses caused by the reflection of depolarized radiation from polarizer 4 (Fig.1). Considering only the photoelastic effect, by analogy with γ an expression for γ_1 at $\theta=\theta_{opt}$ may be found (Khazanov, 2000):

$$\gamma_1 = A_1 \xi^2 p^2 / \pi^2 . \tag{23}$$

Deriving (23) we neglected average over cross-section decrease of V due to average heating of MOE (Khazanov et al., 1999). An increase of the laser power from 0 to 400 W decreased the angle of rotation by 2 degrees (Mukhin et al., 2009), which corresponded to a negligible in practice value $\gamma_1 \approx 0.1\%$. However, when FI is placed in vacuum, the average temperature (and hence γ_1) increases much higher (VIRGO-Collaboration, 2008). In this case good thermal contact of MOE with magnets housing and/or thermal stabilization of the MOE by the Peltier element should be implemented to keep γ_1 negligible.

2.3 Amplitude and phase distortions

The depolarization ratio γ and power losses during the first pass γ_1 are generally the main but not the only parameters of the FI. The output radiation $\mathbf{E}_{out}$ has also spatial (amplitude and phase) distortions. Depending on particular FI applications, the output beam may be a beam at point D^+, a beam at point D^-, or both (Fig. 1). Below we shall assume the first, most frequently used case. For quantitative description of the spatial distortions we shall use

$$\gamma_s = 1 - \left| \int_0^{2\pi} d\varphi \int_0^{\infty} E_{out} E_{ref}^* r dr \right|^2 \Bigg/ \left(\int_0^{2\pi} d\varphi \int_0^{\infty} \left| E_{out} \right|^2 r dr \cdot \int_0^{2\pi} d\varphi \int_0^{\infty} \left| E_{ref} \right|^2 r dr \right), \tag{24}$$

i.e. the difference from unity of the overlapping integral of E_{out} and the reference field E_{ref} that is the field in the absence of thermal effects. To determine analytical expressions for γ_s we shall apply the formalism of the Jones polarization matrices as above. In case of weak polarization distortions (18) and weak phase distortions, i.e. $kL(n(r)-n(0))<<1$, we obtain

$$\gamma_s = \gamma_a + \gamma_i , \tag{25}$$

where

$$\gamma_a = p^2 A_1 / \pi^2 \qquad \gamma_i = p_i^2 A_3 / 4 \tag{26}$$

$$p_i = \frac{L}{\lambda} \frac{\alpha_0 P}{\kappa} P_0 . \tag{27}$$

Values of all γ are summarized in Table 3. Let us discuss the results obtained. First of all, it is important to note that γ_s (as well as γ and γ_1) does not depend on r_0 and is proportional to the square of P_0. Two physical effects contribute to γ_s: isotropic thermal lens (γ_i) and anisotropic distortions (γ_a) due to depolarization. The latter contribution is attributed to the distortions non-uniformity over the cross-section resulting in appearance of amplitude and phase distortions in the beam after propagation through the polarizer (e.g., Maltese cross, astigmatism). Taking into account polarization losses at the first pass γ_1, the total power loss in spatial and polarization mode after the first pass through the FI is $\gamma_{total}=\gamma_1+\gamma_a+\gamma_i$.

		Traditional FI Fig. 1	FI with $\lambda/2$ Fig. 9a	FI with 67.5^0 rotator Fig. 9b
depolarization ratio γ (isolation degree is $1/\gamma$)		$p^2 A_1 / \pi^2$	$p^4 \frac{8A_2}{\pi^4} \xi^2 (b^2 - a^2)$ $\xi>1.3$	$p^4 \frac{6a^2 A_2}{\pi^4} \left(1 + \frac{2}{3}\xi^2 + \xi^4\right)$
no thermal lens compensation	polarization losses γ_1	$p^2 A_1 \xi^2 / \pi^2$	$\frac{p^2 A_1}{\pi^2} \left(2 - \frac{\pi}{2}\right)(\xi^2 + 1)$ $\xi>1.3$	$p^2 A_1 (2 - \sqrt{2}) / \pi^2$
	anisotropic losses γ_a	$p^2 A_1 / \pi^2$	$0(p^4)$	$p^2 A_1 (2 - \sqrt{2}) \xi^2 / \pi^2$
	isotropic losses γ_i	$p_i^2 A_3 / 4$	$p_i^2 A_3 / 4$	$p_i^2 A_3 / 4$
telescope compensation	γ_{1TC}	γ_1	γ_1	γ_1
	γ_{aTC}	γ_a	γ_a	γ_a
	γ_{iTC}	$p_i^2 A_4 / 4$	$p_i^2 A_4 / 4$	$p_i^2 A_4 / 4$
adaptive compensation	γ_{1AC}	$\gamma_1 + A_1 p_{CG}^2 / 8$	$\gamma_1 + A_1 p_{CG}^2 / 8$	$\gamma_1 + A_1 p_{CG}^2 / 8$
	γ_{aAC}	$\gamma_1 + \frac{A_1}{8} p_{CG}^2 + \frac{A_1 \xi}{\pi\sqrt{8}} p p_{CG}$		
	γ_{iAC}	0	0	0

Table 3. Depolarization and power losses after the first pass through FI.

Note that the parameter p_i (27) is analogous to the parameter p (11) accurate within replacement of thermo-optical constants: Q (6) characterizing anisotropic distortions by P (2) characterizing isotropic distortions. Isotropic losses γ_i are determined only by parameter p_i, while p determines isolation degree $1/\gamma$ as well as losses γ_a and γ_i induced by anisotropy of the photoelastic effect.

Since the temperature distribution is not parabolic the thermal lens is aberrational. Such a lens can be represented as a sum of a parabolic lens with focus F and an aberrator that does not introduce any geometrical divergence. Using the method of moments an expression for F can be obtained (Poteomkin & Khazanov, 2005):

$$F = \frac{2A_0}{p_i} k r_0^2 , \tag{28}$$

where A_0 is given in Table 2.

2.4 The influence of beam shape

Above we have discussed thermal distortions of a Gaussian beam. Since a laser beam induces (being a heat source) and simultaneously reads distortions, the value of self-action may depend significantly on the transverse distribution of the intensity. The results obtained can be generalized for an arbitrary axially symmetric beam (Khazanov et al., 2002b), including a super-Gaussian beam with power P_0 and intensity

$$I(r) = P_0 \exp\left(-\frac{r^{2m}}{r_0^{2m}}\right) \cdot \left(\pi r_0^2 \int_0^\infty \exp(-y^m) dy\right)^{-1} . \tag{29}$$

At m=1 the beam is Gaussian, and at $m=\infty$ the beam turns into a flat-top one. Repeating the procedure described in sections 2.2, 2.3 for the laser beam (29) instead of (15), one can show that expressions for the depolarization ratio γ (19-22), for losses in polarization γ_1 (23), and spatial γ_s (25-26) mode during the first pass, and for F (28) are valid at any m, if A_i are replaced by $A_i(m)$, expressions for which are given in Table 2. All equations below are for a Gaussian beam, but they are valid for a super-Gaussian beam after this replacement.
Note that with increasing m the value of $A_{1,3}(m)$ decreases. This means that a flat-top beam is optimal for decreasing the influence of all thermal effects, whereas a Gaussian beam has the strongest self-action.

2.5 Selection of magneto-optical medium

In high-power lasers, magneto-optical materials are chosen taking into account specific features of different nonlinear effects. As a result, figures of merit were introduced: the larger the figure of merit, the better the medium. From the point of view of power losses due to absorption, such a figure of merit is the V/α_0 ratio (Robinson, 1964). From the point of view of self-focusing in pulse lasers, this is parameter VW_{cr} (Zarubina et al., 1997) for thermal self-focusing and VP_{cr} (Malshakov et al., 1997) for electronic Kerr self-focusing.
As has been shown in sections 2.2 and 2.3, all thermal effects are determined by p_i and p. Taking into account that $L\sim1/V$ we obtain figures of merit μ_i and μ:

$$\mu_i = \frac{V\kappa}{\alpha_0 P}\ , \qquad \mu = \frac{V\kappa}{\alpha_0 Q}\ . \tag{30}$$

According to (22, 8) the [001] orientation is better than [111]. In (Khazanov et al., 2002b) it was shown that [001] is the best orientation.
The absorption coefficient α_0 at 1064nm wavelength in TGG can vary by several times from sample to sample, see Table 1, where values of V and κ are also included. The most likely value of κ lies in the range 4-5W/Km. Direct measurements of ξ, P and Q were not done because of difficulty in measuring the photoelastic coefficients p_{ij}. The results of measurements by means of techniques based on thermal effects are shown in Table 1.
As can be seen from expressions (30), and from Table 1, the TGG crystal has a considerable advantage over all glasses due to its high thermal conductivity. At the same time both Q and P can be effectively controlled in glasses by changing their content. For instance, among laser glasses there is a quartz neodymium glass having $Q=0.2\cdot10^{-7}K^{-1}$ (Demskaya & Prokhorova, 1983). If a magneto-optical glass with such a Q were created, its figure of merit μ would be better than in TGG.
Two other terbium garnets have V 35% higher than TGG: TAG (Ganschow et al., 1999; Rubinstein et al., 1964; Geho et al., 2005) and TSAG (Yoshikawa et al., 2002). Verdet constants of $LiTb(MoO_4)_2$ (Guo et al., 2009) and $NaTb(WO_4)_2$ (Liu et al., 2008) are even higher. However, the figures of merit μ_i and μ of all these crystals are unknown up to now. Besides, their diameters are a few mm only.
The greatest disadvantage of TGG is also a relatively small aperture (<30mm), whereas glasses can have a diameter as large as 300 mm. In (Khazanov, 2003; Khazanov, 2004) we proposed to use TGG polycrystalline ceramics in FIs. The first samples of TGG ceramics were made by Dr. A.Ikesue (Japan) in 2003, see Fig. 4, and the first experimental study was done in (Yasuhara et al., 2007). Also, ceramics may be made of other garnets and oxides:

TAG or TSAG (high V and κ) and highly (up to 20%) Nd-doped YAG, Y_2O_5, Sc_2O_5, Lu_2O_5 (low α_0 and high κ). We forecast that the use of FIs in lasers with high average power will expand considerably within the next few years due to the emergence of ceramics. In (Kagan & Khazanov, 2004) we studied specificity of thermal effects in magneto-optical ceramics and showed that figures of merit for ceramics are the same as for a single crystal with [111]-orientation.

Fig. 4. A photograph of the first TGG ceramics samples made by A.Ikesue (Japan) in 2003.

3. Compensation of thermal effects in Faraday Isolators

3.1 Compensation of thermal lens in Faraday Isolators

The temperature distribution in the MOE and, consequently, the distribution of phase of an aberrated laser beam are almost parabolic. Therefore, a great portion of the phase distortions can be compensated by means of an ordinary lens or a telescope (Khazanov, 2000) shown by a dashed line in Fig. 5. Hereinafter we shall call this method "telescopic compensation" and indicate corresponding losses by subscript "TC".

In (Mansell et al., 2001; Mueller et al., 2002) an adaptive method (subscript "AC") for compensating the thermal lens was suggested and experimentally studied. A compensating glass was placed before polarizer 1 (dotted line in Fig. 5). Parameters of the compensating glass were chosen so that the thermal lens had the same focus as in FI but opposite (typically negative) sign. In (Mueller et al., 2002) it was shown numerically that the influence of diffraction can be insignificant. In this case the isotropic losses were totally compensated: $\gamma_{iAC}=0$.

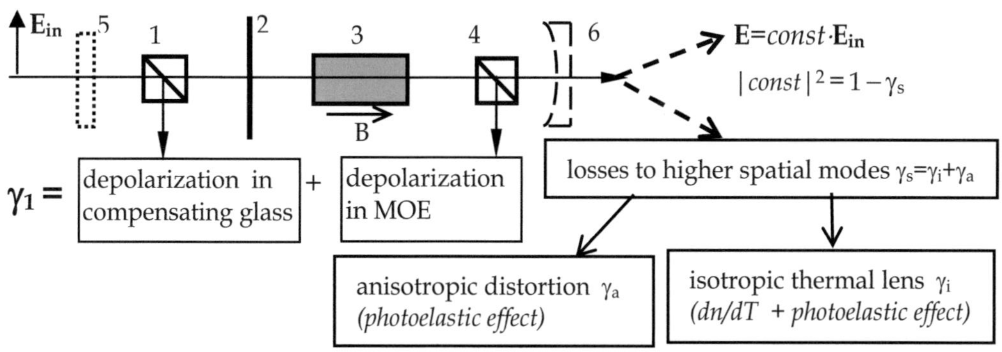

Fig. 5. Power losses during first propagation through FI. 1,4 - polarizers; 2 - $\lambda/2$ plate; 3 - MOE, 5 - compensating glass, 6 - compensating lens or telescope.

The adaptive approach has two certain advantages over the telescopic one: there is no need in adjustment when laser power is changed, and the accuracy of compensation is higher. However, a considerable disadvantage of the adaptive method is that the photoelastic effect in the compensating glass leads to additional (besides isotropic) distortions and, consequently, to losses in the spatial polarization mode. Thus, the above two methods can compensate only for γ_i. The telescopic method is less efficient but does not lead to increase of losses γ_1 and γ_a. The adaptive method totally compensates for γ_i but increases losses γ_1 and γ_a because of the photoelastic effect in the compensating glass, which has not been considered in (Mueller et al., 2002).

In (Khazanov et al., 2004) a detailed theoretical analysis of the two compensation methods is presented. Analytical expressions for all γ are summarized in Table 3. These formulas are valid when $\theta=\theta_{opt}$, condition (18) is obeyed, and phase distortions are weak. Parameter p_{CG} is defined by (11), with all material constants for compensating glass.

As one can see from Table 3, the losses associated with isotropic thermal lens γ_i can be reduced by the telescopic method by $A_3/A_4 \cong 15$ times, as was shown in (Khazanov, 2000). In (Mueller et al., 2002; Mansell et al., 2001) this value appeared to be twice as small as it should be because of mistake made in the calculations.

For a super-Gaussian beam (29), all the formulas in Table 3 are valid if A_i is replaced by $A_i(m)$, expressions for which are given in Table 2. The $A_3(m)/A_4(m)$ ratio grows with increasing m, i.e. the compensation of isotropic thermal lens induced by a super-Gaussian beam is more efficient. In particular, for a flat-top beam the isotropic thermal lens can be totally compensated – $A_4(\infty)=0$. This has a simple physical explanation: at a uniform heat release there is a strictly parabolic temperature distribution in a rod.

A key parameter of the compensating glass is the P_{CG}/Q_{CG} ratio, and the higher this ratio, the better the glass (Khazanov et al., 2004). Specifically, an increase of γ_1 and γ_a by glass is rather small if $P_{CG}/Q_{CG}>10$ and may be neglected if $P_{CG}/Q_{CG}>50$. The thermal lens averaged for two polarizations was almost totally compensated by means of FK51 Schott glass in (Khazanov et al., 2004): the difference of the phase from a constant was reduced from 0.9 to 0.02 radian. At the same time, the astigmatism of the resulting lens was very large, because of the small ratio $P_{FK51}/Q_{FK51}=2.8$.

The photoelastic effect can be totally compensated by using gel instead of glass, as it is done for compensation of thermal lens in laser amplifiers (Roth et al., 2004). Another approach is to use a crystal with natural birefringence, in which the thermally induced birefringence may be neglected. Examples are YLF, KDP, DKDP, LiCAF.

A 5.5-mm-thick DKDP crystal was successfully used in (Zelenogorsky et al., 2007). 2D phase maps are shown in Fig. 6. It was demonstrated in experiment that for 45 W laser power the compensation allows reducing power losses in Gaussian mode γ_s from 26% to 0.5%. Calculations have shown that losses can be reduced to a level of 4.7%, even for a laser power of 150 W. At present, a DKDP crystal seems to be the best choice for adaptive thermal lens compensation and is widely used in FI in high average power lasers.

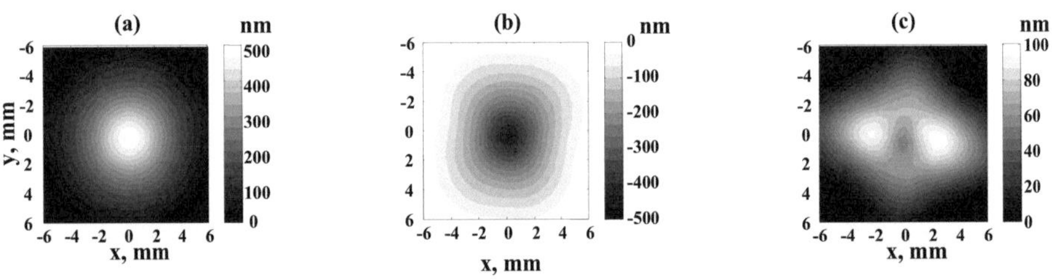

Fig. 6. Thermally induced phase map for FI (a), DKDP (b), and both FI and DKDP (c).

3.2 Depolarization compensation in FI with one magneto-optical element

The idea (Andreev et al., 2002) consists in creating a phase plate, in which all phase incursions that a beam assumes in the MOE are subtracted. For this, the phase plate should have the same transverse distribution of eigen polarizations and phase shift as in the MOE, except that the phase shift is opposite in sign. In this case, the radiation, having successively passed through these two elements, keeps its initial polarization unaltered. If the phase plate is reciprocal, then the non-reciprocal properties of the FI (rotation of polarization by 90^0 during two passes) are maintained.

In a MOE made of glass (ξ=1) or a cubic crystal with the [111] orientation one can derive from (3-7) that $\Psi=\varphi$ and δ_l does not depend on φ and near the beam axis $\delta_l \sim r^2$ (this is not so for crystals with other orientations, but we shall not consider such cases in this section). When a plane wave propagates in crystalline quartz at an angle $\Phi << 1$ relative to the optical axis, there is also superposition of linear and circular birefringences. The phase shift of linear birefringence $\delta_{ql} \sim \Phi^2$. At propagation of a converging or diverging beam (Fig. 7) the angle Φ is proportional to r, i.e., $\delta_{ql} \sim r^2$, and the inclination angle of eigen polarization of the linear birefringence is equal to φ. Therefore, if directions of polarization rotation in the MOE and quartz rotator (QR) are opposite (i.e., $\delta_c = -\delta_{qc}$), then the compensation of depolarization after successive passes through the MOE and QR is possible.

A formula for depolarization ratio γ_q of FI in Fig. 7 was derived in (Andreev et al., 2002). γ_q depends on laser power P_0 and at optimal power has minimum γ_{qmin}:

$$\gamma_{q\min} = A_5(m)(1+2\xi)^2 p^2 / (9\pi^2) . \quad (31)$$

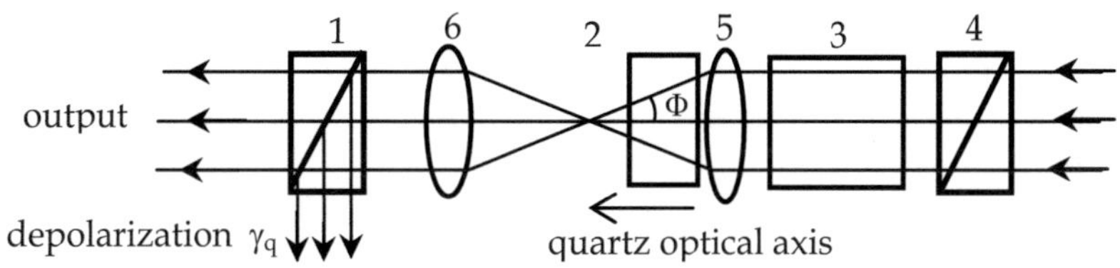

Fig. 7. Depolarization compensation in FI: 1,4 – polarizers, 2 – 45^0 QR, 3 – MOE, 5 and 6 – lenses of telescope. MOE and QR rotate in opposite directions.

Figure 8 shows experimental and theoretical plots for $\gamma(P_0)$ and $\gamma_q(P_0)$. At low power, the thermal effects are small, and the traditional design provides better isolation: $\gamma<\gamma_q$. When power is increased, γ_q decreases reaching its minimum value γ_{qmin}. Without compensation the theory is in good agreement with experiment at high powers, when γ is much greater than the "cold" depolarization ratio $\gamma_{cold}=2.5 \cdot 10^{-4}$. At the focal length of lens 5 f=125mm, the experimental value of γ_{qmin} is considerably greater than the theoretical prediction because γ_{cold} is 2.5 times as great as the theoretical value of γ_{qmin} for this case. At high powers, $\gamma_{cold} << \gamma_q$ and theoretical and experimental values of γ_q coincide. At f=88mm experimental values of γ_q are higher than the theoretical ones, but the difference is not crucial.

Of major interest from a practical standpoint is the γ/γ_{qmin} ratio which shows by how many times the isolation degree can be more when the design in Fig. 7 is used. From (22, 31), with (8) taken into account, for the [111] orientation we obtain $\gamma/\gamma_{qmin}=A_1(m)/A_5(m)$; and this ratio increases with increasing m, see Table 2. In experiment (see Fig. 8) $A_1(1)/A_5(1)=8$ instead of the theoretically predicted 11.5. This difference is attributed to some ellipticity of the beam and non-ideal coaxiality of lens 3 and the beam.

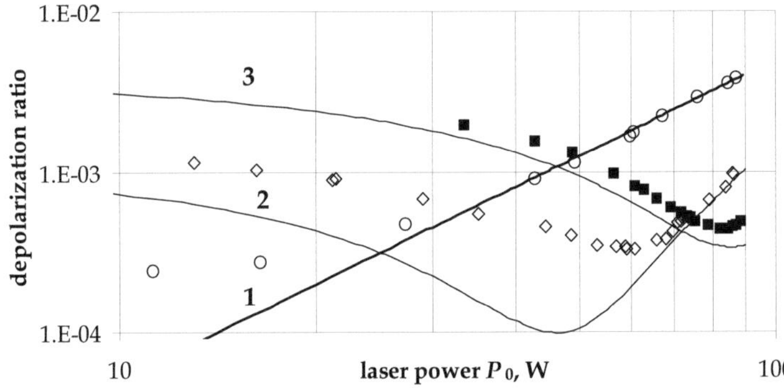

Fig. 8. Depolarization ratio in traditional FI design $\gamma(P_0)$ (curve 1, circles) and design in Fig. 7 $\gamma_q(P_0)$ at f=125mm (curve 2, rhombs) and f=88mm (curve 3, squares) (Andreev et al., 2002).

The main advantage of the design in Fig. 7 is a possibility to upgrade a standard commercial FI without re-assembling its magnetic system. The designs comprising two magneto-optical elements require a special magnetic system but they are more efficient and useful in practice, as will be discussed in the next section.

3.3 Depolarization compensation in Faraday isolators with two magneto-optical elements

It is well known that a 90^0 polarization rotator placed between two identical phase plates with linear eigen polarizations provides total compensation of birefringence (Scott & de Wit, 1971). If there is also circular birefringence in these plates, this statement is valid only when the directions of rotation of the polarization plane in the plates are different. It is however unacceptable for FI, because in this case it loses nonreciprocal properties. Nevertheless, replacing one 45^0 MOE by two 22.5^0 MOEs with a reciprocal optical element between them (a $\lambda/2$ plate or a QR as shown in Fig. 9) compensates depolarization (Khazanov, 1999).

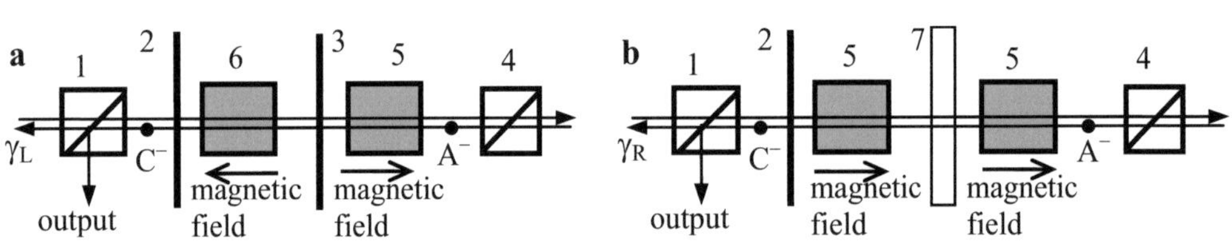

Fig. 9. FI designs with a $\lambda/2$ plate (a) and QR (b) (Khazanov, 1999). 1,4 - polarizers, 2,3 - $\lambda/2$ plates, 5 - 22.5^0 MOE rotating clockwise, 6 - 22.5^0 MOE rotating anticlockwise, 7 - 67.5^0 QR.

Let us find γ_L and γ_R for the novel designs (subscripts "L" and "R" will denote FI designs illustrated in Fig. 9a and 9b, respectively). The field **E** at a point C^- can be easily found:

$$\mathbf{E}_L(C^-) = \mathbf{L}_2(\beta_L+\pi/8)\mathbf{F}(\delta_c=-\pi/4, \delta_l/2)\mathbf{L}_2(\beta_L)\mathbf{F}(\delta_c=\pi/4, \delta_l/2)\mathbf{E}(A^-), \quad (32)$$

$$\mathbf{E}_R(C^-) = \mathbf{L}_2(\beta_R/2+3\pi/8)\mathbf{F}(\delta_c=\pi/4, \delta_l/2)\mathbf{R}(\beta_R)\mathbf{F}(\delta_c=\pi/4, \delta_l/2)\mathbf{E}(A^-)\,, \tag{33}$$

where **F** and $\mathbf{L}_2$ are defined by (1, 17), $\mathbf{R}(\beta_R)$ is the matrix of QR (matrix of rotation by an angle β_R). We assume that the phase shift of the linear birefringence in each MOE is $\delta_l/2$, i.e. δ_l is the phase incursion for an entire pass through FI for all designs in Figs. 1 and 9. In the approximation (18) the substitution (32, 33) into (14) instead of $\mathbf{E}_C$ yields the expressions for $\gamma_{L,R}$. From these expressions it can be seen that at

$$\beta_L=\beta_{optL}=\pi/8+j\pi/2 \qquad \beta_R=\beta_{optR}=3\pi/8+j\pi \tag{34}$$

(j is an integer) $\gamma_{L,R}$ become proportional to the fourth power of δ_l, whereas γ (22) is proportional to the second power of δ_l. Taking into account (18), this indicates that for (34) $\gamma_{L,R} \ll \gamma$, i.e. the isolation degree increases considerably in the two novel designs (Fig. 9) in comparison with the traditional one (Fig. 1). If (34) is valid,

$$\gamma_L(\theta) = p^4 A_2 \pi^{-4}\left[6a^2\left(1+\frac{2}{3}\xi^2+\xi^4\right)+8b^2\xi^2+6b^2\left(1-\xi^2\right)^2\sin^2\theta_4-12ab\left(\xi^4-1\right)\sin\theta_4\right], \tag{35}$$

$$\gamma_R = p^4 6a^2 A_2 \pi^{-4}\left(1+2\xi^2/3+\xi^4\right), \tag{36}$$

where $\theta_4=4\theta-\pi/4$, $a=(\pi-2\sqrt{2})/8$, $b=(2-\sqrt{2})/4$. Note that γ_R does not depend on angle θ at all, whereas γ_L depends on it. By varying angle θ (in practice by rotating both MOEs around the *z* axis) it is possible to find θ_{opt} and minimal value of the depolarization ratio γ_L:

$$\theta_{optL} = \frac{\pi}{16}+\frac{1}{4}\mathrm{Re}\left\{\arcsin\left[\frac{a}{b}\cdot\frac{\xi^2+1}{\xi^2-1}\right]\right\} \tag{37}$$

$$\gamma_L = \gamma_L(\theta=\theta_{optL}) = \begin{cases} p^4 8A_2\pi^{-4}\xi^2\left(b^2-a^2\right) & \xi>1.315 \quad \xi<0.760 \\ p^4 2A_2\pi^{-4}\left[3(a+b)^2+2\xi^2(a^2-b^2)+3\xi^4(a-b)^2\right] & 1\le\xi<1.315 \\ p^4 2A_2\pi^{-4}\left[3(a-b)^2+2\xi^2(a^2-b^2)+3\xi^4(a+b)^2\right] & 0.760<\xi\le 1 \end{cases} \tag{38}$$

Thus, $\gamma_{L,R}$ are determined, like γ in the traditional design, only by two parameters: p and ξ. The polarization distortions which a beam acquires when passing through the first MOE, are compensated, though not totally as in laser amplifiers (Scott & de Wit, 1971), but rather partially yet efficiently when passing through the second MOE. For a super-Gaussian beam (29) one can easily show that formulas (34-38) remain valid at any m, if A_2 is replaced by $A_2(m)$.

Formulas (36,38) are valid if condition (18) is obeyed. In a general case, analytical expressions cannot be obtained. Numerical integration showed that β_{optR} and β_{optL}, like at $\delta_l \ll 1$, are determined by expressions (34). Figure 10 presents the plots of $\gamma(p)$ for all the three designs. A considerable decrease of the depolarization ratio in both novel designs persists even when the condition (18) is disobeyed. It can be seen from Fig. 10 that the design with QR (Fig. 9a) provides better isolation degree than the design with a $\lambda/2$ plate (Fig. 9b). At the same time, we should note an important advantage of the design with a $\lambda/2$ plate: the different direction of polarization rotation in the MOEs and, consequently, different direction of the magnetic field. With an appropriate arrangement of the magnetic system, this reduces the total length of MOEs L (Shiraishi et al., 1986) and leads to an additional decrease in γ_L, which is proportional to L^4.

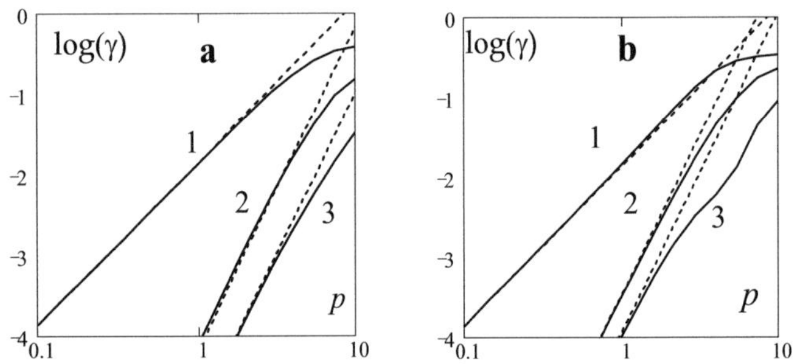

Fig. 10. Numerical (solid curves) and analytical (22, 36, 38) (dashed) plots of $\gamma(p)$ for the traditional FI design in Fig. 1 (1), $\gamma_L(p)$ for FI design with a $\lambda/2$ plate in Fig. 9a (2), and $\gamma_R(p)$ for FI design with QR in Fig 9b (3). (a) glass (b) TGG crystal with the [001] orientation, and (c) TGG crystal with the [111] orientation.

It can be seen from (36) that for TGG (ξ=2.25) γ_R is almost equal for the [001] and [111] orientations; any other orientation is worse (Khazanov et al., 2002a). At the same time, the [111] orientation does not require any mutual alignment of two MOEs. Thus, for the design in Fig. 9b the [111] orientation is more practical.

Power losses during the first passes through the FI designs in Fig. 9 can be calculated using the procedure described in section 2.3 (Khazanov, 2000). The results are summarized in Table 3. As can be expected, γ_i is the same for all three designs. With regard to minimization (γ_a+γ_1), both novel designs are slightly better than the traditional one.

The efficiency of depolarization compensation in the novel designs was first experimentally confirmed for MOEs made of glass (Khazanov et al., 2000). Experimental results for TGG-based FI with QR are summarized in Fig. 11. The first test (rhombs, total TGG crystals length 22mm, diameter 11mm) showed excellent compensation of thermal effects and 45dB isolation at up to 90W power. The most powerful experiments (circles, total TGG crystals

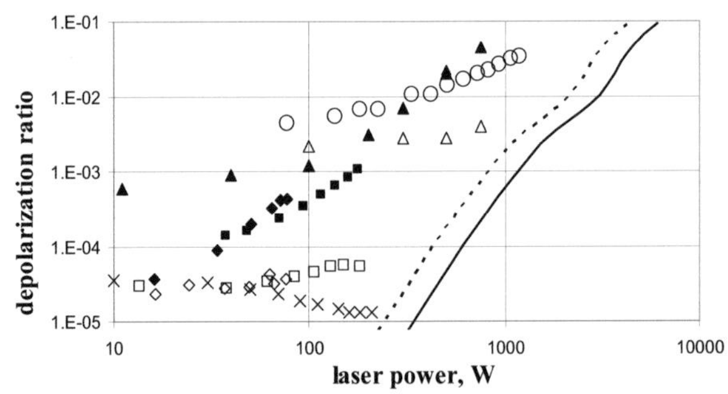

Fig. 11. Experimental plots of $\gamma(P_0)$ (filled symbol) and $\gamma_R(P_0)$ (open symbol) for different FIs: (Andreev et al., 2000b) (rhombs), (Nicklaus et al., 2006) (circles), (Voytovich et al., 2007) (squares, triangles), and unpublished data (crosses). Theoretical curves correspond to $\gamma_R(P_0)$ for squares (solid) and triangles (dashed).

length 15mm, aperture 4x8mm) showed depolarization much worse than the theoretical predictions due to bad crystal quality and non-optimized alignment. Three 20mm-diameter FIs with total TGG crystals length 18mm (triangles, squares, and crosses) showed at least an order of magnitude compensation of depolarization. Isolation degrees (24dB, 42dB, and 49dB) are mostly defined by cold depolarization. Theoretical curves show that these FIs provide more than 20 dB isolation at laser power up to 3kW. Nowadays the FIs with QR (Fig. 9b) are widely used in high average power lasers.

4. Suppression of thermal effects in Faraday Isolators

4.1 Cryogenic Faraday isolators

In (Zheleznov et al., 2006) we suggested cooling FI to liquid nitrogen temperature to improve the high average power characteristics. The Verdet constant and the magnetic field grow when temperature decreases, so the length of the MOE becomes shorter and hence the isolation degree $1/\gamma$ increases. Also, thermo-optical properties of TGG and magneto-optical glasses are improved at nitrogen temperature. All this aspects allow drastically suppressing all thermal effects. Let us discuss it in details.

It is known (Zarubina et al., 1987; Barnes & Petway, 1992; Davis & Bunch, 1984) that the Verdet constant of TGG and magneto-optical glasses depends on temperature according to the law V=const/T. Recent studies (Yasuhara et al., 2007) confirm this dependence for TGG-ceramics too. Therefore, cooling MOE to 77K will make it possible to reduce its length almost by a factor of 4.

Magnetic field also grows when the magnetic system is cooled. But the most frequently used Nd-Fe-B magnet showed second-order phase transition at T>135K; consequently, magnetic field at 77K depends on cooling speed (Zheleznov et al., 2007). In that paper we also showed that there is no phase transition in samarium-cobalt alloy (Sm-Co) magnets and that magnetic field at 77K is lager than at 300K by a factor of 1.2 and does not depend on cooling speed. So, in cryogenic FIs the Sm-Co magnets may be more efficient, even though they are weaker than Nd-Fe-B.

Temperature dependence of the depolarization ratio γ and thermal lens in TGG and magneto-optical glass MOC-04 were measured in (Zheleznov et al., 2006). As temperature was reduced to 102K the focal length F increased by a factor of 2.7 (Fig. 12a), allowing us to expect a 3.6 times increase at 77K. Formulas (27,28) show that this corresponds to a 3.6 times decrease of parameter $\alpha P/\kappa$. Reduction of this parameter for MOC-04 glass is about a factor of 1.8.

At 86K, γ in TGG decreased 8-fold, see Fig. 12b. At 77K, we can expect a decrease by a factor more than 9. Therefore, a decrease of parameter $\alpha Q/\kappa$ by a factor of 3 follows from expressions (22, 11). This TGG crystal was grown from a melt. The thermal conductivity of such crystals in the temperature range from 77K to 300K is almost constant (Slack & Oliver, 1971), so the measured decrease of γ is related to decreasing αQ.

Let us estimate the maximum average power of the FI cooled to 77K. According to (22, 11), at a given value of γ the laser power P_0 is proportional to $\kappa/\alpha QL$. Since parameter αQ is about three times lower and the growth of the Verdet constant and the magnetic field allow a 5-fold reduction of the crystal length L, P_0 will be increased by a factor of 15. Use of flux-grown TGG crystals allows further doubling of the laser power because the thermal conductivity of such crystals is twice as great at 77K (Slack & Oliver, 1971). This allows using traditional FI (Fig. 1) at the laser power up to 10kW.

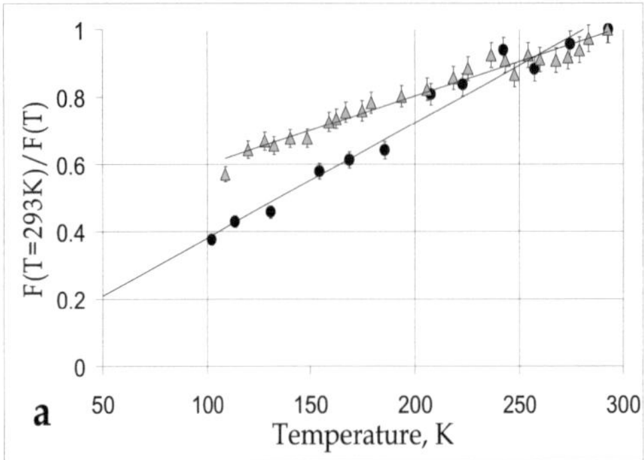

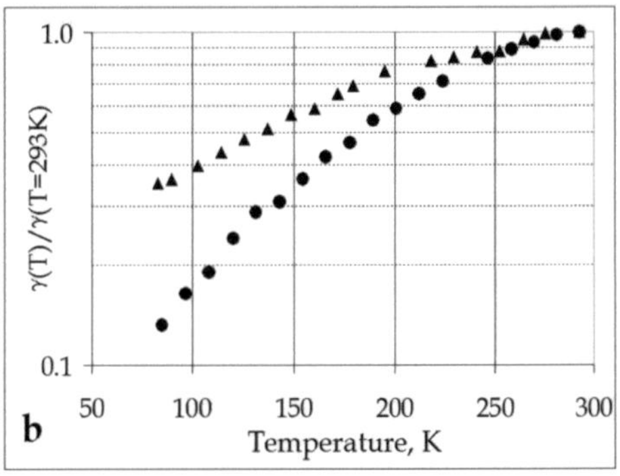

Fig. 12. Temperature dependence of the focal length of thermal lens F (a) and depolarization γ (b) in the TGG with the [001] orientation (circles, $F(T$=293K)=191cm, $\gamma(T$=293K)=5.2·10^{-3}) and in glass MOC-04 (triangles, $F(T$=293K)=581cm, $\gamma(T$=293K)=5.1·10^{-3}).

Traditional (Fig. 1) FI based on [111]-oriented TGG and FIs with QR (Fig. 9b) enjoy also the effect of ξ decreased by a factor of 1.9 (see Fig. 13b), i.e. from 2.25 to 1.2. According to (8,22) and (36) this increases maximum power by a factor of 1.6 in both cases. Bearing in mind that FIs with QR at room temperature can operate at 1-2kW, we obtain that at T=77K the maximum power will be tens of kW. Since the P/Q ratio is little dependent on temperature (Fig. 13a), all the above considerations are valid for both, the isolation and the thermal lens.

Note that in cryogenic FIs the temperature dependence of Verdet constant becomes more important because of increasing $V^{-1}dV/dT$ and deceasing L; compare two terms in (21). In order to suppress the dV/dT effect it is promising to cool TGG(s) through optical surfaces by means of a sapphire, for example (the characteristic length of the TGG crystal(s) in cryogenic FIs is about 3mm). Advantages of such disk geometry (see section 4.3) are huge reduction of depolarization caused by both, temperature dependence of Verdet constant and photoelastic effect. Our recent study showed that FI cooling is a very promising technique, which had been successfully applied in laser amplifiers (Ripin et al., 2004).

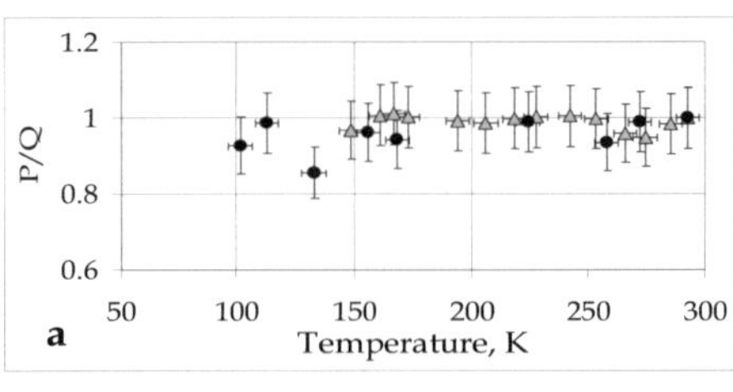

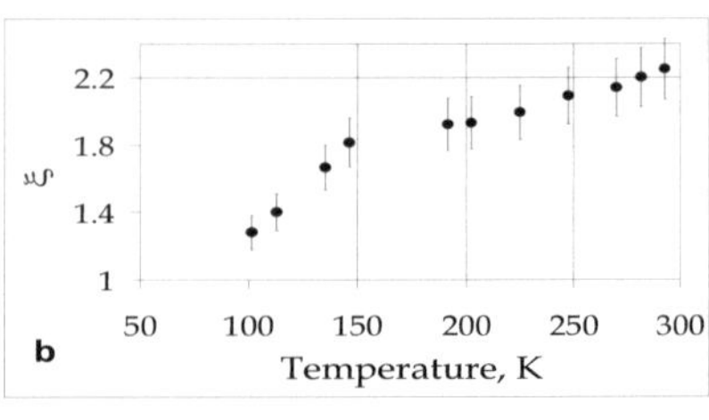

Fig. 13. Temperature dependence of the P/Q ratio (a) and parameter of optical anisotropy ξ (b) for TGG (circles) and glass MOC-04 (triangles).

4.2 Increase of magnetic field of permanent magnets

An obvious way to increase laser power keeping the same isolation degree is to increase magnetic filed B, and hence to decrease MOE length L. In (Zheleznov et al., 2007) we describe experiments with magnetic systems based on superconducting solenoids with B=5T. Such a strong field opens one more approach to suppress the thermal effects: use new magneto-optical media with lower absorption and higher thermal conductivity, for example

YAG and GGG crystals with or without doping. Additional advantage of these crystals is huge increase of their thermal conductivity at 77K, see section 4.1.

Increasing of B is also possible for a magnetic system based on constant magnets, even though it is a complex task. First, an increase of B implies shortening of the MOE, which imposes additional requirements to transverse homogeneity of B. Second, an increase of B demands building-up a mass of magnets and reinforces the mutual demagnetizing action of the neighboring magnets. Moreover, when the required (calculated) distribution of magnetization vector is achieved, one has to overcome technological difficulties at the stage of production and assembling.

In (Gauthier et al., 1986; Shiraishi et al., 1986) magnetic systems with a magnetic field up to 1T were designed. Typically a magnetic system is a set of axially and radially magnetized rings (Geho et al., 2005). The calculations demonstrated that such a system made it possible to form a magnetic field up to 1.7 T (Mukhin et al., 2009). In that paper we also proposed to replace part of the magnets by a magnetic conductor, see Fig. 14. The total energy of the field of magnets decreases, but careful selection of the shape and position of the magnetic conductor allows a local increase of field intensity in TGG.

The magnetic conductor (made of steel) consisted of two parts: the external part that was a screen (Fig. 14d) closing magnetic lines of force of the poles of radially magnetized rings, and the internal part - a pole terminal (Fig. 14e) concentrating magnetic field lines near the TGG. The magnets were made of the Nd-Fe-B alloy: radially magnetized rings with residual induction B_{r1}=12 kGauss and coercitive magnetization force H_{c1}=13 kOe and axially magnetized ring (B_{r2}=10 kGauss, H_{c2}=27 kOe). The magnetic conductor increased the field in the center of the magnetic system from 1.7 T to 2.1 T. Such a strong magnetic field enabled creating a 13mm-diameter FI with the TGG length of only 10.3 mm at modest mass and size of the magnetic system (diameter 132 mm, length 140 mm, mass 12 kg).

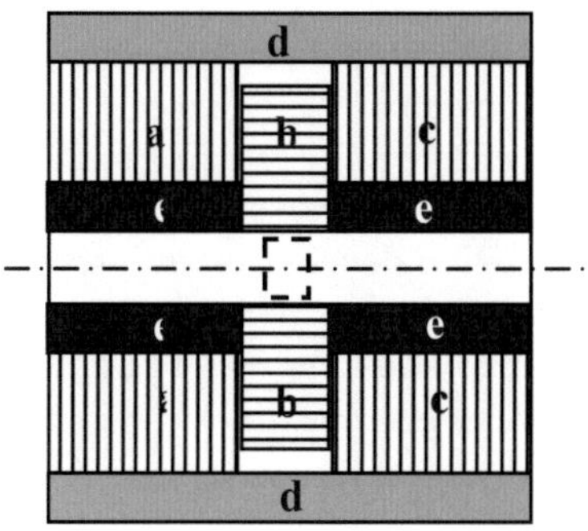

Fig. 14. Design of a magnetic system with B=2.1T: **a, c,** (**b**) are radially (axially) magnetized rings; **d,e** are magnetic conductor. Dashed rectangle shows TGG crystal.

In practice γ is determined by both, heat effects and "cold" depolarization. The latter depends on the TGG quality and magnetic field inhomogeneity. Note that the magnetic conductor also decreases inhomogeneity down to <0.3%, which corresponds to cold depolarization 50dB. In terms of maximal operating power, our FI (Mukhin et al., 2009) is 3…5 times better than the FI manufactured by leading companies and provides the isolation degree of ~30 dB at average laser power of ~400 W, see Fig. 15. Such a high power is ensured by a homogeneous magnetic field of 2.1 T (and, hence, a short TGG), as well as by the [001] orientation of a TGG crystal instead of [111], see (22, 8).

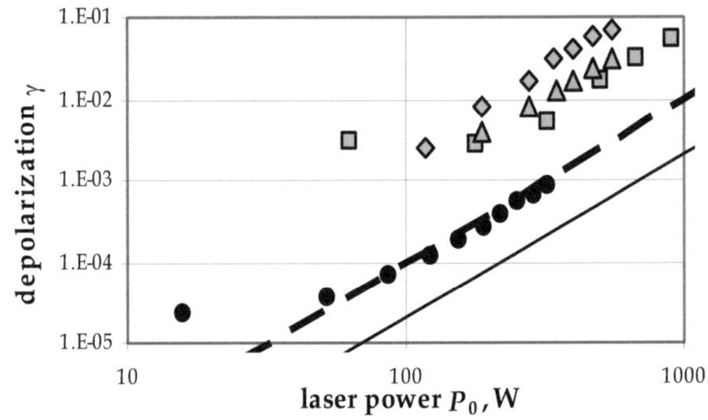

Fig. 15. γ as a function of P_0 for FI with B=2.1T (Mukhin et al., 2009) (circles, dashed line is theory for $\alpha=2.5\cdot10^{-3}$cm^{-1}), and for commercial FIs (Nicklaus et al., 2006) produced by Litton (rhombs), Linos (triangles), and EOT (squares). Solid line is theory for FI with B=2.5T and $\alpha=1.5\cdot10^{-3}$cm^{-1}.

If still stronger magnets are used, it will be possible to make TGG still shorter and to additionally increase admissible optical power. The estimates demonstrated a feasibility of producing FI with TGG 13 mm in diameter and 7 mm long using magnets with the following parameters: B_{r1}=14 kGauss, H_{c1}=16 кOe; B_{r2}=11.2 kGauss, H_{c2}=35 kOe. Recent experiments (Palashov et al., 2009) showed a possibility to reach B=2.5T and to shorten TGG to 8mm. If TGG with small absorption ($\alpha=1.5\cdot10^{-3}$ cm^{-1}) is used, the FI provides 27 dB isolation degree at laser power 1 kW, see Fig. 15. According to (28, 27), the thermal lens focal length in this FI will be 19m at 1kW (we assume the Gaussian beam radius r_0 to be 2.5 mm). The lens may be compensated as discussed in section 3.1.

4.3 Faraday isolators based on slab and disc geometries

Above we considered rod MOE geometry. Use of either slabs or several thin discs (Fig. 16) is more attractive for the thermal effects suppression. In this section we shall study these geometries both for traditional (Fig. 1) and novel (Fig. 9) FI designs. We shall focus on two main questions: how the depolarization depends on the aspect ratio of slab or disc, and what yield in maximum power the novel designs provide in comparison with the traditional one.

For slab geometry we shall use the following assumptions. The laser beam has a uniform intensity distribution over slab aperture and linear polarization along the x axis. Heat is removed only through horizontal surfaces, see Fig. 16a. The ratio of the slab thickness t to the slab width w is small: $R_s=w/t$<<1. Under these conditions, eigen polarizations of the birefringence induced by the photoelastic effect are oriented along the x and y axes, i.e. the matrix of MOE is described by formula (1) at Ψ=0. An expression for the phase shift δ_l is found in (Dianov, 1971):

$$\delta_l = \pi \frac{p}{R_s}\left(\frac{1}{6} - \frac{2y^2}{t^2}\right). \tag{39}$$

It is evident from (39) and the transition rule (8) that the [001] orientation provides a lower depolarization ratio. For the case (18) we derived the following formulas using the same mathematical procedure as for rod geometry (Khazanov, 2004):

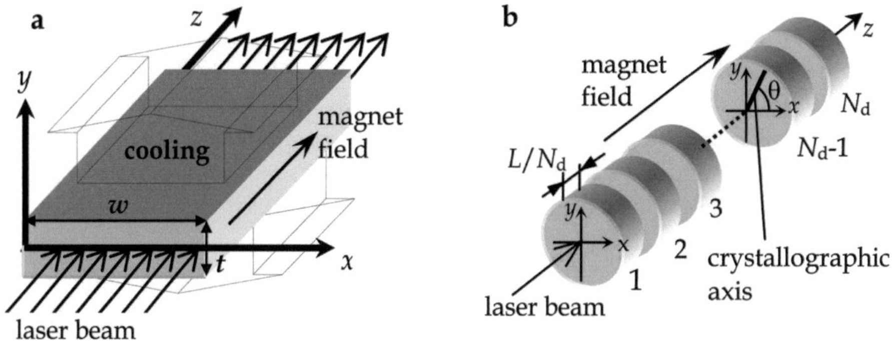

Fig. 16. Use of slabs (a) or discs (b) in Faraday isolators.

$$\gamma_{slab} = R_s^{-2} p^2 / 45 \quad \gamma_{slabL} = R_s^{-4} p^4 \left(\pi - 4\sqrt{2} + 2\right)^2 / 3780 \quad \gamma_{slabR} = R_s^{-4} p^4 \left(\pi - 2\sqrt{2}\right)^2 / 3780 . \qquad (40)$$

Numerical calculations made for an arbitrary aspect ratio R_s show that formulas (40) are valid at $R_s>3$. In practice, as a rule, $R_s>3$. Figure 17a presents the plot for $\gamma(p)$, both for rods and slabs. In practice, necessary values of the depolarization ratio usually lie within 10^{-4}–10^{-2}, where analytical formulas (40) are very accurate as seen from Fig. 17a.

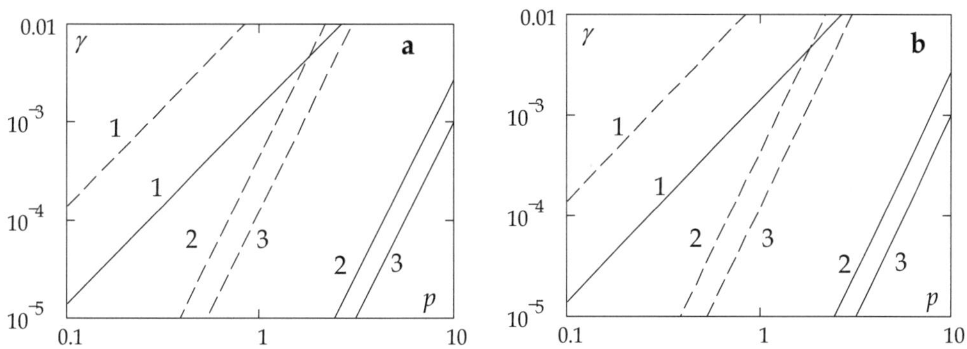

Fig. 17. The plots (solid lines) of $\gamma_{slab}(p)$ for TGG-based FI with slab, R_s=4 (a) and $\gamma_{disc}(p)$ for glass-based FI with discs, R_d=1.8 (b). 1 - FI design in Fig. 1; 2 – FI design in Fig. 9a; 3 – FI design in Fig 9b. Dashed lines show γ as a function of p for rods.

For all the FI designs shown in Figs. 1 and 9, $\gamma_{slab} \ll \gamma$. The depolarization ratio γ_{slab} is inversely proportional to the second power of the aspect ratio R_s for the traditional design, and to the fourth power of R_s for novel designs of FI. The depolarization ratio in novel designs is much lower than in the traditional one.

For TGG, ξ=2.25 and as follows from (40, 8), the [001] orientation provides a depolarization ratio that is 3.4 (11.3) times as small as that with the [111] orientation for the traditional (novel) designs. Estimations show that the FI presented in Fig. 9b with slabs made of TGG with aspect ratio R_s=5 at laser power 10kW provides an isolation degree of 30 dB. Slab-based FIs are used in high power slab lasers.

The idea of using thin discs consists in intensive cooling of optical surfaces (Fig. 16b). In this case the radial temperature gradient (the source of depolarization) will be considerably lower. At the same value of the magnetic field, the total length of magneto-optical elements L should be the same as in the rod geometry, i.e., N_d discs having length $h=L/N_d$ each should be used. Different relations between h, r_0 and disc radius R were considered in detail in (Mukhin & Khazanov, 2004), where it was shown that the analytical expressions obtained for the simplest case of thin ($h<<r_0$) and wide-aperture ($r_0<<R$) disc are valid if $R_d>3$, where $R_d= r_0/h$ is disc aspect ratio. In this case for $\theta=\theta_{opt}$ we have

$$\gamma_{disc}=\frac{A_6}{9\pi^2}(1-\nu)^2\frac{p^2}{R_d^4}\qquad \gamma_{discL}=\frac{8A_7}{(3\pi)^4}(1-\nu)^4\xi^2\left(2a^2+b^2\right)\frac{p^4}{R_d^8}\qquad \gamma_{discR}=\frac{2a^2A_7}{(3\pi)^4}(1-\nu)^4\left(3+2\xi^2+3\xi^4\right)\frac{p^4}{R_d^8}\quad(41)$$

As shown in (Mukhin & Khazanov, 2004), depolarization induced by temperature dependence of Verdet constant is less than γ_{disc} by four orders of magnitude. It is seen from (41) that when the disc thickness is reduced or the beam radius is increased, γ_{disc} drops in all cases. However, creation of thin discs involves engineering problems, see details in (Mukhin & Khazanov, 2004; Yasuhara et al., 2005). Therefore, it is reasonable to use the disc geometry for wide-aperture FIs. Such devices cannot be made of a TGG single crystal. However, magneto-optical glasses and TGG ceramics can be readily employed for this purpose. Figure 17b presents plots of $\gamma_{disc}(p)$ for rods and for discs made of magneto-optical glass (ξ=1).
Thus, γ_{disc} is inversely proportional to the fourth power of the aspect ratio R_d for the traditional design (Fig. 1), and to the eighth power of R_d for novel designs (Fig. 9). When discs are used instead of rods, the depolarization ratio can be reduced considerably if the disc thickness is less than the beam radius.
Estimations based on constants for magneto-optical glass (Table 1) show that when R_d=2.5 the FI presented in Fig. 9b provides an isolation degree of 35 dB at 5kW laser power. Use of TGG single crystal or TGG-ceramics allows using even higher power.
The disc geometry has two major disadvantages: optical surfaces have to be cooled, making the design more complicated and leading to wavefront distortions, and a large number of Fresnel back reflections; see details in (Mukhin & Khazanov, 2004).

5. Conclusion

Let us summarize the main results and discuss prospects for the future studies.
Thermal self-action of laser radiation in FIs leads to degradation of isolation degree $1/\gamma$ and to power losses in the initial spatial and polarization mode during the first pass through a FI. These losses comprise three components (Fig. 5): polarization losses γ_1, losses induced by isotropic thermal lens γ_i, and losses associated with anisotropic amplitude-phase distortions γ_a. The isotropic thermal lens can be compensated, i.e. γ_i can be decreased by an ordinary negative lens or using an adaptive method with negative dn/dT compensator (DKDP crystal is the best choice). In the first case, γ_i can be reduced by a factor of 15. The adaptive method totally nulls γ_i, but increases γ_1 and γ_a. Formulas for all γ are summarized in Table 3.
The influence of the temperature dependence of the Verdet constant on the isolation degree $1/\gamma$ may be neglected as compared to the influence of the photoelastic effect. The FI design

comprising two magneto-optical elements (Fig. 9) and the design with a crystalline quartz in telescope (Fig. 7) provide depolarization compensation. The latter design has an advantage of using standard commercial FI. The most suitable for high power lasers is the FI design shown in Fig. 9b, which can provide 30dB isolation at an average power up to a few kWs.

Neither isolation degree nor power losses in initial Gaussian beam depend on beam radius, but they depend on beam shape. A flat-top shaped beam is optimal, whereas a Gaussian beam has the strongest self-action, see Table 2.

Slab and disk geometries (Fig. 16) allow increasing laser power in FI up to at least 10kW due to optimized thermal management. Note that the analysis presented in section 4.3 for a slab was made only for the [001] and [111] crystal orientations and only for incident polarization parallel to the slab edge. Therefore, additional four parameters appear (inclination angle of incident polarization and three Euler angles). By varying these parameters one can further increase the isolation degree.

Recently, magnetic field of constant magnets in FI was increased from 1T to 2T, thus shortening TGG to 10mm. Further increase up to 3T will probably be possible in the future. In spite of the practical disadvantages of the superconductive solenoids, they provide magnetic field up to 10T (TGG crystal is not cooled). Another approach is cooling the whole FI to 77K, because TGG figures of merit (30) improve drastically due to the Verdet constant and thermo-optical properties increase. In superconductive solenoids or at cooling to 77K the TGG crystal length is 2-3mm only. This allows implementing disc geometry with cooling through optical surfaces by means of sapphire, for example. As a result, in both cases FIs operating at giant average power of tens of kilowatts is possible.

Taking into account figures of merit of magneto-optical medium (30), the TGG crystal is the best choice for high average power lasers. For the FI shown in Fig. 1, the [001] orientation is the best (and [111] is the worst) of all possible orientations. For the FI design shown in Fig. 9b, isolation degree is the same for the [001] and [111] orientations, but the latter is more practical.

From the point of view of magneto-optical glass melting, the possibilities are far from being exhausted. The most promising idea is creation of athermal glass (similarly to available laser glasses), i.e. glass with low values of *P* and *Q* and as a result with figures of merit more than the TGG ones. New magneto-optical crystals, e.g., $LiTb(MoO_4)_2$ and $NaTb(WO_4)_2$ seem to be promising for the future developments.

In the nearest future we forecast wide usage of ceramics (TGG, TAG, TSAG, Nd-doped YAG, Y_2O_5, Sc_2O_5, Lu_2O_5, and others) due to its high thermal conductivity in comparison with glass, large aperture in comparison with single crystal, and low absorption.

Thus, even though up to now Faraday isolators were experimentally tested at average laser power of only 1kW, the 10-kilowatt barrier can be not only reached in the nearest future but also successfully overcome.

6. References

Andreev, N.; Babin, A.; Zarubina, T.; Kiselev, A.; Palashov, O.; Khazanov, E. & Shaveleov, O. (2000a). Thermooptical constant of magneto-active glasses. *Journal of Optical Technology*, Vol.67, No.6, (556-558), 1070-9762

Andreev, N.F.; Palashov, O.V.; Poteomkin, A.K.; Sergeev, A.M.; Khazanov, E.A. & Reitze, D.H. (2000b). A 45-dB Faraday isolator for 100-W average radiation power. *Quantum Electronics*, Vol.30, No.12, (1107-1108), 1063-7818

Andreev, N.F.; Katin, E.V.; Palashov, O.V.; Poteomkin, A.K.; Reitze, D.; Sergeev, A.M. & Khazanov, E.A. (2002). The use of crystalline quartz for compensation for thermally induced depolarization in Faraday isolators. *Quantum Electronics*, Vol.32, No.1, (91-94), 1063-7818

Barnes, N.P. & Petway, L.P. (1992). Variation of the Verdet constant with temperature of TGG. *Journal of the Optical Society of America B*, Vol.9, No.10, (1912-1915), 0740-3224

Carr, I.D. & Hanna, D.C. (1985). Performance of a Nd:YAG oscillator/amplifier with phase-conjugation via stimulated Brillouin scattering. *Applied Physics B*, Vol.36, No.2, (83-92), 0946-2171

Chen, X.; Lavorel, B.; Boquillon, J.P.; Saint-Loup, R. & Jannin, M. (1998). Letter: Optical rotary power at the resonance of the terbium $^7F_6 \rightarrow ^5D_4$ line in terbium gallium garnet. *Solid-State Electronics*, Vol.42, No.9, (1765-1766), 0038-1101

Chen, X.; Galemezuk, R.; Salce, B.; Lavorel, B.; Akir, C. & Rajaonah, L. (1999). Long-transient conoscopic pattern technique. *Solid State Communications*, Vol.110, No.8, (431-434), 0038-1098

Davis, J.A. & Bunch, R.M. (1984). Temperature dependence of the Faraday rotation of Hoya FR-5 glass. *Applied Optics*, Vol.23, No.4, (633-636), 0003-6935

Demskaya, E.L. & Prokhorova, T.I. (1983). Investigation of properties of high-silica glass Nd_2O_3. *Fizika i Himiya Plasmi*, Vol.9, No.5, (554-560)

Denman, C.A. & Libby, S.I. (1999). Birefringence compensation using a single Nd:YAG rod. *Proceedings of Advanced Solid State Lasers*, pp. 608-612

Dianov, E.M. (1971). Thermal distortion of laser cavity in case of rectangular garnet slab. *Kratkiye Soobsheniya po Fisike*, Vol.8, (67-75), 1068-3356

Ganschow, S.; Klimm, D.; Reiche, P. & Uecker, R. (1999). On the crystallization of terbium aluminium garnet *Crystal Research and Technology*, Vol.34, No.5-6, (615-619), 0232-1300

Gauthier, D.J.; Narum, P. & Boyd, R.W. (1986). Simple, compact, high-performance permanent-magnet Faraday isolator. *Optics Letters*, Vol.11, No.10, (623-625), 0146-9592

Geho, M.; Takagi, T.; Chiku, S. & Fujii, T. (2005). Development of optical isolators for visible light using terbium aluminum garnet ($Tb_3Al_5O_{12}$) single crystals. *Japanese Journal of Applied Physics, Part 1*, Vol.44, No.7A, (4967-4970), 0021-4922

Gelikonov, V.M.; Gusovskii, D.D.; Leonov, V.I. & Novikov, M.A. (1987). Birefringence compensation in single-mode optical fibers. *Sov. Tech. Phys. Lett*, Vol.13, No.7, (322-323),

Giuliani, G. & Ristori, P. (1980). Polarization flip cavities: a new approach to laser resonators. *Optics Communications*, Vol.35, No.1, (109-112), 0030-4018

Guo, F.; Ru, J.; Li, H.; Zhuang, N.; Zhao, B. & Chen, J. (2009). Growth and magneto-optical properties of $LiTb(MoO_4)_2$ crystal. *Applied Physics B*, Vol.94, No.3, (437-441), 0946-2171

Ivanov, I.; Bulkanov, A.; Khazanov, E.; Mukhin, I.B.; Palashov, O.V.; Tsvetkov, V. & Popov, P. (2009). Terbium gallium garnet for high average power Faraday isolators: modern aspects of growing and characterization. *Proceedings of CLEO /EUROPE-EQEC 2009,* pp. CE.P.12 MON, Munich, Germany

Jiang, Y.; Myers, M.J. & Rhonenhouse, D. (1992). High Verdet constant Faraday rotator glasses. *SPIE Proceedings,* Vol.1761, (268-272), 9780819409348, Damage to Space Optics, and Properties and Characteristics of Optical Glass

Kagan, M.A. & Khazanov, E.A. (2004). Thermally induced birefringence in Faraday devices made from terbium gallium garnet-polycrystalline ceramics. *Applied Optics,* Vol.43, No.32, (6030-6039), 0003-6935

Kaminskii, A.A.; Eichler, H.J.; Reiche, P. & Uecker, R. (2005). SRS risk potential in Faraday rotator $Tb_3Ga_5O_{12}$ crystals for high-peak power laser. *Laser Physics Letters,* Vol.2, No.10, (489-492), 1612-2011

Khazanov, E.A. (1999). Compensation of thermally induced polarization distortions in Faraday isolators. *Quantum Electronics,* Vol.29, No.1, (59-64), 1063-7818

Khazanov, E.A.; Kulagin, O.V.; Yoshida, S.; Tanner, D. & Reitze, D. (1999). Investigation of self-induced depolarization of laser radiation in terbium gallium garnet. *IEEE Journal of Quantum Electronics,* Vol.35, No.8, (1116-1122), 0018-9197

Khazanov, E.A. (2000). Characteristic features of the operation of different designs of the Faraday isolator for a high average laser-radiation power. *Quantum Electronics,* Vol.30, No.2, (147-151), 1063-7818

Khazanov, E.; Andreev, N.; Babin, A.; Kiselev, A.; Palashov, O. & Reitze, D. (2000). Suppression of self-induced depolarization of high-power laser radiation in glass-based Faraday isolators. *Journal of the Optical Society of America B,* Vol.17, No.1, (99-102), 0740-3224

Khazanov, E.A. (2001). A new Faraday rotator for high average power lasers. *Quantum Electronics,* Vol.31, No.4, (351-356), 1063-7818

Khazanov, E.; Andreev, N.; Palashov, O.; Poteomkin, A.; Sergeev, A.; Mehl, O. & Reitze, D. (2002a). Effect of terbium gallium garnet crystal orientation on the isolation ratio of a Faraday isolator at high average power. *Applied Optics,* Vol.41, No.3, (483-492), 0003-6935

Khazanov, E.A.; Anastasiyev, A.A.; Andreev, N.F.; Voytovich, A. & Palashov, O.V. (2002b). Compensation of birefringence in active elements with a novel Faraday mirror operating at high average power. *Applied Optics,* Vol.41, No.15, (2947-2954), 0003-6935

Khazanov, E. (2003). Investigation of Faraday isolator and Faraday mirror designs for multi-kilowatt power lasers. *SPIE Proceedings,* Vol.4968, (115-126), 9780819447685, Solid State Lasers XII, San Jose, California

Khazanov, E. (2004). Slab-based Faraday isolators and Faraday mirrors for 10kW average laser power. *Applied Optics,* Vol.43, No.9, (1907-1913), 0003-6935

Khazanov, E.A.; Andreev, N.F.; Mal'shakov, A.N.; Palashov, O.V.; Poteomkin, A.K.; Sergeev, A.M.; Shaykin, A.A.; Zelenogorsky, V.V.; Ivanov, I.; Amin, R.S.; Mueller, G.; Tanner, D.B. & Reitze, D.H. (2004). Compensation of thermally induced modal distortions in Faraday isolators. *IEEE Journal of Quantum Electronics*, Vol.40, No.10, (1500-1510), 0018-9197

Liu, J.; Guo, F.; Zhao, B.; Zhuang, N.; Chen, Y.; Gao, Z. & Chen, J. (2008). Growth and magneto-optical properties of $NaTb(WO_4)_2$. *Journal of Crystal Growth*, Vol.310, No.10, (2613-2616), 0022-0248

Malshakov, A.N.; Pasmanik, G. & Poteomkin, A.K. (1997). Comparative characteristics of magneto-optical materials. *Applied Optics*, Vol.36, No.25, (6403-6410), 0003-6935

Mansell, J.D.; Hennawi, J.; Gustafson, E.K.; Fejer, M.M.; Byer, R.L.; Clubley, D.; Yoshida, S. & Reitze, D.H. (2001). Evaluating the effect of transmissive optic thermal lensing on laser beam quality with a Shack-Hartmann wave-front sensor. *Applied Optics*, Vol.40, No.3, (366-374), 0003-6935

Mueller, G.; Amin, R.S.; Guagliardo, D.; McFeron, D.; Lundock, R.; Reitze, D.H. & Tanner, D.B. (2002). Method for compensation of thermally induced modal distortions in the input optical components of gravitational wave interferometers. *Classical and Quantum Gravity*, Vol.19, (1793-1801), 0264-9381

Mukhin, I.B. & Khazanov, E.A. (2004). Use of thin discs in Faraday isolators for high-average-power lasers. *Quantum Electronics*, Vol.34, No.10, (973-978), 1063-7818

Mukhin, I.B.; Voitovich, A.V.; Palashov, O.V. & Khazanov, E.A. (2009). 2.1 tesla permanent - magnet Faraday isolator for subkilowatt average power lasers. *Optics Communications*, Vol.282, (1969-1972), 0030-4018

Nicklaus, K.; Daniels, M.; Hohn, R. & Hoffmann, D. (2006). Optical isolator for unpolarized laser radiation at multi-kilowatt average power. *Proceedings of Advanced Solid-State Photonics*, pp. MB7, Incline Village, Nevada, USA

Palashov, O.V.; Voitovich, A.V.; Mukhin, I.B. & Khazanov, E.A. (2009). Faraday isolator with 2.5 tesla magnet field for high power lasers. *Proceedings of CLEO /EUROPE-EQEC 2009*, pp. CA1.6 MON, Munich, Germany

Poteomkin, A.K. & Khazanov, E.A. (2005). Calculation of the laser-beam M2 factor by the method of moments. *Quantum Electronics*, Vol.35, No.11, (1042-1044), 1063-7818

Raja, M.Y.A.; Allen, D. & Sisk, W. (1995). Room-temperature inverse Faraday effect in terbium gallium garnet. *Applied Physics Letters*, Vol.67, No.15, (2123-2125), 0003-6951

Ripin, D.J.; Ochoa, J.R.; Aggarwal, R.L. & Fan, T.Y. (2004). 165-W cryogenically cooled Yb:YAG laser. *Optics Letters*, Vol.29, No.18, (2154-2156), 0146-9592

Robinson, C.C. (1964). The Faraday rotation of diamagnetic glasses from 0.334 micrometer to 1.9 micrometer. *Applied Optics*, Vol.3, No.10, (1163-1166), 0003-6935

Roth, M.S.; Wyss, E.W.; Graf, T. & Weber, H.P. (2004). End-pumped Nd:YAG laser with self-adaptive compensation of the thermal lens. *IEEE Journal of Quantum Electronics*, Vol.40, No.12, (1700-1703), 0018-9197

Rubinstein, C.B.; Uitert, L.G.V. & Grodkiewicz, W.H. (1964). Magneto-optical properties of rare earth (III) aluminum garnets. *Journal of Applied Physics*, Vol.35, No.10, (3069-3070), 0021-8979

Scott, W.C. & de Wit, M. (1971). Birefringence compensation and TEM_{00} mode enhancement in a Nd:YAG laser. *Applied Physics Letters*, Vol.18, No.1, (3-4), 0003-6951

Shiraishi, K.; Tajima, F. & Kawakami, S. (1986). Compact Faraday rotator for an optical isolator using magnets arranged with alternating polarities. *Optics Letters*, Vol.11, No.2, (82-84), 0146-9592

Slack, G.A. & Oliver, D.W. (1971). Thermal conductivity of garnets and phonon scattering by rare-earth ions. *Physical Review B*, Vol.4, No.2, (592-609), 1098-0121

Soms, L.N. & Tarasov, A.A. (1979). Thermal deformation in color-center laser active elements. 1.Theory. *Soviet Journal of Quantum Electronics*, Vol.9, No.12, (1506-1508), 0049-1748

Tabor, M.J. & Chen, F.S. (1969). Electromagnetic propagation through materials possessing both Faraday rotation and birefringence: experiments with ytterbium orthoferrite. *Journal of Applied Physics*, Vol.40, No.7, (2760-2765), 0021-8979

VIRGO-Collaboration. (2008). In-vacuum optical isolation changes by heating in a Faraday isolator. *Applied Optics*, Vol.47, No.31, (5853 5861), 0003 6935

Voytovich, A.V.; Katin, E.V.; Mukhin, I.B.; Palashov, O.V. & Khazanov, E.A. (2007). Wide-aperture Faraday isolator for kilowatt average radiation powers. *Quantum Electronics*, Vol.37, No.5, (471-474), 1063-7818

Wynands, R.; Diedrich, F.; Meschede, D. & Telle, H.R. (1992). A compact tunable 60-dB Faraday optical isolator for the near infrared. *Review of Scientific Instruments*, Vol.63, No.12, (5586-5590), 0034-6748

Yasuhara, R.; Yamanaka, M.; Norimatsu, T.; Izawa, Y.; Kawashima, T.; Ikegawa, T.; Matsumoto, O.; Sekine, T.; Kurita, T.; Kan, H. & Furukawa, H. (2005). Design and analysis on face-cooled disk Faraday rotator under high average power lasers. *Proceedings of Advanced Solid-State Photonics*. pp. MB43, Vienna, Austria

Yasuhara, R.; Tokita, S.; Kawanaka, J.; Kawashima, T.; Kan, H.; Yagi, H.; Nozawa, H.; Yanagitani, T.; Fujimoto, Y.; Yoshida, H. & Nakatsuka, M. (2007). Cryogenic temperature characteristics of Verdet constant on terbium gallium garnet ceramics. *Optics Express*, Vol.15, No.18, (11255-11261), 1094-4087

Yoshikawa, A.; Kagamitani, Y.; Pawlak, D.A.; Sato, H.; Machidab, H. & Fukudaa, T. (2002). Czochralski growth of $Tb_3Sc_2Al_3O_{12}$ single crystal for Faraday rotator *Materials Research Bulletin*, Vol.37, No.1, (1-10), 0025-5408

Zarubina, T.V.; Kim, T.A.; Petrovskiy, G.T.; Smirnova, L.A. & Edel'man, I.S. (1987). Temperature dependence and dispersion of Faraday effect in glass based on oxide of terbium and cerium. *Optiko-mechanicheskaya Promyshlennost'*, Vol.11, (33-45), 1070-9762

Zarubina, T.V. & Petrovsky, G.T. (1992). Magnetooptical glasses made in Russia. *Opticheskii Zhurnal*, Vol.59, No.11, (48-52), 1070-9762

Zarubina, T.V.; Mal'shakov, A.N.; Pasmanik, G.A. & Poteomkin, A.K. (1997). Comparative characteristics of magnetooptical glasses. *Opticheskii Zhurnal*, Vol.64, No.11, (67-71), 1070-9762

Zarubina, T.V. (2000). Private communication.

Zelenogorsky, V.; Palashov, O. & Khazanov, E. (2007). Adaptive compensation of thermally induced phase aberrations in Faraday isolators by means of a DKDP crystal. *Optics Communications*, Vol.278, No.1, (8-13), 0030-4018

Zheleznov, D.S.; Voitovich, A.V.; Mukhin, I.B.; Palashov, O.V. & Khazanov, E.A. (2006). Considerable reduction of thermooptical distortions in Faraday isolators cooled to 77 K. *Quantum Electronics*, Vol.36, No.4, (383-388), 1063-7818

Zheleznov, D.S.; Khazanov, E.A.; Mukhin, I.B.; Palashov, O.V. & Voytovich, A.V. (2007). Faraday rotators with short magneto-optical elements for 50-kW laser power. *IEEE Journal of Quantum Electronics*, Vol.43, No.6, (451-457), 0018-9197

Faraday isolator with slab magnetooptical elements

Faraday isolator with strong magnet field

Faraday isolator for vacuum application

4

Numerical Simulation of High-Power Operation of 2 μm Co-doped Tm,Ho Solid-State Lasers

O. A. Louchev[1], Y. Urata[1], M. Yumoto[2], N. Saito[2] and S. Wada[2]
[1]*Megaopto Co. Ltd., RIKEN Cooperation Center W414, 2-1 Hirosawa, Wako, Saitama 351-0106,*
[2]*Solid-State Optical Science Research Unit, RIKEN, 2-1 Hirosawa, Wako, Saitama 351-0198,*
[1,2]*Japan*

1. Introduction

The development of solid state 2 μm Tm, Ho lasers has remained a topic of particular interest for many years, due to a number of possible applications such as coherent laser radar, atmospheric sensing, a possible pump source for an optical parametric oscillator operating in the mid-infrared region, and medicine. In recent years significant progress in understanding the basic phenomena underlying Tm,Ho laser operation as well as in developing high power lasers has been achieved (Tyminski *et al.*, 1989; French *et al.*, 1992; Petrin *et al.*, 1992; Jani *et al.*, 1991; Jani *et al.*, 1995; Yu *et al.*, 1998; Alpat'ev *et al.*, 1998; Alpat'ev *et al.*, 1993; Barnes *et al.*, 1996; Lee *et al.*, 1996; Rustad & Stenersen, 1996; Bruneau, 1998; Bourdet & Lescroart, 1999; Jackson & King, 1998; Sudesh & Asai, 2003; Sato *et al.*, 2004; Walsh *et al.* 2004; Galzerano *et al.*, 2004; Izawa *et al.*, 2000; Yu *et al.*, 2006; Zhang *et al.* 2006; Louchev *et al.* 2007 and 2008) and recently reviewed by (Eichhorn, 2008) and (Walsh, 2009). Nevertheless significant effort is still necessary for defining optimized parameters and achieving high lasing efficiency for high power operation associated with significant temperature increase and related effects of pulse energy inhibition and thermal lensing destroying laser operation (Akhmanov *et al.* 1972).

In this work we review main computational and numerical optimization results of a side pumped Tm,Ho:YLF laser for producing MW-power giant pulses with durations of several hundred nanoseconds using the coupled thermo-optical model (Louchev *et al.*, 2007; Louchev *et al.*, 2008). In particular, section 2 focuses on optical model which integrates 8-level rate dynamics model with the oscillator energy equation. In section 3 we discuss the possibility of G-pulse optimization by doping concentration and by Q-switching control. Section 4 focuses on coupling optical and rate dynamics model with the heat transfer computations. In section 5 we demonstrate main effects in water cooled laser operation. In section 6 we compare main results with available experimental data. In section 7 we outline main conclusions.

2. Pulse generation model

Our study is based on 8-level (see Fig. 1 for excited level scheme) rate dynamics model (Barnes et al. 1996; Walsh *et al.*, 2004) describing the population dynamics in Tm,Ho lasers by the following set of equations:

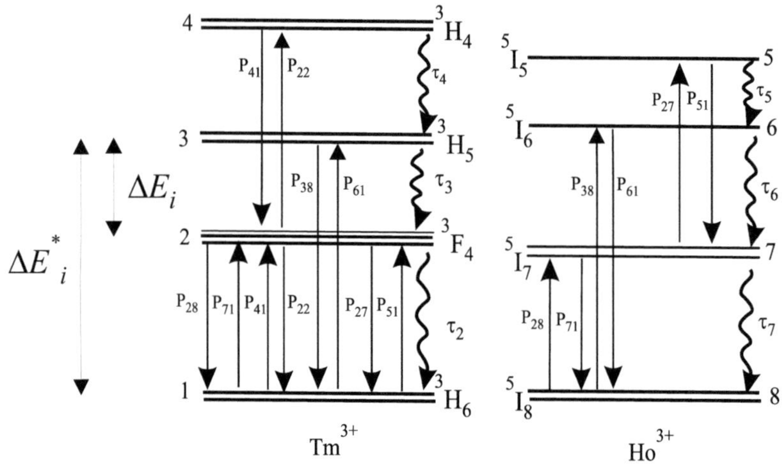

Fig. 1. Energy transfer processes in Tm, Ho doped materials.

$$\frac{dn_1}{dt} = -R_p(t,z,r) + \frac{n_2}{\tau_2} + p_{28}n_2n_8 - p_{71}n_7n_1 - p_{41}n_4n_1 + p_{22}n_2^2, \tag{1}$$
$$+ p_{27}n_2n_7 - p_{51}n_5n_1 - p_{61}n_6n_1 + p_{38}n_3n_8$$

$$\frac{dn_2}{dt} = -\frac{n_2}{\tau_2} + \frac{n_3}{\tau_3} - p_{28}n_2n_8 + p_{71}n_7n_1 + 2p_{41}n_4n_1 - 2p_{22}n_2^2, \tag{2}$$
$$- p_{27}n_2n_7 + p_{51}n_5n_1$$

$$\frac{dn_3}{dt} = -\frac{n_3}{\tau_3} + \frac{n_4}{\tau_4} + p_{61}n_6n_1 - p_{38}n_3n_8' \tag{3}$$

$$\frac{dn_4}{dt} = R_p(t,z,r) - \frac{n_4}{\tau_4} - p_{41}n_4n_1 + p_{22}n_2^2' \tag{4}$$

$$\frac{dn_5}{dt} = -\frac{n_5}{\tau_5} + p_{27}n_7n_2 - p_{51}n_5n_1, \tag{5}$$

$$\frac{dn_6}{dt} = -\frac{n_6}{\tau_6} + \frac{n_5}{\tau_5} - p_{61}n_6n_1 + p_{38}n_8n_3, \tag{6}$$

For the upper laser state ($^5I_7$):

$$\frac{dn_7}{dt} = -\frac{n_7}{\tau_7} + \frac{n_6}{\tau_6} + p_{28}n_2n_8 - p_{71}n_7n_1 - p_{27}n_2n_7 + p_{51}n_5n_1 - \frac{c\sigma_{se}}{\eta}(f_7n_7 - f_8n_8)\varphi(t,\mathbf{r}), \tag{7}$$

For the lower laser state ($^5I_8$):

$$\frac{dn_8}{dt} = \frac{n_7}{\tau_7} - p_{28}n_2n_8 + p_{71}n_7n_1 + p_{61}n_6n_1 - p_{38}n_3n_8 + \frac{c\sigma_{se}}{\eta}(f_7n_7 - f_8n_8)\varphi(t,\mathbf{r}), \tag{8}$$

where n_i are the state concentrations, p_{ij} are the probabilities of the optical transitions (Walsh *et al.*, 2004), η is the refractive index, τ_i are the state lifetimes, $R_p(t,z,r)$ is the pumping source space and time distribution, σ_{se} is the stimulated emission cross-section, f_i are the Boltzmann populations factors and $\phi(t,\mathbf{r})$ is the local laser photon density with dimensionality of $1/m^3$.

The local laser photon density is represented via the product of (i) the total number of photons depending on t and (ii) the normalized space distribution function as $\phi(t,\mathbf{r}) = \Phi_0(t)\phi_0(r,z)$. The resulting equation for total photon number, $\Phi_0(t)$, inside the oscillator cavity is given by the differential equation (Risk, 1988; Rustad & Stenersen, 1996; Walsh *et al.*, 2004; Černý & Burns, 2005):

$$\frac{d\Phi_0(t)}{dt} = \frac{\Phi_0(t)c\sigma_{se}}{\eta}\iint_{V_{cr}}(f_7 n_7 - f_8 n_8)\phi_0(r,z)dV - \frac{\Phi_0(t)}{\tau_c} + \frac{\varepsilon}{\tau_7}\iint_{V_{cr}} n_7 dV, \tag{9}$$

which includes the integrals of the stimulated and spontaneous emissions over the crystal volume, V_{cr}, and where τ_c is the cavity time, and $\varepsilon \approx 10^{-7}$-$10^{-8}$ is the coefficient which takes into account the proportion of photons spontaneously emitted within the solid angle of resonator mirrors, thereby initiating the development of the laser beam.
For the operating resonator the cavity time is:

$$\tau_c^{-1} = \frac{c}{2L_{opt}}\left[-\ln R_l - \ln(1-T_{out}) + \beta + 2\sum_i \alpha_i l_i\right], \tag{10}$$

where $L_{opt} = L_{cav} + (\eta - 1)L_{cr}$ is the characteristic optical length, L_{cav} is the cavity length and L_{cr} is the crystal length; R_l is the back mirror reflectance, T_{out} is the output mirror transmittance and β is the parameter used in our simulations for the optical loss associated with the active Q-switching: β=0 for the open resonator and $\beta >> -\ln R_1 - \ln(1-T_{out})$ for the closed resonator.
The value of $2\sum_i \alpha_i l_i$ in Eq. (10) represents the sum of absorption losses in different elements along which laser beam propagates inside the cavity. This value should be small to ensure the feasibility of the laser operation. Otherwise the significant part of the laser energy generated by the crystal is lost.
For the case of ≈500 ns pulse generation the cavity length L_{cav} >> L_{cr} and the laser beam radius inside the crystal may be considered to be constant. The spatial distribution of photons inside the operating crystal is simplified as:

$$\phi_0(r,z) = \frac{2}{\pi w_0^2 L_{cav}}\exp\left(\frac{-2r^2}{w_0^2}\right), \tag{11}$$

where w_0 is the beam waist size of TEM_{00} mode defined by the resonator parameters (for instance, for the simplest case of the confocal spherical resonator one has $w_0 = \sqrt{L_{cav}\lambda_l/2\pi}$).
The solution of the rates equation together with main oscillator equation gives the value of the output power, (W), as:

$$J(t) = \Phi_0(t)\frac{ch\nu_{las}}{2L_{opt}}\ln\frac{1}{1-T_{out}}, \tag{12}$$

We consider Tm,Ho:YLF operation side pumped by 780 nm LD radiation. For 6 % Tm doped YLF one finds for the absorption coefficient $\alpha=\sigma_a N_{Tm}\approx 2.8$ cm^{-1}. Thus, a 2 mm diameter YLF crystal is able to absorb (1-exp(-2αd))≈0.67-fraction of the incident beam flux in the case of the double-pass pumping scheme providing high uniformity of the absorbed flux over the crystal volume. High incident fluxes are able, in fact, to deplete the $^3H_6$-level in Tm^{3+} and to inhibit absorption (Rustad & Stenersen, 1996; Černý & Burns, 2005). However, when the concentration of the $^3H_6$-level does not fall below 0.9 of the initial Tm-concentration, the related variations of absorption coefficient do not exceed 5 %, and in making simulation one can use for R_p:

$$R_p(t) \approx \frac{\eta_p \eta_a Q_p}{\pi d^2 L_{cr} h \nu_p \Delta t_p}, \tag{13}$$

where $\eta_a = (1-\rho)\left[1-\exp(-2\alpha d)\right]$ is the absorption efficiency of the pumping radiation, ρ is the reflection factor of pumping radiation, Q_p is the pumping pulse energy, Δt_p is the pumping pulse duration, $\eta_p\approx 1.3$ is the pumping quantum efficiency for Tm,Ho:YLF.

3. Giant pulse optimization

Model parameters used for Tm,Ho:YLF in the simulations are fully specified by Walsh *et al.* (2004). Other parameters used in the simulations reported below are: crystal length L_{cr}=2 cm, crystal diameter d=2 mm, cavity length L_{cav}=1 m, output transmittance T=0.05, reflectance R_1=0.98, and beam waist radius w_0=0.85 mm. The simulations reveal two main effects important for efficiency optimization for producing nanosecond G-pulses.

Fig. 2 shows a simulation results for 6 % Tm doping and two values of Ho doping, 0.4 and 1.0 %, for the case of normal pulse operation (without Q-switch) demonstrating a series of relaxation oscillations. The pumping (780 nm laser diode) time for both modes is 0.5 ms and the total absorbed energy $\eta_a Q_p$=1.8 J. Fig.2 (a) shows the power oscillations for the cases of Ho = 0.4 and 1.0 % respectively, whereas Fig. 2 (b) shows the output pulse energies versus time. The final pulse energy for the 0.4 % Ho case (≈0.9 J), is calculated to be higher than that of 1.0 % Ho case (≈0.8 J). However, Fig. 2 (a) shows that the power of the first relaxation spike is higher for the case of 1.0 % Ho. Thus in G-pulse generation the 1.0 % Ho case is able to yield a higher output energy per pulse as compared with the 0.4 % Ho case.

Fig. 3 shows calculations of (a) G-pulse powers and (b) energies versus time for 0.4 % and 1.0 % Ho cases. The calculations assume that Q-switch operation is open by the end of the 0.5 ms pumping period. This figure shows that 1.0 % Ho doping can provide higher energy output per pulse as compared with 0.4 % Ho doping due to significantly larger level of the inverted population achieved after the pumping (See. Fig. 4).

Fig. 5 compares the resulting pulse energy for the normal pulse and for the G-pulse as a function of Ho-concentration for the given 6 % Tm doping. These figures confirm that for the normal pulse 0.4 % Ho doping provides the highest output energy whereas for G-pulse 1.0-1.2 % looks to be optimal.

Let us now focus on the time optimization of G-pulse operation. The above results on G-pulse generation assumed 0.5 ms pumping followed by immediate Q-switch opening. However, Fig. 4 suggests that such Q-switch opening is premature because the inverted population density does not achieve its maximum at the end of pumping.

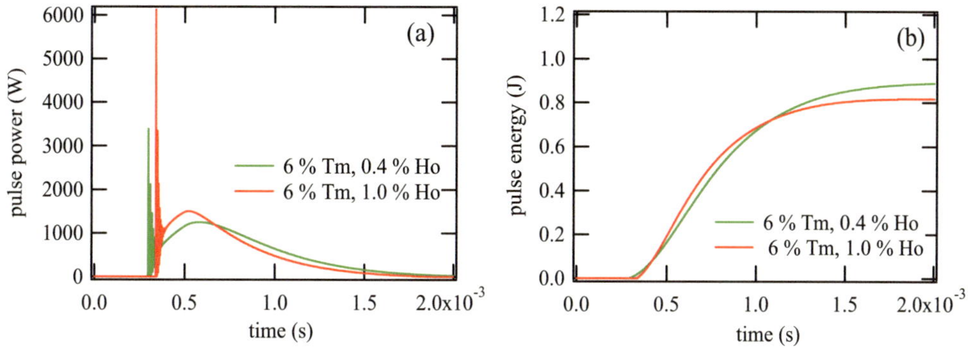

Fig. 2. Pulse generation in normal mode: (a) pulse power versus time and (b) pulse energy versus time for two Tm, Ho concentrations.

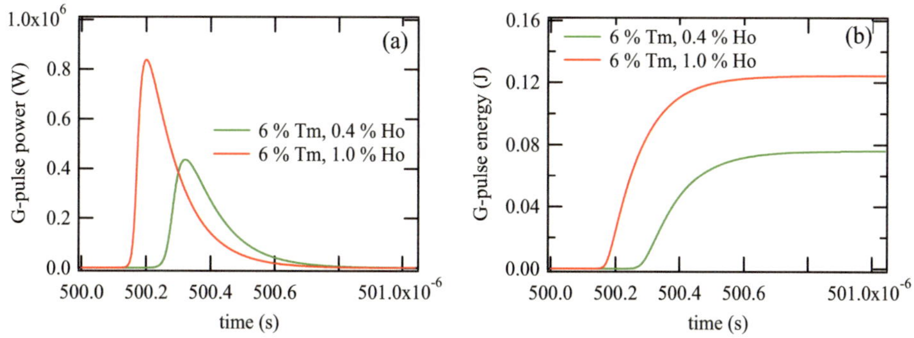

Fig. 3. G-pulse generation: (a) pulse power versus time and (b) pulse energy versus time.

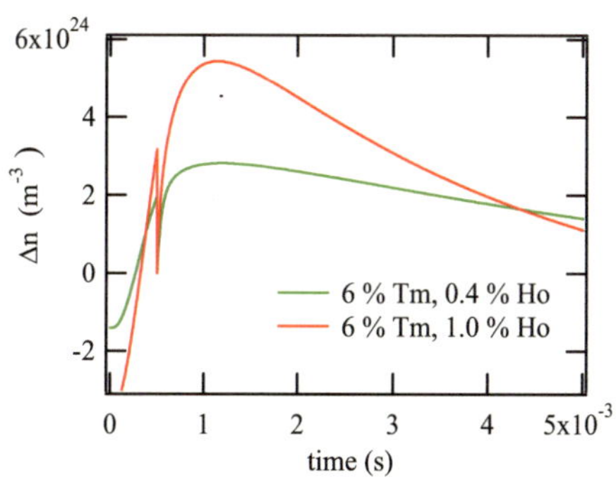

Fig. 4. Inversion population dynamics in G-pulse generation mode.

Now we provide more details to emphasize the importance of proper time control. In particular, Fig. 6 (a) shows the dependence of G-pulse power as function of the time for three characteristic cases: (i) 0.5 ms pumping followed by Q-switch opening (green), giving a pulse energy of ≈0.08 J, (ii) 1.9 ms pumping followed by Q-switch opening (blue) giving a pulse energy of ≈0.11 J and (iii) 0.5 ms pumping followed by Q-switch opening at 1.2 ms (red) giving a pulse energy of ≈0.12 J. Fig. 6 (b) shows the behavior of the inversion population, $\Delta n = f_7 n_7 - f_8 n_8$, for all three cases considered and provides the background

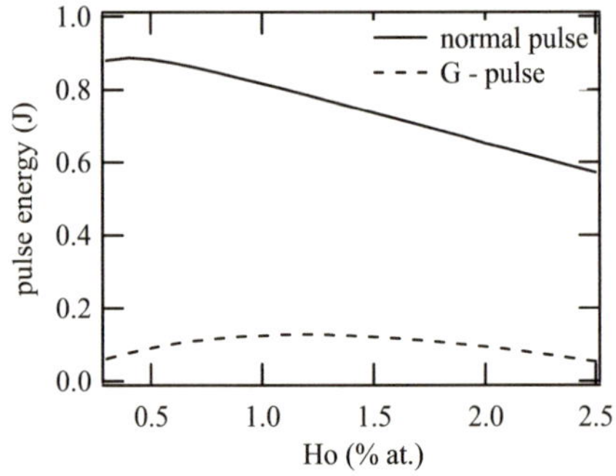

Fig. 5. Comparison of output pulse energy in normal (solid) and G-pulse (dashed) mode.

information for understanding time optimization issues. That is, in the first case the Q-switch is open prior to the moment when maximal population inversion in the Ho^{3+} is achieved and the resulting output G-pulse energy is small. After the pulse is finished and the Q-switch is closed, the excited Tm^{3+} state $^3F_4$ continues to transfer excitation towards the Ho^{3+} $^5I_8$. In this case a very significant part of the excitation "stored" initially in Tm^{3+} is transferred to Ho^{3+} after the G-pulse generation, and later on uselessly dissipated through spontaneous radiation and non-radiative transitions. In the second case of the significantly increased pumping time 1.9 ms with simultaneous opening of Q-switch a significantly higher generated pulse energy is achieved, due to significantly higher electronic excitation energy stored on the Ho^{3+} $^5I_7$ state by the end of the pumping pulse. The third case corresponds to the situation when a relatively short pumping time of 0.5 ms is used together with 0.7 ms delay in opening the Q-switch. This case gives maximal energy output of ≈0.12 J in G-pulse generation because the Q-switch opens exactly at the moment when the concentration on the lasing Ho^{3+} $^5I_7$ state and the value of Δn reach their maximum, allowing maximal energy extraction by the laser pulse.

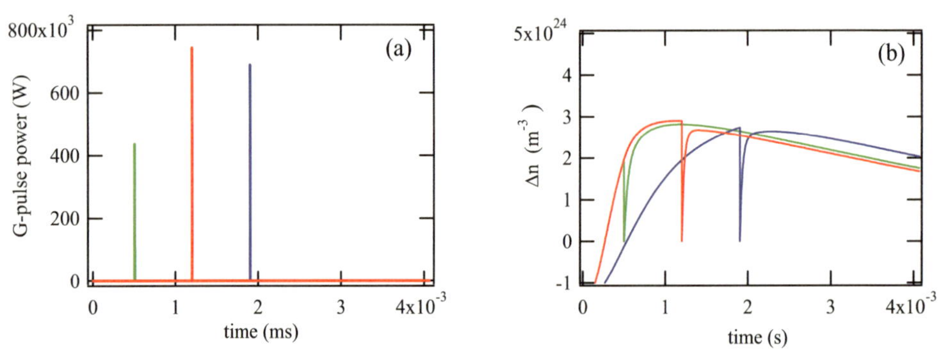

Fig. 6. Optimization of G-pulse generation: (a) G-pulse powers versus time and (b) corresponding population inversion densities versus time for three different scenarios.

Fig. 7 shows final results on G-pulse energy optimization: (a) for the case of extended LD pumping within the range 0.5 - 10 ms followed by immediate Q-switch opening and (b) 0.5 ms LD pumping with delayed Q-switch opening. Energy maxima exist in both cases: at ≈1.9 ms for (a) and at ≈1.2 ms for (b). However, case (b) provides the higher absolute maximum.

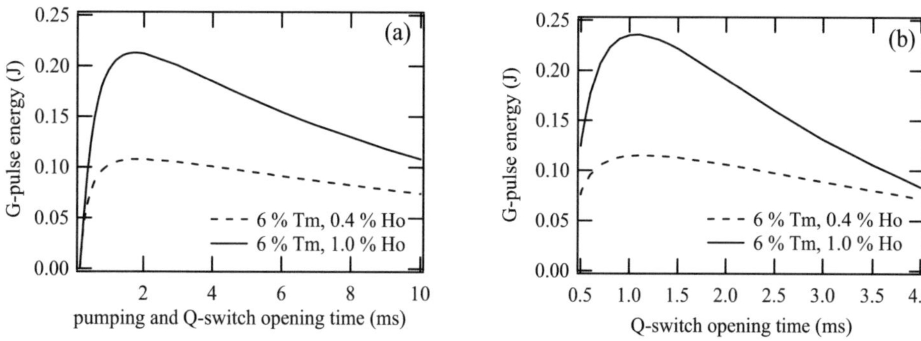

Fig. 7. Optimization of operation times: (a) Q-switch is open by the end of pumping and (b) pumping time is 0.5 ms and Q-switch is open after a delay.

It is worth noting that the delay between the pumping period and Q-switch opening shown above to optimize the energy pulse is associated with the characteristic time of the electronic transfer from the upper exited state $^3H_4$ to $^3F_4$ and, later on, towards $^5I_7$. Our simulation based on the above model shows that the excitation transfer from the $^3F_4$ state towards the $^5I_7$ state happens without any significant deceleration whereas significant delay is associated with the $^3H_4 \rightarrow {}^3F_4$ radiation transfer process, i.e. the rise in concentration at the Tm3+ $^3H_4$ state, defined by the pumping period used, and the depletion of this level, defined by the two-photon process $p_{41}n_4n_1$ and requiring ≈0.7 ms. In the case of extended pumping of several ms this delay does not play a very significant role, in contrast with short pumping times. The decrease in the delivered G-pulse energy shown in both cases for long times is associated with the increasing role of spontaneous radiation loss from both $^3F_4$ and $^5I_7$ states which have characteristic lifetimes of 16 and 15 ms respectively.
To finalize the discussion we would like to note that a ≈0.7 ms delay between pumping and achieving the maximal population inversion density in a Tm,Ho:YLF laser found in this study is experimentally supported by the observation of the amplification signal in a Tm,Ho:YLF crystal pumped by a 780 nm radiation pulse of 65 μs duration (Bruneau *et al.*, 1998). Data recorded for the YLF case shows the amplification signal maximum at 0.6-0.7 ms after the start of the pumping. However, the data for Tm,Ho:YAG operation does not show any significant delay between pumping period and maximal amplification signal.

4. Thermal model

The heat absorbed inside the crystal leads to a temperature increase over the crystal volume. For high power operation this temperature shift is able to change the local values of the Boltzmann population factors of the upper and lower lasing levels:

$$f_i(t,\mathbf{r}) = \frac{g_i \exp[-E_i/k_B T(t,\mathbf{r})]}{\sum_j g_j \exp[-E_j/k_B T(t,\mathbf{r})]}, \tag{14}$$

where k_B is the Boltzmann constant, g_i is the degeneracy of the *i*-level, and $T(t,\mathbf{r})$ is the local temperature.
Generally, the crystal is heated via lattice vibrations due to non-radiative decay of electrons from all levels involved in the excitations. The local heat source is defined by:

$$q_{cr}(t,\mathbf{r}) = \sum_{i=2}^{7} \Delta E_i n_i / \tau_{inr}, \tag{15}$$

where ΔE_i is the energy difference between the *i*-manifold and the next lower manifold into which the electron makes the transition (Fig. 1) and τ_{inr} are the non-radiative times inversely proportional to the non-radiative transition probabilities.

In order to avoid difficulties in defining the probabilities of non-radiative transitions, an estimate of the heat source can be made via the difference between the pumped energy and the energy of stimulated and spontaneous radiation leaving the crystal (Bruneau *et al*, 1998). This approach is mainly used for the CW mode or as an averaged estimate for high-repetition pulsed mode. However, we (Louchev et al., 2008) use this approach for normal or Q-switched mode operation by introducing a modification which takes into account the rate, $\sum_{i=2}^{7} \Delta E_i^* dn_i / dt$, at which the pumped energy is stored inside Tm^{3+} and Ho^{3+} ions as:

$$q_{cr}(t,\mathbf{r}) = R_p(t) h\nu_p - c\sigma_{se}\eta^{-1} h\nu_l (f_7 n_7 - f_8 n_8)\phi(t,\mathbf{r}) - \sum_{i=2}^{7} \Delta E_i n_i / \tau_{ir} - \sum_{i=2}^{7} \Delta E_i^* dn_i / dt, \tag{16}$$

where in addition to ΔE_i we introduce the energy difference between the *i*-manifold and the ground state ΔE_i^* (Fig. 1), ν_p is the pumping radiation frequency, ν_l is the lasing frequency and τ_{ir} are the corresponding radiative times (Walsh *et al.*, 2004).

The calculation of Eq. (16) for the Tm,Ho:YLF laser reveals several effects significant for energy extraction by lasing pulse. First, Fig. 8 shows (a) G-pulse power and (b) the energy and (c) optical loss balance integrated over the crystal volume versus time. In fact, Fig. 8 (b) reveals a very significant extension of the heat release period as compared with the pumping period. In particular, it shows that the heat is released inside the crystal over a period of ≈10 ms, whereas the pumping period is 0.5 ms during which only ≈30 % of heat is released. A two-time lower resulting temperature increase is achieved in crystal prior to G-pulse generation (1.2 ms) as follows from an estimate neglecting the thermal conductivity effect:

$$\Delta T_{cr}(t,\mathbf{r}) \approx \frac{1}{\rho c} \int_0^t q_{cr}(t,\mathbf{r}) dt \cdot \tag{17}$$

Second, Fig. 8 (b) also shows that in the final energy balance ≈0.12 J corresponds to the G-pulse energy, ≈0.75 J corresponds to the heat released inside the crystal and ≈0.84 J corresponds to the energy lost by spontaneous emission. Thus, about 43 % of the pumped energy is directly converted into heat. We should note that the estimates of heat release based on 2-level rate equations treat this value as the difference between the pumped energy and the optical energy of the laser pulse and the spontaneous emission from two levels, $^3F_4$ and $^5I_7$ (Bruneau *et al*, 1998). The energy spontaneously emitted by other levels, i.e. $^3H_5$, $^3H_4$, $^5I_5$ and $^5I_6$, are implicitly included into the heat released inside the crystal (Bruneau *et al*, 1998). The 8-level model used here shows that the contribution of the $^5I_5$ and $^5I_6$ levels into the spontaneous emission loss is negligibly small, whereas the contribution of $^3H_4$ and $^3H_5$ appears to be quite significant, ≈0.1 and ≈0.4 J, respectively. Adding these values to the heat release of ≈0.75 J gives ≈70 %, similar to the result from the 2-level model (Bruneau *et al*, 1998).

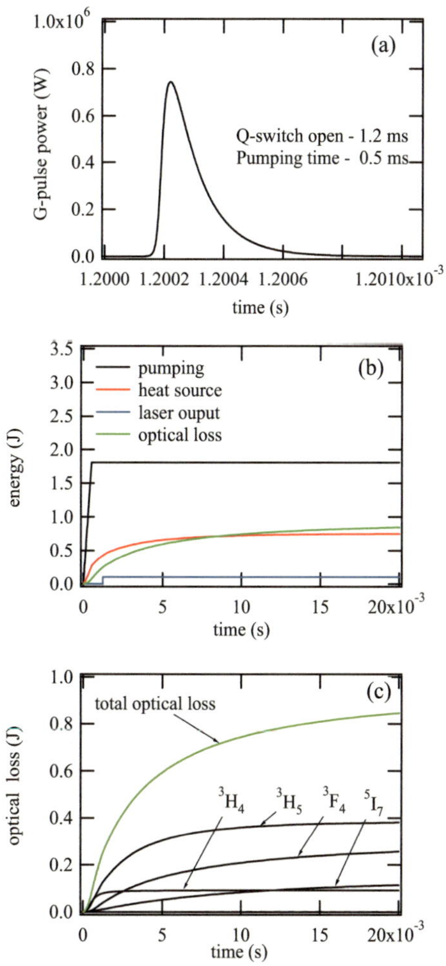

Fig. 8. Pulse power, energy balance versus time during laser operation: (a) G-pulse power with corresponding (b) energy balance and (c) optical loss by spontaneous radiation from different levels.

Thus, only ≈43 % of the pumped energy is released directly as heat inside the crystal, whereas ≈45 % is spontaneously emitted radiation from the crystal at wavelengths: λ_2=1.93 μm, λ_3=4.32 μm, λ_4=2.46 μm and λ_7=2.07 μm. These wavelengths are within the transparency range of the crystal and are therefore able to leave the crystal. Fig. 8 (c) shows the values of the total optical loss, $\sum_{i=2}^{7} \Delta E_i n_i / \tau_{ir}$, and also the losses emitted from all levels, $\Delta E_i n_i / \tau_{ir}$ integrated over the crystal volume. These radiation fluxes leaving the crystal are absorbed by the water flow typically used for crystal cooling. The water absorption coefficients for these wavelengths are given by Wieliczka *et al.*, 1989: α_2=124 cm^{-1}, α_3=300 cm^{-1},

α_4=63.5 cm^{-1} and α_7=31 cm^{-1}. That is, the spontaneously emitted fluxes leaving the crystal are absorbed within lengths of $\approx \alpha_i^{-1}$, i.e. within 80, 33, 157 and 320 μm from the surface, respectively. The absorption of these fluxes in the vicinity of the crystal surface can significantly inhibit heat dissipation from the crystal. The heat transfer to the water flow depends on the Reynolds number, *Re*, defining the level of the flow turbulency dependent on the water flow rate through the channel inside which the operating crystal is set up. Numerical estimates show that for the typical coaxial crystal in a tube water channel geometry and typical flow rates, the value of the heat transfer coefficient is h=10^3-10^5 W/m^2 K (Koechner, 2006). The main thermal resistance to the heat flow from the crystal surface is due to the thermal boundary layer, δ_T, within which the heat conductance dominates over the convective transport. The estimate of δ_T follows from the equivalency of heat fluxes at crystal-water syrface:

$$-k_{cr}\partial T_{cr}/\partial r\big|_{sur} = -k_w \partial T_w/\partial r\big|_{sur} = h(T_{cr}\big|_{sur} - T_{w\infty}),$$

where $k_{cr}\approx 6$ and $k_w\approx 0.6$ W/m K are the thermal conductivity of crystal and water, respectively. That is, using $\partial T_w/\partial r\big|_{sur} \approx -(T_{cr}\big|_{sur} - T_{w\infty})/\delta_T$ one finally obtains for h=10^3-10^5 W/m^{2}K:

$$\delta_T \approx k_w / h \approx 6\text{-}600\ \mu\text{m}. \quad (18)$$

Thus, the spontaneous IR fluxes are absorbed by water within a distance where the heat transfer is dominated by the thermal conductivity. Hence, the absorption of these fluxes is able to significantly inhibit the heat dissipation from the crystal. In order to consider the thermal effect we simulate the complex heat transfer non-steady state, two-dimensional problem by coupling the above optical model with the heat generation and heat transport through the operating crystal, and the water boundary layer inside which the absorption of spontaneously emitted IR radiation takes place. The radially symmetric temperature distribution inside the cylindrical crystal, $T_{cr}(t,\mathbf{r})$, and the thermal boundary layer in water, $T_w(t,\mathbf{r})$, are defined by:

$$\rho_i C_i \partial T_i/\partial t = \text{div}(k_i \text{grad} T_i) + q_i(t,\text{r}), \quad (19)$$

for crystal (*i=cr*) and water (*i=w*) with the boundary condition $T_w = T_{w\infty}$ at $r = R_0 + \delta_T$, where $\delta_T = R_0\left[\exp(k_w/R_0 h) - 1\right]$ takes into account the radial curvature.
Heat source density inside the crystal is defined by Eq. (16) whereas the heat source density due to the absorption of spontaneously emitted IR fluxes in water is defined by:

$$q_w(t,r) = \frac{R_0}{r}\sum_i J_{0i}(t)\alpha_i \exp\left[-\alpha_i(r - R_0)\right], \quad (20)$$

where $J_{0i}(t)$ are the IR flux densities isotropically leaving the crystal given by:

$$J_{0i}(t) = \frac{1}{S_{cr}}\int_V \Delta E_i n_i(t,\text{r})/\tau_{ir} dV \cdot \quad (21)$$

The effect of IR radiation absorption is negligibly small for h>10^5 W/m^2 K, when $\delta_T << \alpha_i^{-1}$. However, for h≈10^4 W/m^2 K (δ_T≈60 μm) this effect is very significant, and can lead to the onset of an inverted temperature distribution inside the crystal when the temperature inside the boundary layer is higher than that inside the crystal.

5. Coupled thermo-optical modeling

In this section we consider coupled optical and thermal model results obtained for G-pulse generation by a Tm,Ho:YLF laser aimed to produce ≈0.1 J pulses of 100-500 ns duration. In particular, we simulate here 20-50 Hz active Q-switched laser operation side-pumped by 0.5 ms LD pulses of ≈785 nm wavelength. In Fig. 9 we show G-pulse simulation results for a 2 cm long 2 mm diameter crystal placed inside a 1 m long cavity with a 0.85 mm radius waist. The Q-switch is opened after a 0.5 ms pumping period with a delay of 1.2 ms to ensure that the G-pulse generation starts after the maximal possible gain is achieved.
In particular, in Fig. 9 we show the results of G-pulse stabilization for h=104 W/m^2 K for three different cases: (a) 2 mm diameter crystal with L_{cr}=2 cm, pumping energy 1.8 J and f=20 Hz operation, (b) 4 mm diameter crystal with L_{cr}=2 cm, pumping energy 4.0 J and f=20 Hz operation, and (c) 4 mm diameter crystal with L_{cr}=2 cm, pumping energy 4.0 J, and f=50 Hz operation.

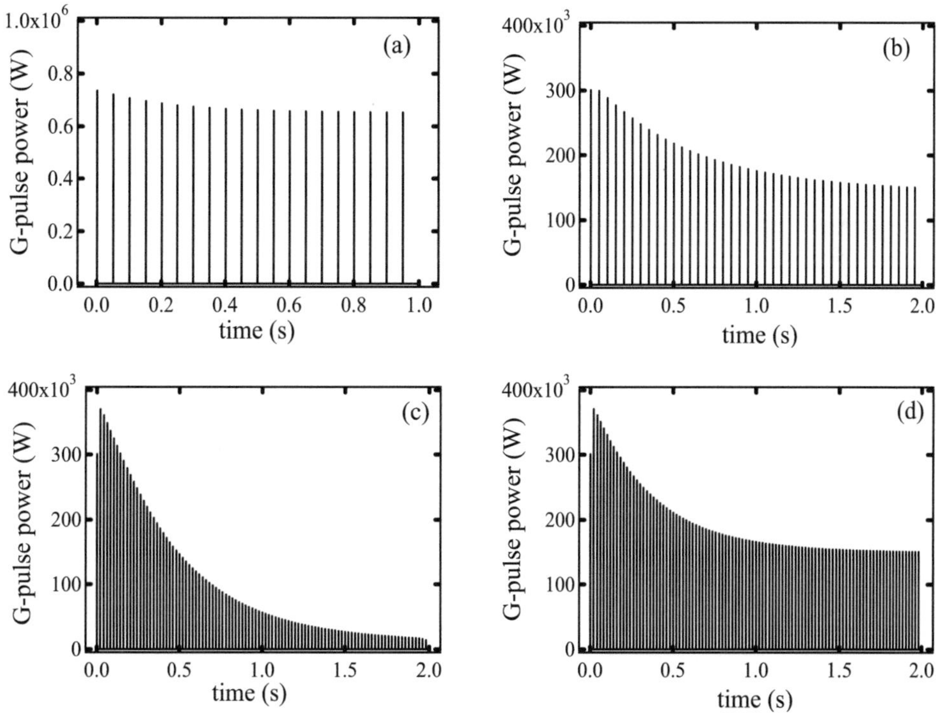

Fig. 9. G-pulse power stabilization versus time for (a) 2 mm diameter crystal, 1.8 J pumping, f=20 Hz and h=10^4 W/m^2 K, (b) 4 mm diameter crystal, 4.0 J pumping, f=20 Hz and h=10^4 W/m^2 K, (c) 4 mm diameter crystal, 4.0 J pumping, f=50 Hz and h=10^4 W/m^2 K and (d) 4 mm diameter crystal, 4.0 J pumping, f=50 Hz and h=10^5 W/m^2 K.

Additionally in Fig.9 (d) we show the pulse power for the case of enhanced heat transfer with h=10^5 W/m^2 K for 4 mm diameter crystal with L_{cr}=2 cm, pumping energy 4.0 J and f =50 Hz operation. In this case the spontaneous radiation leaving the crystal is absorbed outside of the boundary layer, $\delta_T \approx k_w / h \approx 6$ μm.

The case (a) corresponds to ≈0.12 J pulse generation. In 20 Hz repetition rate operation the initial pulse power for this case is decreased by ≈11 %. The case (b) corresponds to ≈0.13 J pulse generation. In this case, after stabilization in 20 Hz repetition rate operation the G-pulse looses about 40 % of its power and energy. In case (c) for 50 Hz repetition rate operation the G-pulse looses about 95 % of its initial power. In case (d) when heat transfer is enhanced the pulse energy decrease is about 60 %. One peculiar point worth noting here is that in 50 Hz operation the second pulse has a 20 % higher pulse energy than the first pulse (see Fig. 9 (c) and (d)). This is due to the inversion population remaining after the first pulse during 20 ms and summarized with the effect of the second pumping pulse.
The temperature evolution and stabilization related to these cases are shown in Fig. 10 (a-d) where the temperature is shown at the crystal axis and surface. The maximal temperatures are achieved at the end of ≈5 ms period and not at the end of pumping period. In particular, Fig. 10 (a), (b) and (c) show that at the initial stage the temperature at the crystal surface is higher than that at the crystal axis. This is due to the high optical fluxes spontaneously emitted from the crystal and heating water near the crystal surface. However, with time and operation stabilization the temperature inside the crystal becomes higher than that at the surface, and the heat from inside the crystal dissipates to the water. In the case (d) the temperature at the surface does not exceed that at the crystal axis even in the initial stage because for this case the spontaneous radiation is absorbed outside of the boundary layer ($\delta_T \approx 6$ μm$<<\alpha^{-1}$).

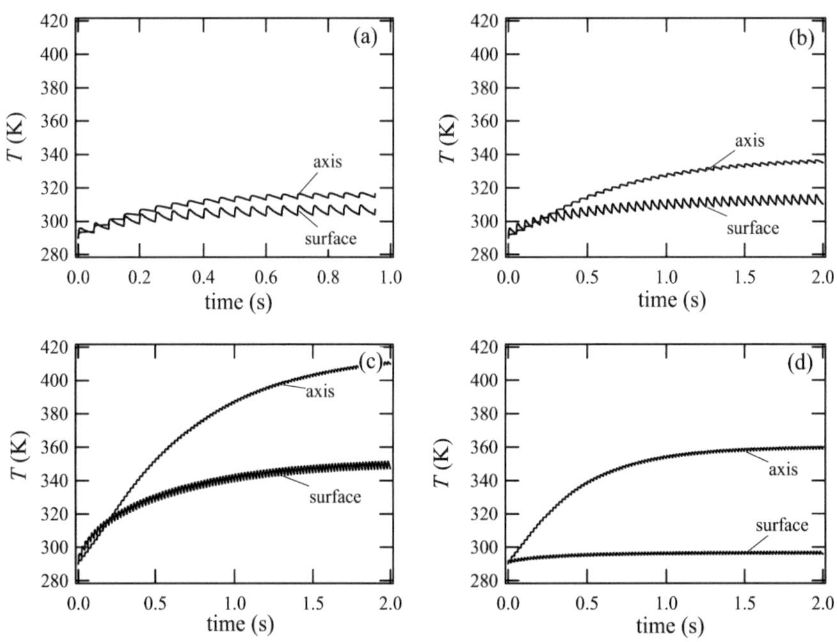

Fig. 10. Temperature evolution at crystal axis and surface for the cases defined in Fig. 9.

Fig. 11 shows the temperature distributions versus radius for the case of 4 J pumping of the crystal 4 mm in diameter and L_{cr}=20 mm for f=20 Hz operation and h=10^4 W/m^2 K. The temperature distribution are given (a) for the first pulse and (b) after pulse energy and temperature stabilization is achieved. The temperature distributions are given immediately after the pumping, 1.2 ms, (solid) and after the full pulse period of 50 ms (dashed). Fig. 11 (a) shows that for the first pulse the temperature in the boundary layer is higher than that in crystal due to the effect of absorption of IR radiation inside the boundary layer. However, after the pulse energy and temperature stabilization shown in Fig. 11 (b) this effect does not look very strong. However, in fact, this effect averaged over the pulse period remains to be very significant in resulting temperature field. After the stabilization, the temperature amplitude inside the crystal over one period decreases to the level of ≈1 K.

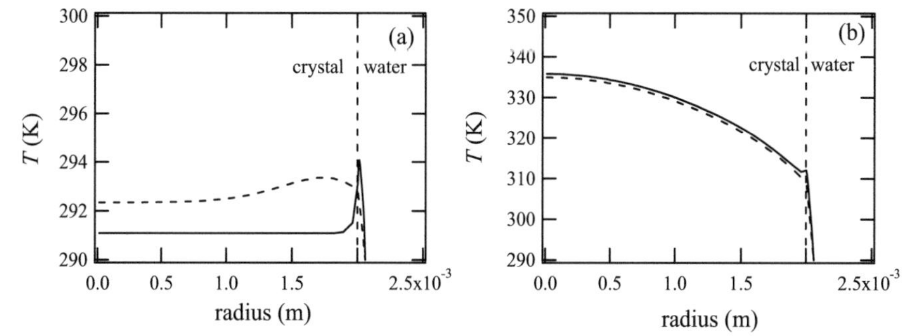

Fig. 11. Temperature distributions versus radius for 1.2 ms (solid) and 50 ms (dashed) for f=20 Hz operation: (a) the first pulse and (b) after pulse energy stabilization.

Let us discuss the above obtained simulation results within the framework of simplistic estimates. First, the effect of heat dissipation inhibition due to the radiation fluxes absorbed within the boundary layer shown by the above simulation can be demonstrated by an analytically resolvable model using the plane steady-state boundary layer approximation ($\delta_T << R_0$):

$$k_w d^2 T_w / dr^2 + \sum_i \bar{J}_{0i}\, \alpha_i \exp(-\alpha_i r) = 0\,, \tag{22}$$

with the mean radiation fluxes leaving the crystal surface given by :

$$\bar{J}_{0i} = \frac{\nu_p}{S_{cr}} \int_t dt \int_V \Delta E_i n_i / \tau_{ir} dV\,, \tag{23}$$

where $V_{cr}=\pi R_0^2 L_{cr}$ is the crystal volume, $S_{cr}\approx 2\pi R_0 L_{cr}$ is the crystal surface through which the spontaneous radiation leaves the crystal and the boundary condition is written at r=0 as $T_w = T_s$ and at $r = \delta_T = k_w / h$ as $T_w = T_w^*$.

After resolving this simplified model one can obtain the following analytical expression for the modified heat transfer coefficient:

$$h^* = -\frac{k_w}{T_s - T_w} \frac{\partial T_w}{\partial r}\bigg|_{r=R_0} = h - \frac{1}{T_s - T_w} \sum_i J_{oi} \left\{ 1 - \frac{h}{\alpha_i k_w} \left[1 - \exp\left(-\frac{\alpha_i k_w}{h} \right) \right] \right\}, \tag{24}$$

which takes into account the effect of spontaneous radiation absorbed within the thermal boundary layer on the heat transfer to the water.

One can see that this estimate gives automatically $h^* = h$ for the case when absorption does not take place within the boundary layer, or primarily occurs outside it. The estimate for the value of h^*/h defined by Eq. (24) is shown in Fig. 12 for the possible range of h=10^3-10^5 W/m^2 K and for IR fluxes corresponding to the simulations given in Fig.9 (a) and averaged over the characteristic time of $\Delta t = \nu^{-1}$=50 ms. In particular, Fig. 12 shows that even for this simplified treatment this effect appears to be very significant. That is, for low values of h<10^4 W/m^2 K the effective heat transfer coefficient falls rapidly and for h=10^3 W/m^2 K the value of h^* tends to zero. Eq. (24) also shows that h^* can become negative (for low h values or high radiation fluxes) meaning that the water layer near the crystal surface can have a temperature that exceeds that of the crystal surface (as shown in simulation on Fig. 11). For a typical value of h=10^4 W/m^2 K (δ_T≈60 μm) Eq. (24) gives $h^* \approx 0.55\,h$ whereas for h=10^5 W/m^2 K (δ_T≈6 μm) the effect decreases giving $h^* \approx 0.94\,h$. The effect of radiation absorption in water becomes negligibly small for high values of h>10^5 W/m^2 K, i.e. when the thickness of the boundary layer becomes smaller than a typical absorption length and all radiation is absorbed outside of the boundary layer.

Second, we have to note here that after onset of operation stabilization the mean temperature distribution across the crystal radius (averaged over one period) is well approximated by modified analytical quasi-steady state approximation (Koechner, 2006):

$$T(r) = T_w^* + \frac{q_{av} R_0}{2h^*} + \frac{q_{av}}{2k_{cr}}\left(R_0^2 - r^2\right), \qquad (25)$$

where for q_{av} we use the averaged heat source density defined by the integral over the characteristic pulse period, $\Delta t = \nu^{-1}$, and the crystal volume as:

$$q_{av} = \frac{1}{V_{cr}\Delta t}\int_{\Delta t}\int_{V_{cr}} q_{cr}(\mathbf{r},t)\,dt\,dV \cdot \qquad (26)$$

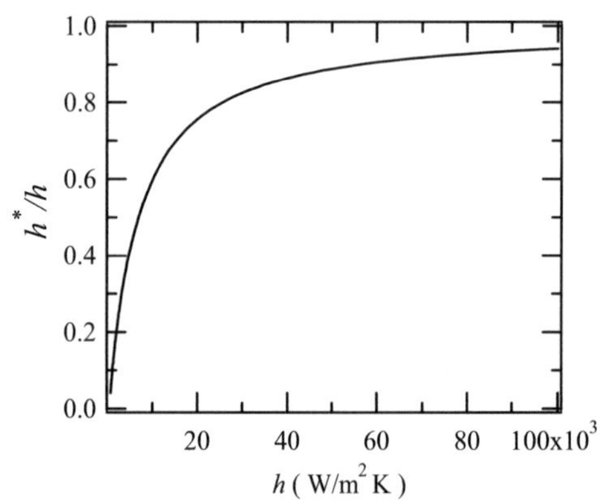

Fig. 12. The influence of absorbed infrared radiation spontaneously emitted from Tm-Ho laser on the heat transfer coefficient from the surface of the operating crystal to the water for 2 mm diameter crystal, 1.8 J pumping and f=20 Hz.

Equation (25) explains well our simulation results given in Fig.9 (c) and (d), and Fig. 10 (c) and (d) which show that the absolute temperature level achieved during laser operation can be decreased by more intensive heat removal (by using higher water flow) or by using lower temperature water (in our simulations T_w^*=290 K). In effect, Eq. (25) shows that intensification of the heat transfer, i.e. increase in h^*, allows one to decrease the temperature of the crystal surface to that of water T_w^* (making $q_{av}R_0/2h^* \to 0$). However, the temperature difference installed across the crystal radius remains the same (≈30 K) because it is defined by the value of $q_{av}R_0^2/2k_{cr}$, and depends only on the crystal radius, crystal heat conductance and the heat source density inside the crystal.
High temperature gradient across the crystal radius is known to lead to thermal lensing effects and output laser beam distortions (Akhmanov *et al.;* 1979). Thus, in order to inhibit the temperature gradient across the radius and onset of thermal lenzing one should (i) decrease the pumping energy per crystal volume unit and (ii) to decrease the crystal radius as shown in Figs. 9 (a) and 10 (a) where significantly lower temperature gradient for 2 mm diameter crystal are found as compared with 4 mm crystal diameter cases shown on Figs. 9 and 10 (b) - (d).
Finally, we have to note that Tm,Ho:YAG has lower temperature gradients across the radius due to two factors. First, YAG has larger heat conductance (11 W/m K) as compared with YLF (6 W/mK). Second, Tm.Ho:YAG has 2 times higher Tm $^3H_4$ absorption cross-section (σ_a=7x10^{-21} cm^2) as compared with Tm,Ho:YLF (σ_a =3.4x10^{-21} cm^2) which allows one to use lower crystal radii with the same absorption efficiency. That is, for 6 % Tm doped YLF and YAG one finds α_{YLF}≈2.82 cm^{-1} and α_{YAG}≈5.75 cm^{-1}. Thus, 2 mm diameter crystal is able to absorb (1-exp(-αd))-fraction of the incident beam, i.e. 43 % in case of YLF and 70 % in case of YAG. For double passing scheme 2 mm YAG absorbs 90 % of the side-pumped incident beam whereas YLF absorbs only 68 %. In case of YLF 90% absorption is achieved in crystals of 4 mm in diameter. It should be noted that increase in the crystal radius increases the pumping density and also heat source density. However, 4-times increase in the heat source density in 2mm YAG crystal as compared with 4 mm YLF is exactly compensated by 4-time decrease in the temperature gradient across the radius shown by Eq. (25).

6. Model verification

6.1 Absorption losses

As noted in section 2 the absorption loss $2\sum \alpha_i l_i$ (in Eq. (10)) taking place inside the cavity is able to decrease significantly and even completely inhibit laser output. In particular, Fig. 13 shows the typical dependence of G-pulse energy from Tm,Ho:YLF laser versus the absorption loss inside the cavity. The absorption loss can be due to several factors. First, for the case of 2 μm lasers this effect can be due to the water vapor present inside the cavity. In effect, the absorption of water for ≈2.1 μm is α≈31 cm^{-1} (Wieliczka, *et al.*, 1989). The absorption coefficient is defined by the absorption cross-section and water molecules density $\alpha = \sigma\, n$. Thus, the absorption cross-section for water is $\sigma = \alpha / n_w = \alpha \rho_w / m$≈9.3x10^{-22} cm^2 (where n_w is the density of molecules in water, ρ_w is the water density and m is the molecular mass of water). For the room temperature 24^0 C the equilibrium water pressure P≈3x10^3 Pa. Thus, for the humidity 50 % the vapor density is

$n_v=P_v/2k_BT\approx 0.36\text{x}10^{18}$ cm^{-3}. Respectively, the value of the absorption coefficient is $\alpha = \sigma\, n_v \approx 3.3\text{x}10^{-4}$ cm^{-1} and for the cavity l=300 cm in length one gets $2\sum_i \alpha_i l_i \approx 0.2$, which suggests that the air humidity can completely inhibit laser operation. However, for a short cavity length of 30 cm $2\sum_i \alpha_i l_i \approx 0.02$ and G-pulse energy loss is ≈25 %.

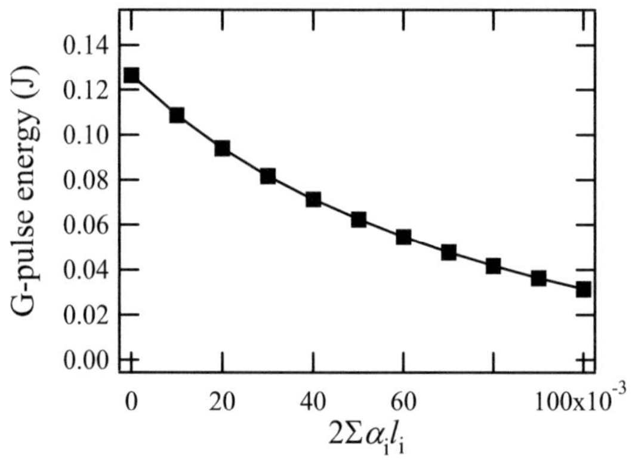

Fig. 13. The effect of absorption loss inside the cavity.

Second, the absorption loss can take place inside the sections of the crystal usually left undoped for the mechanical assembly of the crystal inside water cooling camera, and also inside the Q-switch. For instance, in near IR region undoped YAG has linear absorption coefficient $\alpha\approx 1.5\text{x}10^{-3}$ cm^{-1}. Assuming similar absorption for ≈2.1 μm radiation one finds that for 4 cm of undoped YAG sections $2\sum_i \alpha_i l_i \approx 0.012$ and about 10% loss in G-pulse.

Third, the effect of two-photon absorption which may have a serious effect on G-pulse generation when the laser power attains MW level. In particular, in general case the absorption loss inside optical medium is defined by the following equation including linear and non-linear (two-photon) absorption as:

$$dI / dz = -\alpha I - \beta I^2, \tag{27}$$

where the value β is known to be within the range of 10^{-11}-10^{-9} cm/W for wide range of optical materials (Dmitriev *et al.*, 1997).
This equation suggests that the effect of two-photon absorption is significant when $\alpha \approx \beta I$. In particular, for the undoped YAG sections this takes place when $\beta I\approx\alpha\approx 0.0015$cm^{-1}. Let us estimate the possible range for βI assuming that the target level of pulse energy should attain the level of $Q\approx 0.1$ J with pulse width of $\tau_p\approx 500$ ns. This gives us maximal power inside the cavity of $P=Q/\tau_p$=0.1 J/ 500 ns= 2x10^5 W. The maximal intensity in the laser beam will be $I = P/\pi\, r_0^2 \approx 0.65\text{x}10^7$ W/cm^2 (where $r_0\approx 0.1$ cm is the laser beam radius). Thus, one obtains for $\beta I \approx 10^{-11}$-10^{-9} cm/W x 0.65x10^7 W/cm^2=6.5x(10^{-5} -10^{-3}) cm^{-1}, which shows that the effect of two-photon absorption may be similar and even higher than that of linear absorption $\alpha\approx 1.5\text{x}10^{-3}$ cm^{-1}. Moreover, in experiments with shorter cavity and shorter pulse width this effect may become even more significant. In particular, 500 ns pulses correspond to the cavity length of 3 m. In case of 1 m cavity the pulse duration shortens to the level of ≈100-200 ns, and the maximal power achieves the level of $P\approx 1$ MW. Thus, for increased intensity

one gets $\beta I \approx 3.25\text{x}(10^{-4} - 10^{-2})$ cm^{-1},which shows that the two-photon absorption may be an order of magnitude higher than that of linear absorption, $\alpha \approx 1.5\text{x}10^{-3}$ cm^{-1}.
Additionally one should pay attention to the possible interference of two-photon absorption loss with the output transmittance T_{out}. The above computational model shows that the increase in T_{out} shortens G-pulse width tending to increase its peak power. In fact, the possible involvement of the two-photon absorption can inhibit this effect decreasing G-pulse energy. We should also note that the simulation of the effect of two-photon absorption requires formal correction of Eq. (8) by introduction of an additional loss term $\propto \beta\Phi_0^2$.

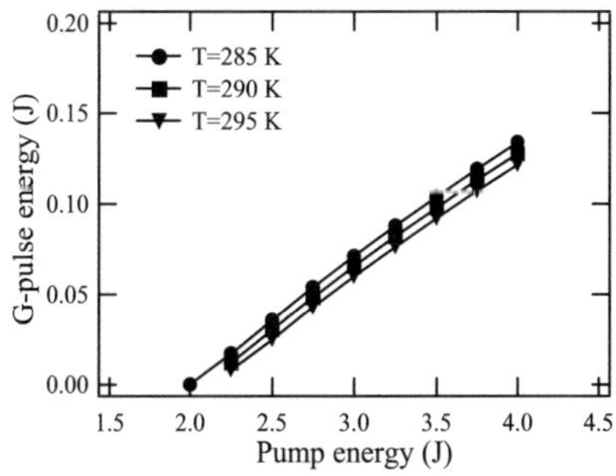

Fig. 14. G-pulse energy versus pumping energy for the experimental conditions by Yu *et al.* 1998.

6.2 Tm,Ho:YLF

For this case the rate dynamics equation and related parameters are fully specified by Walsh *et al.* 2004. The model was verified by using the set of experimental condition reported by Yu *et al.*, 1998 for the development of Q-switched 125 mJ Tm,Ho:YLF laser using ≈790 nm LD pulse of 1.2 ms duration for side-pumping of 2 cm long 4 mm diameter crystal rod set up in 3.3 m cavity with T_{out}=0.18 for the output mirror. Results of our simulations performed for this particular case are given in Fig. 14 where the resulting G- pulse energy is given versus the pumped energy for T=285, 290 and 295 K. The simulations show a good agreement with the experimental data (Yu *et al.*, 1998). In particular, 4 J pumping gives the output energy of about ≈0.14 J for T=285 K and ≈0.13 J for T=290 K, correspondingly. The experiment shows ≈0.125 J for T=292 K (Yu *et al.*, 1998). The pulse generation threshold is computationally found to be near the pumping energy of 2 J which is also close to the value of the extrapolated experimental data (Yu *et al.*, 1998). Additionally, the simulation data given in Fig. 14 shows that the G-pulse energy strongly depends on the operating temperature. For instance, for 4 J pump the temperature increase from 285 K to 295 K is found to decrease G-pulse energy on ≈15%.
In Fig. 15 we show (a) G-pulse power versus time for several pumping energies, (b) G-pulse power versus time for various photon injection seeder power and (c) G-pulse width versus pumping energy. The pulse width shows decrease with the increase of the pumping energy similar to the experimental data. However, the values of pulse width found in our simulation is about 2 times higher that those of experimental study (Yu *et al.*, 1998). The seeder does not significantly change the pulse width, and shortens the time between the moment of Q-switch opening and G-pulse generation.

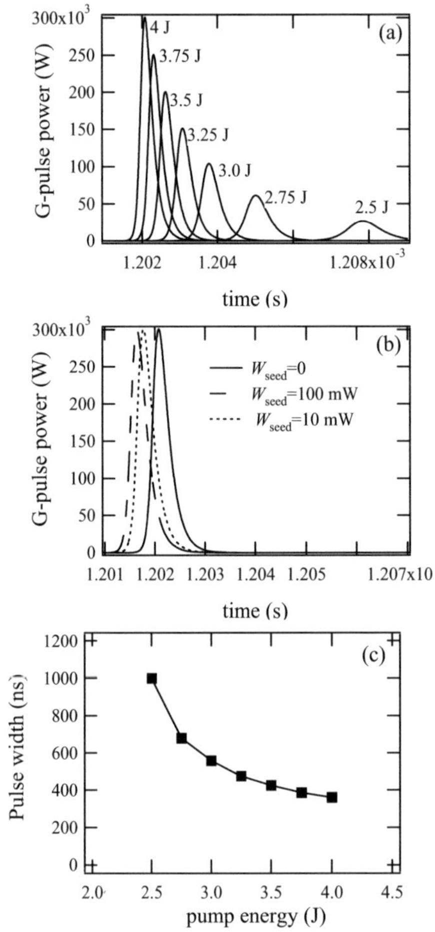

Fig. 15. G-pulse power versus time for (a) for different pump energy, (b) constant pump energy 4 J and photon seeder powers, and (c) pulse width versus the pump energy for experimental conditions by Yu *et al*. 1998.

6.3 Tm,Ho:YAG

In the case of Tm,Ho:YAG serious simplifications can be done allowing model reduction to two rate equations describing the electron density at the excited levels $^3F_4$ and $^5I_7$, and ground state levels $^3H_6$ and $^5I_8$ assuming that the excitation is rapidly transferred from $^3H_4$ to $^3F_4$ level. In particular, this model based on 4-level rate dynamics model (Bruneau, *et al.* 1998) is given by the set of equations:

$$\frac{dn_2}{dt} = R_p(t) - \frac{n_2}{\tau_2} - p_{28}n_2n_8 + p_{71}n_7n_1, \tag{28}$$

$$\frac{dn_7}{dt} = -\frac{n_7}{\tau_7} - p_{27}n_2n_7 + p_{28}n_2n_8 - p_{71}n_7n_1 - \frac{c\sigma_{se}}{\eta}(f_7n_7 - f_8n_8)\phi_0(t,\mathbf{r}), \tag{29}$$

where $n_1 \approx n_{Tm} - n_2$, $n_8 \approx n_{Ho} - n_7$, p_{28}=31.7x10^{-17} cm^3/s, p_{71} =0.09p_{28}, p_{27}=7.24x10^{-17} cm^3/s are the probabilities of the optical transitions, τ_2=11 ms and τ_7=15.6 ms are the level lifetimes, σ_{se}=9.2x10^{-20} cm^2 is the stimulated emission cross-section, f_7=0.104 and f_7=0.018 are the Boltzmann populations factors for T=20 ^{0}C.

The experimentally measured and simulated normal pulse energy versus pumping energy is shown in Fig. 16 (a) and (b). The experiments were done using two crystal set of Tm (6%), Ho (0.4 %): ceramic YAG of 3 mm in diameter and 22 mm long doped zone (within the rod of 70 mm in length). Output cavity transmission is T_{out}=0.15, and cavity length is L_{cav}=40 cm. The temperature and humidity inside the dehumidified box of 30 cm in length was 24.8 ^{o}C and H=0.029 (2.9 %), correspondingly. The temperature and humidity outside the dehumidified box was 25.2 ^{o}C and H=0.42 (42 %), correspondingly. The total loss inside the cavity was estimated for laser beam waist is 1.5 mm and loss inside the undoped YAG sections as $2\sum_i \alpha_i l_i$ =0.034.

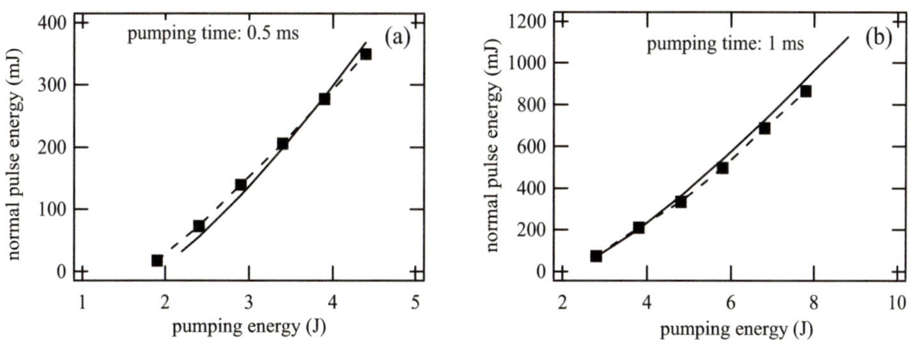

Fig. 16. Normal pulse energy versus pumping energy: solid line - computational, broken line - experiments: (a) pumping time 0.5 ms and (b) pumping time 1.0 ms.

In Fig. 17 we show experimentally measured and simulated normal pulse power versus time made for the same crystal set, cavity length of 100 cm with the temperature of 22 ^{0}C and humidity H=0.36 (36 %) (T_{out}=0.15). Pumping time is 1 ms. Fig. 17 (a) shows very good agreement of the experimental and simulated pulse power versus time. However, the careful comparison shows several discrepancies which may be attributed to a difficulties in defining precise experimental conditions. Nevertheless, all time scales of pulse generation in Fig. 17 (a) and (b), including relaxation period of ≈0.2 ms, agree well with each other.

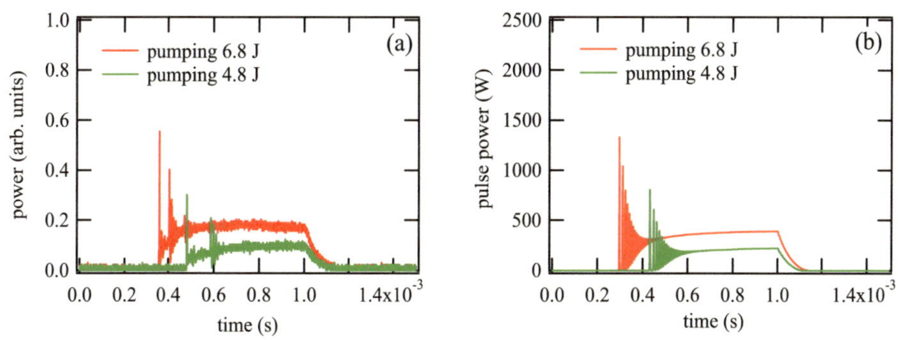

Fig. 17. Normal pulse power versus time (a) experiment and (b) simulation.

It is worth noting that experimental and simulated results show output power stabilization after several dozens of relaxation spikes agreeing with previous analytical derivations by Elder and Payne, 1995.

7. Summary and conclusions

Complex thermo-optical model for simulation and optimization of 2 μm co-doped Tm,Ho solid state lasers is developed based on (i) 8-levels rate dynamics model for the excitation transfer to Ho^{3+} ions from LD pumped Tm^{3+} ions and (ii) the oscillator energy equation with TEM_{00} distribution for stimulated photons inside the laser cavity. This optical model is rigourosly coupled with two-dimensional time dependent heat transfer model including absorption, heat release and heat transfer inside the operating crystal, as well as thermal effect of spontaneously emitted infrared radiation fluxes. In the case of water cooled laser operation these IR radiations fluxes are shown to be fully absorbed within the water boundary layer of several hundreds microns and to produce a significant inhibition of heat dissipation from the operating crystal. In particular, in case of water cooled operation only ≈40 % of the pumped energy is transformed into the heat directly inside the crystal, whereas ≈45 % is IR radiation spontaneously emitted by $^3H_4$, $^3H_5$, $^3F_4$ and $^5I_7$ levels and fully absorbed by water within the thermal boundary layer. This model is used in simulations for a wide range of parameters for 780 nm LD side-pumped co-doped Tm,Ho:YLF and YAG lasers. Two significant effects are found for the particular case of Q-switched high repetition (20-50 Hz) Tm,Ho:YLF laser operation with output pulse energy ≈0.1 J and ≈500 ns pulse duration. First, for obtaining a giant (G) pulse from a 2.06 μm solid state Tm,Ho:YLF laser by the active Q-switching technique, the optimal Ho concentration will be higher than that used in normal operation. In simulations of 500 ns G-pulse generation maximal efficiency occurred at 6 % Tm and 1.0 % Ho, in contrast with 0.4% Ho found to be optimal for the normal pulse generation. Maximal energy output from Tm,Ho:YLF lasers can be achieved by incorporating a delay of about 0.7 ms between 0.5 ms 780 nm LD pulsed pumping and the start of Q-switched G-pulse operation. Secondly, simulation suggests that the ≈0. 7 ms delay after 0.5 ms LD pumping pulse should be used for retrieving the maximum of the pumped energy. Finally, heat transfer effects are analyzed in detail showing output pulse energy decrease for high repetition rates due to temperature influence on the upper and lower level population factors.

8. Acknowledgements

This work was supported by the National Institute of Information and Communications Technology (Japan). We would also like to thank Dr. J. Hester from the Australian Nuclear Science and Technology Organization for numerous valuable comments.

9. References

Akhmanov, S. A.; Khokhlov, R. V. & Sukhorukov, A. P., *Self-focusing, self-defocusing and self-modulation of laser beams*, Laser Handbook, V.2. Eds. F. T. Arecchi, E. O. Schulz-Dubois, North-Holland, Amsterdam 1972. 978-0720402131

Alpat'ev, A. N.; Smirnov, V. A. & Shcherbakov, I. A. (1998). Relaxation oscillations of the radiation from a 2-μm holmium laser with a Cr,Tm,Ho:YSGG crystal. *Quantum Electronics,* Vol. 28, No. 2, 143-146, 1063-7818(print); 1468-4799(online)

Barnes, N. P.; Filer, E. D.; Morison, C. A. & Lee, C. J. (1996). Ho:Tm Lasers: Theoretical. *IEEE Journal of Quantum Electronics,* Vol. 32, No. 1, 92-103, 0018-9197

Bourdet, G. L. & Lescroart, G. (1999). Theoretical modeling and design of a Tm, Ho: $YLiF_4$ microchip laser. *Applied Optics,* Vol. 38, No. 15, 3275-3281, 0003-6935(print); 1539-4522(online)

Bruneau, D.; Delmonte, S. & Pelon, J. (1998). Modeling of Tm, Ho: YAG and Tm, Ho: YLF 2-μm lasers and calculation of extractable energies. *Applied Optics,* Vol. 37, No. 36, 8406-8419, 0003-6935(print); 1539-4522 (online)

Černý, P. & Burns, D. (2005). Modeling and experimental investigation of a diode-pumped Tm:$YAlO_3$ laser with *a*- and *b*- cut crystal orientations. *IEEE Journal of Selected Topics in Quantum Electronics,* Vol. 11, No. 3, 674-681. 1077-260X

Dmitriev, V. G.; Gurzadyan, G. G. & Nikogosyan, D. N. (1997). *Handbook of Nonlinear Optical Crystals 2nd Edition,* Springer-Verlag, 3-540-61275-0, Berlin

Eichhorn, M. (2008). Quasi-three-level solid-state lasers in the near and mid infrared based on trivalent rare earth ions. *Applied Physics B,* Vol. 93, No. 2-3, 269-316, 0946-2171 (print); 1432-0649 (online)

Elder, I. F. & Payne, M. J. P. (1995). Lasing in diode-pumped Tm:YAP, Tm,Ho:YAP and Tm,Ho,YLF. *Optical Communications,* Vol. 145, No. 1-6, 329-339, 0030-4018

French, V. A.; Petrin, R. R.; Powell, R. C. & Kokta, M. (1992). Energy-transfer processes in Y3Al5O12:Tm, Ho. *Physical Review B,* Vol. 46, No. 13, 8018-8026, 1098-0121(print); 1550-235X(online); 1538-4489(CD-Rom)

Galzerano, G.; Sani, E.; Toncelli, A.; Della Valle, D.; Taccheo, S.; Tonelli, M. & Laporta, P. (2004). Widely tunable continuous-wave diode-pumped 2-μm Tm-Ho:KYF_4 laser. *Optics Letters,* Vol. 29, No. 7, 715-717, 0146-9592(print); 1539-4794(online)

Izawa, J.; Nakajima, H.; Hara, H.; & Arimoto, Y. (2000). Comparison of lasing performance of Tm,Ho:YLF lasers by use of single and double cavities. *Applied Optics,* Vol. 39, No. 15, 2418-2421, 0003-6935(print); 1539-4522(online)

Jackson, S. D. & King, T. A. (1998). CW operation of a 1.064-μm pumped Tm-Ho-doped silica fiber laser. *IEEE Journal of Quantum Electronics,* Vol. 34, No. 9, 1578-1587, 0018-9197

Jani, M. G.; Reeves, R. J.; Powell, R. C.; Quarles, G. J. & Esterovitz, L. (1991). Alexandrite-laser excitation of a Tm:Ho:$Y_3Al_5O_{12}$ laser. *Journal of the Optical Society of America B,* Vol. 8, No. 4, 741-746, 0740-3224(print); 1520-8540(online)

Jani, M. G.; Naranjo, F. L.; Barnes, N. P.; Murray, K. E. & Lockard, G. E. (1995). Diode-pumped long-pulse-length Ho:Tm:$YLiF_4$ laser at 10 Hz. *Optics Letters,* Vol. 20, No. 8, 872-874, 0146-9592(print); 1539-4794(online)

Koechner, W. (2006). *Solid-State Laser Engineering 6th Edition,* Springer, 978-0387290942, New York

Lee, C. J.; Han, G. & Barnes, N. P. (1996). Ho:Tm Lasers: Experiments. *IEEE Journal of Quantum Electronics,* Vol. 32, No. 1, 104-111, 0018-9197

Louchev, O. A.; Urata, Y. & Wada, S. (2007). Numerical simulation and optimization of Q-switched 2 μm Tm, Ho:YLF laser. *Optics Express,* Vol. 15, No. 7, 3940-3947, 1094-4087

Louchev, O. A.; Urata, Y.; Saito, N; & Wada, S. (2007). Computational model for operation of 2 μm co-doped Tm, Ho solid-state lasers. *Optics Express,* Vol. 15, No. 19, 11903-11912, 1094-4087

Louchev, O. A.; Urata Y.; Yumoto, M.; Saito, N.; & Wada, S. (2008). Thermo-optical modeling of high power operation of 2 μm co-doped Tm,Ho solid-state lasers. *Journal of Applied Physics,* Vol. 104, No. 3, 033114-033114-9, 0021-8979(printed); 1089-7550(online)

Petrin, R. R.; Jani, M. G.; Powell, R. C. & Kokta, M. (1992). Spectral dynamics of laser-pumped $Y_3Al_5O_{12}$:Tm, Ho lasers. *Optical Materials,* Vol. 1, No. 2, 111-124, 0925-3467

Risk, V. P. (1988). Modeling of longitudinally pumped solid-state lasers exhibiting reabsorption losses. *Journal of the Optical Society of America B,* Vol. 5, No. 7, 1412-1423, 0740-3224(print); 1520-8540(online)

Rustad, G. & Stenersen, K. (1996). Modeling of laser-pumped Tm and Ho lasers accounting for upconversion and ground-state depletion. *IEEE Journal of Quantum Electronics,* Vol. 32, No. 9, 1645-1656, 0018-9197

Sato, A.; Asai, K. & Mizutani, K. (2004). Lasing characteristics and optimizations of diode-side-pumped Tm,Ho:$GdVO_4$ laser. *Optics Letters,* Vol. 29, No. 8, 836-838, 0146-9592(print); 1539-4794(online)

Sudesh, V. & Asai, K. (2003). Spectroscopic and diode-pumped-laser properties of Tm,Ho:YLF; Tm,Ho:LuLF; and Tm,Ho:LuAG crystals: a comparative study. *Journal of the Optical Society of America B,* Vol. 20, No. 9, 1829-1837, 0740-3224(print); 1520-8540(online)

Tyminski, J. K.; Franich, D. M. & Kokta, M. (1989). Gain dynamics of Tm,Ho:YAG pumped in near infrared. *Journal of Applied Physics,* Vol. 65, No. 8, 3181-3188, 0021-8979(printed); 1089-7550(online)

Walsh, B. M. (2009). Review of Tm and Ho materials; spectroscopy and lasers. *Laser Physics,* Vol. 19, No. 4, 855-866, 1054-660X(print); 1555-661(online)

Walsh, B. M.; Barnes, N. P.; Petros, M.; Yu, J. & Singh, U. N. (2004). Spectroscopy and modeling of solid state lanthanide lasers: application to trivalent Tm^{3+} and Ho^{3+} in $YLiF_4$ and $LuLiF_4$. *Journal of Applied Physics,* Vol. 95, No. 7, 3255-3271, 0021-8979(print); 1089-7550(online)

Wieliczka, D. M.; Weng, S. & Querry, M. R. (1989). Wedge shaped cell for highly absorbent liquids: infrared optical constants of water. *Applied Optics,* Vol. 28, No. 9, 1714-1719, 0003-6935(print); 1539-4522(online)

Yu, J.; Singh, U. N.; Barnes, P. & Petros, M. (1998). 125-mJ diode-pumped injection-seeded Ho:Tm:YLF laser. *Optics Letters,* Vol. 23, No. 10, 780-782, 0146-9592(print); 1539-4794 (online)

Yu, J.; Trieu, B. C.; Modlin, E. A.; Singh, U. N.; Kavaya, M. J.; Chen, S.; Bai, Y.; Petzar, P. J. & Petros, M. (2006). 1 J/pulse Q-switched 2 μm solid-state laser. *Optics Letters,* Vol. 31, No. 4, 462-464, 0146-9592(print); 1539-4794(online)

Zhang, X.; Ju, Y. & Wang, Y. (2006). Theoretical and experimental investigation of actively Q-switched Tm,Ho:YLF lasers. *Optics Express,* Vol. 14, No. 17, 7745-7750, 1094-4087

5

Actively Q-switched Thulium Lasers

Jan K. Jabczynski[1], Lukasz Gorajek[1], Waldemar Zendzian[1],
Jacek Kwiatkowski[1], Helena Jelinkova[2], Jan Sulc[2] and Michal Nemec[2]
[1]*Military University of Technology, Warsaw,*
[2]*Czech Technical University, Prague,*
[1]*Poland*
[2]*Czech Republic*

1. Introduction

The near and mid infrared range of optical spectrum (1500 – 5000 nm) has attracted great attention over the last decades, mainly because of a numerous set of absorption / emission lines being "finger prints" of specific chemical and biological compounds (see e.g. (Sorokina & Vodopyanov, 2003)). One of the most reliable techniques for stand alone detection and identification of such agents is laser spectroscopy, for which the coherent, highly monochromatic, tuneable laser sources are required. Research on mid infrared laser sources destined for long range detection is one of the most active areas of solid state photonics nowadays (see e.g. (Godard, 2007) (Mirov et al., 2007) etc.) (Schellhorn, et al., 2007) (Eichhorn, 2008)). One of the necessary requirements for such application is high spatial coherence of laser output beam (parameter M^2 near 1). Among the most perspective types of lasers operating in a mid infrared range (above 2.2 µm-wavelength) are quantum cascade lasers, Cr^{2+}:$A_{II}B_{VI}$ lasers and optical parametric generators. For shorter wavelengths laser diodes and erbium (1600 - 1700 nm), thulium (1800 - 2000 nm), holmium (2050-2200 nm) solid state lasers realised in bulk/crystalline/ or fibre technology are available. The main properties differentiating such groups of lasers are peak power and pulse energy. It has to be pointed out, that for such a wavelength range, damage thresholds of optical elements are severely decreased, comparing to 1-µm wavelength, because of the presence of water vapour and OH groups, having a wide absorption peak just in this wavelength range. The physics of laser action in semiconductor lasers prohibits the high peak power and high energy operation. Moreover, the high CW power 1D or 2D laser diodes stacks characterise low spatial coherence and high beam volume, thus they can be used mainly as optical pumps for other lasers, e.g. holmium or erbium ones.

For short infrared range (< 2 µm wavelength) the Tm or Ho–doped fibre lasers sources operating in a CW regime seem to be the most perspective choice (Gapontsev et al., 2007) (McComb et al., 2009). The pulsed operation of fibre lasers, with energies of tens to hundreds of microJoules, can be realized applying Q-switching technique (Eichhorn & Jackson, 2008) (Eichhorn & Jackson, 2009) or gain-switched method. The best result of 2-µm Q-switched Tm:fibre lasers from the point of view of pulse energy (0.6 mJ for 10 Hz of rep. rate) were obtained (Barnes & De Young, 2009). Mid infrared fibre lasers suffer from lack of good quality active glasses and problems with high peak power / high energy operation for

longer wavelengths. However the promising result namely 10 W at 2.78-µm-wavelength has been demonstrated recently in the ZBLAN fibre laser (Zhu & Jain, 2007).
Solid state lasers based on bulk crystalline gain media offer much higher peak powers and pulse energies compared to fibre ones. For the shortest wavelengths (1600-1700 nm) the best configuration is an erbium doped crystalline laser pumped by erbium fibre laser (Clarkson et al., 2006) or directly by laser diode (Setzler et al., 2005) (Kudryashov et al., 2009), (Eichhorn, 2008). For 1800-2000 nm wavelength range the thulium doped crystalline laser pumped by laser diodes seems to be the best option (Honea et al.,1997) (Budni et al., 2000), (Dergachev et al., 2002) (So et al, 2006), (Schellhorn, 2008), (Schellhorn et al., 2009). Intensive research has been devoted to thulium doped gain media (see e.g. (Huber et al.,1988) (Payne et al., 1992) (Lisiecki et al.,2006)) and methods of Q-switching. Passive Q-switching by means of holmium or chromium doped saturable absorbers as well as special types of quantum dots structures (Gaponenko et al., 2008) were examined, however simple, active Q-switching by means of acousto-optic modulators seems to be the better choice because of matured technology, very low insertion losses and relatively high damage thresholds.
The aim of this work was the theoretical analysis of such laser types and its experimental verification for one type of gain medium namely thulium doped yttrium, lithium fluoride (Tm:YLF).

2. Theoretical analysis of actively Q-switched thulium lasers

2.1 Model of quasi- three -level laser for CW pumping

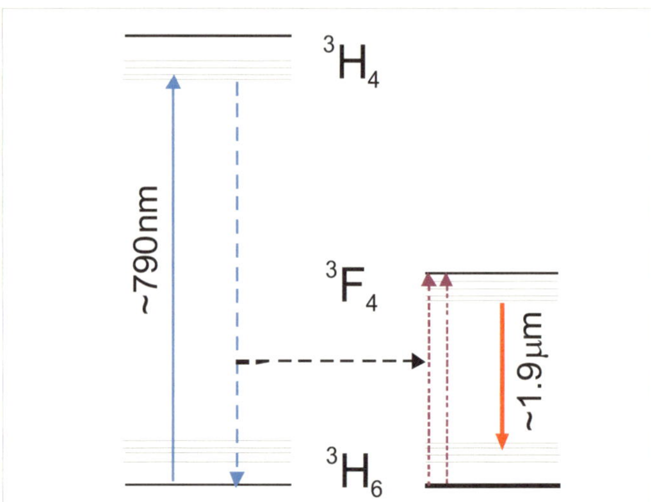

Fig. 1. Scheme of energetic levels of Tm^{3+} gain medium.

The following analysis can be applied to any quasi-three-level laser for which the lower laser levels are partly occupied at room temperature (see e.g. (Beach, 1996) (Bourdet, 2000) (Bourdet, 2001) (Lim & Izawa, 2002)). Typical configuration of Tm^{3+} lasers levels (see Fig. 1) consists of at least three manifolds : $^3H_6$ - pumping and laser lower levels manifold, $^3H_4$ – pumping, upper levels manifold and $^3F_4$ - laser upper level manifold. The pumping process between $^3H_6 \rightarrow {}^3H_4$. corresponds to wavelength ~ 790 nm which is avaiable for typical AlGaAs laser diodes. Excitation of upper laser levels $^3F_4$ is a result of an efficent cross-relaxation process $^3H_4 + ^3H_6 \rightarrow 2 \times ^3F_4$. In the following analysis we assumed, that this two-stage pumping process is characterised by the quantum efficiency equal to 2. Further, we will consider only relaxation/excitation processes (typical for quasi-III-level lasers) between two manifolds $^3H_6$ - lower pump/laser manifold and $^3F_4$ - upper pump/laser manifold. The analysis given in this paper will be followed by the model developed by (Bourdet, 2000).

Primarily, let us define processes of pumping between lower pumping level of number $N_{p,lower}$ - and the upper pumping level of number $N_{p,upper}$ ($N_{p,lower} \rightarrow N_{p,upper}$). Similarly, the laser transition has to be defined between the pair of levels $N_{l,upper}$ and $N_{l,lower}$ ($N_{l,upper} \rightarrow N_{l,lower}$). It was assumed, that it is valid plane wave approximation and both pump and laser beams have the same radius W and area $A_m=A_p=\pi W^2/2$. For such assumptions the analytical formulae describing the quasi-three-level laser can be derived (see details (Bourdet, 2000)).

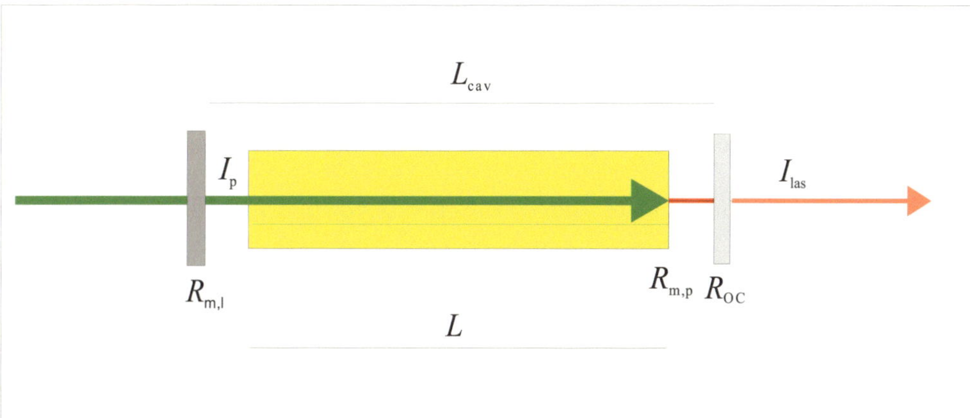

Fig. 2. Scheme of pumping and laser action in the Bourdet's model.

The final expression for relative laser output intensity I_{las} in dependence on incident relative pump intensity I_p and cavity and gain medium parameters is the following:

$$I_{las} = \left(1 - R_{OC}\right)\sqrt{R_{m,l}}\,\frac{\gamma_0 L\left(\dfrac{I_p\left(1-\Gamma\right)\left(1+R_{m,p}\Gamma\right)}{\alpha_0 L} - f_l\right) + \ln\sqrt{R_{OC}R_{m,l}}}{\left(1-\sqrt{R_{m,l}R_{OC}}\right)\left(\sqrt{R_{m,l}}+\sqrt{R_{OC}}\right)}, \tag{1}$$

where: L - rod length, $R_{m,l}$, R_{OC} - rear and output mirror reflectivity at laser wavelength, $R_{m,p}$ - rod reflectivity at pump wavelength. The function Γ (corresponding to the transmission of pump beam for laser action) is given by:

$$\Gamma = \left(R_{m,l}R_{OC}\right)^{\alpha_0/2\gamma_0} \exp\left[-\alpha_0\left(f_p - f_l\right)L\right], \tag{2}$$

where:

$$\alpha_0 = \sigma_p N_0\left(f_{N_{p,lower}} + f_{N_{p,upper}}\right) \;,\; \gamma_0 = \sigma_l N_0\left(f_{N_{l,lower}} + f_{N_{l,upper}}\right),$$
$$f_p = \frac{f_{N_{p,lower}}}{f_{N_{p,lower}} + f_{N_{p,upper}}} \;,\; f_l = \frac{f_{N_{l,lower}}}{f_{N_{l,lower}} + f_{N_{l,upper}}}, \tag{3}$$

N_0 - concentration of active ions, σ_p, σ_l - absorption/emission cross sections at pump/emission wavelengths, respectively, f_N - Boltzmann's occupation factor of N-th level. To calculate the final output laser power P_{out} and the incident pump power P_p the relative intensities I_{las}, I_p have to be multiplied by the mode/pump area $A_m = A_p = A = \pi W^2/2$ and emission/ absorption saturation densities $I_{sat,l}$, $I_{sat,p}$ as follows:

$$P_{out} = A\, I_{las} I_{sat,l} \quad ; \quad P_p = A\, I_p I_{sat,p} , \tag{4}$$

where:

$$I_{sat,l} = \frac{h\nu_l}{\left(f_{N_{l,lower}} + f_{N_{l,upper}}\right)\tau_u \sigma_l} \quad , \quad I_{sat,p} = \frac{h\nu_p}{\left(f_{N_{p,lower}} + f_{N_{p,upper}}\right)\tau_u \sigma_p} \quad , \tag{5}$$

τ_u- life time of upper laser level , $\nu_p = c/\lambda_p$, $\nu_l = c/\lambda_l$ - pump / laser frequencies, λ_p , λ_l- pump laser wavelengths, c- light speed, h -Planck constant.

Let us summarise the main properties of the quasi-three-level laser:

- strong dependence on temperature via Boltzmann's occupation factors f_N occuring in all formulae
- occurrence of additional, reabsorption losses per roundtrip defined by:

$$\gamma_{reab} = 2\gamma_0 f_l L \tag{6}$$

- minimal pump power required for achieving gain medium transparency:

$$P_{p,min} = \beta_{min} A_p I_{sat,p} \quad , \quad \beta_{min} = \frac{f_l}{f_p - f_l} \quad . \tag{7}$$

Moreover, in contrast to the typical case of a four-level laser, the absorption efficiency is different in laser and non-laser conditions. As a result of laser transition the occupations of upper and lower laser level are "clamped" to those at threshold and are dependent only on coupling conditions, losses etc.

For the case of laser action (above threshold) the absorption efficiency is given by:

$$\eta_{abs,las} = \frac{I_{p,abs}}{I_p} = (1-\Gamma)(1+R_{m,p}\Gamma) \quad . \tag{8}$$

As a rule, $\eta_{abs,las}$ is higher compared to non-laser conditions and does not depend on pump power.

In the case of non-lasing conditions the main effect of the diminishing of absorption efficiency consists in saturated absorption. Therefore, the absorption efficiency can be defined in two alternative ways:

$$\eta_{abs,non-las} = (1-\Gamma_{lambertW})(1+R_{m,p}\Gamma_{lambertW}) \quad , \quad \Gamma_{lambertW} = \frac{LambertW\left(I_p \exp(-\alpha_0 f_p L)\right)}{I_p} , \tag{9}$$

where $LambertW(z)$ denotes LambertW function (see e.g. (Barry et al, 2000), (Grace, et al, 2001), (Jabczyński et al, 2003)) defined as follows:

$$LambertW(z)\exp\left[LambertW(z)\right]=z \quad . \tag{10}$$

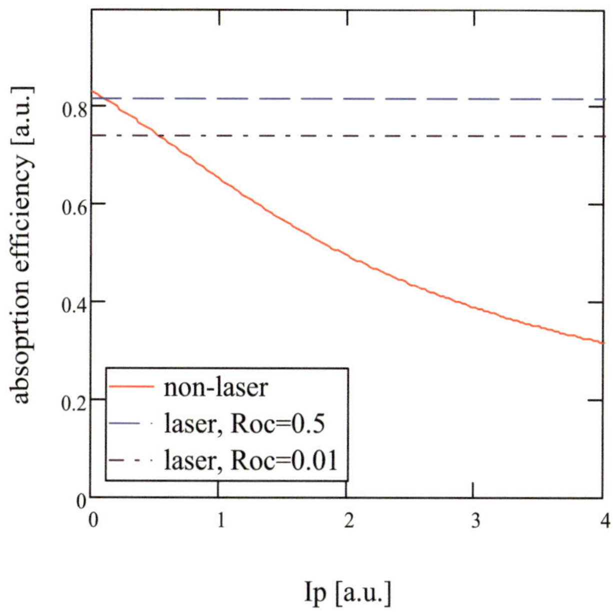

Fig. 5. Absorption efficiency vs. relative pump power; non lasing - continuous curve, lasing R_{oc}=0.5 dotted curve, lasing R_{oc}=0.01 - dash-dot curve.

An alternative way to determine absorption efficiency in non-laser conditions is to find a numerical solution for the pump transmission function Γ of the following equation:

$$\frac{\alpha_o f_p L + \ln\Gamma}{(1-\Gamma)(1+R_{m,p}\Gamma)} = I_p \quad , \tag{11}$$

and substituting the obtained value of Γ to formula (8). The results of absorption efficiency calculation for non-laser and laser conditions are shown in Fig. 5.
Below threshold absorption efficiency decreases with pump intensity. In a laser condition above threshold, absorption efficiency does not depend on pump power and it can even jump to higher values depending on output coupler reflectivity as was shown in Fig. 5. In our case of an actively Q-switched laser the non-lasing condition has to be assumed during the pumping process.

2.1.1 Optimisation of laser parameters

Analysing the formula (1), it is quite easy to find the condition for the optimum length of gain medium. Please note, that the small signal gain (assuming $R_{m,p}$ = 0) can be defined as follows:

$$g_{ss} = \gamma_0\left(\frac{I_p(1-\Gamma)}{\alpha_0 L} - f_l\right). \tag{12}$$

Thus, assuming constant I_p, R_{oc}, $R_{m,p}$, N_o and temperature, after differentiating (12) with respect to L, the final formula on optimal crystal length can be found:

$$L_{\mathrm{opt}} = \frac{\beta_{\min}}{\alpha_0 f_1} \ln\left[\left(R_{m,p} R_{OC}\right)^{\frac{1}{2\gamma_0}} \left(\frac{\beta_{\min}}{I_{\mathrm{p}}}\right)^{\frac{1}{\alpha_0}} \right]. \tag{13}$$

The example of calculation of optimal crystal length dependent on pump intensity is shown in Fig. 6.

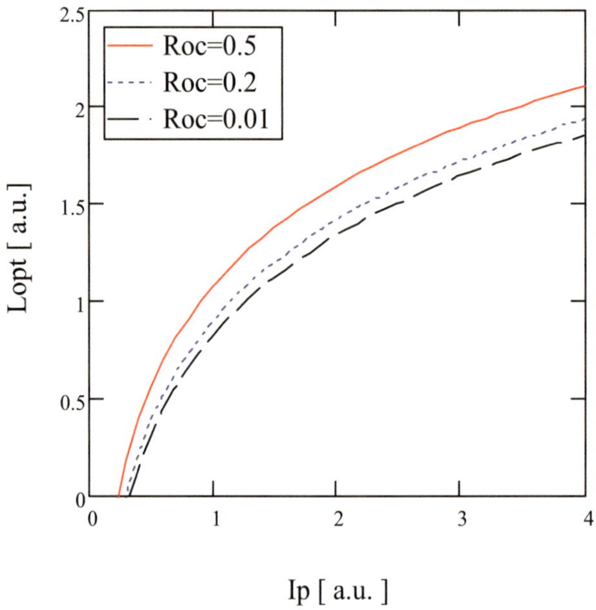

Fig. 6. Optimal crystal length vs. relative pump intensity for different output coupler reflectivities R_{oc}, $R_{m,l}$=0.99.

Please note, that output power for a given pump intensity depends also on output coupler reflectivity and cavity losses. The typical characteristics of output power intensity vs. pump power intensity are shown in Fig. 7.

As was shown in Fig. 6, 7, to obtain the higher output power with increased pump power density the gain medium length and out-coupling losses have to be increased. Please note that threshold pump power is also the function of gain medium length via reabsorption losses.

In the process of optimisation of quasi-three-level laser parameters, the starting point is the available pump power. We have to find the best combination of pump area, output coupler transmission, gain medium length to obtain the maximum output power. Further, the cavity design (curvature of mirrors, distances) should satisfy the best mode matching condition i.e. the fundamental laser mode area has to be comparable to the pump area to minimize the aperture/absorption losses occurring in an un-pumped region. The above presented formulae can give only preliminary indications, all these rules have to be verified experimentally.

It should be noted, that with increase in pump power density, the temperature of the rod increases, causing an increase in reabsorption losses and additional cavity losses as a result of stress induced aberrations. Thus, it is a trade of between requirements for efficient pumping and thermal limitations of a quasi-three-level laser. In practice, output power of such a laser is limited by thermo-optic effects and thermal fracture (see e.g. (So et al, 2006)), so the proper choice of pump geometry, size and dopant level of gain medium is fundamental to mitigate these effects.

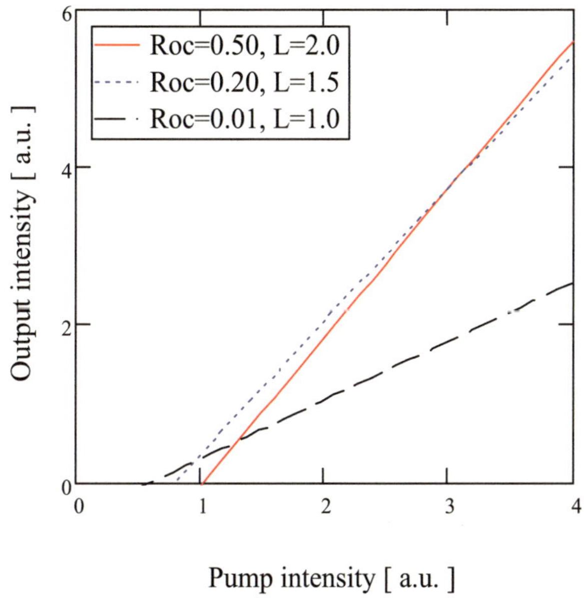

Fig. 7. Output intensity vs. pump intensity.

2.2 The thermo-optic limitations of thulium lasers

For lasers operating in a free running regime the maximum pump power with respect to fracture limit is determined by the doping level and mechanical toughness of the crystal (see e.g. (Koechner, 1996), (Chen, 1999), (So et al, 2006)). Much below fracture limit the operating parameters of any type of diode pumped lasers are governed by the temperature increase inside the gain medium. The temperature profile induced by absorption of a focused pumping beam inside the gain medium of a quasi-three-level laser manifests by means the following effects:

- changes in absorption efficiency and available net inversion,
- reabsorption losses due to temperature dependent occupations of lower laser level,
- paraxial thermal lensing,
- higher order thermal aberration,
- stress induced birefringence and depolarisation losses
- thermal fracture for higher heat loads.

According to the model derived for end pumped lasers (Chen, 1999) the maximum absorbed pump power is given by:

$$P_{abs,lim} = \frac{4\pi R_T}{\alpha_{abs}\eta_h}, \tag{14}$$

where: η_h - heat conversion efficiency, α_{abs} - absorption coefficient, R_T - thermal shock parameter defined, as follows:

$$R_T = \frac{K_c \sigma_{max,fracture}}{\alpha_T E}, \tag{15}$$

where : $\sigma_{max,fracture}$ – fracture limit stress, E – Young's modulus, K_c- thermal conductivity, α_T – linear thermal expansion coefficient.

Assuming top-hat distribution of the pump beam, maximum temperature increase on the z- distance on the axis of the laser rod is given by (Chen, 1999):

$$\Delta T_{max}(z) = \left(1 + \ln\left(\frac{(a/W_{p,0})^2}{1+((z-z_w)/z_p)^2}\right)\right)\frac{\eta_h \alpha_{abs} \exp(-\alpha_{abs} z)}{4\pi K_c (1-\exp(-\alpha_{abs} L))} P_{abs}, \tag{16}$$

where : a – rod radius, $W_{p,0}$ – radius of pump beam in the waist, z_w – waist distance to the rod facet, z_p – Rayleigh range of pump beam, P_{abs}- absorbed pump power. To obtain average temperature in the rod, $\Delta T_{max}(z)$ has to be integrated along the rod.

Maximal thermally induced stress inside the rod occurs on rod facet (z=0). Assuming a linear thermal elasticity model, the maximum stress can be estimated by the following formulae:

$$\sigma_{max} = \alpha_T \Delta T_{max}(0) E \quad . \tag{17}$$

For top-hat heat source distribution inside the cylindrical rod we have a parabolic profile of temperature resulting in optical path difference of a parabolic shape. In the framework of a paraxial thermal lensing model, the thermo-optic optical power induced inside the rod is given by:

	Tm:YAG	Tm:YAlO$_3$	Tm:YLF	Tm:YVO$_4$
Absoprtion peak wavelength [nm]	785	795	792	801
absorption coeff. α_{abs} [cm^{-1}] @1%Tm	1.05	0.9/0.8	0.36	7
Refractive index n	1.81	1.929/1.943	1.47/1.448	1.98/2.16
Temperature dispersion dn/dT[10^{-6}/K]	7.3	9.7	-4.4/-2	8.5/3.9
thermal expansion coefficient α_T[10^{-6}/K]	7.5	10.8	13.3	4.4/11.4
thermal conductivity K_c [W/m/K]	13	11	7.2	11
Young modulus E [GPa]	310	220	85	77
fracture stress $\sigma_{max,fracture}$ [kG/mm^2]	16	16	4	4
Thermal shock parameter R_T [W/cm]	6.5	5.5	1.7	6.1
β_T [mm^2/W/m]	1.17	1.96	0.47	1.71
f_T [mm] @ I_h=1 W/mm^2	85.4	50.8	212.8	58.5

Table 1. Mechanical and optical parameters of 4 typical Tm doped gain media.

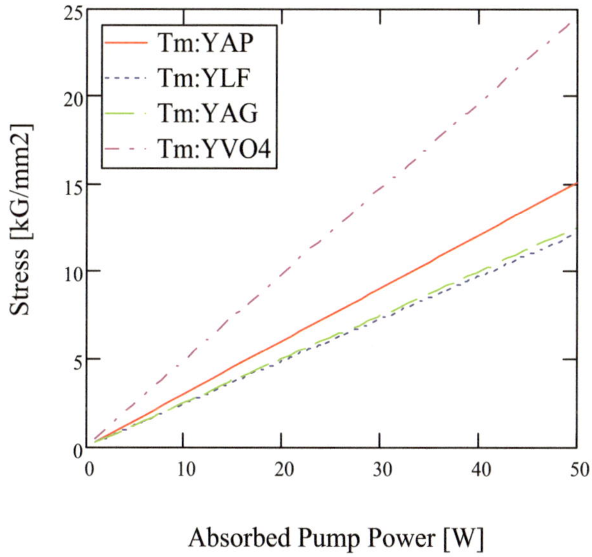

Fig. 8. Maximal stress vs. absorbed pump power for 4 Tm crystals; $W_{p,0}$=0.5mm.

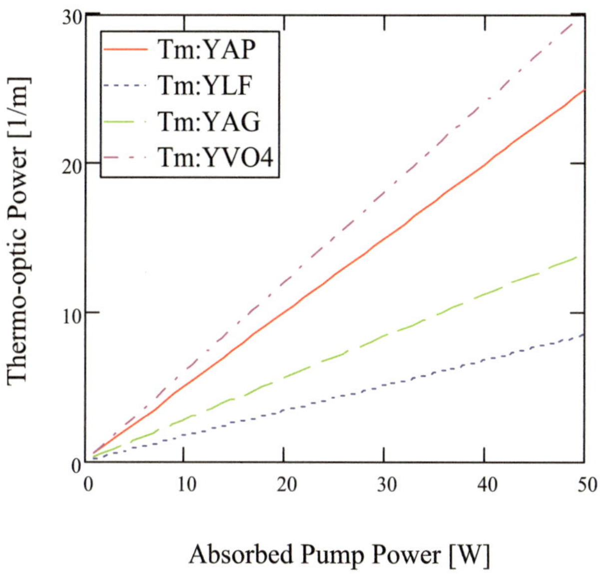

Fig. 9. Thermo-optic power vs. absorbed pump power for 4 Tm crystals; $W_{p,0}$=0.5mm.

$$M_T = f_T^{-1} = \beta_T I_h \quad , \tag{18}$$

where: β_T – thermal lensing parameter, I_h – heat power density defined as follows:

$$\beta_T = \frac{\frac{dn}{dT} + (n-1)(1+\nu)\alpha_T}{K_c} \quad , \quad I_h = \frac{\eta_h P_{abs}}{A_p} \quad , \tag{19}$$

where: ν- Poisson's ration, A_p - averaged pump area, n - refractive index, dn/dT – temperature dispersion of refractive index.
To help in the choice of a proper gain medium for the thulium laser, the mechanical and optical parameters of four Tm-doped media were collected in Table 1.
We have calculated the maximum stress and thermo-optical power vs. absorbed pump power for these 4 gain media (see Fig. 8, 9 respectively).

2.3 Model of actively Q-switched Tm laser

2.3.1 Degnan's analytical model

To analyse the regime of active, periodic Q-switching we can modify the classical models of Q-switching (see e.g. (Degnan, 1989), (Koechner, 1996), (Eichhorn, 2008)). The model of rate equations for a quasi-three-level laser is defined by two functions: Φ – relative intensity of the laser field, and X - modified relative occupation (further called inversion) of the upper laser level:

$$X = X_u - f_l, \tag{20}$$

where f_l –relative occupation of the lower laser level defined by the formula (3).
The set of rate equations consist of two first order differential equations with respect to relative time $\tau = ct/2L_{cav}$ as follows:

$$\begin{cases} \dfrac{d\Phi}{d\tau} = \Phi\left(2\gamma_0 LX - \rho_{cav}\right) \\ \dfrac{dX}{d\tau} = -2\gamma_0 L_{cav}\Phi X - \dfrac{X}{\tau_{rel,u}} + \eta_{abs}\dfrac{I_p}{I_{p,0}} + \dfrac{f_l}{\tau_{rel,u}} + \dots \end{cases}, \tag{21}$$

where: L_{cav} – length of cavity , $\tau_{rel,u} = c\tau_u/2L_{cav}$ – relative upper level lifetime, I_p – relative pump density,

$$I_{p,0} = \alpha_0 L / \eta_{quant}, \tag{22}$$

$$\rho_{cav} = -\ln\left(R_{m,1}R_{OC}\right) + \rho_{on,off}, \tag{23}$$

η_{abs}- absorption efficiency, $\rho_{on,off}$ – losses of active Q-switch, different in pumping and pulsing intervals,

$$\rho_{on,off} = \begin{cases} \rho_{qsw,low} & \text{for} \quad \mathrm{mod}(t, t_{rep}) < t_{off} \\ \rho_{qsw,high} & \text{for} \quad t_{off} < \mathrm{mod}(t, t_{rep}) < t_{on} \end{cases}. \tag{24}$$

The pumping process is periodic with a repetition rate of $f_{rep} = 1/t_{rep}$. In pumping time $t_{on} \approx t_{rep}$, losses of q-switch $\rho_{qsw,high}$ have to be sufficiently high to prevent laser oscillation. In a short interval of duration $t_{off} << t_{on}$, t_{rep} the cavity losses are small and the pulse starts to build up.
For the steady state (after a few initial pulses) the initial inversion X_i before the pulse and final inversion X_f tend to values defined only by pump density, cavity parameters and the repetition period.
We can define threshold inversion for low losses of Q-switch as follows:

$$X_{thr} = \frac{-\ln(R_{m,1}R_{OC}) + \rho_{qsw,low}}{2\gamma_0 L}. \tag{25}$$

To obtain effective Q-switching, pump intensity and duration should be sufficiently high to achieve initial inversion X_{in} a much higher value than X_{thr}:

$$X_{thr} < X_{in} < \frac{-\ln(R_{m,1}R_{OC}) + \rho_{qsw,high}}{2\gamma_0 L}. \tag{26}$$

In pumping time the laser intensity disappears ($\Phi = 0$) and the second equation can be solved analytically, assuming, that absorption efficiency does not change during pumping and other non-linear processes (up-conversion, amplified spontaneous emission) can be neglected. The formula for inversion is the following:

$$X(\tau) = X_\infty - (X_\infty - X(0))\exp(-\tau/\tau_{rel,u}), \tag{27}$$

where: X_∞- inversion achieved for time $\tau >> \tau_{rel,u}$ defined as follows:

$$X_\infty = \tau_{rel,u}\eta_{abs}(I_p)\frac{I_p}{I_{p,0}} + f_1, \tag{28}$$

$X(0) = X_f$ - inversion at the starting point of pumping corresponding to the final inversion after pulse generation. The examples of temporal dependence of inversion for different repetition rates are shown in Fig. 10.

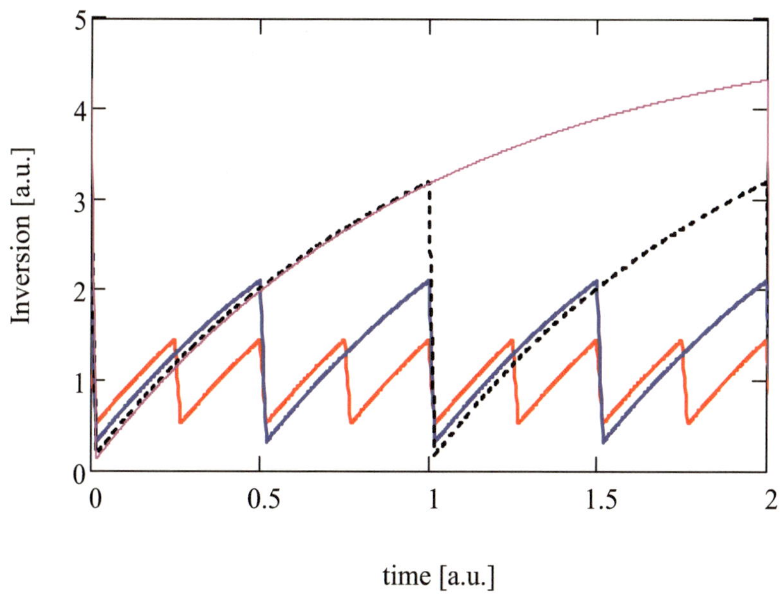

Fig. 10. Changes in inversion vs. time for different repetition periods.

During the pulse build up interval the pumping and relaxation components can be neglected in a second rate equation. Thus, we can find the analytical solution of rate equations in the form of a transcendent equation on initial and final inversions X_i, X_f:

$$\frac{X_f}{X_i} = \exp\left(-\frac{X_i - X_f}{X_{thr}}\right). \tag{29}$$

Using (27), initial inversion can be written as:

$$X_i = X_\infty \frac{1 - \exp\left(-\tau_{rep} / \tau_{rel,u}\right)}{1 - \exp\left(-\tau_{rep} / \tau_{rel,u}\right)\exp\left(-(X_i - X_f) / X_{thr}\right)}. \tag{30}$$

To determine the initial inversion including the cross-relaxation and up-conversion processes the more complicated model of a quasi-three-level laser (Rustad & Stenersen, 1996) can be used.

The pulse duration, pulse energy, and peak power can be determined knowing the initial and final inversions. The conditions of optimisation of an actively Q-switched laser were found (Degnan, 1989). Let us introduce the un-dimensional parameter z defined as follows:

$$z = X_i(I_p, \tau_{rep}) / X_{thr}. \tag{31}$$

The optimal reflectivity $R_{OC,opt}$ of the output coupler can be determined as follows:

$$R_{OC,opt} = \exp\left[-\rho_{pas}\left(\frac{z - 1 - \ln z}{\ln z}\right)\right], \tag{32}$$

where ρ_{pas} – passive cavity losses defined as follows:

$$\rho_{pas} = -\ln\left(R_{m,1}\right) + \rho_{qsw,low}. \tag{33}$$

Let us notice, that $R_{OC,opt}$ depends on the pump density and repetition period.

For optimal out-coupling losses the formulae on pulse duration t_{pulse}, extraction efficiency η_{extr} and output energy E_{out} are the following:

$$t_{pulse} = \frac{2L_{cav}}{c\rho_{pas}}\left(\frac{\ln(z)}{z\left[1 - a(z)(1 - \ln(a))\right]}\right), \quad a(z) = \frac{z - 1}{z\ln(z)}, \tag{34}$$

$$\eta_{extr} = 1 - \frac{1 + \ln(z)}{z}, \tag{35}$$

$$E_{out} = E_0\left(z - 1 - \ln(z)\right), \tag{36}$$

where:

$$E_0 \approx \frac{h\nu_1 A_m \rho_{pas}}{2\sigma_1}. \tag{37}$$

Such an elegant, simple analytical model can give mainly qualitative information about the properties of the Q-switched laser. It should be underlined the role of parameter z equal to the ratio of small signal gain to passive losses. The quality of laser output is proportional to parameter z which could be determined for any type of laser (including high repetition rate Q-switched fibre lasers as well). The short pulses and high peak powers, evidencing effective Q-switching, occur for z > 10.

However, comparing Degnan's analytical model with experiments, quite large differences can be found especially for high pumping rate and long repetition periods. It should be noted, that in this model all non-linear processes are neglected, whose role is growing with increase in inversion. Moreover, with an increase in heat load the several temperature dependent processes (thermally induced aberration, stress, lowering of gain etc.) significantly diminish the performance of the laser.

2.3.2 Numerical modelling of an actively Q-switched Tm:YLF laser.

Comparing results of theoretical analysis with experiments on the Q-switched Tm:YLF laser, it was found, that it is necessary to improve the model adding the thermo-optic and cavity effects. Therefore, we have developed a numerical model (Gorajek, et al, 2009) introducing additional losses $\gamma_{add}(T)$ occurring as a result of the absorption of laser mode wings on unpumped regions of a quasi-three-level medium. In the particular case of our oscillator, the laser scheme for "cold" cavity is nearly half confocal (g_1=1, g_2 = 0.5) for vanished thermal lensing). With increase in negative thermal lensing (typical for Tm:YLF crystal) the mode size at the crystal increases until reaching the stability limit of the cavity (g_1 ~ 2 , g_2 =0.5). However, also, even out of stability range, the spatially limited mode is formed due to diffraction on the gain diaphragm induced by the pumping beam, which could be treated as a kind of "imaginary" lens (Grace, et al , 2001) (Jabczyński et al, 2003). Such a laser mode has a "gaussian" like profile and much higher diffraction losses which depend (via temperature increase) on average pump power. To determine the maximal, average temperature in rod and thermal lensing power the analytical formulae (16), (18), (19) were applied. For calculation of the additional losses $\gamma_{add}(T)$ the numerical procedure proposed in (Jabczyński et al, 2003) was used. The results of the calculation of average temperature and pump dependent losses for two cases: quasi-CW pumping (duty factor 10%) and CW pumping are shown in Fig. 11

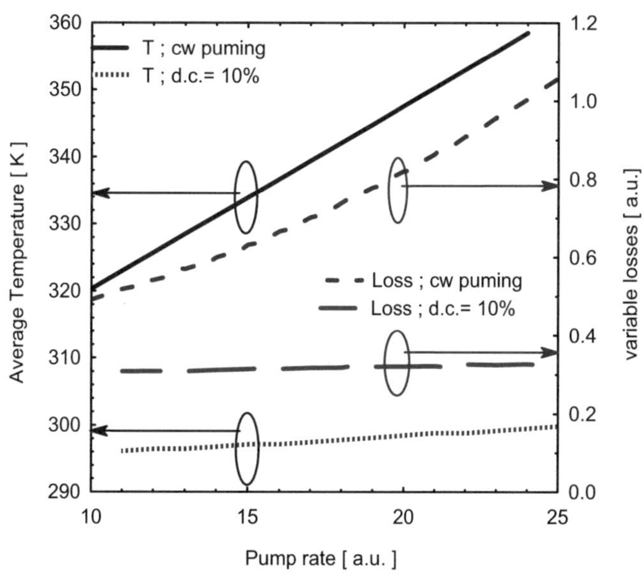

Fig. 11. Average temperature inside rod, pump dependent losses vs. pump rate: CW pumping.

The numerical procedure prepared for analysis of giant pulse formation was divided into two stages. In the first one, before switching off the high Q-switch losses, the internal flux vanishes and only the second equation is numerically solved. Moreover additional non-linear effects causing a drop in inversion can be added. After switching off the Q-switch, the threshold condition is tested and the set of both equations is solved for the giant pulse evolution calculation.

The main difference in analysis of Q-switching in a quasi-three level laser (compared to the four level scheme) consists on the effect of temperature on the available parameters of the giant pulse. Because of the increase in temperature with pump power, the inversion and additional losses significantly influence on the available peak power and pulse duration (see Fig. 12-13).

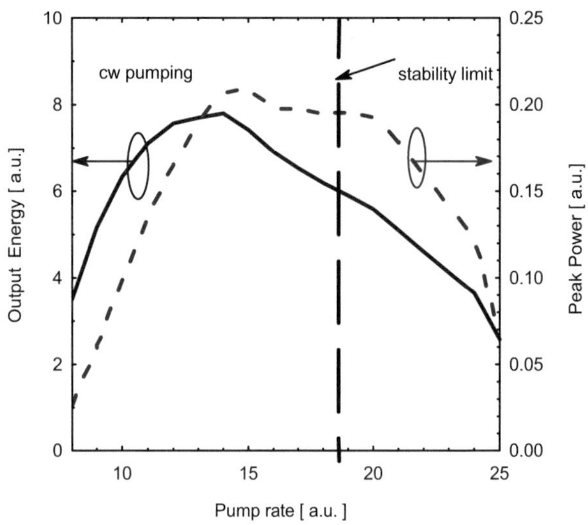

Fig. 12. Average output power, peak power vs. pump rate: CW pumping.

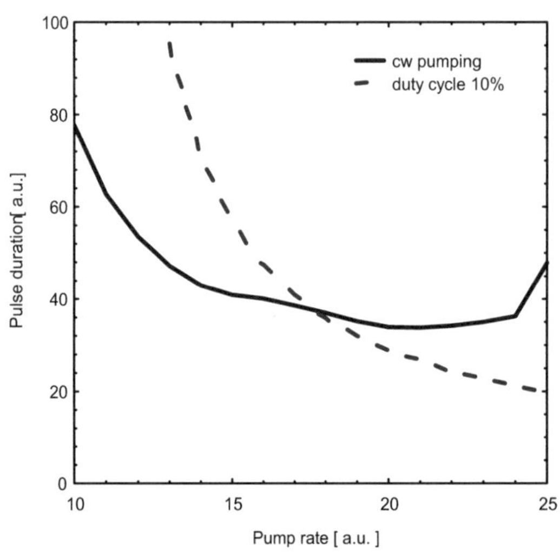

Fig. 13. Pulse duration vs. pump rate; CW pumping continuous curve, pumping duty cycle 10%- dashed curve.

The all above presented effects give the result, that giant pulse is generated for a considerable level of losses dependent on effective average heat power dissipated in the gain medium. Such effects were observed in experiments (see p.3.3 and p.3.4) and we have shown in Fig. 11-13 the results of simulation qualitatively explaining the experiment.

3. Experiments

3.1 Laser head scheme

We have designed the laser head of a diode pumped Tm:YLF laser shown in Fig. 14 destined for tuneable and Q-switched operation (Jabczyński, et al., 2007) (Jabczyński, et al., 2009). The fibre coupled diode of 30-W output power emitting at a 0.792 μm wavelength was deployed as a pump source.

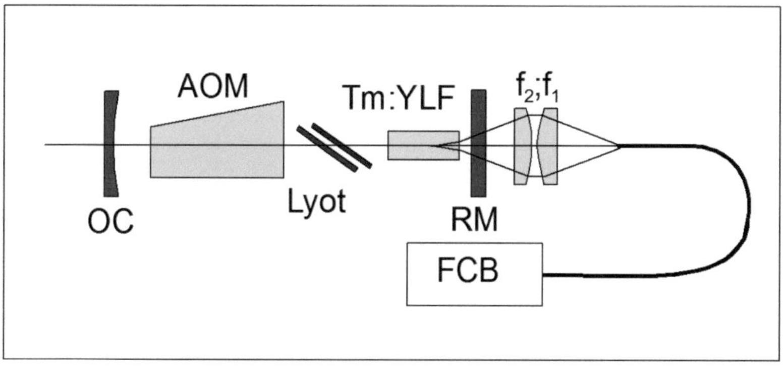

Fig. 14. Optical scheme of a Tm:YLF laser: FCB- fiber coupled laser diode of 30 W, RM- rear flat mirror, OC- output coupler, Tm:YLF- gain crystal, AOM- acousto-optic Q-switch.

The pump beam after passing through a reimaging optic (of 1.5 x magnification) forms inside the gain medium the caustics of waist diameter approximately 0.6 mm and induces a near 60 K increase in temperature for the maximum incident pump power of 26 W.
The uncoated, with 3.5%. doping of thulium YLF rods of 8 and 10 mm size wrapped with indium foil were tested as gain media. The crystals were mounted in a copper heat-sink maintaining 293 K temperature of coolant. The diode wavelength was tuned via the control of thermo-electric cooler voltage to maximise the absorbed pump power, however not higher than 65% absorption efficiency for an 8-mm long rod and 80% for a 10-mm crystal were obtained in the best cases of CW pumping without lasing (see Fig. 15).

3.2 Free running and tunablity characterisation

As the output couplers we have used mirrors of a 0.3 and 0.5 m radii of curvature and 10%, 15%, 20% and 30% transmission at a 1.9-μm wavelength. The energetic characteristics in quasi-CW pumping in dependence on absorbed pump power were presented in Fig. 16.
Above 7-W power with nearly 30% slope efficiency with respect to incident pump power were demonstrated for the best case of quasi-CW pumping for a short, 70-mm resonator. Nearly 5 W were obtained for CW pumping in a short cavity and above 3 W for an elongated 220-mm cavity.

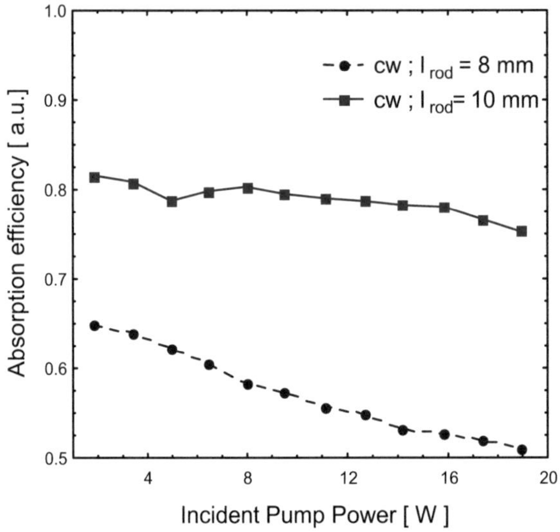

Fig. 15. Absorption efficiency in non-lasing conditions vs. incident pump density for two Tm:YLF rods.

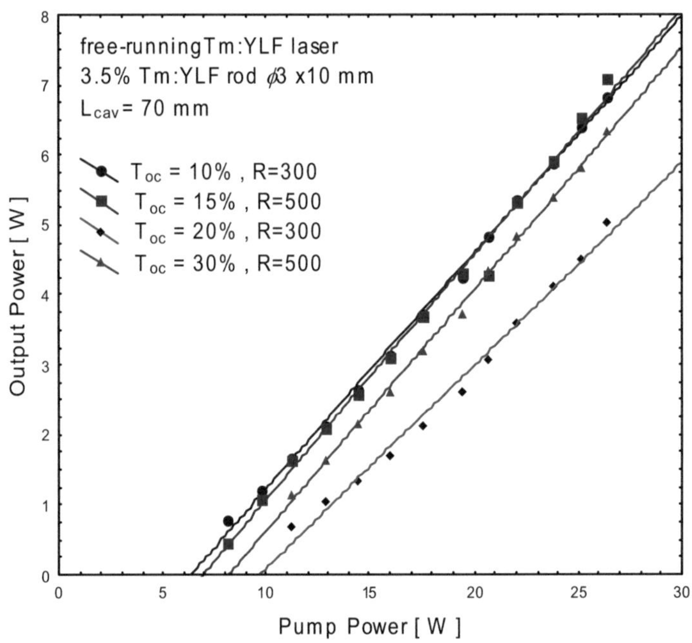

Fig. 16. Output power vs. pump power for quasi-CW pumping (10% duty cycle, 10Hz, 10ms pump duration).

The spectrum of generation was centred at 1908 nm with FWHM of approximately 6 nm (see Fig. 17) for the oscillator without deploying Lyot's filter. In the next step of experiments the cavity was elongated to 220 mm to deploy the Lyot's filter (consisting of 2 quartz plates of 0.92 and 1.84 mm thickness, respectively) and an active Q-switch.

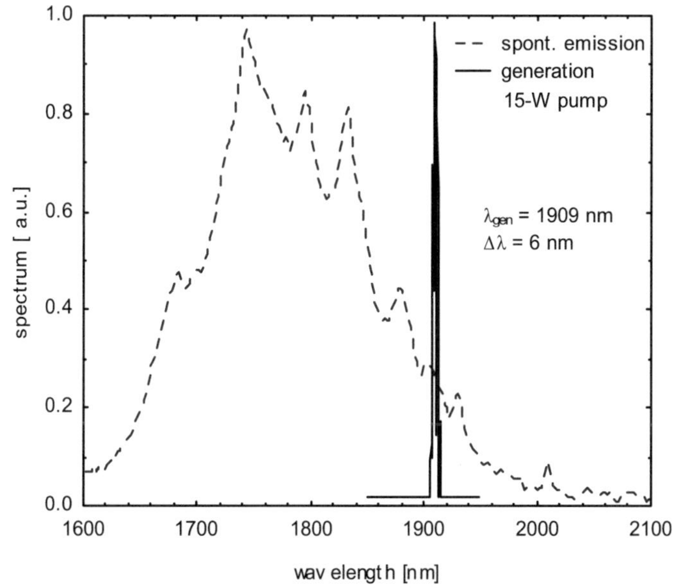

Fig. 17. Spectra of spontaneous emission and free-running of a Tm:YLF laser with 10%- transmission of OC.

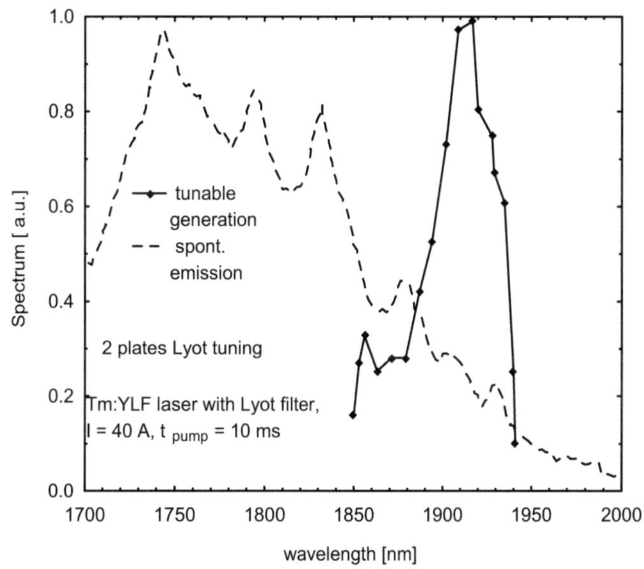

Fig. 18. Spectra of spontaneous emission and tuning of Tm:YLF laser by means of Lyot's filter.

The experiments of tuning were carried out firstly in a free-running regime. The tuning range of 1845-1935 nm was demonstrated for 37° - 47° angle of rotation with respect to the fast axis of quartz plates (see Fig. 18). The linewidth was less than 1 nm. For the Q-switching regime the contrast of a deployed birefringent filter was too low to prevent oscillation on the strongest

1908-nm line. In the last part of characterisation in a free-running regime, we have measured the beam profiles in far field in the focal plane of a 500-focal length lens. The divergence angle was about 4.3 mrad and an estimated parameter $M^2 < 1.3$ for high 20-W incident pump power.

3.3 Q-switching experiments for low duty cycle pumping

For Q-switching we have used a water cooled acousto-optic modulator made of 45-mm long fused silica, operating at a radio frequency of 40.7 MHz with a maximum power of 25 W. In fact, that was the largest element of laser head which determined its size. It was shown in separate experiments that for maximum RF power of the acousto-optic modulator the diffraction efficiency was higher than 80%, diffraction angle was about 7 mrad and the falling edge, i.e. switch off time was about 100 ns.

In the first part of the Q-switching experiments we have estimated the maximum available output energy in free-running for which the acousto-optic modulator can hold off oscillations for a switch on state of RF power. It should be noted that we have used a 220-mm long cavity with Lyot's filter inside introducing additional insertion losses. The laser output was horizontally polarized (perpendicularly to the c-axis of YLF crystal).

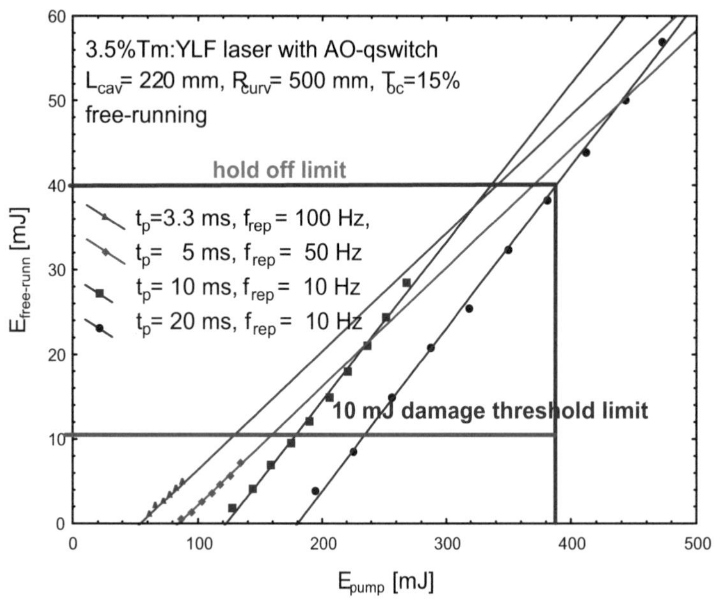

Fig. 19. Available output energy vs. incident pump energy in free running for the state of effective operation of the active Q-switch

As was shown in Fig. 19, for the best case the output energy of 40 mJ (for incident pump energy of 400 mJ) was the upper limit of efficient operation of the Q-switch. However, the real limit of output energy was far lower, because of the damage threshold of the Tm:YLF crystal facet. It was shown, that the output energy above 10 mJ corresponding approximately to 1.5 - 2 GW/cm² of intracavity power density constitutes the upper limit of available pulse energy for the safe operation in a Q-switching regime in the case of our laser head. Thus, we can conclude that a much smaller Q-switch without water cooling will be satisfactory for our purposes.

The results of measurements of pulse duration and peak power for a low duty cycle of 10% (10 Hz of PRF and 10 ms pump duration) were shown in Fig. 20. The shortest pulse of 22-ns duration (see Fig. 21) and 10.5 mJ energy corresponding to 0.45 MW of peak power were demonstrated for the best case of stable output below the risk of damages to laser elements.

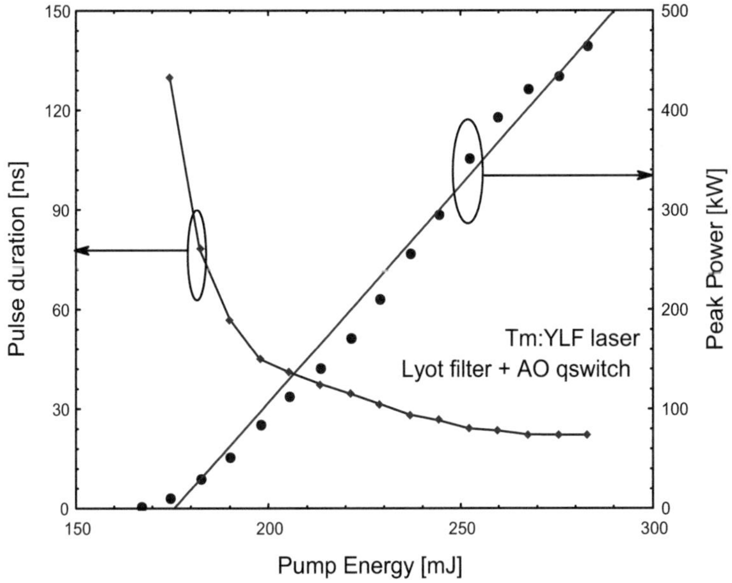

Fig. 20. Pulse duration, peak power vs. pump energy for Q-switching in a low 10% duty cycle pumping regime.

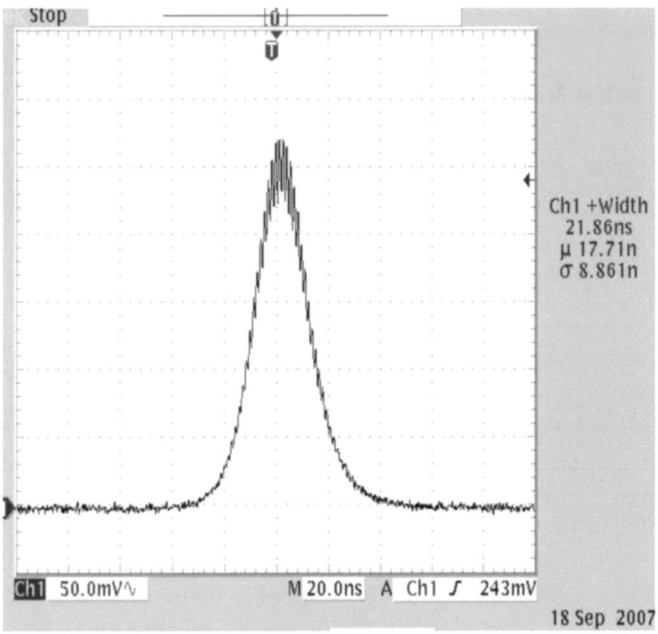

Fig. 21. Oscillogram of the giant pulse of 10.5 mJ of energy.

3.4 Q-switching experiments for CW pumping

For the CW pumping regime the maximum pump power was constituted due to the thermal lensing limit. Because of negative thermal dispersion of Tm:YLF the cavity achieves stability limit for nearly 20-W of incident pump power.

The results of the Q-switching experiments were shown in Fig. 22, 23, and collected in Table 2. Nearly 20% slope efficiency with respect to absorbed pump energy was obtained for high repetition frequency. The experimental results were in agreement with the numerical model presented in p. 2.3.2

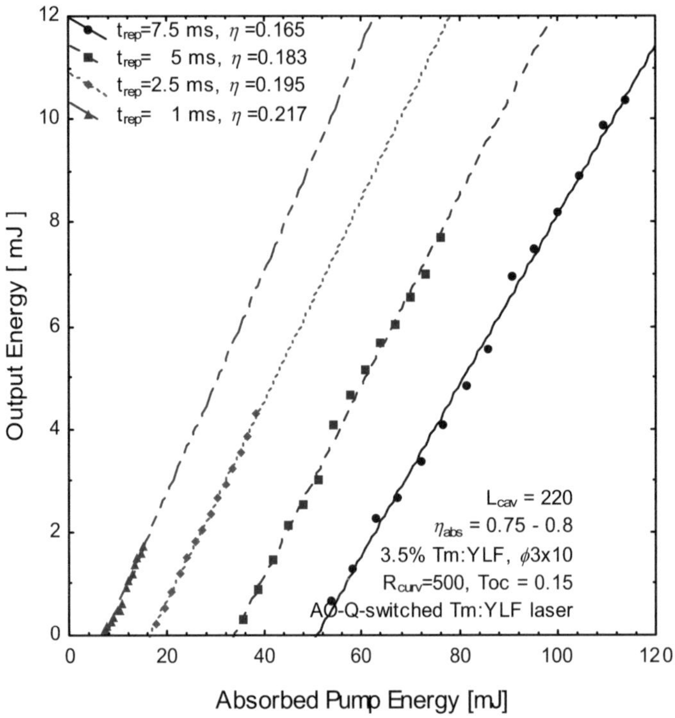

Fig. 22. Output energy vs. absorbed pump energy for different repetition periods.

f_{rep} [Hz]	d.f.	P_{avg} [W]	E_p [mJ]	τ_p [ns]	P_p [kW]
1000	1	1.725	1.725	146	11.8
400	1	1.725	4.31	101	42.7
200	1	1.541	7.7	70	110
133	1	1.38	10.35	46.8	221
10	0.1	0.105	10.5	22	447

Table 2. Results of Q-switching experiments; f_{rep} – pulse repetition frequency, d.f. – duty cycle factor, P_{avg} – average output power, E_p – pulse energy, τ_p – pulse duration, P_p – peak power.

The comparable pulse energies of 10 mJ (last two rows of Table 2) were achieved for both cases of pumping. The much longer pulse duration for a case of CW pumping was caused by the combined effect of an increase in reabsorption and additional diffraction loss (see p. 2.3.2). Please note, that maximum available pulse energy was limited in both cases by reaching the damage thresholds of the rod facet or rear mirror.

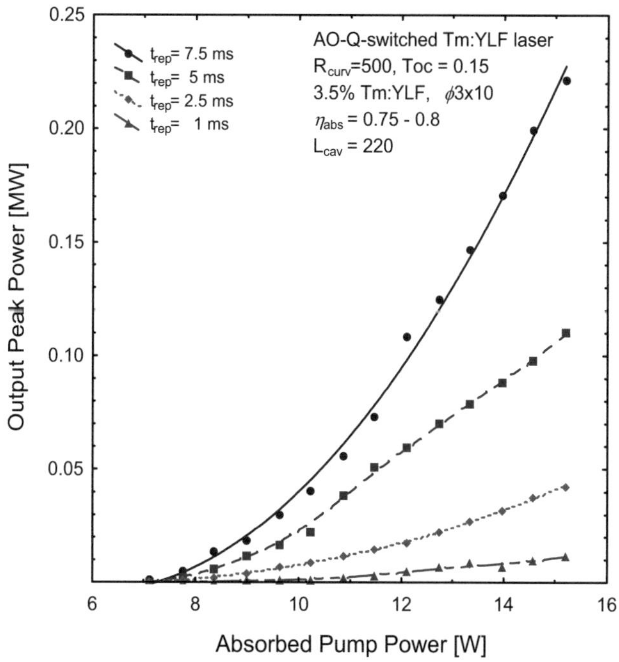

Fig. 23. Output peak power vs. absorbed pump power for different repetition periods.

4. Conclusions

The analytical models of quasi-three-level lasers operating in free running and Q-switching regimes were presented. In both cases appropriate formulae enabling the optimization of such lasers were given and analysed. The numerical model of a quasi-three-level laser operating in a Q-switching regime including additional pump dependent losses, was elaborated to explain the properties of the developed actively Q-switched laser. The main difference in analysis of Q-switching in a quasi-three-level laser (compared to a four-level laser) consists of the effect of temperature on giant pulse parameters. Because of increase in temperature with pump power, the net inversion, additional reabsorption and diffraction losses significantly influence available pulse energy, peak power and pulse duration. The all above mentioned effects result in the fact, that giant pulse is generated for a considerable level of losses dependent on effective average heat power dissipated in the gain medium. The results of numerical modelling were confirmed in the experiments.

To compare models with experiments we have presented the results of investigations of an efficient Tm:YLF laser end-pumped by 30-W fiber coupled laser diode bar. The incident pump density exceeded above 5 times the saturation pump density, thus the drawbacks of the quasi-three-level scheme have been mitigated. We have obtained the best output characteristics (slope and maximum power) for out-coupling losses of 20% evidencing the high roundtrip gain for maximum pump power. Above 7-W of output power for incident

26-W pump power in free running regime was achieved in the best case for a short 70-mm cavity. Above 3 W of output power was demonstrated for CW pumping for an elongated 220-mm cavity. The divergence angle was about 4.3 mrad and estimated parameter $M^2 < 1.3$. To improve the output characteristics in a free running regime, the optimisation of pump size in the gain medium, application of a longer rod and optimised cavity design should be applied.

For the free-running and Q-switching regimes the output spectrum was centred at 1908-nm with linewidth less than 6 nm. For tuning the Lyot's filter consisting of 2 quartz plates was deployed. The tuning range of 1845-1935 nm with less than 1-nm linewidth was demonstrated for the free-running regime. For the Q-switching regime the contrast of a deployed birefringent filter was too low to prevent oscillation on the strongest 1908-nm linewidth.

In the experiments on active Q-switching by means of an acousto-optic modulator, up to 10-mJ output energy was demonstrated. Output energy was limited by damage of the laser elements. Nearly 0.5 MW peak power with pulse durations of 22 ns was achieved for a 10-Hz repetition rate with 10% duty cycle of the pumping regime. The 1.7-W of average power with 12 kW peak power and 1000 Hz repetition rate was demonstrated for the CW pumping regime. The developed laser could constitute the basis for development of the tunable, Q-switched laser source operating at a 2-μm wavelength. Moreover, it could be used as a pump source for Ho:YAG and Cr:ZnSe lasers operating in a gain switching regime for the longer (> 2 μm) wavelengths.

5. Acknowledgments

This work was supported by the Polish Ministry of Science and Higher Education under projects 0T00A00330, NN515 423033, NN515 414834, NN515 345036.

6. References

Barnes, N., & De Young, R. (2009). Tm:germanate Fiber Laser for Planetary Water Vapor Atmospheric Profiling. *The Conference on Lasers and Electro-Optics (CLEO)/The International Quantum Electronics Conference (IQEC) (Optical Society of America, Washington, DC,* 2009 (p. JWA60). Optical Society of America, Washington, DC, 2009

Barry, D., Parlange, J., Li, L., Prommer, H., Cunningham, C., & Stagnitti, F. (2000). Analytical approximation for eal values of the LambertW function. *Math and Computers in Simulation*, Vol. 53, pp. 4-14

Beach, R. (1996). CW Theory of quasi-three-level end-pumped laser oscillators. *Optics Communications*, Vol. 123, pp. 385-393

Bourdet, G. (2001). New evaluation of ytterbium-doped materials for CW laser applications. *Optics Communications*, Vol. 198, pp. 411-417

Bourdet, G. (2000). Theoretical investigation of quasi-three-level longitudinally pumped continuous wave lasers. *Applied Optics*, Vol. 39, pp. 966-971.

Budni, P., Lemons, M., Mosto, J., & Chicklis, E. (2000). High-Power/High-Brightness Diode-Pumped 1.9-μm Thulium and Resonantly Pumped 2.1-μm Holmium Lasers *IEEE J. Sel. Top. Quant. Electron,* Vol. 6, No. 4, pp. 629-634

Chen, Y.-F. (1999). Design Criteria for Concentration Optimization in Scaling Diode End-Pumped lasers to High Powers: Influence of Thermal Fracture. *IEEE Journal of Quantum Electronics,* Vol. 35, pp. 234-239

Clarkson, W., Shen, D., & Sabu, J. (2006). High-power fiber-bulk hybrid lasers. *Proceedings SPIE,* Vol. 6100, pp. 61000A-1-13

Degnan, J. (1989). Theory of optimally coupled Q-switched laser. *IEEE Journal of Quantum Electronics*, Vol. 25, pp. 214-220

Dergachev, A., Wall, K., & Moulton, P. (2002). A CW Side-pumped Tm:YLF Laser. *OSA TOPS, Advanced Solid State Lasers, ed. M.E. Ferman L.R. Marshall,* Vol. 68, pp. 343-346

Eichhorn, M. (2008). First investigations on an Er^{3+}:YAG SSHCL. *Appl. Phys. B*, Vol. 93, pp. 817-822

Eichhorn, M. (2008). High-Power Resonantly Diode-Pumped CW Er^{3+}:YAG Laser. *Appl. Phys. B*, Vol. *93*, pp. 773-778

Eichhorn, M. (2008). Quasi-three-level solid-sate lasers in the near and mid infrared based on trivalent rare earth ions. *Appl Phys B*, Vol. 93, pp. 269-316

Eichhorn, M., & Jackson, S. (2008). High-pulse-energy, actively Q-switched Tm^{3+} -doped silica 2 m fiber laser pumped at 792 nm. *Optics Letters*, Vol. 32, pp. 2780-2782

Eichhorn, M., & Jackson, S. (2009). High-pulse-energy, actively Q-switched Tm^{3+},Ho^{3+}-codoped silica 2 m fiber laser. *Optics Letters*, Vol. 33, pp. 1044-1046

Gaponenko, M., Denisov, I., Kisel, V., Malyarevich, A., Zhilin, A., Onushchenko, A., et al. (2008). Diode-pumped Tm:KY$(WO_4)_2$ laser passively Q-switched with PbS-doped glass. *Appl Phys B*, Vol. 93, pp. 787-791

Gapontsev, D., Platonov, N., Meleshkevich, M., Drozhzhin, A., & Sergeev, V. (2007). 415W single-mode CW thulium fiber laser in all-fiber format. *CLEO Europe, 2007,* paper. CP2-3-THU

Godard, A. (2007). Infrared (2-12 μm) solid-state laser sources: a review. *C.R. Physique*, Vol. 8, pp. 1100-1128

Gorajek, L., Jabczyński, J.K. , Zendzian, W., Kwiatkowski, J., Jelinkova, H., Sulc, J., et al. (2009). High repetition rate, tunable, Q-switched, diode pumped Tm:YLF laser. *Opto-Electronics Rev.,* Vol. 6, pp. 23-35

Grace, E., New, G., & Franch, P. (2001). Simple ABCD matrix treatment for transversely varying saturable gain. *Optics Express*, Vol. 26, pp. 1776-1778

Honea, E., Beach, R., Sutton, S., Speth, J., Mitchell, S., Skidmore, J., et al. (1997). 115-W Tm:YAG Diode-Pumped Solid-State Laser. *IEEE Journal of Quantum Electronics,* Vol. 33, pp. 1592-1600.

Huber, G., Duczynski, E., & Peterman, K. (1988). Laser pumping of Ho—Tm-, Er- doped garnet laser at room temperature. *IEEE Journal of Quantum Electronics*, Vol. 24, pp. 920-923.

Jabczyński, J., Gorajek, L., Zendzian, W., Kwiatkowski, J., Jelinkova, H., Sulc, J., et al. (2009). High repetition rate, high peak power, diode pumped Tm:YLF laser. *Laser Phys. Letters*, Vol. 6, pp. 109-112

Jabczyński, J., Kwiatkowski, J., & Zendzian, W. (2003). Modeling of beam width in passively Q-switched end-pumped laser. *Optics Express*, Vol. 11, pp. .552-559

Jabczyński, J., Zendzian, W., Kwiatkowski, J., Jelinkova, H., Sulc, J., & Nemec, M. (2007). Actively Q-switched diode pumped thulium laser. *Laser Phys. Letters*, Vol. 4, pp. 863-867

Koechner, W. (1996). *Solid-State Laser Engineering.* Springer Verlag, ISBN 3-540-60237-2, Berlin

Kudryashov, I., Katsnelson, A., Ter-Gabrielyan, N., & Dubinskii, M. (2009). Room Temperature Power Scalability of the Diode-Pumped Er:YAG Eye-Safe Laser. *CLEO-Baltimore,* paper CWA2

Lim, C., & Izawa, Y. (2002). Modeling of End-Pumped CW Quasi-Three-Lvele Lasers. *IEEE Journal of Quantum Electronics*, Vol. *38*, pp. 306-311

Lisiecki, R., Solarz, P., Dominiak-Dzik, G., Ryba-Romanowski, W., Sobczyk, M., Cerny, P., et al. (2006). Comparative optical study of thulium-doped YVO_4, $GdVO_4$, and $LuVO_4$ single crystals. *Phys. Rev. B*, Vol. 74, pp. 035103.

McComb, T., Shah, L., Sims, A., Sudesh, V., Szilagyi, J., & Richardson, ,. M. (2009). Tunable Thulium Fiber Laser System for Atmospheric Propagation Experiments. *Conference on Lasers and Electro-Optics (CLEO)/The International Quantum Electronics Conference (IQEC) (Optical Society of America, Washington, DC,* paper CthR5.

Mirov, S., Fedorov, V., Moskalev, I., & Martyshkin, D. (2007). Recent Progress in Transistion-Metal-Dped II-VI Mid-IR Lasers. *IEEE Journal of Selected Topics in Quantum Electronics*, Vol. 13, pp. 810-822

Payne, S., Chase, L., Smith, L., Kway, W., & Krupke, W. (1992). Infrared Cross-Section Measurements for Crystals Doped with Er^{3+}, Tm^{3+}, and Ho^{3+}. *IEEE Journal of Quantum Electronics*, Vol. 28, pp. 2619-2630

Rustad, G., & Stenersen, K. (1996). Modeling of Laser-Pumped Tm and Ho Lasers Accounting for Upconversion and Ground-State Depletion. *IEEE Journal of Quantum Electronics*, *32*, pp. 1645-1655.

Schellhorn, M. (2008). High-power diode-pumped Tm:YLF laser. *Appl. Phys. B*, Vol. 91, pp. pp. 71-74

Schellhorn, M., Ngcobo, S., & Bollig, S. (2009). High-Power Diode-Pumped Tm:YLF slab laser. *Appl. Phys. B*, Vol. 94, pp. 195-198

Schellhorn, M., Eichhorn, M., Kieleck, C., & Hirth, A. (2007). High repetition rate mid-infrared laser source. *C. R. Physique*, Vol. 8, pp. 1151-1161

Setzler, S., Francis, M., Young, Y., Konves, J., & Chicklis, E. (2005). Resonantly pumped eyesafe erbium lasers. *IEEE J. Sel. Top. Quant. Electron*, Vol. 11, pp. 645-657

So, S., MacKenzie, J., Shepherd, D., Clarkson, W., Betterton, J., & Gorton, E. (2006). A power-scaling strategy for longitudinally diode-pumped Tm:YLF lasers. *Appl. Physics B.*, Vol. 84, pp. 389-393

Sorokina, I., & Vodopyanov, K. (2003). *Solid-State Mid-Infrared Laser Sources.*: Springer Verlag, ISBN 3-540-00621, Berlin

Zhu, X., & Jain, R. (2007). 10-W-level diode-pumped compact 2.78 μm ZBLAN fiber laser. *Opt. Letters*, Vol. 32, pp. 26-28.

6

Efficient Intracavity Beam Combining of Multiple Lasers in a Composite Cavity

Ming Lei
Department of Electronic Engineering
Tsinghua University
China

1. Introduction

High power or high energy solid-state lasers are required in many applications, but are limited in beam quality or brightness at high pump level by thermal effects. Beam combining with two lasers is an effective way to solve this problem and has been successfully realized in the past (Sabourdy et al., 2002; Sabourdy et al., 2003; Qinjun et al., 2005; Eckhouse et al., 2005). In order to get higher output energy with good beam quality, researchers often regard the two-channel combined configuration as the elementary laser and combine an even number of elementary lasers in a tree architecture (Sabourdy et al., 2002; Sabourdy et al., 2003; Qinjun et al., 2005). These direct extending schemes can successfully combine 2×N channel lasers into one beam intracavity, but the whole scaling geometry is really complicated and bulk, which bring more difficulties for alignment among multiple branches. Besides all these additions of laser beams are only obtained with spatial Gaussion beams(Sabourdy et al., 2002; Sabourdy et al., 2003; Qinjun et al., 2005), which limits the output power for scaling. Using a planar interferometric coupler, Ishaaya firstly reported intracavity beam addition of transverse multimode laser beam distributions (Ishaaya, et al., 2004), then more than two lasers combination has also been demonstrated (Eckhouse, et al., 2005; Eckhouse, et al., 2006). In these schems, the thick planar interferometric coupler with a high-precision plane is the key component. But it is difficult to fabricate this coupler, further, the intracavity loss will increase when multiple beams being combined. In addition, most of these more than two-channel combining schemes are based on the open-ended configuration (Sabourdy et al., 2002; Sabourdy et al., 2003; Qinjun et al., 2005; Eckhouse et al., 2005; Eckhouse, et al., 2006; Ishaaya, et al., 2004), if the symmetries of many branches are not well guaranteed, the loss will be unavoidably introduced from every open end of the beam splitter or coupler, consequently resulting in instability of the whole composite cavity. Generally speaking, these kinds of cavities are not very easy to implement at present.

Recently, we have presented a new close-ended Four-Mirror Cavity to combine two beams with two gain media intracavity (Ming & Mali, 2007). Base on this, in this letter, we propose a novel and practical composite-cavity, named Six-Mirror Cavity, to combine four beams with four gain media intracavity. This cavity is based on a close-ended configuration, which makes the output very stable, even when multiple channels combining at the high pump level. Also, it is not the direct extending of the two-channel scheme as the conventional strategies, the reduction of two mirrors compared with the scaling scheme shown in Ref (Ming & Mali, 2007), makes the whole scaling configuration simple, compact and easy to implement. Moreover, it is an approach for efficient intra-cavity beam addition of transverse multimode laser beam distributions, possessing considerably more energy than that of Gaussian beam distributions. The whole cavity is composed of several LD pumped laser modules. Compared to end-pumped scheme, the diode-side-pumped configuration has a more excellent scalability to obtain high output energy (Fujikawa et al., 1997). Several side-pumped lasers with slab and rod media geometries were investigated. Slab geometry requires expensive slab-shaped materials, and it is difficult to generate symmetrical beam patterns because of the rectangular cross section of the laser medium (Golla et al., 1995). On the contrary, side-pumped scheme by using rod laser systems can overcome the above-mentioned shortcomings and is especially appropriate for beam combination. Therefore, here we adopt the diode arrays side-pumped rod laser as the basic module and combine four laser modules intracavity with a six-mirror cavity.

2. Six-mirror cavity configuration

The basic configuration for energy addition of four lasers with six-mirror cavity is schematically presented in Fig.1. The cavity is based on close-ended resonator, which is composed of six end mirrors M_1-M_6. M_1-M_5 are flat 100% reflectors at the laser wavelength (1064nm) and M_6 is the output mirror with 80% transmission at 1064nm. Thanks to two 50/50 beam splitters, BS, the lasers produced by each arm combine together into one beam in the end and export from the output coupler M_6.

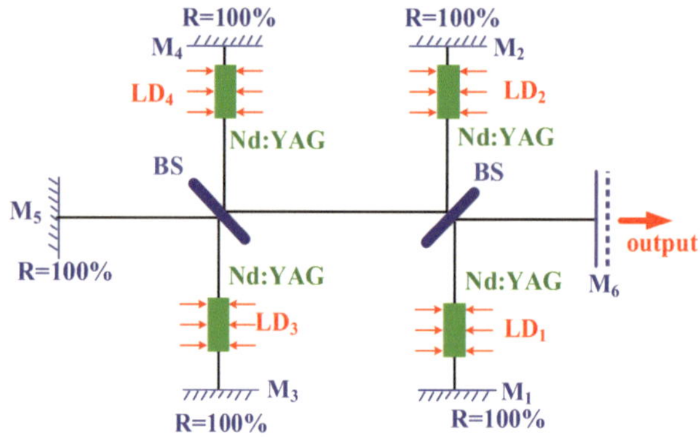

Fig. 1. Schematic of the experimental setup of the six-mirror cavity. BS: beam splitter; LD: laser diode; M_1-M_6: mirrors;

The whole system consists of four amplifying modules, i.e., four laser heads, arranged in the respective branch arm. Fig.2 shows the schematic cross section of the side-pumped Nd:YAG rod laser head. The laser rod (diameter of 5mm, length of 55mm, Nd-doping level of 1.0 at.%) is placed in a glass tube for direct water cooling. Outside the tube, a number of linear LD arrays are located circular-symmetrically and densely around the rod, generating 808nm laser that is directly coupled into the rod. The two end faces of the rod are AR-coated at 1064nm and wedged into 2 degree, which prevents the self-oscillation of the rod. Each pump LD arrays is directly attached to a copper heat sink. The temperature of the pump modules is controlled by the water flow through the copper heat sinks to regulate the temperature of the diode lasers within an accuracy of ± 0.2°C.

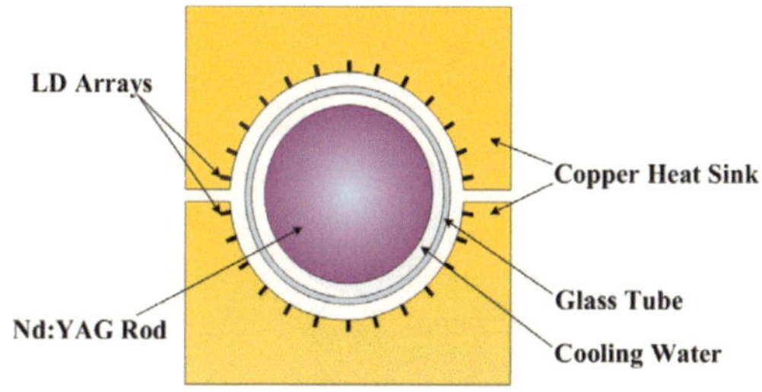

Fig. 2. Schematic cross section of the side-pumped Nd:YAG rod laser head

Every laser head works in a free-running mode and the LD energy supply provides 240μs electric pulse and 1Hz repetition rate. 1μs pulse synchronization has been set among the four channels with the outer-trigger, so that the laser beams produced by every laser heads can be combined temporally and spatially at the same time.

3. Experimental results and analysis

We use EPM2000 two-channel joulemeter/power meter and J50HR energy probe (Molectron, Inc.) to measure the output energy and a laser beam analyzer (Spiricon M^2-200) to detect the beam quality and the intensity distribution for the combined laser and the four individual lasers. As is illustrated in Fig.3, the output energies of the six-mirror cavity when only one LD arrays(LD_1,LD_2,LD_3 or LD_4) is pumping, and the combined energy when four LD arrays are pumping simultaneously are shown. A single beam multimode output exceeding 453mJ (165μs duration, 1Hz repetition rate) at 1064nm is obtained when the four laser heads work simultaneously in the six-mirror cavity.
In order to demonstrate the improved brightness of six-mirror cavity, four Fabry-Parot lasers with the cavity length of 31cm (same to the length of l_2+ l_7 in Fig.4(b)) are characterized for reference with the experimental setup shown in Fig.4(a). When the four LD arrays, i.e. four laser heads, work at the maximum pump energy of 1.26J with 70A operating current, the output properties in five cases are listed in Table 1 and the brightness is calculated by the expression (Fan, 2005)

$$B = \frac{E}{\lambda^2 \cdot (M_X^2 \cdot M_y^2)} = \frac{k \cdot E}{M_X^2 \cdot M_y^2} \quad (1)$$

where

$$k = 1/\lambda^2 \quad (2)$$

The increasing output energy and the good combined beam quality are well shown in Table 1 and Fig.4. Using this six-mirror cavity, four independent elementary multimode lasers have been successfully combined into one beam intracavity with the combination efficiency of 90.7% (453.3/(124.7+131.2+115+128.6)=0.907), a rather high value despite the disparity and multimode distribution among the four branch laser heads features. These results can be explained as follows. In the laser cavity, the four elementary lasers are inter-seeds of each other. One laser beam imprint its transverse distubution content on the other three beam distributions. The combined laser tends to operate so that the losses are minimum. Therefore, each of the transverse beam distribution adds with its counterpart in the other three beams and four multimode beam distributions have similar distribution composition.

Consequently four multimode beams combine intracavity successfully and considerably higher output energy is obtained in laser system.

The brightness of the combined laser has been significantly improved more than 3 times compared to single F-P cavity laser. Furthermore, the experiments also show that when the pump energy is fixed, the laser output of the six-mirror cavity is stable with no change in energy or beam quality, which shows that this cavity can withstand environmental perturbations very well.

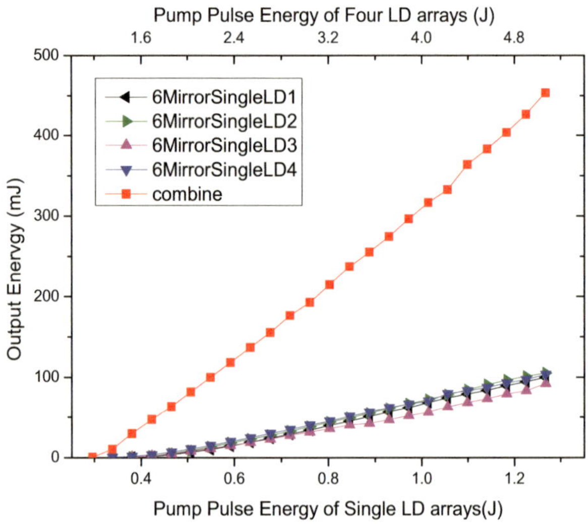

Fig. 3. Dependence of the output energies on the pump energy launched on each LD arrays. The filled triangles with four directions show the output energies of the six-mirror cavity when only one LD arrays(LD1,LD2,LD3 or LD4) is pumping respectively; The filled triangles show the output energies of the six-mirror cavity when LD1~LD4 are pumping simultaneously with the pump energy at the top.

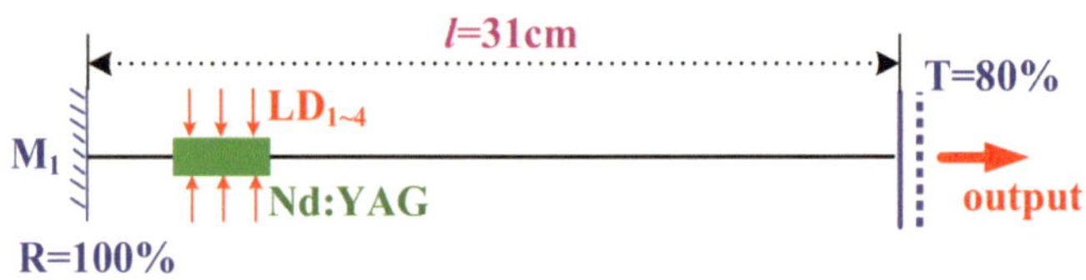

(a)

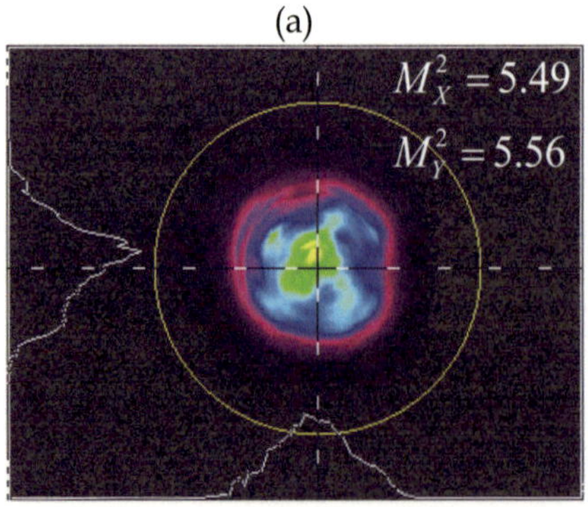

(b)

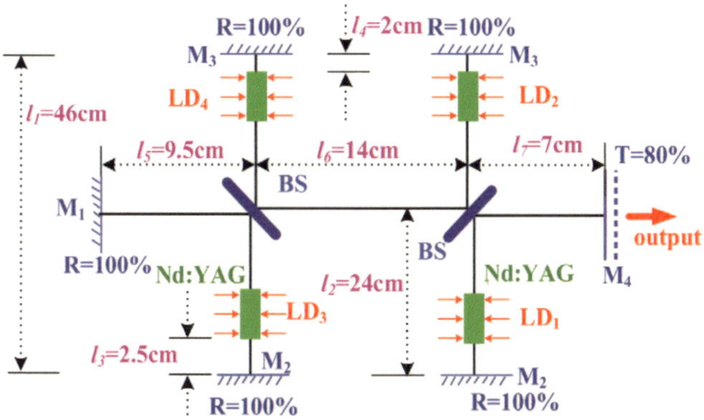

(c)

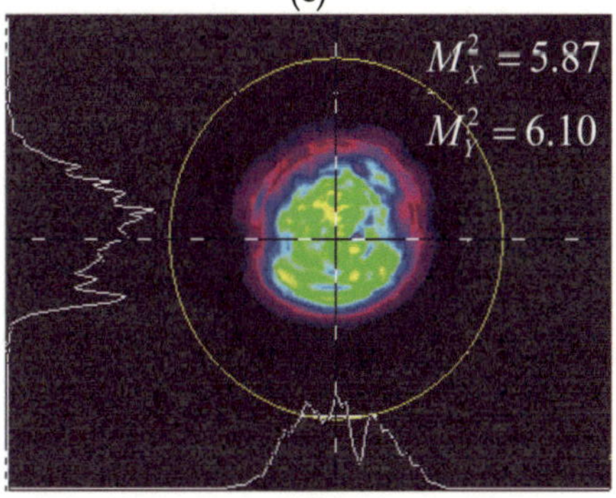

(d)

Fig. 4. The setup of Fabry-Parot cavity and six-mirror cavity and the detected output intensity distributions at the maximum pump energy
(a) The setup of the Fabry-Perot cavity laser
(b) The intensity distributions of the Fabry-Perot cavity laser with individual LD_1 arrays pumping
(c) The setup of six-mirror cavity laser
(d) The intensity distribution of the six-mirror cavity laser with four LD arrays pumping simultaneously

4. Conclusion

A new laser scheme for energy scaling with a composite six-mirror cavity is demonstrated. Unlike the conventional method that directly extends the two-beam addition configuration to 2×N channels geometry in a tree architecture, in this paper, we present a new close-ended six-mirror cavity, which is more compact, stable and practical, to combine four multimode laser beams intracavity. The combined output characteristics including energy, beam quality and the improved brightness of this laser are investigated experimentally. The results demonstrate that four multimode individual Nd:YAG lasers have been combined intracavity successfully with this six-mirror cavity, and a single beam laser output exceeding 453mJ with 165μs duration is achieved, with 90.7% combining efficiency. The brightness of the combined laser has been significantly improved compared to single F-P cavity laser. In conclusion, the use of this six-mirror cavity provides a novel approach for the efficient improvement of brightness and energy scaling by combination of multiple lasers.

Parameter	Pump current	Pump Energy	Output energy	Beam quality		Calculated Brightness $mJ/(m^2 \cdot ster)$
				M_X^2	M_Y^2	
F-P1	70A	1.26 J	124.7 mJ	5.49	5.56	4.08k
F-P2		1.26 J	131.2 mJ	5.62	5.97	3.91k
F-P3		1.26 J	115 mJ	5.24	5.66	3.87k
F-P4		1.26 J	128.6 mJ	5.37	5.78	4.14k
Six-mirror combined	70A	1.26J+1.26J +1.26J+1.26J	453.3 mJ	**5.87**	**6.10**	**12.66k**

Table 1. The output properties in five cases at the maximum pump energy
k: coefficient. F-P1: The Fabry-Perot cavity laser with LD_1 arrays pumping
F-P2: The Fabry-Perot cavity laser with LD_2 arrays pumping
F-P3: The Fabry-Perot cavity laser with LD_3 arrays pumping
F-P4: The Fabry-Perot cavity laser with LD_4 arrays pumping
Six-mirror combined: The Six-mirror cavity laser with LD_1, LD_2, LD_3 and LD_4 arrays pumping simultaneously

5. References

Eckhouse, V.; Ishaaya, A. A. & Shimshi, L. (2005). Imposing a Gaussian distribution in multichannel laser resonators, IEEE Journal of Quantum Electronics, Vol. 41, pp. 686-693, 2005.

Eckhouse, V.; Ishaaya, A. A. & Shimshi, L. (2006). Intracavity coherent addition of 16 laser distributions, Optics Letters, Vol. 31, pp. 350-352, 2006.

Fan, T. Y. (2005). Laser beam combining for high-power, high-radiance sources, IEEE Journal of Selected Topics in Quantum Electronics, Vol. 11, pp. 567-577, 2005.

Fujikawa, S.; Kojima, T. & Yasui, K. (1997). High-power and high-efficiency operation of a CW-diode-side-pumped Nd:YAG rod laser, IEEE Journal of Selected Topics in Quantum Electronics, Vol. 3, pp. 40-44, 1997.

Golla, D.; Knoke, S. & Schone, W. (1995). 300-W cw diode-laser side-pumped Nd:YAG rod laser, Optics Letters, Vol. 20, pp. 1148-1150, 1995.

Ishaaya, A. A.; Shimshi, L. & Davidson, N. (2004). Coherent addition of spatially incoherent light beams, Optics Express, Vol. 12, pp. 4929-4934, 2004.

Ming, L. & Mali, G. (2007). Experimental investigation of laser power addition with composite four-mirror cavity, Laser Physics Letters, Vol. 4, pp. 16-19, 2007.

Qinjun, P.; Zhipei, S. & Yahui, C. (2005). Efficient improvement of laser beam quality by coherent combining in an improved Michelson cavity, Optics Letters, Vol. 30, pp. 1485-1487, 2005.

Sabourdy, D.; Kermene, V. & Desfarges-Berthelemot, A. (2002). Coherent combining of two Nd:YAG lasers in a Vernier-Michelson-type cavity, Applied Physics B: Lasers and Optics, Vol. 75, pp. 503-507, 2002.

Sabourdy, D.; Kermene V. & Desfarges-Berthelemot, A. (2003). Efficient coherent combining of widely tunable fiber lasers, Optics Express, Vol. 11, pp. 87-97, 2003.

7

Compact, High Brightness and High Repetition Rate Side-Diode-Pumped Yb:YAG Laser

Mikhail A. Yakshin, Viktor A. Fromzel, and Coorg R. Prasad
Science and Engineering Services, Inc.
USA

1. Introduction

Efficient, compact, high average power (100 W and higher) and brightness Q-switched solid-state lasers capable of operating at high pulse repetition frequencies (PRF) of 10 kHz and higher are required for many applications such as material processing, frequency conversion, remote sensing, etc. These lasers represent a scaling up of nearly an order of magnitude over the current generation of diode-pumped solid state lasers. To achieve this level of performance, it is essential to provide a high pump power density in the laser medium, to reduce thermal loads and gradients in active medium and to obtain a good laser beam quality and brightness. Thermal effects in a laser gain medium are generally the main limiting factors for power scaling of diode-pumped solid-state lasers when near diffraction limited output beam is required. Diode-pumped Yb:YAG lasers are a very attractive alternative to the lasers utilizing classical laser material such as Nd:YAG for reducing thermal effects and for scaling the Q-switched output power to the desired level. Yb:YAG has nearly four times less heat generation during lasing than comparable Nd:YAG laser systems [Bibeau et al., 1998, Honea et al., 2000, Rutherford et al., 2001, Goodno et al., 2001] due to a much smaller quantum defect in Yb^{3+}. However, there are two shortcomings with Yb:YAG crystals related to a quasi-three-level nature of its laser transition. The first shortcoming is significant reabsorption at the laser wavelength preventing many laser configurations from being effective. However, recent advances in the development of diffused bonded composite YAG crystals have made it possible to diminish reabsorption losses and achieve a high brightness output. Another drawback is the relatively high level of the laser threshold pump power, which is noticeably, higher than in Nd:YAG lasers. But the last disadvantage is not too important for high average output power lasers, which usually are pumped significantly above threshold. Along with the choice of the gain medium, the important parameters to consider in the design of high power and brightness solid state lasers are the architectures chosen for the diode pumping scheme, laser resonator layout for thermal lensing compensation, energy extraction and cooling of the laser crystal. All of these play critical roles in average power scaling especially when a good quality laser beam is needed. A number of different approaches have been tried by other investigators for developing high power Yb:YAG lasers. Conventional rod lasers allow scaling to high average powers [Bibeau et al., 1998, Honea et al., 2000]. But obtaining a good beam quality at high average power is a difficult task due to considerable stress-induced birefringent and a strong thermal lensing in laser rods. Nevertheless, efficient birefringent compensation in an end-pumped Nd:YAG rod laser with CW output power of 114 W and a beam quality value of M^2 = 1.05 has been demonstrated [Frede et al., 2004]. Another approach to development of high power rod lasers was recently demonstrated in cryogenically (~ 77- 100 K) cooled Yb:YAG rod lasers, where 165 W CW output power in near-diffraction-limited (M^2 = 1.02) beam with optical-to-optical efficiency of 76%[Ripin et al., 2004] and even 300 W average power with the M^2 ~ 1.2 and 64% optical-to-optical efficiency has been obtained [Ripin et al., 2005]. But the disadvantage of this laser design is the expensive and rather impractical cryogenic technique.

Traditional zigzag slab geometry, which is known as face pumping [Kane et al., 1984], is more promising than straight-through geometries for scaling to high average power levels while maintaining good beam quality [Koechner, 1999]. However, practical use of slab lasers have been limited by the low laser efficiency that is typically seen in side-pumped slab lasers and by the complexity in engineering a robust, high-power zigzag slab laser system. Modified zigzag-slab laser designs employing conduction cooling and pumping geometry called edge-pumping have been also developed for high power CW and Q-switched Yb:YAG lasers [Rutherford, 2001]. The edge-pumping zigzag slab design eliminates the complexity of the cooling-pumping interface design of the conventional slab lasers, but achieving TEM_{00} mode operation of this laser at high levels of pumping meets is also difficult. Another slab laser design is the end-pumped zigzag slab laser architecture [Goodno et al., 2001]. The slab is pumped from each end by laser diode bars using a lens duct. The diode light is injected through the special coating on the TIR face of the crystal, undergoes TIR reflection from the 45° input face and is guided down the length of the slab. By using a Yb:YAG composite slab pumped from each end by a 700-W laser diode bar stack and an image-inverting stable resonator, 215 W of CW power with linear polarization and average M^2 beam quality of ~ 1.5 was obtained. A constraint of this design for high average power scaling is the considerable energy concentration (pump and intra-cavity fluence) on the small input face of the slab. The thin disk laser design [Karszewski et al., 1998, Stewen et al., 2000] is another approach to high average power lasers. In this laser design, the pump light from a bundle of fiber coupled diodes is imaged onto the center of the thin crystal disk by means of spherical or parabolic mirrors. The pump light not absorbed in the crystal after the first double pass is repeatedly re-imaged onto the crystal. A CW multimode output power of 1070 W at 1030 nm with 48% optical efficiency has been reported in an 8% doped Yb:YAG disk by using of a 16-fold pass through the crystal. This laser design is promising but its disadvantages are a very complicated multi-element optical pumping scheme and a relatively low storage energy because of the small pumped volume of the Yb:YAG crystal that limits the power scaling potential in the Q-switched regime.

In this chapter, we describe the development of compact, side-diode-pumped, Q-switched, TEM_{00}-mode Yb:YAG lasers producing from 65 to 120 W of output power at 10-30 kHz PRF with very high (> 30%) efficiency.

2. Design of the high-average power Yb:YAG laser

Design of the laser configuration, which controls the pumping, cooling and energy extraction, plays a critical role in average power scaling of lasers when a good quality laser beam is required. Achieving high average power in a TEM_{00} output from a Q-switched Yb:YAG laser became possible due to:

1. Proper thermal design of all optical and mechanical components in the laser to: ensure effective heat dissipation from laser crystal using diamond heat spreader plate and limit thermal optical effects, avoid damage to critical components due to overheating;
2. Utilization of flexible multi-pass side-diode-pumping schemes along with a composite laser crystal with low (3%) Yb doped Yb:YAG sandwiched between undoped YAG to uniformly pump the gain region of the laser crystal, and thereby obtain good TEM_{00} mode quality, and efficiently couple pump light from 940 nm pump diodes;
3. Optical thermal lensing compensation to maintain mode stability;
4. Maintaining the energy fluence within the resonator at levels below optical damage threshold;

Below we consider this in details.

2.1 Yb:YAG composite crystal design

A 3% doped Yb:YAG composite crystal with overall dimensions of 3(H) x 10(W) x 21.6(L) mm (doped region is 1.5 x 10 x 10.8 mm) with diffusion bonded clear YAG ends and top plate was conductively cooled from its back and front faces (10 x 21.6 mm) and pumped through the front face by a 250 W CW laser diode bar stack at 940 nm. Figure 1 shows side and top view of this composite Yb:YAG crystal. A back side of the Yb:YAG composite crystal (10 x 21.6 mm) was attached to a copper heat sink through a thin highly thermo-conductive spacer, presumably thin diamond plate. Pumping of the crystal is performed from the top face (10 x 21.6 mm) in multi-pass pumping scheme. To ensure several passes of the pumping beam, a highly reflected coating at 940 nm was introduced on the back side of the crystal. The size of pumping spot on the Yb:YAG crystal was 1.0 x 10.8 mm. Extension of

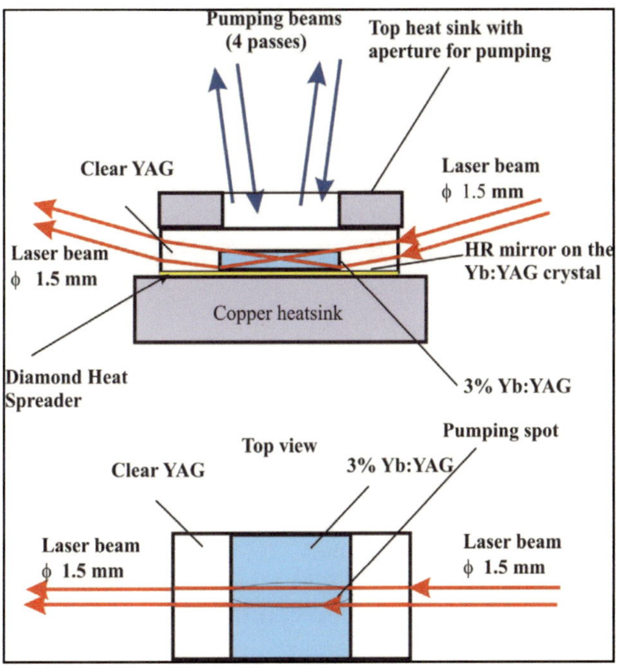

Fig. 1. Side and top views of the composite Yb:YAG crystal. The laser beam executes TIR from the cooled bottom face.

the doped Yb:YAG section of the crystal beyond the pumped width of 1.0 mm to 10 mm provides effective suppression of a spontaneous emission amplification (ASE), that is important for Q-switched operation of the laser.

For scaling up average power of the lasers, proper thermal design of active elements is very important. In order to estimate thermal load and heat distribution in side-pumped composite Yb:YAG crystal showed in Figure 1, we performed numerical calculations using a finite element code (NISA / HEAT III program). It was assumed that the absorbed pump energy in the crystal has uniform distribution over entire pumped volume. The reason for this assumption was the fact that absorption of the pump light at wavelength of 940 nm for one straight pass through the crystal thickness (1.5 mm) was measured to be only ~ 20% from incident beam intensity. Because diode pump beam performed several (4-6) passes through the crystal, the total absorbed pump energy can be considered uniformly distributed. Maximum heat power generated in the crystal is taken as 20 W (~ 10 % of total absorbed pump power).

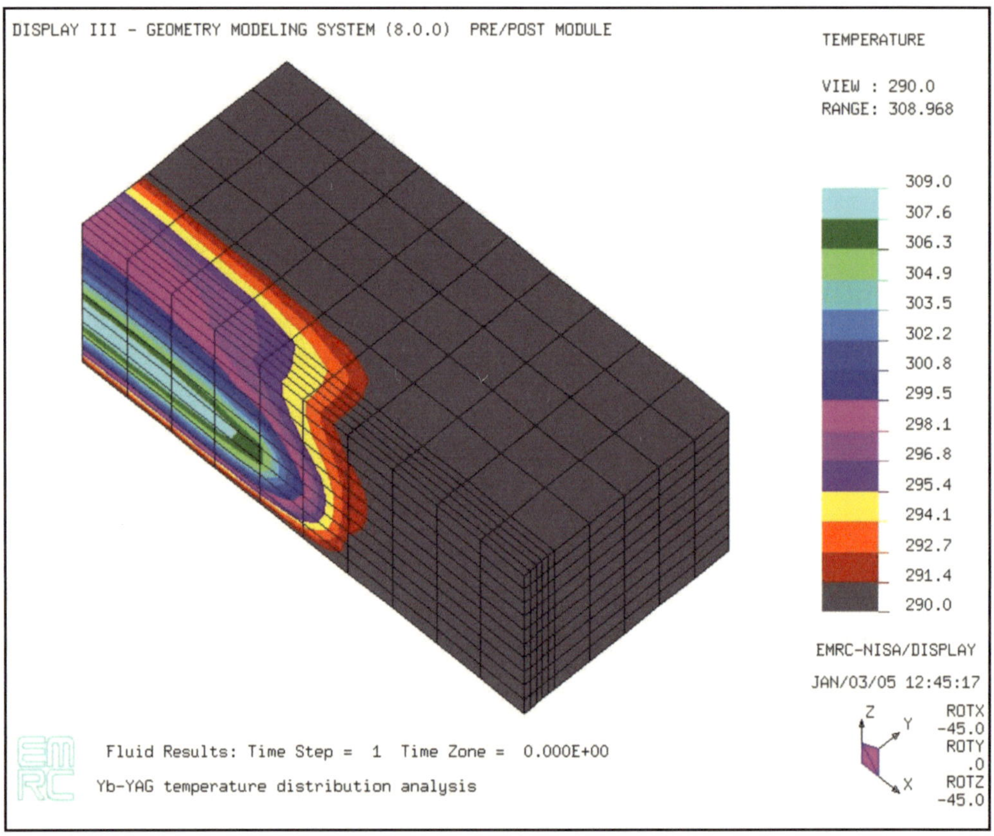

Fig. 2. Temperature distribution in composite 3%Yb:YAG crystal of 3 x 10 x 21.6 mm with homogeneous absorbed pump light area of 1.5 x 1.5 x 10.8 mm, which is cooled from the bottom and from the top faces, except window of 0.22 x 13 mm. Maximum heat power generation in crystal is 20 W. Because of symmetry, only one fourth of the crystal is shown.

Figure 2 shows calculated temperature distribution in the composite crystal with dimension of 3 mm x 10 mm x 21.6 mm, when heat power generation in the crystal is equal to 15 W (because of symmetry, only one fourth of the crystal is shown). Crystal in Figure 2 is cooled from the bottom and from the top faces (21.6 x 10 mm), except small window of 3 x 14 mm on the top face for pump beam passage. Heat transfer process on the boundaries between crystal and cooling plates assumed to be much faster than inside the crystal. In this case, maximum temperature gradient ΔT between the cooled bottom face and the hottest point in the crystal is $\Delta T \sim 19°C$ (see Figure 2). The hottest region of the crystal locates inside top clear YAG part of this composite crystal approximately on 1 mm in depth from upper surface. Because heat transfer speed on the crystal boundaries assumed to be very fast, the temperature of the crystal rises only in limited region around upper window for passing of pumping.

The temperature gradient in the crystal can be divided on two parts: linear growing of the temperature and parabolic temperature distribution. The first part is responsible for appearance of optical wedge in the crystal, while the second part leads to thermal lensing. Knowledge of the temperature gradient arising along the horizontal (along axis Z in Figure 2) and the vertical (along axis Y in Figure 2) planes of the crystal allows to calculate expected focal distance $F_{h,,v}$ of the thermal lens in these directions:

$$F_{h,v} = \frac{t_{h,v}^2}{8 \cdot \Delta T_p \cdot (P_t \mp Q_t) \cdot l} \tag{1}$$

where $t_{h,v}$ is characteristic width of the Yb:YAG crystal in the horizontal (h) or vertical (v) direction, respectively; l is the length of the crystal; ΔT_p is the maximum parabolic temperature gradient along the corresponding width t_{hv}, and P_t and Q_t are the thermo-optical coefficients of the YAG crystal (P_t = 87x 10^{-7} K^{-1}, Q_t = 17x 10^{-7} K^{-1}) [Durmanov et al., 2001].

Using equation (1) and calculated temperature distribution in the crystal, expected focal distance of thermal lens at maximum pump level can be estimated as $F_h \sim 36$ cm in the horizontal plane of the crystal (along the axis Z) and $F_v \sim 22$ cm in the vertical plane (along the axis Y). However, direct measurements of optical power of astigmatic thermal lenses arising in these crystals (see below) showed that thermal lenses in this Yb:YAG crystal are occurred to be nearly twice stronger than calculated ones. Perhaps, additional heating of the crystal, which exceeds 10% taking into account in the calculation, stems from absorption of the scattering pump light, reabsorption of fluorescence and "non-active" losses in the Yb:YAG crystals (usually ~ 3 - 4 x 10^{-3} cm^{-1}) at laser wavelength.

2.2 Multi-pass pumping scheme

An efficient optical pumping scheme is one of the key elements of the laser design. To improve laser pumping efficiency, we used multi-pass pumping scheme shown in Figure 3, which provided from 6 to 8 passes of the pump beam through the Yb:YAG crystal. This pumping scheme to use the same principle of repeatable passing of the pumping beam through the crystal as that of using in think disk lasers [Karszewski et al., 1998, . Stewen et al., 20001] but our scheme is much simpler. In this multi-pass scheme, a plano-concave spherical focusing lens L_1 (focal length F_1 = 75 - 100 mm depends on diode stack beam divergence) placed in front of the laser diode stack at its focal distance from the pumped Yb:YAG crystal center and a concave mirror M_2 (HR @ 940 nm) with radius of curvature R_2 equal to chosen distance between this mirror and Yb:YAG crystal (available, for example, R_2

= F_1, but not absolutely necessary) placed next to the lens. Second curved mirror M_3 with the same radius of curvature ($R_2 = R_3$) was placed on the opposite side of the diode focusing lens, and its distance from flat reflector M_1 was also taken equal to the mirror radius of curvature.

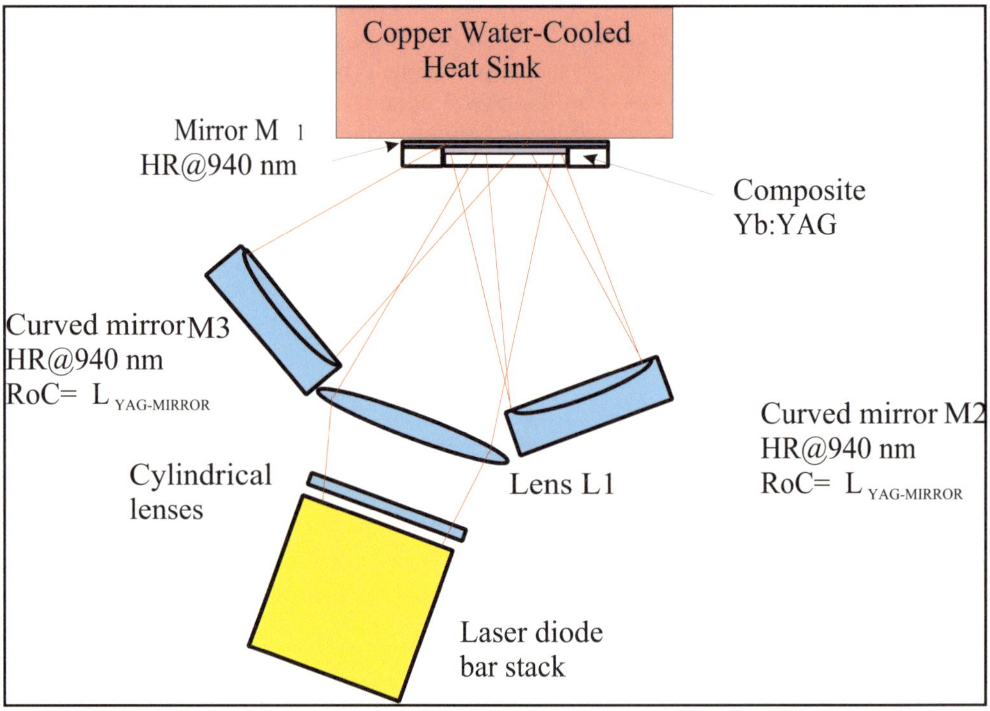

Fig. 3. Multi-pass pumping scheme of Yb:YAG laser

The pumping beam from the laser diode bar stack is focused on the Yb:YAG crystal by a plano-concave spherical focusing lens L_1. The beam is incident on the crystal at a small angle θ_1 (in our experiments, this angle was taken ~ 7°) to the crystal surface. After the first transverse pass through the crystal, the pumping beam is reflected back by a flat mirror M_1 (HR @ λ = 940 nm) located directly on the Yb:YAG crystal. Then it makes the second pass through the crystal, and continues to pass towards the concave mirror M_2 (HR @ λ = 940 nm). This curved mirror is also tilted at some angle θ_2, and thus allows the pump beam to retrace its pass for the third and forth transverse pass through the crystal. The mirror M_2 reflects the pump beam not exactly back as the incident beam is coming, i.e. back to the diode, but instead directs it to the second curved mirror M_3, which in its turn retraces the beam back. As a result, pumping beam performs additional two passes through the crystal. The number of pump beam passes through the crystal can be increased to 8 and even more, if to use a lens with appropriate focal distance in front of Yb:YAG crystal.

Results of comparison of effectiveness of pumping schemes providing two, four and six passes of the pump beam through the crystal are shown in Figure 4. In these experiments, the output pulse energy of the same laser utilizing one or another pumping scheme was investigated. A low PRF (13 Hz) and a simple resonator consisted of a flat output coupler with reflectivity of 0.9 and a curved HR mirror (Roc = 1 m) were used. The length of resonator was ~ 7 cm. Lower curve shows output pulse energy of the laser, when the pumping beam performed only two passes through the Yb:YAG crystal (mirrors M_2 and M_3 were blocked). Middle curve represents output of the same laser, when four-pass pumping scheme was used (only mirror M_3 was blocked). Upper curve in Figure 4 corresponds to the laser utilizing six-pass pumping scheme. It can be seen from comparison of the lower and the middle curves in Figure 4 that efficiency of the laser increases nearly twice, when pumping beam performs four passes through the Yb:YAG crystal compared with two passes through the same crystal. Utilizing additional two passes of the pumping beam through the crystal, i.e. six-pass pumping scheme, compared with four-pass pumping scheme gives another increase in the laser efficiency of ~ 15%.

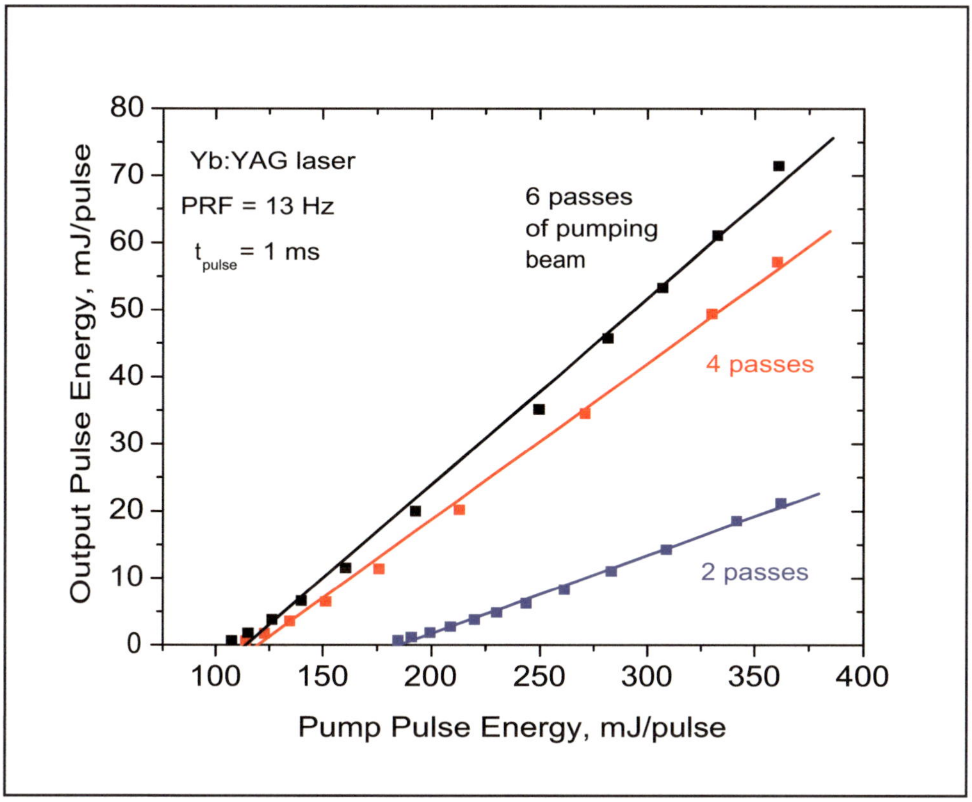

Fig. 4. Output performance of Yb:YAG laser with two-, four-and six-passes of pump beam through the crystal

As it was pointed, the pumped spot size formed by lens L_1 on the Yb:YAG crystal depends on focal distance of this lens and beam divergence of the pumping diode stack in both directions. Varying focal distance of the focusing lens L_1 in front of the diode stack, the size of the pumping spot on the crystal can be optimized. In case of using of the lens L_1 in front of the diode stack with focal distance of 100 mm and mirrors M_2 and M_3 with radius of curvature of 75 mm, the measured pumping beam intensity distribution in the Yb:YAG crystal (for two passes) is shown in Figure 5. It is seen that the pump beam spot has elliptical shape with dimensions of ~ 9.5 mm and ~ 1.1 mm ($1/e^2$ intensity level) in the horizontal and the vertical direction, respectively, and pumping beam size stays nearly unchangeable, when the distance between focusing lens and crystal is varied on ~ 8 mm. The length of the pumping spot on the crystal increases to 10.8 mm, when all six pumping beam passes are accomplished, while in the vertical plane spot stayed unchangeable.

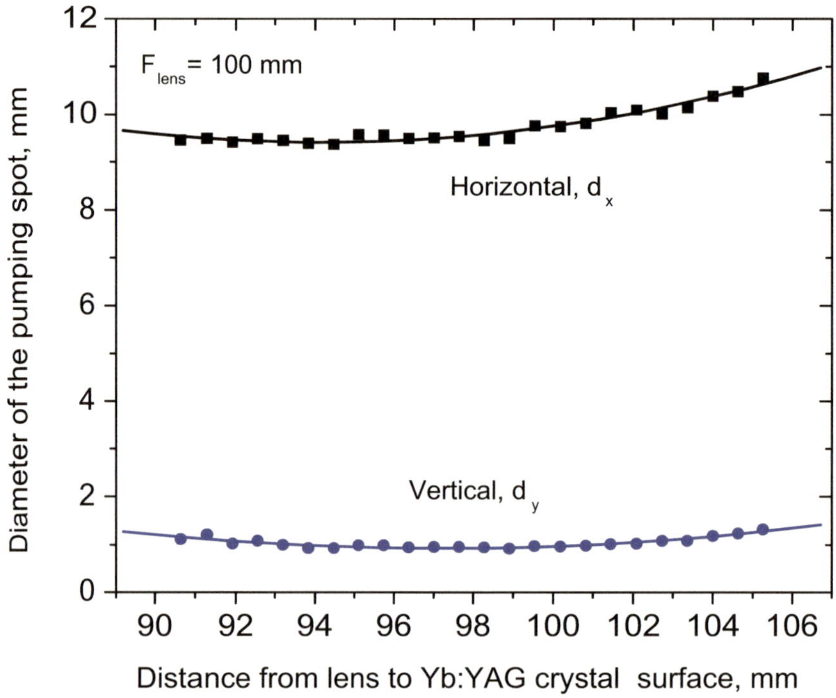

Fig. 5. Measured size of the pumping spot in the composite 3% Yb:YAG crystal in the horizontal and the vertical plane ($1/e^2$ level).

2.3 Yb:YAG single gain module design

To simplify laser design and alignment, Yb:YAG crystal in cooling mount, laser diode bar stack, focusing lens and two HR curved mirrors placed in adjustable mounts have been assembled on the individual base plate forming Yb:YAG gain module. Every such module can be aligned and optimized for the best performance individually in a simple flat-flat resonator.

Figure 6 shows the picture of the assembled gain module. On the right side of the picture, a copper water-cooled mount containing Yb:YAG composite crystal is seen. The crystal with dimensions of 3 x10 x21.6 mm is conductively cooled from the back and the front surfaces. To ensure an efficient heat transfer process between crystal and water-cooled copper heat sink, an optically polished 0.3 mm CVD diamond plate was used as an intermediate heat transfer layer and Yb:YAG crystal was in a direct contact with CVD plate. A 100 μm indium foil used between the front water-cooled heat sink and Yb:YAG crystal surface. On the left edge of the base plate, a pump laser diode stack is mounted on the adjustable water-cooled manifold. The lens in front of the pump diode stack serves to focus the pumping diode beam on the Yb:YAG crystal. A distance between focusing lens and Yb:YAG crystal is adjusted in order to ensure the pumping spot on the Yb:YAG crystal to be elliptical in shape with dimensions of ~ 9.5 mm and ~ 1.1 mm (at $1/e^2$ intensity level) in the horizontal and the vertical direction, respectively. Gain module plate also includes two adjustable optical-mechanical mounts for two concave mirrors (HR @ 940 nm, radius of curvature F_2 = 75 mm) placed next to the both sides of the focusing lens and used in multi-pass pumping scheme for retracing residual pumping beam back to the Yb:YAG crystal. In the laser resonator, gain module can be rotated on any required angle regarding the axis of resonator without disturbing a crystal pumping system. It is very convenient for building a thermally-compensated multi-component laser resonator.

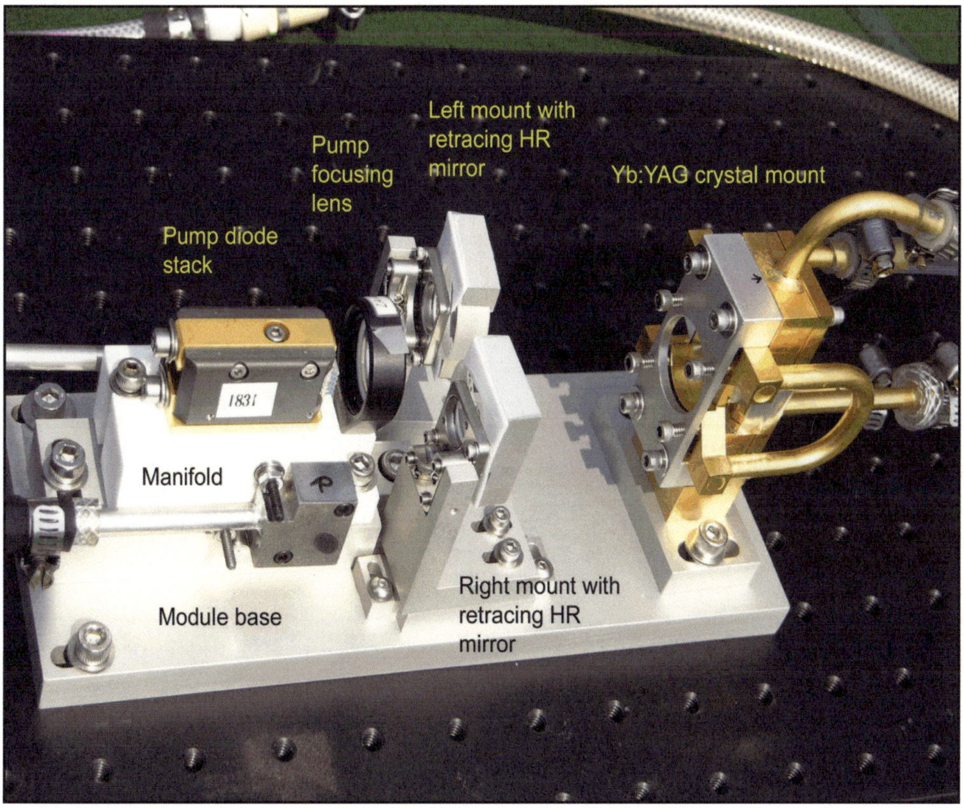

Fig. 6. View of the assembled Yb:YAG gain module

2.4 Measurement of thermal lenses induced in a single Yb:YAG gain module

Accurate measurement of thermal lens induced in our diode-pumped composite Yb:YAG crystal were performed with side-diode pumped gain module shown in Figure 6 and containing a 3% doped composite Yb:YAG crystal. For this purpose, a collimated beam of the probe laser passing through the crystal was used. In front of the crystal, a piece of flat wire mesh having 200 x 200 cells/inch (0.021 mm) was inserted. CCD camera located after the crystal showed a distinct image of this wire mesh along the whole cross-section of the crystal. When thermal lens was induced in the crystal, initial image was distorted and images of adjacent wire mesh nodes became closer each other because of focusing action of thermal lens. Scanning CCD camera along direction of the probe beam, one can find such position of the camera, where images of adjacent wire mesh nodes are merged in one line. Distance from correspondent principal plane of the crystal to the CCD camera location, where such junction of node images took place, apparently, has to be equal to the focus distance of the measured thermal lens. For astigmatic thermal lens, such merge of images occurs on different distances for the vertical and horizontal directions.

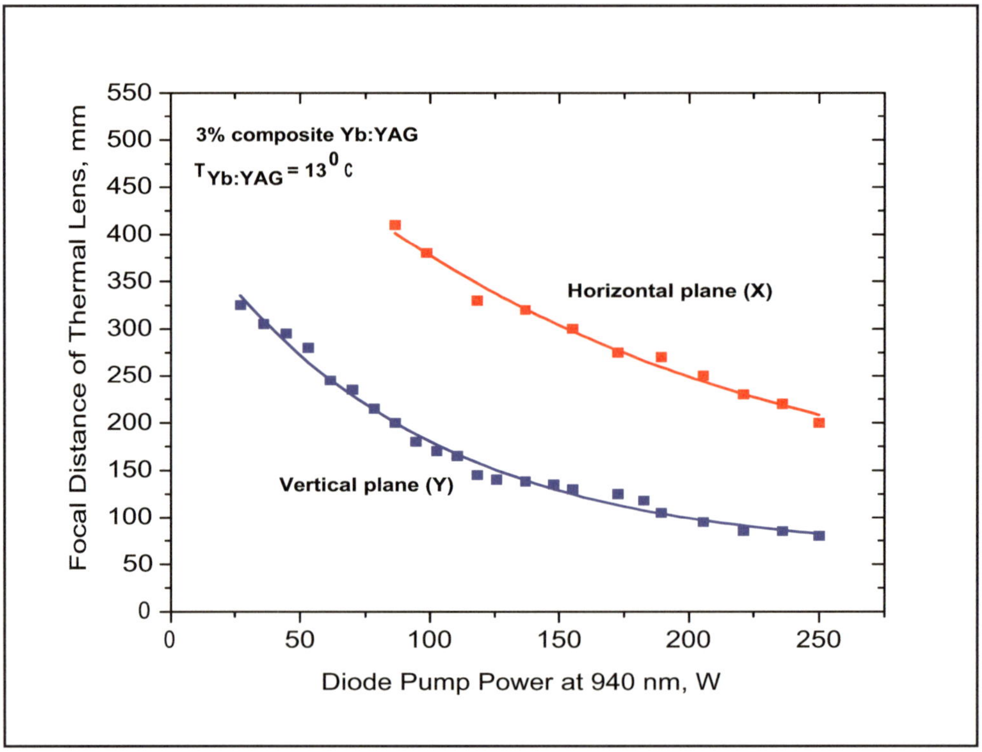

Fig. 7. Measured focal distances of thermally induced lens in side-diode-pumped composite 3% Yb:YAG crystal mounted in the gain module shown in Figure 6.

Figure 7 shows the measured focal distances of thermally induced lenses in the vertical and the horizontal plane of the Yb:YAG crystal (plane XZ and plane XY in Figure 2, respectively) plotted as a function of incident on the crystal pump power. It was found that induced positive thermal lens in this composite crystal is considerable astigmatic and its measured focal power approximately twice bigger than that estimated from thermo-analysis on assumption that 75% of incident pump power is absorbed in Yb:YAG crystal and only ~ 10 % of this pump power is converted in heat in the crystal. At maximum available CW pumping of 250 W, the focal distance of the induced lens in this 3% composite Yb:YAG crystal was measured to be ~ 8 cm and ~ 20 cm in the vertical and horizontal planes, respectively. The difference between the measured and calculated focal powers of thermal lens can be explained by the fact that real heating of the crystal was much higher than that of taking in calculations (10% of absorbed pump power) and maybe also because of possible arising of additional "electronic" lens due to the excitation of Yb^{3+} ions [Antipov et al., 2006]. With practical point of view, such astigmatic and very short focal distance lens induced in Yb:YAG crystal creates considerable obstacles for development of stable TEM_{00} -mode laser resonator.

3. Performance of the Yb:YAG laser

3.1 TEM_{00} - mode CW performance of one gain module Yb:YAG laser

Experimental investigation of the TEM_{00}-mode operated laser with one gain module and thermally-compensated flat-curved resonator was performed at first in CW regime of operation. It can be noted that the method of compensation of the positive thermally induced lens by means of inserting intra-cavity negative lens or appropriate convex resonator mirrors (or both) can provide TEM_{00}-mode operation only for relatively narrow range of pumping. Parameters of our resonator were calculated in assumption that induced lens compensation will take place around maximum level of pumping - ~ 220 - 250 W.

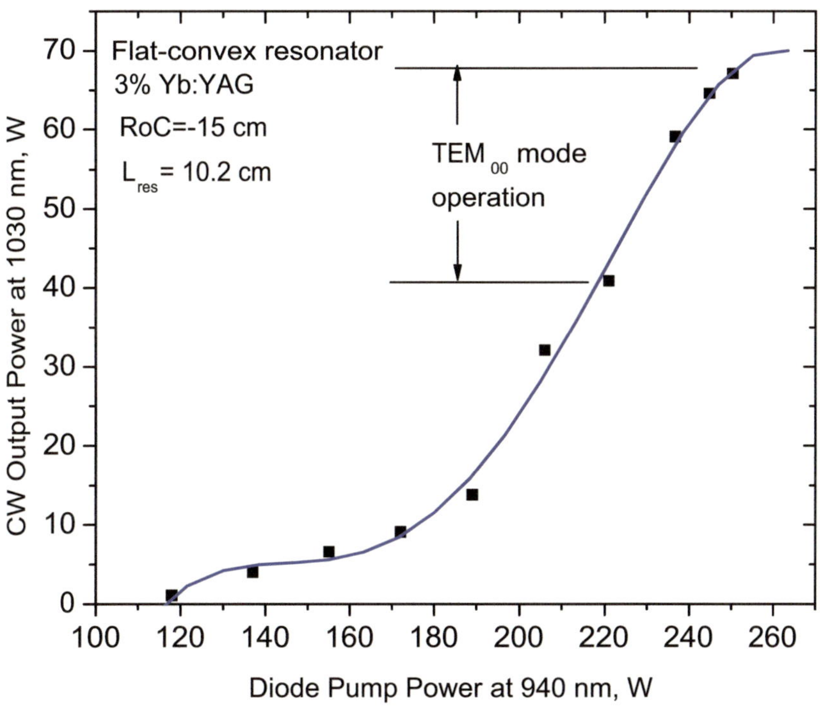

Fig. 8. Output power of Yb:YAG laser with thermally compensated resonator and one composite 3% Yb:YAG crystal in CW regime depends on incident diode pump power.

TEM_{00}-mode operation of this laser in CW regime have been achieved for several tested resonator configurations both containing negative intra-cavity cylindrical lenses and convex mirrors without lenses. The maximum TEM_{00}-mode output power of the laser for all investigated resonator configurations was in range of 40-67 W. It was found that sensitivity of the resonator to optical alignment increases and TEM_{00}-mode laser output reduces when negative cylindrical lens are used for thermal lens compensation compared with the case of using a convex mirror for the same purposes. It should be noticed that we tried to obtain exact compensation of thermal lens only in the horizontal plane, where laser aperture was larger (~ 1.5 mm). Compensation of the thermal lens in the vertical plane, where it was stronger than that of in the horizontal plane, was not exact especially at maximum available pump power of 250 W, however, laser beam profile in this plane always was very close to TEM_{00} - mode. Figure 8 shows the CW TEM_{00} - mode output of the laser obtained with such simple flat-convex thermally compensated resonator. Resonator was formed by convex HR mirror with radius of curvature of -15 cm placed on ~ 42 mm from the crystal center, and flat output coupler ($R_{1.03}$ = 0.75). The geometrical length of resonator was 10.2 cm. TEM_{00}-mode operation of this laser is observed in the range of pump power of ~ 220-250 W. Increasing of resonator length more than 12.5 cm caused drop in output power and appearance of TEM_{01} and higher order modes. Obviously, that compensation of the thermal lens in the vertical plane became insufficient with increasing the resonator length larger than 12.5 cm. Maximum achievable CW output power of this Yb:YAG laser with thermally compensated resonator in TEM_{00}-mode regime was about 67 W. In multimode regime, maximum CW output power in comparable conditions was ~ 78 W. Figure 9 shows laser beam intensity distribution when laser operates in TEM_{00} mode regime with thermal lens compensated resonator and diode pump power of 250 W. The measured TEM_{00} mode profile shows that the laser output beam is slightly elliptical M^2 ~ 1.3 and ~ 1.5 in the horizontal and vertical directions respectively.

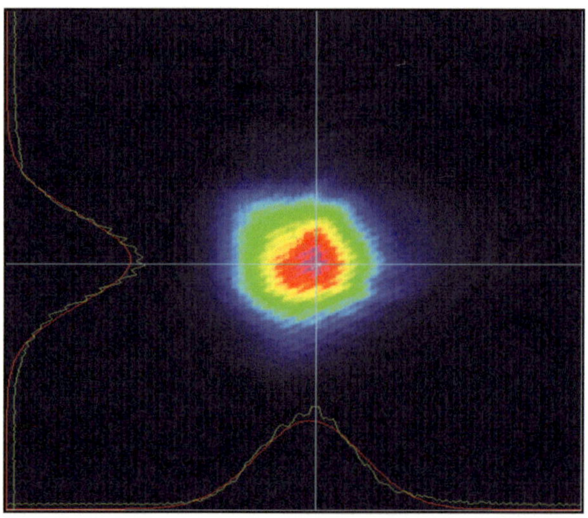

Fig. 9. TEM_{00}-mode beam profile of the CW operated Yb:YAG laser with thermally compensated resonator. Pump power is 250 W, output power is 67.1 W.

3.2 Q-switched performance of one-gain-module Yb:YAG laser

For Q-switched TEM_{00}-mode operation of the Yb:YAG laser with thermally compensated resonator described above, the length of resonator was increased to 130 mm and an acousto-optical Q-switch (NEOS Technology, model QS27-4S-S) was inserted in the cavity. At the same time, optical length of resonator was only 115 mm, i.e., stayed in the limits defined above for a pure TEM_{00}-mode operation of the laser in thermally compensated resonator. With chosen method compensation of thermally induced lens by means of using not intra-cavity negative lens but resonator mirror with appropriate negative optical power (a convex high reflective mirror with radius of curvature of -15 cm), the laser design looked very compact and simple. A transmission of the flat output coupler was taken of 50% (instead of $R_{1.03}$ = 0.75 for CW operation). This transmission was found to be the optimal for Q-switched operation of this laser providing the maximum output and reduction of the intra-cavity flux much below the damage threshold of all optical components of the laser. Figure 10 shows a photo of this bread-board Q-switched laser containing a single Yb:YAG gain module. The back HR concave mirror (R_{oc} = - 15 cm) is separated on 44 mm from the center of Yb:YAG crystal and mounted on a massive base to improve mechanical stability of the laser resonator.

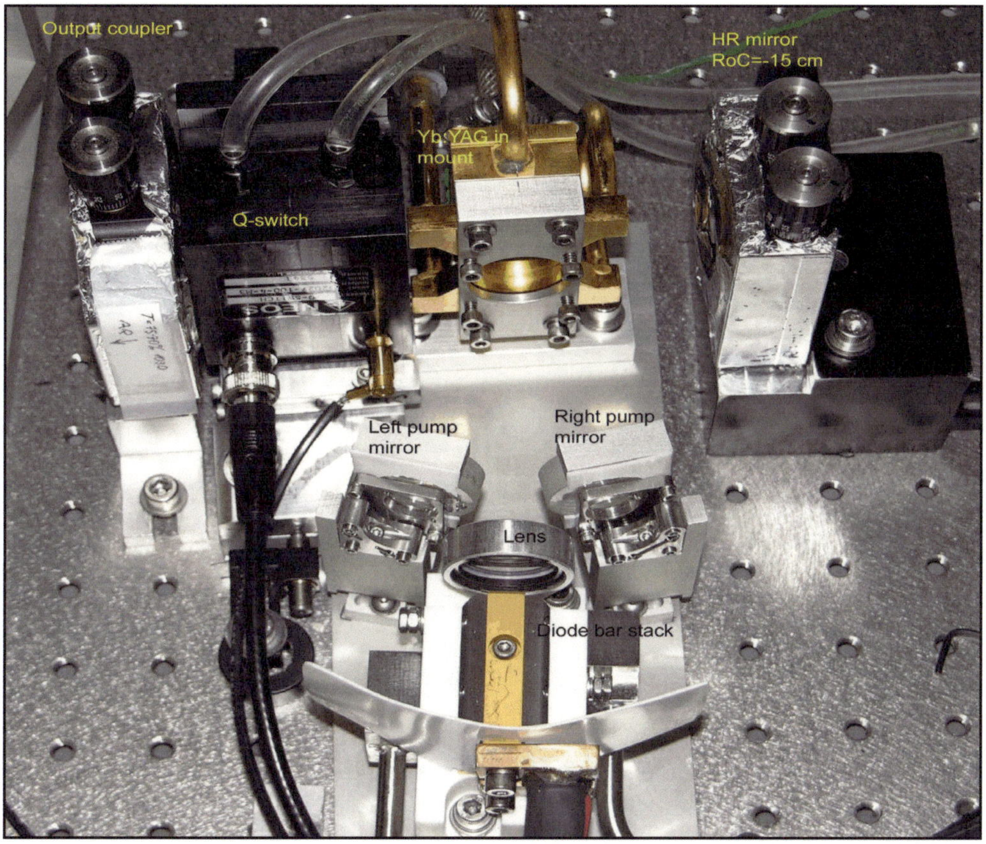

Fig. 10. View of assembled one-gain-module Yb:YAG Q-switched TEM_{00}-mode oscillator.

Figure 8 shows average output power of this one-gain-module laser depends on pump power emitted by diode stack in CW and Q-switched regime at PRF of 10 kHz. It can be seen that in both regimes output power of the laser was nearly the same. The maximum Q-switched output at TEM_{00} mode was ~ 56 W at 250 W of pump power.

Figure 11 shows average output power of this one-gain-module laser depends on pump power emitted by diode stack in CW and Q-switched regime at PRF of 10 kHz. It can be seen that in both regimes output power of the laser was nearly the same. The maximum Q-switched output at TEM_{00} mode was ~ 56 W at 250 W of pump power. Compared to maximum TEM_{00}-mode output of 67 W for similar laser reported earlier, there is ~ 15% reduction of the output for the assembling unit. This reduction is caused, most probably, by slight reduction of absorbed diode pump power because of different spectral width of radiation emitted by different diode bar stacks used in both cases. Some increase in the resonator length due to insertion of Q-switch, in its turn, deteriorates compensation of the thermally induced lens in Yb:YAG crystal in the vertical direction introducing additional diffraction losses. Pulse duration of the laser in Q-switched regime (PRF 10 kHz) gradually reduces from ~40 ns at pump power of 150 W to ~ 20 ns at maximum pump power of 250 W. The TEM_{00}-mode output beam profile in Q-switched regime was slightly elliptical, - bigger in the vertical plane and smaller in the horizontal direction.

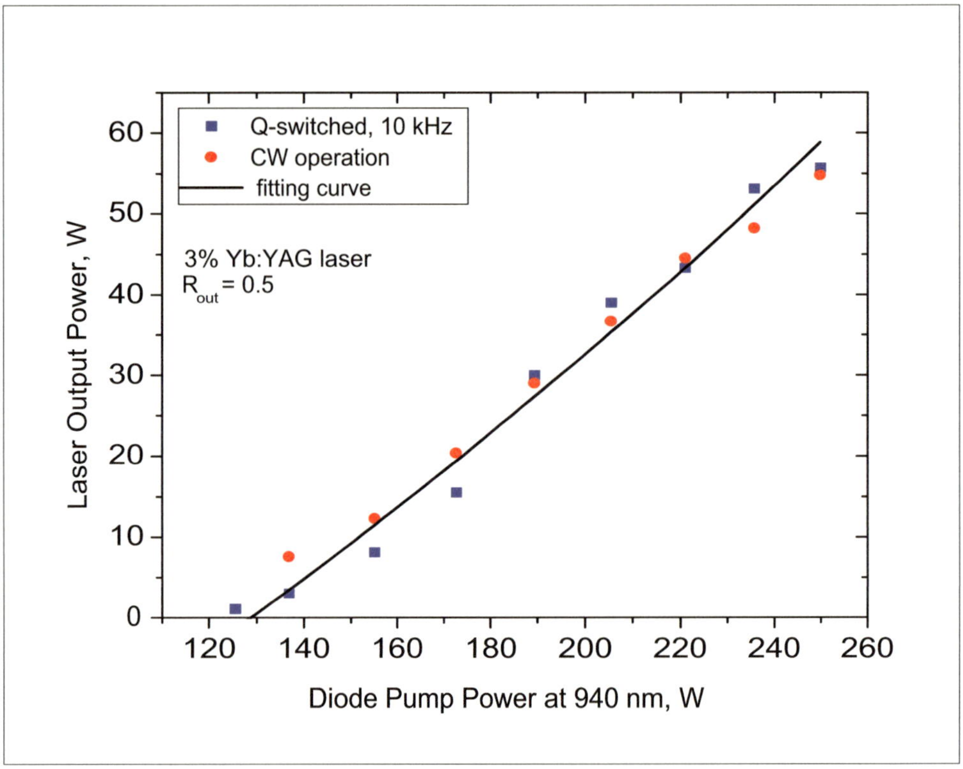

Fig. 11. Output power of assembled one-gain-module Yb:YAG laser with AO Q-switch and compensated resonator in CW and Q-switched regime at PRF of 10 kHz depends on diode pump power.

Thermally induced lens in Yb:YAG crystal permanently changed with increasing of laser pump power. Simultaneously some optical wedge grew in the crystal. That is why resonator of the laser become slightly misaligned with growing of pump power and required to be corrected (usually by means of a small turning of the back HR mirror). As a consequence of that, "cold" and "hot" alignments of resonator were different. During the experiments it was also found that switching of the laser from CW regime to Q-switched regime was accompanied as well by misalignment of laser resonator because of appearance of optical wedge in acousto-optical Q-switch, when RF power was applied. Transient process of reaching an equilibrium temperature distribution for Q-switch fused silica crystal usually takes ~ 2 -3 minutes because of a big size of this crystal (12 x 30 x 50 mm). Besides, drop in laser output, which was caused by misalignment of resonator, immediately increased heating of the Yb:YAG crystal and produced quantitative and qualitative changes in thermally induced lens and optical wedge in Yb:YAG crystal. All of described above thermal processes created some difficulties for development of safe procedure of tuning on and off the laser and maximization of its pumping, especially in Q-switched regime.

3.3 Performance of the Yb:YAG laser with oscillator- double-pass-amplifier optical schematic

To increase of an average output power of this Q-switched Yb:YAG laser to 100 W, a new laser design employing Power Oscillator- Double-Pass-Amplifier optical schematic was accomplished. The other similar gain module is used as a Double-Pass-Amplifier. Figure 12 shows an optical schematic of this laser. The laser is consisted of two identical side-diode-pumped gain modules. One of them was placed in a simple flat-convex resonator similar to thermally compensated for TEM_{00}-mode resonator used in described above One-Gain-Module Q-switched Yb:YAG laser and served as a powerful master oscillator. The second Yb:YAG gain module is used as a double-pass amplifier. To fulfill two passes of the laser beam through the amplifier, a straight-through and TIR bouncing passes of the incident beam through the Yb:YAG crystal were carried out. Redirection of the laser beam back into amplifier after the first pass was performed by two HR flat turning mirrors. Advantage of using such laser system design is not only in obtaining of a good laser beam quality, but also in reducing of intra-cavity fluence compared with possible design of this laser, when two Yb:YAG gain modules are placed in a common resonator. A distance between two Yb:YAG crystals (in oscillator and in amplifier) and between two successive beam passes through the amplifier is approximately equal to double focal distance of the thermal lens induced in every Yb:YAG crystal in the vertical plane. It means that the laser beam diameter in the vertical plane is re-imaged after every passage it through the crystal. In the horizontal plane,

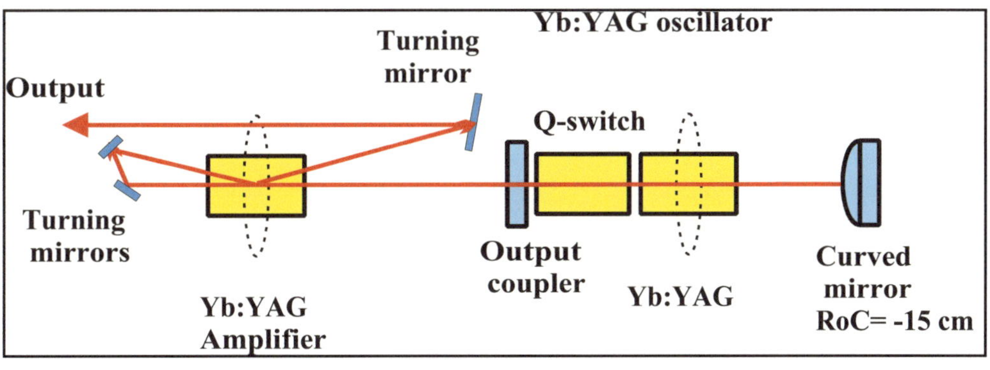

Fig. 12. Optical schematic of two-gain-module oscillator-double-pass amplifier Q-switched Yb:YAG laser.

where initial divergence of the beam was ~ 2.8 x 10^{-3} rad and focal distance of thermal lens in the Yb:YAG crystal was much larger (~ 20 cm)), laser beam diameter after second pass through the amplifier was equal to ~ 1.2 mm and it convergences with an small angle of ~ 3 x 10^{-3} rad. The third turning mirror is used for redirection of the laser beam forward after the second pass through the amplifier. Output laser beam as a whole has TEM_{00} mode profile but different divergence in the horizontal and the vertical planes. To estimate expected performance of the double-pass amplifier, we performed calculations of a small signal gain, α, of this amplifier depends on diode pump power using Franz-Nodvik analysis [Franz & Nodvik, 1963], which takes into account gain saturation on signal fluence. At maximum pump power of 250 W, the average small signal gain was estimated to be ~ 0.55 cm^{-1}. For both laser beam passes, the calculated input beam fluence was below the saturated fluence, which is ~ 9.3 J/cm^2 for Yb:YAG, amounting of ~ 0.44 J/cm^2 for the first pass and ~ 0.73 J/cm2 for the second pass with initially incident average power of 50 W at 10 kHz PRF. The total amplification of Yb:YAG gain module pumped at maximum pump power of 250 W was calculated to be ~ 1.66 and ~ 1.45 for the first and the second passes of the laser beam through the crystal, respectively, when average incident on the amplifier laser Q-switched power was 50 W at PRF of 10 kHz. The total passive energy losses of the laser beam between these two passes were taken as 10%. In these conditions, an average laser output power obtaining from amplifier is expected to be ~ 83 - 85 W after the first pass, and ~ 120-125 W after the second pass.

Figure 13 shows results of experimental measurements of the Q-switched average output power of the described above laser system consisting of Powerful Oscillator and Two-Pass Amplifier at 10 kHz depends on pump diode current (upper curve). Lower curve shows Q-switched average output power of oscillator alone with the same pump diode current. TEM_{00}-mode operation of the oscillator was obtained with high diode current of 70 to 80 Amps (pump power of 220 to 250 W). Maximum Q-switched average power at PRF 10 kHz from oscillator incident on the amplifier was ~ 55 W. After first pass through the amplifier, this average power was increased to ~ 90 W, that is, the maximum amplification for one pass was ~ 1.6. It can be seen that experimental measured value of amplification for the first pass of the laser beam through the Yb:YAG amplifier was very close to the calculated value. After the second pass, 115 W of Q- switched TEM_{00} mode average output power at PRF of 10 kHz have been obtained from this laser system (coefficient of amplification was ~ 1.3). At maximum pumping, Q-switched pulse duration was measured to be ~ 24 ns at 10 kHz and increased to 42 ns at 20 kHz. The TEM_{00}-mode profile of the output beam of this laser was elliptical with larger angular divergence in the vertical direction (where thermal lens in Yb:YAG crystals is stronger). Measured laser output beam quality (after amplifier) showed that in spite of the laser beam is slightly astigmatic but it has nearly TEM_{00} mode profile in both directions - with M^2x = 1.2 and M^2y = 1.33 in the horizontal and in the vertical plane, respectively. In multi-mode operation, laser beam quality of this laser was also comparatively not far from cited above values - measured beam quality was M^2x = 1.7 and M^2y = 1.9, respectively.

4. Conclusion

A compact, high average power, high brightness, diode-pumped Q-switched Yb:YAG laser have been developed. This laser utilized a composite laser crystal with doped Yb:YAG and

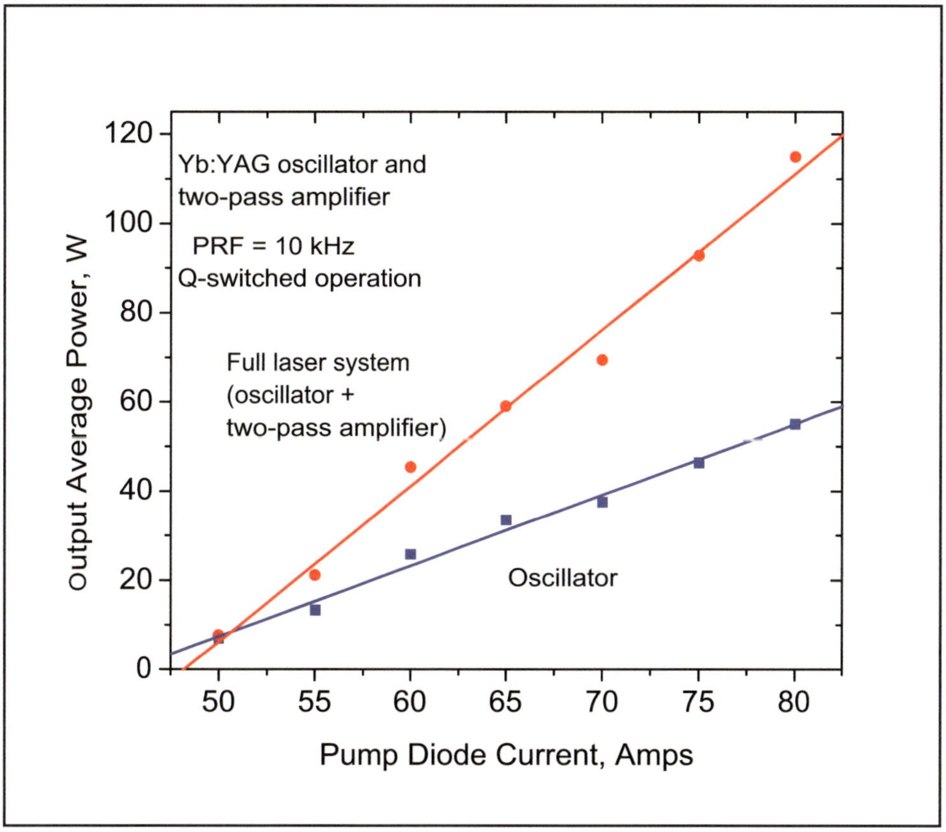

Fig. 13. Output performance of Q-switched Yb:YAG laser employing Powerful Oscillator-Two-Pass Amplifier scheme (upper curve) at PRF of 10 kHz. Performance of the oscillator alone is also shown (lower curve).

clear YAG crystals and an innovative, six-pass side-diode-pumping scheme to couple the 940 nm diode pump beam through a low Yb concentration doped (3.0 atomic %) YAG laser crystal to achieve excellent pump distribution and coupling (> 95%) efficiency. It was capable of delivering 115 W of average output power with a high overall efficiency (> 23% optical), in a nearly TEM_{00} mode (M^2 ~ 1.2) at pulse repetition frequency (PRF) of 10 kHz and pulse duration of ~ 24 ns.

5. Acknowledgements

This research was supported by Air Force Research Laboratory, Kirtland Air Force Base, NM award FA9451-04-C-0152.

6. References

Antipov O.L., Bredikin D.V., Eremeykin O.N., Savikin A.P., Ivakin E.V., Sukhadolau A.V. (2006). "Electronic mechanism for refractive-index changes in intensively pumped Yb:YAG laser crystals", *Optics Letters*, 31, p. 763-765.

Bibeau, C.; Beach R.J., Mitchell, S.C., Emanuel, M.A., Skidmore, J.A., Ebbers, C.A., Sutton, S.B., and Jancatis, K.S., (1998). "High-average-power 1-µm performance and frequency conversion of a diode-end-pumped Yb:YAG laser," *IEEE J. Quantum Electron*, 34, p. 2010.

Durmanov S. T.; Novoselov V.G., Rudnitsky Y.P., Smirnov G.V., Lubimov V.V., Malashko Ya. I., (2001). "Thermo-optic lens in the active element of Nd:YAG lasers and the study of the possibilities of its compensation," *Proceeding of the International Conference on Lasers'2000*, December, 2000 (V.J.Corcoran & T.A.Corcoran, eds.), STS Press, McLean, VA, p.766.

Franz, L.M.; and Nodvik, J.S., (963). "Theory of pulse propagation in a laser amplifier,", 34, p.2346.

Frede, M.; Wilhelm R., Brendel M., Fallnich C., Seifert F., Willke B., and Danzmann K., 2004, "High power fundamental mode Nd:YAG laser with efficient birefringent compensation", *Optics Express*, 12, p.3581.

Goodno G.D.; Palese S., Harkenrider J., and Injeyan H., (2001). "Yb:YAG power oscillator with high brightness and linear polarization," *Optics Letters*, 26, pp.1672-1674.

Honea, E.C.; Beach, R.J., Mitchell, S.C., Skidmore, J.A., Emanuel, M.A., Sutton, S.B., and Payne, S.A., (2000). "High-power dual-rod Yb:YAG laser," *Optics Letters*, 25, pp. 805-807.

Kane, T.J., J.M. Eggleston, and R.L. Byer, (1984). *IEEE J.Quantum Electron.*, QE-20, p. 289.

Karszewski M.; Brauch U., Contag K., Erhard S., Giesen A., Johannsen I., Stewen C., and Voss A., (1998) "100 W TEM operation of Yb:YAG thin disk laser with high efficiency,"*OSA TOPS Vol.*19 *Advanced Solid State Lasers*, Walter R. Bosenberg and Martin M.Fejer (eds.), Optical Society of America, p.296.

Koechner, W, (1999). *Solid-State Laser Engineering*, 5th ed., Springer-Verlag, Berlin.

Ripin D.J.; Ochoa J.R., Aggarwal R.L., and Fan T.Y., (2004). "165-W cryogenically cooled Yb:YAG laser", *Optics Letters*, 19, pp.2154-2156.

Ripin D.J.; Ochoa J.R., Aggarwal R.L., and Fan T.Y., (2005). "300-W cryogenically cooled Yb:YAG laser", *IEEE J. Quantum Electron.*, QE-41, pp. 1274-1277.

Rutherford, T.S.; Tulloch, W.M., Sinha, S., and Byer, R.L., (2001). "Yb:YAG and Nd:YAG edge-pumped slab lasers," *Optics Letters*, 26, pp. 986-988.

Stewen C.; Contag K., Larionov M., Giesen A., and Hugel H., (2000). *IEEE, J. Sel. Top. Quantum Electron*, 6, p. 650.

8

In-line Typed High-Precision Polarization Lidar for Disaster Prevention

Tatsuo Shiina
Graduate school of Advanced Integration Science,
Chiba University
Japan

1. Introduction

Prediction of local and steep weather change such as heavy rain and lightning strike are urgently needed for disaster prevention. In Japan, Automated Meteorological Data Acquisition System (AMeDAS) has been operational throughout the country, with a mesh size of 20 km. Lightning Location Positioning (LLP) system, Lightning Position and Tracking System (LPATS), and SAFIR (Surveillance et Alerte Foudre par Interférometrie Radioélectrique) have been also used for lightning detection. (MacGorman & Taylor, 1989; Théry, 2001; Hayashi, 2006; Matsudo et al, 2007; Hauf et al, 2007) All of these weather observation systems, however, are comprehensive and necessary to analyze the multipoint data for the weather forecast in a certain area.

Heavy rain and lightning strike locally occur during a short time period. They are often broken out at low-altitude and mainly caused by ice-crystals in cloud. Such local and steep weather change should be observed and be distinguished at a point observation. Radar (Radio Detection and Ranging) is a solution to catch weather change, but it can only detect rain droplets and cannot detect cloud particles which are precursors of heavy rain. To predict the heavy rain and lightning strike, low-altitude atmosphere and cloud should be observed. Furthermore, the flow of ice-crystals should be monitored.

Lidar (LIght Detection And Ranging) can observe the atmosphere and cloud. It is stand-alone system and can distinguish ice-crystals from water droplets. On the other hand, Lidar technique is unsuited for observation in the near range because the laser beam will be dangerous for human eye.

To develop the disaster prediction system for local weather change, we apply the lidar technique. The system was designed with emphasis on the following characteristics:

Stand-alone
Compactness
Eye-safety

To realize such a system, in-line optics, or common optics for transmitter optics and receiver optics, was adopted. It is usual technique for microscope optics, while the pulsed energy on lidar is too high that no one could accomplish its development since the first proposal. (Measures, 1984) Micro Pulse Lidar (MPL) broke though the problem and built up the new generation in lidar field. (Spinhirne, 1993, 1994 & 2001; Grund & Sandberg, 1996; Lee et al,

1996; Welton, 2001; Hwang et al, 2002) Our first developed lidar is also MPL. By installing the in-line optics, near range detection and narrow Field Of View (FOV) observation were realized simultaneously.
After the success of our in-line MPL, the next approach for in-line lidar for high precision polarization measurement was conducted. While it is not eye-safe, it is capable of near range detection and narrow FOV observation.
Though the development of the unique lidars, we brushed up our skill and built up the stand-alone local weather prediction system for disaster prevention.
This chapter introduces the lidar theory, in-line optics, initial and next approach of the original lidar development for low-altitude atmosphere and cloud.

2. Outlook

The goal of our research is the prediction of local weather for prevention of disasters such as heavy rain and lightning strikes. To achieve this, monitoring of the flow of ice-crystals in the cloud is effective. Local and steep weather changes such as heavy rain and lightning strikes occur in the low altitude atmosphere. At first our group developed a compact in-line type micro pulse lidar (MPL) system which can detect local and fast changes in low-altitude clouds and monitor the flow of ice-crystals in clouds by measuring the depolarization effect. (Shiina, 2002, 2005a, & 2005b) Thanks to the in-line optics, the system can detect near range echoes with the narrow field of view of 0.1 mrad. It can also distinguish the ice-crystals from sphere particles by examining the depolarization. The laser pulse energy is in the order of micro joule, and the system is eye-safe. We have succeeded in the prediction of heavy rain, while the prediction of lightning strikes is yet to be accomplished, because the correlation between the ice crystal flow and lightning discharge is indirect. In order to establish direct correlation between the lightning discharge and the transmitted lidar beam, we considered the use of the Faraday effect, by which the electromagnetic pulse due to the lightning discharge rotates the polarization plane of the propagating beam in the ionized atmosphere. By measuring the rotation angle of the polarization plane of the propagating beam, the electron density, discharge current and the lightning position can be estimated. We aim to develop the new concept lidar for lightning discharge detection.
Numerical analysis using a cloud-to-ground discharge model showed that the rotation angle for visible light is <1 degree. (Shiina, 2008a) A laboratory experiment using an impulse voltage generator and a discharge chamber (with a gap length in the order of cm) was conducted, and the experimental results were in agreement with the analysis results. (Shiina, in press) Another laboratory experiment using a larger impulse voltage generator (with a gap length in the order of m) was conducted, and the rotation of the polarization plane of the beam propagating through the immediate vicinity of the discharge path was confirmed. (Shiina, 2008b; Fukuchi, in press)
In order to detect a small rotation angle of the polarization plane, we developed a new concept lidar system for high precision polarization measurement. The isolation between the orthogonal polarizations must be improved up to >30 dB, which was realized by a specially designed polarization- independent optical circulator composed of Glan laser prisms and highly transparent Faraday rotators. Thanks to the in-line optics, the echoes could be detected from near range. We examined the echoes in the viewpoint of accuracy of estimation of rotation angle of the polarization plane and system stability. (Shiina, 2007a)

3. First approach – lidar theory and depolarization measurement -

3.1 Three types of lidar optics

There are three types of fundamental structures of lidar optics.(Shiina, 2005b) A biaxial type shown in Fig.3-1(a) has separate optics for the optical transmitter and the optical echo receiver, and the laser beam is transmitted parallel to the receiver axis. A coaxial type shown in Fig.3-1(b) transmits the laser beam along the receiver axis. An in-line type shown in Fig.3-1(c) has a common optics for the transmitter and the receiver, i.e. the telescope is used for both transmission and reception.

The biaxial and coaxial types have the demerit of having a blind area (non-detectable range of lidar echo) at near range. For the biaxial type, an easy way to shorten the blind area is to broaden the receiver's field of view (FOV), but this increases the background noise and decrease the dynamic range of lidar echo. Another method is to set the path of the transmitted laser beam at a slant angle against the receiver axis. However, in this configuration, the laser beam deviates from the receiver's FOV beyond a certain distance.

In the coaxial type, the laser path and the receiver axis are collinear. This necessitates the implementation of a transmitter mirror (usually located on the back side of the secondary mirror). Therefore, the shadow of the transmitter mirror (secondary mirror) creates a blind area at near range. Furthermore, a pinhole (spatial filter) is usually inserted at the focal point of the receiver telescope in order to narrow the receiver's FOV to eliminate the background light. The lidar echo from near range is out of focus and is blocked by the pinhole, resulting in the blind area.

In contrast, by using in-line typed optics, the laser path is always overlapped with the receiver's FOV even if it is narrow of the order of 0.1 mrad. As the beam size of the transmitted laser beam can enlarged up to the telescope aperture, there is no blind area of near range lidar echoes. There exists the technical problem of separating the transmitted beam and the lidar echoes.

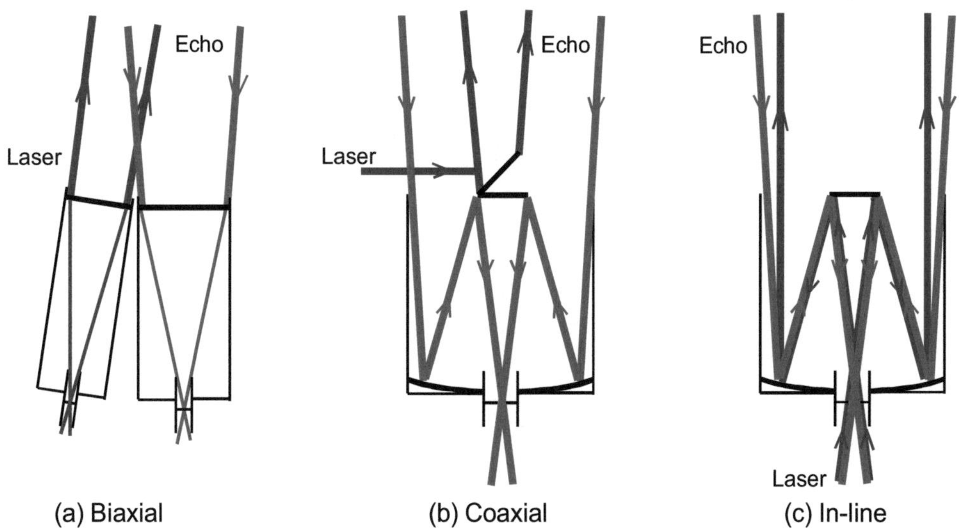

Fig. 3-1. Three types of lidar optics.

3.2 Lidar equation

The lidar echo characteristics were evaluated by calculating the lidar echo power *Pr(L)* and its signal-to-noise ratio *SNR(L)*.(Measures, 1984; Weitkamp, 2005; Fujii & Fukuchi, 2005)

$$Pr(L) = P_0 K A_r \frac{c\tau}{2} \beta(L) Y(L) T(L)^2 / L^2 + Pb$$
$$T(L) = \exp(-\int_0^L \alpha(l) dl) \quad (3\text{-}1)$$

$$SNR(L) = \frac{\sqrt{M}\sqrt{\eta \Delta t / h\nu}\,\mathrm{Pr}(L)}{\sqrt{\mu}\sqrt{\mathrm{Pr}(L) + Pd}} \quad (3\text{-}2)$$

where, P_0 is the transmitting power, K is the system optical efficiency, A_r is the receiver's area, c is the speed of light, τ is the pulse width, $\beta(L)$ is the backscattering cross section, $Y(L)$ is the geometrical form factor, $T(L)$ is the transmittance, Pb is the background light power, $\alpha(L)$ is the atmospheric extinction coefficient, M is the number of signal summation, η is the detector's quantum efficiency, μ is the detector's noise factor, h is Planck's constant, υ is the light frequency, Pd is the equivalent dark current power.

The collimated beam is transmitted to the atmosphere, scattered in all directions, and a small part of the scattered light is collected by the receiver. The fraction of the scattered light returning to the receiver is proportional to the solid angle subtended by the receiver aperture, so the lidar echo intensity is inversely proportional to the square of the distance (from the location of the lidar to the point at which the scattering occurs). The transmitted beam and the backscattered echo are attenuated by the atmospheric transmittance, i.e., due to the atmospheric extinction coefficient. The geometrical form factor *Y(L)* is an overlap function between the receiver's FOV and the transmitting beam. It is determined by the specifications of the transmitted laser beam, the receiving telescope, and the field stop aperture. (Halldorsson & J. Langerholc, 1978; Harms, 1979; Sugimoto et al, 1990)

3.3 Simulation

The lidar echo was estimated for the three type lidars: the coaxial type lidar, the improved coaxial type lidar, and the in-line type lidar with an annular beam. The specifications are summarized in Shiina, 2005b. The coaxial type and the improved type lidars have been put to practical use. The transmitting beam was fired from the center of the telescope, generally on the secondary mirror. The improved coaxial lidar was designed for near range echo measurement. The divergence of the transmitting beam and the receiver's FOV were larger than those of the coaxial type lidar. A photomultiplier (PMT) was used as the detector in the case of the improved coaxial type lidar, while and avalanche photodiode (APD) was used for other lidar types. The coaxial type lidar is the typical compact lidar. The intensity of the transmitting beam was nearly equal in these models, while their intensity distributions and aperture sizes were different.

At first, the geometrical form factor *Y(L)* was examined. The result is shown in Fig. 3-2(a). The *Y(L)* of the improved coaxial type lidar has the fastest rise of the three. This is because the beam divergence and the receiver's FOV were large. The *Y(L)* of the in-line type lidar rises slowly and needs a longer distance to reach unity. The important point, however, is the receiver's efficiency at near distance for the lidar echo estimation. The coaxial type lidar cannot detect the echo within 0.2 km in this condition. The *Y(L)* of the in-line type lidar never becomes 0 and is higher than that of the other two type lidars at the nearest distance. This effect is due to the similarity between the intensity distribution of the annular beam and the receiver's efficiency distribution in the near range. The lidar echo variations along the propagated distance are shown in Fig. 3.2(b). The atmospheric extinction coefficient was α=1.70x10^{-4}/m and the backscattering coefficient was β=3.4x10^{-6}/m. The usual lidar echo variation along the propagated distance has a peak such as that of the coaxial type and the improved coaxial type lidar. The peak is caused by the relation between the geometrical form factor *Y(L)* and the lidar echo intensity *P(L)*, which both depend on the propagated distance. On the other hand, the echo variation of the in-line type lidar has no peak, i.e., the lidar echo is maximum at the nearest distance and decreases monotonously along the propagated distance. This is mainly due to the efficiency of the geometrical form factor in the near range. Although the improved coaxial type lidar had the high signal-to-noise ratio in the near range, the ratio decreases rapidly along the further propagated distance in comparison with those of the coaxial and the in-line type lidar. The improved coaxial type lidar needs the distance of more than a few kilometers to obtain the ideal echo variation, in which the lidar echo is inversely proportional to the square of the propagated distance. The in-line type lidar approximately has the ideal echo variation from the nearest distance. This facilitates the analysis of the near range echo, because there is no need for compensation for the variation of the factor *Y(L)*. The result also indicates that the in-line type lidar can detect any obstructions (human, building, etc.) without blind range.

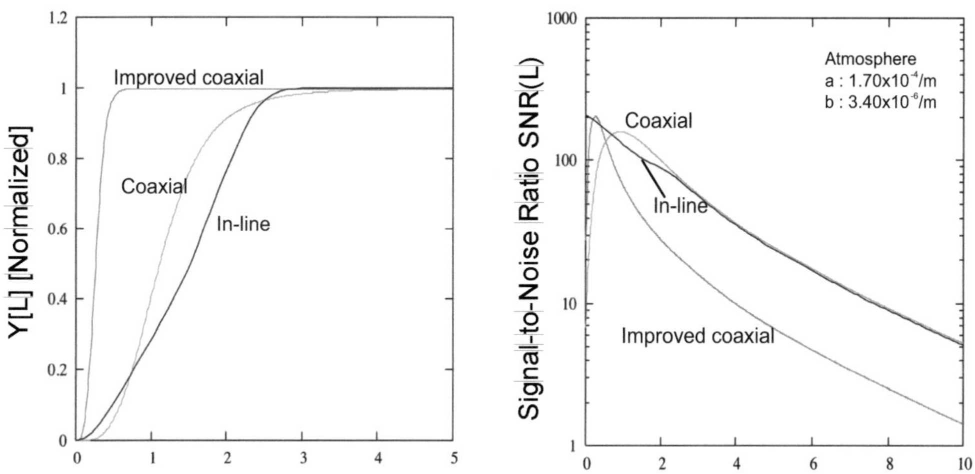

Fig. 3-2. Geometric form factor Y(L) and Signal-to-Noise ratio SNR(L) with the three lidars.

3.4 Depolarization measurement

Small particles suspended in the atmosphere include aerosols, raindrops, cloud particles, ice-crystals. Rayleigh scattering and Mie scattering are commonly used for lidar observation. Cloud particles, which we intend to measure, have a diameter of about 5 µm, and cause Mie scattering. (Pal & Carswell, 1973; Ryan et al, 1979; Sassen, 1991; Zaccanti et al, 1993; Gai et al, 1996) The polarization plane of incident bream and backscattering echo against a cloud particle and an ice-crystal are illustrated in Fig. 3-3. In the scattering, the backscattering echoes keep the incident polarization plane when the beam hits the cloud particles (spherical particles). On the contrary, non-spherical particles such as ice-crystals change the incident polarization. It is known as depolarization effect. It is caused by the difference of Fresnel's refraction coefficient between the parallel and the orthogonal polarization. When the beam hits the plane surface of ice-crystal perpendicularly, the depolarization never occurs. When the beam hits the crystal at an angle, one can distinguish the ice-crystals from the spherical particles such as cloud particles by examining the orthogonal polarization components of the lidar echoes.

The spherical particles, however, cause depolarization in the case of multiple scattering. (Sassen & Petrilla, 1986; Kerscher, 1995; Bruscaglioni et al, 1995) To reduce the contribution of multiple scattering, it is effective to narrow the receiver's FOV. The biaxial and coaxial lidars with narrow FOV have a broad blind area, because it a certain distance is needed for the transmitted beam to enter into the receiver's FOV (the overlap function has a slow rise with respect to distance). In contrast, the in-line type lidar can have a narrow FOV with no blind area.

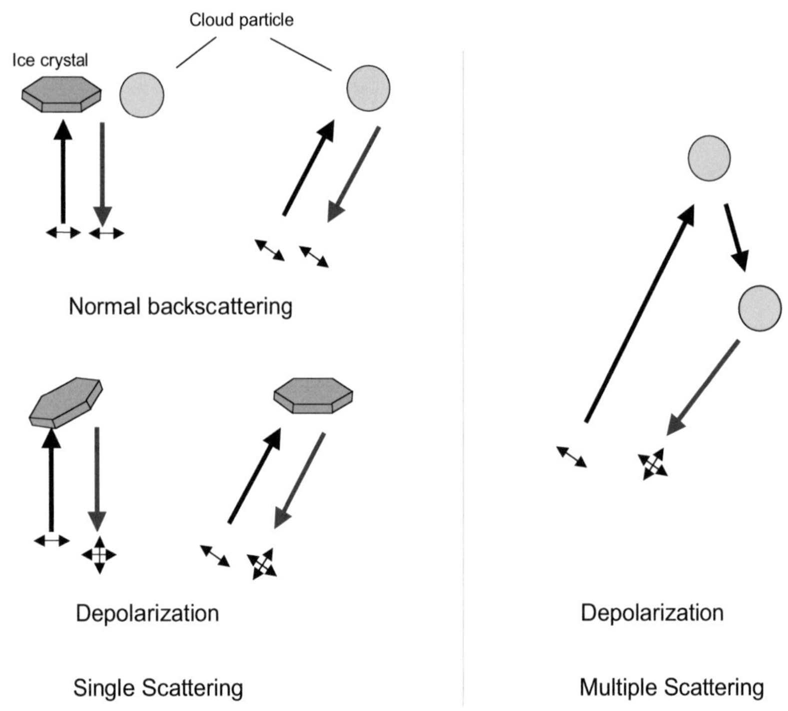

Fig. 3-3. Normal backscattering and depolarization by ice-crystals and cloud particles.

4. First Approach – In-line MPL-

4.1 Apparatus

The setup of the in-line MPL is illustrated in Fig. 4-1. The optical circulator and the pair of Axicon prisms were installed into the lidar optics. (Shiina, 2005a) The laser source is a LD pumped YLF laser, of wavelength 1.047μm, pulse width 5 ns, output energy 80μJ, and

maximum pulse repetition frequency 50 kHz. The transmitted beam is collimated by the beam expander just behind the laser head and passes though the optical circulator.
The optical circulator makes allows measurement of the orthogonally polarized echoes with two detectors, that is, the same polarized component (p-component) as the transmitting beam and the orthogonal one(s-component).(Shiina, 2002) The s-component echo is reflected at the polarized beam splitter (PBS2) and goes to the detector APD(s), while the p-component echo passes through the circulator optics and goes to the detector APD(p) (See Fig.4-2). The insertion and isolation characteristics of the optical circulator are summarized in Table 4-1. The insertion loss as the transmitter is 2dB and the isolation between the orthogonal polarized echoes is about 20dB. As the insertion loss of the Faraday rotator was large, the sensitivity of the s-component echo became twice of that of the p-component one. It is convenient to estimate the depolarization effect by comparing the weak s-component echo with the p-component one. All optical elements have small tilts (1.5 degrees) against the optical axis. The laser head also has a tilt of 3 degrees against the optical axis. As the in-line MPL could not reject the directly reflected light from the lidar optics perfectly, the reflected light influenced the lidar echo signal (ringing or saturation of the lidar receiver's electrical circuit, etc.). Pinholes were inserted in front of the detectors (APDs) in order to prevent the light directly reflected from the in-line optics from entering the detectors. Its aperture was determined in combination with the tilts of the in-line optics.
In the in-line type lidar, the same telescope is used for transmission and reception. When the transmitted beam passes through the telescope, part of the beam is reflected by the secondary mirror, obstructing the transmitted beam and causing a large noise signal upon returning to the detector. Therefore, a pair of Axicon prisms was installed to prevent this reflection. The prism pair creates an annular beam, whose hole overlaps the secondary mirror of the reflecting telescope. The insertion loss of the transducer was 0.932 dB. The insertion loss of the whole lidar optics was about 3 dB. The total energy of the output beam was 8 kW. The eyepiece focuses the transmitting annular beam at the field stop aperture (FSA). The FSA is a pinhole of 650 μm diameter. At the focal point, the annular beam changes its beam shape into the nearly non-diffractive beam, of which spot size is less than 4μm diameter. (Belanger & Rioux, 1978; Durnin & Miceli, 1987; Indebetouw, 1989; Scott & McArdle, 1992; Arrimoto et al, 1992; Doskolovich et al, 1993; Kono, 1995 & 1996; Soifer et al, 1997) The beam then passes though the FSA without any obstruction. The annular beam finally goes to the telescope. A Schmidt-Cassegrain reflecting telescope was used for the transmitter and receiver. The annular beam was expanded up to the telescope aperture (304 mm diameter). The quality of the transmitting beam was verified by the identification between the annular beam enlarged by the telescope and that in front of the eyepiece. The light directly reflected from the in-line optics was suppressed to below 0.1 mW. The FOV of the telescope is 0.1 mrad, determined by the focal length of the telescope and the aperture size of the FSA. The divergence of the transmitting beam can be controlled by the beam expander. Though the divergence does not depend on the telescope's FOV, it is basically less than the FOV.
The detectors are NIR enhanced Si-APDs (PerkinElmer 30954E). They have responsivity of 36 A/W at wavelength 1064 nm and at the operating voltage. The responsivity rises up over 500 A/W near the breakdown voltage. The voltage over the breakdown is in the region of the Geiger mode. In our system, the APDs were used in the analog mode between the operating voltage and the near breakdown voltage. All optics including the laser head and the detectors were mounted on the telescope tube. It allows measurement of lidar echoes

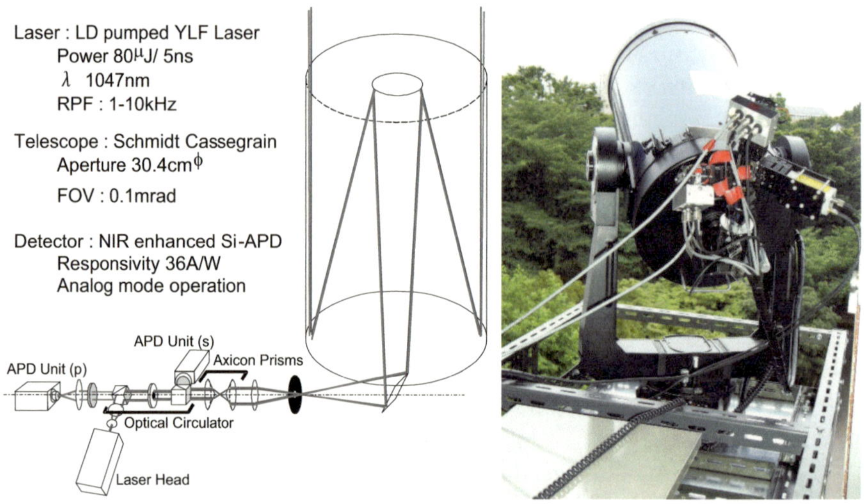

Fig. 4-1. Optics of In-line typed micro pulse lidar.

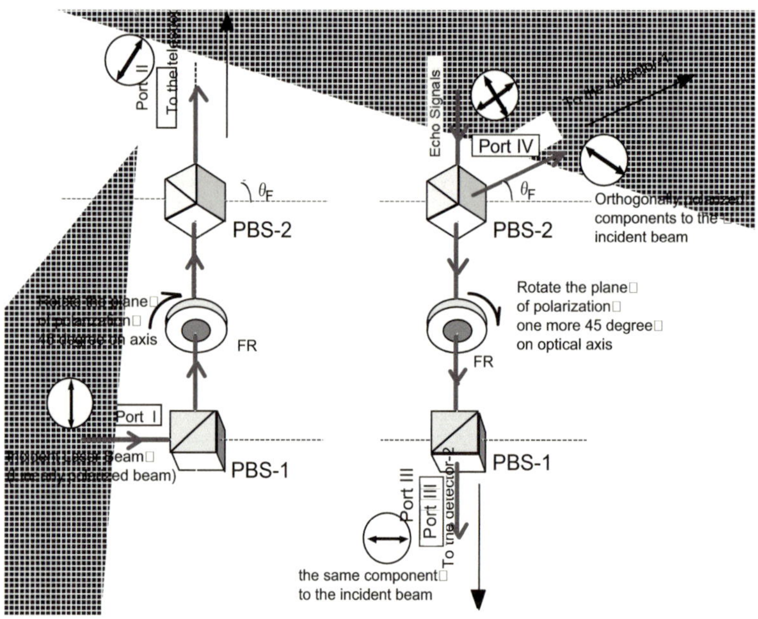

(a) transmitter (b) Receiver

Fig. 4-2. The optical circulator for in-line type lidar.

	Port I	Port II	Port III	Port IV
Transmitter	- (Incident TEM00 Mode)	2.08	>60.0	>60.0
Receiver	-	- (Parallel component to the incident beam)	3.27	21.5
	-	- (Normal component to the incident beam)	17.2	0.50

Table 4-1. Insertion and isolation characteristics of the Optical Circulator in Decibels.

from the near distance with the narrow FOV. The incident beam energy into an human eye was 5.34 J/s. It is lower than the maximum permissible exposure (MPE) value of 11.4 J/s in the same condition. (American National Standards Institute, 1986) The in-line MPL keeps eye-safety even in front of the system.

4.2 Near range observation

Figure 4-3 shows the observation results of ice-clouds during 22:40-4:20 local time (JST) on January 24, 2009. Lidar echoes of p- and s- components were captured simultaneously. The observation angle was 4 degrees from zenith. In General, ice-crystals in the sky are parallel to the ground.(Sassen, 1991) Since the angle of incidence was too small, echoes observed from upper clouds at the altitude of 1200-2000 m (A) (B) contained only p-component, while sparse and intensive s-component echoes were widely captured from cloud at the altitude of 600-1800 m (C) during 22:40-3:00. Especially strong echo was detected in s-component echo during 2:40-3:00 (D). In this period, strong wind was also observed in that period. Since ice-crystals leaned against the zenith direction while falling, ice-crystal echoes were captured in s-component.

The downdraft could be also recognized by lowering of the cloud base. Although snowfall or downfall could not be observed on the ground, the ice-crystals would have melted away during descent because the ground temperature was about 4 degrees.(Sassen, 1991) Indeed, both of the p- and s-component echoes could not been detected near the ground. The in-line typed MPL with the narrow FOV and depolarization measurement can examine dynamics of ice-crystals in low-altitude clouds.

Next, measurement of heavy rain and thundercloud are shown in Fig.4-4. The eve of the heavy rain was observed in the low altitude atmosphere at July 4, 2006, as shown in Fig.4-4(a). The cloud base was around 1200 m and above. At 17:00, strong echoes (p-component) appeared at very low altitude (100-600 m). The echoes reached near ground, while it did not grow up to the cloud base. The downfall was not detected on the ground. Although the droplets caused the strong echoes could not distinguished by visual contacts, heavy rain started just after the observation. The system could detected the steep change in low-altitude and local area. Low altitude thundercloud was also observed as a trial at the evening on July 14, 2006, as shown in Fig.4-4(b). The cloud base came close to the ground; about 400 m. Strong echoes appeared in the cloud. The cloud echoes usually come from the cloud base, and decrease exponentially over a characteristic distance of about 300 m. The result of thundercloud did not show such a exponential change, but showed strong signals in the cloud. It is not easy to clear the casual correlation between the thundercloud echoes and thunder activity. To capture the precursors of the lightning and predict the lightning strike, we must catch more direct sign from the lightning.

5. Next step – Direct detection of lightning discharge-

5.1 Faraday effect

The interaction between the light and the atmosphere is caused by scattering (Mie, Rayleigh, Raman), and also by the magneto-optical effect and electro-optical effect. The magneto-optical effect (Faraday effect) is associated with the lightning discharge. It has been reported as an optical measurement method for magnetic confinement fusion reactor. (Kawahata & Okajima, 2000) The polarization plane of a beam propagating parallel to the magnetic flux is rotated in a partially ionized atmosphere (plasma) (Fig.5-1). The rotation angle is

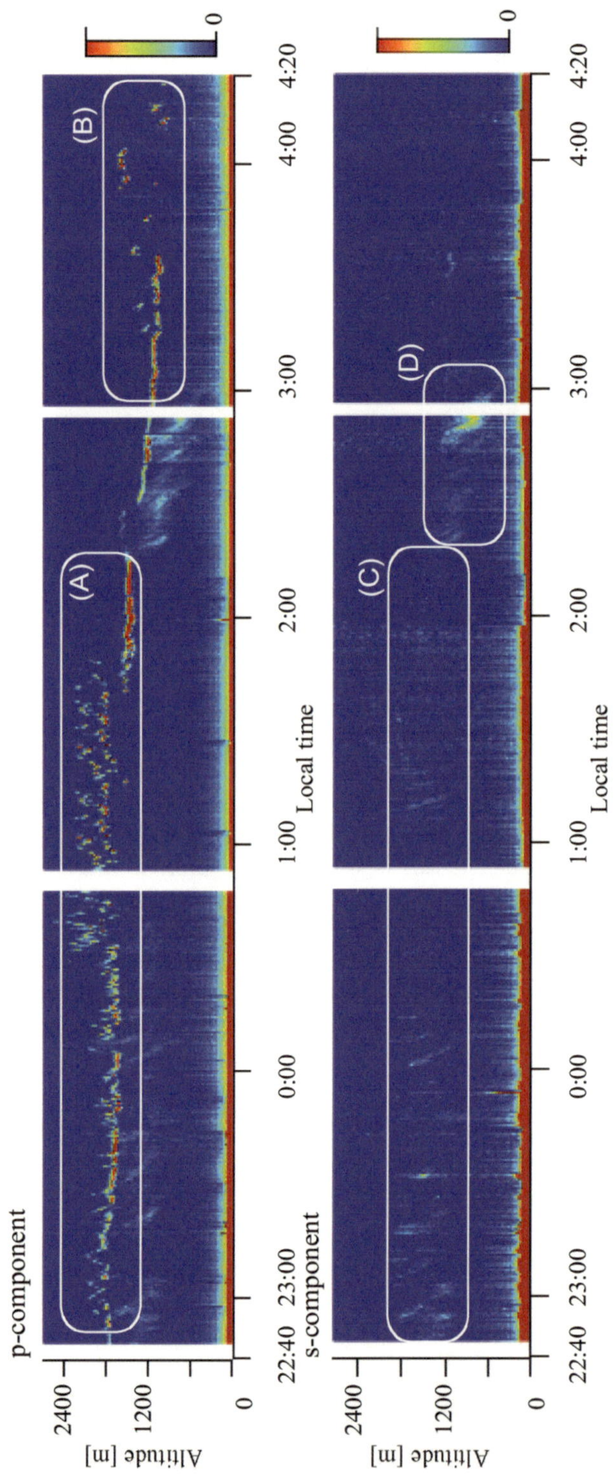

Fig. 4-3. long term ice-cloud measurement. January 24, 2009. Temp. 3.7deg Hum. 70% Cloudy

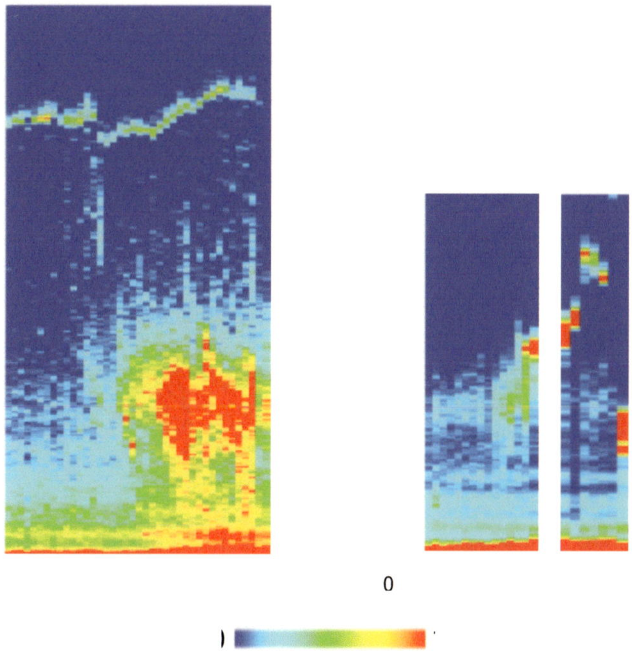

Fig. 4-4. lidar echoes (p-components) in bad weather condition. (a) July 4, 2006. Temp. 28deg. Hum. 52%. Before heavy rain (b) July 14, 2006, Temp. 32deg. Hum. 50% Thunder cloud

proportional to the product of the ionization electron density n_e and the magnetic flux density B along the beam propagation path. The linearly polarized beam can be regarded as a combination of the clockwise and the counterclockwise circularly polarized beams. The refractive indices of the ionized atmosphere for each circularly polarized beam are as follows.

$$n_{\pm} = \left(1 - \frac{\omega_{pe}^2}{\omega^2}\frac{\omega}{\omega \pm \omega_{ce}}\right)^{1/2}$$
$$\omega_{pe} = \sqrt{\frac{e^2 n_e}{\varepsilon_0 m_e}} \qquad \omega_{ce} = \frac{eB}{m_e} \tag{5-1}$$

where ω_{pe}, ω_{ce} are the plasma and electron cyclotron frequencies, respectively, e is the fundamental charge, m_e is the electron mass, and ε_0 is the permittivity of free space. Therefore, the rotation angle of polarization of the beam propagated at distance L (=L_1~L_2) is obtained as follows.

$$\begin{aligned}\delta &= \frac{\pi}{\lambda}\int_{L_1}^{L_2}(n_+ - n_-)dl \\ &= 2.62\times 10^{-13}\lambda^2 \int_{L_1}^{L_2} n_e B dl\end{aligned} \tag{5-2}$$

where λ is wavelength of the propagating beam. Since δ is proportional to λ^2, the rotation angle for visible light is small. Therefore, the polarization angle rotation must be measured with high accuracy in order to detect lightning discharge.
When the Faraday effect is applied to lightning measurement, the atmosphere needs to be partially ionized, and the magnetic flux due to the lightning discharge must exist. Cloud-to-cloud discharge, which causes 20-30 times continuous discharge, satisfies those conditions. (Franzblau & Popp, 1989; Franzblau, 1991; Stith et al, 1999; Society of Atmospheric Electricity of Japan, 2003)

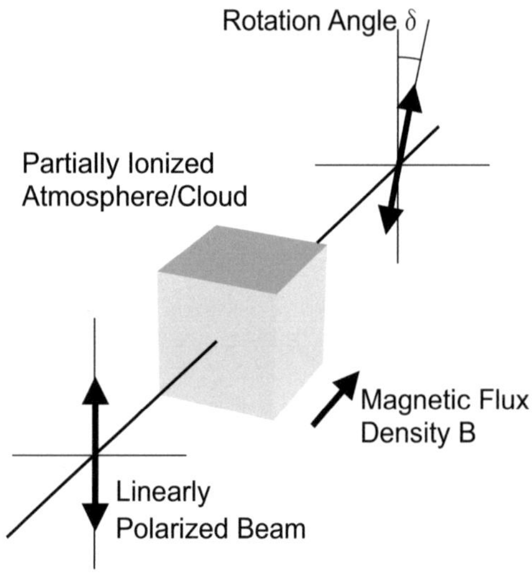

Fig. 5-1. Faraday effect.

5.2 New concept lidar

The analysis and experimental results have shown that the rotation angle of polarization plane of the propagating beam is less than 1 degree, so that the mutually perpendicular polarization components must be measured with a sensitivity and accuracy of >30 dB in order to detect lightning discharges. The rotation of the polarization plane only occurs in a nearly perfectly ionized atmosphere, so the signal cannot be detected unless the transmitted beam intersects the discharge path. On the other hand, the shock wave (variation in the neutral gas density) generated by the discharge can be detected over a broader range, while it causes no rotation of the polarization plane. (Fukuchi, 2005) This was confirmed by high voltage discharge experiment in the next section.
Therefore, the scenario of the lightning detection using the lidar system is designed as follows. At first the system roughly scans the sky in the direction in which the occurrence of a cloud-to-cloud lightning discharge is likely. If a shock wave is detected, the 3-demensional lightning position is estimated. Next, by scanning the neighborhood of the lightning position with higher precision to intersect the propagating beam and the lightning discharge path, the rotation angle of the polarization plane is measured. In general, lower area (bottom) of clouds will be scanned, as the beam penetrates only a few hundred meters in clouds. The distribution of the discharge location and its change will lead to the prediction of lightning strike.

The lidar system must be capable of measurement at near range with a narrow field of view in order to eliminate the effects of multiple scattering. The use of in-line optics is effective in meeting this requirement. The system must also have scanning capability to search the cloud-to-cloud lightning discharge. The concept of the lidar system for lightning detection is shown in Fig.5-2. For the detection of the small rotation angle, differential detection should be used.

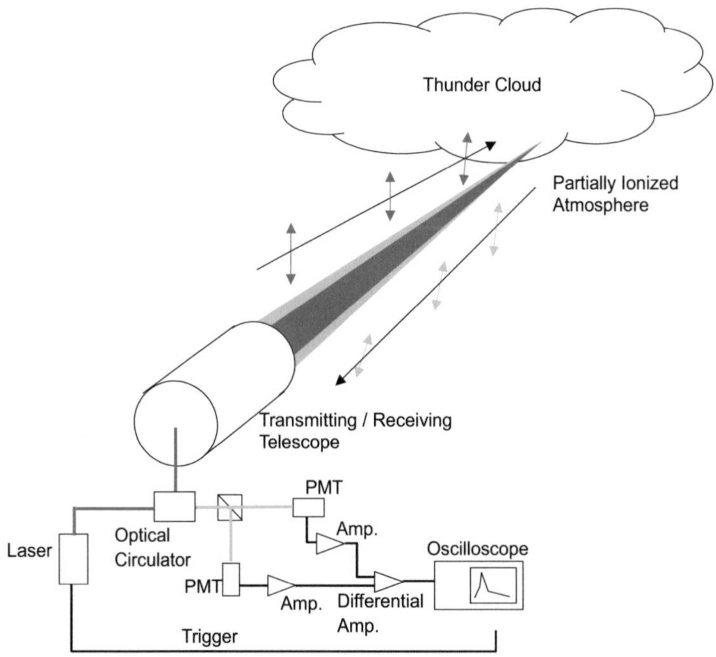

Fig. 5-2. Concept of lidar lightning detection.

6. Demonstration –Ground based experiment-

6.1 Apparatus

Figure 6-1 shows the experimental setup of the high-voltage discharge experiment. (Shiina, 2008b; Fukuchi, in press) The experiment was conducted in a high voltage experiment hall using an impulse voltage generator (HAEFELY SGΔA1600-80). The discharge gap between the needle electrodes was 0-2 m and the charging voltage was >1000 kV. A voltage divider and Rogowski coil were used to measure the charge voltage and the discharge current. The laser beam was transmitted near the discharge path. The polarization plane of the linearly polarized laser beam was so adjusted by a half wave plate (HWP) that its photon flux was equally divided into the two mutually orthogonal polarization components. The intensities of the orthogonal polarization components were detected by photodiodes (PDs) with amplifiers. A differential amplifier was also installed in the receiver circuit to detect the small rotation angle of the polarization plane. To eliminate electromagnetic noise caused by the discharge, the laser power supply and the receiver circuit were placed inside copper boxes. Signal cables were also shielded by wire mesh. The specifications of the discharge equipment and the optical detection system are summarized in Table 6-1. The differential output was detected only when the polarization plane was rotated by the Faraday effect.

The position of the propagating beam could be adjusted with respect to the discharge path and the discharge terminals.
The rotation angle of the polarization plane was estimated from the intensities of the orthogonal polarization components or the differential output by eq.(6-1). I_p and I_s are the intensities of the orthogonal polarization components.

$$\delta = \frac{\pi}{4} - \tan^{-1}(\sqrt{I_s} / \sqrt{I_p}) = \tan^{-1}\left(\frac{\sqrt{I_p} - \sqrt{I_S}}{\sqrt{I_P} + \sqrt{I_S}}\right) \quad (6\text{-}1)$$

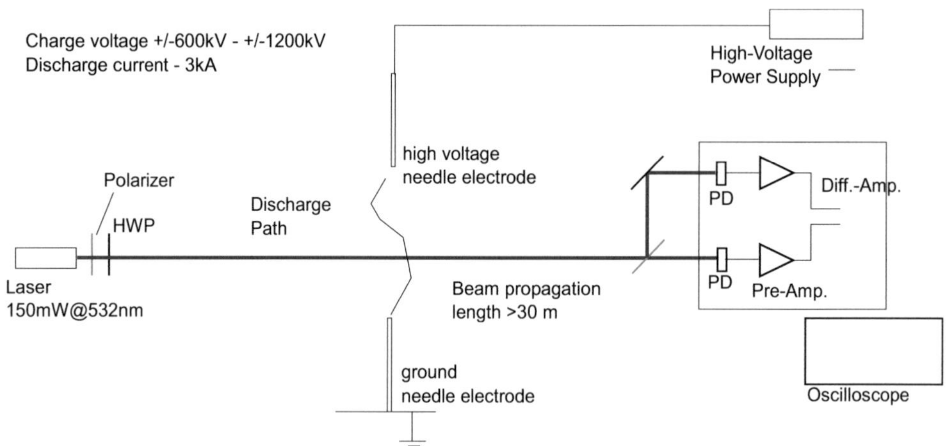

Fig. 6-1. Experimental setup of the high voltage discharge experiment.

When |*Ip-Is*|<<*Ip, Is*, the rotation angle is approximated by the following equation.

$$\delta = \frac{I_p - I_s}{I_p + I_s} \quad (6\text{-}2)$$

The estimation of the rotation angle is illustrated in Fig. 6-2. The polarity of the rotation angle indicates the spatial relation between the beam and the discharge path.

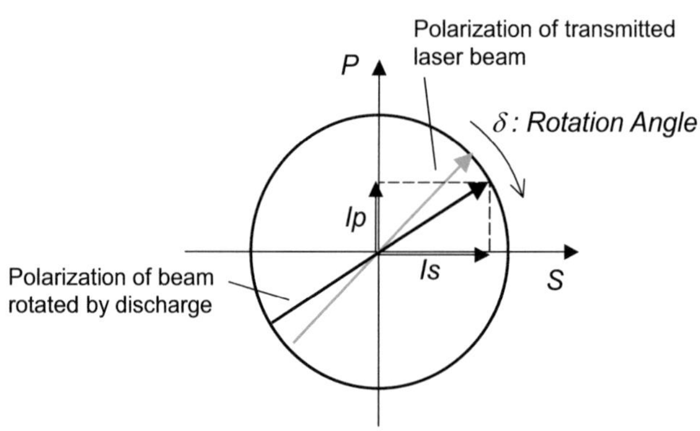

Fig. 6-2. Differential detection.

Discharge equipment	
Manufacturer, model	HAEFELY SG ΔA1600-80
Maximum charging voltage	+/-1600 kV
Discharge waveform	Lightning Impulse
Electrodes	needles
Discharge gap length	0-2 m
Beam/Receiver	
Light source	Nd:YAG green laser λ=532 nm, CW
Power	150 mW
Detector	Photodiode + Amplifier
detection	Differential detection

Table 6-1. Specifications of the discharge equipment and the optical detection system

6.2 Rotation angle detection

6.2.1 Detection of shock waves

Lightning discharge generates shock waves, which accompany variations in the air density and cause fluctuations of the propagating beam.(Fukuchi, 2005) Signals due to the shock waves are shown in Fig. 6-3. The discharge gap between the needle terminals was 77 cm and the charge voltage was -1200 kV. The propagating beam passed 4 cm below and 3 cm to the left of the high voltage needle electrode. In this case, the rotation angle was not detected because of the spatial separation between the discharge path and the beam. The air density variation accompanying the shock wave does not contribute to the Faraday effect, so the differential output is zero. In Fig. 6-3, the shock wave appeared 30 μs after the discharge trigger, so the distance between the discharge path and the propagating beam was calculated as 1 cm. In the experiment, we confirmed that the shock wave could be detected at a few hundred μs after the discharge trigger. Therefore, the shock wave signal can be used an indicator to locate the discharge location.

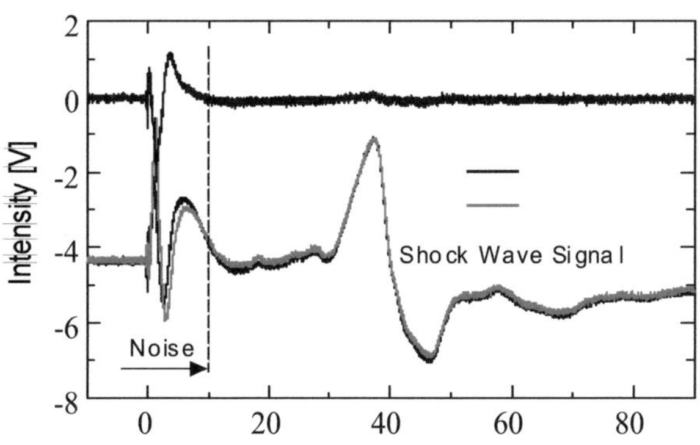

Fig. 6-3. Detection of shock wave.

6.2.2 Detection of polarization rotation angle

The differential output signals corresponding to the rotation angle of the polarization plane are shown in Fig.6-4. The typical discharge current is also shown. The discharge gap between the needle terminals was 77 cm and the charging voltage was +/-1200 kV. The propagating beam passed 2 cm under the high voltage needle electrode. The waveform before 10 μs could not be evaluated because of the electromagnetic noise due to the discharge. The differential outputs in the case of positive discharge (+1200 kV) and negative discharge (-1200 kV) showed opposite polarity. The output signals had the same response time as the discharge current. The rotation angle evaluated using eq. (6-1) was δ=0.53 degrees for positive polarity and δ=0.50 degrees for negative polarity. The dynamic range of >30 dB of the differential amplifier enabled detection of the small rotation angle. The results were in agreement with the results of numerical analysis and preliminary experiment using short gap discharge.

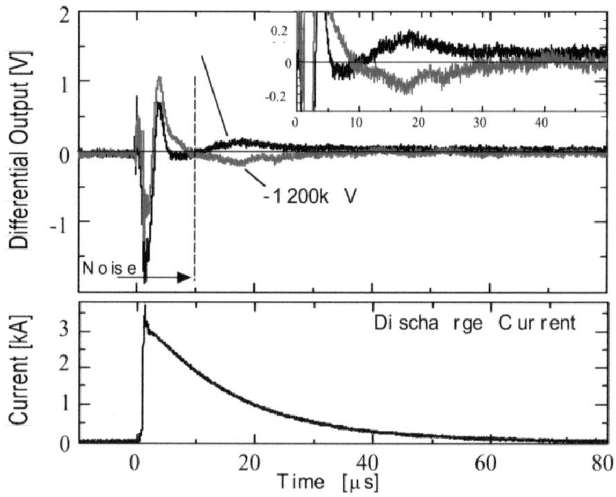

Fig. 6-4. Obtained waveform showing rotation angle detection for positive and negative polarities.

To suppress the electromagnetic noise, the receiver optics and electrical circuits were put in a shielded room. Due to spatial limitations caused by the introduction of the shield room, the position of the propagating beam was changed to 30 cm above the ground needle electrode from 2cm below the high voltage needle electrode. The electron density does not change significantly in the discharge path on arc or spark discharge. The discharge gap length was extended from 77 cm to 100 cm. This caused the shot-to-shot fluctuations of the discharge path in the extended discharge gap. The rotation angle depends on the distance between the discharge path and the propagating beam. Figure 6-5 shows the results of the experiment. The discharge gap between the needle terminals was 100 cm and the charge voltages were +1200 kV. Fig. 6-5(a) shows the differential output signals. The influence of the electromagnetic noise on the waveform decreased in comparison with the former experiment. Photographs of the discharge path in Fig. 6-5(b) were obtained simultaneously with the waveforms in Fig. 6-5(a). The position of the propagating laser beam is also indicated. The separation distance between the beam center and the discharge path was <2 cm for (A) and >6 cm for (B). The rotation angle was estimated as 0.54 degrees in case (A). The existence of the differential output is dependent on the distance between the discharge path and the beam. The output signal thus appeared when the probing beam was located within 2 cm apart from the discharge path, where the atmosphere was nearly perfectly ionized (n_e~10^{25} m^{-3}).

The present sensitivity of the rotation angle of the polarization plane is <1 degree. It is sufficient to detect the rotation angle only in a perfectly ionized atmosphere (n_e~10^{25} m^{-3}). The rotation angle can be detected only if the transmitting beam crosses the neighborhood of the discharge path. On the other hand, the shock wave does not rotate the polarization plane, and can be detected over a broader spatial region. Therefore, the observation algorithm for lidar application is designed as follows. At first, the lidar system roughly scans the observation region. When a shock wave is detected, the lightning position is estimated. Next, the neighborhood of the lightning position is scanned with higher spatial resolution, and the rotation angle of the polarization plane is measured. The discharge current, magnetic flux density, and ionized density of atmosphere are estimated. The distribution of the ionized atmosphere and its change will lead to the prediction of lightning strike.

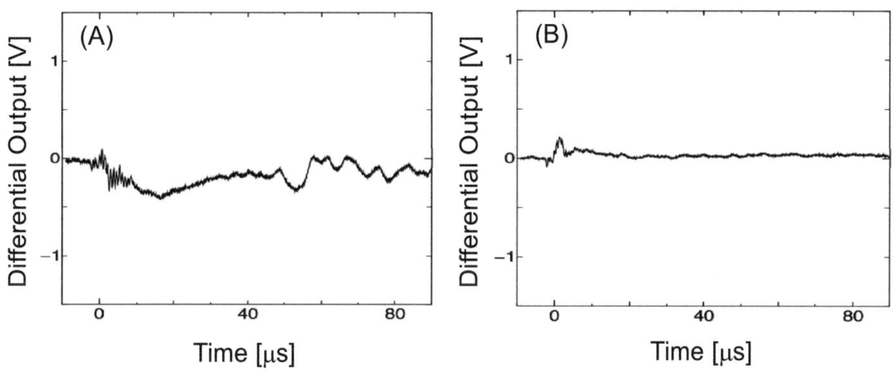

(a)Differential outputs of rotation angles

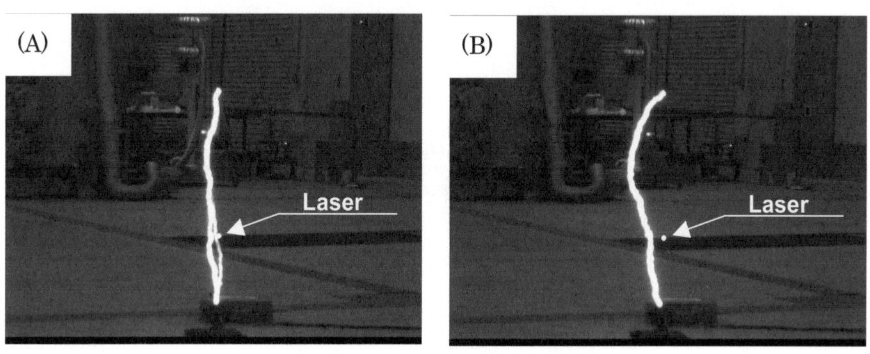

(b)Snapshots of discharge path

Fig. 6-5. Discharge experiment with electromagnetic shield room.

7. High precision polarization lidar

7.1 System setup

The lidar system was developed under the concept of the above lidar design. (Shiina, 2007a & 2008c) A schematic diagram is shown in Fig. 7-1, a photograph is shown in Fig. 7-2, and the specifications are summarized in Table 7-1. The optical circulator and a pair of Axicon prisms were installed into the lidar optics to realize the in-line optics. All optical components were selected to realize the high polarization extinction ratio and the high-power light source.

The laser source is a second harmonic Nd:YAG laser of wavelength 532 nm and pulse energy 200mJ. The polarization plane of the beam is balanced by a half wave plate (HWP) so the intensities of the parallel (*p*-) and orthogonal (*s*-) component beams are equal. The controlled beam passes through the specially designed polarization independent optical circulator. The beam changes its wave shape to the annular by a pair of Axicon prisms to expand its beam size up to the telescope diameter and to prevent the second mirror of the telescope from blocking the beam. All of the optics including the Axicon prisms had small tilts at the flat surface and AR coatings in every surface because the directly reflected light goes back to the detectors. Nevertheless, gated photomultiplier tubes (PMTs) are used for detection. The gate function stops the PMT operation until the outgoing beam exits the lidar optics. This protects the PMTs from reflections of the high power laser pulse from optical components in the in-line optics. The time delay between the beam firing and the start of the gate function is 0.2 μs. In other words, the system can detect the lidar echo signals from the near range of >30 m.

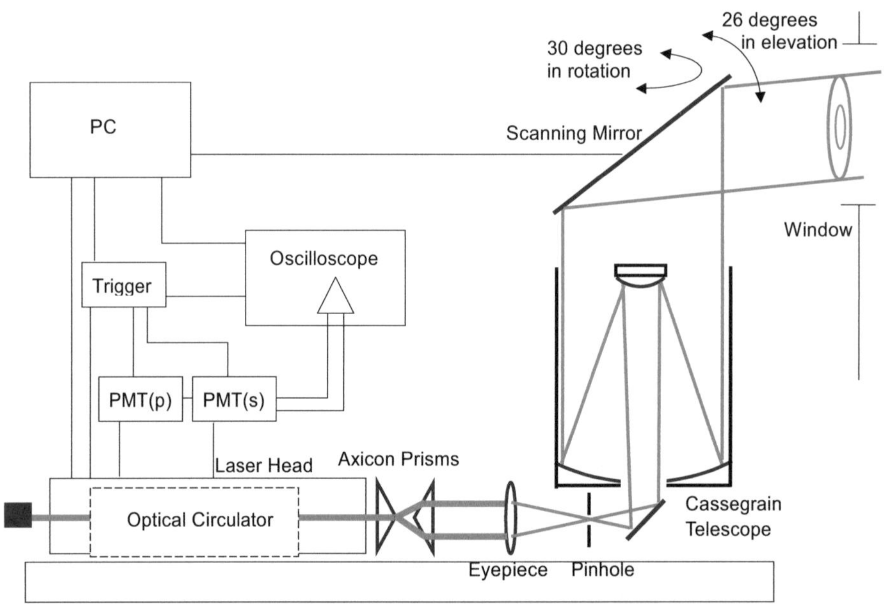

Fig. 7-1. Systematic diagram of high-precision polarization lidar system.

The scanning mirror was installed into the lidar system. The scanning area of the observation was limited to 26 degrees in elevation and 30 degrees in azimuth because of the constraint of the installation site.

To detect the small rotation angle of < 1 degree, the polarization-independent optical circulator for high power green laser was developed, as shown in Fig. 7-3. To avoid damage from the strong incident laser pulse, half wave plates (HWP), Gran laser prisms (GLP), Faraday rotators (FR), and mirrors (M) were chosen based on high threshold for optical tolerance, high transmittance (high reflectance for mirrors), and high extinction ratio for the polarization. The linearly polarized beam is divided equally at GLP1 into orthogonal polarization beams, which go through each path and are joined together at GLP2. The combined beam is transmitted into the Axicon prisms. When the lidar echo re-enters the optics, the parallel (*p*-) polarization echo to the incident beam is detected at GLP3. The echo

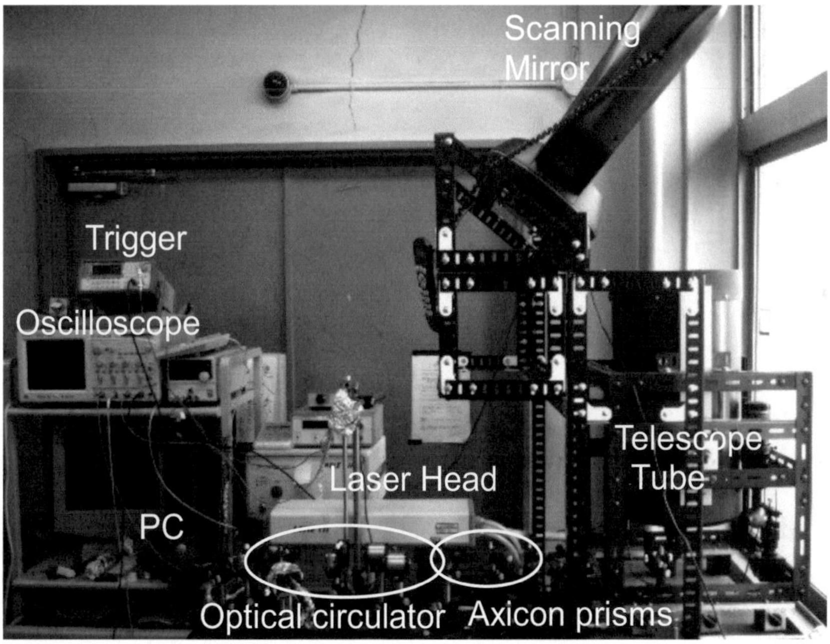

Fig. 7-2. Snapshot of high precision polarization lidar system.

Laser	*Nd:YAG* laser (Lotis TII)	
	Pulse Energy	200 mJ
	Wavelength	532 nm
	Repetition rate	10 Hz
	Beam Diameter	27 cmϕ
Telescope	Shmidt-Cassegrain *(Celestron NextarGPS)*	
	Aperture	28 cmϕ
	Field of View	0.177 mrad
Detector	PMT with gate function (2 ports for p- and s-polarizations) *(Hamamatsu K.K.)*	
	Filter bandwidth	3 nm
Observation	Range	0-20 km(max)

Table. 7-1. Lidar Specifications.

is picked up in the direction perpendicular to the illustration. The orthogonal (*s*-) polarization echo goes back into the circulator and is detected at GLP1. The isolation and the insertion loss of the optical circulator are summarized in Table 7-2. The transmission efficiency of the laser beam was 1.05 dB (79%) on average, which is acceptable considering the transmittances of the optical components. It is also confirmed that the GLP had a sufficiently high extinction ratio of >30 dB for the polarization.

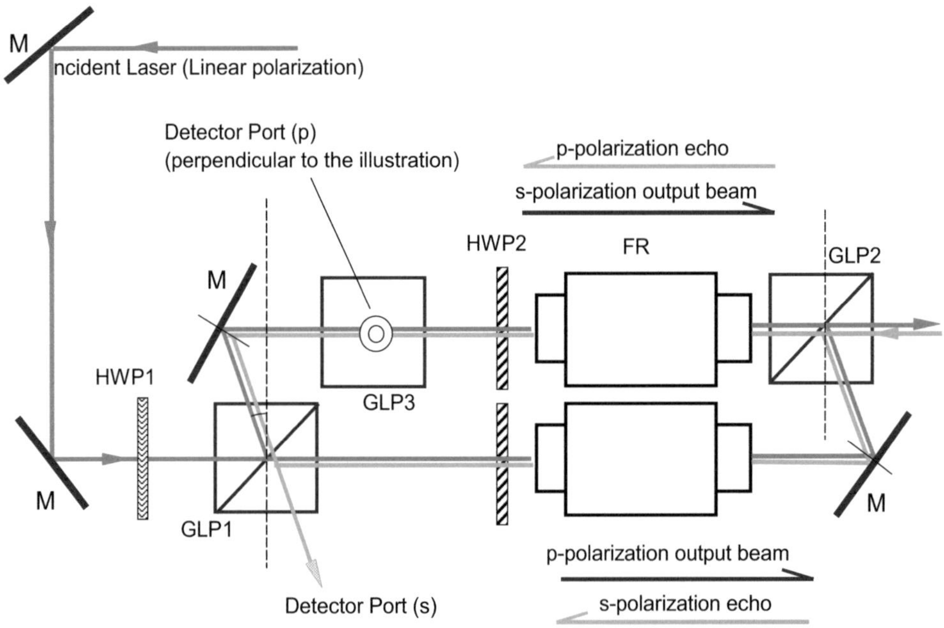

Fig. 7-3. Polarization-independent optical circulator.

Mode	Pol.	Insertion loss [dB]	Isolation	
			p-pol. echo [dB]	*s*-pol. echo [dB]
Transmitter	*p*	1.01	>39	>39
	s	1.08		
Receiver	*p*	1.25	-	>35.9
	s	1.65	>35.4	-

Table 7-2. Performance of polarization independent optical circulator.

7.2 Measurement

Here, echo characteristics of the high precision polarization lidar are explained in perspective of measurement range, intensity balance between p- and s-polarization components, and accuracy. Thanks to the in-line optics, the echo can be detected from near range of 30-50 m. Since the near range echo is large, the receiver's trigger is delayed by the gate function of PMT to restrict the current, especially in the case of far range measurement. If the PMT output current is too large, the balance between p- and s-polarizations is disrupted. In the following figures, the delay was adjusted adequately. Echoes were obtained from 330 m in Fig. 7-4, and from 820m in Figs. 7-5-7-7. We have confirmed that the delay of more than 3.5-4.0 μs (the lidar echo distance of 500-600 m) is acceptable to obtain the balanced echoes.

At first, the calculated accuracy of rotation angle of the polarization plane was checked by using a hard target. Figure 7-4 shows the p- and s-component echoes from a lightning rod located at a distance of 480 m from the lidar. As this distance is not large enough to eliminate the large echoes, p- and s-polarization echoes were not equal especially in near range of <400m. The echoes were summed over 1024 shots. As the lightning rod is a cylinder, the incident beam is depolarized if it was hit on a decline. Here we evaluated the depolarization as the rotation angle to check the accuracy of the lidar echoes. By using the derivation from Equation (6-1), the rotation angle was calculated as -0.497 degrees. The negative sign indicates the clockwise rotation of polarization plane in Fig. 6-1, that is, s-polarization component was larger than p-component. This result indicates the ability for the detection of the small rotation angle of 1 degree.

Figure 7-5 shows the balance and the accuracy of the echo intensities in the high precision polarization lidar by the atmospheric fluctuations. The figure shows the difference between p- and s-component echoes summed over 4096 shots. The difference at <1 km is large, while that at far distance becomes small, of the order of <100 μV. This difference indicates the accuracy of rotation angle of +/-0.35. That is, the atmospheric fluctuation is restricted enough to identify the rotation due to a lightning discharge. The accuracy, however, is obtained under summation over a large number of shots. The lightning discharge has the duration in the order of a 10-100 μs. See Fig.6-4). Although the cloud-to-cloud discharge continues a few dozens times, the summation should be finished within the period.

The low-altitude cloud observation is shown in Fig. 7-6. The vertical axis is the range-corrected signal ($P_L L^2$) shown in logarithmic scale. The water cloud was detected at 2.34 km ahead. Although p- and s-component echoes from the cloud base were equal intensity, the difference became large at the inside of the cloud. The incident beam penetrated up to 500-600m. The effect of multiple scattering becomes large when the beam penetrates inside cloud. The high precision polarization lidar has the narrow FOV of 0.177 mrad, which successfully eliminated multiple scattering inside the cloud at about 300m. However, the contribution of multiple scattering cannot be ignored for echoes from deeper locations

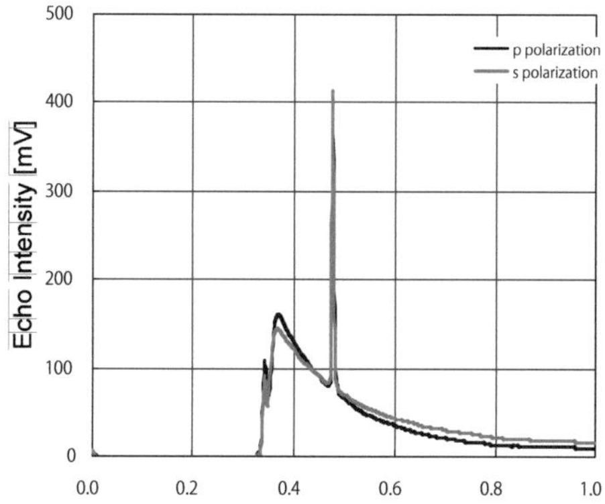

20:00() 24.7.2009, Temp. 25.1 deg, Hum. 62%

Fig. 7-4. Hard target detection.

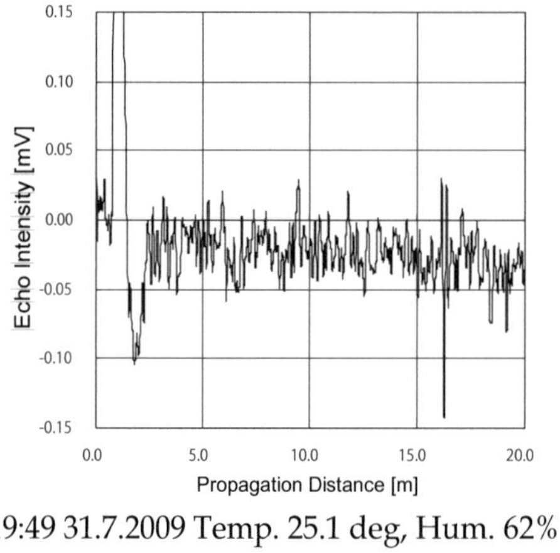

Fig. 7-5. Accuracy check for long propagation distance.

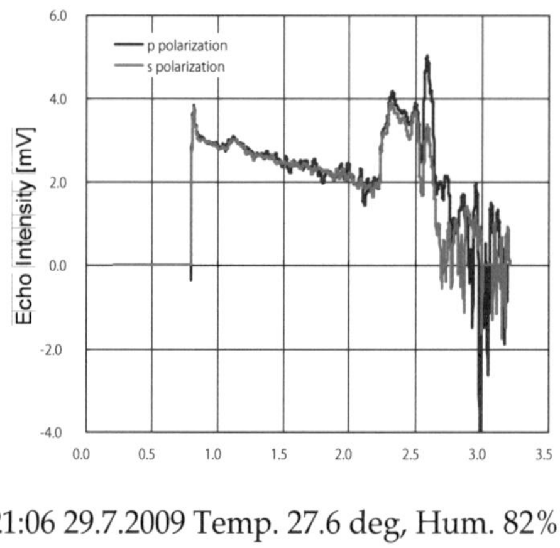

Fig. 7-6. Cloud observation.

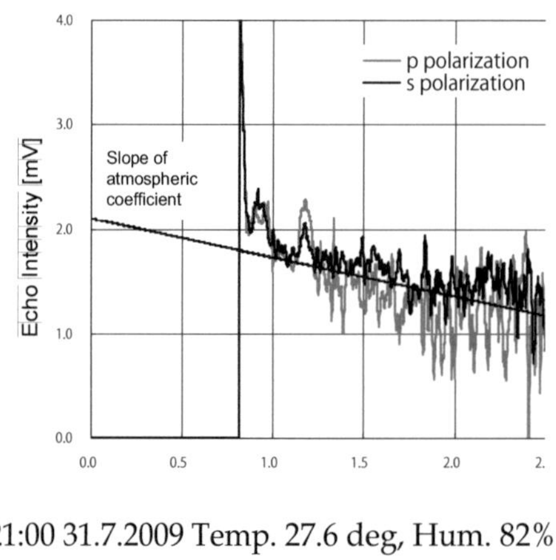

Fig. 7-7. Estimation of atmospheric coefficient.

inside clouds. The lidar echoes were also examined in the viewpoint of the atmospheric extinction coefficient. The atmospheric extinction coefficient is derived from the slope of the range-corrected signal, shown in Fig. 7-7. The p- and s-component echoes were well balanced. Although the small peak at 1.2km was a thin cloud, there are no influence to the backward echoes. The visibility calculated by the atmospheric coefficient was well coincide with the actual visibility. The near range echo also has the enough accuracy to evaluate the atmospheric characteristics.

8. Summary

Lidars for local weather prediction for prevention of disasters such as heavy rain and lightning strike were developed. As in-line optics were adopted to the in-line MPL and the high precision polarization lidar, the near range measurement could be accomplished with the narrow FOV. Optical circulators were also developed originally not to only separate echoes from the transmitting beam, but also to detect the orthogonal polarization echoes. The polarization extinction ratio between p- and s-polarization echoes was about 20 dB in the in-line MPL. The system can distinguish the ice-crystals from spherical particles stably in long period. The extinction ratio was improved to more than 30dB in the high precision polarization lidar. This improvement realized the measurement of Faraday effect caused by lightning discharge.

The current approach led to application of lidar to detection of hazardous gases. A mini Raman lidar with in-line optics to detect the hydrogen gas leak in near range is currently under development. A mini lidar to monitor the closed space atmosphere such as a factory and an exhibition hall is also under development.

The lidar studies made progress to the another field too. We found that the annular beam used in the in-line lidar optics transforms its intensity distribution into that of the non-diffractive beam through the propagation and that the transformed beam has the tolerant characteristics in the atmospheric fluctuation.(Shiina, 2007b) Now The technique tries to apply to penetrate the longer distance or to monitor the deeper area in the dense scattering media.

The near range detectable in-line lidar is counted on continued outstanding success to the various application by adjusting its size and specifications.

9. Acknowledgement

The author would like to express his thanks to Tetsuo Fukuchi of Central Research Institute of Electrical Power Industry and Kazuo Noguchi of Chiba Institute of Technology for their supports in developing the lidars and discussing the data analyses. These studies were funded by the Grant-in-Aid for Young Scientists (A) and (B), Mitutoyo Association for Science and Technology, and Kansai Research Foundation for technology promotion.

10. Reference

American National Standards Institute, "American national standard for the safe use of lasers", ANSI Z136.1-1986, pp.1-96, 1986.

C. J. Grund and S. P. Sandberg, "Depolarization and Backscatter Lidar for Unattended Operation", Advanced in Atmospheric Remote Sensing with Lidar, A. Ansmann, R.

Neuber, P. Rairoux, and U. Wandinger eds.(Springer-Verlag, Berlin, 1996), pp.3-6, 1996

C. Théry, "Evaluation of LPATS data using VHF interferometric observations of lightning flashes during the Eulinox experiment", Atmospheric Research, Vol. 56, Issues 1-4, 397-409, 2001

C. Weitkamp, *Lidar Range-Resolved Optical Remote Sensing of the Atmosphere*, Springer, 2005

D. R. MacGorman, W. L. Taylor, "Positive cloud-to-ground lightning detection by a direction-finder network", Journal of Geophysical Research, Vol. 94, No. D11, 13,313–13,318, 1989

E. Franzblau and C. J. Popp, "Nitrogen oxides produced from lightning", J. Geophys. Res., Vol.94, No.D8, pp.11.089-11.104, 1989

E. Franzblau, "Electrical discharges involving the formation of NO, NO_2, HNO_3, andO_3" , J. Geophys. Res., Vol.96, No.D12, pp.22.337-22.345, 1991

E. J. Welton, J. R. Campbell, J. D. Spinhirne,a nd V. S. Scott, "Global monitoring of clouds and aerosols using a network of micro-pulse lidar systems", in Lidar Remote Sensing for Industry and Environmental Monitoring, U. N. Singh, T. Itabe, N. Sugimoto eds., Proc. SPIE, 4153, pp.151-158, 2001.

G. Indebetouw, "Nondiffracting Optical Fields: Some Remarks on their Analysis and Synthesis", J. Opt. Soc. Am. A, Vol. 6, No. 1, pp.150-152, 1989

G. Scott and N. McArdle, "Efficient Generation of Nearly Diffraction-Free Beams using an Axicon", Opt. Engin., Vol. 31, No. 12, pp.2640-2643, 1992

G. Zaccanti, P. Bruscaglioni, M. Gurioli, and P. Sansoni, "Laboratory simulations of lidar returns from clouds : experimental and numerical results", App. Opt., Vol. 32, No. 9, pp.1590-1597, 1993

H. S. Lee, I. H. Hwang, J. D. Spinhirne, V. Stanley Scott, "Micro Pulse Lidar for Aerosol and Cloud Measurement", Advanced in Atmospheric Remote Sensing with Lidar, A. ansmann, R. Neuber, P. Rairoux, and U. Wandinger eds.(Springer-Verlag, Berlin, 1996), pp.7-10, 1996

I. H. Hwang, M. K. Nam, B. Ranganayakamma, and H. S. Lee, "A Compact Eye-safe Dual Wavelength Lidar and Application in Biological Aerosol Detection", Lidar Remoet Sensing in Atmospheric and Earth Sciences Pert I, L. Bissonnette, G. Roy, and G. Vallee eds.(Defence R&D Canada, 2002), pp. 205-208, 2002

J. D. Spinhirne, "Micro Pulse Lidar", IEEE trans. Geosc. Rem. Sens., Vol. 31, No. 1, pp.48-55, 1993.

J. D. Spinhirne, "Micro Pulse Lidar Systems and Applications", Proceedings of 17th International Laser Radar Conference, pp.162-165, 1994.

J. D. Spinhirne, E. J. Welton, J. R. Campbell, "Measurement of the Vertical Distribution of Aerosol by Globally Distributed MP Lidar Network Sites", IEEE IGARSS 2001Meeting, Sydney, July 2001.

J. Durnin and J. J. Miceli, Jr., "Diffraction-Free Beams", Phys. Rev. Lett., Vol. 58, No. 15, pp.1499-1501, 1987

J. Harms, "Lidar return signals for coaxial and noncoaxial systems with central obstruction", Appl. Opt., Vol. 18, No. 10, pp.1559-1566, 1979

J. S. Ryan, S. R. Pal, and A. I. Carswell, "Laser backscattering from dense water-droplet clouds", J. Opt. Soc. Am., Vol. 69, No. 1, pp.60-67, 1979

J. Stith, J. Dye, B. Ridley, P. Laroche, E. Defer, K. Baumann, G. Hubler, R. Zerr, and M. Venticinque, "NO signature from lightning flashes", J. Geophys. Res., Vol.104, No.D13, pp.16.081-16.089, 1999

K. Kawahata and S. Okajima, "Interferometry and Polarimetry -Principle of Interferometry and Polarimetry-", Japan Society of Plasma Science and Nuclear Fusion Research, Vol.76, No.9, pp.845-847, 2000(written in Japanese)

K. Kono, Y. Mitarai, and T. Minemoto, "New Super-Resolution Optics with Double-Concave-Cone Lens for Optical Disk Memories", J. Opt. Mem. Neural Networks, Vol. 5, No. 4, pp.279-285, 1996

K. Kono, M. Irie, and T. Minemonto, "Generation of Nearly Diffraction-Free Beam Using a New Optical System", Opt. Rev., Vol. 4, No. 3, pp.423-428, 1997

K. Sassen and R. L. Petrilla, "Lidar depolarization from multiple scattering in marine stratus clouds", App. Opt., Vol. 25, No. 9, pp.1450-1458, 1986

K. Sassen, "The polarization lidar technique for cloud research: A review and current assessment", Bulletin American Meteorological Society, Vol.72, No.12, pp.1848-1866, 1991

L. L. Doskolovich, S. N. Khonina, V. V. Kotlyar, I. V. Nikolsky, V. A. Soifer, and G. V. Uspleniev, "Focusators into a Ring", Opt. Quant. Elect., Vol. 25, pp.801-814, 1993

M. Kerscher, W. Krichbaumer, M. Noormohammadian, and U. G. Oppel, "Polarized Multiply Scattered LIDAR Signals", Opt. Rev., Vol. 2, No. 4, pp.304-307, 1995

M. Gai, M. Gurioli, P. Bruscaglioni, A. Ismaelli, and G. Zaccanti, "Laboratory simulations of lidar returns from clouds", App. Opt., Vol. 35, No. 27, pp.5435-5442, 1996

N. Sugimoto, I. Matsui, and Y. Sasano, "Design of lidar transmitter-receiver optics for lower atmospheric observations: geometrical form factor in lidar equation", Jpn. J. Opt., Vol.19, No.10, pp. 687-693, 1990

P. Belanger and M. Rioux, "Ring Pattern of a Lens-Axicon Doublet Illuminated by a Gaussian Beam", Appl. Opt., Vol. 17, No. 7, pp.1080-1086, 1978

P. Bruscaglioni, A. Ismaelli, G. Zaccanti, M. Gai, and M. Gurioli, "Polarization of Lidar Returns from Water Clouds : Calculations and Laboratory Scaled Measurement", Opt. Rev., Vol. 2, No. 4, pp.312-318, 1995

R. Arimoto, C. Saloma, T. Tanaka, and S. Kawata, "Imaging Properties of Axicon in a scanning optical system", Appl. Opt., Vol. 31, No. 31, pp.6653-6657, 1992

R. M. Measures, *Laser Remote Sensing*, John Wiley & Son, New York, 1984.

S. Hayashi, "Numerical Simulation of Electrical Space Charge Density and Lightning by using a 3-Dimensional Cloud-Resolving Model", Scientific Online Letters on the Atmosphere, Vol. 2, p.124-127, 2006

S. R. Pal and A. I. Carswell, "Polarization properties of lidar backscattering from clouds", Appl. Opt., Vol. 12, No. 7, pp.1530-1535, 1973

Society of Atmospheric Electricity of Japan, Atmospheric Electricity, Chapter 2, Corona publishing Tokyo, 2003 (Written in Japanese)

T. Halldorsson and J. Langerholc, "Geometrical form factors for the lidar function", Appl. Opt., Vol. 17, No. 2 pp. 240-244, 1978

T. Hauf, U. Finke, S. Keyn, O. Kreyer, "Comparison of a SAFIR lightning detection network in northern Germany to the operational BLIDS network", Journal of Geophysical Research 112(d18): D18114, 2007

T. Fujii and T. Fukuchi, *Laser Remote Sensing*, Taylor & Francis, 2005

T. Fukuchi, K. Nemoto, K. Matsumoto, and Y. Hosono, "Visualization of High-speed Phenomena using an Acousto-optic Laser Deflector", IEEJ Transactions on Fundamentals and Materials, Vol.125-A, No.2, pp.113-118, 2005 (written in Japanese)
T. Fukuchi and T. Shiina, "Measurement of rotation of polarization plane of laser radiation propagating though impulse discharge in air", IEEJ Transactions on Electrical and Electric Engineering, in press.
T. Shiina, E. Minami, M. Ito, and Y. Okamura, "Optical Circulator for In-line Type Compact Lidar", Applied Optics, Vol. 41, No. 19, pp.3900-3905, 2002
T. Shiina, K. Yoshida, M. Ito, and Y. Okamura, "In-line type micro pulse lidar with annular beam -Experiment-", Applied Optics, Vol.44, No.34, pp.7407-7413, 2005a
T. Shiina, K. Yoshida, M. Ito, and Y. Okamura, "In-line type micro pulse lidar with annular beam -Theoretical Approach-", Applied Optics, Vol.44, No.34, pp.7467-7473, 2005b
T. Shiina, T. Honda, and T. Fukuchi, "High-precision polarization lidar - lidar in-line optics - ", CLEO Pacific Rim 2007 proceedings, pp.1499-1500, 2007a
T. Shiina, M. Ito, and Y. Okamura, "Long range propagaion characteristics of annular beam", Optics Communications, Vol.279, pp.159-167, 2007b
T. Shiina, T. Honda, and T. Fukuchi, "Evaluation of polarization angle rotation of propagating light in a partially ionized atmosphere under discharge conditions", Electrical Engineering in Japan, Vol.163, No.4, pp.1-7, 2008a
T. Shiina, T. Honda, and T. Fukuchi, "Optical measurement of high-voltage discharge in air for lidar lightning detection", APLS The Review of Laser Engineering Supplemental Volume 2008, Vol.36, pp.1279-1282, 2008b
T. Shiina, Masakazu Miyamoto, Dai Umaki, Kazuo Noguchi and Tetsuo Fukuchi, "Fundamental Measurement by In-line typed high-precision polarization lidar", SPIE Asia-Pacific Remote Sensing 2008 proceedings, Vol.7153, pp.71530B-1 - 71530B-8, 2008c
T. Shiina, T. Honda, and T. Fukuchi, "Measurement of polarization rotation of propagating light in a partially ionized atmosphere under discharge conditions", Electrical Engineering in Japan, in press.
V. Soifer, V. Kotlyar, and L. Doskolovich, *Iterative Methods for Diffractive Optical Elements Computation*, Tatlor & Francis, 1997
Y. Matsudo, T. Suzuki, M. Hayakawa, K. Yamashita, Y. Ando, K. Michimoto, V. Korepanov, "Characteristics of Japanese winter sprites and their parent lightning as estimated by VHF lightning and ELF transients", Journal of atmospheric and solar-terrestrial physics, vol.69, no.12, 1431-1446, 2007

9

Precision Dimensional Metrology based on a Femtosecond Pulse Laser

Jonghan Jin[1,2] and Seung-Woo Kim[2]
[1]*Center for Length and Time, Division of Physical Metrology, Korea Research Institute of Standards and Science (KRISS)*
[2]*Ultrafast Optics for Ultraprecision Technology Group, KAIST*
Republic of Korea

1. Introduction

Metre is defined as the path traveled by light in the vacuum during the time interval of 1/299 792 458 s. The optical interferometer allows a direct realization of metre because it obtains the displacement based on wavelength of a light source in use which is corresponding to the period of interference signal. Due to the periodicity of interference signal, the distance can be determined by accumulating the phase continuously to avoid the 2π ambiguity problem while moving the target. Conventional laser interferometer systems have been adopted this relative displacement measurement technique for simple layout and high measurement accuracy.

Recently the use of femtosecond pulse lasers (fs pulse laser) has been exploded because of its wide spectral bandwidth, short pulse duration, high frequency stability and ultra-strong peak power in precision spectroscopy, time-resolved measurement, and micro/nano fabrication. A fs pulse laser has more than 10^5 longitudinal modes in the wide spectral bandwidth of several hundred nm in wavelength. The longitudinal modes of a fs pulse laser, the optical comb can be described by two measurable parameters; repetition rate and carrier-offset frequency. A repetition rate, equal spacing between longitudinal modes is determined by cavity length, and a carrier-offset frequency is caused by dispersion in the cavity. Under stabilization of the repetition rate and a carrier-offset frequency, longitudinal modes are able to be employed as a scale on the optical frequency ruler with the traceability to the frequency standard, cesium atomic clock.

Optical frequency generators were suggested and realized to generate a desired well-defined wavelength by locking an external tunable working laser to a wanted longitudinal mode of the optical comb or extracting a frequency component directly from the optical comb with optical filtering and amplification stages. Optical frequency generators can be used as a novel light source for precision dimensional metrology due to wide optical frequency selection with the high frequency stability.

In this chapter, the basic principles of a fs pulse laser and optical frequency generators will be introduced. And novel measurement techniques using optical frequency generators will be described in standard calibration task and absolute distance measurement for both fundamental research and industrial use.

2. Basic principles of precision dimensional metrology

2.1 Optical comb of a fs pulse laser

Precision measurement of optical frequency in the range of several hundred THz has lots of practical difficulties because the bandwidth of photo-detectors only can reach up to several GHz. Simply it can be determined by beat notes with well-known optical frequencies like absorption lines of atoms or molecules in the radio frequency regime, which are detectable. And the other method, frequency chain, was realized which could connect from radio frequency regime to optical frequency regime with numerous laser sources and radio /micro frequency generators, which were stabilized and locked to the frequency standard by beat signals of them in series. The construction and arrangement of components for this frequency chain should be changed according to the optical frequency we want to measure. That makes it is not attractive as general optical frequency measuring technique in terms of efficiency and practicality.

The advent of a fs pulse laser could open the new era for precision optical frequency metrology. The absolute optical frequency measurement technique was suggested based on the optical comb of a fs pulse laser, which could emit a pulse train using a mode-locking technique. Since the optical comb has lots of optical frequency modes, it can be employed as a scale on the frequency ruler under stabilization. Prof. Hänch in MPQ suggested this idea, and verified the maintenance of phase coherence between frequency modes of the optical comb experimentally. However, it could not be used because the spectral bandwidth is not wide enough for covering an octave. Dr. Hall in NIST realized the wide-spectral optical comb firstly with the aid of a photonic crystal fiber, which could induce the non-linear effect highly. That allows a direct optical frequency measurement with the traceability to the frequency standard. And it also has been used in the field of precision spectroscopy.

Laser based on the optical cavity can have lots of longitudinal modes with the frequency mode spacing of $c/2L_c$, where c is speed of light and L_c represents the length of optical cavity. When spectral bandwidth of an amplifying medium is broad enough to have several longitudinal modes, it can be operated as a multi-mode emission. Typical a monochromatic laser is designed to produce a single frequency by shortening length of the optical cavity, which leads wide frequency mode spacing. In the case of a fs pulse laser, the Ti:Sapphire has very wide emission bandwidth of 650 to 1100 nm with the absorption bandwidth of 400 to 600 nm. Even if the spectral bandwidth is only 50 nm with frequency mode spacing of 80 MHz at a center wavelength of 800 nm, number of longitudinal modes can be reached to 3×10^5. Though the modes are oscillating independently, the phases of whole modes can be made same based on strong non-linear effect, Kerr lens effect.

Mode-locking can be achieved by Kerr lens effect in the amplifying medium with a slit. When strong light propagates into a medium, Kerr lens effect can lead change of refractive index of the medium according to the optical intensity of an incident light. Therefore, the refractive index of the medium, n, can be expressed as

$$n = n_0 + n_2 \cdot I \qquad (2\text{-}1)$$

where n_0 is the linear refractive index, n_2 is the second-order nonliner refractive index, and I is intensity of the incident light. Since the plane wave has Gaussian-shaped intensity profile spatially, the high intensity area near an optical axis suffers the high refractive index relatively. That makes self-focusing of the strong light as shown in figure 2-1(a). In order to generate a pulse train, the difference gains for continuous waves and pulsed waves are induced by two methods; One is the adjusting the beam size of a pumping laser to have more gain in the only high intensity area as shown in figure 2-1 (b). The other is removal of the continuous wave by a slit in figure 2-1 (c). That is, the short pulse has stronger optical intensity than a continuous wave, which is caused by Kerr lens effect in the amplifying medium. By designing the optimal cavity the short pulse will be activate with high gain, and the continuous light will be suppressed. In the begging of generation of a short pulse, the small mounts of shock or impact should be needed to give intensity variation for inducing non-linear effect efficiently.

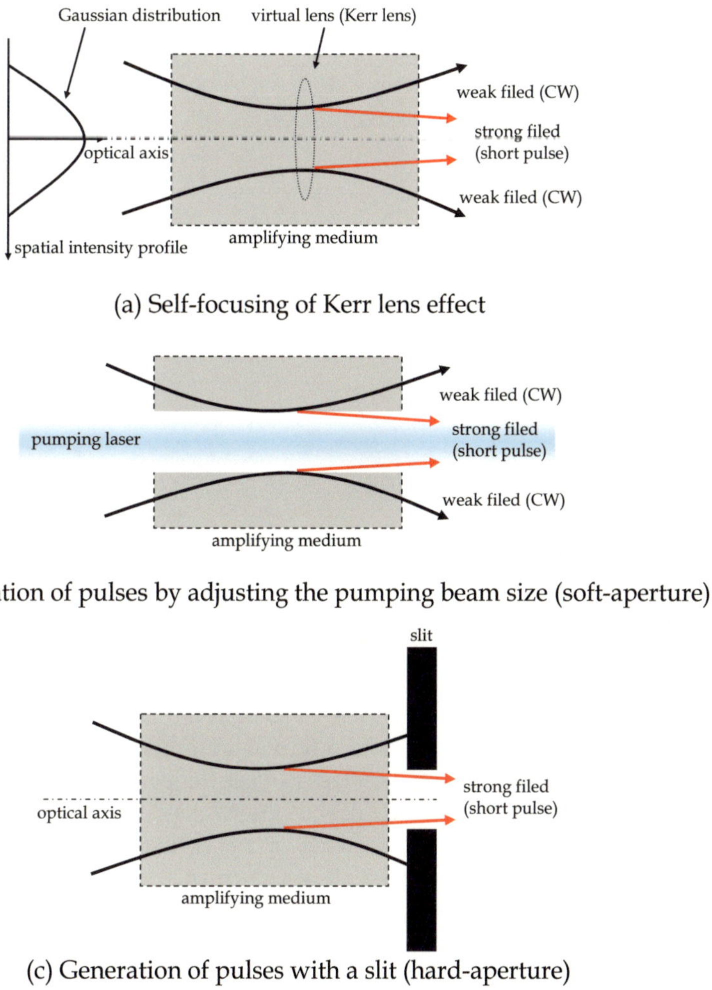

Fig. 2-1. Basics of Mode-locking; Kerr lens effects and cavity design for introducing differential gain

Figure 2-2 shows the generation of pulse train when the modes are phase-locked according to the number of participating frequency modes. The pulse duration can be shortened by employing numerous frequency modes in the mode-locking process, it can be achieved several fs or less in time domain. Typically a commercialized Ti:Sapphire fs pulse laser has the spectral bandwidth of more than 100 nm in wavelength (There are approximately 10^5 or more longitudinal modes.) and the pulse duration of less than 10 fs.

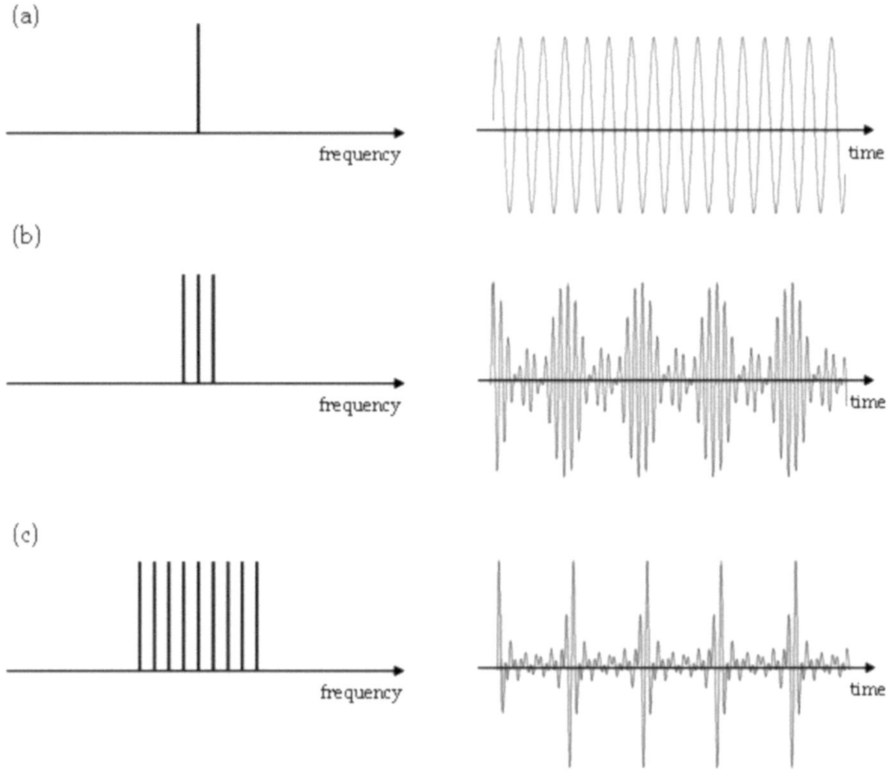

Fig. 2-2. Relationship between frequency and time-domain; (a) single frequency, (b) three frequencies, (c) nine frequencies with same initial phase values.

Figure 2-3(a) shows the optical comb, which can be described by repetition rate, f_r and carrier-offset frequency, f_o. Repetition rate, mode spacing in the frequency domain, can be monitored by photo-detector easily, and adjusted by translating of the end mirror of the cavity to change the cavity length. The carrier-offset frequency caused by dispersion in the cavity is defined as frequency shift or offset of the whole optical comb from the zero in the frequency domain. It can be measured by a self-referencing f-2f interferometer, and controlled by prism pairs, diffraction gratings, or chirped mirrors in order to compensate the dispersion. Figure 2-3(b) shows pulse train of a fs pulse laser in time domain. The time interval of pulses, T, is defined by the reciprocal of the repetition rate, f_r. The shift of the carrier signal from the envelope peak because of the pulse-to-pulse phase shift, $\Delta\Phi$, is caused by carrier-offset frequency, f_o.

Even if the optical comb has a wide spectral bandwidth, that is not sufficient for f-2f interferometer, which is needed the octave-spanned spectrum. The fs pulse laser delivers into a photonic crystal fiber for octave-spanning, then both frequency, f, and doubled frequency, 2f, components are obtained. The broadening of the spectrum can be occurred by non-linear phenomena such as four-wave mixing generation, self phase modulation, and stimulated Raman scattering. The broadened spectrum divided into two parts, low

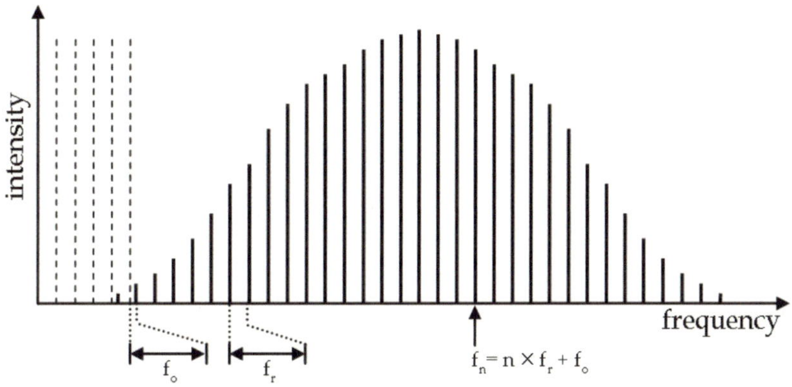

(a) Optical comb of a fs pulse laser; repetition rate (f_r), carrier-offset frequency (f_o), n^{th} longitudinal mode (f_n)

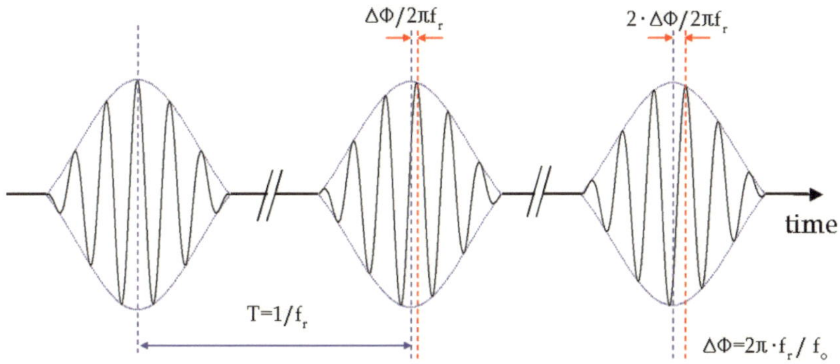

(b) Pulse train of a fs pulse laser; time interval between pulses (T), pulse-to-pulse phase shift ($\Delta\Phi$)

Fig. 2-3. Optical comb of a fs pulse laser in spectral domain and time domain.

frequency part (f) and high frequency part (2f) by a dichromatic mirror. Here, the nth frequency of longitudinal modes, f_n, can be expressed by the simple equation of

$$f_n = n \times f_r + f_o \tag{2-2}$$

The low frequency part (f_n) is frequency-doubled, $2f_n$, by a second harmonic generator, then makes the beat note with high frequency part (f_{2n}). It gives the carrier-offset frequency, f_o, which is given as

$$2f_n - f_{2n} = 2(n \times f_r + f_o) - (2n \times f_r + f_o) = f_o \tag{2-3}$$

Therefore, all longitudinal modes of the optical comb can be stabilized precisely by locking two parameters, f_r and f_o, to the atomic clock in the radio frequency regime. For this locking process, phase-locked loop (PLL) was adopted. PLL generates periodic output signal with different duty ratio, which depends on the frequency and phase differences of two inputs based on XOR logic. The periodic output signal can convert to DC signal by a low pass filter, then the DC signal acts as input of voltage-controlled oscillator to coincide frequencies and phases of two inputs. Figure 2-4 shows a schematic of control loop for repetition rate and carrier-offset frequency stabilization of the fs pulse laser.

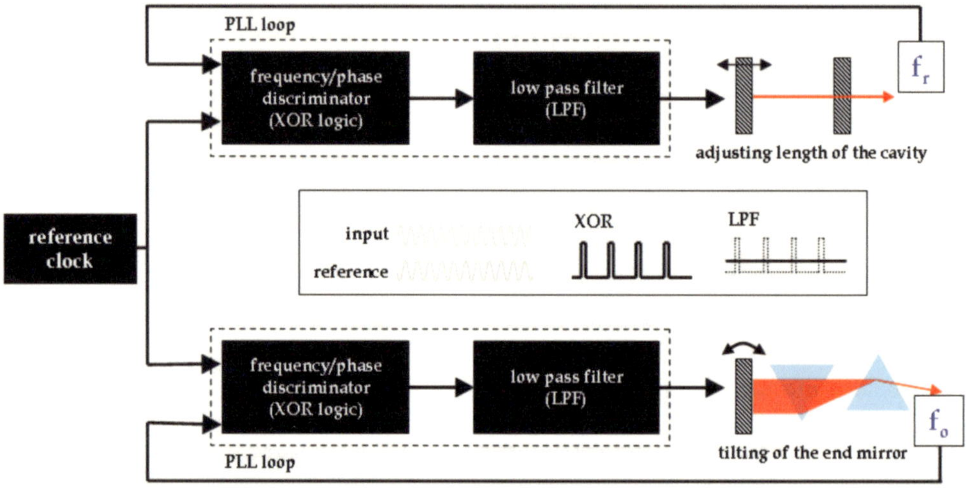

Fig. 2-4. Schematic of control loop for fs pulse laser stabilization

Since the optical frequency of more than several hundreds THz is too fast to detect directly, the optical comb has attractive advantages in the field of ultra-precision optical frequency measurement. Since the optical comb acts as a precision scale on an optical frequency ruler, the optical frequency, f can be obtained by simply adding or subtracting the beat note, f_b to the nearest frequency mode, f_n. It can be given as

$$f = f_n \pm f_b = (n \cdot f_r + f_o) \pm f_b \qquad (2\text{-}4)$$

By modulating repetition rate or carrier-offset frequency slightly, the sign of beat note can be determined. And more highly accurate optical frequency measurements will be achieved with the advent of ultra-stable optical clocks in the near future.

2.2 Realization of optical frequency generators

Optical frequency generator is an ultra-stable tunable light source, which is able to produce or to extract desired optical frequencies (wavelengths) from the optical comb of a fs pulse laser. Because optical interferometer allows direct realization of length standard based on the wavelength of a light source in use, one or more well-defined wavelengths should be secured. Typically the stabilized wavelength can be obtained by absorption lines or transition lines of atoms or molecules. For multi-wavelength applications it will be a complicated system with different stabilities at desired wavelengths. However, optical frequency generator is a simple solution to produce numerous well-defined wavelengths which are traceable to an unique reference clock of time standard.

In order to realize the optical frequency generator, Extra-cavity laser diode (ECLD), one of useful light sources for wide wavelength tuning range of ~ 20 nm with narrow line-width, was adopted as a working laser as shown in figure 2-5. However, it is not proper to apply it for precision dimensional metrology due to ignorance of absolute frequency and unsatisfied frequency stability. By integrating the ECLD with the stabilized optical comb, the optical frequency generator can be realized as shown in figure 2-6. The ECLD can tune the wavelength easily by tilting the end mirror using a DC-motor and a PZT coarsely. The input current and the temperature control of the ECLD allow precise wavelength tuning with the resolution of several Hz.

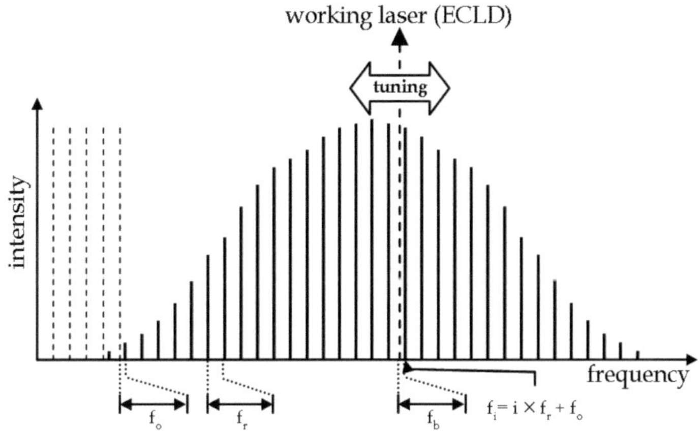

Fig. 2-5. Basic concept of an optical frequency generator with a working laser

A working laser, ECLD consists of a diode laser and an extra cavity; One side of diode laser was anti-reflection coated, the other side was high reflective coated for the end mirror of an external cavity. The diode laser is temperature controlled with the resolution of less than 1 mK by an attached thermometer and an active cooling pad. The light emitted from the diode laser is collimated and goes to the diffraction grating. Then the diffracted light is propagated to the end mirror of external cavity, which can be angle-adjustable using attached DC motor and PZT. Since the angle of the end mirror can be read by an angle sensor precisely, the wavelength of ECLD can be tuned with the resolution of less than 0.02 nm.

The wavelength tuning of ECLD by the DC motor can achieve the resolution of 0.02 nm with speed of 8 nm/s in the range of 765 to 781 nm. The wavelength can be tuned precisely using PZT within only the range of 0.15 nm (75 GHz) due to the short travel of PZT. This control loop has the bandwidth of several hundreds Hz, which is limited by resonance frequency of the mechanical tilting end mirror.

When a target wavelength is inputted, the wavelength of a working laser is tuned coarsely by DC motor and PZT. Because of the short tuning range of PZT, it should be within the small deviation from a target wavelength, 0.15 nm. After that the fine tuning will be done by modulating the input current under temperature stabilization according to the beat note between a nearest comb mode and ECLD, f_b. The current control can be achieve the control rate of 1 MHz. Since the wave-meter has the measurement uncertainty of 30 MHz, it can determine only integer part (i) of the nearest comb mode (f_i) with known values of repetition rate (f_r) and carrier-offset frequency (f_o). And the control loop operates based on a phase-locked loop (PLL) technique which is referred to an atomic clock. The output power of the optical frequency generator is 5 to 20 mW with linewidth of less than 300 kHz at 50 ms, which is corresponding to the coherence length of more than 1 km.

Figure 2-7 shows coarse wavelength tuning performance using the DC motor and the PZT attached on the ECLD. The tuning range is from 775 nm to 775.00025 nm with the step of 0.00005 nm in vacuum wavelength. The frequency stability at each step is about 10^{-8}, which

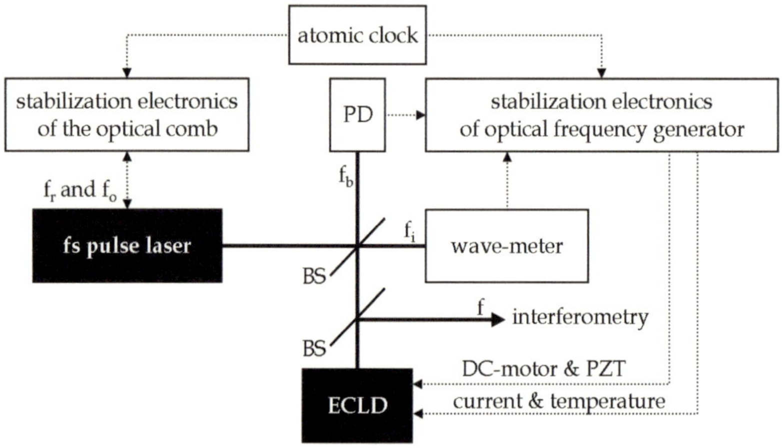

Fig. 2-6. Schematic of the optical frequency generator with wavelength tunable light source; PD(photo-detector), solid lines for optical delivery and dotted lines for electric signals

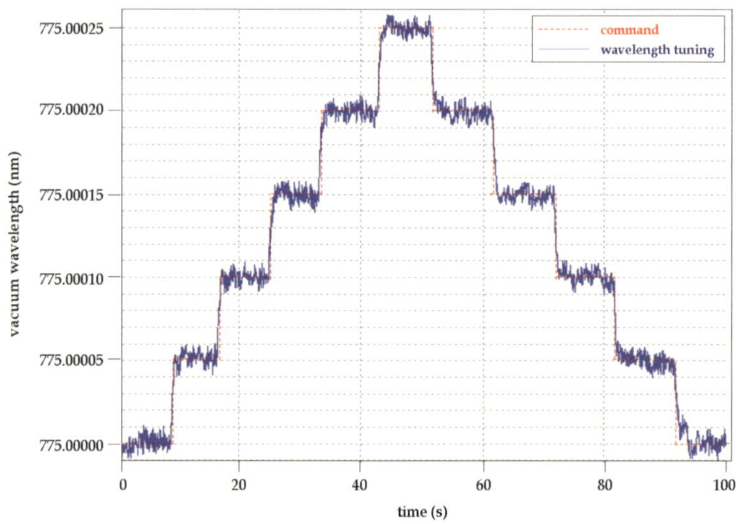

Fig. 2-7. Coarse wavelength tuning performance of an optical frequency generator; solid line is wavelength of ECLD and dotted line is command for coarse tuning.

is frequency deviation of 4 MHz (peak-to-valley value) at 387 THz in frequency. This coarse tuning can be done within 1 s at each step. And then finally the stability of wavelength can be achieved to 1.3×10^{-10} at 10 s by locking beat note, f_b to the rubidium reference clock. This type of the optical frequency generator, however, has a drawback which is a degradation of frequency stability due to the mechanical control and resonance of the ECLD. The other trial for obtaining better frequency stability is the direct extraction of desired wavelengths from the optical comb as shown in figure 2-8. The extraction process consists of coarse and fine filtering. For coarse extraction, the wide tuning range is necessary to cover the whole spectral range of a fs pulse laser even if it has relatively broad filtering bandwidth. In order to extract a single mode from the optical comb, fine filtering should have the filtering bandwidth smaller than a repetition rate. The optical power of the extracted mode is very weak, thus it should be amplified for applications.

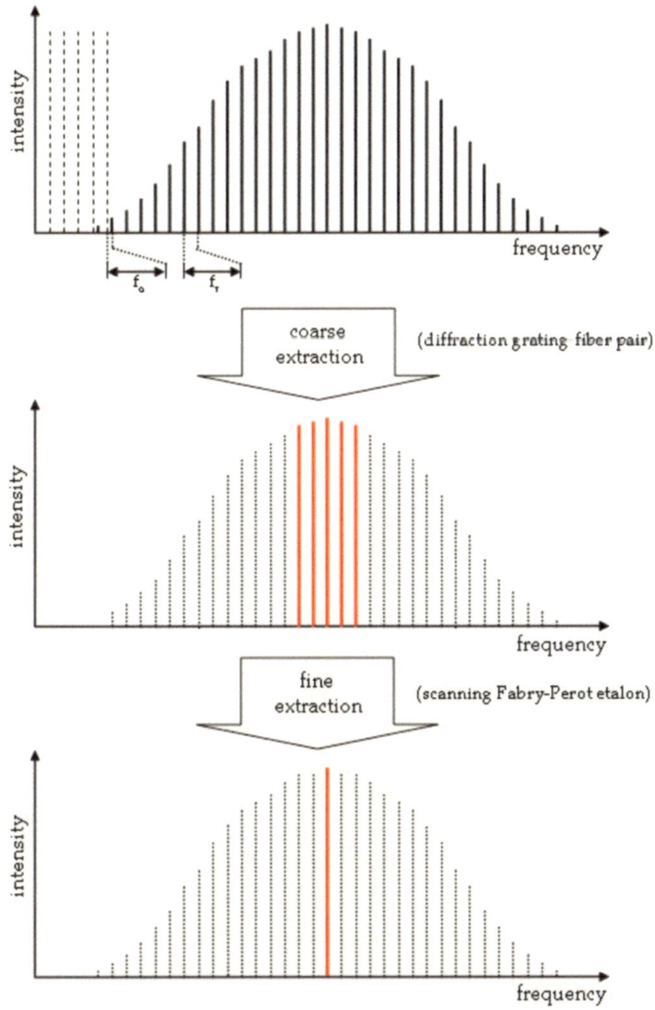

Fig. 2-8. Basic concept of optical frequency generator based on one-mode extraction; dotted longitudinal modes filtered out.

Figure 2-9 shows the optical layout and a schematic of the control system. The desired wavelength can be extracted by diffraction grating-fiber pair coarsely and scanning Fabry-Perot etalon (FPE) finely. By tilting the angle of the diffraction grating or translating the focusing lens of optical fiber, the optical comb can be filtered roughly with the resolution of several hundreds MHz. The FPE can select a desired mode by adjusting the length of the etalon with the resolution of several tens MHz, which is enough to resolve an individual comb mode. However, the selected mode is too week to detect directly, consequently the injection locking technique is adopted for optical amplification. Optical power of ~ 10 mW can be achieved with the amplification factor of 10^6. To evaluate the frequency stability, the selected mode is frequency-shifted by AOM and then interfered with the original optical comb.

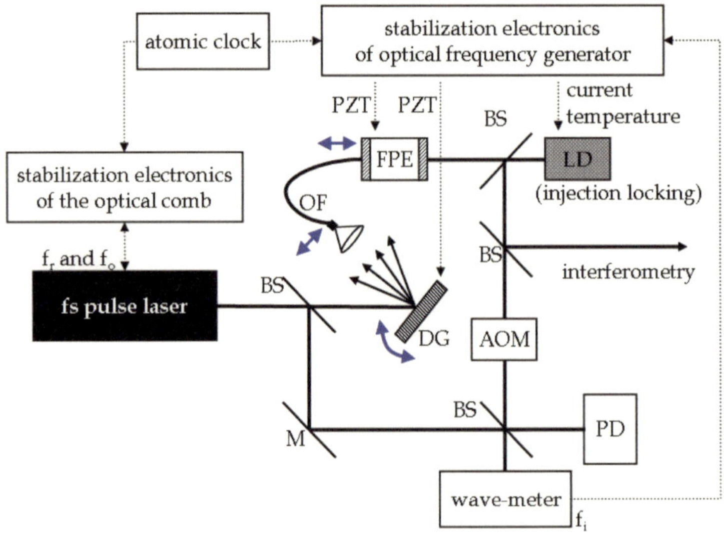

Fig. 2-9. Schematics of optical frequency generator based on direct extraction; DG(diffraction grating), OF(optical fiber), BS(beam splitter), M(mirror), AOM(acoustic-optic modulator), FPE(Fabry-Perot etalon), PD(photo-detector), LD(laser diode)

	optical frequency generator with a working laser	optical frequency generator based on a mode extraction
tuning range	~ 20 nm	10 ~ 15 nm
wavelength stability	$10^{-10} \sim 10^{-11}$	10^{-12} (same as the reference clock)
optical power	20 mW	10 mW
control time	< 1 s	< 1 s

Table 2-1. Summary of the performance of optical frequency generators

Figure 2-10 shows the tuning performance with the step of 5 mode spacing of ~ 400 MHz at a repetition rate of 80 MHz. The angle of diffraction grating was adjusted near 373221494.8 MHz, and then the length of the Fabry-Perot etalon was controlled precisely. The frequency

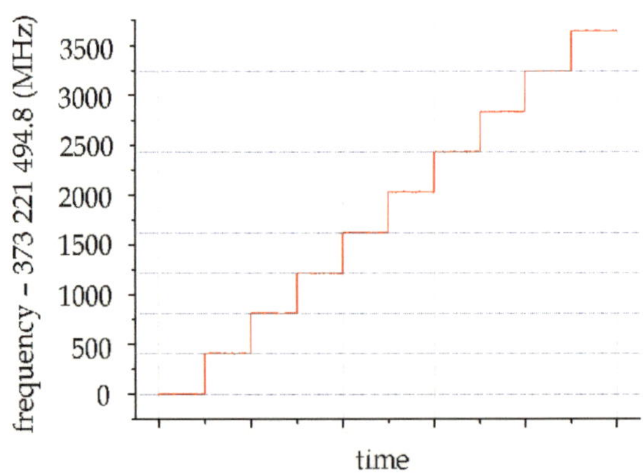

Fig. 2-10. Tuning performance of optical frequency generator based on one-mode extraction

stability of an extracted mode can be achieved 10^{-12} at 10 s, which is the same level as the reference clock. Table 2-1 shows the summary of the performance of optical frequency generators.

2.3 Basic principles of optical interferometer

Optical interference is the superposition of two or more lights that result in a new pattern, which depends on the optical path differences of rays. Since the interference pattern is very sensitive to the optical path difference, the displacement can be determined with sub-wavelength resolution. Figure 2-11(a) shows Michelson's interferometer, the most well-known optical layout, used in precision dimensional metrology. Light emitted from a light source is divided into two paths, reference arm and measurement arm. The reflected lights from both arms recombine and make interference patterns, which is modulated by optical path difference, z. As the interference intensity, I(z), is a sinusoidal signal, it can be simply expressed by

$$I(z) = I_o (1 + \Gamma) \cos (2\pi/\lambda \cdot 2z) \tag{2-5}$$

where I_o is background intensity, Γ is visibility, λ is wavelength, and z is optical path difference. The background intensity is proportional to the intensity of the light source, and visibility can be determined from degree of coherence of a light source. The brightest intensity of the interference pattern can be observed at $z = \lambda/4 \cdot n$ (n=0, 2, 4, 6, ...), and the darkest intensity is at $z = \lambda/4 \cdot n$ (n=1, 3, 5, ...). Because of its periodic pattern, the optical path difference can be determined within less than a half wavelength uniquely. In order to get over the 2π-ambiguity problem, the interference pattern should be counted continuously during the motion of the target mirror. This relative displacement measurement technique is one of the most precise metrological tools, and it can achieve measurement resolution of less than 1 nm with phase detection resolution of 1/1000, which is corresponding to $\lambda/2000$.

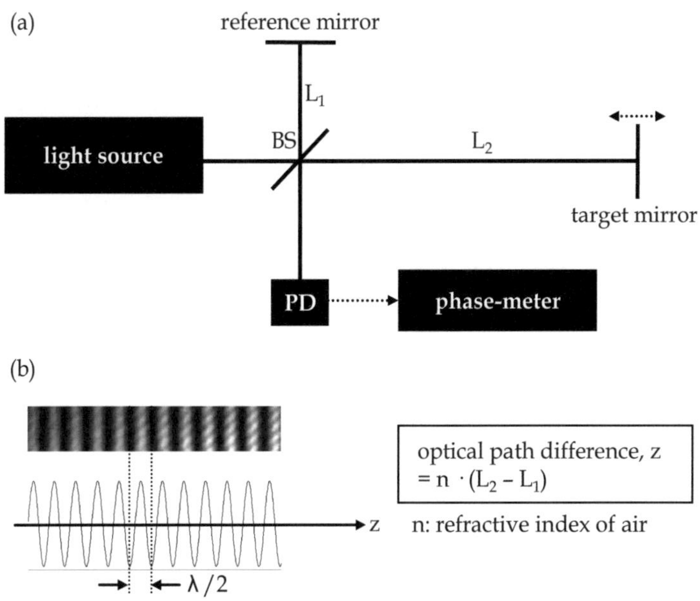

Fig. 2-11. Relative displacement measurement. (a) Michelson's interferometer setup, (b) Interference pattern caused by the optical path difference; BS(beam splitter), PD(photo-detector)

However, it has practical difficulties in the long-range measurement or large step measurement because the target mirror should be installed for translating from the initial position to the final position continuously without any interruption.
Multi-wavelength interferometer can solve this practical difficulty since it provides numbers of phase values obtained from several wavelengths. The distance, L, is given in the form of simultaneous equations as

$$L = \lambda_1/2 \cdot (m_1+f_1) = \lambda_2/2 \cdot (m_2+f_2) = \ldots = \lambda_N/2 \cdot (m_N+f_N) \quad (2\text{-}6)$$

where the subscript N indicates the total number of individual wavelengths in use. As all m_i ($i=1,2,3,\ldots,N$) should be positive integer numbers, a unique solution of L can be determined by solving equation (2-6) numerically in association with a proper estimation for the feasible range of L. In doing that, the required minimum number of wavelengths increases when the extent of the unknown range of L increases. As a general rule, four equations are found to be enough when a good assumption of L is available with an error of less than ± 1 mm.
The combined standard uncertainty of the measured distance, u(L) comes from the uncertainty of air wavelength, u(λ), and uncertainty for phase detection, u(f). The combined standard uncertainty is given as

$$u(L) = [(u(\lambda)/\lambda)^2 L^2 + (\lambda \cdot u(f))^2]^{1/2} \quad (2\text{-}7)$$

The uncertainty of air wavelength, u(λ)/λ, comes from stability of the vacuum wavelength of the light source and uncertainty related to compensation of a refractive index of air. Commercialized laser for metrological use has a stability of vacuum wavelength of more than 10^{-8}. The uncertainty related to refractive index of air can be up to 10^{-7} in the typical environment controlled laboratory by monitoring the environment parameters such as temperature, relative humidity, pressure, etc. Since relative term in equation (2-7) is proportional to the distance, L, thus it can be neglected for short measurement range. For long-range measurement, this term will be dominant. However, for the measurement in the vacuum condition or in space, this term will be disappeared.
The uncertainty of phase detection, u(f), is ~ 1/100 for commercialized phase-meter. The second term in the right side of equation (2-7) is proportional to the wavelength in use. Because the optical wavelength for optical interferometry is several hundreds nm, this term will be several nm. When the shorter wavelength is used or the electronics of phase-meter is improved, this term will be decreased.
Wavelength sweeping interferometer is also one of good candidates of absolute distance measurement, which can determine distance directly without any motions of a target mirror. Instead of moving the target mirror, wavelength of the light source will be swept continuously from λ_1 to λ_2 to obtain distance, L. During wavelength sweeping, changes of the interference pattern, Δm and Δf, are detected according to equation (2-8).

$$L = \lambda_1/2 \cdot (m_1+f_1) = \lambda_2/2 \cdot (m_1+\Delta m+f_1+\Delta f) \quad (2\text{-}8)$$

where Δm, Δf are changes in the integer part and the excess fraction part, respectively. After removing unknowns, m_1 and f_1, length, L, can be expressed as equation (2-9) in terms of a synthetic wavelength, λ_s, simply.

$$L = \lambda_1\lambda_2/2|\lambda_1-\lambda_2| \cdot (\Delta m+\Delta f) = \lambda_s/2 \cdot (\Delta m+\Delta f) \quad (2\text{-}9)$$

Since the synthetic wavelength, λ_s, is much longer than optical wavelengths, λ_1 and λ_2, the phase unambiguity range can be extended. For example, when the wavelengths of λ_1 and λ_2 are 630 nm and 630.01 nm respectively, the synthetic wavelength becomes 39.690630 mm. Wavelength sweeping interferometer has advantages; it does not need multiple stabilized light sources and the initial estimation of L. However, the combined standard uncertainty of wavelength sweeping interferometer is worse than that of the multi-wavelength interferometer because of long synthetic wavelength, λ_s. The combined standard uncertainty can be derived as equation (2-10), which is similar to equation (2-7). The length dependent term of equation (2-10) is related to the stability of synthetic wavelength, which is ~ 10^{-6}. The second term of the right side related to the phase detection will be dominant even if the uncertainty of phase detection, u(f), is extremely small due to the long synthetic wavelength. For example, this term will be approximately 0.4 mm when the synthetic wavelength is 40 mm and the uncertainty of phase detection is 1/100.

$$u(L) = [(u(\lambda_s)/\lambda_s)^2 L^2 + (\lambda_s \cdot u(f))^2]^{1/2} \quad (2\text{-}10)$$

3. Precision dimensional metrology using the optical frequency generators

3.1 Precision length calibration of gauge blocks

Gauge block is one of the widely-used length standards in industry. In order to retain a traceability chain from the definition of metre, gauge blocks should be calibrated based on the well-defined wavelengths. For this task, lots of national metrology institutes have been installed and operated optical interferometer systems for gauge block calibration.

Figure 3-1 shows gauge block calibration system installed in KRISS (Korea research institute of standards and science). The system consists of three major parts; light sources, interferometer part, and environment monitoring part. Light source part contains three different lasers, which are stabilized HeNe laser (633 nm), frequency doubled Nd:YAG laser (532 nm), and Rb-stabilized laser (532 nm). The light in use is selected by mechanical shutters, and then is delivered to the interferometer part via an optical fiber. The delivered light is collimated and divided into two paths; one goes to the reference mirror and the other goes to the gauge block attached to the platen. The reflected lights interfere with each other, and are observed by the CCD camera. In order to compensate the refractive index of air using Edlen's formula, environment parameters such as temperature, pressure, relative humidity and CO_2 contents in the air are monitored. And the interferometer part is isolated by thermal insulators at standard temperature of 20 °C.

Optical frequency generator with a working laser can be a newly developed light source for gauge block calibration. The wavelength generation range of the optical frequency generator is from 765 nm to 781 nm. The coarse tuning using DC-motor attached on the mirror of the extra cavity has the resolution of 0.02 nm with the maximum speed of 8 nm/s. And fine tuning range using PZT attached on the mirror is 0.15 nm (75 GHz) with control bandwidth of 2 kHz. The optical power of the optical frequency generator is about 10 mW, which is sufficient for length metrology. The lind-width of the working laser is 300 kHz at 50ms, which allows the coherence length of more than 1 km. For gauge block calibration the optical frequency generator produces four different wavelengths (776.99983 nm, 777.99925 nm, 779.99727 nm, and 781.00002 nm) instead of using several stabilized light sources such as HeNe laser, Nd:YAG laser, etc. Since this concept gives the traceability to the time standard directly, the direct connection between length and time standards can be realized.

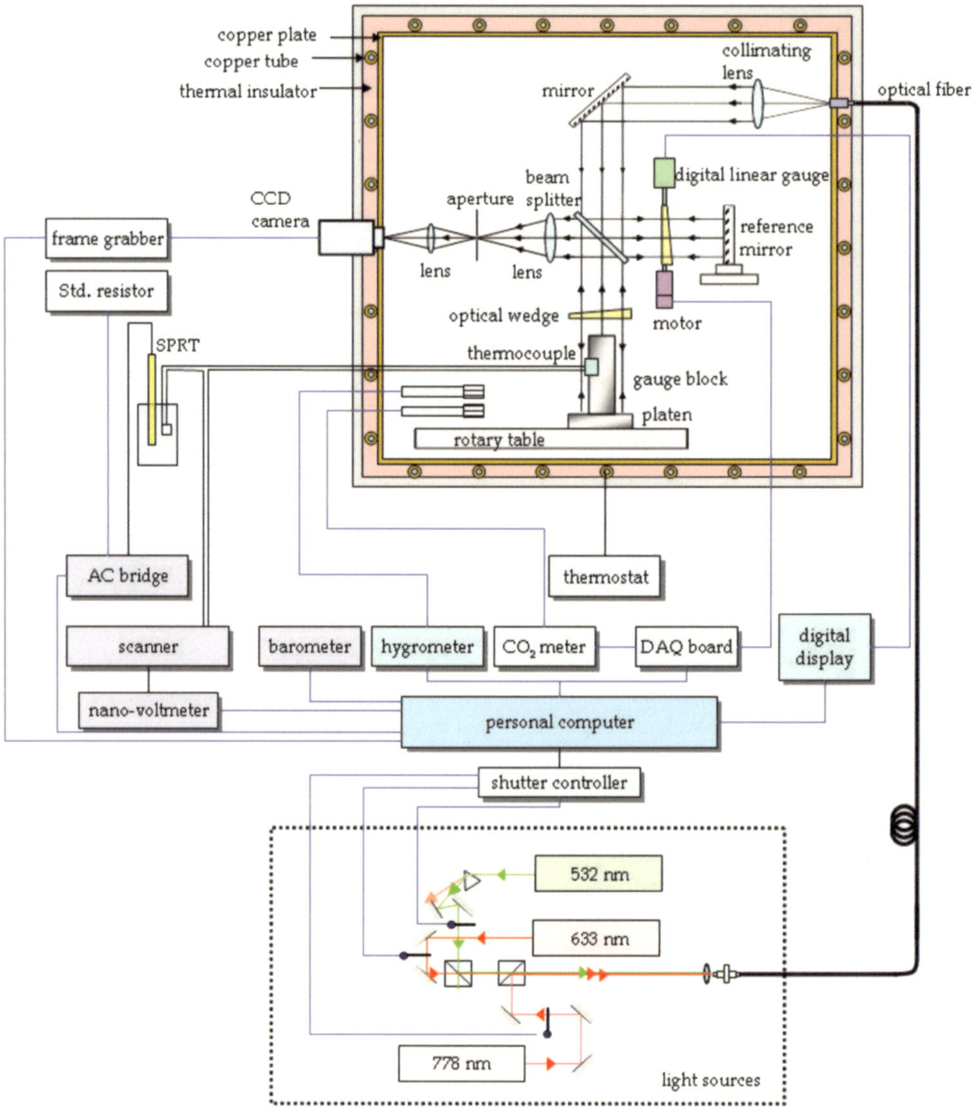

Fig. 3-1. Layout of gauge block calibration system in KRISS

Figure 3-2 shows the optical layout of the gauge block calibration system based on the optical frequency generator. The interferometer setup is basically same as the previous gauge block calibration system in KRISS. The light emitted by optical frequency generator is collimated by a double-let lens with the diameter of 50 mm, and then goes to the plate type beam splitter through a tuning mirror. The separated lights propagate both a gauge block attached on a base plate and reference mirror, and then reflect. The interference pattern can be observed by 2D CCD camera. The environmental parameters (pressure, temperature, relative humidity, and CO_2 contents in the air) are monitored to compensate the refractive index of the air during the measurements.

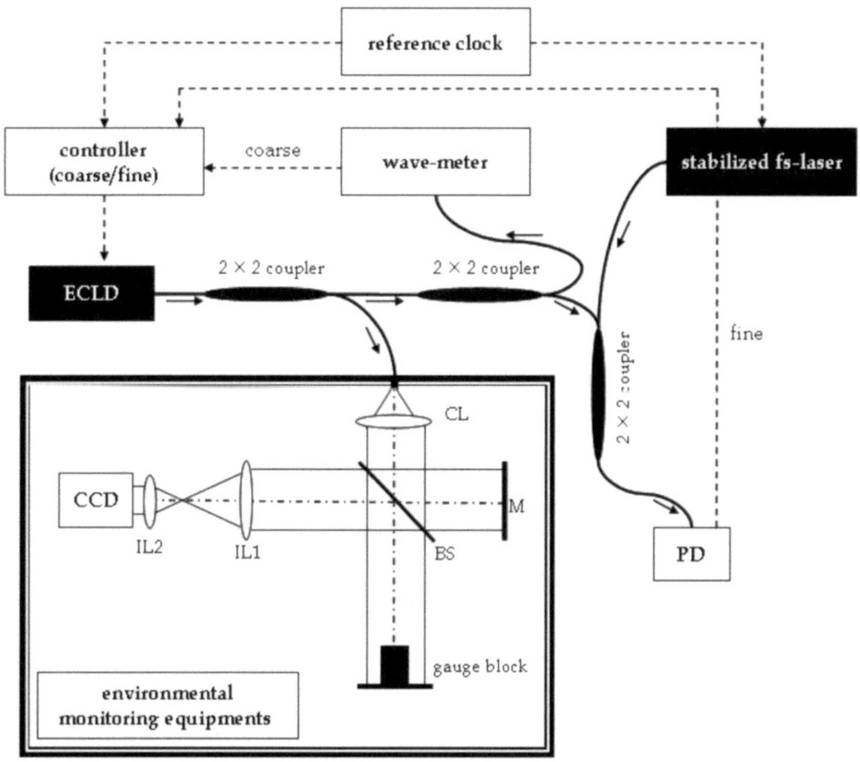

Fig. 3-2. Optical layout of the gauge block calibration system based on the optical frequency generator; CL (collimation lens), M (mirror), BS (beam splitter), IL (imaging lens), PD (photo-detector), ECLD (external cavity laser diode)

The length of the gauge block can be determined by phase difference between top surface of the gauge block and the base plate. Fourier-transform method was adopted for phase determination. The three different lines are chosen; one (G) is at the center of the top surface of the gauge block and the other lines (B_a and B_b) are at the base plate with same offset from the chosen line at the top surface of the gauge block. In order to eliminate the DC component, the slopes of lines are removed by least square method. And then these are Fourier-transformed with Hanning window after zero-padding to improve the resolution in the spectral domain. The initial phase of each line can be extracted at the peak of the amplitude in the spectral domain. The phase of the base plate can be determined by averaging the phase values of B_a and B_b to eliminate the tilting effect of the gauge block and the base plate. Excess fraction parts in equation (2-6) for each wavelength can be given by phase difference between the phase at the center of top surface of the gauge block (Φ_G) and the average phase of the base plate. Figure 3-3 shows the overall calculation procedure of the excess fraction part based on Fourier-transform method.
The measurement result of 25 mm gauge block is summarized in Table 3-1. The excess fraction parts were obtained by Fourier-transform method from interference patterns as described before. To cancel out the random vibration effect, 20 interference patterns were

captured repeatedly at each wavelength. The standard deviation of the excess fraction was less than 5/100. The mean value of the calculated length of the gauge block was 24.999890 mm when the refractive index of air and temperature of gauge block were corrected.

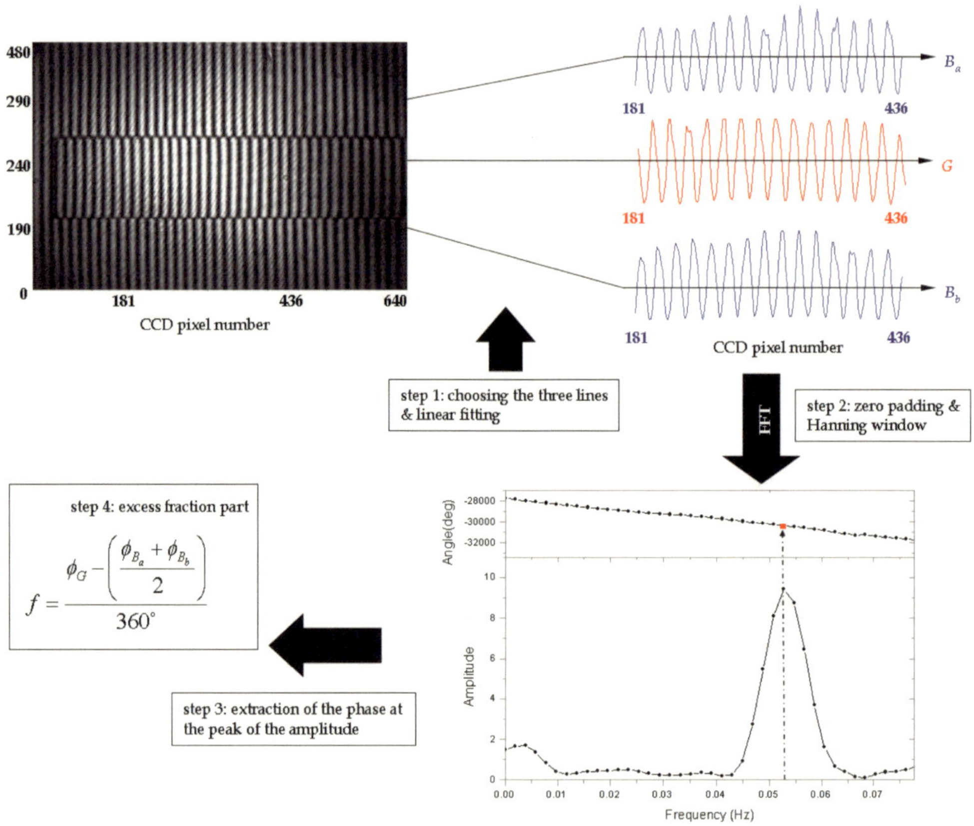

Fig. 3-3. Calculation procedure of the excess fraction part based on Fourier-transform method

State	Measured and computed data 1	2	3	4
Vacuum wavelength (nm)	776.99983	777.99925	779.99727	781.00002
Excess fraction, f	0.9483	0.2591	0.5773	0.2430
Deviation of f	0.0010	0.0014	0.0010	0.0047
Integer part, m	64366	64284	64119	64037
Absolute distance, L (nm)	24999890.2	24999888.9	24999890.2	24999888.7
Stability of measured excess fraction		0.0035		
Theraml expansion coefficient of gauge block		8.4×10^{-6} /K		
Refractive index of air		1.00027050		

Absolute distance, L, finally determined by averaging the computed values: 24.999890 mm
(The refractive index of air and temperature of gauge block are corrected.)

Table 3-1. Measurement result of 25 mm gauge block.

The length of gauge blocks based on the optical interferometry, L, can be expressed as

$$\begin{aligned} L &= \mathrm{mean}[\lambda_i/2\cdot(m_i+f_i)] - \mathrm{mean}[L_0\cdot(n_i-1)] - L_0\alpha(T_{GB}-20) + L_\Phi + L_W + L_E + L_G \\ &= L_{fit} + L_n + L_T + L_\Phi + L_W + L_E + L_G \end{aligned} \quad (3\text{-}1)$$

where, i is integer number (i=1, 2, 3, … N), N indicates the total number of individual wavelengths in use, mean[A_i] is the mean value of A_i, L_0 is nominal value of the gauge block, n is refractive index of air, α is thermal expansion coefficient of the gauge block, T_{GB} is temperature of the gauge block, L_{fit} is the measured length based on the optical interferometry, L_n, L_T, L_Φ, L_W, L_E, and L_G are the lengths corresponded to the compensation of refractive index of air, correction of thermal expansion to the standard temperature of 20 °C, phase compensation for surface roughness of the gauge block and the base plate, wringing process, optical components errors, and geometrical errors such as parallelism, flatness, respectively.

The uncertainty of determination of the excess fractions at four difference wavelengths (777 nm, 778 nm, 780 nm, and 781 nm) is 0.02, which is caused by repeatability of phase detection and stability of the measured excess fraction, 0.0035. And the uncertainty of the wavelengths in use is 1.9×10^{-10}, which comes from the frequency stabilities of the repetition rate, carrier-offset frequency, and beat note between the optical comb and a working laser. Therefore, the uncertainty of the gauge block calibration interferometer becomes as

$$u(L_{fit}) = [\,(7.8\ \mathrm{nm})^2 + (1.9 \times 10^{-10}\cdot L_0)^2\,]^{1/2} \quad (3\text{-}2)$$

The air wavelength can be obtained by dividing the vacuum wavelength into the refractive index of air. The refractive index of air can be determined by updated Edlen's formula with the environmental parameters of temperature, pressure, relative humidity, and CO_2 contents of air. The uncertainty of measuring environmental parameters comes from equipment measuring accuracy and deviation of parameters during the experiments. The uncertainty of temperature is 5.2 mK when the deviation of temperature is 1.5 mK with measurement uncertainty of thermometer, 5 mK (k=1). The uncertainty of relative humidity is given as 1.0 % R.H. at 20 % R.H. caused by equipment accuracy of 1.0 % R.H. and deviation of 0.01 % R.H. The uncertainty of pressure is 3.0 Pa at 1 atm when the deviation is 0.45 Pa with measurement accuracy of a manometer, ±0.003 %. The uncertainty of CO_2 contents in air is estimated 10.0 ppm at 450 ppm caused by equipment accuracy of ±1 % and deviation of 5.5 ppm. The uncertainty of update Edlen formula itself is 10^{-8}. These uncertainty factors give the uncertainty of air wavelength as

$$u(L_n) = 1.4 \times 10^{-8}\cdot L_0 \quad (3\text{-}3)$$

Since the length of gauge block is defined at the temperature of 20 °C, the thermal expansion of the gauge block caused by temperature difference should be compensated. The material of gauge block and base plate is chrome carbide, which has the thermal expansion coefficient of 8.4×10^{-6} /K. The uncertainty of thermal expansion coefficient is $1.0/\sqrt{3} \times 10^{-6}$ /K, and the uncertainty of the temperature of the gauge block is 15 mK. The uncertainty of the compensation of thermal expansion could be expressed as

$$u(L_T) = 1.3 \times 10^{-7} \cdot L_0 \qquad (3\text{-}4)$$

The uncertainty of phase compensation for surface roughness of the gauge block and the base plate comes from the optical constant differences caused by different materials and surface roughness of gauge block and base plate. The term related to optical constant differences can be ignored by adopting the same material for the gauge block and the base plate. The uncertainty of surface roughness is estimated 5 nm by mechanical or optical profilers.
The uncertainty of wringing effect is 6.9 nm, which can be determined by wringing the same gauge block on the base plate repeatedly. The uncertainty of optical components can be estimated by wave-front errors of each components, λ/10~ λ/20. However, it's very difficult to extract optical system error from the each, which can be determined by measuring amount of bending of interference patterns generated by the installed Twyman-Green interferometer with two flat mirrors. It is estimated as λ/15 within 2/5 of full measurement area, which is only used for this calibration task. Therefore, the uncertainty of optical

Uncertainty sources	Contributions	Numerical values for L_0=25 mm
Gauge block interferometer	$[(7.8\ \text{nm})^2 + (1.9\times10^{-10}\cdot L_0)^2]^{1/2}$	7.8 nm
Excess fraction	0.02	
Wavelength in use	1.9×10^{-10}	
Repetition rate	1.3×10^{-12}	
Carrier-offset frequency	8.1×10^{-13}	
Beat note	1.9×10^{-10}	
Estimating the refractive index of air	$1.4\times10^{-8}\ L_0$	0.4 nm
Air temperature	5.2 mK	
Air pressure	3 Pa	
Air humidity	0.6 % R.H.	
CO_2 content in air	6.0 ppm	
Edlen's formula	10^{-8}	
Thermal expansion of gauge block	$1.3\times10^{-7}\ L_0$	3.3 nm
Thermal expansion coefficient	5.8×10^{-7} /°C	
Surface temperature of gauge block	15 mK	
Wave-front error of the optical components of the gauge block interferometer	12 nm	12 nm
Wringing of gauge block	6.9 nm	6.9 nm
Surface roughness of gauge block	5.0 nm	5.0 nm

* L_0 denotes the nominal length of a gauge block whose length is given in meter.

Table 3-2. Uncertainty evaluation of absolute calibration of gauge blocks.

components becomes 12 nm. The uncertainty of geometrical errors is 0.6 nm, which can be given according to ISO defined grade of gauge blocks.
Table 3-2 shows the summary of the uncertainty evaluation of the absolute calibration of gauge blocks. The combined standard uncertainty is 17 nm (k=1) for the 25 mm gauge block.
This research was a meaningful work because it was a first realization of length and time connection. Even if the standard was well established in national metrological institutes, the traceability chain should be coupled well for maintaining the nation's industrial qualify. Therefore, the calibration of gauge block is an important task to keep the traceability chain from the standard because the gauge block is one of the practical length standards, and widely used in the industry. As the time standard has the higher stability and uncertainty, the dimensional metrology will have better uncertainty by linking length to time standards.

3.2 Absolute distance measurement

Absolute distance measurement can determine the length with a single operation based on optical interferometry. Based on the two absolute distance measurement techniques described in section 2.3, multi-wavelength interferometer and wavelength sweeping interferometer, the distance can be obtained with the optical frequency generator; Four different wavelengths, λ_1, λ_2 , λ_3 and λ_4 ,are chosen as 780.206901 nm, 780. 203668 nm, 779.953524 nm and 770.204349 nm, respectively for multi-wavelength interferometer principle, and the wavelength between λ_1 and λ_2 is scanned continuously for wavelength sweeping interferometer as shown in figure 3-4. During the wavelength scanning for rough estimation of the distance, the change of phase in terms of the integer part and the excess fraction part in equation (2-9), Δm and Δf, are 13 and -0.06940, respectively. That leads to an estimation of the distance as 1195.205502 nm along with a synthetic wavelength, λ_s, of 184.853 mm. The optical heterodyne technique is adopted for real-time phase detection, and it is realized that the optical frequency is slightly shifted by the two acousto-optic modulators, AOM1 and AOM2, in series as shown in figure 3-5. The avalanche photo-detector, APD1, monitors the reference signal as $A \cdot \cos[2\pi\Delta f \cdot t]$, and the other avalanche photo-detector, APD2, obtains the measurement signal as $B \cdot \cos[2\pi\Delta f \cdot t+\Phi]$, where A and B denote amplitude, Δf is frequency difference caused by two AOMs, phase shift,Φ, represents $4\pi f \cdot L$. The commercialized phase-meter can detect the phase shift, Φ, with the bandwidth of more than 1 kHz. From the value of phase shift, the distance, L can be extracted with the uncertainty of phase-meter accuracy, $0.0016 \cdot \lambda_s/2$. And the uncertainty of the measured excess fractions is $0.0012 \cdot \lambda_s/2$. Therefore, the uncertainty of the phase detection is estimated as 258.8 μm.
The uncertainty of the synthetic wavelength comes from the uncertainty of wavelength in use. It is evaluated as $2.0 \times 10^{-6} \cdot L$, which leads 2.4 μm at the distance of 1.195 m. And the uncertainty for refractive index of air is only 17 nm, which can be determined by $1.4 \times 10^{-8} \cdot L$ at L=1.2 m. Therefore, the estimation distance of 1195.205502 nm has the measurement uncertainty of 258.8 μm, which can be calculated according to the equation (2-10). The uncertainty of phase detection is the most contributed uncertainty component in this measurement. It is important note that the estimation value is a good assumption for multi-wavelength interferometer.

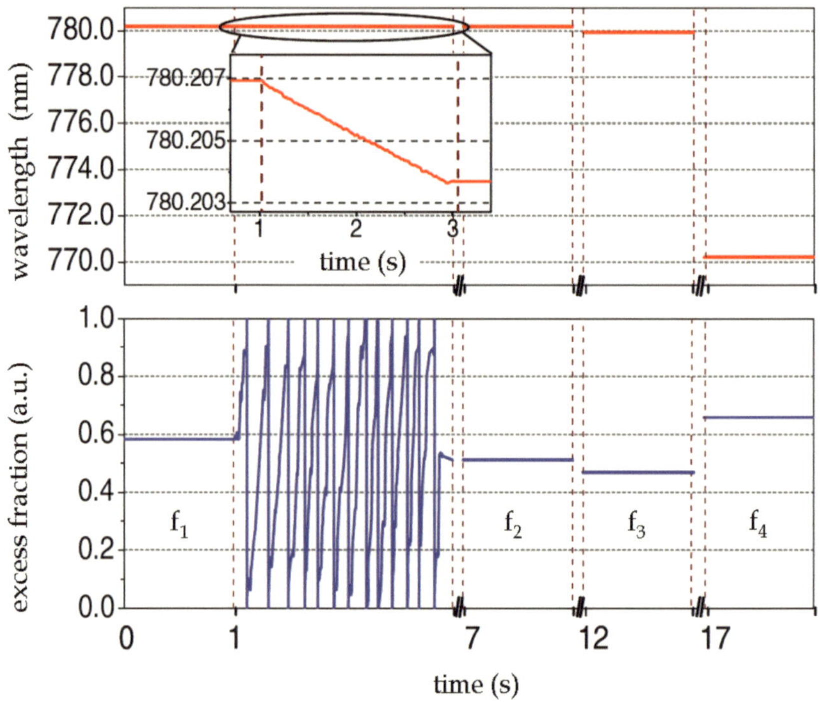

Fig. 3-4. Wavelength generation profile and the changes of phase in terms of excess fraction for the absolute distance measurement.

State	Measured and computed data 1	2	3	4
Vacuum wavelength (nm)	780.206961	780.203668	779.953524	770.204349
Refractive index of air	1.000263346	1.000263345	1.000263347	1.000263410
Air wavelength (nm)	780.001551	779.998260	779.748180	770.001523
Excess fraction, f	0.5817	0.5123	0.4688	0.6578
Deviation of f	0.0007	0.0002	0.0011	0.0005
Integer part, m	3064834	3064847	3065830	3104637
Absolute distance, L (nm)	1195.287863	1195.287863	1195.287865	1195.287862
Initial estimation of L by means of sweeping wavelength from state 1 to 2: 1195.205502 mm				
Absolute distance, L, finally determined by averaging the computed values: 1195.287863 mm				

Table 3-3. summary of experimental data for the absolute distance of ~ 1.2 m

The uncertainty of the optical frequency generator comes from the uncertainties of repetition rate, carrier-offset frequency, and beat note as mentioned in the section 3.1. However, the uncertainty is improved by the modified electric control loop with high-performance filters. It is estimated as $5.9 \times 10^{-12} \cdot L$, which is 0.01 nm at L=1.195 m. The uncertainty of refractive

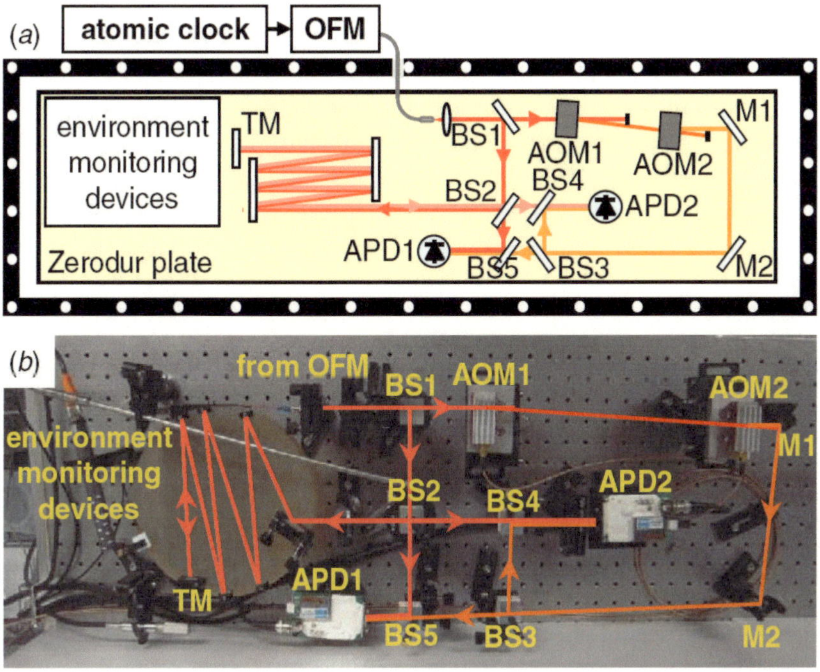

Fig. 3-5. Interferometer setup for absolute distance measurement. (a) schematic of the interferometer, (b) photo of the interferometer; AOM(acousto-optic modulator), APD(avalanche photo-detector), BS(beam splitter), M(mirror), TM(termination mirror)

index of air is same as the previous task, which is given as 17.1 nm at L=1.195 m. The uncertainty of the multi-wavelength interferometer comes from the uncertainties of phase-meter accuracy, measured excess fractions, and the exact fraction method. It can be estimated as 0.0061 $\cdot \lambda/2$, which is 2.3 nm. The combined standard uncertainty can be expressed as

$$u(L) = [\ (2.3\ \text{nm})^2 + (1.4 \times 10^{-8} \cdot L)^2\ ]^{1/2} \qquad (3\text{-}5)$$

At the measured distance of ~1.195 m, the combined standard uncertainty is 17.1 nm (*k*=1). Table 3-4 shows the uncertainty evaluation of absolute length measurement based on the optical frequency generator.

This absolute distance measurement has various applications; one is for precision positioning of space fleet. Several satellites make specific arrangements for obtaining large aperture, which gives high-quality images of the universe. The missions are for finding earth-like planets, discovering the fundamentals of the creation, and detecting the gravitational waves. Since this operation will be done in space, the uncertainty of the refractive index of air, the most dominant uncertainty, can be eliminated. The combined standard uncertainty for the precision positioning in space is proportional to the measuring distance, L, simply. For example, when the distance between satellites is 1 km, the combined standard uncertainty is estimated only as 6.4 nm according to the table 3-4. The other is for positioning and control of ultra-precision stages in the field of semi-conductor and flat-panel display. The travel of these stages is several meters in the well-stabilized environmental conditions. Generally the relative displacement measurement technique has been used for these applications, but it has drawbacks; since it can not give the absolute values, the stage should be moved from a reference point continuously to obtain the absolute coordinates. Because it is occurred some drifts due to the environment variation and mis-reading of interference signals, homing (setting a reference point) should be done periodically. Absolute distance measurement will be a good alternative measurement solution for these applications.

Uncertainty sources	Contributions	Numerical values for L=~1.195 m
Optical frequency generator	5.9×10^{-12} L	0.01 nm
Repetition rate	1.3×10^{-12}	
Carrier-offset frequency	2.2×10^{-12}	
Beat note	5.8×10^{-12}	
Estimating the refractive index of air	1.4×10^{-8} L	17.1 nm
Air temperature	5.2 mK	
Air pressure	3 Pa	
Air humidity	0.6 % R.H.	
CO_2 content in air	6.0 ppm	
Edlen's formula	10^{-8}	
Multi-wavelength interferometer	$0.0061 \times \lambda/2$	2.3 nm
Phase-meter accuracy	$0.0016 \times \lambda/2$	
Measured excess fractions	$0.0012 \times \lambda/2$	
Multi-wavelength algorithm	$0.0058 \times \lambda/2$	
Combined standard uncertainty (k=1)	$[(2.3\ \text{nm})^2+(1.4 \times 10^{-8}\ \text{L})^2]^{1/2}$	17.1 nm

Table 3-4. Unceratainty evaluation of absolute distance measurement

4. Summary and future works

Dimensional metrology has a long history from the ancient times. The long sticks or bars have been used with precision scales on that for measuring length. In the 18th century the light emitted from lamps was suggested and realized as a tool for precision dimensional metrology by using interference phenomenon. In 1960s the advent of laser opened a new way for precision length and profile measurement techniques, which leads today's optical interferometry using long coherence length. Recently the fs pulse laser has been in spotlight because of its short pulse duration and the wide spectral bandwidth, which allows precision optical frequency metrology and spectroscopy with the traceability to the frequency standard.

In this chapter, the fs pulse laser was adopted as a light source for precision dimensional metrology. The optical comb of the fs pulase laser was acted as scales in the optical frequency regime. Based on that, the two types of optical frequency generators, which could produce a desired single wavelength with high stability, were suggested and realized. One was constructed with a working laser, which could tune the wavelength within the range of 20 nm. Due to low control rate caused by mechanical resonance, it had degradation in terms of wavelength stability and line-width. The other was based on direct extraction of a single mode from the optical comb with the help of optical filtering components, a diffraction grating-optical fiber pairs and scanning Fabry-Perot etalon. However, the extracted mode should be amplified for dimensional measurement because it was too weak to use. The injection locking technique was adopted for this amplification process. This type of the optical frequency generator could maintain the wavelength stability, which was same as the reference clock. If the several diode lasers with different emission wavelengths will be locked to the optical comb at the same time, the distance can be measured in real-time by shortening the wavelength tuning procedure.

Absolute distance measurement can determine the distance with a single operation without 2π ambiguity problem. It could be realized by using the optical frequency generator, which could produce individual wavelengths and scan the wavelength smoothly. By adopting the optical frequency generator as a light source the absolute length calibration of gauge blocks could be realized with the combined standard uncertainty of 17 nm for 25 mm gauge blocks. Four individual wavelengths were chosen for multi-wavelength interferometer, the phases at each wavelength were obtained by Fourier-transform method. This standard work gave the direct coupling between length and time because the optical comb of a fs pulse laser was stabilized to the atomic reference clock, frequency standard. Now KRISS is expanding this scheme to the long range gauge block of more than 1 m.

The absolute distance of 1.195 m was measured using the optical frequency generator. To obtain the estimated value of the distance, wavelength sweeping interferometer principle was applied. With the estimated value, the absolute distance was determined by measured phases at four different wavelengths. The phases were obtained by optical heterodyne technique in real-time. The combined standard uncertainty of the measured distance of 1.195 m was 17.1 nm. In this case, the combined standard uncertainty will be decreased dramatically in the vacuum condition because the most dominant uncertainty was uncertainty of refractive index of air. That means it will a good solution for satellite fleets for space missions or precision stages operating in the vacuum. And a fs pulse laser based on optical fiber allows more simple and reliable measurements in industry because of its easy handling and low cost.

5. Acknowledgements

These research works were accomplished at KAIST (principal investigator: Prof. Seung-Woo Kim, swk@kaist.ac.kr). Now Dr. Jonghan Jin is working as a senior researcher at Centre for Length and Time, Division of Physical Metrology, Korea Research Institute of Standards and Science (KRISS), and the present e-mail address is jonghan@kriss.re.kr.

6. References

Jin, J.; Kim, Y. -J.; Kim, Y.; Kim, S. –W. & Kang, C. –S. (2006) Absolute length calibration of gauge blocks using the optical comb of a femtosecond pulse laser, *Optics Express*, Vol. 14, No. 13, pp. 5568-5974, ISSN 1094-4087

Jin, J.; Kim, Y. -J.; Kim, Y. & Kim, S. –W. (2007) Absolute distance measurement using the optical comb of a femtosecond pulse laser, *International Journal of Precision Engineering and Manufacturing*, Vol. 8, No. 7, pp. 22-26, ISSN 1229-8557

Kim, Y. -J.; Jin, J.; Kim, Y. ; Hyun, S. & Kim, S. –W. (2008) A wide-range optical frequency generator based on the frequency comb of a femtosecond laser, *Optics Express*, Vol. 16, No. 1, pp. 258-264, ISSN 1094-4087

Hyun, S.; Kim, Y. -J.; Kim, Y. ; Jin, J. & Kim, S. –W. (2009) Absolute length measurement with the frequency comb of a femtosecond laser, *Measurement Science and Technology*, Vol. 20, pp. 095302-1-6, ISSN 0957-0233

Minoshima, K. & Matsumoto, H. (2000) High-accuracy measurement of 240-m distance in an optical tunnel by use of a compact femtosecond laser, *Applied Optics*, Vol. 39, No. 30, pp. 5512-5517, ISSN 0003-6935

Yamaoka, Y.; Minoshima, K. & Matsumoto, H. (2002) Direct measurement of the group refractive index of air with interferometry between adjacent femtosecond pulses, *Applied Optics*, Vol. 41, No. 21, pp. 4318-4324, ISSN 0003-6935

Bitou, Y.; Schibli, T. R.; Minoshima, K. (2006) Accurate wide-range displacement measurement using tunable diode laser and optical frequency comb generator, *Optics Express*, Vol. 14, No. 2, pp. 644-654, ISSN 1094-4087

Schibli, T. R.; Minoshima, K.; Bitou, Y.; Hong, F. –L.; Bitou, Y.; Onae, A. & Matsumoto, H. (2006) Displacement metrology with sub-pm resolution in air based on a fs-comb wavelength synthesizer, *Optics Express*, Vol. 14, No. 13, pp.5984-5993, ISSN 1094-4087

Bitou, Y. & Seta K. (2000) Gauge block measurement using a wavelength scanning interferometer, *Japanese Journal of Applied Physics*, Vol. 39, pp. 6084-6088, ISSN 0021-4922

Schibli, T. R.; Minoshima, K.; Hong, F. –L.; Inaba, H.; Onae, A.; Matsumoto, H.; Hartl, I.; Fermann, M. E. (2004) Frequency metrology with a turnkey all-fiber system, *Optics Letters*, Vol. 29, No. 21, pp.2267-2469, ISSN 0146-9592

Lay, O. P.; Dubovitsky, S.; Peters, R. D. & Burger, J. P (2003) MSTAR: a submicrometer absolute metrology system, *Optics Letter*, Vol. 28, pp. 890-892, ISSN 0146-9592

Walsh, C. J. (1987) Measurement of absolute distances to 25 m by multiwavelength CO_2 laser interferometry, *Applied Optics*, Vol. 26, No. 9, pp. 1680-1687, ISSN 0003-6935

Bien, F.; Camac, M,; Caulfield, H. J. & Ezekiel, S. (1981) Absolute distance measurements by variable wavelength interferometry, *Applied Optics*, Vol. 20, No. 3, pp. 400-403, ISSN 0003-6935

Dändliker, R.; Thalmann, R. & Rregué (1988) Two-wavelength laser interferometry using super heterodyne detection, *Optics Letters*, Vol. 13, No. 5, pp. 339-341, ISSN 0146-9592

Kubota, T.; Nara, M. & Yoshino, T. (1987) Interferometer for measuring displacement and distance, *Optics Letters*, Vol. 12, No. 5, pp. 310-312, ISSN 0146-9592

Jost, J. D.; Hall, J. L. & Ye, J. (2002) Continuously tunable, precise, single frequency optical signal generator, *Optics Express*, Vol. 10, pp. 515-520, ISSN 1094-4087

Birch, K. P. & Downs, M. J. (1993) An updated Edlen equation for the refractive index of air, *Metrologia*, Vol. 34, pp. 479-493, ISSN 0026-1394

Schuhler, N.; Salvadé, Y.; Lévêque, S.; Dändliker, R & Holzwarth, R. (2006) Frequency-comb-referenced two-wavelength source for absolute distance measurement, *Optics Letters*, Vol. 31, No. 21, pp. 3101-3103, ISSN 0146-9592

Slavadé, Y.; Schuhler, N.; Lévêque, S. & Floch, S. L. (2008) High-accuracy absolute distance measurement using frequency comb referenced multiwavelength source, *Applied Optics*, Vol. 47, No. 14, pp. 2715-2720, ISSN 0003-6935

Coddington, I; Swann, W. C.; Nenadovic, L. & Newbury, N. R. (2009) Rapid and precise absolute distance measurement at long range, *Nature Photonics*, Vol. 3, pp. 351-356, ISSN 1749-4885

Jin, J.; Kim, Y. -J.; Kim, Y. ; Hyun, S. & Kim, S. –W. (2008) Absolute distance measurement using the frequency comb of a femtosecond pulse laser, *Proceedings of the European Society of Precision Engineering and Nanotechnology (EUSPEN) International conference*, O7.2, Zurich, 05/2008, EUSPEN, Cranfield

Jin, J.; Kim, Y. -J.; Kim, Y. ; Hyun, S. & Kim, S. –W. (2007) Precision length metrology using and optical frequency generator, *Proceedings of the Asian Society of Precision Engineering and Nanotechnology (ASPEN) 2007*, pp. 39-41, Kwangju, 11/2007, KSPE, Seoul (invited)

Jin, J.; Kim, Y. -J.; Kim, Y. ; Hyun, S. & Kim, S. –W. (2007) Precision length metrology using and optical frequency synthesizer, *Proceedings of the Conference on Laser and Electro-Optics-Pacific Rim (CLEO-PR)*, pp. 1443-1444, Seoul, 08/2007, OSA, Seoul (invited)

Jin, J.; Kim, Y. -J.; Kim, Y. & Kim, S. –W. (2006) Absolute length metrology using a femtosecond pulse laser, *Proceedings of the 2nd International Conference on Positioning Technology*, pp. 119-121, Daejeon, 10/2006, KSPE, Seoul

Jin, J.; Kim, Y. -J.; Kim, Y.; Kim, S. –W. & Kang, C. -S (2006) Absolute length calibration of gauge blocks using optical comb of a femtosecond pulse laser, *Proceedings of SPIE Optics & Photonics*, pp. 6292O-1, San-diego, 08/2006, SPIE, Bellingham

Jin, J.; Kim, Y. -J.; Kim, Y.; Kim, S. –W. & Kang, C. -S (2006) Absolute length calibration of gauge blocks using optical comb of a femtosecond pulse laser, *Proceedings of the European Society of Precision Engineering and Nanotechnology (EUSPEN) International conference*, pp. 410-414, Baden, 05/2006, EUSPEN, Cranfield

Kim, S. –W.; Oh, J. S.; Jin, J.; Joo, K. N. & Kim, Y. -J. (2005) New precision dimensional metrology using femtosecond pulse lasers, *Proceedings of the European Society of Precision Engineering and Nanotechnology (EUSPEN) International conference*, pp. 135-138, Montpellier, 05/2005, EUSPEN, Cranfield

Kim, S. –W.; Joo, K. N.; Jin, J. & Kim, Y. -J. (2005) Absolute distance measurement using femtosecond laser, *Proceedings of SPIE*, pp. 58580N-1-8, Munich, 06/2005, SPIE, Bellingham

Ye, J. & Cundiff, S. T. (2005). *Femtosecond optical frequency comb: Principle, operation, and applications,* Springer, ISBN 0-387-23790-9, New York

Rulliére, C. (1998). *Femtosecond laser pulses; Principles and experiments,* Springer, ISBN 3-540-63663-3, Berlin Heidelberg

10

Micro-Solid-State Laser for Ignition of Automobile Engines

Masaki Tsunekane, Takayuki Inohara, Kenji Kanehara and Takunori Taira

Institute for Molecular Science, Nippon Soken, Inc.

Japan

1. Introduction

Recently in consideration of the problem of protecting the global environment and preserving fossil resources, the research and development of new clean vehicles driven by clean energy sources, such as electricity, fuel cell, etc., has been progressing worldwide. However it is difficult to replace all conventional gasoline vehicles to clean vehicles immediately, because they still have several hurdles to get over, costs of the clean vehicles and the energy sources, range between refuelling, the availability of refuelling or recharging stations, vehicle performance, fuel cell lifetime, etc. Therefore the improvement of the efficiency of conventional internal combustion gasoline engines, and the reductions of CO_2 and harmful pollutant emissions have become more important today.

A laser has been discussed widely as one of the promising alternatives for an ignition source of the next generation of efficient internal combustion engines (Hickling & Smith, 1974; Dale et al., 1997; Phuoc, 2006). Laser ignition can change the concept of ignition innovatively and has many advantages over conventional electric spark plug ignition. Figure 1 shows the schematics of combustion engines ignited by (a) an electric spark plug and a laser (b), (c).

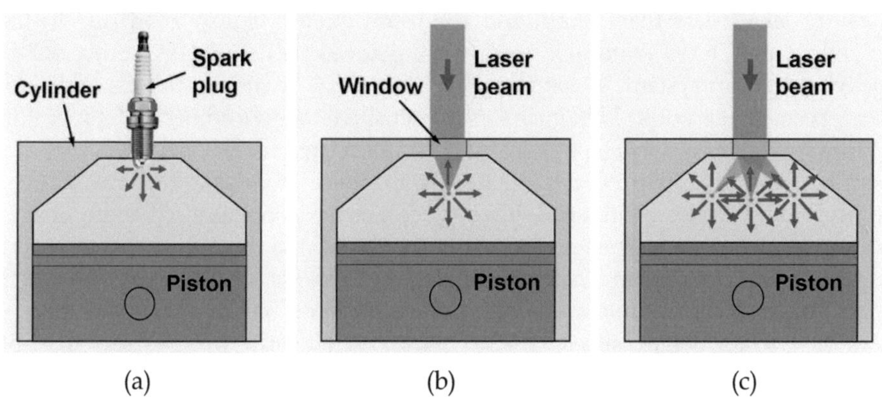

Fig. 1. Schematics of the combustion engines ignited by (a) a spark plug and (b), (c) a laser. (c) shows multipoint ignition.

Using a laser, the ignition plasma may be located anywhere within the combustion chamber because laser ignition doesn't need electrodes. Optimal positioning of ignition apart from the cold cylinder wall allows the combustion flame front to expand rapidly and uniformly in the chamber and thus increases the efficiency as seen in (b). In addition, laser ignition has great potential for simultaneous, spatial multipoint ignition within a chamber as shown in (c). This shortens combustion time dramatically and improves the output and efficiency of engines effectively (Phuoc, 2000; Morsy et al., 2001). Further a laser can ignite leaner or high-pressure mixtures that are difficult to be ignited by a conventional electric spark plug (Weinrotter et al., 2005). A laser igniter is also expected to have a longer lifetime than a spark plug due to the absence of electrodes.

One of the major difficulties of laser ignition for actual applications, especially for automobiles is the dimension of the lasers. For breakdown in fuel mixtures, light intensities in the order of 100GW/cm^2 are necessary at the focal point of ignition. Then lasers with pulse energies higher than 10mJ, beam quality factors, M^2 of lower than 3 and pulse durations of shorter than 10ns have been used for combustion experiments. But the commercially available laser heads have table top size due to the complexities of the laser cavity and the cooling system.

In reductions of system size and costs, a multiplexing fiber optics delivery system seems to be ideal and practical for laser ignition of multicylinder engines. But it is difficult to deliver ignition light through fibers to each cylinder of an engine directly, because the optical damage threshold of fibers is still several orders of magnitude less than the peak energy levels required by laser ignition at present (Joshi et al., 2007). The fundamental problem of a fiber is attributed to the need to deliver relatively high power pulses with sufficient beam quality to focus the output light to the intensity required for breakdown.

2. Characteristics of passively Q-switched laser

A passively Q-switched solid-state laser, especially a Nd:YAG/Cr^{4+}:YAG laser end-pumped by a fiber-coupled laser diode (LD) has been proposed as a promising ignition laser recently (Kofler et al., 2007). It has a simple structure, only two functional optical elements and no external power for optical switching is necessary hence the dimension of the laser head can be reduced. In addition, a short pulse operation less than 1ns is easily obtained by reduction of the cavity length less than 10mm and the beam quality is also good due to the soft aperture effect of a Cr:YAG saturable absorber (Zaykowski & Dill III, 1994; Sakai, 2008). The fiber delivered pump system allows not only further size reduction but also reliable laser operation, because the pump LD which is very sensitive to environmental temperature can be positioned at relatively stable place inside a car apart from the hot engine.

In generally, passively Q-switched lasers have large pulse-to-pulse energy fluctuations and large timing jitters under continuous-wave (CW) pumped operations (Huang et al., 1999; Tang et al., 2003) due to thermal and mechanical instabilities. Such fluctuations and jitters have strongly restricted the applications of passively Q-switched lasers. On the other hand, operation frequency of igniters of internal combustion engines is less than 60Hz, corresponding to an engine speed of 7200 rpm, and the duty cycle is less than 5% for automobiles. In such a low frequency, quasi-continuous-wave (QCW) pumped operations with a low duty cycle, passively Q-switched lasers are expected to operate stably due to initialization of the thermal and mechanical conditions during pulses.

The characteristics of passively Q-switched lasers have been analyzed in detail and various optimum design criteria have been presented (Szabo & Stein, 1965; Degnan, 1995; Xiao & Bass, 1997; Zhang et al., 1997; Chen et al., 2001; Pavel, 2001; Patel & Beach, 2001). But there are still several discrepancies in the theoretical calculations and the experimental results especially for output energy and efficiency. We think that the main cause is uncertainness of size of the laser mode. It is not easy to estimate the actual laser mode size and the beam quality accurately, because the aperture formed in the saturable absorber of a Cr:YAG crystal has a complex spatial distribution of transmission and it changes dynamically (Zabkar et al., 2008).

In this paper we demonstrated the optimum design of a high-brightness (high peak power and high beam quality), passively Q-switched micro-solid-state laser for ignition of engines. The performance of the micro-laser including fluctuations of pulse-to-pulse energy and timing jitters in QCW pumped operations was evaluated in detail. The combustion experiments in a static test chamber and a dynamic real automobile engine ignited by the micro-laser were demonstrated and discussed compared to a conventional spark plug. From the results, we could confirm that a high-brightness laser could dramatically reduce the minimum ignition energy and we also found that multi-pulse (pulse-train) ignition was effective to improve ignition possibility for leaner mixtures. Finally we successfully demonstrated the prototype laser igniter which had the same dimension as a spark plug including all optics for ignition.

3. Performance of micro-solid-state laser for ignition

3.1 Performance of passively Q-switched micro-laser module

Figure 2 shows (a) a schematic drawing and (b) a photograph of a passively Q-switched micro-laser module (Tsunekane et al., 2008). An active medium of a 1.1at.% Nd doped YAG crystal (crystal orientation of <111>, Sumitomo Metal Mining Co., Ltd.) with a length of 4mm is longitudinally pumped by a fiber-coupled, conductive-cooled, 120W (peak) QCW 808nm laser diode (JENOPTIK laserdiode GmbH). The core diameter of the fiber is 0.6mm with N.A. of 0.22. The pump light from the fiber was collimated by a lens set to have a diameter of 1.1mm in the active medium. Antireflection (<0.2%) and high-reflection (>99.8%) coatings at 808nm and 1064nm, respectively, were deposited on the pumped surface of the Nd:YAG crystal. High-reflection (>90%) and antireflection (<0.2%) coatings at 808nm and 1064nm, respectively, were deposited on the other, intra-cavity surface of the

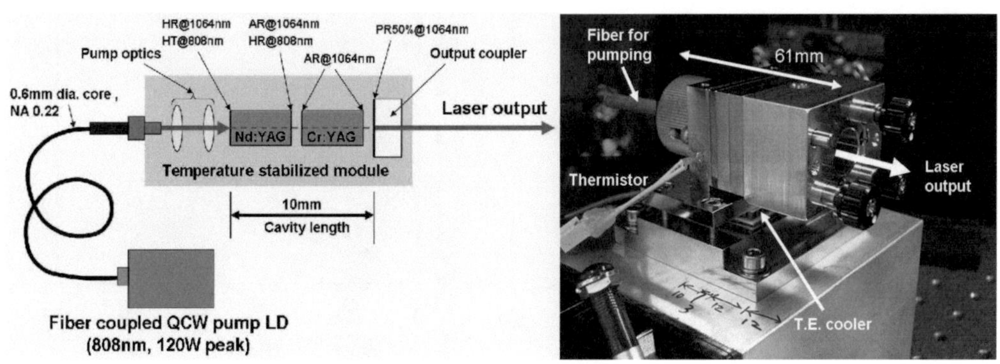

(a) (b)

Fig. 2. (a) Schematic drawing and (b) photograph of the passively Q-switched micro-laser module.

crystal. The high-reflection coating at 808nm makes efficient pump absorption possible by a round trip path of the pump light and it can also prevent a closely situated Cr:YAG crystal from pump-induced breaching (Zaykowski & Wilson, Jr., 2003; Jaspan et al., 2004). Antireflection (<0.2%) coatings at 1064nm were deposited on both surfaces of the saturable absorber of a Cr^{4+}:YAG crystal with a length of ~4mm (crystal orientation of <100>, Scientific Materials Corp.). The output coupler is flat with a reflectivity of 50% at 1064nm. The cavity length is 10mm. These optical elements were aligned carefully with an output coupler and fixed in the conductive-cooled, temperature-stabilized module (40mm-width × 28mm-height × 61mm-length) as shown in the figure. The module does not include focusing optics of the output beam for breakdown. In the following experiments, the pump energy was controlled by changing the pump duration with the peak pump power maintained at 120W. The maximum pump duration is 500μs limited by the diode. The repetition rate was 10Hz constant.

Figure 3 (a) shows the output energy of the passively Q-switched micro-laser as a function of the initial transmission of a Cr:YAG crystal. The output of a passively Q-switched laser forms a pulse train, which is well known. The closed circles and the solid line denote the experimental values and the calculation of pulse energy (energy per pulse), respectively, and the open circles denote the experimental values of the total output energy (sum of pulse energies) at a pump duration of 500μs. The pulse energy increases to 4.3mJ as an initial transmission of the Cr:YAG crystal decreases to 15%. On the other hand, the total output energy decreases from 25 to 12mJ as an initial transmission decreases from 80% to 15%, because the pulse-to-pulse interval becomes longer and then the number of laser pulses decreases even though the pulse energy increases. The decrease in total output energy simply means the decrease in efficiency of the laser.

Figure 3 (b) shows the pulse width as a function of the initial transmission of a Cr:YAG crystal. The pulse width was measured by a 10GHz InGaAs detector (ET-3500, Electro-Optics Technology, Inc,) with a 12GHz oscilloscope (DSO81204B, Agilent Technology). The closed circles and broken line denote the experimental values and the calculation of the pulse width, respectively. The pulse width decreases as the initial transmission decreases. The shortest pulse width of 300ps was obtained at an initial transmission of 15%.

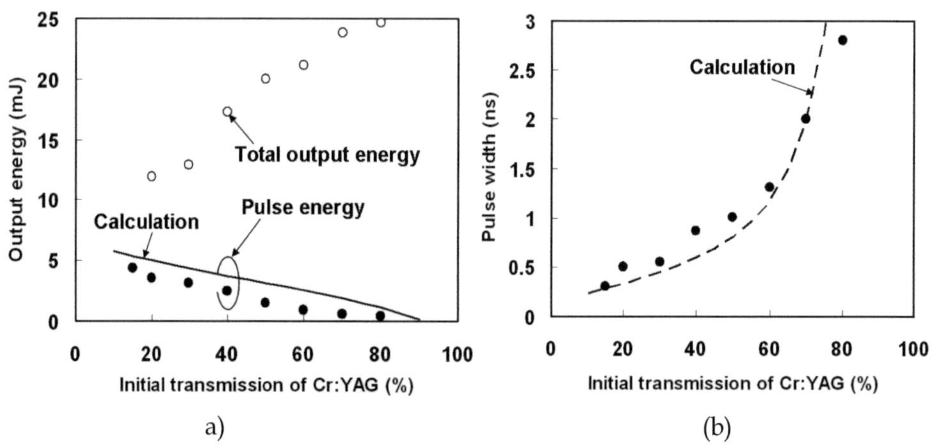

Fig. 3. (a) Output energy and (b) pulse width of the passively Q-switched micro-laser as a function of the initial transmission of Cr:YAG.

In the theoretical calculations shown in Fig.3, we assumed the ground-state absorption cross section of Cr^{4+}:YAG as σ_{SA}=2×10^{-18} cm^2 and the excited-state absorption as σ_{ESA}=5×10^{-19} cm^2. These are very important parameters but vary greatly in previous reports hence we used the averaged values in recent reports (Burshtein et al., 1998; Xiao et al., 1999; Feldman et al., 2003). In the calculation of pulse energy, we also assumed that the laser mode has the same size as the pump beam. Theoretically the output pulse energy is proportional to the area of the laser mode, so it is understood that the actual laser mode size is smaller by 10% or more than the pump beam due to the aperture effect of the saturable absorber of a Cr:YAG crystal as shown in Fig.3(a). As the initial transmission is higher, the discrepancy is larger. On the other hand, the calculation of pulse width agrees well with the experimental results as seen in Fig.3 (b), because the calculation of pulse width has no relation to the size of the modes.
Though the highest pulse energy and shortest pulse width were obtained at an initial Cr:YAG transmission of 15%, optical damage was observed at the output coupler and then the beam quality was degraded. The beam quality was also degraded at initial transmissions of higher than 70%, because the aperture effect of a Cr:YAG crystal is weak.
To use a laser for ignition, optical intensities of the order of 100GW/cm^2 are necessary at the focal point for breakdown. From our experimental observations, stable breakdown in air was observed at a pulse energy larger than 1.5mJ and a pulse width less than 1ns using an aspheric focus lens with a focal length of 10mm. In addition to the pulse energy, the total output energy is also necessary to ignite fuel-lean mixtures as discussed later. However as seen in Fig.3(a) the pulse energy and the total output energy are in the conflicting relation. Therefore we selected 30% as an optimum initial transmission of a Cr:YAG saturable absorber in our laser configurations. The laser performances were tested in detail and finally it was applied to the combustion experiments in the optimum condition.
Figure 4 shows the laser output energy and optical-to-optical conversion efficiency as a function of pump duration at an initial Cr:YAG transmission of 30%. As a unique characteristic of passively Q-switched lasers, the output pulse energy is constant until the following pulse is generated, then the output characteristic changes to the shape of stairs. The interval of pulse generation is almost constant at 100μs. The output energy obtained was 2.7mJ per pulse and totally 11.7mJ (sum of 4 pulses) was obtained at a pump duration of 500μs. The optical-to-optical conversion efficiency changes largely and periodically by the pump duration and the maximum efficiency of 19% was obtained at the durations of pulse generation.

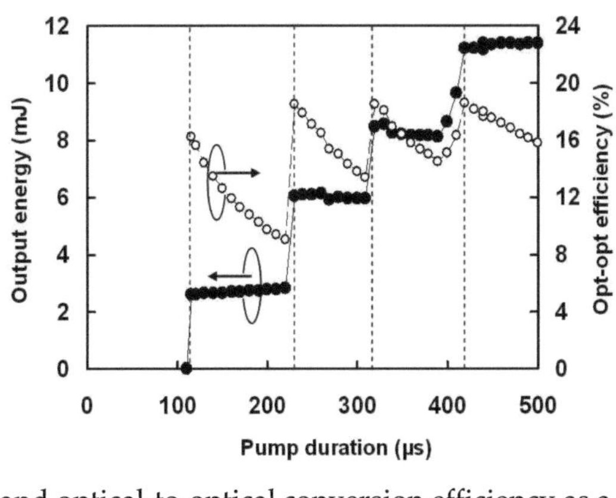

Fig. 4. Output energy and optical-to-optical conversion efficiency as a function of pump duration at an initial transmission of 30%.

Figure 5 shows the fluctuation of the total output energy as a function of pump duration which was estimated statistically from 500 consecutive pulses. The perpendicular doted lines show the pump durations at which the number of pulses increases as shown in Fig.4. It is understood that the fluctuation increases by increase in the number of pulses and at the duration when the number of pulses changes. The fluctuation is still less than 100μJ (3%).

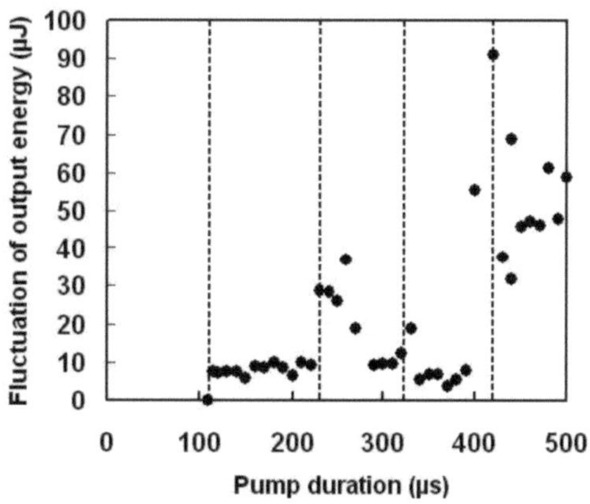

Fig. 5. Fluctuation of the total output energy as a function of pump duration which was estimated statistically from 500 consecutive pulses.

Figure 6 shows (a) the delay times of the each laser pulse from the standup of the pump LD current and (b) the jitters (standard deviation) of the delay times estimated statistically from 500 consecutive pulses as a function of pump duration. As seen in Fig.6 (a), the delay times are not dependent on the pump duration. The pulse interval is constant around 100μs and is equal to the interval of pulse generation. On the other hand, the jitter of the delay time strongly depends on the pump duration and also on the number of pulses. The first pulse has a small jitter for 200ns or less, and it is not dependent on pump duration, but the pulses generated later have a large jitter of 1μs or more, and the jitter changes sharply with the

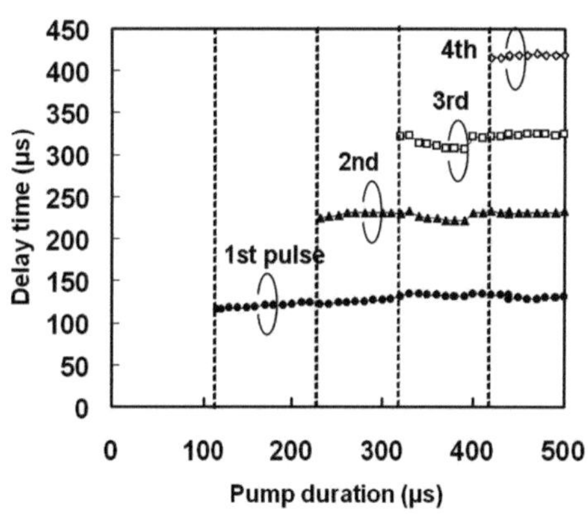

a)

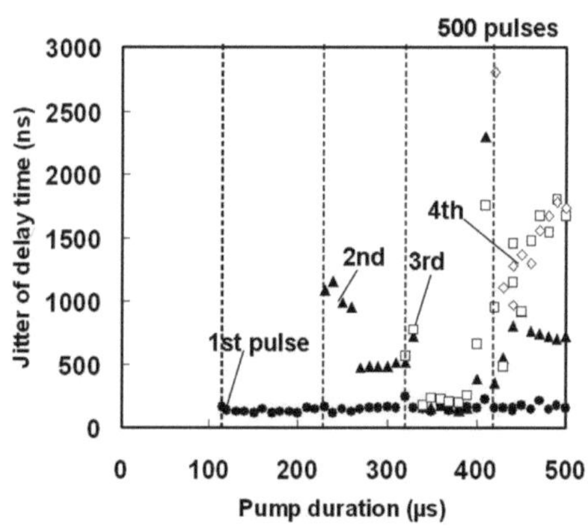

(b)

Fig. 6. (a) Delay times of the each laser pulse from the standup of the pump LD current and (b) jitters (standard deviation) of the delay times estimated statistically from 500 consecutive pulses, as a function of pump duration pump duration as seen in Fig.6 (b). Thermal lens and distortion which grow during QCW excitation in the Nd:YAG crystal strongly influence oscillation timing of a laser pulse. Then the jitter becomes large in the pulse generated later. However it should be mentioned that the jitter of the micro-laser is still 0.5% or less, and are much smaller than CW-pumped passively Q-switched lasers which have pulse-to-pulse jitters of 10% or more. This is because the oscillation condition is thermally initialized by low repetition rate, QCW pumping. It was confirmed that the fluctuation and jitter of the passively Q-switched micro-laser in QCW operations were within the tolerance limits for actual automobile applications.

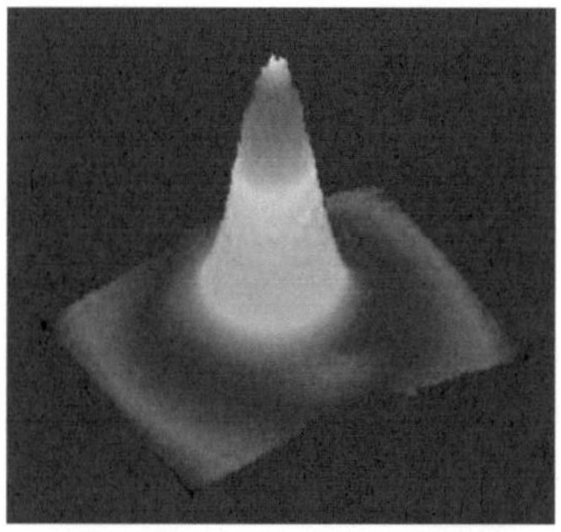

Fig. 7. 3D intensity profile of the output beam (first pulse) at a pump duration of 130μs.

Figure 7 shows the 3D intensity profile of the output beam (first pulse) measured by Beam Star-FX33D (Ophir) at a pump duration of 130μs. The M^2 value was calculated as 1.2. The beam profile does not change largely by changing the repetition rate from 1 to 100Hz. The pulse width was 600ps as measured in Fig.3(b) the brightness of the micro-laser was calculated as 0.3PW/sr-cm^2 which is one order of higher than our previous report (Sakai, 2008). To confirm the aperture effect of a Cr:YAG saturable absorber, we measured the beam quality of a micro-laser without the Cr:YAG crystal. The cavity length of 8mm and the output coupler with a reflectivity of 10% were employed to simulate the similar laser mode size in a cavity and threshold (cavity loss) as the passively Q-switched laser. The M^2 value of the simulated laser was measured to be 5 or more. Therefore it is understood that the saturable absorber works effectively as an aperture in the laser cavity. In our optimum designed micro-laser, each pulse of multipulse train has almost the same energy, pulse width and beam quality, hence each pulse can generate breakdown plasma independently in a fuel-mixture.

3.2 High temperature operation of passively Q-switched laser

For the practical use of laser igniters, stable operation is required at temperatures of up to 150˚C. Thus, we tested the operation of a passively Q-switched Nd:YAG/Cr:YAG laser at high temperature. First, we studied the temperature dependence of the transmission of a Cr:YAG saturable absorber at a wavelength of 1064nm. Three Cr:YAG crystals with different crystal orientations of <100>, <110> and <111> were prepared. The <100> is popular for passive Q-switching, while the <110> is used for polarization stabilized operation. They have the same initial transmission of 30% at 1064nm and at room temperature. Antireflection (<0.2%) coatings at 1064nm were performed to both surfaces of all the samples. The temperatures of the crystals were controlled by a themo-electric heater. As incident light sources, we used a commercial CW Nd:YVO_4 laser (MIL-1064-100-5,

Broadband, Inc.) for measurement of the initial transmissions and the passively Q-switched micro-laser we developed for measurement of the saturated transmissions of the Cr:YAG crystals. The polarization of the incident beams to the crystals was linear and rotated by a quarter wave plate in front of the crystals.

Figure 8 shows the initial transmissions of the Cr:YAG crystals as a function of crystal temperature. The incident beam had a power of 5mW CW and a diameter of 2mm in the crystals. Slight, 5% increases of initial transmissions were observed with increase in crystal temperatures from 25 to 150°C for all the crystals.

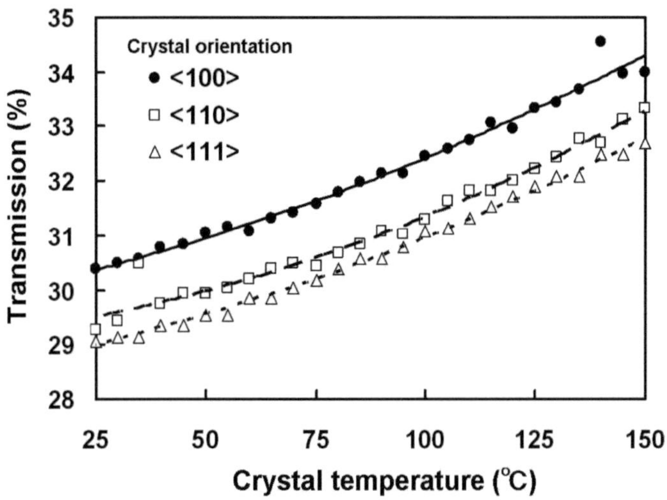

Fig. 8. Initial transmissions of the Cr:YAG crystals with three different orientations as a function of crystal temperature

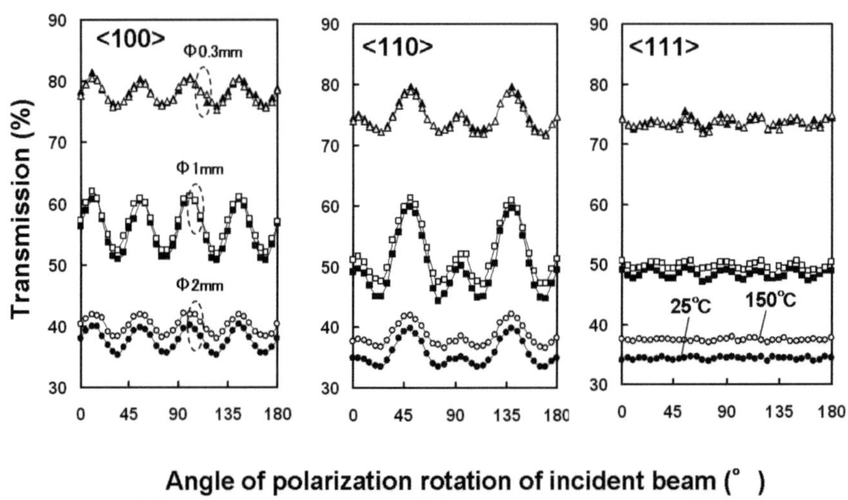

(a) (b) (c)

Fig. 9. Saturated transmissions of Cr:YAG crystals at 25°C and 150°C with crystal orientations of (a) <100>, (b) <110>, and (c) <111> as a function of the angle of incident beam polarization rotation.

Figure 9 shows the transition of the saturated transmissions of the Cr:YAG crystals as a function of the angle of incident beam polarization rotation at temperatures of 25 and 150˚C and at three different incident beam diameters of 0.3, 1 and 2 mm. The crystal orientations in (a), (b), and (c) are <100>, <110>, and <111>, respectively. The incident Q-switched pulse had an energy of 2mJ constant and a width of 600ps. As seen in these figures, it is understood that the saturated transmissions at the highest beam intensity with a diameter of 0.3mm are the same for both temperatures and the situation is independent on a crystal orientation. Therefore it is expected that the performance of a Cr:YAG saturable absorber dose not change even at 150°C.

Figure 10 (a) shows the input-output characteristics of a Nd:YAG laser without a Cr:YAG crystal at various crystal temperatures of up to 150°C. The temperature-controlled Nd:YAG crystal was longitudinally pumped by a fiber-coupled LD with QCW operation (peak pump power 120W, repetition rate 1Hz). The cavity length was 20mm and the reflectivity of the flat output coupler was reduced to 10% to simulate a cavity loss by a Cr:YAG crystal. As seen in the figure, as the temperature of the Nd:YAG crystal increases from room temperature to 150°C, the threshold increases by 60% and the slope efficiency decreases.

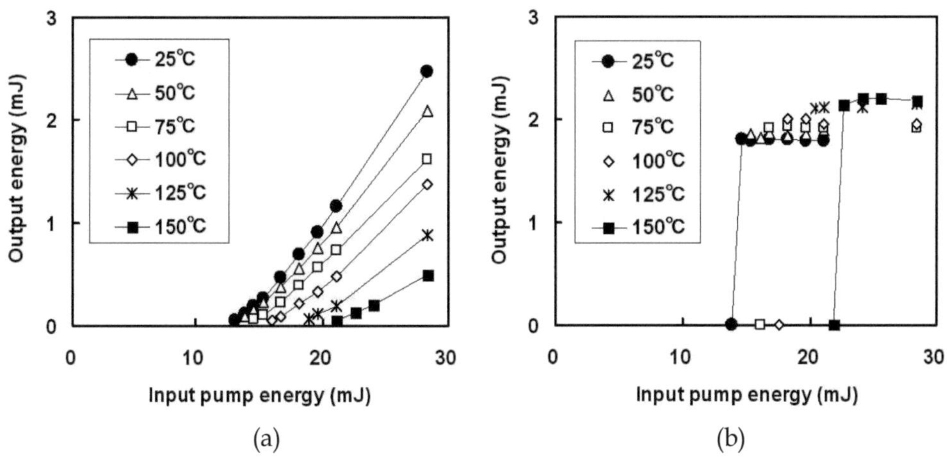

Fig. 10. (a) Input-output characteristics of a Nd:YAG laser without a Cr:YAG crystal at various crystal temperatures of up to 150°C, and (b) input-output characteristics of a Nd:YAG/Cr:YAG passively Q-switched laser at the same crystal temperatures.

Figure 10 (b) shows the input-output characteristics of a Nd:YAG/Cr:YAG passively Q-switched laser at various crystal temperatures. The temperatures of the two crystals are the same, and the pump and cavity layouts are the same as those of the Nd:YAG laser shown in Fig.10 (a), but the reflectivity of the output coupler is 50%. The initial transmission of the Cr:YAG crystal is 30% at room temperature. The experimental data at 20°C and 150°C are connected with a line only so that the results may be understood clearly. The thresholds of the passively Q-switched laser at 20°C and 150°C are almost equivalent to those of the Nd:YAG laser at the same temperatures shown in Fig.10 (a), and we can understand that the characteristics of the Cr:YAG/Nd:YAG passively Q-switched laser at 150°C are mainly decided by that of a Nd:YAG crystal (Tsunekane & Taira, 2009). It should be also emphasized that the output pulse energy increases slightly even as the temperature of the crystals increases to 150°C. From the results, we can further confirm that the Nd:YAG/Cr:YAG passively Q-switched micro-laser is a suitable light source for laser ignition.

4. Ignition in constant volume chamber

The laser ignitions for stoichiometric to fuel-lean C_3H_8/air mixtures by a high-brightness passively Q-switched micro-laser were studied and compared with a conventional electric spark plug experimentally in a constant-volume (~200cm^3) chamber at room temperature (25°C) and atmospheric pressure (100kPa). The C_3H_8 gas and air were introduced in the combustion chamber after mixing it thoroughly in specified ratios within another chamber beforehand. The ignition experiment was conducted after the flow of the combustible mixture in the chamber settled. The chamber is equipped with three windows. Two of them are laterally opposite to each other for flow visualization and one is for introduction of a laser light or a spark plug (compatible window). The combustion processes were observed by Schlieren photography (shadowgraph) which can visualize a slight refractive-index gradient of transparent media. Highly uniform, incoherent light from a Xenon lamp was introduced to the chamber and the 2D images of the transmitted light through the windows were detected by a high-speed video camera (25000 fps) synchronized with a laser pulse or an electric trigger.

Figure 11 shows the Schlieren photographs of the early stage of ignition and subsequent combustion ignited by (a) a micro-laser and (b) a spark plug in stoichiometric mixture. The air fuel ratio (A/F) is 15.3. The total energy from the laser is ~9mJ (3 pulses) and the input electric energy to the plug is 35mJ. The laser beam is focused inside the chamber through the window by an aspheric lens with a focal length of 10mm so as to have the same ignition position as the spark plug. Then both images show the same position with the same scale in the chamber. The numbers indicate time after a laser shot or an electric trigger on. As shown in the lowermost figures of Fig.11 (a) and (b), the cross-section area of the flame kernel generated by the laser is 3 times larger than the spark plug at 6ms after ignition, even though the ignition energy of the laser is 1/3 (Tsunekane et al., 2008). Therefore it is confirmed that laser ignition effectively accelerates the flame kernel growth due to the absence of quenching effect by electrodes.

The combustions in the fuel-lean mixtures where the ratio of air increased were observed using the same constant-volume chamber. The A/Fs of the mixtures were changed from 15.2 (stoichiometric) to 18.1. The numbers of laser pulses (pulse train) were also changed

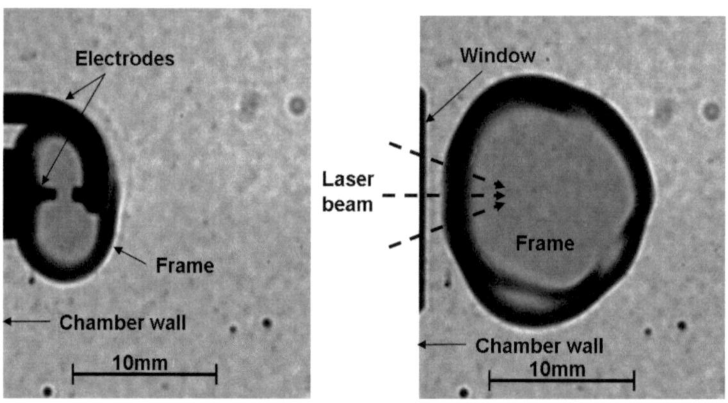

(a) (b)

Fig. 11. Schlieren photographs for early stage of ignition in a constant-volume chamber ignited by (a) a micro-laser and (b) a spark plug in a stoichiometric mixture.

Ignition probability (%)

C_3H_8+Air		Laser					Spark Plug
ER	A/F	1 pulse (3.0mJ)	2 pulses (5.6mJ)	3 pulses (8.1mJ)	4 pulses (10.6mJ)	5 pulses (13.0mJ)	
1.00	15.2	100	100	100	100	100	100
0.97	15.7	0	50	100	100	100	33
0.93	16.3	–	0	100	60	100	0
0.88	17.2	–	–	22	0	100	–
0.84	18.1	–	–	–	0	0	–

Table 1. Ignition probabilities for complete combustion in a stoichiometric and fuel-lean mixtures in a constant-volume chamber estimated from the repetitive ignition.

from 1 to 5 by changing the pump duration to understand the effect of the total optical energy on combustion. Table 1 summarizes the ignition probabilities for complete combustion estimated from the repetitive ignition experiments in each condition. The horizontal lines show the numbers of laser pulses and the total optical energies, and the right end shows the experimental results from a spark plug. The vertical lines show the A/Fs. The values become high toward the bottom which means the mixtures become leaner. ER is equivalent ratio, the value that broke the A/F (15.2) of the stoichiometric mixture by the A/F of the specific mixture. From this table, it can be seen that the ignition probability of a leaner mixture improves with increase in the number of laser pulses. Then it is understood that higher total ignition energy is necessary for a leaner mixture. One hundred percent ignition is accomplished in the fuel-lean mixture of A/F of 17.2 by 5 laser pulses (5-pulse train) with the total optical energy of 13mJ. Such a multiple pulse ignition is advantageous for actual applications compared to conventional laser ignition by a single big pulse with an optical energy of more than 10mJ. The peak intensity of a laser pulse can be reduced by maintaining total ignition energy, and the optical damage to coatings on optics can be avoided. Moreover, the number of pump LDs can be reduced. It has the advantage not only of miniaturization of the ignition system but also of low price.

From these experimental results of laser ignition by a micro-laser, it should be mentioned that in a stoichiometric mixture, 100% ignition was accomplished by a single pulse with an optical energy of 3mJ. From the Schlieren photography measurements, it was observed that in the lean mixtures, the first laser pulse (3mJ) was still enough for breakdown and flame kernel formation, but the growth of the flamee kernel was slow compared to that in the stoichiometric mixture and disappeared quickly if the following pulses were not injected.

In the case of spark plug ignition shown in the right end, the ignition probability is below 100% even in an A/F of 15.7, which is slightly leaner than in a stoichiometric mixture. Schlieren photography also demonstrated that the growth of the flamee kernel was slow and the time during the boundary of the flame kernel in contact with the electrodes was long, and therefore it could be understood that the quenching effect by electrodes of a spark plug was more significant in lean mixtures.

5. Ignition in real automobile engine

Finally, ignition tests for a real automobile engine were performed. We used a commercial engine of 1AZ-FSE (TOYOTA Motor Corp.), which is a 2.0 L, straight-4 piston engine with a gasoline direct injection system. Figure 12 shows the optical layout for laser ignition. The micro-laser module was fixed not on the engine directly but on a metal frame closely positioned on the upper side of the engine and the output beam was carefully aligned by using three mirrors to the optical axis of the focal lens fixed in the spark plug hole. The optical path from the module to the focal lens was around 830mm. A transmission lens with a focal length of 300mm was used to control the beam size. The focal lens had a focal length of 10mm. The ignition point of a laser was set to be the same point as a spark plug by tuning the height of the focal lens. Thus this experiment was not optimized for laser ignition. In this experiment, three of the four cylinders (from #1 to #3) were ignited by conventional spark plugs and the #4 cylinder was ignited by a laser. Each ignition timing was carefully controlled and optimized. The repletion rates of the igniters were 13.3Hz corresponding to an engine speed of 1600rpm.

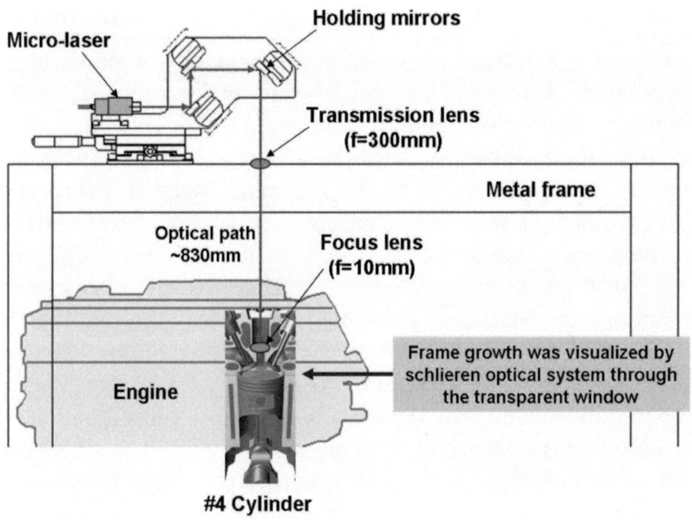

Fig. 12. Optical layout of laser ignition experiments for a real engine

The experimental setup for visualization of combustion with reflective Schlieren photography is schematically shown in Fig.13. The collimated and highly homogenized light from a Xenon lamp was introduced into the combustion chamber through a transparent window which was formed at the side wall of the cylinder head of #4 and was reflected back at a mirror, which was situated inside the cylinder wall at the opposite side of the window. The reflected light was imaged by a high-speed CCD camera (25000fps). Laser beam was introduced into the chamber through a normal spark plug hole.

Schlieren photographies of the early stage of ignition and subsequent combustion in a real engine are summarized in Fig.14. In the micro-laser ignitions, combustion processed images of three different pulse-trains, i.e., single-pulse, two-pulse and four-pulse trains, are demonstrated. The right side shows the image of a conventional spark plug ignition. Total optical energies into the chamber are also shown in brackets, but the energy from the micro-laser decreases to 85% due to optical loss at the three alignment mirrors. In this figure, the

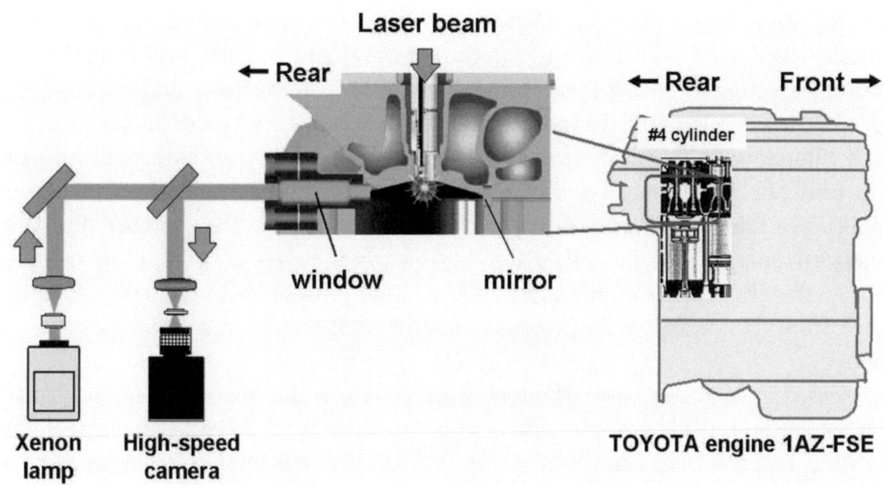

Fig. 13. Experimental setup for visualization of combustion with reflective Schlieren photography

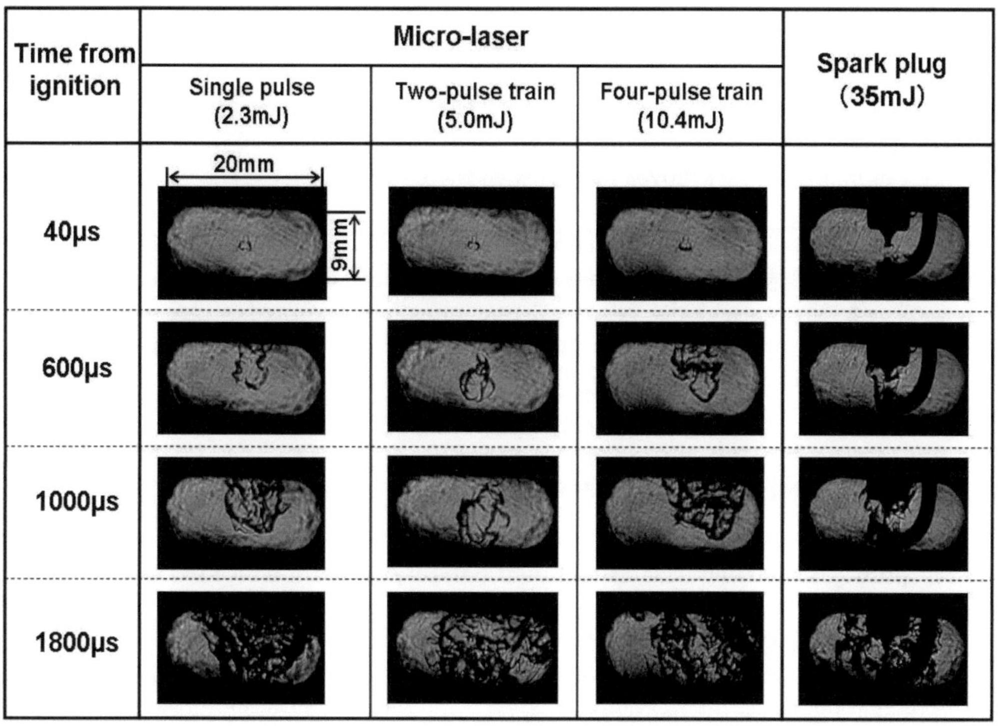

Fig. 14. Schlieren photographies of the early stage of ignitions by a micro-laser and a spark plug and the subsequent combustions in a real engine

A/F is 14.5, which is a stoichiometric mixture of gasoline and air. It should be emphasized that a single laser pulse with an energy of 2.3mJ can successfully ignite a real engine. We think this will be the lowest energy ever reported for laser igntion of a real automobile engine. High-brightness, passively Q-switched micro-laser can reduce the ignition energy dramatically compared with previous ignition lasers (Kroupa, 2009) and a spark plug. The operation of an engine ignited by a single laser pulse is quite stable and no miss-ignitions appeared during our experiment for several tens of minutes. However, look carefully at the Schlieren photographs in Fig.14 at the early stage until 1000μs, and you will find that the shadows of the flames by single and two-pulse trains are weak compared with those of four-pulse train. We think flame growth is perturbed by the flow in the chamber due to lack of total ignition energy. On the other hand, such flow helps a spark plug escape from quenching effect by the electrodes and then the flame grows effectively with a high ignition energy of 35mJ, in contrast with previous experimental results in a static constant-volume chamber.

We also tested in lean mixtures. The lean limit (A/F), where the combustion was slightly unstable, of single-pulse, two-pulse and four-pulse were 16, 17.6, and 18.8, respectively, but a spark plug had a higher lean limit of 20~21. We think the total optical energy of a laser needs up to 20mJ, where the lean limit of laser ignition will be comparable with that of spark plug ignition. Recently the maximum total output energy of 21mJ was obtained with an optical-to-optical conversion efficiency of 23% from the same micro-laser module by increasing of the peak power of fiber-coupled pump diode to 180W.

From the Schlieren photography, it is understood that the flame-growing processes of laser and electric spark plug ignitions at the early stage are quite different. In a spark plug, the flame generated at the gap of the electrodes forms an eddy structure and stays around the outer electrode, and grows continuously and stably to a large flame. On the other hand, the flame generated by a laser moves randomly in the free space of the chamber by turbulent flow during the growth.

The contamination and damage of an optical window in the combustion chamber are well-known serious problems of laser ignition (Ranner et al., 2007). In our experiments, Al_2O_3 was used as a window material. No visual contaminations and no damages were observed on the window surface after the combustion experiments for several hours, but long-duration tests under various engine conditions are necessary for choice and optimization of practical windows.

6. Real spark plug size micro-laser module

In Fig.15, we show the first prototype micro-laser module which has the same dimensions as a spark plug. This module includes not only pumping optics from a fiber to a solid-state material but also beam expanding and focusing optics for ignition. The laser igniter has the same optical design and the same performance as the experimental module in Fig.2, and it is physically possible to ignite a real engine by installing it instead of a spark plug to a plug hole. For real operation on an engine, however, the mechanical design inside the module should be improved to sustain the high temperature (up to 150°C) and vibration of a real engine. In this report, we used conventional single crystals, but we think YAG ceramics are promising actual light sources for laser ignition, because they have higher uniformity and stress resistance, and are suitable for mass-production. In addition, if a composite structure of Nd:YAG/Cr:YAG is possible, then the compact and rugged, monolithic laser cavity will be made (Feng, 2004; Taira, 2007).

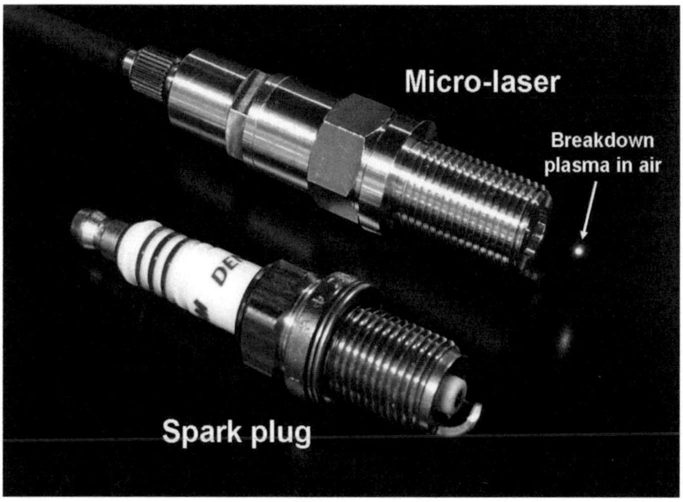

Fig. 15. Prototype spark plug size micro-laser head and a conventional spark plug

7. Conclusion

A high-brightness, passively Q-switched Nd:YAG/Cr:YAG micro-laser was developed and optimized for ignition of engines. The output energies of 2.7mJ per pulse and 11.7mJ in total (four-pulse train) were obtained at a pump duration of 500μs with an optical-to-optical conversion efficiency of 19%. A pulse duration of 600ps and an M^2 value of 1.2 were obtained and the brightness of the micro-laser was calculated as 0.3PW/sr-cm^2. The optical power intensity at the focal point of ignition was calculated as 5TW/cm^2. The fluctuations of the total output energies and jitters of the delay time of a laser oscillation were less than 100μJ (3%) and 0.5%, respectively. We further confirmed that the output pulse energy of a passively Q-switched Nd:YAG/Cr:YAG laser did not change even though the temperature of the crystals increases to 150°C. The enhanced combustion by the micro-laser ignition was successfully demonstrated in a constant-volume chamber with room temperature and with atmospheric pressure. The cross-section area of a flame kernel generated by the micro-laser was 3 times larger than a spark plug at 6ms after ignition in a stoichiometric mixture (A/F 15.2) of C_3H_8/air. The effective laser ignition for lean mixtures was also accomplished by a multiple pulse (pulse train) of the micro-laser. Ignition of 100% was successfully demonstrated by a five-pulse train in a lean mixture of an A/F 17.2, where spark plug ignition failed. Finally, ignition tests for a real automobile engine were performed. A single laser pulse with an energy of 2.3mJ could ignite and drive the engine stably. It will be lowest energy ever reported for laser ignition of a real automobile engine. We can confirm that an optimally designed, high-brightness, passively Q-switched micro-laser reduced the ignition energy dramatically compared with previous ignition lasers and a spark plug and the dimension of the laser head can be reduced to real spark plug size.

8. Acknowledgment

We are grateful to Japan Science and Technology Agency (JST) for financial support on Practical Application Research of "Micro solid-state laser for internal-combustion engine control" promoted by JST Innovation Plaza Tokai (2006-2009). We thank Dr. Fujikawa of Toyota Central R&D Labs., Inc. for technical discussion of combustion and Mr. Mizutani of the equipment development division of IMS for fabrication of the micro-laser modules.

9. References

Hickling, R. & Smith, W. R. (1974). Combustion bomb tests of laser ignition, SAE Paper no.740114

Dale, J. D.; Checkel, M. D. & Smy, P. R. (1997). Application of high energy ignition systems to engines, Prog. Energy Combust. Sci. vol.23, pp379-398

Phuoc, T. X. (2006). Laser-induced spark ignition fundamental and applications, Opt. Lasers Eng. vol.44, pp351-397

Phuoc, T. X. (2000). Single-point versus multi-point laser ignition: experimental measurements of combustion times and pressures, Combustion and Flame vol.122, pp.508-512

Morsy, M. H. ; Ko, Y. S., Chung, S. H. & Cho, P. (2001). Laser-induced two-point ignition of premixture with a single-shot laser, Combustion and Flame vol.125, pp724-727

Weinrotter, M.; Kopecek, H., Tesch, M., Wintner, E., Lackner, M. & Winter, F. (2005). Laser ignition of ultra-lean methane/hydrogen/air mixtures at high temperature and pressure, Exp.Themal and Fluid Sci., vol.29, pp.569-577

Joshi, S.; Yalin, A. P. & Galvanauskas, A. (2007). Use of hollow core fibers, fiber lasers, and photonic crystal fibers for spark delivery and laser ignition in gases, Appl. Opt., vol.46, pp.4057-4064

Kofler, H.; Tauer, J., Tartar, G., Iskra, K., Klausner, J., Herdin, G. & Wintner, E. (2007). An innovative solid-state laser for engine ignition, Laser Phys. Lett. vol.4, pp322-327

Zaykowski, J. J. &Dill III, C. (1994). Diode-pumped passively Q-switched picosecond microship lasers, Opt. Lett. vol.19, pp.1427-1429

Sakai, H.; Kan, H. & Taira, T. (2008). >1MW peak power single-mode high-brightness passively Q-switched Nd^{3+}:YAG microchip laser, Opt.Exp.vol.16, pp.19891-19899

Huang, S-L.; Tsui, T-Y., Wang, C-H. & Kao, F-J. (1999). Timing jitter reduction of a passively Q-switched laser, Jpn. J. Appl. Phys. vol.38, pp.L239-241

Tang, D. Y.; Ng, S. P., Qin, L. J. & Meng, X. L. (2003). Deterministic chaos in a diode-pumped Nd:YAG laser passively Q switched by a Cr^{4+}:YAG crystal, Opt. Lett., vol.28. pp.325-327

Szabo, A. & Stein, R. A. (1965). Theory of laser giant pulsing by a saturable absorber, J. Appl. Phys. vol.36, 1562-1566

Degnan, J.J. (1995). Optimization of passively Q-switched lasers, IEEE J. Quantum Electron., vol.31, pp.1890-1901, Nov.

Xiao, G. & Bass, M. (1997). A generalized model for passively Q-switched lasers including excited state absorption in the saturable absorber, IEEE J. Quantum Electron., vol.33, pp.41-44, Jan.

Zhang, X.; Zhao, S., Wang, Q., Zhang, Q., Sun, L. & Zhang, S. (1997). Optimization of Cr^{4+}-doped saturable-absorber Q-switched lasers, IEEE J. Quantum Electron., vol.33, pp.2286-2294

Chen, Y. F.; Lan, Y. P. & Chang, H. L. (2001). Analytical model for design criteria of passively Q-switched lasers, IEEE J. Quantum Electron., vol.37, pp.462-468, March

Pavel, N.; Saikawa, J., Kurimura, S. & Taira, T. (2001). High average power diode end-pumped composite Nd:YAG laser passively Q-switched by Cr:YAG saturable absorber, Jpn. J. Appl. Phys. vol.40, pp.1253-1259, March

Patel, F. D. & Beach, R. J. (2001). New formalism for the analysis of passively Q-switched laser systems, IEEE J. Quantum Electron., vol.37, pp.707-715, May

Zabkar, J.; Marincek, M. & Zgonik, M. (2008). Mode competition during the pulse formation in passively Q-switched Nd:YAG lasers, IEEE J. Quantum Electron., vol.44, pp.312-318, Apr.

Tsunekane, M.; Inohara, T., Ando, A., Kanehara, K. & Taira, T. (2008). High peak power, passively Q-switched Cr:YAG/Nd:YAG micro-laser for ignition of engines, in Proc. Advanced Solid State Photonics, OSA Tech. Dig., MB4, Jan.

Zaykowski, J. J. & Wilson Jr., A. L. (2003). Pump-induced bleaching of the saturable absorber in short-pulse Nd:YAG/Cr^{4+}:YAG passively Q-switched microchip lasers, IEEE J. Quantum Electron., vol.39, pp.1588-1593, Dec.

Jaspan, M. A. ; Welford, D. & Russell, J. A. (2004). Passively Q-switched microlaser performance in the presence of pump induced bleaching of the saturable absorber, Appl. Opt., vol.43, pp.2555-2560, Apr.

Burshtein, Z.; Blau, P., Kalisky, Y., Shimony, Y. & Kokta, M. R. (1998). Excited-state absorption studies of Cr^{4+} ions in several garnet host crystals, IEEE J. Quantum Electron., vol.34, pp.292-299, Feb.

Xiao, G.; Lim, J. H., Yang, S., Stryland, E. V., Bass, M. & Weichman, L. (1999). Z-scan measurement of the ground and excited state absorption cross sections of Cr^{4+} in yttrium aluminum garnet, IEEE J. Quantum Electron., vol.35, pp.1086-1091, July

Feldman, R.; Shimony, Y. & Burshtein, Z. (2003). Dynamics of chromium ion valence transformations in Cr,Ca:YAG crystals used as laser gain and passive Q-switching media, Opt. Mat., vol.24, pp.333-344

Tsunekane, M.; Inohara, T., Ando, A., Kanehara, K. & Taira, T. (2008). Compact, high peak power, passively Q-switched micro-laser for ignition of engines, in Proc. Conf. Laser Electro-Optics., OSA Tech. Dig. CFJ4

Tsunekane, M.; Inohara, T., Ando, A., Kanehara, K. & Taira, T. (2008). Compact and High-brightness Passively Q-switched Cr:YAG/Nd:YAG Laser for Ignition of Engines, in 3rd EPS-QEOD Europhoton conf. Tech. Dig. FRoB.2, Sep.

Feldman, R.; Shimony, Y. & Burshtein, Z. (2003). Passive Q-switching in Nd:YAG/Cr:YAG monolithic microhip laser, Opt. Mat., vol.24, pp.393-399

Graham-Rowe, D. (2008). Lasers for engine ignition, nature photonics vol.2, pp.515-518, Sep.

Feng, F.; Lu, J., Takaichi, K., Ueda, K., Yagi, H., Yanagitani, T. & Kaminskii, A. A. (2004). Passively Q-switched ceramic Nd^{3+}:YAG/Cr^{4+}:YAG lasers, Appl. Opt., vol.43, pp.2944-2947, May

Taira, T. (2007). RE^{3+}-ion-deped YAG ceramic lasers (invited), IEEE J. Sel. Top. Quantum Electron., vol.13, pp.798-809, May

Kroupa, G.; Franz, G. & Winkelhofer, E. (2009). Novel miniaturized high-energy Nd-YAG laser for spark ignition in internal combustion engines, Opt. Eng. vol.48, no.1, pp.014202-1 to 5, Jan.

Tsunekane, M. & Taira, T. (2009). High temperature operation of passively Q-switched, Cr:YAG/Nd:YAG micro-laser for ignition of engines, in Proc. Conf. Laser Electro-Optics/Europe-EQEC., Tech. Dig. CA.P.30, June

Ranner, H.; Tewari, P. K., Kofler, H., Lackner, M., Wintner, E., Agarwal, A. K. & Winter, F. (2007). Laser cleaning of optical windows in internal combustion engines, Opt. Eng. vol.46, no.10, pp.104301-1 to 8, Oct.

11

High Gain Solid-State Amplifiers for Picosecond Pulses

Antonio Agnesi and Federico Pirzio
Laboratorio Sorgenti Laser
Dipartimento di Elettronica dell'Università di Pavia
Via Ferrata 1 - 27100 Pavia
Italy

1. Introduction

Picosecond solid-state lasers are attractive for many industrial and scientific applications, such as precision material processing (Dausinger et al., 2003; Breitling et al., 2004), nonlinear optics (McCarthy & Hanna 1993; Ruffling et al., 2001; Sun et al., 2007) and laser spectroscopy (Mani et al., 2001). In contrast with traditional laser processing performed with 10-100 ns multi-kHz sources, picosecond laser-matter interaction is basically non-thermal, relying on multi-photon ionisation and photo-ablation processes that allow cleaner and much higher spatial definition in laser marking, drilling and cutting. Femtosecond pulses would perform even better in principle, but at the expense of a significant increase in complexity of the laser system that most often is unwelcome in industrial environments.

Furthermore, the multi-kW peak-power levels allowed by cw mode-locked picosecond lasers with average power of at least few watts are already sufficient to produce efficient frequency conversion by harmonic, sum- or difference-mixing and parametric generation.

Semiconductor saturable absorber mirrors (SESAMs) are widely employed for the passive mode-locking of picosecond solid-state lasers (Keller, 2003). SESAMs are very effective and highly reliable when used in low-power oscillators; however, when employed in high-power oscillators, they require a special design as their thermal management becomes a very important issue (Burns et al., 2000; Neuhaus et al., 2008). Indeed, the intense intracavity radiation of this particular operating regime may induce significant optical and thermo-mechanic stress effects, leading to rapid degradation of their performance.

An alternative (and not new) approach to powerful cw picosecond sources is to use a master-oscillator power-amplifier (MOPA) system, in which a seed from a low-power, robust picosecond laser is amplified to the required average power levels through extra-cavity diode-pumped amplifiers.

Some recent results have pointed out the great potential of this approach (Snell et al., 2000; Agnesi et al., 2006a; Nawata et al., 2006; Farrell & Damzen, 2007; McDonagh et al., 2007). To our knowledge, the most powerful cw picosecond source reported to date was a Nd:YVO_4 MOPA system longitudinally-pumped with 216 W at 888 nm, where a 60-W cw picosecond mode-locked laser was amplified to 111 W, with 53% amplifier extraction efficiency (McDonagh et al., 2007). Though the master oscillator of this example could be well

classified as "high-power", the effectiveness and the convenience of the power amplification to reach very high power levels is clear.
It is also worth noticing that fibre laser technology is rapidly approaching maturity also in the field of high-power ultrafast laser applications (Fermann & Hartl, 2009), though reliable large-area, photonic fibres sustaining picosecond pulses with energy >1 μJ are still subject of extensive research. Most likely, this will take some time before full commercial exploitation. Indeed, all-fibre amplification of femtosecond pulses, stretched to nanosecond or sub-nanosecond time duration, is definitely easier than direct amplification of pulses only few picosecond long. However, robust picosecond master oscillators, passively mode-locked by SESAMs (Okhotnikov et al., 2003) or other techniques (Porta et al., 1998), can be successfully realised with readily available fibre laser components. These considerations suggest an attractive approach to picosecond MOPAs consisting in a compact rugged picosecond fibre oscillator and a powerful diode-pumped solid-state amplifier with mode size properly scaled in order to avoid damage.
Grazing-incidence side-pumped Nd:YVO_4 slabs allow efficient power extraction owing to the very high single-pass gain achievable in such configuration (Bernard & Alcock, 1993; Damzen et al., 2001). This article reviews picosecond MOPA systems employing this particular amplification technique. Simple numerical models are presented and applied to the design of MOPAs, as well as for the interpretation of their performance and limitations.
A four-pass amplification setup including a photo-refractive phase-conjugating mirror was first reported to yield 12.8 W, nearly diffraction-limited, 8.7-ps pulses (Ojima et al., 2005) starting from a 100-MHz, 290-mW commercial cw mode-locked Nd:YVO_4 master oscillator. An improved and more powerful setup also employing phase-conjugating mirror delivered up to 25 W (Nawata et al., 2006).
A simpler setup with either a single- or double-pass slab yielded as much as 8.4 W with 7.5-ps pulses at 150 MHz (30% amplifier extraction efficiency from 28-W pump power), using a 50-mW seed oscillator (Agnesi et al., 2006a).
Other applications require instead intense picosecond pulses from compact diode-pumped solid-state laser systems at lower repetition rates ($\lesssim$1 MHz). Usually, intra- or extra-cavity pulse-picking from a diode-pumped low-power picosecond oscillator at ~ 100-MHz repetition rate is used to seed a regenerative amplifier. By this means, the pulse energy is increased from nanojoules to within a range from few microjoules to few millijoules, depending on the operating frequency (Siebold et al., 2004; Kleinbauer et al., 2005; Killi et al., 2005).
For this aim, too, grazing-incidence bounce amplifiers provide an interesting alternative, since their gain of ≈ 30 dB/pass allows efficient energy extraction in just two or three passes, thus avoiding the higher complexity of the former schemes and requiring only a versatile, extra-cavity acousto-optic pulse-picker. This needs to be appropriately synchronised to the mode-locked train, as well as to the amplifier pump pulse (if the amplifier is not cw-pumped).
In the first demonstration of such an amplifying configuration, the high-frequency picosecond seeder output was sampled synchronously, and the selected pulse (with energy < 1 nJ) was injected into a two-stage amplifier, yielding an output energy up to 10 μJ at 100 Hz (Agnesi et al., 2006b). A remarkable 1-MHz high repetition frequency and pulse energy up to 76 μJ were later reported by Nawata *et al.*, (Nawata et al., 2007) who employed a more complex cw-pumped phase-conjugation setup with a double-pass grazing-incidence slab. Repetition frequency as high as 4 MHz was also reported for an extra-cavity multi-pass

$Nd:YVO_4$ amplifier (Gerhard et al., 2008), yielding 80 μJ at low frequency and 1.8 μJ at 4 MHz. Effective material processing results were demonstrated with such a laser system.
However, the highest pulse energy was achieved with a refined qcw-pumped single-pass two-slabs amplifier design (Agnesi et al., 2008a), yielding as much as 210 μJ, up to 1 kHz repetition rate, and 11-ps time duration pulses.
Highly efficient harmonic conversion to 532 nm and 266 nm was readily observed, owing to the ≈ 20-MW peak power of the amplified picosecond pulses.
Travelling-wave parametric generation spanning the ranges 770-1020 nm (signal), 1110-1720 nm (idler) was also demonstrated (Agnesi et al., 2006b).
Particular applications such as photocatode injection (Will et al., 2005), low-threshold parametric generation (Agnesi et al., 1993; Butterworth et al., 1996) and implementation of pulse-format and wavelength typical of free-electron-lasers (Edwards et al., 2002) by all-solid-state laser technology require instead amplification of bunches of picosecond pulses.
Again, side-pumped high-gain bounce amplifiers can be successfully employed to increase the pulse energy of pulse trains as long as ~ 1 μs. A remarkable example of such a laser source delivering trains of ~ 2500 pulses of 12-ps time duration, 5-GHz repetition rate at 1064 nm, and train energy of 250 mJ, was reported recently (Agnesi et al., 2008b). This review is organised as follows. Section 2 gives the theoretical background for understanding and modelling grazing-incidence slab amplifiers; Sections 3-5 review some representative experimental results achieved by our research group with slab amplifiers operated in several regimes, as well as interesting frequency conversion applications; in Section 6 we finally summarise our results and trace few conclusions.

2. Numerical model

In this Section we review simple, yet effective, numerical models for grazing-incidence class of amplifiers, for several operating regimes such as cw and pulsed up to multi-kHz repetition rate. Limitations due to the finite amplifying bandwidth are discussed. These models will be applied to the interpretation of experimental results reviewed in the next Sections. More generally, their use extends to a wider class of amplifier, including, for example, cascaded systems and bounce amplifiers based on different laser materials, provided the pump absorption depth is of the order of ~ 1 mm or less, and the integrated single-pass gain reaches useful levels.

2.1 Operation in cw regime

The model is based on standard cw amplifier theory (Koechner, 2006). Let us introduce a reference system for bounce beam propagation in the active material slab shown in Fig. 1. The length and the thickness of the slab are L and W, respectively.
A particularly useful transverse local frame for the seed beam is $\xi - y$. Assuming small grazing angles, i.e. $\theta \ll 1$ rad, we may approximate $x(s, \xi) = \theta L/2 + \xi - s\theta$. The propagation inside the amplifier occurs with a nearly-constant beam cross section in order to optimise the overlap efficiency, therefore the gain can be calculated along the longitudinal s coordinate as in a ray-tracing approximation:

$$\frac{dI(\xi, y)}{ds} = \sigma n(x, y) I(\xi, y) \qquad (1)$$

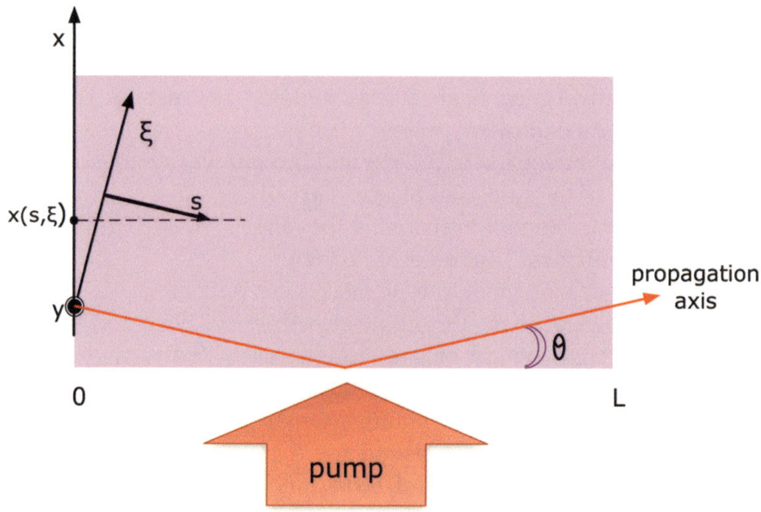

Fig. 1. Model of the slab amplifier (seen from above): the beam enters from the left side with a grazing angle θ. The beam cross section is most easily described by coordinates (ξ,y).

The saturated population inversion density is

$$n(x,y) = \frac{R_p(x,y)\tau}{1 + I(\xi,y)/I_s} \tag{2}$$

and the pump rate is given by

$$R_p(x,y) = \frac{\lambda_P}{\lambda_L}\frac{\alpha_p}{WL}\frac{P_{inc}\Theta(y)e^{-\alpha_p x}}{h\nu} \tag{3}$$

The step function Θ is defined as: $\Theta(y) = 1$ for $|y| < W/2$, $\Theta(y) = 0$ for $|y| > W/2$; σ is the stimulated emission cross section at the laser frequency $\nu = c/\lambda_L$; τ is the fluorescence time; $I_s = h\nu/(\sigma\tau)$ is the laser saturation intensity; α_P is the saturated absorption coefficient at the pump wavelength λ_P, corresponding to the incident pump intensity $I_P = P_{inc}/(WL)$ (P_{inc} is the power transmitted through the pump face and absorbed by the crystal), $\alpha_P = \alpha_{P0}/(1 + I_P/I_{SP})$ and I_{SP} is the pump absorption saturation intensity (Bermudez et al., 2002).
Inserting Eqs. (2) and (3) into Eq. (1) and integrating, one obtains

$$\ln\left[\frac{I_o(\xi,y)}{I_i(\xi,y)}\right] + \frac{I_o(\xi,y) - I_i(\xi,y)}{I_s} = \frac{\lambda_P}{\lambda_L}\frac{\alpha_P P_{inc}\Theta(y)}{WLI_s}\int_0^L ds\,\exp\left[-\alpha_p\left|x(\xi,s)\right|\right] \tag{4}$$

The double-pass model can be readily derived from Eqs. (1-3), requiring that the forward-travelling intensity equals the backward-travelling intensity at $s = L$. We assume that, after the first pass, the beam retraces its path backward through the amplifier. Eventually, we end up with two equations, one for the single-pass ($q = 1$) and one for the double-pass amplifier ($q = 1/2$):

$$q\ln\left[\frac{I_o(\xi,y)}{I_i(\xi,y)}\right]+\frac{I_o(\xi,y)-I_i(\xi,y)}{I_S}=\frac{\lambda_P}{\lambda_L}\frac{\alpha_P P_{inc}\Theta(y)}{WLI_S}\psi(\xi) \tag{5}$$

where

$$\psi(\xi)=\frac{2}{\alpha_P\theta}\left[1-\cosh(\alpha_P\xi)\exp\left(-\frac{\alpha_P\theta L}{2}\right)\right] \tag{6}$$

Notice that the small-signal gain coefficient is proportional to the right-hand side of Eq. (5): it is assumed to have a flat-top profile in the vertical direction while it has a smooth symmetric profile in the horizontal ξ-direction, with a peak on axis. This is a consequence of the bounce occurring at the mid-point of the slab. Moreover, thermal distortions that usually accompany the pump deposition are averaged accordingly, yielding the same symmetry as for the gain distribution along the ξ axis. Thermal lensing retains the usual distribution along the y axis.

As anticipated, an arbitrary seed beam profile can now be traced along the amplifier with Eqs. (5) and (6).

A particularly handy approximation, that speeds up the calculations significantly, consists in assuming an average gain coefficient over the horizontal seed diameter:

$$\overline{\psi}=\frac{2}{w_s}\int_0^{w_s/2}\psi(\xi)d\xi=\frac{2}{\alpha_P\theta}\left[1-\frac{2}{\alpha_P w_s}\sinh(\alpha_P w_s/2)\exp\left(-\alpha_P\theta L/2\right)\right] \tag{7}$$

Rather than the transverse intensity distribution of the rays with coordinates (ξ,y), the spatially-averaged Eq. (5) yields now a single output intensity, given the seed cross section $w_S \times W$ and the input intensity:

$$q\ln\left[\frac{I_o}{I_i}\right]+\frac{I_o-I_i}{I_S}=\frac{\lambda_P}{\lambda_L}\frac{\alpha_P P_{inc}}{WLI_S}\overline{\psi} \tag{8}$$

It is worth noticing that the crystal end faces limit the effective width of the input seed. The beam width may be comparable with the distance between the centre of the beam and the edge of the slab input face, i.e. $\theta L/2$. On the other hand, when increasing the angle θ there is a possibility that the beam width becomes comparable with its distance from the other edge of the slab, of width D. Therefore, the effective beam diameter to be amplified is the minimum out of the quantities w_S, θL and $2(D-\theta L/2)$.

2.2 Amplification in pulsed, low-frequency regime ($f \ll 1/\tau$)

In pulsed amplification regime, usually a pulse-picker selects a single pulse from a cw mode-locked laser and sends it to the amplifier which was previously pumped to achieve an appropriate gain level. Assuming pump pulse duration T, and averaging the spatial gain distribution as in Eq. (7), the pump rate equation yields the small-signal exponential gain:

$$\frac{dn}{dt}=R_P-\frac{n}{\tau} \tag{9}$$

$$g_0 = \left\langle \int_0^L \sigma n \, ds \right\rangle = \frac{2}{w_s} \int_0^{w_s/2} d\xi \left[\int_0^L \sigma n(\xi, s) \, ds \right] = \sigma R_p \tau \left(1 - e^{-T/\tau}\right) \overline{\psi} \tag{10}$$

$$g_0 = \frac{\lambda_P}{\lambda_L} \frac{E_{inc}}{WLF_S} \frac{\tau}{T} \left(1 - e^{-T/\tau}\right) \overline{\psi} \tag{11}$$

We may now use the Franz-Nodvik amplifier model (Franz & Nodvik, 1963) to calculate the output fluence F_o, given the input fluence F_i and the amplifier saturation fluence $F_S = h\nu/\sigma$:

$$F_o = F_S \ln\left[1 + e^{g_0} \left(e^{F_i/F_S} - 1\right)\right] \tag{12}$$

So far we have considered amplification in a single slab. However, the output fluence from the first slab may be used as the input for a second pass or to the next slab, and so on, allowing in general different beam sizes and grazing angles to be considered for each stage of the amplifier chain.

2.3 Amplification in pulsed, high-frequency regime ($f \gtrsim 1/\tau$)

This operating regime is mostly employed in laser systems for industrial applications, where high processing speed is extremely important, even at the expense of some pulse energy reduction. In this case a cw pump laser is chosen owing to duty-cycle limitations of more powerful qcw laser diode arrays employed at lower frequency (typically $< 1/\tau$).

The amplifier gain (or the inversion population) between two amplified pulses is still restored according to Eq. (9), starting from the gain g_f just after the amplified pulse and reaching g_i just before the next seed pulse enters the amplifier.

The Franz-Nodvik model can still be used, provided we use the right initial gain, which, in turns, depends on the pulse repetition frequency (Koechner, 2006):

$$F_o = F_S \ln\left[1 + \left(e^{F_i/F_S} - 1\right) e^{g_i}\right] \tag{13}$$

$$g_i = g_\infty - \left(g_\infty - g_f\right) e^{-\frac{1}{f\tau}} \tag{14}$$

$$F_o = F_i + \left(g_i - g_f\right) F_S \tag{15}$$

We notice that, according to Koechner's notation, $g_\infty = \lim_{f \to 0} g_i(f)$ hence we identify $g_\infty = g_0$ as given in Eq. (11).

Again, Eqs. (13-15) apply to single-pass amplification, but sequential application of the same model readily accounts for multi-pass schemes or amplifier chains.

2.4 Bandwidth limitation and gain narrowing

The most important of limitations arising from the finite bandwidth of the amplifier concern single- or multi-pulse amplification in pulsed regime, when small-signal gain is very high, say > 1000. This situation is commonly encountered in regenerative and multi-pass amplifiers (Walker et al., 1994; Le Blanc et al., 1996). However, single-pass amplification in a couple of grazing-incidence Nd:YVO_4 slabs readily yields 60 dB overall small-signal gain, leading to severe pulse broadening if too short seed pulses are injected.

In order to predict this effect quantitatively, we developed a numerical model based on Franz-Nodvik equations, adding gain filtering (of lorentzian form with fwhm = $\Delta\nu_g$) on each of the M slices into which the amplifier is sectioned. The basic assumption is that the gain depletion in each slice must be small enough (< 1%), so that we may apply the exact gain-filter model:

$$\hat{u}_{n+1}(\nu) = \hat{u}^{(u)}_{n+1}(\nu)\exp\left[\frac{(g_{n+1}+g_n)}{2}\frac{\gamma(\nu)}{2}\right] \tag{16}$$

$$\gamma(\nu) = \frac{1}{1+\left(2\nu/\Delta\nu_g\right)^2} \tag{17}$$

$$u_n(t) = \int_{-\infty}^{+\infty}\hat{u}_n(\nu)e^{-i2\pi\nu t}\,d\nu \tag{18}$$

where $\hat{u}^{(u)}_{n+1}(\nu)$ is the Fourier transform of the amplified (unfiltered) field $u^{(u)}_{n+1}(t)$ at the end of the (n+1)-th section, according to the Franz-Nodvik model:

$$G_{n+1} = \frac{e^{E_n/E_s}}{e^{E_n/E_s}-1+1/G^{(0)}_n} \tag{19}$$

$$E_n = \int_{-\infty}^{+\infty}\left|u_n^2(t)\right|dt \tag{20}$$

$$G^{(0)}_n = G_0^{1/M} \tag{21}$$

$$g_n = \ln(G_n) \tag{22}$$

Here G_0 is the total small-signal gain, and g_n is the result of the previous step. The condition for accurate computation is that

$$\frac{g_n - g_{n+1}}{g_n} \ll 1 \tag{23}$$

Therefore M is chosen accordingly. We notice that computation time might be conveniently reduced provided one chose a logarithmic distribution of amplifier sections lengths, since most severe saturation occurs at the end of the amplifier. However, in order to model amplifier gains as high as 60 dB, we found that M = 500 yields sufficiently accurate results with only few seconds of computation time on a 2.4-GHz laptop.

Let us consider typical data for a pulsed grazing-incidence amplifier: $\Delta\nu_g$ = 212 GHz (*a*-cut Nd:YVO_4, fluorescence fwhm = 0.8 nm (Zayhowski & Harrison, 1997)), F_S = 125 mJ/cm^2 and E_S = 188 µJ. We calculated the output energy and pulse width as a function of small-signal gain, input energy and seed pulse width. The results are summarised in Figs. 2-4.

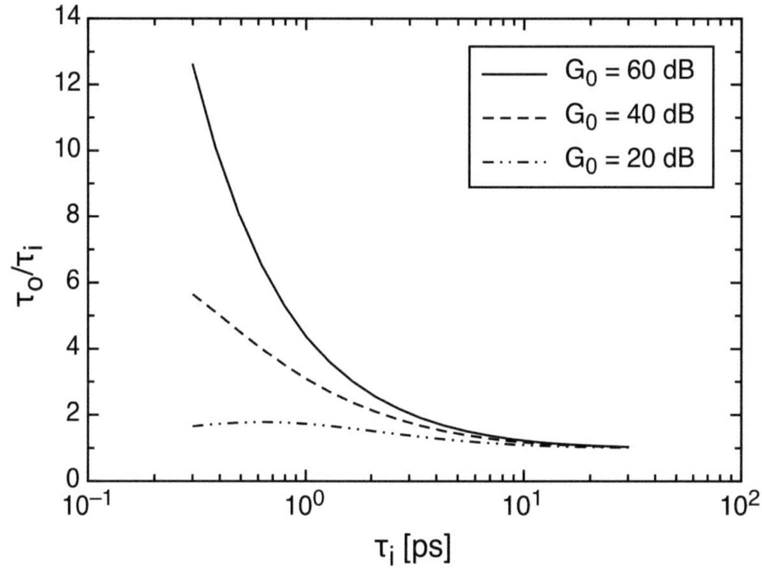

Fig. 2. Output pulse width broadening as a function of the seed pulse duration. Three small-signal gain levels are considered. Injected seed energy of 1 nJ is typical of most picosecond seeders.

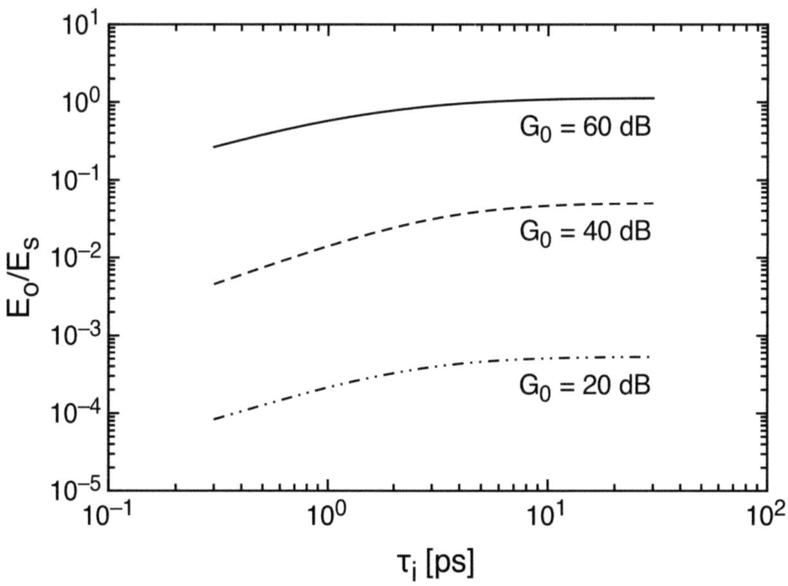

Fig. 3. Output energy as a function of the seed pulse width, for three small-signal gain levels. Injected seed energy is 1 nJ.

In conclusion, we note that seed pulse duration exceeding four times the minimum width set by the amplifier bandwidth (≈ 1.5 ps in this example, assuming *sech*^{2} pulse shape) produces no significant broadening after amplification even at an unsaturated gain of 60 dB (Fig. 2). The gain narrowing effect is most clearly seen at shorter seed pulse widths, as expected. Notice that at relatively low gain values, G_0 < 20 dB, the output pulse width reaches a maximum broadening as the seed pulse shortens, then approaches the seed duration provided it is short enough. The physical explanation is that the gain modulation is not as strong as with higher gain, so that only a relatively small central region of the seed spectrum is amplified but the overall non-amplified spectrum energy dominates, leading to an output pulse width still approaching that of the seed pulse.

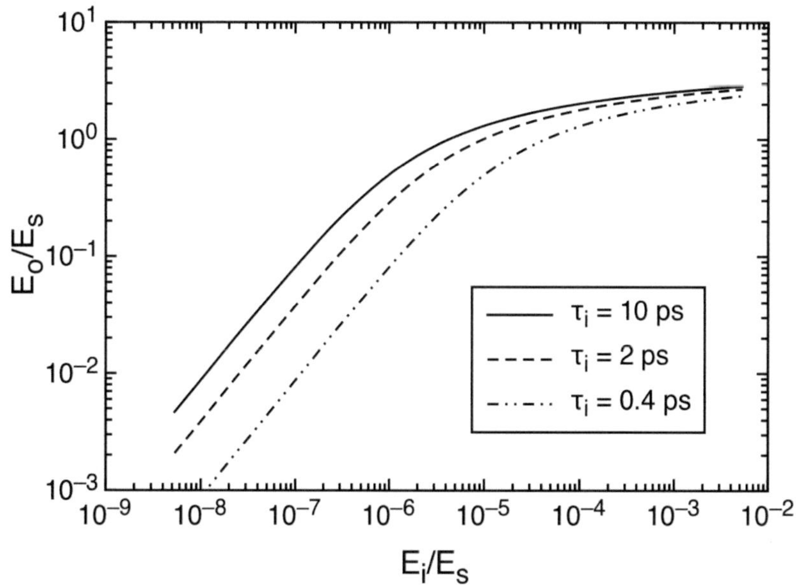

Fig. 4. Output energy as a function of the seed energy (both normalised to the saturation energy); three different seed pulse width have been considered.

As far as efficient energy extraction is of concern, Fig. 3 shows that with the given (typical) amplifier parameters a small-signal gain G_0 = 50-60 dB is required to reach saturation starting from the typical seed energy level ≈ 1 nJ. It is remarkable that, whilst this is usually achieved with either regenerative or extra-cavity multi-pass setups, typical gains of qcw grazing-incidence modules ≈ 30 dB/slab yield such gain levels with only two slabs or a single slab in two-passes, if only one is careful enough to suppress ASE.

Fig. 4 shows that gain saturation in the Nd:YVO_4 slab is optimised with seed pulse width > 6 ps: this means that, owing to gain narrowing, all the seed spectral energy is used for the effective amplification bandwidth.

2.5 Temporal distortions from amplifier saturation

This effect is most relevant when amplifying relatively long pulse trains, typically with an envelope duration > 50 ns for G_0 ~ 60-70 dB. It originates from the strong amplification experienced by the leading edge of the train envelope, which depletes the gain reducing the amplification of the trailing edge of the envelope.

The distortion can be readily calculated from the Franz-Nodvik model, given the power envelope $P(t)$ of the pulse train, the small-signal gain G_0 and the saturation energy E_S:

$$G(t) = \frac{e^{E_i(t)/E_s}}{e^{E_i(t)/E_s} - 1 + 1/G_0} \tag{24}$$

$$E_i(t) = \int_{-\infty}^{t} P_i(t')dt' \tag{25}$$

$$P_o(t) = G(t)P_i(t) \tag{26}$$

Therefore, the envelope modulation waveform required for pre-compensation of the saturation distortion can be easily computed, in order to yield a nearly flat-top output pulse from the amplifier. Owing to the strong amplifier saturation, the output energy reduction due to the pre-compensation is not very penalizing, notwithstanding the significant energy reduction of the modulated seed. This will be discussed in more detail in Section 5.

3. Amplification of a cw picosecond laser

Here we discuss the simplest setup that one can conceive, i.e. a single grazing incidence slab, side-pumped by a single diode array, for amplification of a low-power cw picosecond laser. In particular, the seeder is a Nd:YVO_4 oscillator (Fig. 5), longitudinally pumped by a 1-W laser diode at 808 nm. The laser is passively mode-locked with a SESAM, generating a 50-mW average power train of 6.8-ps, linearly polarized pulses at a 150-MHz repetition rate in each of the two diffraction-limited output beams.

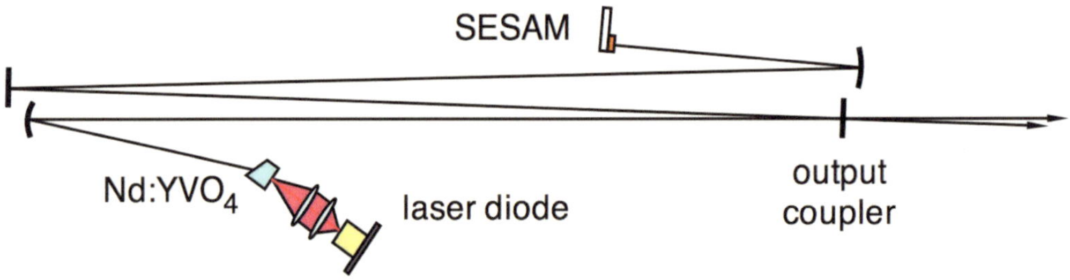

Fig. 5. Representative layout of a cw diode-pumped picosecond oscillator.

A 4 x 2 x 16 mm^3, *a*-cut, 6°-wedged, 1%-doped Nd:YVO_4 slab is employed as amplification head, pumped by a TE-polarised 10-mm x 1-µm laser diode array, emitting at 808 nm and collimated along the fast-axis by a 0.9-mm focal length microlens. This laser diode is actively cooled and temperature controlled with a thermoelectric cooler and emits a beam of maximum power of 32 W. The amplifier crystal is antireflection (AR) coated at 1064 nm on the input and output faces and simply polished and uncoated on the pumped side face. The Fresnel loss of about 14% of the incident pump power limits the absorbed pump power at 28 W. The 4 x 16 mm^2 slab faces are placed in contact with a water-cooled heat exchanger by thin indium foils. A proper choice of the water temperature set point is crucial in optimising the amplifier performances. The reason is twofold: i) minimisation of the thermal stress inside the slab and ii) reduction of the thermal red-shift of the fluorescence bandwidth of $Nd:YVO_4$, that may reduce the amplifier gain and performance. In this case the optimum operating temperature is 8°C. The collimated beam emitted by the pump diode is polarisation-rotated by a half-wave plate and focused into the $Nd:YVO_4$ slab with a 15-mm focal length cylindrical lens.

In order to estimate the vertical thickness of the gain sheet, the seed beam is collimated (0.9-mm half-width at $1/e^2$) and sent to the amplifier without any focusing optics. In these conditions, because of the strong gain shaping of the pump beam in the high-gain amplifier and of the great mismatch between the seed diameter and the gain-sheet thickness, the beam divergence along the y-direction is nearly pump-insensitive. Considering a flat-top y-profile for the output beam leaving the amplifier, the thickness W of the gain-sheet can be readily estimated by measuring the full vertical divergence angle $\theta_y \approx 2\lambda_L/W$. In this example it turns out $W \approx 70$ µm. Notwithstanding the clearly non optimised beam size, the seed beam yields up to 3.5 W in a single pass through the amplifier for an optimum internal grazing-incidence angle of ≈ 3°.

The setup for the cw amplification experiments is shown in Fig. 6.

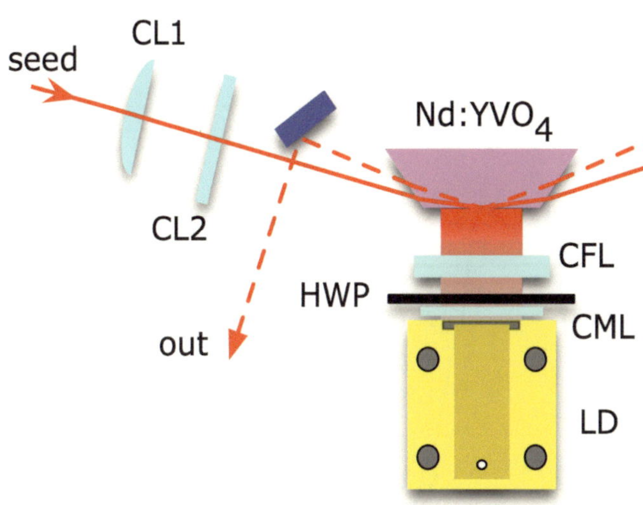

Fig. 6. Setup for the cw amplifier. CL1: 250-mm focal cylindrical focusing lens; CL2: 100-mm focal cylindrical focusing lens; SL1: 200-mm focal spherical lens; P: right-angle prism; CFL: 15-mm focal cylindrical pump focusing lens; HWP: half-wave plate; CML: collimation microlens (0.9-mm focal length).

Being the thickness W of the gain sheet set by the pump focusing lens and limited by the residual smile of the laser diode, the depth of the gain sheet in the horizontal plane is basically determined by the pump absorption depth, $1/\alpha_P \approx 0.5$ mm, given the typical pump spectrum width (≈ 2-3 nm) and the 1%-doping concentration of $Nd:YVO_4$. This generates an elliptical transverse gain profile that suggests the need for a different seed focusing in the horizontal and vertical directions. As shown in Fig. 6, the cylindrical lens CL1 focuses the oscillator beam down to $w_x \approx 100$ µm in the horizontal plane, whereas a cylindrical lens CL2 focuses the seed beam in the vertical plane to $w_y \approx 36$ µm. In order to choose the seed waist inside the amplifier, both the thickness of the pumped region and the Rayleigh range of the focused beam should be taken into account. In particular the pumped region length along the seed propagation direction should be shorter than two times the Rayleigh range of the focused beam in order to take advantage of the entire length of the active region. These considerations set a minimum acceptable dimension of ≈ 30 μm for the focused seed beam waist. Moreover, the grazing incidence angle has to be chosen as small as possible in order to maximise the gain while avoiding clipping effects.

With 34-mW seed injection the amplifier yields as much as 6.1 W after a single pass for the 28-W absorbed pump power, corresponding to a 22% optical-to-optical efficiency. The beam quality of the seed ($M^2 = 1.1$) is only slightly degraded to $M_x^2 = 1.3$ and $M_y^2 = 1.2$ at the maximum amplification level.

Fig. 7 shows the autocorrelation traces of both seed and amplified pulses and the correspondent optical spectra. The pulse duration after amplification is only slightly increased and the effect of spectral gain narrowing is quite evident. We notice that, owing to dishomogeneous broadening due to spatial hole burning (Flood et al., 1995), longitudinally-pumped picosecond lasers with gain at the end usually show excess bandwidth with respect to the measured pulsewidth. Therefore, in this case gain narrowing improves the pulsewidth × bandwidth product from 0.68 for the seed to 0.42 for the amplified pulses, approaching the Fourier limit of 0.32 for a *sech*^{2} pulse shape.

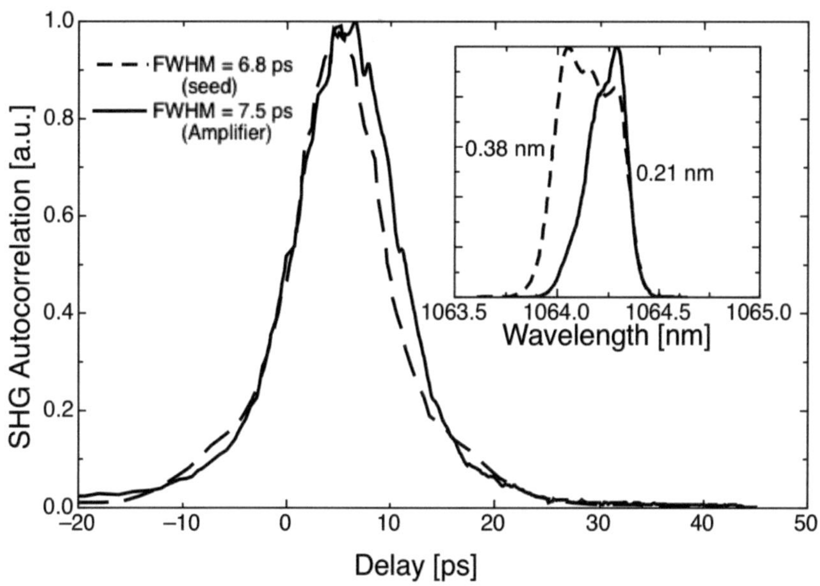

Fig. 7. Pulse autocorrelation of seed (dashed curve) and amplified (continuous curve) beam. Inset: corresponding optical spectra for seed and amplified pulses.

Double-pass amplification can be conveniently realised by re-imaging the single-pass output with a spherical lens SL1 (200-mm focal length) and a right-angle antireflection-coated prism P (see Fig. 5), thus maintaining the same spot size as in the first pass, but with a slightly increased grazing angle of ≈ 4°. Beam extraction occurs before the lens CL2.

The first consequence of realising a return path in the amplifier is the immediate growth of the amplified spontaneous emission (ASE) background. When the pump diode beam is focused for single-pass amplification, addition of the second pass boosts the ASE to > 1 W in this setup. In order to reduce the gain in the amplifier and hence the amount of ASE, the pump diode focusing must be relaxed, for example replacing the 15-mm cylindrical lens with a 20-mm lens.

This increases W to ≈ 95 µm. In these conditions ASE background reduces to < 200 mW and as much as 8.4 W are obtained at 28-W absorbed pump power after second pass amplification. The corresponding optical-to-optical efficiency increases to ≈ 30%.

Owing to a relatively small horizontal beam width compared to the horizontal transverse gain profile dimension, off-axis thermal aberrations in the critical horizontal plane are reasonably low. Therefore beam quality is well conserved also after the second pass ($M_x^2 = 1.4$ and $M_y^2 = 1.3$). Results similar to those reported in Fig. 7 are obtained for both pulse autocorrelation and spectrum.

Several different configurations were investigated, in single and double pass and also employing a single 300-mm focal spherical lens to focus the seed into the Nd:YVO$_4$ slab instead of the cylindrical optics shown in Fig. 6. Both the amplifier gain as a function of the injected seed power and the output power as a function of the absorbed pump power are summarised in Figs. 8 and 9.

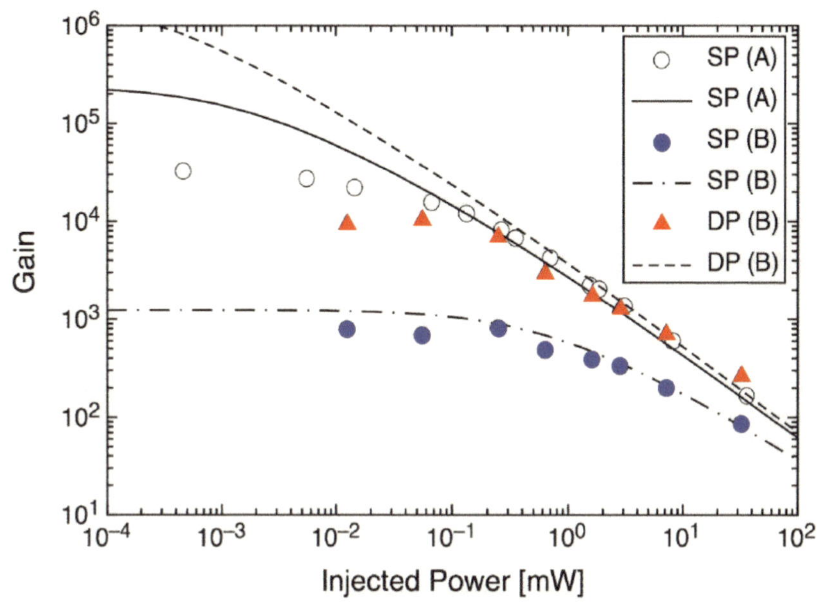

Fig. 8. Gain curves for the single- and double-pass amplifier. (A) Setup with pump focusing with 15-mm focal and seed focusing with cylindrical lenses of focal lengths f_x = 250 mm and f_y = 100 mm. (B) Setup with pump focusing with 20-mm focal and seed focusing with 300-mm focal spherical lens. SP single pass, DP double pass. Continuous curves: numerical results. Points: data.

Both in single or double pass setup, a small signal gain higher than 40 dB for the lowest input power is obtained. The model employed for fitting the experimental data is described in Section 2.1. The following physical parameters of the laser crystal can been used to fit the experimental data: $I_S = 2$ kW/cm², $\alpha_{P0} = 24$ cm⁻¹, $I_{SA} = 3.61$ kW/cm². The unsaturated absorption coefficient has been calculated assuming the absorption spectrum averaged over the laser diode emission spectrum. The best-fit saturation intensity value I_S yields a product $\sigma\tau$ about 30% smaller than the one usually reported in literature. Such a behaviour, beside intrinsic model approximations, can be explained by a reduction of the fluorescence lifetime due to up-conversion effects and by a reduction of the effective emission cross section due to the mismatch between the spectrum peaks of seed and amplified pulses (see the inset of Fig. 7).

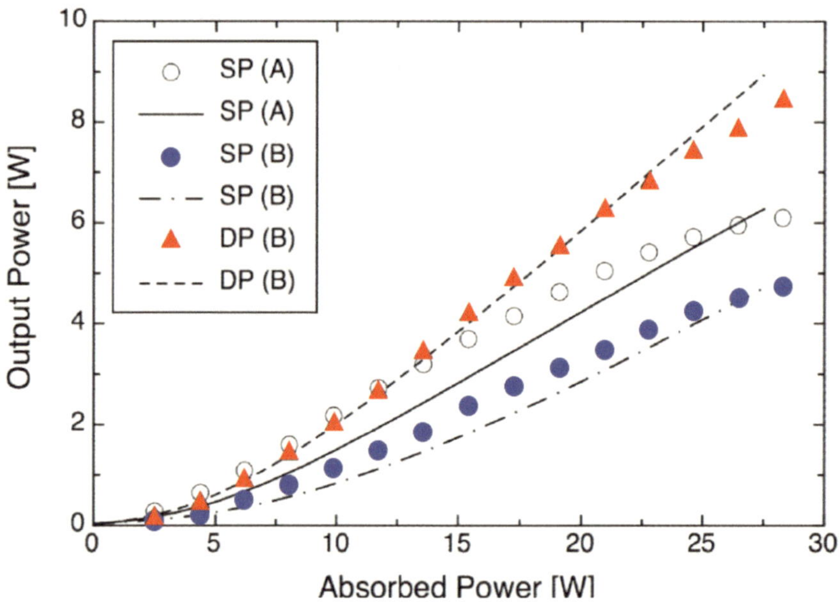

Fig. 9. Output power as a function of the absorbed pump power for the single- and double-pass amplifier. (A) Setup with pump focusing with 15-mm focal and seed focusing with cylindrical lenses of focal lengths $f_x = 250$ mm and $f_y = 100$ mm. (B) Setup with pump focusing with 20-mm focal and seed focusing with cylindrical lenses $f_x = 250$ mm and $f_y = 100$ mm. SP single pass, DP double pass. Continuous curves: numerical results. Points: data.

As clearly shown by Fig. 8, the amplifier behaviour tends to deviate from the model for decreasing injected power levels, suggesting that the maximum available gain is lower than that predicted. For very small injected power, gain saturates around $\approx 10^4$ owing to ASE, while gain saturation with higher input seed power levels recovers a fair agreement between predictions and experiments, since stored energy is more effectively extracted from the amplified beam rather than being wasted into wide-angle ASE.

4. Single-pulse amplification up to 1 kHz repetition rate

A typical, representative layout of a laser system for single picosecond pulse amplification after pulse-picking is shown in Fig. 10.

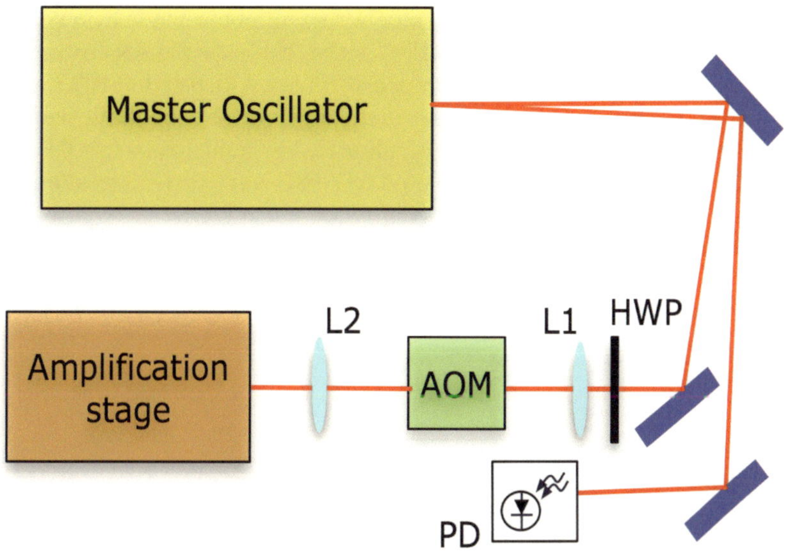

Fig. 10. Layout of a laser system for single pulse amplification. HWP: half-wave plate; L1: focusing spherical lens; L2: collimating spherical lens; AOM: acousto-optic modulator; PD: photodiode.

The master oscillator in this case is still similar to that used for cw experiments, providing two output beams each carrying out about 70-mW average power, 7.7-ps long pulses but with a reduced repetition rate ≈ 48-MHz. This last parameter is important since well-separated pulses are easier to pick up with optoelectronic devices.

The repetition frequency downscaling scheme for single pulse amplification employs an acousto-optic modulator (AOM) to pick up single pulses from the cw mode-locked oscillator. The major benefit of using AOM with respect to faster electro-optic modulators for pulse picking is the simpler and easier to customise driving electronics. Moreover AOMs are cheaper and do not require fast switching of high voltages that generates electromagnetic noise difficult to suppress. The pulse train sampled by the pulse picker has programmable repetition frequency, basically upper-limited by the duty cycle of the high-power diode arrays pumping the amplifier stage, that is typically 10%. In case of vanadate slab amplifiers, since the fluorescence time is ≈ 100 μs, the maximum operating frequency for qcw laser diodes bars is fixed at 1 kHz.

Given a temporal separation between two adjacent master oscillator pulses larger than the minimum deflection window allowed by the AOM, the key issues is to be able to drive the pulse picker synchronously with respect to the laser pulses. The correct timing can be realised conveniently by employing one of the two beam outputs from the master oscillator to generate a clock signal intrinsically synchronous to the pulse train for the pulse-picker driving electronics. In this example, a time separation between two adjacent pulses of about 20 ns allows a suppression ratio between selected and adjacent pulses better than 2% with a deflection efficiency exceeding 40%. If higher suppression ratio were an issue, a faster pulse picker, two acousto-optic pulse pickers in series or an oscillator running at a lower repetition rate might be employed.

The amplification stage setup is shown in detail in Fig. 11. The seed emerging from the pulse picker is injected into a grazing-incidence single-pass amplifier made by a couple of 4 x 2 x 16 mm^3, *a*-cut, 5°-wedged, 1%-doped Nd:YVO_4 slabs. The slabs are AR-coated at 1064 nm at the input and output faces, while the pumped sides are AR-coated at 808 nm. Each slab is pumped by a 150-W peak power qcw laser diode array with an emitting size of 10 mm x 1 µm. The radiation emitted by each diode is collimated by a microlens to ≈ 0.9-mm thickness x 10-mm width stripe and polarisation rotated by a half-wave plate to be aligned with the *c*-axis of the vanadate slabs. Notice that, owing to the much higher peak power of the diode arrays, tight focusing into the slab is no more required as in the cw amplifier of Section 3.

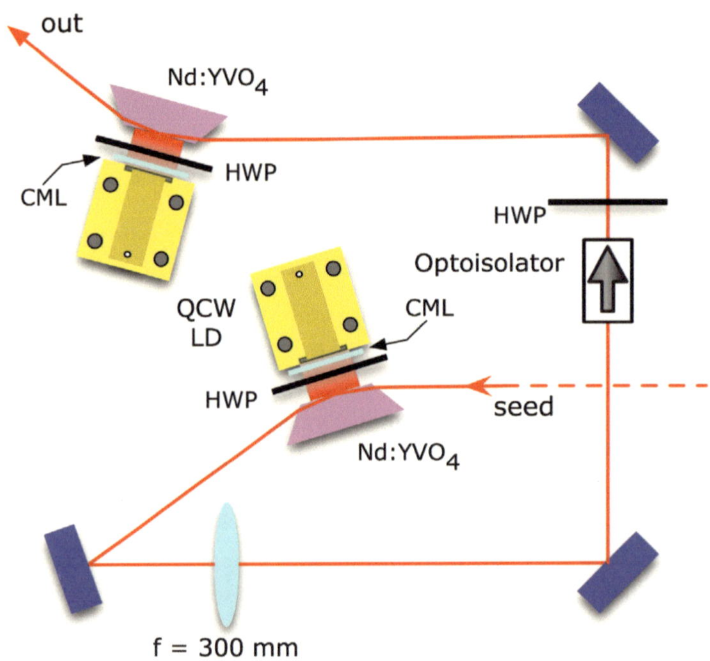

Fig. 11. Setup for the single-pulse amplification 2-slabs module. HWP: half-wave plate; CML: collimation microlens (0.9-mm focal).

In order to optimise the energy storage in the amplifiers the pump pulse duration is set to about 100 µs, slightly longer than the fluorescence time of the 1%-doped Nd:YVO_4 slab. The pulse picking occurs on the trailing edge of the pump current pulses, when the maximum gain is stored in the amplifiers medium.

Pulse duration is slightly increased after amplification. As shown in Fig. 12, fwhm 7.7-ps long seed pulses are slightly stretched to 11 ps long. The 0.37-nm wide pulse spectrum is substantially preserved during amplification, improving the quality parameter pulsewidth × bandwidth.

In order to fully exploit the potential of high-gain amplifier modules, besides the optimisations concerning the grazing angle as well as the matching between the seed beam waist and the gain layer, the main issue is to be able to contrast successfully those effects, such as amplified spontaneous emission (ASE), that compete with the amplifying beam in depleting the population inversion in the gain media. Since the ASE generated in the one

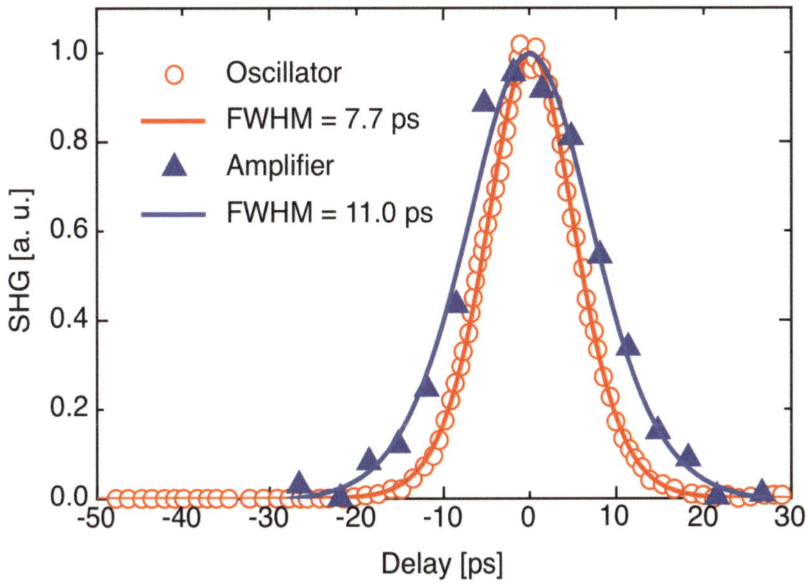

Fig. 12. Pulse autocorrelation of the oscillator (circles) and amplified (triangles) beam. Also shown are the best-fit traces corresponding to *sech*^{2} intensity envelope.

slab can seed the other, the two amplifiers must be kept sufficiently separated, about 30 cm in the final setup, in order to reduce the solid angle under which the first slab is seen by the second and consequently reduce the reciprocal ASE seeding. This solution yields significant performance improvements with respect to more closely separated slabs only few centimetres apart (Agnesi et al., 2006b). Furthermore, an optical isolator is inserted between the two amplifiers to prevent self oscillations and back-injection in the direction of the master oscillator. A spherical lens (300-mm focal length) placed at about 250 mm from the second slab re-collimates the spatial mode before entering the second amplifier.

A small-signal gain as high as $\approx 10^6$ ($\approx$ 60 dB) is measured for this double-slab module. Injecting pulses with energy of about 0.4 nJ, yields amplified pulses with energy as high as 210 μJ with weak gain saturation. Fig. 13 shows the energy extraction curve of the amplifier. The model for experimental data fitting is described in Section 2.2. The following best-fit parameters values can be employed: saturation fluence F_S = 133 mJ/cm^2 (correspondent effective emission cross-section $\sigma \approx 15 \times 10^{-19}$ cm^2), absorption coefficient α_P = 22 cm^{-1}, gain thickness W = 1.2 mm (flat-top approximation). For the seed beam it is assumed an equivalent flat-top profile diameter w_s = 0.6 and 0.8 mm in the first and second slab respectively, while the grazing incidence angle θ is 1.6° in both slabs, according to a direct measurement of the deviation angle at the slabs output for the configuration yielding the highest pulse energy. The calculated angular sensitivity of the second amplifier is shown in Fig. 14. The model confirms that the slab amplifiers are operated near the optimum predicted angle.

Moreover, the numerical model suggests some useful optimisation criteria for further improvements of the system performances. First of all, in order to fully exploit the gain available in the first slab it is necessary to reduce the transverse dimensions of the seed beam in order to match its confocal parameter with the length of the pumped region. Such a relatively small beam size allows to explore smaller grazing incidence angles without

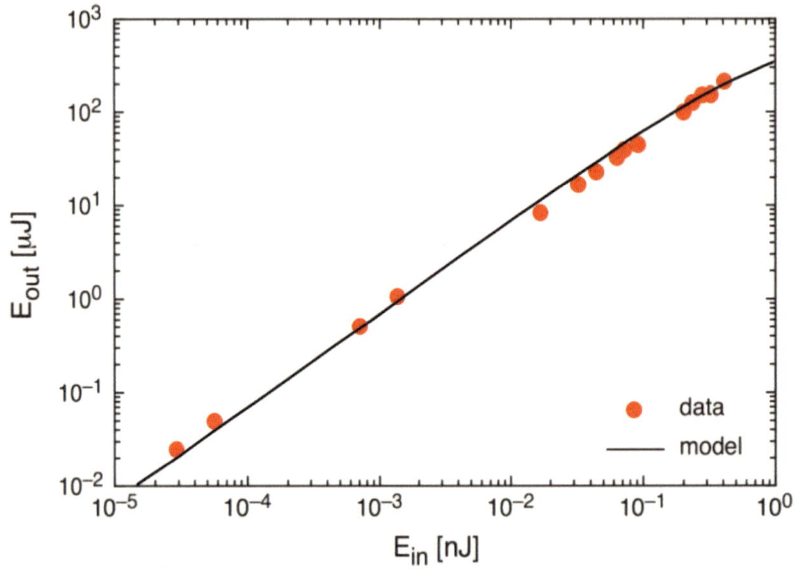

Fig. 13. Energy extraction curve of the amplifier (experiment and numerical model).

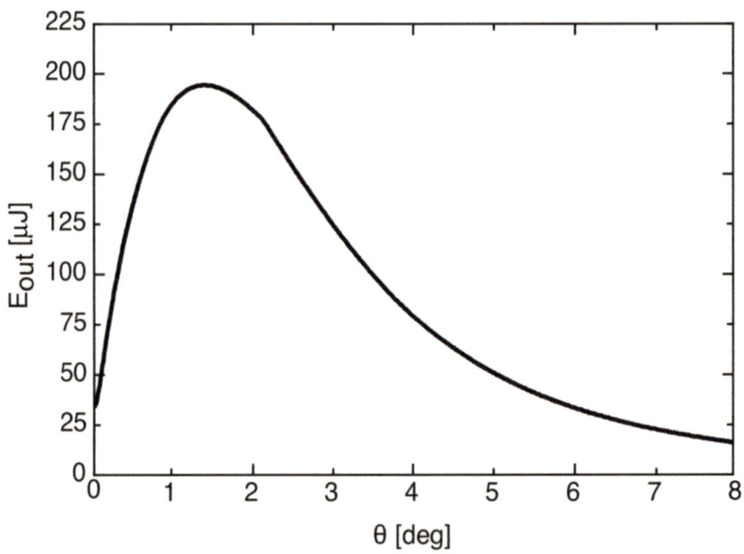

Fig. 14. Calculated energy extraction curve of the second amplifier as a function of the grazing incidence internal angle θ corresponding to a full pump power and maximum seeding injection.

clipping the beam, hence maximising the small-signal gain in the first unsaturated amplifier. Beam clipping should be carefully avoided, since it produces scattering and beam distortions that can substantially limit amplifiers performance. In the second slab the beam transverse dimensions and the grazing incidence angle should be increased in order to reduce non-linear effects and damage issues. The upper limit to input beam waist is given in this case by the gain region thickness W. According to these guidelines, output energy is predicted to increase up to 0.5 mJ with very small grazing incidence angles ($\approx$ 0.3°) in the first slab and beam radius $\approx$ 0.5 mm in the second slab.

It is worth noticing that for an accurate measurement of the output pulse energy it is necessary to remove the contribution of both ASE and undesired background pulses. In fact, since the static extinction ratio of the AOM is $\approx$ 1:2000, this value can be considered the attenuation factor for pulses far away from the selected one, while pulses coming immediately before and after the selected one will surely experience an even lower extinction ratio. All of these contributions can become significant after amplification and affect a direct measurement carried out with a power meter. However, the post-amplification pulse contrast will be significantly increased by second harmonic generation (SHG), since the integrated contribution of the small background pulses will be practically negligible and the only significant contribution will come from the main pulse. Following these considerations, the amplified pulse energy at the fundamental wavelength can be inferred from a measurement of both SHG conversion efficiency and second harmonic pulse energy. With this setup it was obtained up to 160 µJ at 532 nm with a conversion efficiency of 75% in an angularly phase-matched 3 x 3 x 8 mm^3, type-I LiB_3O_5 (LBO) crystal without any focusing, owing to the high peak intensity level ($\approx$ 4.8 GW/cm^2) of the nearly diffraction-limited amplified picosecond pulses. Fourth harmonic generation was also performed, employing the SHG pulse as the pump for a 3 x 3 x 7 mm^3 β-BaB_2O_4 (BBO) crystal. As much as 64 µJ at 266 nm were generated with 40% conversion efficiency.

The high peak power makes such laser sources ideal for simple, travelling wave optical parametric generation (OPG). With only 10 µJ at 1064 nm, 55%-efficiency SHG in a 15-mm LBO crystal yields $\approx$ 8-ps pulses at 532 nm for pumping an OPG, consisting in a 15-mm type-II $KTiOPO_4$ (KTP) for phase-matching in the xz plane. Tight focusing to $\approx$ 40-µm spot radius yields signal generation in the range 770-1020 nm (idler range was $\approx$ 1110-1720 nm) with angle tuning, with small efficiency $\approx$ 1%. This OPG might be readily amplified by a second stage pumped by a more energetic pulse at 532 nm, with much higher conversion efficiency and energy.

5. Amplification of high repetition rate pulse trains

Some applications requires amplification of bunches of pulses instead of single pulses or continuous trains. The repetition rate of the micro-pulses contained in the bunches is set by the master oscillator, while temporal shape, duration and repetition frequency of the micro-pulses envelope (macro-pulse) can be settled by a suitable pulse-picking device. High small-signal single-pass gain diode side-pumped grazing incidence amplifier modules can be employed also for macro-pulse amplification, owing to their good performances in terms of gain, beam quality and micro-pulse duration preservation, their relatively simple optical arrangement and their cost effectiveness.

The main issue, peculiar of this application, is related to the severe temporal distortions experienced by the envelope of the pulse burst during amplification, due to gain saturation and extremely high gain values. This effect is strongly dependent on the macro-pulse duration and in the end limits the possibility to obtain rectangular-shaped amplified envelopes for macro-pulses longer than few μs or even few hundreds of ns, depending on the amplifiers saturation level. In order to overcome amplification distortions it is necessary to inject in the amplifier chain a conveniently shaped macro-pulse with a temporal envelope designed to compensate saturation effects (Butterworth et al., 1996). This can be done by modulating the radio-frequency driving signal of the AOM pulse picker. An example of rectangular-shaped and custom-shaped macro-pulses is shown in Fig. 15.

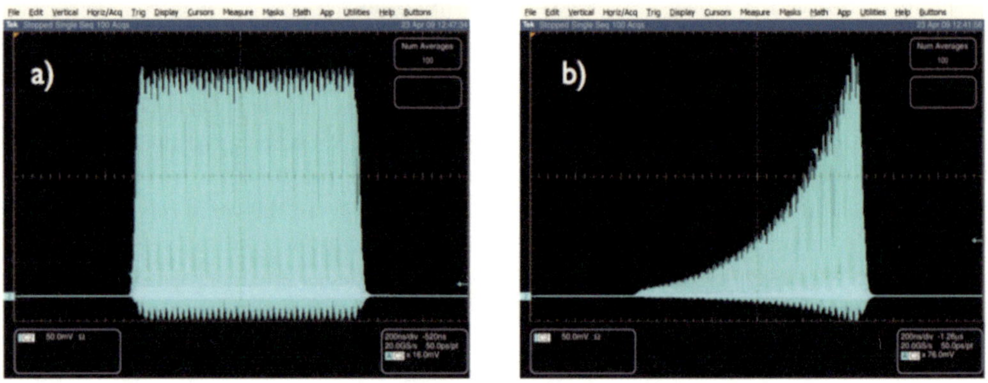

Fig. 15. Examples of two different deflected waveforms. a) Rectangular-shaped macro-pulse envelope; b) Macro-pulse envelope reshaped for amplification distortions compensation. Note that the energy of the modulated envelope is reduced to ~ 1/3.

In the example reported here, the master oscillator for pulse train amplification experiments is an high-frequency passively mode-locked Nd:YVO_4 oscillator (Agnesi et al., 2005). The ≈ 3-cm long V-folded resonator is pumped by a 1-W cw laser diode emitting at 808 nm, yielding ≈ 20-mW average power, ≈ 5-GHz repetition rate, 6-ps long mode-locking pulses.
The AOM pulse-picker yields typically 500-ns long macro-pulses containing approximately 2500 picosecond micro-pulses from the continuous train (any macro-pulse length > 50 ns can be chosen). The macro-pulses are amplified employing three Nd:YVO_4 slabs, as shown in Fig. 16.
Each 14 x 4 x 2 mm^3, *a*-cut, 1% doped vanadate slab is side-pumped by a pulsed 150-W peak power laser diode array. A couple of optical isolators are employed to prevent self-lasing and dangerous, high-intensity back-injections in the direction of the master oscillator. The deflected laser beam emerging from the acousto-optic modulator needs to be carefully collimated to a vertical dimension optimally matched to the pumped layer, while the path between the three amplification heads is kept sufficiently long in order to reduce the amount of ASE generated in the first amplifier and injected into the others. All these optimisations contribute to achieve a single-pass gain of ≈ 70 dB. As shown in Fig. 17, pulse duration increases from 6 ps to 8.9 ps after amplification.
A maximum output energy of about 2.5 mJ per macro-pulse is obtained after the three stage amplifier. The pulse spectrum after amplification is significantly narrowed: starting from a master oscillator 0.4-nm wide spectrum (fwhm), it shrinks to ≈ 0.19 nm at the output of the diode pumped amplifiers. Beam quality is well preserved ($M^2 \approx 1.2$) as Fig. 18 clearly suggests.

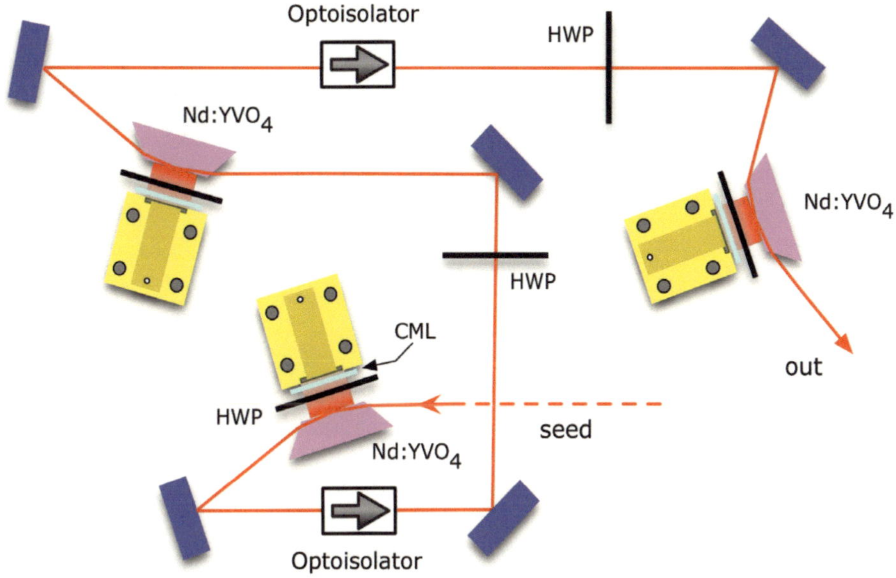

Fig. 16. Diode pumped amplifiers setup for pulse train amplification. Each laser diode beam is polarisation-rotated by a half-wave plate (HWP) and collimated by a 0.9-mm collimation microlens (CML).

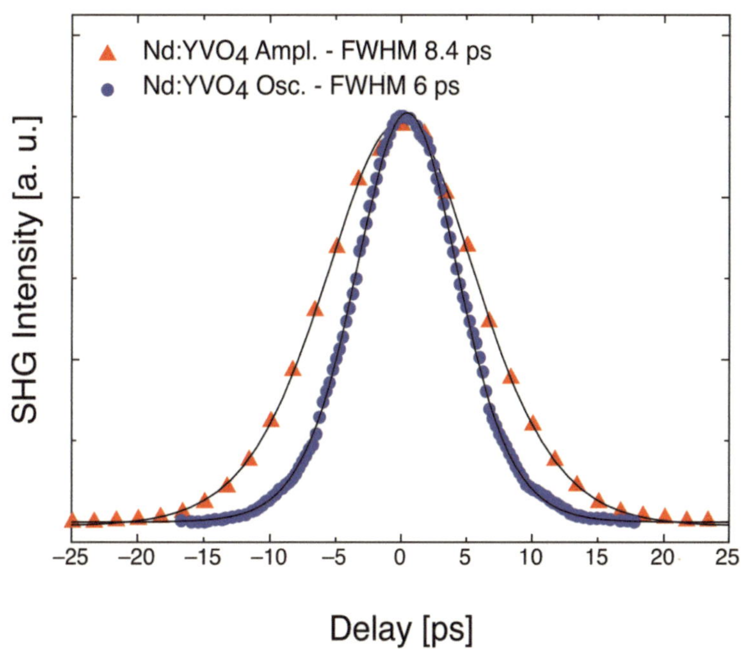

Fig. 17. Autocorrelation traces of both master oscillator (circles) and amplified pulses (triangles). Also shown the $sech^2$ experimental data fitting.

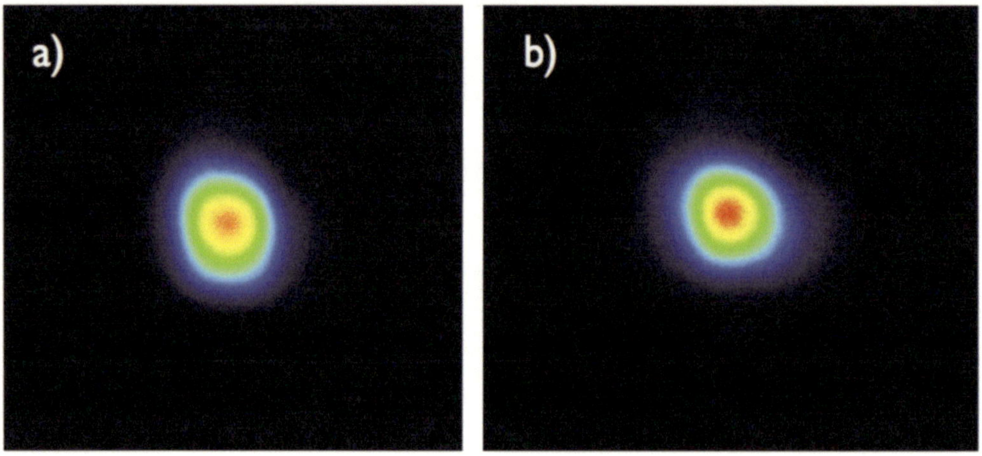

Fig. 18. Laser beam profile before amplification (a) and at the amplifier output (b).

Macro-pulse repetition rates up to 1 kHz can be chosen, owing to the duty cycle limitation of qcw diode arrays discussed previously.

A particular high-energy application required that the macro-pulses emerging from the diode pumped amplification stage was further amplified in a couple of 12-cm long, 6-mm diameter, flash-lamp pumped Nd:YAG rods.

At the maximum pump level, energy as high as 300 mJ was obtained for a 500-ns macro-pulse containing approximatively 2500, 12-ps long micro-pulses. The good beam quality ($M^2 = 1.5$) and peak power > 10 MW of the single pulses in the amplified burst allowed efficient nonlinear frequency conversion.

SHG efficiency as high as 60% at 532 nm was achieved in a 16-mm long type-I LBO crystal. The second harmonic output beam was used to synchronously pump an optical parametric oscillator (OPO) plane-plane cavity. The nonlinear medium was a 12-mm long AR-coated KTP crystal cut for type II phase-matching. Both the OPO mirrors and the crystal coatings were designed for singly-resonant oscillation around 800 nm, even though much broader tuning behaviour might be obtained in principle, with a suitable choice of mirror coatings. At the maximum pump energy at 532 nm, with a 40% transmitting output coupler the OPO generated about 60-mJ at 800 nm, in 500-ns long, rectangular-shaped macro-pulses.

6. Conclusions

Effective amplification of diode-pumped picosecond laser oscillators employing grazing-incidence side-pumped slab modules has been reviewed. Numerical models, useful for the design and the investigation of the amplifier characteristics, both in terms of energy extraction and pulse dynamics, have been illustrated. Some particularly representative examples of such amplification techniques, under various operating regimes, have been reported and discussed, comparing their results with the numerical model and putting them in perspective against other state of the art amplification schemes.

The most distinctive features of grazing-incidence amplifiers can be summarised as follows:

- extremely high gain that allows considerable setup simplification;
- strong gain shaping allowing beam quality preservation along the fast axis of the diode array, notwithstanding the strong thermal lensing in that direction;

- compactness, modularity and easy scalability.

It has been pointed out that picosecond laser amplification might be now take advantage of simple, rugged cw mode-locked fibre lasers, adding DPSS amplifiers such as grazing-incidence slabs. This is especially interesting for high-energy picosecond pulse amplification at multi-kHz up to MHz repetition rate.

Higher power and energy levels can readily be obtained by adding pump diodes or even diode stacks, instead of using a single diode array to pump the bounce face of the slab (Minassian et al., 2005).

Applications of this amplification technique can also be attractive for other laser materials such as Yb^{3+}-doped crystals (mostly Yb:YAG), Nd:glass, Nd:YAG and Nd:YLF. It is required that the absorption depth be small enough, and the absorbed pump power sufficiently high to produce a population inversion density yielding an integrated gain ≈ 20-30 dB per pass, in order to exploit all the useful features of the grazing-incidence slab. Particularly attractive are wide-band materials such as Nd:glasses that might allow high-energy femtosecond amplification with much simpler setups than usually reported in literature, at least using powerful qcw diode arrays at relatively low frequency (< 1 kHz). Quasi-three level materials such as Yb^{3+}-doped crystals also allow sub-picosecond and femtosecond amplification, however their re-absorption losses have to be taken into account properly and the pump geometry carefully designed to maximise efficiency. On the other hand, they are definitely more attractive than glasses for power upscaling, owing to their superior thermal and mechanical properties.

7. Acknowledgement

We acknowledge the contribution of Bright Solutions, Srl, for supporting and discussion. This research received funding from the European Community's Seventh Framework Programme FP7/2007-2011 under grant agreement n◦ 224042.

8. References

Agnesi, A.; Reali, G. C.; Kubecek, V.; Kumazaki, S.; Takagi, Y. & Yoshihara, K. (1993). β-barium borate and lithium triborate picosecond parametric oscillators pumped by a frequency-tripled passive negative-feedback mode-locked Nd:YAG laser. *Journal of the Optical Society of America B*, Vol. 10, No. 11, (November 1993) pp. 2211-2217, ISSN: 0740-3224

Agnesi, A.; Pirzio, F.; Tomaselli, A.; Reali, G. & Braggio, C. (2005). Multi-GHz tunable-repetition-rate mode-locked Nd:GdVO$_4$ laser. *Optics Express*, Vol. 13, No. 14, (July 2005) pp. 5302-5307, ISSN: 1094-4087

Agnesi, A.; Carrà, L.; Pirzio, F.; Reali, G.; Tomaselli, A.; Scarpa, D. & Vacchi, C. (2006a). Amplification of a low-power picosecond Nd:YVO$_4$ laser by a diode-laser side-pumped grazing-incidence slab amplifier. *IEEE Journal of Quantum Electronics*, Vol. 42, No. 8, (August 2006) pp. 772-776, ISSN: 0018-9197

Agnesi, A.; Carrà, L.; Pirzio, F.; Scarpa, D.; Tomaselli, A.; Reali, G.; Vacchi, C. & Braggio, C. (2006b). High-gain diode-pumped amplifier for generation of microjoule-level picosecond pulses. *Optics Express*, Vol. 14, No. 20, (October 2006) pp. 9244-9249, ISSN: 1094-4087

Agnesi, A.; Carrà, L.; Dallocchio, P.; Pirzio, F.; Reali, G.; Tomaselli, A.; Scarpa, D. & Vacchi, C. (2008a). 210-μJ picosecond pulses from a quasi-cw Nd:YVO_4 grazing-incidence two-stage slab amplifier package. *IEEE Journal of Quantum Electronics*, Vol. 44, No. 10, (October 2008) pp. 952-957, ISSN: 0018-9197

Agnesi, A.; Braggio, C.; Carrà, L.; Pirzio, F.; Lodo, S.; Messineo, G.; Scarpa, D.; Tomaselli, A.; Reali, G. & Vacchi, C. (2008b). Laser system generating 250-mJ bunches of 5-GHz repetition rate, 12-ps pulses. *Optics Express*, Vol. 16, No. 20, (September 2008) pp. 15811-15815, ISSN: 1094-4087

Bermudez, J. C. G.; Pinto-Robledo, V. J.; Kir'yanov, A.V. & Damzen M. J. (2002). The thermo-lensing effect in a grazing incidence, diode-side-pumped Nd:YVO_4 laser. *Optics Communications*, Vol. 210, No. 1-2, (September 2002) pp. 75-82, ISSN: 0030-4018

Bernard, J. E. & Alcock, A. J. (1993). High-efficiency diode-pumped Nd:YVO_4 slab laser. *Optics Letters*, Vol. 18, No. 12, (June 1993) pp. 968-970, ISSN: 0146-9592

Breitling, D.; Ruf, A. & Dausinger, F. (2004). Fundamental aspects in machining of metals with short and ultrashort laser pulses, Proceedings of SPIE Vol. 5339 Photon processing in microelectronics and photonics III, pp. 49-63, ISBN 9780819452474, San Jose CA (USA), January 2004, SPIE, Bellingham WA (USA)

Burns, D.; Hetterich, M.; Ferguson, A. I.; Bente, E.; Dawson, M. D.; Davies, J. I. & Bland, S. W. (2000). High-average power (>20-W) Nd:YVO_4 lasers mode locked by strain-compensated saturable Bragg reflectors. *Journal of the Optical Society of America B*, Vol. 17, No. 6, (June 2000) pp. 919-926, ISSN: 0740-3224

Butterworth, S. D.; Clarkson, W. A.; Moore, N.; Friel, G. J. & Hanna, D. C. (1996). High-power quasi-cw laser pulses via high-gain diode-pumped bulk amplifiers. *Optics Communications*, Vol. 131, No. 1-3, (October 1996) pp. 84-88, ISSN: 0030-4018

Damzen, M. J.; Trew, M.; Rosas, E. & Crofts G.J. (2001). Continuous-wave Nd:YVO_4 grazing-incidence laser with 22.5 W output power and 64% conversion efficiency. *Optics Communications*, Vol. 196, No. 1-6, (September 2001) pp. 237-241, ISSN: 0030-4018

Dausinger, F.; Hügel, H. & Konov, V. (2003). Micro-machining with ultrashort laser pulses: from basic understanding to technical applications, Proceedings of SPIE Vol. 5147 ALT'02 International Conference on Advanced Laser Technologies, pp. 106-115, ISBN 9780819450173, Adelboden (Switzerland), November 2003, SPIE, Bellingham WA (USA)

Edwards, G. S.; Hutson, M. S.; Hauger, S.; Kozub, J. A.; Shen, J.-H.; Shieh, C.; Topadze, K. & Joos, K. (2002). Comparison of OPO and Mark-III FEL for tissue ablation at 6.45 μm, Proceedings of SPIE Vol. 4633 Commercial and biomedical applications of ultrafast and free-electron lasers, pp. 194-200, ISBN 9780819443724, San Jose CA (USA), April 2002, SPIE, Bellingham WA (USA)

Farrell, D. J. & Damzen, M. J. (2007). High power scaling of a passively modelocked oscillator in a bounce geometry. *Optics Express*, Vol. 15, No. 8, (April 2007) pp. 4781-4786, ISSN: 1094-4087

Fermann, M. E. & Hartl, I. (2009). Ultrafast fiber laser technology. *IEEE Journal of Selected Topics in Quantum Electronics*, Vol. 15, No. 1, (January 2009) pp. 191-206, ISSN: 1077-260X

Flood, C. J.; Walker, D. R. & Van Driel, H. M. (1995). Effect of spatial hole burning in a mode-locked diode end-pumped Nd:YAG laser. *Optics Letters*, Vol. 20, No. 1, (January 1995) pp. 58-60, ISSN: 0146-9592

Frantz L. M. & Nodvik, J. S. (1963). Theory of pulse propagation in a laser amplifier. *Journal of Applied Physics*, Vol. 34, No. 8, (August 1963) pp. 2346-2349, ISSN: 0021-8979

Gerhard, C.; Druon, F.; Blandin, P.; Hanna, M.; Balembois, F.; Georges, P. & Falcoz, F. (2008). Efficient versatile-repetition rate picosecond source for material processing applications. *Applied Optics*, Vol. 47, No. 7, (March 2008) pp. 967-974, ISSN: 0003-6935

Keller U. (2003). Recent developments in compact ultrafast lasers. *Nature*, Vol. 424, (August 2003) pp. 831-838, ISSN: 0028-0836

Killi, A.; Dörring, J.; Morgner, U.; Lederer, M. J.; Frei, J. & Kopf, D. (2005). High speed electro-optical cavity dumping of mode-locked laser oscillators. *Optics Express*, Vol. 13, No. 6, (March 2005) pp. 1916-1922, ISSN: 1094-4087

Kleinbauer, J.; Knappe, R. & Wallenstein, R. (2005). 13-W picosecond $Nd:GdVO_4$ regenerative amplifier with 200-kHz repetition rate. *Applied Physics B*, Vol. 81, No. 2-3, (July 2005) pp. 163-166, ISSN: 0946-2171

Koechner, W. (2006). *Solid-State Laser Engineering*, Springer, ISBN 9780387290942, Berlin

Le Blanc, C.; Curley, P. & Salin, F. (1996). Gain-narrowing and gain-shifting of ultra-short pulses in Ti:sapphire amplifiers. *Optics Communications*, Vo. 131, No. 4-6, (November 1996) pp. 391-398, ISSN: 0030-4018

Mani, A. A.; Dreesel, L.; Hollander, P.; Humbert, C.; Caudano, Y.; Thiry, P. A. & Peremans, A. (2001). Pumping picosecond optical parametric oscillators by a pulsed Nd:YAG laser mode locked using a nonlinear mirror. *Applied Physics Letters*, Vol. 79, No. 13, (September 2001) pp. 1945-1947, ISSN: 0003-6951

McCarthy, M. J. & Hanna, D. C. (1993). All-solid-state synchronously pumped optical parametric oscillator. *Journal of the Optical Society of America B*, Vol. 10, No. 11, (November 1993) pp. 2180-2189, ISSN: 0740-3224

McDonagh, L.; Wallenstein, R. & Nebel, A. (2007). 111 W, 110 MHz repetition-rate, passively mode-locked TEM_{00} $Nd:YVO_4$ master oscillator power amplifier pumped at 888 nm. *Optics Letters*, Vol. 32, No. 10, (May 2007) pp. 1259-1261, ISSN: 0146-9592

Minassian, A.; Thompson, A.; Smith, G. & Damzen, M. J. (2005). High-power scaling (>100 W) of a diode-pumped TEM_{00} $Nd:GdVO_4$ laser system. *IEEE Journal of Selected Topics in Quantum Electronics*, Vol. 11, No. 3, (May/June 2005) pp. 621-625, ISSN: 1077-260X

Nawata, K.; Ojima, Y.; Okida, M.; Ogawa, T.& Omatsu T. (2006). Power scaling of a picosecond $Nd:YVO_4$ master-oscillator power-amplifier with a phase-conjugate mirror. *Optics Express*, Vol. 14, No. 22, (October 2006) pp. 10657-10662, ISSN: 1094-4087

Nawata, K.; Okida, M.; Furuki, K. & Omatsu, T. (2007). MW ps pulse generation at sub-MHz repetition rate from a phase-conjugate $Nd:YVO_4$ bounce amplifier. *Optics Express*, Vol. 15, No. 15, (July 2007) pp. 9123-9128, ISSN: 1094-4087

Neuhaus, J.; Kleinbauer, J.; Killi, A.; Weiler, S.; Sutter, D. & Dekorsy, T. (2008). Passively mode-locked Yb:YAG thin-disk laser with pulse energies exceeding 13 μJ by use of an active multipass geometry. *Optics Letters*, Vol. 33, No. 7, (April 2008) pp. 726-728, ISSN: 0146-9592

Ojima, Y.; Nawata, K. & Omatsu T. (2005). Over 10-W picosecond diffraction-limited output from a $Nd:YVO_4$ slab amplifier with a phase conjugate mirror. *Optics Express*, Vol. 13, No. 22, (October 2005) pp. 8993-8998,, ISSN: 1094-4087

Okhotnikov, O. G.; Gomes, L. A.; Xiang, N.; Jouhti, T.; Chin, A. K.; Singh, R. & Grudinin A. B.(2003). 980-nm Picosecond Fiber Laser. *IEEE Photonics Technology Letters*, Vol. 15, No. 11, (November 2003) pp. 1519-1521, ISSN: 1041-1135

Porta, J.; Grudinin, A. B.; Chen, Z. J.; Minelly, J. D. & Traynor, N. J. (1998). Environmentally stable picosecond ytterbium fiber laser with a broad tuning range. *Optics Letters*, Vol. 23, No. 8, (April 1998) pp. 615-617, ISSN: 0146-9592

Rufling, B.; Nebel, A. & Wallenstein, R. (2001). High-power picosecond LiB_3O_5 optical parametric oscillators tunable in the blue spectral range. *Applied Physics B*, Vol. 72, No. 2, (January 2001) pp. 137-149, ISSN: 0946-2171

Siebold, M.; Hornung, M.; Hein, J.; Paunescu, G.; Sauerbrey, R.; Bergmann, T. & Hollemann, G. (2004). A high-average power diode-pumped Nd:YVO_4 regenerative laser amplifier for picosecond-pulses. *Applied Physics B*, Vol. 78, No. 3-4, (February 2004) pp. 287-290, ISSN: 0946-2171

Snell, K. J.; Lee D.; Wall, K. F. & Moulton, P. F. (2000). Diode-pumped, high-power cw and modelocked Nd:YLF lasers, OSA Trends in Optics and Photonics Vol. 34, pp. 55-59, ISBN 1557526281, Davos (Switzerland), February 2000, Optical Society of America, Washington, D.C.

Sun, Z.; Ghotbi, M. & Ebrahimzadeh, M. (2007). Widely tunable picosecond optical parametric generation and amplification in BiB_3O_6. *Optics Express*, Vol. 15, No. 7, (April 2007) pp. 4139-4147, ISSN: 1094-4087

Walker, D. R.; Flood, C. J.; van Driel, H. M.; Greiner, U. J. & Klingerberg, H. H. (1994). High power diode-pumped Nd:YAG regenerative amplifier for picosecond pulses. *Applied Physics Letters*, Vol. 65, No. 16, (October 1994) pp. 1992-1994

Will, I.; Koss, G. & Templin, I. (2005). The upgraded photocatode laser of the TESLA Test Facility. *Nuclear Instruments and Methods in Physics Research. Section A*, Vol. 541, No. 3, (April 2005) 467-477, ISSN: 0168-9002

Zayhowski, J. J. & Harrison, J. (1997). Miniature solid-state lasers, in: *Handbook of Photonics*, Gupta, M. C., (Ed.), pp. 326-392, CRC Press, ISBN 0849389097, Boca Raton FL (USA)

12

Dynamics of Continuously Pumped Solid-State Regenerative Amplifiers

Mikhail Grishin and Andrejus Michailovas
Institute of Physics & EKSPLA uab
Lithuania

1. Introduction

Regenerative amplifiers are extensively used for amplifying pulses generated by mode-locked oscillators (Koechner, 2006). This is a powerful technique providing several orders of magnitude gain, virtually unlimited by amplified spontaneous emission, (the well known nemesis for multi-pass amplifiers) (Forget et al., 2002). Such uniquely high gain is achieved due to multiple passes of optical pulse through the gain medium. Multiple passes are organized by placing the gain media in an optical resonator. The number of round trips is typically controlled by an electro-optic switch (Nickel et al., 2005) [occasionally with acousto-optic modulator (Norris, 1992)]. The optical switch also provides quality control of the optical cavity, suppressing lasing and reducing the time period when parasitic amplification of spontaneous emission takes place. In addition, the optical cavity provides "filtering" of spatial mode (there is no spatial imperfection accumulation during multiple passes of amplifying pulse). Consequently, the possibility exists to obtain perfect beam quality. At the same time the stable resonator does not allow expanding mode diameter too much (Magni, 1986), restricting capabilities for amplifying optical pulses to very high intensity - when it is required this job is placed to subsequent high aperture amplifiers (Siebold et al., 2008). An auxiliary technique, vitally important to amplify femtosecond pulses to high energies, is chirped-pulse amplification (the stretcher-compressor scheme) (Strickland & Mourou, 1985; Mourou & Umstadter, 1992).

In respect to the gain, the regenerative amplifiers probably have only one competitor - fiber amplifiers (Fermann et al., 2002; Liu H. et al., 2008). These amplifiers as well as all lasers based on optical fibers exhibit impressive progress over the last decade (Jeong et al., 2004). However, the extremely small mode diameter and large medium length intrinsic for optical fibers lead to significant influence of detrimental nonlinear effects. These distort the amplifying signal's optical spectrum eventually resulting in serious limitation of peak power so that 0.7 MW before compression is one of the best achievements (Röser et al., 2007). Regenerative amplifiers are commonly designed with bulk materials allowing larger mode diameter that moves away the peak power limit well above tens of megawatts (Kleinbauer et al., 2008).

Most frequently used material for amplification of femtosecond pulses (and thoroughly dominating below 100 fs) is titanium doped sapphire. Broad gain bandwidth, exceptionally good lasing properties and opto-mechanical characteristics place this medium in such a special position. High pulse energy and high average output power have been achieved by using Ti:saphire amplifiers (Walker et al., 1999; Matsushima et al., 2006). The disadvantages of this gain medium are related to possible means of pumping. First, corresponding absorption lines are located in the green spectrum range, where suitable laser diodes with reasonable power are not available. Second, the short gain relaxation time (3.2 μs) requires pulsed pumping with high pulse energy (usually with Q-switched frequency-doubled neodymium lasers) in order to store substantial population inversion and consequently to obtain high output energy. We should remark here that the issues described in present chapter are not valid for Ti:saphire regenerative amplifiers since we use approximations which suitable only for long lifetime media. From the other hand due to very short lifetime the Ti:saphire amplifiers do not exhibit extraordinary dynamic properties on which we are mainly focusing in the chapter.

Another family of popular laser materials is ytterbium doped crystals and glasses. Their wide spectrum supports amplification of ultrashort pulses (however not as short as Ti:sapphire supports). Moreover ytterbium materials allow direct laser diode pumping with intrinsically small quantum defect (typically 10% or even less). The latter enhances overall power efficiency and reduce heat generation in active elements, that in turn alleviates thermal effects which inhibit average power increase. Long upper-level lifetime of ytterbium ions virtually in all the crystals and glasses allows good capacity of stored energy under convenient continuous laser diode pumping. What somewhat challenges operation with ytterbium materials is relatively low gain (typically less than 10% per pass) which originated from peculiar to this materials small stimulated emission cross section. From the other hand regenerative amplification is indeed a way of efficient energy extraction at low gain; just special attention should be given to reduce inrtracavity losses (Biswal et. al., 1998). An alternate way to improve stimulated emission features is cryogenically cooled active elements (Kawanaka et al., 2003) although this method is bulky and, as a rule, it narrows gain bandwidth. Then special attention should be placed to host crystal selection not to limit development towards shorter pulses (Pugžlys et al., 2009).

Ytterbium doped media are able to withstand very intense optical pumping without detriment to exited-state population which in other materials can be limited by quenching effects (e.g. exited state absorption or up-conversion). This favorable property permits use of active elements in thin disc geometry. In particular, Yb:YAG thin disk lasers are scalable to very high average power and to high pulse energies (Speiser & Giesen, 2008). Extremely short optical pass within the thin disc reduces nonlinear effects (in essence the Kerr effect) allowing high peak power pulses even without using stretcher-compressor technique (Kleinbauer et al., 2008).

The amplifiers based on neodymium gain media have their specific advantages. High stimulated emission cross section simplifies system design and reduces requirements for optical components. Good energy storage capabilities allow high output energy and power efficient operation when pumping with laser diodes. Neodymium laser materials are well suited for picosecond pulse durations and are competitive for moderate average power. Systems based on Nd:YVO_4 and Nd:$GdVO_4$ crystals routinely produce more than 10 W of output power (Kleinbauer et al., 2005; Clubley et al., 2008).

The regenerative amplifiers of solid state lasers designed for scientific applications usually operate at low or moderate repetition rates (not exceeding several kHz). Presently, there is rising demand for high repetition rate ultrashort-pulse solid state lasers for material micro-processing (Meijera et al., 2002). On the other hand, a new generation of fast electro-optic switches became available such as Pockels cells based on β-barium borate along with improved high-voltage electronics (Nickel et al., 2005; Siebold et al., 2004). As a result, picosecond and femtosecond lasers with repetition rate of the order of 100's kHz have come onto the market (Raciukaitis et al., 2006). Regenerative amplifiers are an important part of most ultrafast industrial solid state laser systems. Both high system efficiency and stable output parameters over a wide range of pulse repetition rates are essential for this actively developing field. For creation of power-efficient systems, neodymium and ytterbium laser gain media which may be directly pumped by laser diodes is advantageous. Long lifetime of the upper laser level typical of both these ions supports accumulation of substantial population inversion under continuous laser diode pumping. However, this long inversion lifetime may also cause stability problems at high repetition rates. Continuously pumped regenerative amplifiers demonstrate peculiar pulse amplification dynamics when the pulse repetition period becomes comparable or shorter than the gain relaxation time (Müller et al., 2003). Period doubling bifurcations develop generating periodically alternating energy pulses or even sequences of pulses having chaotic energy distribution.

Complex dynamic behavior is well known phenomenon in laser physics (Haken, 1975). Generally, nonlinear differential equations describing laser dynamics tend to have unstable solutions containing multi-stabilities and bifurcations when a number of independent variables representing system states are equal or more than three (Lorenz, 1963). By no means complete list of laser systems exhibiting complicated dynamics includes Q-switched gas lasers (Arecchi et al., 1982) passively Q-switched solid state lasers (Tang et al., 2003), optically injected solid state lasers (Valling et al., 2005). The specifics of high repetition rate regenerative amplifiers is such that their operation can be described with two differential equations, but periodic disturbance caused by release of the amplified pulse complicates the system behavior. Unlike many lasers which have been created specially to study dynamic phenomena and chaotic behavior, dynamics of regenerative amplifiers needs to be understood from a more utilitarian position in order to comprehensively optimize real systems. To date, only a few articles have been dedicated to this phenomenon despite its critical influence on the performance of regenerative amplifiers. Period doubling regime passing to chaotic operation has been observed for a system based on ytterbium doped glass and the role of bifurcations has been investigated both theoretically and experimentally (Dörring et al., 2004). However, one of the important parameters, the seed pulse energy, was left beyond the scope, and so applicability of the obtained results was restricted. The experiments were confined to studying cavity dumping of the Q-switched laser, an approximately equivalent system to the regenerative amplifier seeded by extremely low pulse energy. Our recent theoretical work has presented a generalized picture of stability features of a continuously pumped high repetition rate regenerative amplifier based on laser media with long relaxation time. The regions exhibiting different system behavior have been mapped in the space of non-dimensional control parameters: repetition rate and round trip number (Grishin et al., 2007). Additionally this analysis revealed the importance of the seed pulse energy and demonstrated that increase in the seed energy helps in eliminating the instabilities. Comprehensive utilization of these theoretical results has promoted top performance obtained from multi-kilohertz Yb:YAG disk amplifier (Metzger et al., 2009). Experimental study, performed with Nd:YVO_4 system, has thoroughly confirmed theoretical conclusions and on its basis the concept of regenerative amplifiers optimization has been formulated (Grishin et al., 2008; Grishin et al., 2009).

In the present chapter we summarize information from our already published papers and expand this description on the basis of our recent results. In the first theoretical part of the chapter we present a concept for modeling of continuously pumped solid state regenerative amplifiers. Simplified rate equations in normalized form allow reducing the system analysis to a task which has been thoroughly accomplished in theory of discrete-time dynamical systems (Alligood et al., 1996). Then we focus more attention to the influence of the main governing parameters on dynamics of regenerative amplification and limitations of the practical system performances occurred due to instabilities. We show why seed pulse energy plays such an important role at repetitive regime in contrast to low repetition rates. Influence of parasitic intracavity losses is shown to be a factor which not only decreases energy extraction efficiency (as in all laser systems) but also enhances instability range. We analyze numerically obtained stability diagrams which allow determination of optimal operation regime when maximum output energy can be extracted from the amplifier as stable pulse train. Important laser parameters which influence the system performance indirectly will be considered too. Heating of intracavity optical components depends not only on the range of output average power and the optical components quality but also on the amplification regime. The Kerr nonlinearity which limits system performance for short optical pulses will be evaluated taking into account multiple passes.
The verification of the created model is presented in the experimental part of the chapter. We investigate operation peculiarities of the system consisting of mode-locked master oscillator, the preamplifier and the regenerative amplifier based on $Nd:YVO_4$ crystals and analyze criteria of stable operation limitation. The experimental dependences of the output energy versus round trip number and repetition rate well agree with theoretical data. We demonstrate that increase of the seed pulse energy up to the value predicted by numerical model ensures stable operation within full range of repetition rates. The chapter is concluded by summarizing of presented theoretical and experimental results and suggesting of further investigations.

2. Theory of regenerative amplification in repetitive regime

2.1 Principle of operation

Regenerative amplifier can be regarded as a system in which an optical resonator provides multiple passes, a gain medium is responsible for amplification, and an electro-optic switch serves as a valve in-turn admitting weak input pulse and releasing amplified pulse. At that, spatial properties of the amplified radiation are primarily determined by the optical resonator; the output energy mostly depends on the population inversion stored in the gain medium whereas the amplified pulse duration is imposed by the input seed pulse. A schematic diagram of a conventional solid-state diode-pumped regenerative amplifier is depicted in Fig. 1. This optical layout by no means differs from that for a common Q-switched laser and consequently there are certain similarities in operation (Murray & Lowdermilk, 1980). The resonator quality (Q-factor) is controlled by an electro-optic switch. The electro-optic switch usually consists of a Pockels cell, a quarter-wave plate and a polarizer. An operational cycle of the regenerative amplifier consists of two successive stages: low-Q and high-Q. When voltage is not applied to the Pockels cell the wave plate along with the polarizer provides high intracavity losses (low-Q state of the resonator). Laser action is suppressed by high losses, and the gain medium, being under continuous pumping, accumulates population inversion. The amplification takes place during high-Q stage. It starts when quarter-wave voltage is applied to the Pockels cell and the seed pulse is injected into the resonator. The intracavity losses become minimal and are kept low for some pre-set time while the optical pulse circulating in the resonator is amplifying, simultaneously consuming a certain part of the stored energy. The intracavity pulse energy grows until the gain becomes equal to the resonator losses and then the pulse energy decays. As soon as the intracavity energy reaches a desired level the Pockels cell voltage is switched off. This dumps the amplified pulse out of the cavity as the output pulse. The system returns into the initial state. Then during the next low-Q stage the depleted part of the inversion population is restored by uninterrupted pumping and the cycles iterate.

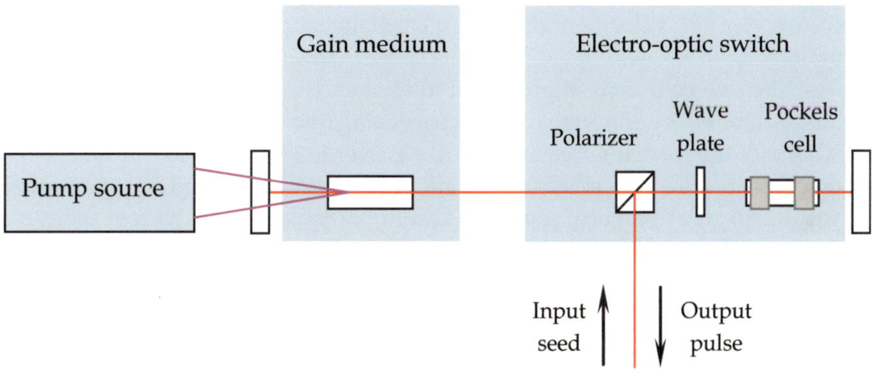

Fig. 1. Optical layout of typical solid-state diode pumped regenerative amplifier.

At higher repetition rates, when the pump stage is comparable or shorter than the gain relaxation time, the operation cycles become interdependent. The equilibrium between population inversion depletion caused by amplification and inversion restoration caused by pumping may become unstable. This often leads to breakdown of the single-energy regime and to generation of periodically alternating high/low energy pulses, or more complicated instability patterns. We will consider this and related phenomena theoretically in the first part of the present chapter.

2.2 Rate equations and basic terms

The process of regenerative amplification is essentially determined by the interaction of the intracavity radiation with the laser medium excited by pumping. The rules of this interaction can be established by using a simple phenomenological notion of a dynamic balance between basic processes: pumping and relaxation, amplification and extinction. Thus the system evolution can be described by coupled differential equations for the population inversion density N and photon number ϕ. We use the space independent rate equations which have been formulated for idealized four-level laser medium with a homogenously broadened line (Svelto, 1998):

$$\frac{dN}{dt} = R_p - \frac{\sigma c}{V}\phi N - \frac{N}{T_1} \tag{1}$$

$$\frac{d\phi}{dt} = \left(\frac{\sigma c L_a A_a}{V} N - \frac{1}{T_c}\right)\phi . \tag{2}$$

The density of population inversion grows due to pumping (proportionally to the pumping rate R_p) and decays with the constant rate inversely proportional to the upper laser level lifetime T_1. This balance is significantly influenced by stimulated emission, the contribution of which depends on the gain medium characteristics and particularly on the stimulated emission cross section σ. The second equation establishes a rule of stimulated photons multiplication. The photon number increase proportionally to the population inversion and at the same time it decreases because of intracavity losses. The losses here are defined in terms of the photon lifetime T_c which determines the rate of decay for the light field in the optical resonator. Geometry of the amplifier is accounted for in terms of the mode volume within the optical resonator (V), the active medium length (L_a) and the beam area in the active medium (A_a). The velocity of light c is a constant coupling temporal and spatial terms of the equations.

Since we use the rate equations approximation for modeling, the range of proposed model applicability is limited by the range of the particular rate equation's validity. Concerning general validity of the rate equation approach, we refer to (Svelto, 1998) for details. Here we just mention that this is a conventional method of laser dynamics study. It gives sufficiently accurate results for most practical purposes and in particular for analysis of multi-pass and regenerative amplification (Murray & Lowdermilk, 1980; Lowdermilk & Murray, 1980)

We shall re-arrange the equations to a form containing macroscopic non-dimensional terms. This is a common way to reduce the number of control parameters in order to provide simplified picture of system behavior without losing information. A primary normalization coefficient is the term proportional to the pump level, $G_0=R_pT_1\sigma L_a$. The physical meaning of G_0 is steady state gain coefficient per pass. Such gain is achieved at the equilibrium of population inversion density ($dN/dt=0$) obtained under constant pumping ($dR_p/dt=0$), constant rate of spontaneous decay (T_1) and in absence of lasing ($\phi=0$). Thus we can proceed to equations for the normalized gain $g=N\sigma L_a/G_0$ and normalized energy $\varepsilon=\phi\sigma/(A_aG_0)$:

$$\frac{dg(\tilde{t})}{d\tilde{t}}=-\varepsilon(\tilde{t})g(\tilde{t})+\frac{1-g(\tilde{t})}{\tau_1} \tag{3}$$

$$\frac{d\varepsilon(\tilde{t})}{d\tilde{t}}=\varepsilon(\tilde{t})\left[g(\tilde{t})-g_t\right]. \tag{4}$$

Additionally, we have modified the time scale by introducing normalized time $\tilde{t}=tG_0/T_0$, a product of the current number of cavity round trips and the steady state gain coefficient. Here the denominator T_0 represents the round trip time. As we will see later, the term $\tilde{t}$ assigns a natural time scale to the high-Q phase of operation. The governing parameter, which defines the amplification period (effective round trip number), will be introduced on this basis. The normalized relaxation time of the gain $\tau_1=T_1G_0/T_0$ is essentially determined by the upper laser level lifetime T_1. A term accounting for parasitic losses of the optical resonator is represented by normalized threshold gain, $g_t=T_0/(T_cG_0)$. This parameter can be also expressed in terms of loss coefficient l as $g_t=-\ln(1-l)/G_0$, by using common threshold criterion of gain and losses equality.

Note that equations formulated with those new variables do not contain pumping characteristics in explicit form. Parameter G_0, proportional to the pumping rate, is hidden in the composition of basic variables. The pump effect (as well as other control parameters effects) is easy to restore when applying modeling results by performing reciprocal transformation from normalized to real parameters. Moreover, further in the theoretical part of this chapter we often omit the words "effective" and "normalized" just for shortening. Although Eqs. (3) and (4) have been obtained in the approximation of a four-level system, the identical equations can be formulated for the more general case of quasi-three-level systems [the initial rate equations and conditions of their applicability can be found in (Svelto, 1998)]. In the latter case the explicit expressions of normalized and effective terms look slightly more complicated but their physical meaning remains unchanged.

2.3 Consideration of a single operation cycle

Below in this section we define relations between basic system parameters deriving the rate equations for separate stages of regenerative amplifier operation. First we consider the low-Q phase. Since during this phase of operation amplification is suppressed while pumping takes place, it is also called pump stage. Inasmuch as there is no lasing during this stage (i.e. ε=0), a set of Eqs. (3) and (4) transforms into a single equation:

$$\frac{dg(t^{\sim})}{dt^{\sim}} = \frac{1-g(t^{\sim})}{\tau_1}. \tag{5}$$

Initial conditions specify the gain at the beginning of the pump phase, g_{pi}. Taking into account that $t^{\sim}/\tau_1=t/T_1$, we can find the relation between the initial gain g_{pi} and the final gain g_{pf} for a certain pump phase duration T:

$$g_{pf} = 1-\left(1-g_{pi}\right)\exp\left(-\frac{T}{T_1}\right) \tag{6}$$

In the next section we will often use diminished form of the equation (6): $g_{pf} = \hat{g}_p(g_{pi}, T/T_1)$, where the function $\hat{g}_p$ just establishes a rule of the gain transformation $g_{pi} \rightarrow g_{pf}$ for the pump stage.

Then let us proceed to the high-Q phase. The equations for this operation stage can be simplified as well. We remind here that we explore laser media having long relaxation time, the case functionally important for diode-pumped systems. Since the buildup time of the optical pulse is usually short in comparison with the pump phase duration and the upper laser level lifetime, the population inversion change due to pumping and relaxation processes is much smaller than the inversion depletion caused by amplification. Hence, the terms containing spontaneous decay and optical pumping can be neglected in Eq. (3). Also, as we have assumed negligible pump contribution during high-Q phase, this stage of operation can be called in a more common and more informative manner as the amplification stage.

The next assumption presumes low intra-cavity losses ($g_t \rightarrow 0$). This approximation substantially simplifies the basics of the presented method. The simplification appears not only due to existence of the analytical solution for the rate equations but also, more importantly, because of the reduced number of parameters governing the system. An influence of the parasitic losses on amplification dynamics will be accounted in section 2.7 after the essence of the theoretical approach has been presented. For the present, we come to the situation at which Eqs. (3) and (4), when describing amplification stage, reduce to the following:

$$\frac{dg(t^{\sim})}{dt^{\sim}} = -\varepsilon(t^{\sim})g(t^{\sim}) \tag{7}$$

$$\frac{d\varepsilon(t^{\sim})}{dt^{\sim}} = \varepsilon(t^{\sim})g(t^{\sim}) . \tag{8}$$

Initial conditions specify the system state at the beginning of the amplification phase: the initial gain g_{ai} and the energy of the input pulse from which the amplification starts – the seed energy, ε_s. Also it is natural to constrain our consideration to the case of low seed energy with respect to the stored energy ($\varepsilon_s << g_{ai}$ in terms of normalized parameters). As a consequence of these assumptions, the solutions of coupled Eqs. (7) and (8) can be found in analytic form. Such solutions obviously describe temporal evolution of the gain and intracavity energy. One can also find other physical sense, more convenient for further consideration. In case of fixed amplification phase duration, temporal evolution is terminated at the moment $t^{\sim}=\tau$. Then the solutions also express how the system parameters on the amplification phase completion (final gain g_{af} and output pulse energy ε_f) depend on the initial conditions and governing parameters:

$$g(t^{\sim}) = \frac{g_{ai}}{1+\frac{\varepsilon_s}{g_{ai}}\exp\left(g_{ai}t^{\sim}\right)} \Rightarrow g_{af} = \frac{g_{ai}}{1+\frac{\varepsilon_s}{g_{ai}}\exp\left(g_{ai}\tau\right)} \tag{9}$$

$$\varepsilon(t^{\sim}) = \frac{g_{ai}}{1+\frac{g_{ai}}{\varepsilon_s}\exp\left(-g_{ai}t^{\sim}\right)} \Rightarrow \varepsilon_f = \frac{g_{ai}}{1+\frac{g_{ai}}{\varepsilon_s}\exp\left(-g_{ai}\tau\right)} . \tag{10}$$

Here the terms τ and ε_s are actual parameters controlling the amplification, whereas the initial gain is a variable coupling the equations. The normalized amplification stage duration τ is a product of cavity round trip number and steady state gain G_0. Further we will call this term in a more comprehensible manner an "effective round trip number". Essentially, the term τ also represents total multi-pass small signal gain factor for the amplifier. The Equations (9) and (10) can be presented in diminished form: $g_{af} = \hat{g}_a(g_{ai}, \varepsilon_s, \tau)$ and $\varepsilon_f = \hat{\varepsilon}(g_{ai}, \varepsilon_s, \tau)$, simplifying further operation with these formulas. Now, as the basic terms have been introduced, we can summarize a rule for subscript notations. The pump and amplification operation phases are designated with p and a subscripts respectively; the initial and final states are notated with i and f subscripts correspondingly.

2.4 Coupling of successive cycles. Discrete-time dynamical system approach

Evaluation of the output energy is a trivial task for low repetition rates, i.e. when the pump phase duration significantly exceeds the inversion relaxation time, $T>>T_1$. The gain in this case reaches saturation before the amplification phase starts, that is the initial gain g_{ai} tends to unity. Consequently, the output energy versus round trip number is strictly defined by Eq. (10) alone. In general case, and particularly at high pulse repetition rate the initial conditions for the current operation stage depend on the previous system state. Hence, the initial gain for each cycle depends not only on operation parameters but also on the system pre-history. In order to determine the gain at the beginning of amplification we shall relate final and initial states of successive operation cycles.

Now we proceed from evaluation of the gain and pulse energy within single operation stages to a description of those stages as a whole by analyzing the temporal evolution of the boundary values. We assign initial and final gains and pulse energy as variables defining the system state. The term $g_{ai}(1)$ is introduced as the initial gain of the amplification phase for the first cycle of operation. This stage finishes with the final gain denoted as $g_{af}(1)$. The subsequent pump phase of the current cycle obviously begins from the same gain value, $g_{pi}(1)=g_{af}(1)$. Similarly, the gain evolution continuity should be taken into account for coupling of all operation cycles. There is a boundary relation $g_{af}(k)=g_{pi}(k)$ within the cycle number k and for subsequent cycles: $g_{pf}(k)=g_{ai}(k+1)$. The legend of the gain evolution in a discrete time scale can be presented as follows:

$$...g_{ai}(k) \to g_{af}(k) = g_{pi}(k) \to g_{pf}(k) = g_{ai}(k+1) \to g_{af}(k+1) = g_{pi}(k+1) \to g_{pf}(k+1)... \quad (11)$$

The corresponding time boundary points of neighboring operation stages can be described as $t_{ai}(k)=(k-1)(\tau T_0/G_0+T)$ and $t_{af}(k)=k\,\tau T_0/G_0+(k-1)T$. Note that in an assumption of short amplification phase, the duration of complete cycle (dumping period) is approximately equal to the pump phase duration T. Hence the term $(T/T_1)^{-1}$ represents normalized pulse repetition rate of the regenerative amplifier (also called dumping frequency).

Analogous transition to the discrete time scale can be applied to energy evolution. However, unlike continuity of the gain evolution, the build-up of intra-cavity energy $\varepsilon_s \to \varepsilon_f(k)$ interrupts at the end of the amplification phase of each cycle at the moment of pulse dumping and then it begins again from the constant level which corresponds to the seed pulse energy ε_s. Hence the term $\varepsilon_f(k)$, determining the output energy, does not depend of its own pre-history. It is dependent on the gain and can be found from Eq. (10) for any operation cycle provided that the gain is known. Therefore, the gain becomes the only independent variable that needs to be analyzed.

We can consider the expressions obtained earlier [$\hat{g}_p$ for the pump phase and $\hat{g}_a$ for the amplification phase explicitly represented by Eq. (6) and Eq. (9)] as the rules of the system state updating. These solutions of the rate equations serve as transformation rules that take the current state as input and update it by producing a new state. This new output state becomes the input for the next stage of operation. Then it is possible to combine amplification and pump phases within certain operation cycle and to form a joint gain transformation rule. We introduce $\hat{g}_\Sigma$ as the composition of functions $\hat{g}_a$ and $\hat{g}_p$ exhibiting the gain transformation rule for the complete cycle, $\hat{g}_\Sigma = \hat{g}_p(\hat{g}_a)$. Then using expressions of the inner functions we can present an explicit form of the function $\hat{g}_\Sigma$:

$$\hat{g}_\Sigma(g_{ai}) = 1 - \left[1 - \frac{g_{ai}}{1 + \frac{\varepsilon_s}{g_{ai}}\exp(g_{ai}\tau)}\right]\exp\left(-\frac{T}{T_1}\right). \quad (12)$$

Thus, we have reduced the regenerative amplification to the evolution of single variable (system state, g_{ai}) in a discrete time scale; and also we have found a rule of this variable updating. The basic properties of this updating function fit to the mathematical definition of so called maps [functions whose domain (input) space and range (output) space are the same]. Then the regenerative amplification can be described by using the theory of one-dimensional discrete-time dynamical systems (one-dimensional maps) (Alligood et al., 1996). The sequence of the system states $g_{ai}(1)$, $g_{ai}(2)$, … $g_{ai}(k)$ … is called an orbit in terms of this theory. The orbits can be calculated by using a recurrent formula determining the subsequent state of the system in terms of the present state: $g_{ai}(k+1) = \hat{g}_{\Sigma}[g_{ai}(k)]$.

It is obvious that in a regular single-energy regime the gain depletion during the amplification phase should be compensated by restoring the population inversion during the pump phase. In terms of states evolution, the initial gain of the amplification stage eventually should iterate, i.e. there is a certain gain value (designated as g_1) such that the subsequent gains stabilize upon reaching that value, $g_{ai}(k+1)=g_{ai}(k)=g_1$. Consequently, the system eigenstate satisfying the condition $g_1 = \hat{g}_{\Sigma}(g_1)$ should exist. The solution of this equation is known as a fixed point in the discrete-time dynamical system theory. Being exactly equal to the fixed point, the system state reproduces itself after each cycle that leads to operating in a regular manner. However, requirement of technical feasibility of such a regime establishes a more strict condition to be fulfilled. The system should return to the fixed point after some perturbation has occurred, in other terms, more common for theory of dynamical systems, the fixed point should be attracting. Thus, study of regenerative amplification is reduced to the analysis of conditions of the fixed point existence and its stability characterization.

Since the equation determining the fixed points is transcendental we start analysis of the system state evolution with the graphical illustration. For more intuitive presentation, the fixed point existence condition [$g_1 = \hat{g}_{\Sigma}(g_1)$ with the explicit form of $\hat{g}_{\Sigma}$ given by Eq. (12)] is rearranged into the following form:

$$1-(1-g_1)\exp\left(\frac{T}{T_1}\right) = \frac{g_1}{1+\frac{\varepsilon_s}{g_1}\exp(g_1\tau)}. \tag{13}$$

The right-hand part of this expression represents the gain transformation function during amplification, $\hat{g}_a$ [see Eq. (9)]. The left-hand part may be regarded as an inverse function of the gain recovery during the pump phase [transformed Eq. (6) gives $g_{pi} = \hat{g}_p^{-1}(g_{pf})$]. Since gain continuity implies equality of the boundary states ($g_{ai}=g_{pf}$ and $g_{af}=g_{pi}$) we can combine both curves g_{af} versus g_{ai} and g_{pi} versus g_{pf} on a common diagram (Fig. 2). A space of system states, defined by this means, can give an intuitively simple but strict and fruitful picture of the system state evolution. The intersection of those curves, having clear physical meaning, gives solution of Eq. (13), i.e. it determines the fixed point of the system. It is important that the intersection of these curves always exists and it is always single for any set of control parameters. Really, the amplification stage curve is single-peaked, it begins from zero and always lies under the state space diagonal ($g_{af}=g_{ai}$). The letter is natural because during the amplification stage (provided, as we assumed, negligible pumping contribution) the population inversion is depleting by transforming to the intracavity pulse energy and respectively the gain can only decrease, $g_{af}<g_{ai}$. The pump stage curve is a straight line whose slope depends on the repetition rate. This curve begins in the right upper corner of

the state space [(1, 1) point] and also always locates under the state space diagonal. Note, the state space, due to proper normalizing, has dimensions of (0-1)×(0-1). Thus, these curves cannot help intersecting and they intersect only once. Moreover, since the basic properties of the curves are universal, a fixed point existence and uniqueness is not only the result of mathematical speculations obtained under certain approximations but also the consequence of inherent physical properties of regenerative amplification.

One of two necessary requirements for existence of a stable single-energy regime, namely the fixed point uniqueness, is fulfilled and then the main concern is the fixed point stability study. Figures 2(a) and 2(b) represent diagrams of system state evolution for two typical cases. Figure 2(a) presents the orbit converging into an attracting fixed point. Such a convergence means that the regenerative amplifier eventually (after sufficient number of reiterations, when initial value of the orbit is "forgotten") starts producing regular pulsing. It is intuitively seen that the behavior of the resulting orbit (convergent or non-convergent) depends on the slope of the "amplification" curve in the fixed point with respect to the slope of the "pump" curve. Strictly speaking, the fixed point becomes attracting if the derivative of $\hat{g}_\Sigma$ function in this point satisfies the requirement $|\hat{g}'_\Sigma(g_1)| < 1$ (Alligood et al., 1996). The condition $|\hat{g}'_\Sigma(g_1)| = 1$ represents the transition point between stable and unstable operation. In case of $|\hat{g}'_\Sigma(g_1)| > 1$ the fixed point is repelling, and consequently stable operation becomes unfeasible.

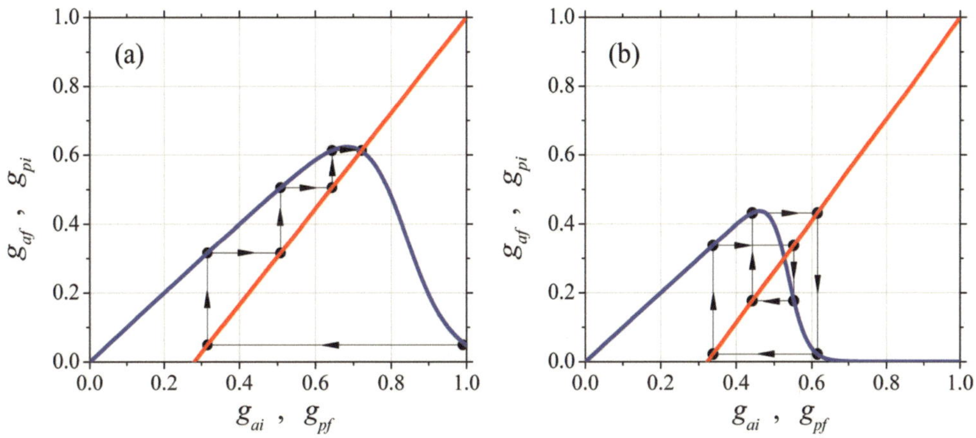

Fig. 2. Graphical presentation of the orbits in state space. The fixed point is the intersection of the "amplification" and "pump" (blue and red) curves. Transition to the stable (attracting) fixed point (a) at ε_s=3×10^{-7}; τ=18.0; T_1/T=3.0. Period-4T orbit (b) at ε_s=10^{-10}; τ=42.0; T_1/T=2.56.

It becomes apparent that in the latter case the system is unable to reproduce its own state after one cycle of operation. However, such a iteration may occur after two or several cycles. The corresponding set of system states is called periodic orbit. An example of periodic orbit is depicted in Fig. 2(b). The condition for existence of the orbit with double period, 2T can be written by introducing an appropriate composite function. Define $\hat{g}^2_\Sigma = \hat{g}_\Sigma(\hat{g}_\Sigma)$ to be the result of applying the map-function $\hat{g}_\Sigma$ to the system state two times. The system state g_2, such that $g_2 = \hat{g}_\Sigma^2(g_2)$, is the fixed point analogue but suited for two successive operation cycles. Generally the orbit with the period of mT exists if there is a system eigenstate g_m satisfying the equation: $g_m = \hat{g}_\Sigma^m(g_m)$. Here the term m is an integer number exhibiting a factor of output pulse repeatability for the corresponding multi-energy regime. If such a regime is realized, the system produces quasi-periodic sequence of the output pulses. The pulses of identical magnitude in this sequence appear each time in a multiplied period equaled to mT. In much the same way as the existence of a fixed point does not ensure stable operation, the existence of a periodic orbit does not in itself mean that the corresponding regime is realizable. Additional analysis of the orbit stability is required, that, similar to the fixed point case, reduces to evaluation of the corresponding map-function derivative. The orbit $g_{ai}(k)$ of period-m is stable provided that $|(\hat{g}_\Sigma^m)'(g_m)| < 1$. Computation of the derivative for this composite function is feasible since its value eventually (at $k \to \infty$) tends to the product of its inner function derivatives at points along the orbit: $(\hat{g}_\Sigma^m)'(g_m) = \hat{g}_\Sigma'[g_{ai}(1)]\hat{g}_\Sigma'[g_{ai}(2)] \cdots \hat{g}_\Sigma'[g_{ai}(k)]$. If the absolute value of the product of the derivatives is larger than one, then periodicity of the orbit becomes unfeasible meaning that the system exhibits chaotic behavior. This is the Lyapunov number criterion of deterministic chaos (Alligood et al., 1996).

We can remind here that the map-function $\hat{g}_\Sigma$ is itself a function of system parameters (τ, ε_s, T/T_1). As one of the governing parameters is varied the corresponding fixed point passes through different states of stability. A pass through the position $|(\hat{g}_\Sigma^m)'[g_{ai}(k)]| = 1$ causes qualitative change of the system operation. Such transitions (e.g. transition from stable to unstable regime at m=1) is usually called a bifurcation. A set of control parameters (operation point) at which the bifurcation occurs is referred to as the bifurcation point. Correspondingly the diagram of the output parameter versus one of the control parameters for the system exhibiting bifurcations is often called a bifurcation diagram. Among many possible types of bifurcations, known for dynamical systems, we have met here the bifurcation of period doubling. This relatively simple type of dynamic behavior is one of the consequences of the fixed point uniqueness. This attribute gives also primary unambiguity of the system behavior. The dynamics of regenerative amplification and output characteristics of the system are determined by the set of control parameters alone in contrast to e.g. bi-stability effects where initial value of the orbit, $g_{ai}(1)$ may also influence the operation. One can imagine the latter as qualitative change of system behavior caused by a way of switching it on (e.g. in practice either one has the pumping source enabled first and then seeding or other way around). The unambiguous relation between control parameters and operation regimes is quite an important property of regenerative amplifiers and our further analysis always implies this property without necessarily mentioning it.

2.5 Diagrams of dynamic regimes in the parameter space

Simplifying assumptions and non-dimensional effective parameters, introduced for the basic rate equation model, reduce the number of independent control parameters of the system to the set of three. These are the normalized repetition rate, T_1/T, amplification phase duration expressed in terms of the effective round trip number τ and normalized seed energy ε_s. Analysis of stability for orbits of the initial gain of amplification phase [$g_{ai}(k)$] at each given control parameter provides thorough picture of regenerative amplifier behavior.

The orbits were calculated by iterating of Eq. (12) in the range of control parameters wide enough to comprehend all the relevant dynamics features: $0.2<(T_1/T)<20$; $10<\tau<110$; $10^{-11}<\varepsilon_s<10^{-3}$. We used as much as 3000 iterations, the sufficient number to be confident that the results are independent of the system state at the beginning of iterating. The orbits were analyzed in two stages. At first, the minimal number of cycles between repeating system states was revealed for each orbit in parameter space. It was performed by direct comparison of the system state sequences with themselves but shifted by a certain cycle number, $g_{ai}(k)$ versus $g_{ai}(k+m)$. In that way the periodic orbits up to m=32, including regular ones (m=1), were identified. Then the Lyapunov number criterion was applied to the residual unidentified orbits. They were separated into two fundamentally different bunches: chaotic and eventually periodic.

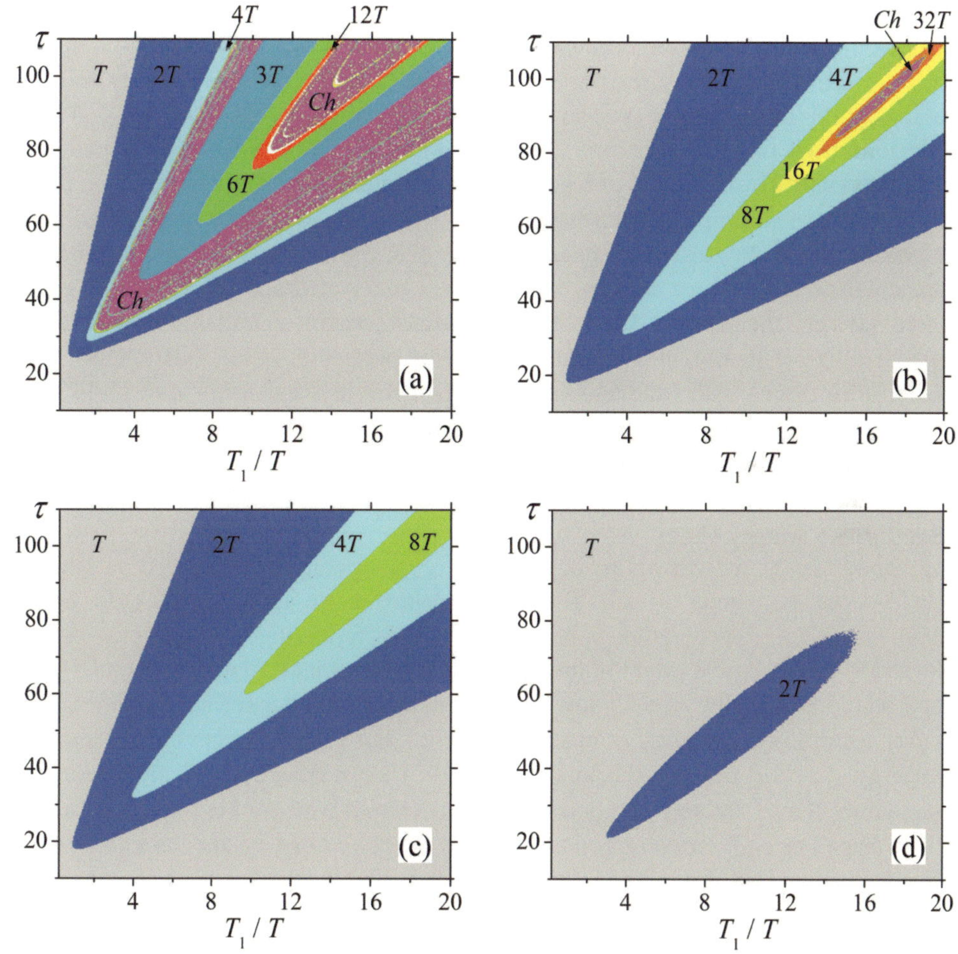

Fig. 3. Diagrams of amplification dynamics in parameter space for different seed energies: $\varepsilon_s=10^{-10}$ (a); $\varepsilon_s=2.5\times10^{-7}$ (b); $\varepsilon_s=3\times10^{-7}$ (c); $\varepsilon_s=1.3\times10^{-4}$ (d).

Thus, the following dynamic regimes were distinguished in three-dimensional space of control parameters in accordance with the orbits properties: (1) The orbits evolving into stable fixed points (m=1) corresponding to the regular system behavior (single pulse energy output, i.e. 1T-regime). (2) Periodic orbits corresponding to multi-energy regimes with repeatability coefficients in the range of $2 \le m \le 32$. (3) Eventually periodic orbits having larger repeatability factor (m>32), for which the m-number itself is not identified. (4) Regime of deterministic chaos in accordance with the Lyapunov number criterion.

The regions of different dynamics are mapped in space of the repetition rate – round trip number (Fig. 3). The major part of the parameter space is occupied by the regions corresponding to the following regimes: single-energy (1T); quasi-periodic with fundamental period of two (2T, 4T, 8T, 16T, and 32T); quasi-periodic with fundamental period of three (3T, 6T, and 12T); and chaotic behavior. These domains are marked with different colors, whereas the rest of the space containing the remaining zones of eventually periodic orbits is left white. The boundaries between adjacent colors (i.e. between different regimes) represent manifolds of bifurcation points in parameter space.

As it is seen, the dynamics turned out multifarious. Chaotic regime ordinarily comes out from the chain of successive period doubling bifurcations: T-2T-4T-8T-16T-32T… The chaotic zone itself has fine structure. Quasi-periodic "windows" with various periods are disseminated in it. The dynamics of regenerative amplification strongly depends on the seed value. The pattern is complex for low seed level ($\varepsilon_s < 10^{-9}$), the parameter space contains more than one clearly distinguishable chaotic regions [Fig. 3(a)]. Quasi-periodic regimes with fundamental period of three are observed between zones of chaotic dynamics. The higher the seed energy, the simpler the instability pattern becomes. Initially, chaotic domain shrinks to ellipse [Fig. 3(b)] and disappears from the parameter space. Furthermore, period doubling bifurcations with fundamental period of two only remain for $\varepsilon_s > 2.52 \times 10^{-7}$. Then the maximum order of bifurcations decreases [Figs. 3(c) and 3(d)] and finally, at $\varepsilon_s > 1.9 \times 10^{-4}$ the system becomes stable in the whole range of control parameters.

2.6 Seed pulse energy effect

The obtained results, demonstrating dependence of the operation on the seed pulse energy, are in essence not quite trivial. This phenomenon is in controversy with intuitive comprehension of regenerative amplification. The following speculations seem to demonstrate the negligible extent of the seed influence or at least to evidence much simpler looking relations. Imagine, initially low seed pulse energy ε_{s1} after certain number of round trips ($\Delta\tau$) is amplified to energy ε_{s2} of several orders of magnitude larger but still much less than energy stored in the gain media, $g_{ai} >> \varepsilon_{s2} >> \varepsilon_{s1}$. Then further amplification should give the same output as if the amplification has initially started with seed energy ε_{s2} because the previous stage ($\varepsilon_{s1} \rightarrow \varepsilon_{s2}$) virtually has not changed the stored energy and, as a consequence, the system gain. This logic leads to an inference that a lack of seed energy can be compensated with additional round trips. Consequently, the operation diagrams have to look identical but shifted in the coordinate of round trip number for different seed values. Accurate computations give absolutely different results, Fig. 3.

Let us consider in details some subtleties which lead to this difference. In time-domain the amplification process looks exactly as simple logic predicts, that can be confirmed with straightforward calculations. Equation 10 gives equal but shifted in time intracavity energy evolutions, $\varepsilon(g_{ai}, \varepsilon_{s1}, \tilde{t}) = \varepsilon(g_{ai}, \varepsilon_{s2}, \tilde{t} - \Delta\tau)$ and this shift, $\Delta\tau$ can be determined as:

$g_{ai}\Delta\tau = \ln\frac{g_{ai}}{\varepsilon_{s1}} - \ln\frac{g_{ai}}{\varepsilon_{s2}}$. This logic is accurate when we are considering amplification phase as "isolated" with given initial gain; this is absolutely true for low repetition rates.

Previously it was commonly accepted that regenerative amplification virtually is independent of the seed energy because only low repetition rates were under consideration. In reality, even for low repetition rate systems the seed energy value should not be too low. However, the reason for that is rather different of what we are describing. Simply, competing parasitic processes of amplification of spontaneous emission always exists in regenerative amplifiers. Thus, the seed energy should be well above the spontaneous emission level in order to get the amplified seed at the output instead of amplified spontaneous emission. However, at low repetition rates the sufficient pulse seed energy is extremely small, e.g. down to 10^{-15} J in accordance to the experiments presented in the classical paper (Murray & Lowdermilk, 1980). Although, some exceptions may take place; they involve special applications which require very clean, high contrast optical pulse, e.g. parametric chirped pulse amplification (Ross et al., 2007).

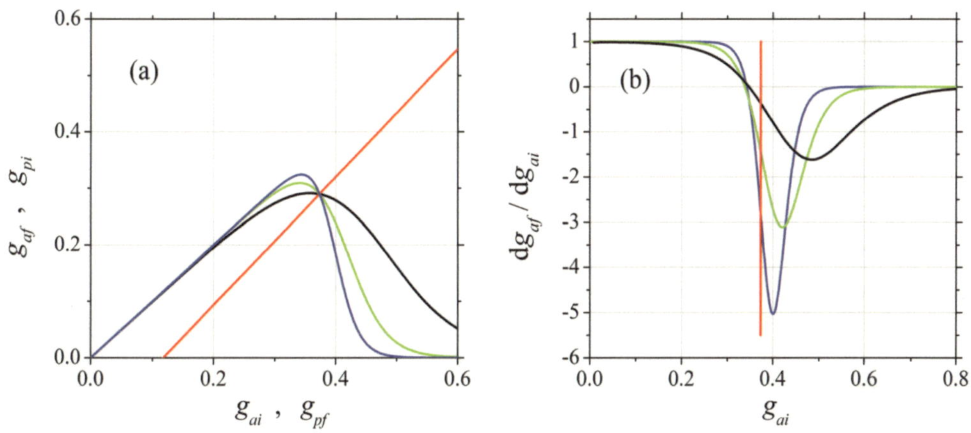

Fig. 4. Typical state space diagrams of regimes having equal fixed points at different seed energies (a) and derivatives of the „amplification" curves (b).

It becomes apparent that at high repetition rates the initial gain depends on preceding operation cycles and can be determined indeed by taking them into account. This procedure is in essence nothing else than the fixed point determination. Let us return to geometric presentation of fixed points in state space. Figure 4(a) represents diagrams of the final gain against initial gain for three seed energy values. Corresponding numbers of round trips have been selected so that the fixed points at certain repetition rate are identical. All the curves intersect in a single point which corresponds to the fixed point for the repetition rate T_1/T=8.0 This means that decrease in seed energy is compensated by increasing of round trips but only in a sense of equal position of the fixed points. However the shapes of curves g_{af} versus g_{ai} are different and their derivatives in the intersection point are dependent on the seed energy [Fig. 4(b)]. Such derivatives indeed determine the regenerative amplification stability as we have described in section 2.4 by referring to the theory of discrete-time dynamical systems.

The identity of the fixed points can be realized by compensation of the seed energy difference by appropriate selection of the round trip number. Output energies corresponding to those fixed points are equal too. However the peculiarity of operation at high repetition rates is such that dynamical system behavior is absolutely different. Thus at high repetition rates the seed pulse energy becomes one of those critical parameters which determine the operation regime of the regenerative amplifier. The specific operation points, which have been analyzed here, can be also found in the diagrams of dynamic regimes (Fig. 3). With regard to stability, they were classified as stable, 2*T*-periodic and chaotic for seed values of 1.3×10^{-4}, 2.5×10^{-7} and 10^{-10} respectively. The dynamic regimes which are in general possible to obtain (by changing the round trip number) within a certain range of seed values are summarized in Table 1.

Existing regimes	Seed value range	
	g_t=0	g_t=0.028
Chaos and "all" periods	$<2.52\times10^{-7}$	$<1.4\times10^{-5}$
T, 2*T*, 4*T*, 8*T*, 16*T*, 32*T*...	2.52×10^{-7} - 2.56×10^{-7}	1.4×10^{-5} - 1.5×10^{-5}
T, 2*T*, 4*T*, 8*T*, 16*T*	2.56×10^{-7} - 2.72×10^{-7}	1.5×10^{-5} - 1.74×10^{-5}
T, 2*T*, 4*T*, 8*T*	2.72×10^{-7} - 3.56×10^{-7}	1.74×10^{-5} - 2.5×10^{-5}
T, 2*T*, 4*T*	3.56×10^{-7} - 1.39×10^{-6}	2.5×10^{-5} - 4.1×10^{-5}
T, 2*T*	1.39×10^{-6} - 1.90×10^{-4}	4.1×10^{-5} - 3.5×10^{-3}
T (stable)	$>1.90\times10^{-4}$	$>3.5\times10^{-3}$

Table 1. Possible regimes versus seed pulse energy range.

2.7 Influence of parasitic intracavity losses on dynamic pattern

The approximation of negligible losses is a good way to present the main ideas for application of the discrete-time dynamics method for regenerative amplification and to understand the dynamic patterns most relevant at high repetition rates. However, this approximation has limited application in practice. The output pulse energy grows in the lossless system monotonically together with amplification phase duration and reaches saturation at the level of $\varepsilon_{max}=1-\exp(-T/T_1)$ that corresponds to full conversion of stored energy (population inversion) to the output pulse energy. Consequently the number of round trips can be increased, without detriment to output energy, to the values high enough for operation behind the bifurcation zone that in turn assures stable operation. Actually, the system is always (i.e. irrespective of losses) stable provided that the population inversion is well depleted during the amplification phase. In this case the initial gain tends to the constant, determined only by the repetition rate, [$g_{af}\rightarrow0 \Rightarrow g_{pi}\rightarrow0$; then from Eq. 6 follows that $g_{ai}=g_{pf}\rightarrow1-\exp(-T/T_1)$]. Consequently, the interdependence of operation cycles vanishes that results in eliminating of immediate cause of unstable behavior. In reality, parasitic losses prevent utilization of this property since because of losses the mode of complete gain depletion becomes inefficient.

Well known efficiency criterion, to dump optical pulse off the resonator at the moment when the current gain has dropped down to the threshold gain ($g_{af}=g_t$), is not applicable to repetitive operation as relating to only "isolated" operation cycles. Power efficiency enhancement takes place at high repetition rates when stored energy is left partially under-depleted ($g_{af}>>g_t$) forming a substantial gain background after several operation cycles. The proportion of gain to losses, which eventually determines extraction efficiency of the stored energy, can be well improved by this means. However incomplete depletion is an origin of operation cycles interdependence therefore in presence of losses the system efficiency in some sense collides with the system stability.

Parasitic losses in laser systems are given by optical components imperfection and diffraction losses of the optical resonator. The latter are objects of resonator geometry optimization. In case of solid-state lasers pumped longitudinally (virtual absence of hard apertures) high order optical aberrations (spherical e.g.) may become the main origin of diffraction losses, especially at high pumping intensities (Liu C. et al. 2008). Among optical components, the electro-optic switch is usually the most critical part; contributions of the Pockels cell and the polarizer to the loss factor surpass the remaining components (Müller et al., 2003). Practically, the level of total parasitic losses can vary in quite a wide range, but the typical value should not exceed a few percent per roundtrip for high-quality, well optimized systems. Here we should remark that the losses, inherent for quasi three-level gain media and related to partially populated ground state, are not dissipative and they do not belong to the parasitic losses which we are considering.

At the account of intracavity losses, general, qualitative pattern of amplification dynamics (fixed points uniqueness, variety of orbits for the repulsive fixed points, significance of the seed pulse energy) remains the same, but naturally the quantitative difference factors in. The intracavity losses of regenerative amplifier not only reduce efficiency (that is natural for lasers), but also substantially interfere in total system stability.

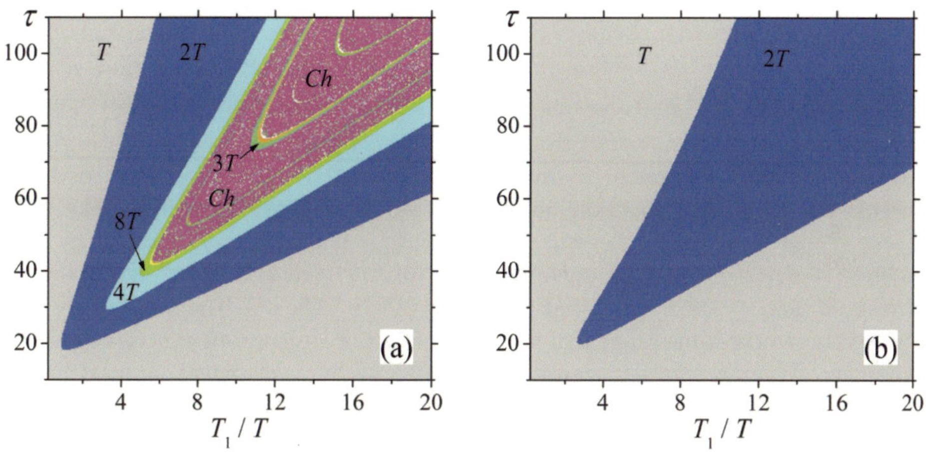

Fig. 5. Diagrams of amplification dynamics for threshold gain g_t=0.028 and seed pulse energies ε_s=3×10^{-7} (a); ε_s=1.90×10^{-4} (b).

The diagrams of regenerative amplification dynamics in parameter space (round trips - repetition rate) for intracavity losses corresponding to the threshold gain g_t=0.028 are presented in Fig. 5. Fixed point calculation and their stability evaluation were performed analogously as described in sections 2.4 and 2.5. The only difference is function $\hat{g}_a$, relating final and initial gains [solutions of Eqs. (4) and (7)], was calculated numerically since analytical solution is unknown in case of nonzero losses. The diagram presented in Fig 5(a) can be compared with that given for the same seed pulse energy ε_s=3×10^{-7} but for zero losses [Fig 4(c)]. The influence of losses results in a more complicated dynamic pattern; the zone of chaotic dynamics evolves and the high order bifurcations shift closer to the tip of the instability zone (towards lower repetition rate and lower round trip number). The second diagram [Fig 5(b)] was calculated for the seed energy ε_s=1.90×10^{-7}, which at zero losses provided stable operation in the whole range of control parameters. Now a period doubling zone (2*T*) has occupied a certain part of the parameter space and noticeably narrowed the range of stable operation.

A more cumulative picture of dynamical regimes is presented in Table 1. One can see some general change for the worse for system stability with respect to the zero-loss case. The decrease in stability caused by losses is not an obvious phenomenon (why not increase?). The reason for that bears similarity to the seed energy effect. The losses decrease pulse energy addition per round trip that can be compensated by increasing of round trip number, but only in a sense of fixed point identity. The derivative magnitude of the gain transformation function, $|\hat{g}'_{\Sigma}(g_1)|$, in this point has changed so that system stability becomes worse as the losses are increasing. This phenomenon is not obvious but the conclusion is straightforward - the parasitic losses should be minimized as much as technically possible not only for efficiency but also for better stability.

2.8 System performance representation in parameter space

For understanding regenerative amplifier performance, not only dynamical regimes but also output pulse energy should be represented in relation to the parameters governing the system. In previous sections we have described in detail the method of fixed points and related orbits determination. Actually, during calculation of the initial gain orbit, $g_{ai}(k)$, the orbit of output energy $\varepsilon_f(k)$ comes out automatically as the second solution of coupled Eqs. (4) and (7), $\varepsilon_f(k) = \hat{\varepsilon}[g_{ai}(k), \varepsilon_s, \tau, g_t]$. If governing parameters are such that operation is unstable then the initial gain varies from cycle to cycle by specific means and consequently the output energy becomes a multi-valued function of governing parameters. Representation of such a function on 2*D* diagram is unfeasible, it looks like a mess. Therefore we present typical dependencies of output energy versus round trip number (bifurcation diagrams) at several repetition rates for the fixed loss factor (corresponding to the threshold gain g_t=0.028) and seed energy ε_s=7.7×10^{-7} (Fig. 6). The selected seed pulse energy belongs to the same range that typical for functionally important class of seed lasers operating in CW mode-locking regime with moderate average power (around hundreds milliwatt). The repetition rates are chosen according to the diagram of dynamic regimes [Fig. 6(a)] so that typical bifurcation diagrams up to 16*T*-regime are demonstrated.

It is seen that the output energy variation in the presence of period doubling is so high that it virtually leaves no opportunity to use such a regime in practical applications. For example at repetition rate T_1/T=4.6 and in 30-60 round trips range the output energy alternates between high and low value so badly that the output pulse train looks almost as at twice less repetition rate [Fig. 6(c)]. It is even more pity that such bad stability often appears in regimes which potentially capable of providing high output energies.

The maximum capability of the system can be determined by calculating the output energy exactly in the fixed point: g_{ai}=$g_1 \Rightarrow \varepsilon_1 = \hat{\varepsilon}(g_1, \varepsilon_s, \tau, g_t)$. This energy is always a single-valued function of the governing parameters regardless of whether the fixed point is attractive or repulsive. In case of a repulsive fixed point (i.e. unstable operation), the corresponding "fixed point" output energy becomes an artificial parameter but it can serve as a convenient reference for evaluating the power efficiency reduction caused by instability effects.

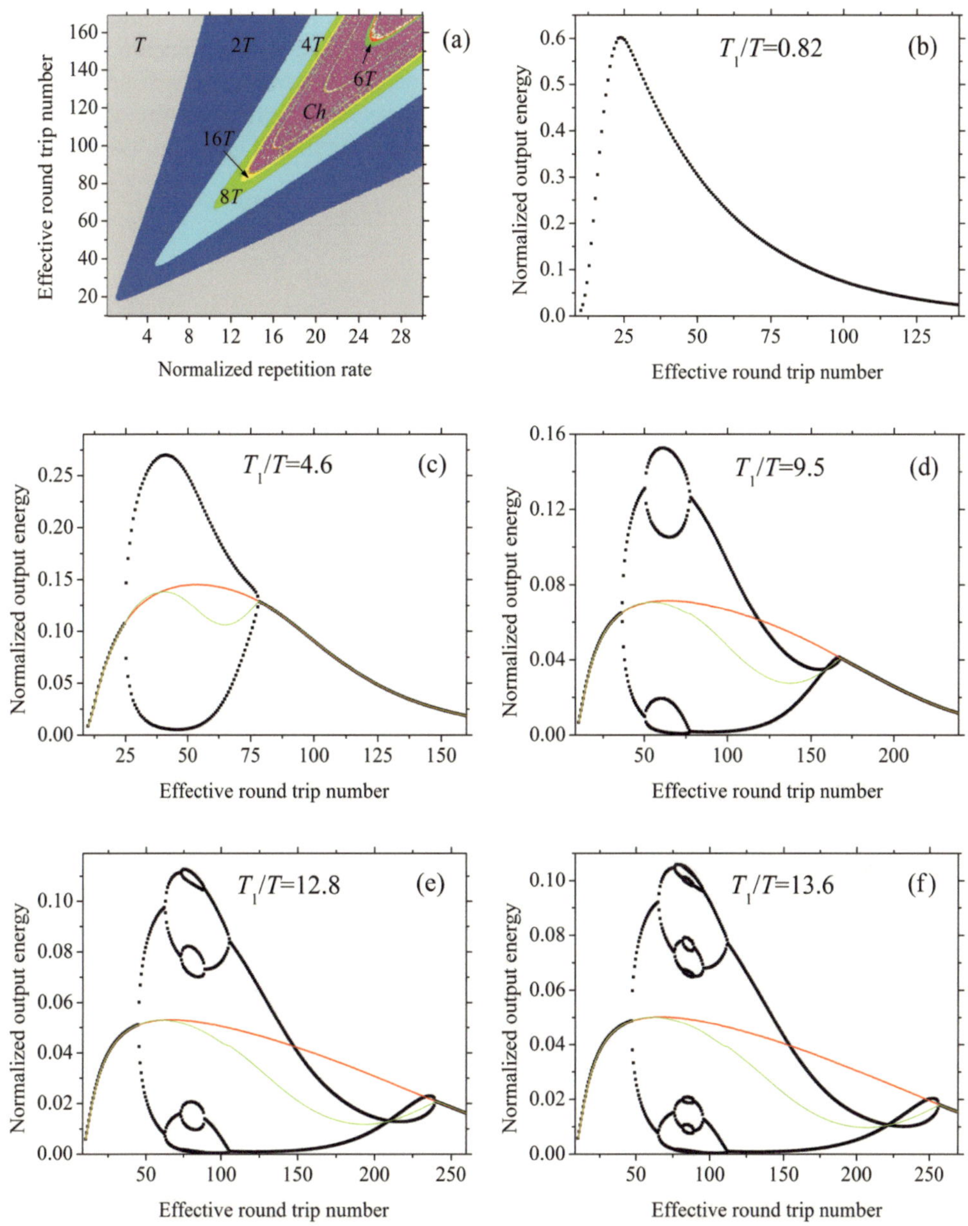

Fig. 6. Diagram of dynamic regimes (a) and corresponding bifurcation diagrams for selected repetition rates (b–f) at ε_s=7.7×10^{-7} and g_t=0.028. Pulse energy, averaged pulse energy and "fixed point" energy correspond to black, green and red lines respectively.

We also determined real (accounting multi-energy nature) output energy averaged over large number of operation cycles $\langle \varepsilon_f(k) \rangle$ (Fig. 6). Interestingly, the real averaged energy is considerably lower than the reference "fixed point" energy in regimes exhibiting pronounced period doubling. The question is "what can the origin of this energy defect be?"

One can suppose that in a multi-energy regime a relatively larger part of the resonator energy circulating is redistributed to the channel of parasitic losses. Some evidence for this explanation is that the same curves calculated in case of zero losses coincide in spite of bifurcations. The only alternate reason is stored energy depletion through spontaneous emission during the pump phase. However, the latter effect becomes dominant when the population inversion is on average high (i.e. at low round trips), that is not the case now.

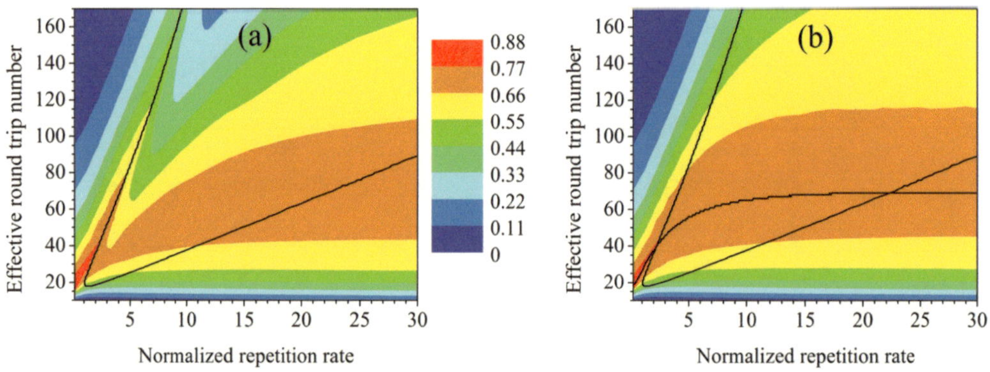

Fig. 7. Efficiency of stored energy extraction with respect to the separatrix (a); the reference ("fixed point") efficiency with respect to the separatrix and τ_{max} curve (b); both diagrams are for ε_s=7.7×10^{-7} and g_t=0.028.

Single-valued functions characterizing output energy allow presentation of an informative picture of system performance in *2D* diagrams. The extraction efficiency of stored population inversion may serve as a more convenient parameter for this purpose. The extraction efficiency can be defined as averaged output energy divided by maximum stored energy available at a given repetition rate, $\langle\varepsilon_f(k)\rangle/\varepsilon_{max}$, where $\varepsilon_{max}=1-\exp(-T_1/T)$. Then the value calculated in the fixed point ($\varepsilon_1/\varepsilon_{max}$) can be regarded as the maximum attainable efficiency at a given round trip number. Typical diagrams of these parameters in the space of repetition rate versus round trip number are presented in Fig. 7. Additional normalization to ε_{max} introduced for already normalized terms is not meshing. Quite the contrary, it simplifies the practical use of theoretical data. The efficiency is the ratio which has the same value for real "dimensional" and normalized parameters.

Apparently, stable single-energy operation is the only suitable regime for routine use of regenerative amplifiers. For completeness sake we can note that one may successfully use certain unstable regimes for specific applications provided that there is comprehensive understanding of the essence of period doubling (Metzger et al., 2009). However this is rather the exception than the common rule. Thus, we can omit a detailed picture of dynamical regimes in order to move towards more pragmatic considerations. It is sufficient to leave only one curve in the parameter space defining the range of operating points in which operation is stable. This curve (further referred to as a separatrix) represents a manifold of the first order bifurcation points (1*T*-2*T* boundary) and actually separates zones of stable and unstable operation in the parameter space. Both real and "fixed point" energies are equal within a range of control parameters providing stable operation. Some drop of real efficiency with respect to that obtained in the assumption of stable operation is observed immediately below the upper branch of the separatrix, in the instability zone [Fig 7(a)]. The cross-sections of this feature are observed in Fig. 6 at several repetition rates and we have already concluded that this "valley" originates from enhanced influence of parasitic loses in the period doubling regime.

The distribution of the "fixed point" extraction efficiency ($\varepsilon_1/\varepsilon_{max}$) in the parameter space contains sufficient information to determine system performance when accompanied with the separatrix curve. This curve confines the space of stable operation that is the range where the "fixed point" data correspond to reality. For system optimization it is important to find operation points potentially providing maximum performance at each given repetition rate. The manifold of such points represents a curve of round trip number versus repetition rate which is mapping the peak value of the "fixed point" extraction efficiency (*2D* distribution in parameter space) and further referred to as τ_{max} curve [Fig 7(b)]. The potential performance becomes real at repetition rates in which the τ_{max} curve is outside of the instability region and when the round trip number is set equal to τ_{max}. If the τ_{max} curve is inside the instability zone then the system capabilities are underexploited. The corresponding range of repetition rates can be called the "critical range". Within the critical range there are two possible positions of the operating point which may provide maximum output in the stable regime - the points along the lower and upper branch of the separatrix [see Fig 7(b)]. According to the diagram, at lower repetition rates in the critical range the upper branch has an advantage from an efficiency point of view. Operation at lesser round trip number (lower branch of the separatrix) becomes preferable at higher repetition rates. Thus optimization of the regenerative amplifier is actually reduced to selection of the round trip number which provides maximum output pulse energy and at the same time allows stable operation for the required repetition rate range(imposed by the system specifications). The corresponding round trip number is logical to call optimum (τ_{opt}). Obviously, τ_{opt} is equal to τ_{max} outside the critical range.

Instructive inference of considered above regenerative amplification properties is that within critical repetition rates the round trip number takes on optimum value either near the lower or near the upper separatrix branch but always at the margin of unstable operation. In general, operation at the margin of stability incurs challenges for robust operation in real systems which undergo technical noises. Even slight changes to control parameters may result in system instability. Therefore reliably stable operation generally requires setting the operating parameters well away from the instability border, but this in turn leads to a reduction of the system performance.

There is an important parameter related to intracavity losses which may influence performance of regenerative amplifiers indirectly: the amount of intracavity energy dissipated during the amplification stage. Accumulated over round trips, the fraction of intracavity energy, dissipated through parasitic losses is subject to the specific operation regime. In particular, multiple passes of the already amplified optical pulse lead to substantial enhancement of energy dissipation. This, in turn, may give unacceptably high heating of intracavity components caused by the absorbed part of the dissipated power. One of the critical components in this respect can be the Pockels cell crystal. It may lose contrast under excessive heating possibly resulting in failure of regenerative amplifier operation. This effect is especially pronounced for systems intended for high power applications. The energy defect, arisen due to parasitic losses (ε_l), can be determined as a product of intracavity energy, integrated over the amplification stage, and loss factor expressed by the threshold gain g_t. Using this definition and Eqs. (4) and (7) we get:

$$\varepsilon_l = g_t \int_0^{\tau} \varepsilon(\tilde{t})d\tilde{t} = \int_0^{\tau} \varepsilon(\tilde{t})g(\tilde{t})d\tilde{t} - \varepsilon_f = g_{ai} - g_{af} - \varepsilon_f , \tag{14}$$

where the integration variable $\tilde{t}$ is the current normalized time. The same result can be obtained from the energy conservation law since during short (as we have initially assumed) amplification stage there are only two energy consumption channels - useful signal and parasitic losses. In case of multi-energy regime the effective lost part of the energy can be found by averaging:

$$\langle \varepsilon_l \rangle = \frac{1}{k} \sum_k \left(g_{ai}(k) - g_{af}(k) - \varepsilon_f(k) \right) \tag{15}$$

The diagram of dissipated energy (also normalized to maximum stored energy ε_{max}), for the set of governing parameters used before, is presented in Fig. 8(a). We can see that the large number of passes, typical for the upper separatrix branch, substantially contributes to this parameter. So the dissipated energy is about 7.5 times higher at the repetition rate when theoretical efficiencies for both branches are equal (at $T_1/T \approx 6.5$). This increase with respect to the lower branch often makes operation at high round trip number, well above τ_{max} point inefficient, despite theoretical preference.

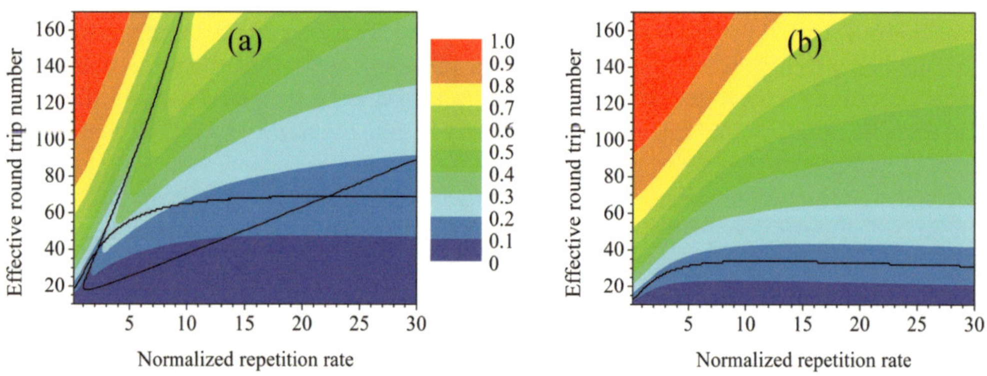

Fig. 8. Dissipated intracavity energy, separatrix and τ_{max} curves for g_t=0.028 and ε_s=7.7×10^{-7} (a); dissipated energy, and corresponding τ_{max} curve for the seed energy ε_s=2.46×10^{-4} (b).

All the issues described in this section are regarded to one certain seed energy ε_s=7.7×10^{-7}. The corresponding diagrams give a typical but single section of multidimensional space of control parameters. However, as we can already conclude from amplification dynamics data presented in sections 2.6 and 2.7, the critical repetition rate range depends on the seed pulse energy. Thus, the picture of regenerative amplification is still incomplete and it is time to proceed to consideration of system optimization taking into account influence of the seed pulse energy.

2.9 Stability diagrams and pulse duration effects

It is possible to present data which allow evaluation of regenerative amplification of different seed pulse energies by considering a single diagram. The necessary premises for doing that have been formulated in the previous sections. The condition where bifurcations are absent in the whole parameter space (Table 1) gives a general understanding of the seed energy influence; however this condition is too strong from a practical point of view. In order to thoroughly utilize power capabilities of the regenerative amplifier at a certain repetition rate, the round trip number should be set equaled to τ_{max}. Stable operation requires having the point τ_{max} outside the instability zone (delimited by the separatrix curve on the parameter space). The inter-positioning of the separatrix and the τ_{max} curve contains sufficient information to determine the critical repetition rate range for a certain seed energy and the possible optimum operation points (τ_{opt}) within this range (which lie, as we have known, along one of the separatrix branches). The diagrams consisting of τ_{max} curves and separatrixes (further referred to as stability diagrams) for selected pulse seed energies are presented in Fig. 9.

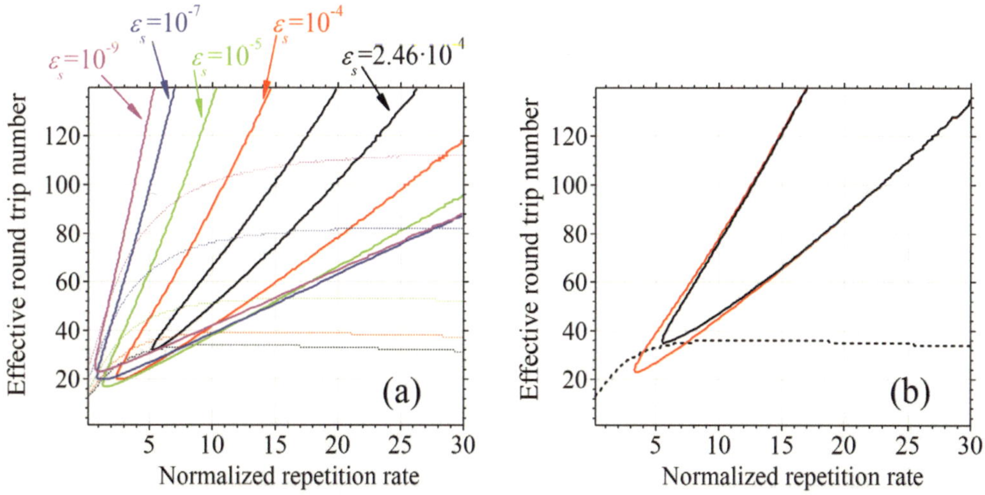

Fig. 9. Stability diagrams for the threshold gain g_t=0.028; the curves couples of the same color (the separatrix and τ_{max} curve) correspond to certain seed energies (a). The separatrixes are shown for the short pulse (black solid line) and long pulse approximations (red solid line) and τ_{max} curve (dash line) at g_t=0.028 and ε_s=1.7×10^{-4} (b).

The approach of stability diagrams forms a more systematic concept of the system behavior and specifically allows estimation of the seed level which may enable one to avoid instability effects at the required pulse repetition rate. Also we can see that the critical range shrinks as the seed energy is increasing. The most critical repetition rate is T/T_1=5.5, the point of the "worst stability", requiring the highest seed energy for optimal operation. Finally one can determine the seed pulse energy at which τ_{max} curve does not pass instability zone at all. For the specific parasitic losses that we consider (g_t=0.028) this energy is equal to 2.46×10^{-4}. Consequently, a regenerative amplifier seeded with pulse energy higher than that value (further referred to as "ample") allows theoretically attainable average power and stable operation over the whole repetition rate range. Also we can mention here another advantage of the "ample" seed operation: As soon as the critical range has disappeared there is no need to operate at the upper separatrix branch and to suffer from large intracavity energy dissipation peculiar to this regime. The diagram of dissipated energy for the ample pulse seed energy, ε_s=2.46×10^{-4} is depicted in Fig. 8(b).

The approach of stability diagrams is a straightforward way of regenerative amplifiers optimization. However in real systems there are specific effects which influence dynamic behavior and output parameters and they cannot be elaborated by using only approximation of simplified rate equations. Nevertheless it is possible to understand some important contributions staying basically within present approach.

The major application of regenerative amplifiers is amplification of short (even more common term is ultrashort) optical pulses. Now we shall consider the influence of pulse duration on regenerative amplification. The theory described above is based on rate equations formulated for idealized four-level system. One of the positions in the definition of that "ideality" is instant depopulation of the lower laser level (which is also called terminal level). In reality we can assume that the lower laser level is virtually unpopulated, only provided that the amplified optical pulse is much longer than the terminal-level lifetime (long pulse approximation). Otherwise, if the length of the optical pulse is much less than the terminal lifetime then the terminal level will remain populated resulting in "faster" decay of population inversion (indeed during amplification of a single pulse). Comprehensive evaluation of actual terminal-level lifetime provided by (Bibeau, et al., 1995) for different neodymium doped laser media gives actual values well exceeding 100 ps. So we can conclude that amplification of pulses shorter than 100ps virtually for all neodymium based systems is more appropriate to analyze within the short pulse approximation that is assuming negligible terminal-level depopulation during singe pulse amplification. This, in fact, constrains applicability of presently described approach and leaves beyond the scope functionally important ultrashort pulses. We can note in advance that net contribution of the terminal-lifetime effect to regenerative amplifier behavior is rather weak at high repetition rates. However it gives some quantitative refinements to the picture presented above.

In order to elaborate regenerative amplification in the short pulse approximation we need to re-calculate the fundamental relations of final and initial gains $g_{af} = \hat{g}_a(g_{ai})$. In this approximation the amplification of a single pulse was regarded as that in truly three-level gain media with initially empty ground state. After single pass amplification the ground state becomes partially populated, but by the second pass it is empty again and consequently the gain defect, which appeared due to "instant" three-levelness, is recovered. The latter is the case since we assume the terminal-level lifetime to be much shorter than the round trip time, T_0. Typical T_0 value is in the range between ten and a few tens of nanoseconds, so this is good approximation for most of neodymium media [actually except some fluoride crystals and glasses in which the neodymium terminal lifetime is of 10 ns order (Bibeau, et al., 1995)]. By this means and by using sequential procedure the basic relations $g_{af} = \hat{g}_a(g_{ai})$ and corresponding output energies were determined. The subsequent procedures (evaluation of the fixed points and their stability analysis) stay unchanged for the short pulse approximation.

We revealed that the influence of a terminal level appears as follows. There is noticeable deviation between the τ_{max} curves at low repetition rates. However both "short pulse" and "long pulse" curves virtually coincide at the repetition rates $T_1/T>1.0$, indeed in the range that we are studying. The separatrix curves practically coincide at low seed energy levels, although filling of the instability zone differs from what have been seen earlier [e.g. in Fig 6(a)] towards not so wide variety of regimes. These peculiarities add little in practical essence. Therefore we have come to nothing more than qualitative description and statement of that fact.

The only noticeable influence of the terminal lifetime was found at high seed energy, approaching to the ample level. The stability diagram, depicted in Fig.9 (b), shows what the difference is. In comparing two separatrixes we can point out that the tip of the "short pulse" separatrix is somewhat shifted towards higher round trip number and higher repetition rate. Such a shrink of the instability zone gives certain improvement of general system stability. The ample seed pulse energy determined for short pulses is almost 1.5 times less than that found within long pulse approximation (ε_s=1.7×10^{-4} against 2.46×10^{-4}). This improvement is rather unexpected result since terminal-level "bottleneck" in some respect hampers the ideal four-level amplification. Thus at least partially and shortly populated terminal level does not act as additional losses as one might imagine.
Another phenomenon, substantial at high intensities and therefore requiring intent attention while ultrashort optical pulses are amplifying, is the Kerr effect. This nonlinearity makes an impact on amplification process by intensity dependent refractive index change in volume intracavity components. The influence occurs in both spatial and spectral domains and commonly is described as Kerr lensing and self-phase modulation. Reduction of this negative influence usually implies decrease of effective optical pulse intensity which can be quantified in terms of so-called *B* integral, in essence representing nonlinear on-axis phase shift (Brawn, 1981). In case of regenerative amplification multiple passes should be taken into account in order to evaluate total *B* integral accumulated during amplification stage:

$$B \approx B_1 \int_0^{\tau} \varepsilon(\tilde{t})d\tilde{t}\,, \qquad (16)$$

where B_1 is single pass *B* integral calculated for the intensity which is equal to the gain media saturation fluence divided by the pulse duration (see the Appendix section for details). The energy integral in this formula exhibits a factor of effective impact of multiple passes. We have already met such an integral when evaluated total lost part of the energy (see Eq. 14) and the value proportional to this factor has been depicted in previous section in Fig. 8. We can see that operation at the upper branch of the separatrix with short pulses is strongly unadvisable since the *B* integral is increased by several times. The cause of that is obviously multiple passes of intense pulses peculiar in a regime behind the instability zone.
That concerns the B_1 value, its reduction, in a condition of fixed pulse duration, simply implies standard methods of the mode area increase and shortening of the volume intracavity components. However these possibilities are rather limited. Even so the thin disc geometry allows tremendous reduction of gain media length (Speiser & Giesen, 2008) but the Pockels cell still can exhibit a real challenge. Among known to date Pockels cell materials only the BBO crystal is suitable for high average power (due to low absorption) and for high repetition rates (thanks to relatively low acoustic ringing). However the transverse electro-optic effect, the only functional for this crystal, permits shortening the optical pass or aperture increasing only in limited extent until driving voltage becomes unacceptably high (Nickel et al., 2005). Thus, since the Kerr effect often restricts the capabilities of regenerative amplifiers it is important at least to select correctly the operation regime in order to minimize its impact. We can note that operation at higher seed energy (ε_s larger than ample seed pulse energy) is beneficial in this respect too. At that condition, operation at maximum output (τ_{max} curve) does not suffer from too high multiple pass factor of the *B* integral [see Fig. 8(b)].

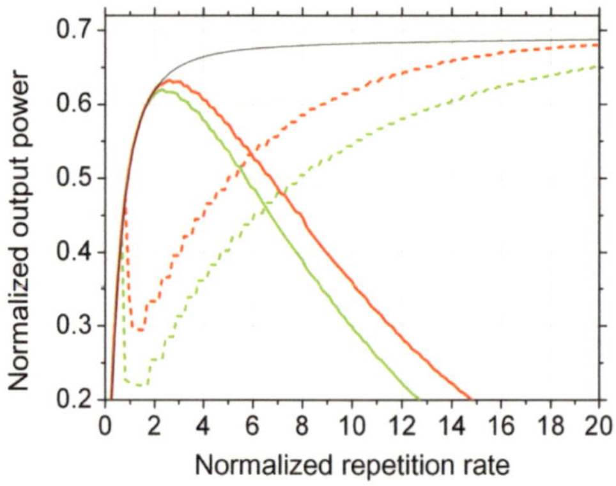

Fig. 10. Theoretical output power versus repetition rate for the threshold gain of 0.028 and normalized seed pulse energy of ε_s=7.7×10^{-7} (red line) and ε_s=7.7×10^{-9} (green line). The data of high and low separatrix branches correspond to solid and dashed lines respectively. The black line is the reference curve of achievable power.

At the end of the theoretical part of the chapter we are giving diagrams of the basic performance parameter - maximum output power which can be achieved in a condition of stability maintaining ($\tau=\tau_{opt}$), Fig. 10. Presented normalized average power is defined as a product of normalized energy and normalized repetition rate. The power curve, calculated for τ_{max} round trip number, can be regarded as the reference curve corresponding to theoretical efficiency limit. Really, in accordance with the τ_{max} definition, the corresponding output power is the highest obtainable average power in assumption of period doubling absence. The calculations show that this power is invariant under the seed energy, although the τ_{max} itself is not. Achieving the calculated maximum average power implies the best possible utilization of the stored pump energy, i.e. it assures the highest power efficiency. Obviously the same top performance we can reach operating at the ample seed energy provided that the round trip number is set equal to τ_{max}. This attribute is the main benefit of the high seed pulse energy at high repetition rates. The power defect with respect to the reference power curve observed for lower seed energies is essentially caused by instability effects. We present curves related to both branches of the separatrix within critical repetition rate range. The data related to the upper branch are not always reliable because as we have shown earlier this regime suffers from indirect effects related to multiple passes of intense optical pulse such as the Kerr effect and excessive heating of optical components.
In concluding this theoretical part, now we are ready to proceed to experimental verification of the ideas developed above.

3. Amplification experiments

3.1 Experimental setup

The amplification experiments were carried out in order to demonstrate basic features of amplification dynamics and to verify theoretical results presented above. The knowledge of potential capabilities and of general limitations makes it possible to provide the best regime selection and deliberate optimization of control parameters. In practicality, this means to maximize extraction of the given stored population inversion as a stable train of output pulses. We leave outside the scope of present consideration optimization of the pump characteristics and the geometry of the optical resonator allowing more power in TEM_{00} mode, as these do not relate directly to the amplification dynamics. The parasitic intracavity losses, although formally a governing parameter, are not an object of consideration; they should simply be reduced as much as technically possible. Since the repetition rate is usually imposed by the specifications it appears as a variable but a given parameter. There are two adjustable parameters which can be used for the system optimization - the number of roundtrips and the seed pulse energy.

We performed experiments with a system based on $Nd:YVO_4$ crystals, a gain medium with truly four-level nature (except terminal-level nuances). The schematic diagram of the experimental setup that we used for investigation of regenerative amplification is shown in Fig. 11. In essence the system consists of the seed source and the regenerative amplifier itself.

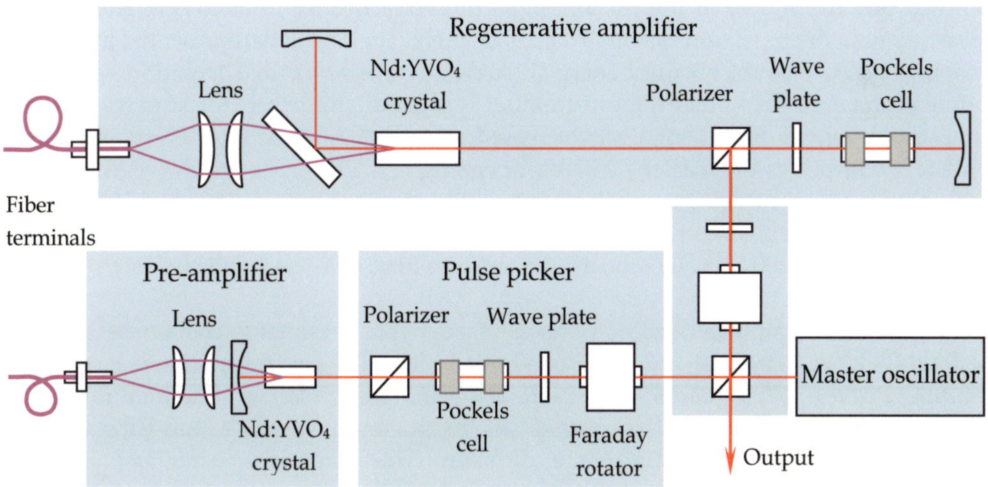

Fig. 11. Schematic diagram of the experimental setup.

The seed pulse source for regenerative amplification experiments was based on a diode-pumped passively mode locked picosecond $Nd:YVO_4$ laser of moderate power. It generated a continuous pulse train with repetition frequency of 82 MHz and average power of 300 mW. The laser was able to produce optical pulses with duration as short as 6 ps. The short pulses were used in experiments where dynamics peculiar to high peak intensities were of interest. The initial investigations were focused on the "pure" dynamics not disturbed by optical nonlinearities. These experiments were carried out with 58 ps duration pulses obtained by installing an etalon in the oscillator cavity (the etalon narrows the bandwidth, thus widening the pulse duration).

A pulse picker was used to select pulses for further amplification and in this way to control the effective repetition frequency of the seed source. This part of the seed source system is important for high repetition rate operation, especially based on high gain laser media like neodymium doped vanadates. If the pulse picker is not used then two negative effects caused by unwanted pulses have place. These pulses continuously pass the optical resonator of the regenerative amplifier during pump stage and go out spatially coinciding with useful output signal. This leads first to reduced pulse contrast and second to parasitic consumption of stored energy. One can decrease the seed pulse energy and compensate for that by increasing number of round trips. This simple approach may often avoid bad influence of the unwanted seed background while operating at low repetition rates. The reduction of the seed energy, as we have already seen (theoretically), is not a good idea when turning operation to high repetition rates. In our setup the pulse picker was an electro-optic switch based on an RTP Pockels cell. Selected pulses formed an input signal for the preamplifier. The remaining pulses of the master oscillator train were directed to the fast photodiode for synchronization of electro optic components of the system including the pulse picker itself.

A double pass Nd:YVO_4 preamplifier installed behind the pulse picker was used to increase the seed pulse energy to required ample energy. High emission cross section of the Nd:YVO_4 crystal make this system efficient at relatively low input average power. Only 2 W of pumping was sufficient to achieve a gain coefficient of more than two orders of magnitude. The seed pulse energy was 3.2 nJ when pumping of the preamplifier was switched off. The energy of the pre-amplified pulse reached 1.1 μJ at 10 kHz and steadily decreased with the repetition rate to 370 nJ at 200 kHz. The calculation presented in the next section will show that the obtained energy is sufficient to ensure stable operation.

Simple estimations show that the preamplifier is a good alternative in comparison with a more straightforward seed source scheme based on a powerful master oscillator. In order to provide 370 nJ pulses the master oscillator operating in a CW mode locking regime with a reasonable repetition frequency of 50 MHz should generate average power of 18.5 W. This way is really prodigal since the useful part of this power is much lower, e.g. only 74 mW even operating at 200 kHz. Obviously the preamplifier is a much more energy-efficient solution.

The regenerative amplifier was comprised of an optical resonator containing the gain medium (Nd:YVO_4 crystals) and an electro-optic switch. The electro-optic switch consisted of a BBO Pockels cell, a quarter-wave plate and a thin-film polarizer. The total multi-pass gain of the regenerative amplifier depends on the number of cavity round trips which is determined by the amplification-stage duration. This important parameter is easily controlled by setting the time interval during which the high voltage is applied to the Pockels cell.

The laser crystal was continuously pumped by the fiber coupled laser diode module with fiber core diameter of 400 μm and numerical aperture of 0.22. Optimal pump power, providing maximum output in TEM_{00} mode, was set to be 44 W. This optimization was performed in CW generation mode. Provided that no voltage was applied to the Pockels cell and the quarter-wave plate was adjusted for maximum output (optimal output coupling conditions), 12.5 W of average power was obtained.

The output radiation was diverted from the input signal path by standard optical circulator based on Faraday rotator. The repetition rate of the system was limited to 200 kHz by electronics driving the electro-optic switches.

3.2 Application of stability diagrams to amplification experiments

Now we can apply the concepts developed in the theoretical part to estimating behavior of a real system. At the beginning the basic system properties should be evaluated in order to perform reciprocal transformation of normalized parameters to dimensional ones corresponding to real operation conditions. The steady state small signal gain, determined by the pump intensity, was found directly as the ratio of the seed energies measured right before and behind the active element of the regenerative amplifier. The parasitic intracavity losses were derived from the specifications of the optical components. The laser characteristics of $Nd:YVO_4$ crystal (emission cross section and gain relaxation time) were taken from (Peterson et al., 2002). We explored amplification of three seed pulse energies differing by about two orders of magnitude: pre-amplified seed, unamplified seed and attenuated seed. This set of input signals covers the functionally important range. The value of the unamplified seed, 3.2 nJ, is of the same order of magnitude as the pulse energy of commonly used moderate power solid-state picosecond lasers. Operation with the seed energy intentionally attenuated to 32 pJ provides opportunity to evaluate typical behavior of the regenerative amplifier seeded with potentially attractive low power sources, e.g. with ultrafast laser diodes, which would substantially reduce system size and complexity. The seed energy obtained with the pre-amplifier was expected to be high enough to reach ample level. These seed energies were measured at the output of the seed formation system. However we observed that during further amplification the seed energy was not completely exploited. Mode mismatching reduces the effective seed energy. In general, it is difficult to avoid mode mismatching between a seed laser and the optical cavity of the regenerative amplifier in both spatial and spectral domains. Spectral mismatching can exist even with identical gain media because of e.g. different temperatures of laser crystals in those devices (Murray & Lowdermilk, 1980).

Parameter	Value
Wavelength	1064 nm
Emission cross section	11.4×10^{-19} cm^2
Gain relaxation time	83 μs
Effective mode diameter in the laser rod	1 mm
Steady state small signal gain	2.94
Seed pulse energies	11 pJ (low); 1.1 nJ (medium); 240 nJ (high)

Table 2. Parameters used for stability diagrams calculation

In order to have appropriate data for theory verification we experimentally estimated the overall level of mismatching. We measured real double pass small signal gain of the whole regenerative amplifier (output/input energy ratio while the Pockels cell was disabled) and compared this value with "pure" steady state gain of the laser element. Thus we determined the effective seed value to be about three times less than the measured one. The primary parameters of the regenerative amplifier which we used for calculations are summarized in Table 2.

Before proceeding to stability diagrams describing our particular experimental conditions, we note that we utilized a linear cavity that establishes double pass through the gain media, while the modeling has implied single pass for one cavity round trip. This factor of two was taken into account so that the y-axis of stability diagrams corresponds to real round trip number, $t/T_0=½\tau/G_0$. The other relations between dimensional and normalized parameters remain unchanged as they have been given in the section 2.2.

The stability diagrams describing amplification experiments are presented in Fig. 12. These data were obtained in the approximation of short pulses (accounting for terminal-level lifetime effect). As we have already known, the system capabilities are completely exploited when the τ_{max} curve is outside the instability region. The diagrams show that at low seed energy (11 pJ) the appropriate repetition rates should be less than 20 kHz. For the medium seed level (1.1 nJ) this range increases to 25 kHz. At higher repetition rates the τ_{max} curve enters the instability zone. It is important that for 240 nJ or higher seed energies the τ_{max} curve does not enter the instability zone in the whole range of repetition rates. So, for our laser system this energy corresponds to the ample seed value, sufficient to eliminate negative features of amplification dynamics, and thus to completely exploit the system capabilities.

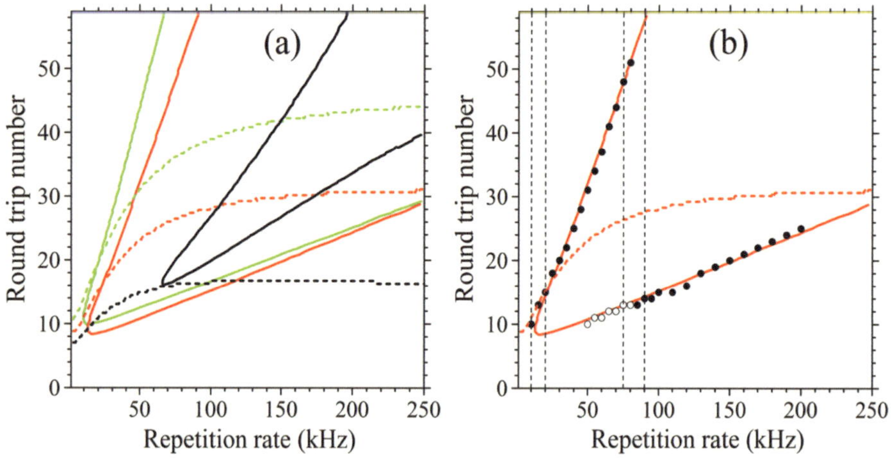

Fig. 12. (a) Separatrix curves (solid lines) and τ_{max} curves (dotted lines) in parameter space. Black, red and green lines correspond to seed pulse energies of 240 nJ, 1.1 nJ and 11 pJ, respectively. (b) Operating point trajectories (vertical dashed lines) and measured number of optimal round trips for the pulse durations of 58 ps (solid circles) and 9 ps (open circles) with respect to stability diagram for the seed energy of 1.1 nJ.

3.3 Experimental bifurcation diagrams

The initial experiments were carried out with medium seed pulse energy (the preamplifier was disabled). Various dynamic regimes, depending on set of control parameters, were observed. As an illustration, Fig. 13 shows oscilloscope screen shots of the output pulse train in typical single-energy and period doubling regimes.

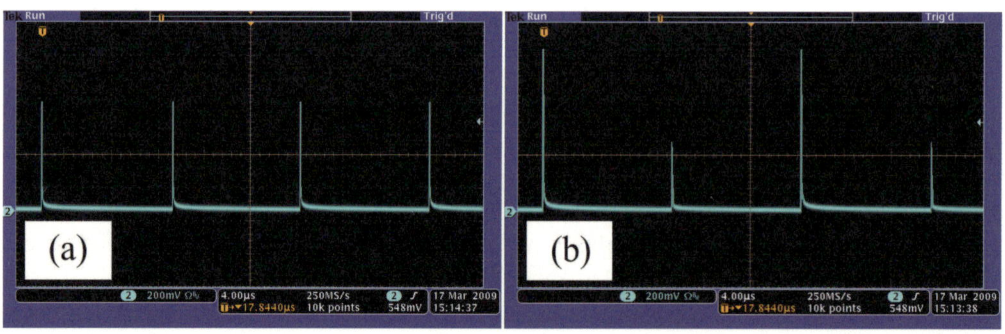

Fig. 13. Screenshots of typical pulse trains at 90 kHz. Stable energy output and the $2T$ period doubling regime were obtained at number of round trips equal to 14 (a) and 16 (b) respectively.

Experimentally obtained diagrams of the average output power and pulse energy versus number of round trips demonstrating system behavior at different repetition rates are presented in Fig. 14. The specific repetition rates were chosen to describe the most relevant cases of regenerative amplifier dynamics in respect of system optimization. The single-peaked dependence inherent to low repetition rates appears at 10 kHz [Fig. 14(a)]. The average power and the pulse energy reach the maximum values simultaneously, when the round trip number is equal to ten. At 20 kHz the situation is different [Fig. 14(b)]. The shape of the energy curve shows that the system undergoes bifurcation in the 9–13 range of the round trip numbers. However, in this case the period doubling does not affect the system performance because the output power reaches its maximum value in a single-energy regime. This repetition rate is still not critical.

Instability effects become more pronounced at higher pulse repetition rates. The period doubling not only breaks the energy stability but also distorts the curve of the average power (as described in section 2.8). This curve now has two explicit peaks [Figs. 14(c) and 15(d)]. The first peak, corresponding to the maximum power, is located in a period doubling zone, whereas the second one is just over the instability edge. The optimal regime is obtained in the vicinity of the bifurcation point. At 75 kHz the optimal number round trips is equal to 48. This point is close to the second power peak, on the right side of the period doubling zone [Fig. 14(c)]. For 90 kHz repetition rate the optimal number round trips is equal to 13 and is situated right before the first bifurcation point [Fig. 14(d)].

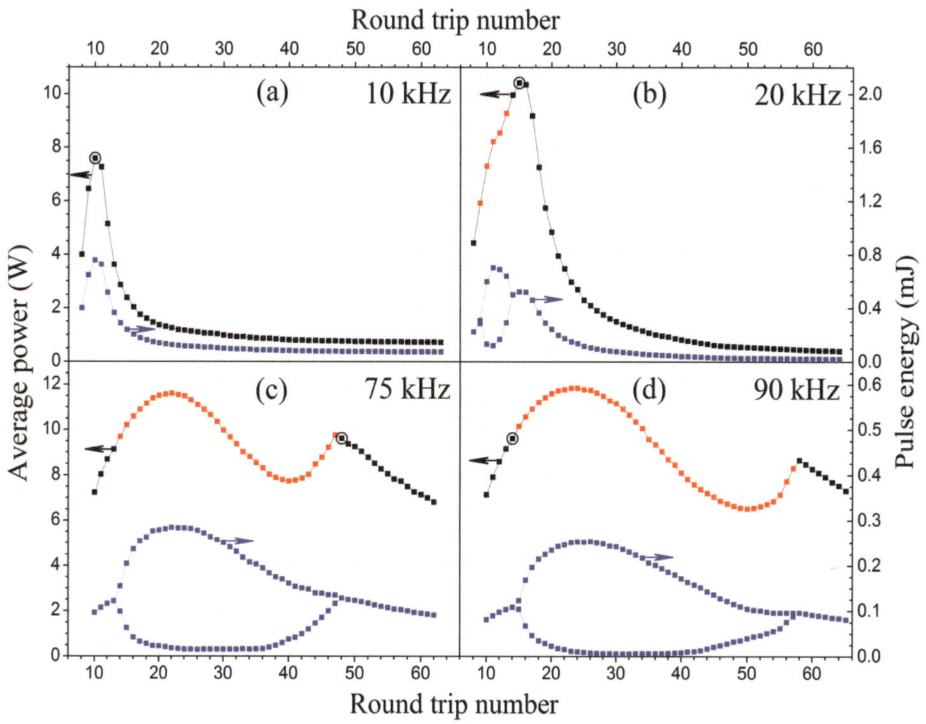

Fig. 14. Experimental average power (black and red dots correspond to stable and unstable regimes respectively) and pulse energy (blue dots) versus number of cavity round trips for the selected repetition rates. The encircled points correspond to the maximum power at stable operation.

The trajectories of the operating points corresponding to variation round trip number at a constant repetition rate are presented in Fig. 12(b). This trajectory at 10 kHz does not pass the instability zone. At 20 kHz the optimal operating point is above the instability zone. Both repetition rates 75 kHz and 90 kHz are critical - the optimal number of round trips is on the stability edge and is rather far from the point of the highest attainable power. The experimentally observed results confirm theoretical predictions that for critical repetition rates: (i) the output energy exhibits unacceptable fluctuations when the amplifier produces the highest average power, (ii) the highest stable pulse energy is reached close to the instability edge.

Some effects caused by nonzero terminal-level lifetime were also observed in the experiment. The presented bifurcation diagrams at 75 kHz and 90 kHz theoretically should exhibit not only $2T$ but also $4T$ regimes [like in Fig. 6(e)], if the long pulse approximation is applied. The regime of $4T$ period doubling was not observed experimentally and it should not exist theoretically, provided that contribution of the terminal-level lifetime is accounted. At the same time, as the theory has predicted, terminal-level lifetime effect does not influence the system performance at such low seed level.

Real deviation from theory is observed at the lowest repetition rate (10 kHz). The output energy decays too fast behind the peak point in comparison with theoretical expectations [see Fig. 6(b)]. This occurred because of the Kerr effect influence was substantial at low repetition rates even for initial experiments performed with relatively long 58 ps pulse duration.

3.4 Performance evaluation and discussion

It has been shown in the theoretical part that variation of only roundtrip number does not solve the stability problem; increase in the seed pulse energy is required in order to avoid bifurcations and corresponding instability at high repetition rates. So, we proceed to experimental verification of the seed energy influence. We compared operation for three cases: "low", "medium", and "ample" pulse seed energy.

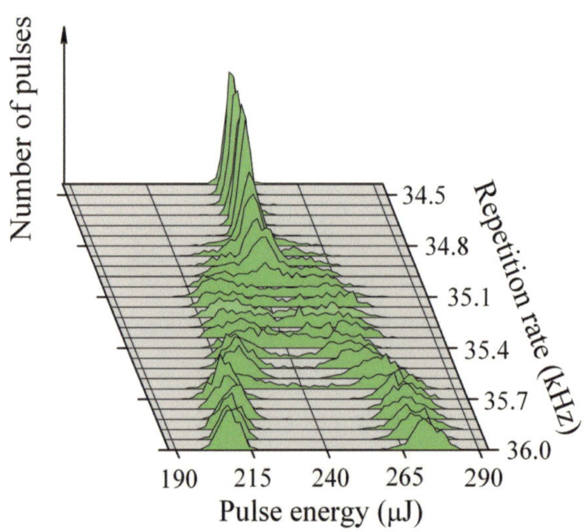

Fig. 15. Experimental energy histogram in the vicinity of the bifurcation point.

The measurements were performed at optimal round trip numbers. They were set for maximum average power while maintaining stable operation for every repetition rate. The formulation of stability criteria in real experimental environments requires certain attention. Experimental discrimination of amplification regimes to some extent suffers from uncertainty because of technical instabilities, not related to fundamental system properties. Technical noises, in essence slight modulation of governing parameters usually with random distribution, limit system stability. The most typical of these are pumping source noises, seed pulse energy fluctuation, synchronization jitter, resonator disturbance by mechanical vibrations and by air flow. In our setup the technical, not disturbed by period doubling, standard deviation of output energy was less than 0.7%. In proximity to a bifurcation point the deviation increased. The diagram of typical transition from single energy to period doubling regime "under magnification" of the repetition rate scale is shown in Fig. 15; the bifurcation point looks rather as a spot than as a point. The uncertainty takes place in the range where the deviation of energy is clearly higher than the technically conditional level but where the two peaks are still not distinguishable. In the repetition rate scale this range does not exceed 1 kHz (typically the value is 0.5 kHz). In the round trips scale measurements with such fine steps are not possible (in contrast to theoretically grounded terms the real experimental round trips are discrete), and besides, the analysis of the energy histogram is a time consuming procedure. So, we used simple phenomenological criterion formulated for the particular setup allowing real time measurements. The operation was considered stable when the standard deviation of the pulse energy did not exceed 1%. Apart from that, special care was taken regarding origin and spectrum of disturbing noises to be sure that they are virtually non-periodic. In particular the pulse picker driving electronics had to be improved in order to eliminate seed train modulation at the frequency equal to half the system repetition rate. We can note that even barely perceptible presence of "resonant" components (sub-harmonics of dumping frequency) in the spectrum of technical noises may tremendously enhance the influence of period doubling and can change the dynamical pattern beyond recognition.

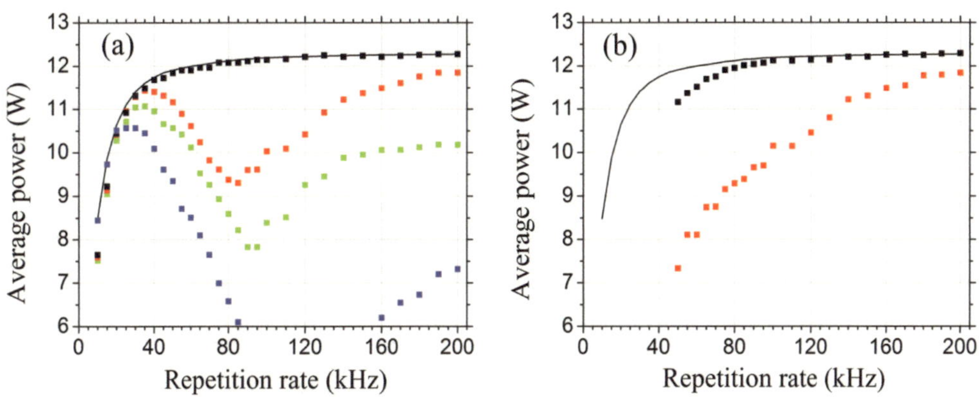

Fig. 16. Experimental output power versus repetition rate for 58 ps (a) and 9 ps (b) pulses. Black, red and green dots correspond to measured seed pulse energies of 700 nJ, 3.2 nJ and 32 pJ, respectively. Theoretical curve of achievable power is solid line in both diagrams. Blue dots in Fig. 16(a) are Q-switch experiment results.

Experimental dependences of the average output power versus repetition rate are presented in Fig. 16(a). The output is virtually independent of the seed level for low repetition rates. At higher rates there is a drop in power for low and medium seed levels in compliance with theoretical notions. The most significant power decrease appears in the 80–95 kHz range, then the output power steadily grows as the repetition rate increases.

Non-monotonic behavior of the power curve originates from the specific location of optimal operation points in the parameter space. Corresponding experimental data with respect to the theoretical stability diagram for medium seed energy is presented in Fig. 12(b). The operating points coincide with the theoretical curve of τ_{max} until the latter enters into the instability zone (at 25 kHz). Then operation at the upper branch of the separatrix becomes optimal. However as the repetition rate increases the stable operating point moves further away from the τ_{max} position, resulting in lower output energy. Consequently, starting from 85 kHz the optimal operation point switches to the lower border of the instability region. In this regime the operating point gradually comes closer to the τ_{max} position, and consequently, the output power steadily increases with increasing repetition rate.

The pre-amplified seed pulse has sufficient energy to maintain stable operation at maximum power, in accordance with ample seed properties. The power curve obtained with the preamplifier shows no signs of downward excursion with respect to the reference curve (the curve of theoretically attainable power corresponding to τ_{max} round trip number). Some slight deviation observed at low repetition rates was assumed to be caused by the Kerr effect. This nonlinearity can cause additional intensity-dependent intracavity losses (more pronounced at higher pulse energies), and so its influence is more pronounced at lower repetition rates. In order to differentiate the Kerr effect influence we performed experiments with nanosecond pulses in the same setup. The seed source was disabled so that the regenerative amplifier was transformed to a Q-switched laser with cavity dumping. Basically this device can be regarded as a regenerative amplifier seeded by spontaneous emission getting into the lasing mode. Thus, the seed energy becomes extremely low and additionally the laser system turns to the nanosecond domain (≈15 ns output pulse length was obtained in our setup). The former gives a large drop in power at critical repetition rates, the latter gives enhanced, with respect to the regular regenerative amplification, output energy at low repetition rate (the Kerr effect is negligible for such long pulses). The nanosecond pulse energy at low repetition rates exceeds the picosecond energy and very well agrees to the theoretical curve [Fig. 16(a)]. This demonstrates that experimental deviations appeared because of the Kerr effect and consequently general validates that the theoretical approach we have developed is wholly satisfactory.

Heating of intracavity components was not observed to affect regenerative amplifier operation in optimal round trip number experiments. However, clear evidence of excessive heating was noticed during recording of bifurcation diagrams in the worse (from this point of view) regime. The beam quality was observed to deteriorate for 90 kHz repetition rate and at round trip numbers between 46 and 56. From the other hand the average power defect reached maximum in this range [see Fig. 14(d)]. As we have concluded theoretically, this defect is indeed due to dissipated power, the absorbed fraction of which is heating intracavity optics.

We also compared performances of the regenerative amplifier seeded by high and medium pulse energy for the functionally important case of shorter optical pulses. Typically about 9 ps pulse duration was obtained at the output of the amplifier seeded by the 6 ps pulse.

This duration is close to the minimum value supported by the gain bandwidth of the Nd:YVO_4 crystal in high total-gain applications such as regenerative amplification. The measurements were constrained to dumping rates above 50 kHz. Nevertheless, the intensities were substantial, and the Kerr effect influence was so strong that it eventually resulted in decrease of the output power [Fig. 16(b)]. The average power obtained with the pre-amplified seed was slightly lower than that theoretically predicted below 80 kHz and the difference reached 6.7% at 50 kHz. However, the comparison of these characteristics with those obtained at a medium seed level shows that the benefit of the preamplifier is even more pronounced in case of shorter pulses. The difference is related to a large decrease of the average power below 85 kHz for the case of the medium seed. This is direct consequence of inefficient operation at the upper separatrix branch in critical repetition rate range under the Kerr effect influence. Experimental optimal operation points for both 58 ps and 9 ps pulses with respect to theoretical stability diagram are represented in Fig. 12(b). The optimal operating points for the short pulse experiments were always settled along the less efficient lower branch of the parameter separatrix. Attempts to operate at the upper branch (optimum for long pulses) resulted in an even larger decrease of the output. In order to quantify this difference we estimated the multi-pass B integral of the system (Eq. 22 in the Appendix). The B integral calculated at 50 kHz repetition rate gave values of 1.3 (acceptable) and 7.6 (problematic) at the transition from low to upper separatrix branch respectively. Thus the high seed energy gives additional advantage at shorter pulses due to a significantly lower value of the optimal number of round trips.

The amplification experiments which were performed with Nd:YVO_4 regenerative amplifier have shown that the developed theoretical approach accurately agrees with experimental data and can be used for practical system-design guidelines.

4. Conclusions

Continuously pumped regenerative amplifiers are subject to energy instability at high pulse repetition rates due to period doubling bifurcation. Theoretical concepts representing a generalized picture of operation features have been in-detail worked out in order to differentiate and understand instability effects. Experimental data for Nd:YVO_4 regenerative amplifier have been presented; and possible techniques for performance optimization have been analyzed. An increase in the seed pulse energy has been demonstrated to improve amplification dynamics. Addition of a preamplifier is shown to be not only a convenient means for regenerative amplification investigation at a wide range of seed energies but also an efficient way to get top performance in practice. The Nd:YVO_4 preamplifier delivered seed energy high enough to provide stable operation at repetition rates up to 200 kHz with average output power near the theoretical limit.

We have not performed appropriate experiments to elaborate amplification dynamics for ytterbium based regenerative amplifiers. However, it would be interesting to verify theoretical predictions for ytterbium doped crystals as for media exhibiting pronounced quasi-three-level behavior. Especially, because similar types of crystals may exhibit different sensitivity to bifurcations in the same setup, as reported by (Buenting et al., 2009) and (Sayinc et al., 2009). In addition, bulk ytterbium doped materials are low gain materials and therefore the preamplifier technique is not so easily applicable to such systems. There is demand to create another method to keep regenerative amplifiers stable at maximum output power, possibly some kind of feedback. Such an idea was formulated by (Dörring et al., 2004) but as far as we know neither theoretical modeling nor practical realization of this approach has been reported so far. Essentially, this can be an attempt to stabilize the inherently unstable balance between pumping and inversion depletion – that is constant forced return of the system state to the originally repulsive fixed point. Actually the initial gain of amplification phase should be kept constant by the feedback. However, straightforward engineering does not work in this case because output energy signal of the current operation cycle does not contain sufficient information to adequately control pump power of the subsequent cycle. So elaboration of this problem can be a goal for further theoretical research and experimental work.

5. Appendix. Multi-pass *B* integral for regenerative amplifier

The conventional quantitative gauge of the Kerr effect in laser systems is the B integral, nonlinear on-axis phase shift which light waves with wavelength λ undergo propagating through the media:

$$B_{sp} = \frac{2\pi}{\lambda} \int n_2(z) I(z) dz . \tag{17}$$

The terms $I(z)$ and $n_2(z)$ are distributions of on-axis intensity and nonlinear refractive index along current coordinate z. In order to evaluate the Kerr effect accumulated during regenerative amplification the integration should be performed over all the roundtrips of the optical cavity (multi-pass B integral). Then the full integration length is a product of the optical cavity pass length and round trip number. In an approximation of relatively low single pass gain (the integral within the gain medium can be replaced with the average) and also for moderate Kerr effect influence (iteration of the intensity profile in the optical resonator is not disturbed too much by self focusing) we can replace the overall integral with a sum of single pass integrals:

$$B \approx \frac{2\pi}{\lambda} \sum_{NRT} \int n_2(z) I_{NRT}(z) dz , \tag{18}$$

where index NRT implies summation over round trips. The sum of integrals is equal to the integral of the sum and also $n_2(z)$ is independent of round trip number function, then we obtain:

$$B \approx \frac{2\pi}{\lambda} \int n_2(z) \sum_{NRT} I_{NRT}(z) dz . \tag{19}$$

In an assumption of Gaussian beam shape for which the peak intensity is equal to $2P/(\pi w^2)$ (the term P is the optical power, the term w is the Gaussian beam radius) we can calculate the intensity for the pulse duration Δt in terms of pulse energy:

$$\sum_{NRT} I_{NRT}(z) \approx \frac{2}{\pi w^2(z) \Delta t} \sum_{NRT} E_{NRT} . \tag{20}$$

Summation of energies can be rewritten as time integration, and then dimensional current energy can be represented in terms of normalized energy which we have introduced in section 2.2:

$$\sum_{NRT} E_{NRT} \approx \frac{1}{T_0} \int_0^{NRT \cdot T_0} E(t)dt = \frac{A_a F_{sat} G_0}{T_0} \int_0^{NRT \cdot T_0} \varepsilon(t)dt = A_a F_{sat} \int_0^{\tau} \varepsilon(t^{\sim})dt^{\sim} . \quad (21)$$

At that, the limit of integration $NRT{\cdot}T_0$ (overall pulse propagation time) is reducing to the effective round trip number τ when the integration variable, current dimensional time t, has been transformed to the normalized time $t^{\sim}$. Also, the laser material properties, responsible for stimulated emission, are combined to a conventional macroscopic term, saturation fluence, the ratio of photon energy and stimulated emission cross section: $F_{sat}=h\omega/\sigma$. The beam area in the active medium obviously can be expressed through the Gaussian beam radius: $A_a=\pi w_a^2$. And finally, using straightforward transformation and normalized energy integral calculated in section 2.8 (Eq. 14), we derive explicit expression of B integral, the diminished form of which has been used in section 2.9 (Eq. 16):

$$B = B_1 \int_0^{\tau} \varepsilon(t^{\sim})dt^{\sim} \text{, where } B_1 = \frac{4\pi}{\lambda} \frac{F_{sat}}{\Delta t} \int n_2(z) \left[\frac{w_a}{w(z)} \right]^2 dz \text{ and } \int_0^{\tau} \varepsilon(t^{\sim})dt^{\sim} = \frac{g_{af} - g_{ai} - \varepsilon_f}{g_t} . \quad (22)$$

We have obtained the multi-pass B integral for evaluation of the Kerr effect in regenerative amplifiers as the product of two factors. The first factor, B_1 represents attributes of the system geometry, material parameters and optical pulse duration. Essentially, this term is single-pass B integral calculated for the Gaussian beam in given optical cavity and for pulse energy fluence equal to the gain medium saturation fluence. The second factor is a function of regenerative amplifier regime represented in normalized terms (the dissipated energy divided by the threshold gain). This is convenient for practical application form in which functional physical contributions are separated.

6. Acknowledgements

The authors wish to acknowledge the technical assistance of Juozas Verseckas from EKSPLA UAB in preparation of the experimental setup, Vidmantas Gulbinas from Institute of Physics and Lucian Hand from Altos Photonics Inc. for fruitful discussions of the manuscript. This work was partially financed by the Eurostars Project E!4335-UPLIT.

7. References

Alligood, K; Sauer, T. & Yorke, J. (1996) *Chaos. An Introduction to Dynamical Systems*, Springer-Verlag, ISBN 0-378-94677-2, New-York

Arecchi, F.; Meucci, R.; Puccioni, G. &. Tredicce, J (1982) Experimental Evidence of Subharmonic Bifurcations, Multistability, and Turbulence in a Q-Switched Gas Laser. *Phys. Rev. Lett.*, Vol. 49, Issue 17 (October 1982), 1217-1220, ISSN 0031-9007

Bibeau, C.; Payne, S.; & Powell, H. (1995) Direct measurements of the terminal laser level lifetime in neodymium-doped crystals and glasses. *J. Opt. Soc. Am. B*, Vol. 12, No. 10 (October 1995), 1981-1992, ISSN 0740-3224

Biswal, S.; Itatani, J.; Nees, J. & Mourou, G. (1998) Efficient energy extraction below the saturation fluence in a low-gain low-loss regenerative chirped-pulse amplifier. *IEEE J. Sel. Top. Quantum Electron.*, Vol. 4, Issue 2 (March 1998), 421-425, ISSN 1077-260X

Brawn, D. (1981) *High Peak Power Nd:Glass Laser Systems*, Springer-Verlag, ISBN 0387105166, New-York

Buenting, U.; Sayinc, H.; Wandt, D.; Morgner, U.; & Kracht, D. (2009) Regenerative thin disk amplifier with combined gain spectra producing 500 μJ sub 200 fs pulses. *Opt. Express*, Vol. 17, No. 10 (May 2009), 8046–8050, ISSN 1094-4087

Clubley, D.; Bell, A. & Friel, G (2008) High average power Nd:YVO based pico-second regenerative amplifier. *Proc. SPIE*, Vol. 6871 (February 2008), 68711D, ISSN 0277-786X

Dörring, J.; Killi, A.; Morgner, U.; Lang, A.; Lederer, M. & Kopf, D. (2004) Period doubling and deterministic chaos in continuously pumped regenerative amplifiers. *Opt. Express*, Vol. 12, No. 8 (April 2004), 1759-1768, ISSN 1094-4087

Fermann, M.; Galvanauskas, A. & Sucha G. (2002) *Ultrafast Lasers: Technology and Applications*, Marcel Dekker, ISBN 0-203-91020-6, New York

Forget, S.; Balembois, F.; Georges, P. & Devilder P. (2002) A new 3D multipass amplifier based on Nd:YAG or Nd:YVO_4 crystals. *Appl. Phys. B*, Vol. 75, No. 4-5 (October 2002), 481-485, ISSN 0946-2171

Grishin, M.; Gulbinas, V. & Michailovas, A. (2007) Dynamics of high repetition rate regenerative amplifiers. *Opt. Express*, Vol. 15, Issue 15 (July 2007), 9434-9443, ISSN 1094-4087

Grishin, M.; Gulbinas, V.; Michailovas, A. & Verseckas, J. (2008) Operation Features of Regenerative Amplifiers at High Repetition Rate, *Technical digest in CD*, paper CFB7, Conference on Lasers and Electro-Optics, CLEO-08, San-Chose, CA, USA, May 4-9, 2008, OSA

Grishin, M.; Gulbinas, V. & Michailovas, A. (2009) Bifurcation suppression for stability improvement in Nd:YVO_4 regenerative amplifier. *Opt. Express*, Vol. 17, Issue 18 (August 2009), 15700–15708, ISSN 1094-4087

Haken, H. (1975). Analogy between higher instabilities in fluids and lasers. *Physics Letters A*, Vol. 53, Issue 1, (May 1975) 77-78, ISSN 0375-9601

Jeong, Y.; Sahu, J.; Payne, D.; & Nilsson, J. (2004) Ytterbium-doped large-core fiber laser with 1.36 kW continuous-wave output power. *Opt. Express*, Vol. 12, Issue 25 (December 2004), 6088-6092, ISSN 1094-4087

Kawanaka, J.; Yamakawa, K.; Nishioka, H. & Ueda, K. (2003) 30-mJ, diode-pumped, chirped-pulse Yb:YLF regenerative amplifier. *Opt. Lett.*, Vol. 28, Issue 21 (November 2003), 2121-2123, ISSN 0146-9592

Kleinbauer, J.; Knappe, R.; & Wallenstein, R. (2005) 13-W picoseconds Nd:$GdVO_4$ regenerative amplifier with 200-kHz repetition rate. *Appl. Phys. B*, Vol. 81, No. 2-3 (July 2005), 163-166, ISSN 0946-2171

Kleinbauer, J.; Eckert, D.; Weiler, S.; & Sutter, D. (2008) 80 W ultrafast CPA-free disk laser, *Proc. SPIE*, Vol. 6871 (February 2008), 68711B, ISSN 0277-786X

Koechner, W. (2006) *Solid-State Laser Engineering*, Springer, ISBN 978-0-387-29094-2, USA

Liu, C.; Riesbeck, T.; Wang, X.; Ge, J.; Xiang, Z.; Chen J. & Eichler, H. (2008) Influence of spherical aberrations on the performance of dynamically stable resonators, *Optics Communications*, Vol. 281, Issue 20 (October, 2008) 5222–5228 ISSN 0030-4018.

Liu, H.; Gao, C.; Tao, J.; Zhao, W.; & Wang, Y. (2008) Compact tunable high power picosecond source based on Yb-doped fiber amplification of gain switch laser diode. *Opt. Express*, Vol. 16, Issue 11 (May 2008), 7888-7893, ISSN 1094-4087

Lorenz, E. (1963) Deterministic nonperiodic flow. *Journal of the Atmospheric Sciences*, Vol. 20, Issue 2 (March 1963), 130-141, ISSN 0022-4928

Lowdermilk, W. & Murray, J. (1980) The multipass amplifier: theory and numerical analysis. *J. Appl. Phys.*, Vol. 51, No. 5 (May 1980), 2436-2444, ISSN 0021-8979

Magni, V. (1986) Resonators for solid-state lasers with large-volume fundamental mode and high alignment stability. *Applied Optics*, Vol. 25, No. 1 (January 1986), 107-118, ISSN 0003-6935

Matsushima, I.; Yashiro, H & Tomie, T. (2006) 10 kHz 40 W Ti:sapphire regenerative ring amplifier. *Opt. Lett.*, Vol. 31, Issue 13 (July 2006), 2066-2068, ISSN 0146-9592

Meijer, J.; Dub, K.; Gillner, A.; Hoffmann, D.; Kovalenko, V.; Masuzawa, T.; Ostendorf, A.; Poprawe, R.; & Schulz, W. (2002) Laser Machining by short and ultrashort pulses, state of the art and new opportunities in the age of the photons. *CIRP Annals - Manufacturing Technology*, Vol. 51, Issue 2 (February 2002), 531-550, ISSN 0007-8506

Metzger, T.; Schwarz, A.; Teisset, C.; Sutter, D.; Killi, A.; Kienberger, R. & Krausz, F. (2009) High-repetition-rate picosecond pump laser based on a Yb:YAG disk amplifier for optical parametric amplification. *Opt. Lett.*, Vol. 34, No. 14 (July 2009), 2123-2125, ISSN 0146-9592

Mourou, G. & Umstadter, D. (1992) Development and Applications of Compact High-Intensity Lasers. *Phys. Fluids B*, Vol. 4, No. 7 (July 1992), 2317-2325, ISSN 0899-8213

Müller, D.; Giesen, A & Hügel, H. (2003) Picosecond thin-disk regenerative amplifier. *Proceedings of SPIE*, Vol. 5120 (November 2003), 281-286, ISSN 0277-786X

Murray, J. & Lowdermilk, W. (1980) Nd:YAG regenerative amplifier. *J. Appl. Phys.*, Vol. 51, No. 7 (July 1980), 3548-3555, ISSN 0021-8979

Nickel, D.; Stolzenburg, C.; Bevertt, A.; Geisen, A.; Haüssermann, J.; Butze, F.; & Leitner, M. (2005) 200 kHz electro-optic switch for ultrafast laser systems. *Rev. Sci. Instrum.*, Vol. 76, No. 3 (March 2005), 033111-033111/7, ISSN 0034-6748

Norris, T. (1992) Femtosecond pulse amplification at 250 kHz with a Ti:sapphire regenerative amplifier and application to continuum generation. *Opt. Lett.*, Vol. 17, Issue. 14 (July 1992), 1009-1011, ISSN 0146-9592

Peterson, R.; Jenssen, H. & Cassanho, A. (2002) Investigation of the spectroscopic properties of Nd:YVO_4. In: *Proc. OSA TOPS, Advanced Solid-State Lasers*, M.E. Fermann and L.R. Marshall, (Ed.), Vol. 68, 294-298, OSA

Pugžlys, A.; Andriukaitis, G.; Baltuška, A.; Su, L.; Xu, J.; Li, H.; Li, R.; Lai, W.; Phua, P.; Marcinkevičius, A.; Fermann, M.; Giniūnas, L.; Danielius, R.; & Ališauskas S. (2009) Multi-mJ, 200-fs, CW-pumped, cryogenically cooled, Yb,Na:CaF_2 amplifier. *Opt. Lett.*, Vol. 34, Issue 13 (July 2009), 2075-2077, ISSN 0146-9592

Raciukaitis, G.; Grishin, M.; Danielius, R.; Pocius, J. & Giniūnas, L. (2006) High repetition rate ps- and fs- DPSS lasers for micromachining. *Congress Proceedings (in CD)*, Vol. 99, Paper M1001, ISBN 0-912035-85-4, International Congress on Applications of Lasers & Electro- Optics, ICALEO 2006, Scottsdale, USA, October 30- November 2, 2006, Laser Institute of America

Röser, F.; Eidam, T.; Rothhardt, J.; Schmidt, O.; Schimpf, D.; Limpert, J. & Tünnermann, A. (2007) Millijoule pulse energy high repetition rate femtosecond fiber chirped-pulse amplification system. *Opt. Lett.*, Vol. 32, Issue 12 (December 2007), 3495-3497, ISSN 0146-9592

Ross, I.; New, G. & Bates, P. (2007) Contrast limitation due to pump noise in an optical parametric chirped pulse amplification system. *Optics Communications*, Vol. 273, Issue 2 (May 2007), 510-514, ISSN 0030-4018

Sayinc, H.; Buenting, U.; Wandt, D.; Neumann, J. & Kracht, D. (2009) Ultrafast high power Yb:KLuW regenerative amplifier. *Opt. Express*, Vol. 17, No. 17 (August 2009), 15068-15071, ISSN 1094-4087

Siebold, M.; Hornung, M.; Hein, J.; Paunescu, G.; Sauerbrey, R.; Bergmann, T. & Hollemann, G. (2004) A high-average-power diode-pumped Nd:YVO_4 regenerative laser amplifier for picosecond-pulses. *Applied Physics B*, Vol. 78, Num. 3-4 (February 2004), 287-290, ISSN 0946-2171

Siebold, M.; Hein, J.; Hornung, M.; Podleska, S.; Kaluza, M.; Bock, S. & Sauerbrey, R. (2008) Diode-pumped lasers for ultra-high peak power. *Appl. Phys. B*, Vol. 90, No. 3-4 (March 2008), 431-437, ISSN 0946-2171

Speiser, J. & Giesen, A. (2008) Scaling of thin disk pulse amplifiers. *Proc. SPIE*, Vol. 6871 (February 2008), 68710J, ISSN 0277-786X

Strickland, D. & Mourou, G. (1985) Compression of amplified chirped optical pulses. *Opt. Comm.*, Vol. 56, Issue 3 (March 1985), 219-221, ISSN 0030-4018.

Svelto, O. (1998) Principles of Lasers, Plenum Press, ISBN 0-306-45748-2, New-York

Tang, D.; Ng, S.; Qin, L.; & Meng, X. (2003) Deterministic chaos in a diode-pumped Nd:YAG laser passively Q switched by a Cr^{4+} YAG crystal. *Opt. Lett.*, Vol. 28, Issue 5 (March 2003), 325-327, ISSN 0146-9592

Valling, S.; Fordell, T. & Lindberg, A. (2005) Experimental and numerical intensity time series of an optically injected solid state laser. *Opt. Commun.*, Vol. 254, Issue 4-6 (October 2005), 282-289, ISSN 0030-4018

Walker, B.; Toth, C.; Fittinghoff, D.; Guo, T.; Kim, D.; Rose-Petruck, C.; Squier, J.; Yamakawa, K.; Wilson, K.; & Barty, B. (1999) A 50 EW/cm^2 Ti:sapphire laser system for studying relativistic light-matter interactions. *Opt. Express*, Vol. 5, Issue 10 (November 1999), 196-202, ISSN 1094-4087

13

Toward TW-Peak-Power Single-Cycle IR Fields for Attosecond Physics and High-Field Science

O. D. Mücke[1], S. Ališauskas[1,2], A. J. Verhoef[1], A. Pugžlys[1], A. Baltuška[1], V. Smilgevičius[2], J. Pocius[3], L. Giniūnas[3], R. Danielius[3], and N. Forget[4]

[1]*Photonics Institute, Vienna University of Technology, Gusshausstrasse 27-387, A-1040 Vienna,*
[2]*Laser Research Center, Vilnius University, Saulėtekio av. 10, LT-10223 Vilnius,*
[3]*Light Conversion Ltd., P/O Box 1485, Saulėtekio av. 10, LT-10223 Vilnius,*
[4]*Fastlite, Bâtiment 403, Ecole Polytechnique, FR-91128 Palaiseau,*
[1]*Austria*
[2,3]*Lithuania*
[4]*France*

1. Introduction

Optical Parametric Chirped-Pulse Amplification (OPCPA) (Dubietis et al., 1992, 2006) has attracted a lot of attention as a promising route toward intensity scaling of few-cycle laser pulses. Intense carrier-envelope phase (CEP)-stable few-cycle laser pulses have numerous intriguing applications in attosecond and high-field science (for a recent review, see Krausz & Ivanov, 2009) including the production of attosecond XUV/soft-X-ray pulses by high-harmonic generation (HHG) (Hentschel et al., 2001; Kienberger et al., 2004; Sansone et al., 2006), tomographic imaging of molecular orbitals (Itatani et al., 2004), and laser-induced electron diffraction (Meckel et al., 2008). A major challenge for using HHG in studies of time-resolved tomography of molecular dissociative states is the low ionization potential I_p of excited molecular states. The resulting competition between state depletion and HHG prevents generation of broad HHG spectra necessary for tomographic reconstruction. One solution are laser sources with high ponderomotive energy $U_p \propto \lambda^2 I$ at moderate intensity level, i.e., infrared (IR) CEP-stable few-cycle high-power laser systems. The present immense interest of the ultrafast community in high-U_p-sources (Fuji et al., 2006; Vozzi et al., 2006, 2007, 2009; Takahashi et al., 2008a, 2008b; Gu et al., 2009; Mücke et al., 2009a, 2009b; Moses et al., 2009) has three major reasons:

(1) These sources open the door to previously inaccessible regimes of light-matter interactions (Blaga et al., 2009) and in particular they allow experimental investigations of the λ-scaling laws of strong-field physics (Tate et al., 2007; Colosimo et al., 2008; Agostini & DiMauro, 2008; Doumy et al., 2009) (Keldysh parameter $\propto\lambda^{-1}$, electron energies $\propto\lambda^2$, HHG cutoff $\propto\lambda^2$, minimum attosecond pulse duration $\propto\lambda^{-1/2}$). In addition, e.g., laser-induced electron diffraction experiments would benefit from IR driving because of the shorter de Broglie electron wavelength and consequently higher spatial resolution (Meckel et al., 2008).

(2) Because of the λ^2-dependence of the HHG cutoff (Sheehy et al., 1999; Shan & Chang, 2001; Gordon & Kärtner, 2005), HHG driven by intense few-cycle IR sources holds great promise for the realization of bright coherent sources in the soft- and potentially even hard-X-ray region (Popmintchev et al., 2009). The achievable HHG photon fluence depends both on the microscopic single-atom response of the gas and on macroscopic effects in the gas target like phase matching and absorption. Currently, both microscopic and macroscopic parts are subject of a heated scientific debate. While the majority of studies (Tate et al., 2007; Schiessl et al., 2007; Colosimo et al., 2008) proclaimed a $\lambda^{-(5\text{-}6)}$-scaling of the single-atom HHG conversion efficiency or even worse (Shiner et al., 2009), Frolov et al. (2009) (also compare Pérez-Hernández et al., 2009) argued that the energy-dependence of the atomic photo-recombination cross-section and irregularities in the photoionization cross-section (e.g., originating from Cooper minima or electron correlation effects) can lead to more favorable HHG efficiency scaling ($\propto\lambda^{-5.3}$ for H, $\propto\lambda^{-4.6}$ for He, and $\propto\lambda^{-3.9}$ for Ar) and in some cases (e.g., in Xe) even to an increase of the HHG efficiency with increasing λ. Concerning macroscopic effects, recent work on phase matching of higher harmonics driven by high-energy IR pulses (Yakovlev et al., 2007; Popmintchev et al., 2008, 2009; Takahashi et al., 2008a, 2008b; Falcão-Filho et al., 2009) indicate the feasibility to compensate the sharp drop of the microscopic HHG efficiency macroscopically by an increased optimal gas pressure for phase matching and a strongly decreasing reabsorption of the generated X-rays at higher photon energies.
(3) The development of high-U_p-sources is also of paramount importance for attosecond photoelectron spectroscopy of solid surfaces and surface-adsorbate systems (Cavalieri et al., 2007; Miaja-Avila et al., 2008). Applying the experimental techniques of attosecond science to solid surfaces is very challenging and significantly more difficult than gas-phase experiments because the solid target is not continuously replaced (like in a gas jet), the target density can not be varied (e.g., to reduce space-charge effects), and detection of ions as well as using them as observable is impossible. Nevertheless, Cavalieri et al. (2007) have recently demonstrated that the attosecond ponderomotive streaking technique can be extended to solid surfaces in a narrow intensity window: space-charge effects originating from above-threshold ionization (ATI) electrons cause a severe background problem if the few-cycle driver pulse intensity is too high. If, on the other hand, the intensity is too low, the ponderomotive streaking is too small to be observed. For a maximum acceptable intensity determined by the experimental conditions (e.g., damage threshold of the solid target, background problems from space-charge effects), IR high-U_p-sources are superior to 0.8-µm Ti:sapphire sources because they exhibit stronger ponderomotive streaking effects.
For many applications in attosecond science, in particular for the generation of isolated attosecond XUV/soft-X-ray pulses, extremely short pulses comprising only one or two light oscillations underneath the field envelope are required. The well-established standard technology for the generation of few-cycle driver pulses is spectral broadening of mJ-level femtosecond pulses from Ti:sapphire amplifier systems in noble gases. Using 5-fs 1-mJ 720-nm driver pulses obtained in this way, coherent X-rays in the keV photon energy range were demonstrated by HHG in He (Seres et al., 2005). At present, the energy throughput of gas-phase broadening schemes, such as the hollow-core fiber compressor (Nisoli et al., 1996, 1997) and filamentation (Hauri et al., 2004, 2007), is limited to 4-5 mJ at 0.8 µm due to ionization losses. In filamentation of intense femtosecond pulses, a self-guiding channel ("filament") is formed due to the dynamical balance between self-focusing by the nonlinear Kerr effect and defocusing by the plasma created by gas ionization. During the nonlinear pulse propagation, the spectrum is significantly broadened by the Kerr nonlinearity and

plasma-induced blue-shifting. Since no waveguiding structure is required in filamentation and therefore coupling losses are eliminated, the filamentation technique permits higher energy throughput than the hollow-core fiber compressor. With IR pulses we expect to surpass the present energy limitation for gas broadening schemes since the critical power of self-focusing scales as λ^2 (Bergé, 2008).

In this chapter we present a hybrid IR OPCPA/filamentation approach to pursue a TW-peak-power single-cycle IR source that will find many applications in attosecond and high-field science. The chapter is organized as follows: In section 2, we provide the theoretical background of optical parametric amplification (OPA) with special emphasis on Type-II phase matching in KTP crystals, as employed in our IR OPCPA. In section 3, a CEP-stable 10-mJ-class four-stage OPCPA system at 1.5 μm is discussed in detail. In section 4, we present experiments on spectral broadening and self-compression of multi-mJ 1.5-μm pulses via filamentation in noble gases.

2. Type-II optical parametric amplification in KTP crystals

In ultrafast parametric amplifiers (Cerullo & De Silvestri, 2003), an intense pump beam of high frequency ω_p is used to amplify a signal and an idler beam with frequencies ω_s and ω_i ($\omega_p > \omega_s > \omega_i$), respectively:

$$\omega_p = \omega_s + \omega_i \,. \tag{1}$$

The signal ω_s lies within $\omega_p/2$ and ω_p, the corresponding idler within $\omega_p/2$ ("degeneracy") and 0. In addition to the energy conservation (1), for an efficient energy transfer in the parametric interaction also the corresponding momenta of the pulses, $k_j = 2\pi n_j/\lambda_j$ (j=p,s,i), must be conserved ("phase matching"):

$$\vec{k}_p = \vec{k}_s + \vec{k}_i \quad \Leftrightarrow \quad \Delta\vec{k} = \vec{k}_p - \vec{k}_s - \vec{k}_i = 0 \,. \tag{2}$$

The phase matching condition (2) cannot be fulfilled in bulk isotropic materials in the normal dispersion region ($n_p > n_s > n_i$). However, phase matching can be achieved in birefringent crystals taking advantage of the different refractive indices of the ordinary (o) and extraordinary (e) beams. Depending on the polarizations of the pump, signal and idler beams, one distinguishes different types of phase matching. For example, Table 1 illustrates three different types of phase matching in the crystallographic XZ-plane in KTP (Liu et al., 2001).

phase-matching type	γ_{sp}	γ_{si}
Type I ($e_s + e_i \to o_p$)	2/3	2
Type II ($e_s + o_i \to o_p$)	2/3	2/3
Type III ($o_s + e_i \to o_p$)	2	2/3

Table 1. OPA phase-matching types in the XZ-plane in the biaxial crystal KTP and corresponding cross-phase modulation (XPM) coupling coefficients γ_{sp} and γ_{si}

In general, phase matching can be achieved by angle-tuning the nonlinear crystal, i.e., by adjusting the polar θ-angle and azimuthal φ-angle, or by changing the crystal temperature.

Type-I OPAs feature ultrabroadband phase matching and gain, especially for operation near the degeneracy point, $\omega_s = \omega_i = \omega_p/2$ (Cerullo & De Silvestri, 2003). For this reason, Type-I phase matching has been extremely popular for the generation of ultrashort pulses and high-energy few-cycle driver pulses for attosecond physics and high-field science. The experiences made with these systems during the past few years, however, have also revealed some severe drawbacks inherent to Type-I phase matching:

- Type-I parametric amplification of CEP-stable two-cycle IR seed pulses obtained from difference-frequency generation (DFG) to the energy level close to 1 mJ has been demonstrated (Fuji et al., 2006; Vozzi et al., 2006, 2007; Gu et al., 2009; Moses et al., 2009). However, the inherently low DFG seed energy causes a sizable superfluorescence background (Fuji et al., 2006; Gu et al., 2009; Moses et al., 2009) that prevents further energy upscaling.
- Type-I OPAs exhibit ultrabroadband phase matching, but vice versa they are also negatively optimized for back-conversion into the pump.
- Even though Type-I OPAs deliver output spectra with FWHM bandwidth in excess of 200 nm, the quality of the resulting compressed pulses most often remains poor due to intrinsically steep slopes of the parametrically amplified spectra.
- Near-degenerate Type-I OPAs have the worst possible quantum defect for the signal wave (i.e., half of the pump photon energy is "lost" with the idler).
- When scaling up the pulse energies to the millijoule-level, (cascaded) four-wave mixing (FWM) can lead to problems due to parasitic self-diffraction (Beržanskis et al., 1995; Varanavičius et al., 1997).

For these reasons, we decided to pursue a hybrid Type-II OPCPA/filamentation scheme to realize a TW-peak-power single-cycle IR source. By employing more narrowband Type-II phase matching, one can optimize the spectral brightness of the seed at the expense of the seed energy, achieve a more uniform saturation across the pulse spectrum, minimize energy back-conversion into the pump, and minimize self-diffraction. Following the pioneering work of Kraemer et al. (2006, 2007), we employ Type-II KTP (1.030/1.064 µm pump, ~1.5 µm signal, ~3.3-3.7 µm idler) because these crystals are transparent for the mid-IR idler wavelength and exhibit a relatively broad bandwidth around 1.5 µm.

$KTiOPO_4$ (KTP) and $KTiOAsO_4$ (KTA) are biaxial crystals permitting Type-II ($e_s + o_i \rightarrow o_p$) phase matching in the crystallographic *XZ*-plane (i.e., φ=0°): the *o*-beam propagates into the *y*-direction, $n_o=n_y$, the *e*-beam parallel to the *XZ*-plane, $n_x<n_e<n_z$, with

$$n_e(\theta) = n_x \frac{\sqrt{1+\tan^2\theta}}{\sqrt{1+(n_x/n_z)^2\tan^2\theta}} . \tag{3}$$

For noncollinear Type-II ($e_s + o_i \rightarrow o_p$) phase matching, where α is the (internal) noncollinearity angle between the pump and signal beams inside the crystal and β the resulting angle between pump and idler satisfying Eq. (2), the analysis by Liu et al. (2001) yields the relations

$$\frac{n_y^{\mathrm{p}}}{\lambda_{\mathrm{p}}} = \frac{n_z^{\mathrm{s}}}{\lambda_{\mathrm{s}}}\cos\alpha + \frac{n_y^{\mathrm{i}}}{\lambda_{\mathrm{i}}}\cos\beta \tag{4}$$

$$\beta = \arcsin\left(\frac{\lambda_{\mathrm{i}}}{\lambda_{\mathrm{p}}}\frac{n_y^{\mathrm{p}}}{n_y^{\mathrm{i}}}\sin\alpha\right) - \alpha \tag{5}$$

from which the optimum phase-matching angle θ_{pm} can be determined

$$\sin\theta_{pm} = \frac{n_z^s}{\frac{n_y^p}{\lambda_p}\cos\alpha - \frac{n_y^i}{\lambda_i}\cos(\alpha+\beta)}\sqrt{\frac{\left(\frac{n_x^s}{\lambda_s}\right)^2 - \left(\frac{n_y^p}{\lambda_p}\cos\alpha - \frac{n_y^i}{\lambda_i}\cos(\alpha+\beta)\right)^2}{\left(n_x^s\right)^2 - \left(n_z^s\right)^2}}. \tag{6}$$

Eq. (5) reflects the angular dispersion of the idler for noncollinear (i.e., $\alpha\neq 0$) phase matching. Using the Sellmeier equations summarized in the Appendix for computing the refractive index components of KTP, the above equations can readily be evaluated. Fig. 1(a) depicts the resulting theoretical angle-tuning curves.

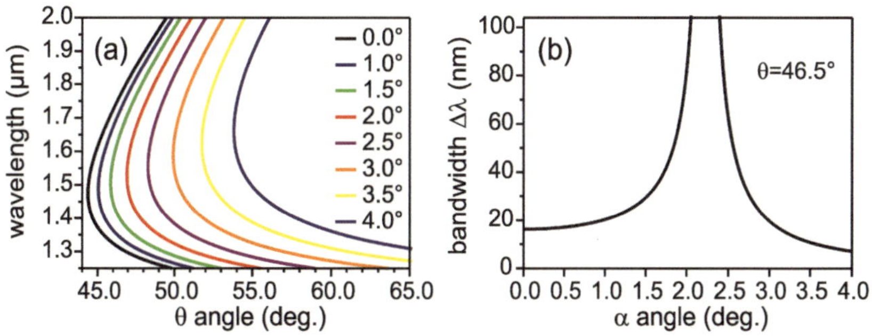

Fig. 1. Noncollinear Type-II ($e_s + o_i \rightarrow o_p$) phase matching in KTP for λ_p=1.064 µm: (a) Signal wavelength versus phase-matching angle θ for various (internal) noncollinearity angles α as indicated. (b) Parametric FWHM phase-matching bandwidth $\Delta\lambda$ [computed from Eq. 8(a)] versus noncollinearity angle α for θ=46.5°, L=10 mm, and λ_s=1.55 µm. Signal-idler group-velocity matching is achieved for $\alpha_{si}\approx$2.2°. At this point, Eq. 8(a) loses its applicability.

The broadest phase-matching bandwidth can be achieved in an OPA, when the group velocities (GV) of signal and idler pulses $v_{s,i}=(\partial k_{s,i}/\partial\omega)^{-1}$ are matched [see Fig. 1(b)]. In case of perfect group-velocity matching, the maximum phase-matching bandwidth is then determined by the mismatch of the group-velocity dispersion (GVD) of signal and idler pulses, $GVD_{s,i}=(\partial^2 k_{s,i}/\partial\omega^2)$. The signal-idler group-velocity mismatch (GVM)

$$\frac{1}{\mathrm{GVM_{si}}} = \frac{1}{v_i\cos(\alpha+\beta)} - \frac{1}{v_s} \tag{7}$$

can be controlled by varying the noncollinearity angle α, and the parametric FWHM gain bandwidth can be calculated from

$$\Delta\lambda = \begin{cases} \dfrac{\lambda_s^2\left|\mathrm{GVM_{si}}\right|}{c_0 L} & \text{for } \dfrac{1}{\mathrm{GVM_{si}}}\neq 0 \quad (8a) \\ \dfrac{0.8\lambda_s^2}{c_0\sqrt{\left|\mathrm{GVDM_{si}}\right|L}} & \text{for } \dfrac{1}{\mathrm{GVM_{si}}} = 0 \quad (8b) \end{cases}$$

with the vacuum velocity of light c_0, crystal length L, and signal-idler GVD mismatch $GVDM_{si}$ (Liu et al. (2001)). The optimum (internal) angle for signal-idler group-velocity matching can also directly be obtained from the expression (Cerullo & De Silvestri, 2003)

$$\alpha_{si} = \arcsin\left(\frac{1 - v_s^2 / v_i^2}{1 + 2v_s n_s \lambda_i / v_i n_i \lambda_s + n_s^2 \lambda_i^2 / n_i^2 \lambda_s^2}\right)^{1/2} . \tag{9}$$

For θ=46.5°, λ_p=1.064 μm and λ_s=1.55 μm, Eq. (9) predicts a value of α_{si} =2.2°.
The parametric gain G can be calculated from the coupled equations describing parametric processes (Cerullo & De Silvestri, 2003). If the OPA is seeded with a small light intensity I_{s0} at the signal wavelength only (i.e., I_{i0}=0) and neglecting pump depletion (i.e., pump intensity I_p=const.), the solution of the coupled equations are

$$I_s(L) = I_{s0}\left[1 + \frac{\Gamma^2}{g^2}\sinh^2(gL)\right] \xrightarrow{\Gamma L \gg 1} I_s(L) \approx \frac{1}{4} I_{s0} \exp(2gL) \tag{10a}$$

$$I_i(L) = I_{s0}\frac{\lambda_s}{\lambda_i}\frac{\Gamma^2}{g^2}\sinh^2(gL) \xrightarrow{\Gamma L \gg 1} I_i(L) \approx \frac{1}{4}\frac{\lambda_s}{\lambda_i} I_{s0} \exp(2gL) \tag{10b}$$

where the formulas on the right-hand side give the limit for large gains. In Eqs. (10) we used the gain coefficients defined as

$$g = \sqrt{\Gamma^2 - (\Delta k / 2)^2} \text{ , with } \Gamma^2 = \frac{8\pi^2 d_{eff}^2}{n_p n_s n_i \lambda_s \lambda_i \varepsilon_0 c_0} I_p \tag{11}$$

with the permittivity of vacuum ε_0. The effective nonlinear optical coefficient d_{eff} of KTP for Type-II phase-matching in the *XZ*-plane is (Dmitriev et al., 1999)

$$d_{eff} = d_{32} \sin\theta \text{ , with } d_{32}(1.064\,\mu\text{m}) = 2.65\,\text{pm/V} . \tag{12}$$

From Eqs. (10), we see that the parametric gain G is given by

$$G = 1 + \frac{\Gamma^2}{g^2}\sinh^2(gL) \xrightarrow{\Gamma L \gg 1} G \approx \frac{1}{4}\exp(2gL) . \tag{13}$$

Fig. 2 shows the theoretical gain spectrum calculated for the experimental conditions in our booster-amplification stage 4 (see Figs. 8 and 9, Table 2). Note, however, that for the high intensities in our OPCPA stage 4, pump depletion cannot be ignored which results in a reduced gain in the experiment.
In our discussion, we so far have neglected spatial effects. However, the energy transfer in the parametric interaction is limited by the Poynting vector walk-off between the pump, signal and idler pulses. For KTP, the (internal) angle between pump and signal beams, for which Poynting vector walk-off is compensated, is given by (Dmitriev et al., 1999)

$$\rho(\theta) = -\arctan[(n_o / n_e(\theta))^2 \tan\theta] + \theta . \tag{14}$$

For a Type-II ($e_s + o_i \rightarrow o_p$) KTP OPA pumped at 1.03 μm or 1.064 μm, the external walk-off compensation angle is 2.2°.

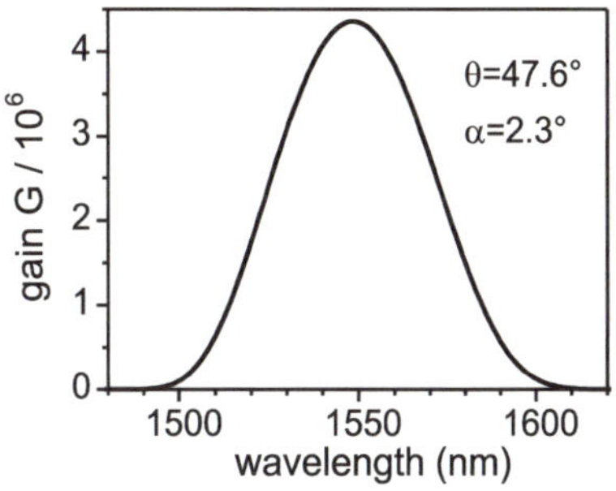

Fig. 2. Gain spectrum (pump depletion neglected) of a Type-II ($e_s + o_i \rightarrow o_p$) KTP OPA pumped at λ_p=1.064 μm. Pump intensity I_p=20 GW/cm^2, θ=47.6°, α=2.3°, and L=10 mm.

3. CEP-stable 10-mJ-class chirped parametric amplifier at 1.5 μm

In this section, we describe in detail a CEP-stable four-stage IR OPCPA system delivering 1.5-μm signal pulses (and corresponding idler pulses at ~3.3-3.7 μm) with up to 12.5 mJ pulse energy before recompression. This IR OPCPA is based on a CEP-stable two-stage OPA seed source discussed in subsection 3.1 and two booster-amplification stages presented in subsection 3.2.

3.1 CEP-stable 4-μJ OPA seed source

The scheme of our two-stage OPA seed source is depicted in Fig. 3.

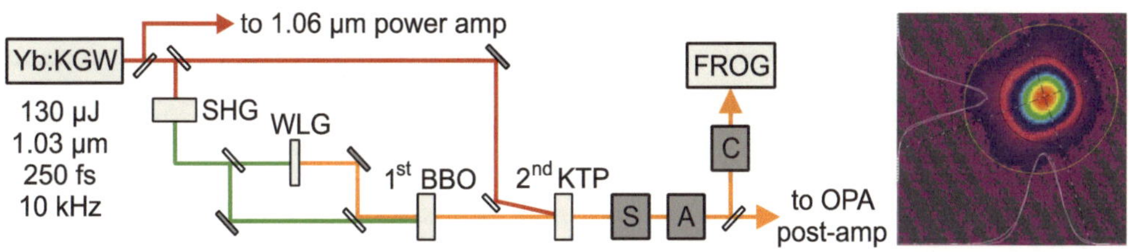

Fig. 3. Scheme of the CEP-stable two-stage 1.5-μm OPA seed source. Yb:KGW, 250 fs Yb:KGW MOPA system; SHG, second-harmonic generation in BBO; WLG, white-light generation in sapphire; S/C, grating stretcher/compressor; A, acousto-optic programmable dispersive filter (DAZZLER). Right panel, far-field beam profile of second-stage signal wave.

In our IR OPCPA scheme (see Figs. 3 and 8), both Yb:KGW and Nd:YAG regenerative amplifiers (RAs) are simultaneously seeded from a single Yb:KGW master oscillator that has a modest FWHM bandwidth of 30 nm centered at 1.04 μm [see Fig. 4(a)]. To seed the Nd:YAG RA, we pick up unused 1064-nm light behind a transmission grating in the pulse stretcher of the Yb:KGW MOPA. The repetition rate of the Yb:KGW diode-pumped solid state (DPSS) MOPA (Pharos, Light Conversion, Ltd.), tunable in the range of 1-100 kHz, was set at 10 kHz as the 500th harmonic of the flash-lamp-pumped Nd:YAG amplifier (Ekspla Ltd.) operating at 20 Hz. In the Nd:YAG RA, an intracavity etalon is used to narrow the pulse bandwidth and make the pulse duration safe for post amplification [see Fig. 4(b)].

The 1.03-μm output from the femtosecond Yb:KGW MOPA is first split into two parts by means of a variable beam splitter (consisting of a half-wave plate and a thin-film polarizer);

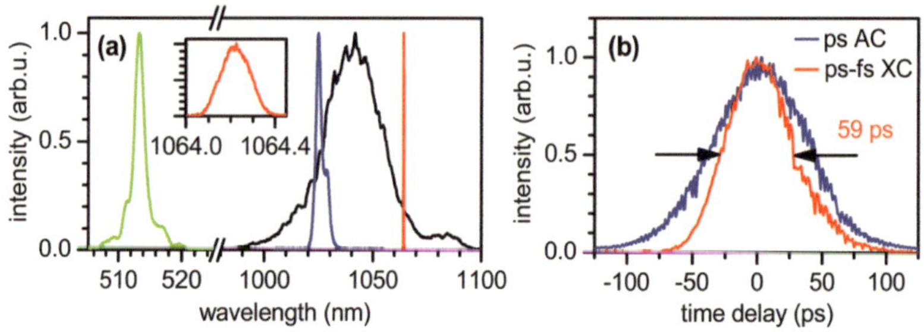

Fig. 4. Optical synchronization of the Yb:KGW and Nd:YAG amplifiers. (a) Laser spectra: Kerr-lens mode-locked Yb:KGW oscillator (black curve), Yb:KGW regenerative amplifier (blue), Nd:YAG with an intracavity 2-mm thick etalon (red), and SHG of Yb:KGW RA (green). (b) Pulse measurement of the ps Nd:YAG amplifier showing an autocorrelation (AC, blue) and a cross-correlation (XC, red) between the Yb:KGW (250 fs) and Nd:YAG.

part one is used for implementing the first OPA stage, part two is used for pumping the second OPA stage. In the first OPA stage, the 1.03-μm pulses are first frequency-doubled in a 1-mm thick Type-I BBO crystal (θ=23.4°, φ=90°). Typically 8.5 μJ of 515-nm pulses are again split by a variable beam splitter into two parts: 1.3 μJ (measured behind a variable aperture used for fine-adjusting the input beam diameter) are focused onto a 10-mm thick sapphire plate using an 87-mm focusing lens. In the sapphire plate, a stable white-light (WL) continuum extending to wavelengths >840 nm [see Fig. 5(a)] is created in a single filament. For a sapphire plate thickness of 10 mm we obtain a stronger and more stable WL seed at >780 nm than for thinner plates of 4–6 mm. The WL continuum, which is recollimated with a 40-mm lens, is used to seed the first OPA stage. The WL seed pulses and the 515-nm pump pulses are combined collinearly [i.e., α=0° to avoid angular dispersion of the idler, compare Eq. (5)] on a dichroic beam splitter and both are focused onto a 4-mm thick Type-I BBO crystal (θ=22.8°, φ=90°) with an f=20 cm spherical mirror to a $1/e^2$ pump beam diameter of 120 μm. In the pump beam, a variable aperture is used to adjust the pump energy to 1.4 μJ (thus, the parametric gain) and to minimize the detrimental effects of amplified spontaneous emission (ASE). By adjusting both the θ-angle of the BBO crystal and time delay between the seed and pump pulses, different wavelength regions can be phase matched for efficient parametric amplification [Fig. 5(a)]. In particular, selecting the ~795 nm wavelength for amplification, this configuration produces CEP-stable idler pulses (Baltuška et al., 2002) at 1.44 μm [see Fig. 5(b)] that we use as a seed in the second OPA stage.

Following the pioneering work of Kraemer et al. (2006, 2007), we employ Type-II KTP crystals (1.03/1.064 μm pump, ~1.5 μm signal, ~3.3-3.7 μm idler) for the subsequent OPA stages 2-4 because these crystals (unlike borate crystals) are transparent for the mid-IR idler wavelength and exhibit a relatively broad bandwidth around 1.5 μm. The CEP-stable idler pulses from the first OPA stage are recollimated with a 10-cm lens and focused onto a 6-mm thick Type-II KTP crystal (θ=45.5°, φ=0°) using a 50-cm lens. The pump beam is focused onto the same KTP crystal using a 100-cm lens under an (external) walk-off compensation angle of 2.2° with respect to the seed beam. For a pump beam diameter of ~550 μm measured at the KTP crystal input face and 69 μJ (measured behind a variable aperture) pump pulses, we achieve 4 μJ signal pulses, i.e., a pump-signal conversion efficiency of ~6%

in the second OPA stage. The pulse-to-pulse intensity fluctuations of the two-stage OPA amounts to 2.5% rms noise, only two times larger than that of the Yb:KGW MOPA pump (1.2% rms). The far-field beam profile of the second-stage signal wave shown in Fig. 3 is nearly Gaussian; the beam propagation factor was determined to be M^2=1.13±0.04 as compared to M^2<1.2 of the Yb:KGW MOPA pump.

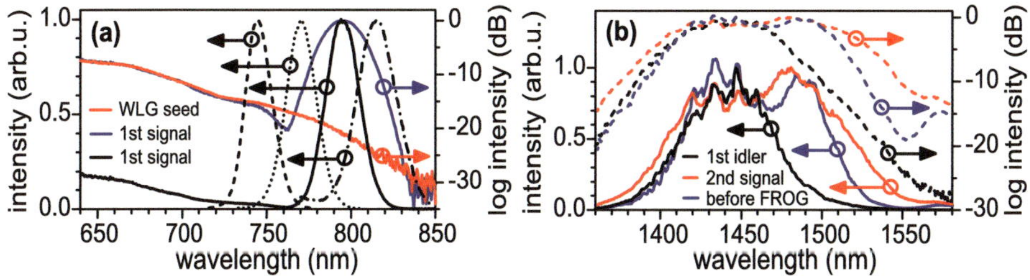

Fig. 5. Spectral properties of two-stage 1.5-μm OPA. (a) Near-IR WL seed (red) and signal spectra (blue and black solid) of the first OPA stage. The dashed and dotted curves indicate tunability of the first-stage signal (and corresponding idler). (b) Infrared passively CEP-stable seed (black), amplified second-stage signal (red), spectrum before FROG setup (blue). The amount of ASE is immeasurable in absence of the WL seed.

The strong nonlinear optical Kerr effect in KTP [nonlinear refractive index coefficient n_2(KTP) = 23.7×10^{-16} cm^2/W as compared to n_2(BBO) = 2.9×10^{-16} cm^2/W (Koechner, 2006)] raises the important question how severe is pump/idler-to-signal cross-phase modulation (XPM) in the second OPA stage and its effect on CEP stability. The nonlinear effects accumulated during the OPA process in a nonlinear crystal of length L can be quantified by introducing a generalized B-integral

$$B=\frac{2\pi n_2}{\lambda_s}\int_0^L dz\left[I_s(z)+\gamma_{sp}I_p(z)+\gamma_{si}I_i(z)\right]. \tag{15}$$

The coefficients γ_{sp} and γ_{si} (see Table 1), which quantify the pump/idler-to-signal coupling, are 2 for parallel polarizations and 2/3 for orthogonal polarizations (Agrawal, 2007). Obviously, Type-II (e_s + o_i → o_p) phase matching as used in our amplification stages 2-4 minimizes the detrimental XPM contribution of pump and idler on the signal wave.

CEP stability of the OPA 2 output was studied by means of inline f-to-$2f$ interferometry: a supercontinuum is generated by focusing the 1.5-μm pulses into a 3-mm thick sapphire plate with a 15-mm lens. After recollimation, the second harmonic of the 1.5-μm pulses is generated in a 0.2-mm thick Type-I BBO crystal (θ=19.8°, φ=90°). With a polarizer, the supercontinuum and SHG are projected onto a common axis and the resulting f-to-$2f$ interferograms from 650 to 790 nm are recorded with a spectrometer (see Fig. 6). The observation of stable interference fringes directly proves CEP stability and negligible influence of XPM on the CEP. The slow CEP drift observed in Fig. 6 is clearly of an environmental origin and can easily be compensated for by feedback stabilization of the interferometer formed by the seed and pump paths in the first OPA stage (see Fig. 3).

In stand-alone applications of our two-stage OPA, the well-behaved spectral phase of the second-stage signal can readily be compensated for by highly reflective chirped mirrors. Here, keeping in mind mainly the development of a >10 mJ four-stage IR OPCPA system, we instead demonstrate that the second-stage output can be stretched to ~40 ps and again recompressed to a sub-40-fs duration using a grating-based stretcher/compressor pair (Kalashnikov et al., 2005) employing 500 grooves/mm 96% efficient gold reflection gratings and an IR high-resolution acousto-optic programmable dispersive filter (DAZZLER) (Verluise et al., 2000).

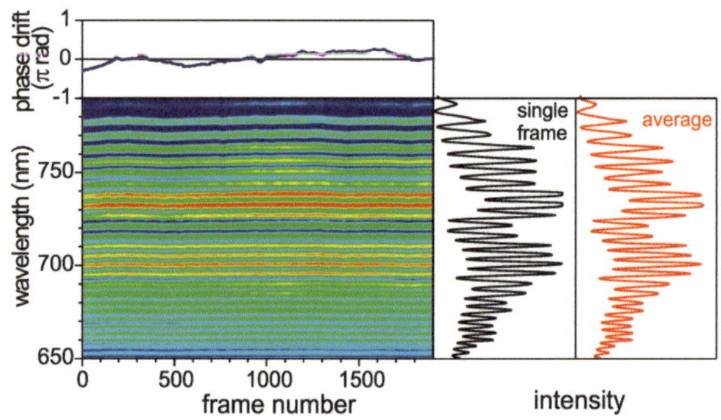

Fig. 6. *f*-to-2*f* interferograms exhibiting stable fringes due to CEP stability. The red and black curves on the right panels indicate the spectra averaged over 1900 frames (2 ms frame exposure time) and a single frame, respectively, on the same intensity axis. The blue curve in the above panel indicates the extracted phase drift of the second-stage signal.

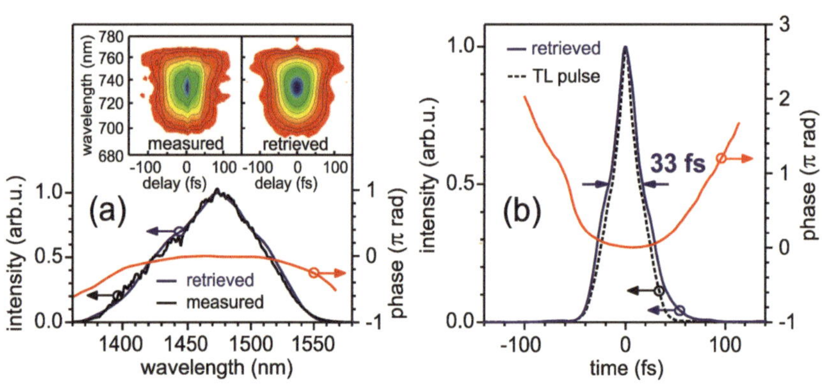

Fig. 7. SHG-FROG characterization of the stretched and recompressed 1.47-µm pulses: (a) Measured spectrum (black curve), retrieved spectral intensity (blue) and phase (red). The insets show the measured and retrieved FROG traces. (b) Retrieved temporal intensity (blue) and phase (red) profile indicating a 33-fs FWHM pulse duration. The transform-limited intensity profile (black dashed) corresponds to a 28-fs duration.

Fig. 7 shows SHG-frequency-resolved optical gating (FROG) measurements of 1.47-µm pulses with a 92-nm FWHM bandwidth from the second OPA stage after stretching to ~40 ps and recompression to a 33-fs FWHM pulse duration.

Our parametric infrared source, easily tunable in the 1.5-1.6 µm range, might find applications in pump-probe experiments on InAs quantum dot semiconductor optical amplifiers (QD SOAs) operating in the 1.55-µm telecommunications wavelength range (Zilkie et al., 2006) or for buried-waveguide writing in semiconductors based on three-photon absorption (Nejadmalayeri et al., 2005).

3.2 Power-amplification to pulse energies exceeding 10 mJ

In the subsequent booster-amplification stages 3 and 4 shown in Fig. 8, the 1.5-μm signal pulses are amplified from the 4-μJ level to energies >10 mJ before recompression.

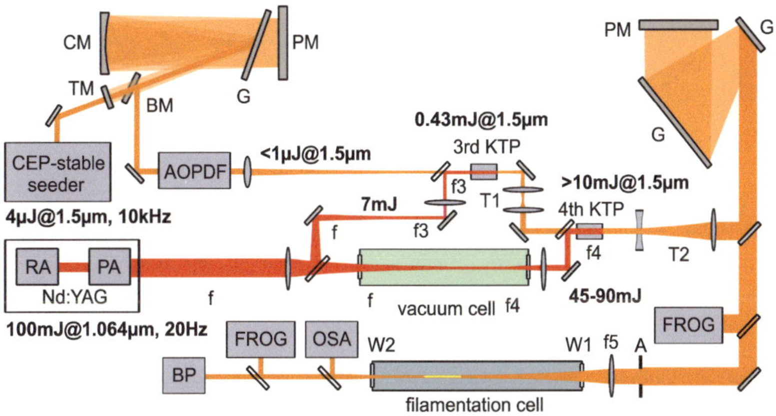

Fig. 8. Scheme of the OPCPA power-amplification stages 3 and 4: G, reflection grating; CM, curved mirror; PM, plane mirror; TM/BM, top/bottom mirror; AOPDF, acousto-optic programmable dispersive filter; RA, regenerative amplifier; PA, double-pass post amplifier; f, f3, f4, f5, lens focal lengths; T1, T2, telescopes; A, iris aperture; W1/W2, input/output windows; OSA, optical spectrum analyzer; BP, beam profiler (pyroelectric 2D array).

As already mentioned in the previous subsection, in order to optimize the energy extraction from the 60-ps long Nd:YAG pump pulses, the passively CEP-stable 1.5-μm pulses from the front-end are temporally stretched to ~40 ps using a grating-based stretcher. In addition, a high-resolution IR DAZZLER is used for higher-order dispersion control. To guarantee a homogeneous pump profile free of hot spots, we relay-image the 10-mm-diameter crystal rod in the Nd:YAG power amplifier onto the 10-mm thick KTP crystals in stages 3 and 4. The measured surface damage threshold of KTP for our pump pulses (21 GW/cm^2) determines a pump spot diameter of 2 mm and 3.1 mm for stages 3 and 4, respectively. Relay-imaging is achieved with three lenses with focal lengths of f=75 cm, f3=10 cm, and f4=35 cm (see Fig. 8). Because of the larger pump intensities in the fourth OPA stage, the focus needs to be placed inside a vacuum cell to avoid a breakthrough in air. The 1.5-μm (seed) pulses are focused onto the third-stage KTP crystal with a 750-mm lens and imaged onto the fourth-stage KTP crystal with telescope T1. The (external) walk-off compensation angle between the pump and seed beams is 2.2°. With this pumping geometry and ~90 mJ pump pulses, we have achieved up to 12.5 mJ signal pulses centered at 1.57 μm with a pump-signal conversion efficiency of ~22% in the final OPCPA stage. To avoid damage to the gold gratings in the OPCPA compressor, we expand the beam diameter of the fourth-stage output by a factor of 5 to 9.5 mm (at the 1/e^2-level) by means of a Galilean beam expander T2.

The spectra of the seed and amplified signal pulses of the power-amplification stages are shown in Fig. 9. In principle, saturating the OPCPA stages permits amplification of pulses with nearly 80 nm bandwidth corresponding to a Fourier limit of ~65 fs. As idler absorption increases above 3.4 μm in KTP, we can achieve higher output powers when tuning the signal center wavelength above 1.55 μm. The SHG-FROG characterization data of 3.5-mJ

1.57-μm pulses with 62 nm bandwidth from the 20-Hz four-stage IR OPCPA (see Fig. 10) indicate a 74.4-fs FWHM pulse duration, close to the transform limit (TL) of 72.6 fs. Ultimately, with further optimization sub-70-fs pulse durations seem in reach by recompressing pulses with bandwidths approaching 80 nm.

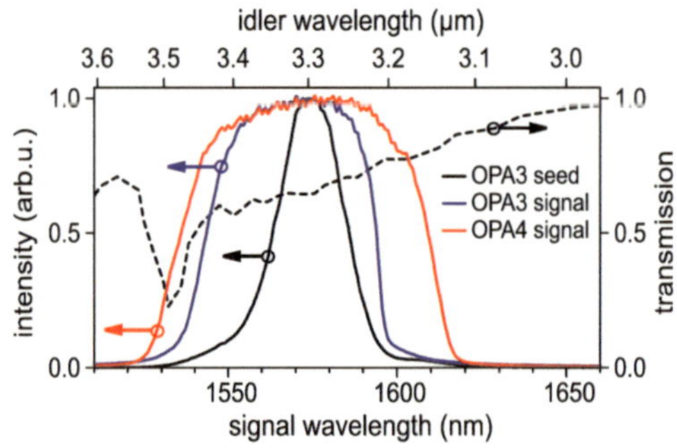

Fig. 9. Spectral properties of the power-amplification OPCPA stages: spectrum of the third-stage seed, amplified signal spectra after stages 3 and 4. The amount of superfluorescence is immeasurable in absence of the WL seed in OPA stage 1. The dashed curve indicates the idler transmission through 10 mm of KTP.

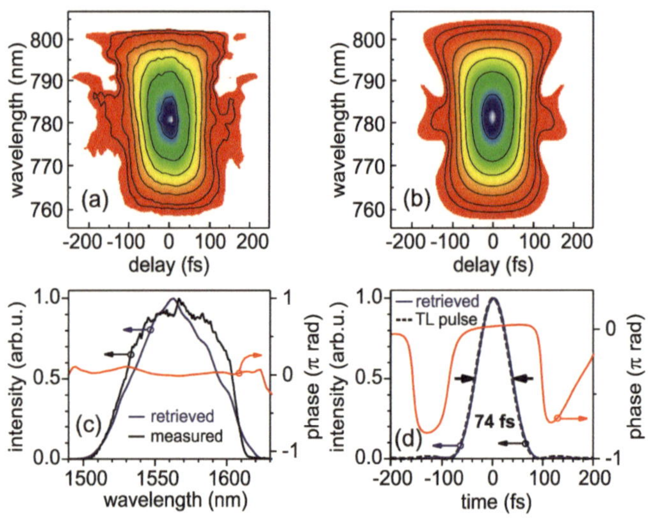

Fig. 10. SHG-FROG characterization of the 20-Hz output from the four-stage IR OPCPA: (a) Measured and (b) retrieved FROG traces. (c) Measured spectrum (back curve), retrieved spectral intensity (blue) and phase (red). (d) Retrieved temporal intensity (blue) and phase (red) profiles exhibiting a 74.4-fs FWHM pulse duration. The TL intensity profile (dashed) corresponds to a 72.6-fs duration.

Concerning the CEP stability of our four-stage IR OPCPA, there are in principle three sources of CEP instabilities:

i. Possible superfluorescence backgrounds can compromise the CEP stability on short time scales. Since superfluorescence is well-suppressed in our IR OPCPA, this effect is expected to be negligible.

ii. On a longer (say, second) time scale, the CEP can drift due to slow environmental relative path-length changes of pump and seed arms in the first OPA stage [see Fig. (3)]. At the moment, we are implementing a feedback stabilization of these path-length drifts to remove slow CEP drifts from the four-stage IR OPCPA output.

iii. As already discussed in subsection 3.1, the CEP stability could in principle also be compromised by amplitude-to-phase-noise conversion due to pump/idler-to-signal XPM in OPA stages 2-4. The importance of this effect can readily be estimated by the generalized B-integral defined in Eq. (15). Here, ignoring the less intense idler wave and assuming a constant pump intensity over the crystal length L=10 mm, Eq. (15) simplifies to

$$B = \frac{2\pi n_2}{\lambda_s} L \gamma_{sp} I_p \,. \tag{16}$$

From this equation, the XPM-induced CEP noise $\Delta\phi_{CE,rms}$ can be estimated assuming 2% rms intensity fluctuations of the Nd:YAG pump amplifier. The results of this estimate summarized in Table 2 indicate that amplitude-to-phase-noise conversion due to XPM is completely negligible in OPA stages 3 and 4. Thus, the output of the four-stage IR OPCPA inherits the excellent short-term CEP stability of the two-stage OPA front-end.

	OPA3	OPA4
pump pulse energy	7 mJ	90 mJ (maximum)
pump diameter	2 mm	3.1 mm
pump intensity	3.7 GW/cm^2	20 GW/cm^2
B	0.02 π	0.11 π
$\Delta\phi_{CE,rms}$	4×10^{-4} π	2.2×10^{-3} π

Table 2. Estimate for XPM-induced CEP noise $\Delta\phi_{CE,rms}$ in OPA stages 3 and 4.

4. Spectral broadening and pulse self-compression via filamentation in noble gases

In this section, we demonstrate single-filament IR supercontinuum generation via femtosecond filamentation in noble gases. Depending on the experimental conditions, two filamentation regimes can be achieved: (i) the filamentation regime without plasma-induced pulse self-recompression, as discussed in section 4.1, and (ii) the self-compression regime, as discussed in section 4.2. Ultimately, since the critical power of self-focusing scales as λ^2, we expect to surpass the current energy limitation [4-5 mJ at 800 nm (Suda et al., 2005; Stibenz et al., 2006)] with the multi-mJ femtosecond pulses obtained from our IR OPCPA. In addition, a promising route for further pulse energy upscaling is the use of circularly/elliptically polarized input pulses (Trisorio & Hauri, 2007; Varela, 2009).

4.1 Multimillijoule filamentation without plasma-induced pulse self-compression

In the filamentation experiments (see Fig. 8), the 1.57-μm OPCPA pulses were focused using a 50-cm lens placed 4 cm in front of the AR-coated input window W1 of a 138-cm long gas cell filled with argon (I_p=15.76 eV) or krypton (I_p=13.99 eV) at the absolute pressure of 4-5 bar.

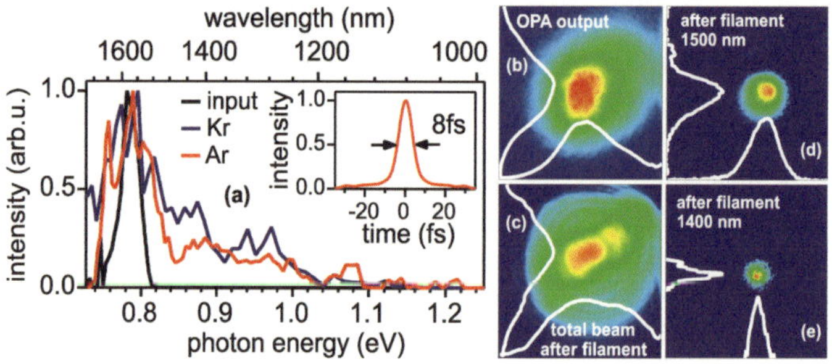

Fig. 11. (a) Filamentation of 1.57-μm pulses in noble gases: Individually normalized input (black curve) and output spectra for filamentation of 0.8-mJ pulses in krypton (blue) and 2.5-mJ pulses in argon (red) at a 5-bar pressure. The inset shows the TL intensity profile computed from the argon output spectrum. (b)-(e) Far-field spatial beam profiles measured with the pyroelectric 2D array: (b) before the OPCPA grating compressor; (c) total (frequency-unresolved) beam profile behind the Ar cell; (d) beam profile at 1500 nm; (e) beam profile at 1400 nm. (d) and (e) are taken at the same camera position as (c) by inserting narrowband filters into the beam. Image size is 12.4 mm×12.4 mm.

In the filamentation regime without plasma-induced pulse self-recompression (see Fig. 11), we generated ~3-mJ 600-nm-wide IR supercontinua of high spatial quality supporting 8-fs pulse durations, which corresponds to less than two optical cycles at 1.5 μm.

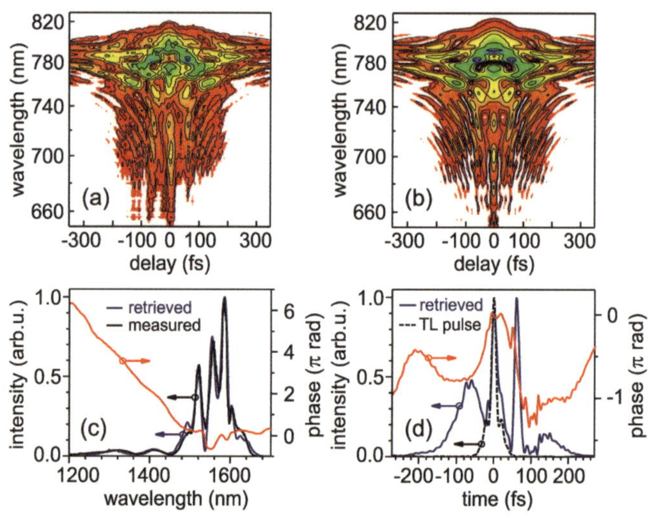

Fig. 12. SHG-FROG characterization of 2.1-mJ filamentation output pulses for argon at 4 bar: (a) Measured and (b) retrieved FROG traces. (c) Measured spectrum (black curve), retrieved spectral intensity (blue) and phase (red). (d) Retrieved temporal intensity (blue) and phase (red) profiles. The TL intensity profile (dashed) corresponds to a 14.6-fs duration.

SHG-FROG data of such spectrally broadened pulses are displayed in Fig. 12. We emphasize that the FROG characterization (Figs. 12 and 13) and corresponding output pulse energy measurements were performed without aperturing the filamentation output beam. In the experiment shown in Fig. 12, the input pulse energy was 3.0 mJ, output energy 2.1 mJ, corresponding to an energy throughput of 68% including the 8% reflection losses on the uncoated 1-mm thick BK7 output window W2 (see Fig. 8). The SHG-FROG data reveal a rather complex spectro-temporal structure. The observed strong nonlinear phase leads to a temporal break-up into two peaks of 20 fs and 15 fs FWHM duration, separated by 60 fs. Since a clean single-filament spatial profile was observed simultaneously, we conclude that the temporal splitting apparently helps to keep the pulse intensity below the break-up threshold of a single filament. A more elaborate time-frequency analysis of these data employing Wigner distributions is presented in section 4.3.

4.2 Self-compression of millijoule infrared pulses

The filamentation regime involving plasma-induced pulse self-compression is particularly attractive for the pursuit of TW-peak-power single-cycle IR sources. Recently, Hauri et al. (2007) demonstrated that filamentation of rather long ~55-fs OPA pulses at 2 µm in a xenon cell allows the generation of self-compressed 17-fs 0.27-mJ pulses. The limited input pulse energy available in that experiment implied the use of xenon as a noble gas with the highest nonlinearity. Detailed numerical investigation of self-compression of 2-µm laser filaments (Bergé, 2008) predicted a number of highly attractive features of femtosecond filamentation at longer carrier wavelength λ: (i) The bandwidth of the generated supercontinuum increases with λ; (ii) for comparable ratios of input power over critical power, P_{in}/P_{crit}, filaments at IR wavelengths have higher pulse energy than near-VIS filaments; (iii) the filament channel extends over longer distances and its waist scales $\propto\lambda$; (iv) self-steepening becomes more pronounced with increasing λ; (v) for gases with moderate ionization potentials (I_p<20 eV, e.g., argon or xenon), the numerical calculations reveal that mid-IR filamentation easily permits self-compression to single-cycle pulse durations, as compared to self-compressed 2-3-cycle durations at visible wavelengths. Self-compression of 2-µm pulses resulted in supercontinua exhibiting a much flatter spectral phase over the full bandwidth as compared to 800-nm pulses. Bergé (2008) also made the important observation that for self-compression of 2-µm pulses, due to nonlinear pulse propagation the shortest achievable pulse duration survives only over shorter distances (~15-20 cm) in the gas medium as compared to the 800-nm case (~60 cm), i.e., the proper choice of the output window position with respect to the filament channel is crucial for observing optimum self-compression.

By lowering the input pulse energy and tuning the gas pressure in the cell, we achieved the regime of pulse self-compression. In the experiment shown in Fig. 13, CEP-stable 2.2-mJ 74.4-fs 1.57-µm input pulses are compressed in a single filament in argon down to a 19.8-fs duration. This represents a temporal compression of the input pulses by a factor of ~4. The output energy was 1.5 mJ, corresponding to the energy throughput of 66%, again including the 8% reflection losses on window W2. The IR supercontinuum with a 130-nm FWHM bandwidth originates from a 12-15-cm long filament. The low-intensity spectral wings of the supercontinuum extend throughout the VIS and are easily visible with the naked eye (see Fig. 14).

As argued above, careful optimization of the propagation distance in the pressurized argon cell behind the filament might lead to the observation of even shorter pulse durations (Bergé, 2008). In addition, the spectral phase is remarkably reproducible on a daily basis which holds potential for further recompression using fixed-dispersion chirped mirrors.

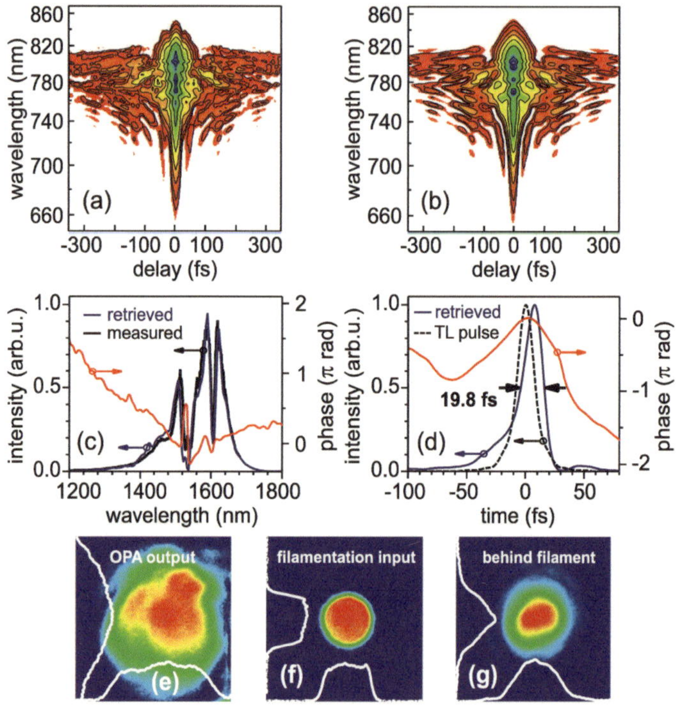

Fig. 13. Self-compression of 1.5-mJ pulses in argon at 5 bar: (a) Measured and (b) retrieved FROG traces. (c) Measured spectrum (black curve), retrieved spectral intensity (blue) and phase (red). (d) Retrieved temporal intensity (blue) and phase (red) profiles indicating a 19.8-fs pulse duration. The TL intensity profile (dashed) corresponds to a 15.9-fs duration. (e)-(g) Far-field spatial beam profiles measured with the pyroelectric 2D array: (e) after OPCPA grating compressor; (f) apertured filamentation input behind the iris aperture A; (g) total beam profile behind the filamentation cell. Image size is 12.4 mm×12.4 mm.

Fig. 14. Photo of the spectrally resolved filamentation output in the self-compression regime.

Concerning the CEP stability of the filamentation output, several femtosecond filamentation experiments performed at various center wavelengths (Hauri et al., 2004, 2007) have clearly demonstrated that the filamentation process preserves the CEP stability of the input pulses. Therefore, we are convinced that the filamentation process does not degrade the CEP stability of the OPCPA input pulses in our experiment. Nevertheless, in order to quantify the quality of CEP stability, we are planning (i) to perform inline *f*-to-2*f* interferometry to the OPCPA output of stages 3 and 4, (ii) quantify possible differential CEP drifts caused in the individual amplification stages 3 and 4 using a TADPOLE-type technique, (iii) and finally perform *f*-to-3*f* interferometry (since a sizable third harmonic is generated inside the filament) to the filamentation output.

4.3 Time-frequency analysis employing Wigner distributions

For the time-frequency analysis, it is very instructive to look at the Wigner distributions

$$W(t,\omega)=\int d\tau\, E^*\left(t-\frac{\tau}{2}\right)E\left(t+\frac{\tau}{2}\right)\exp(i\omega\tau) \tag{17}$$

directly computed from the FROG data (Paye, 1992). When integrating these 2D data over wavelength, one obtains the temporal intensity profile, while integration over time yields the pulse spectrum:

$$I(t)=\int d\omega\, W(t,\omega) \text{ and } \tilde{I}(\omega)=\int dt\, W(t,\omega) \tag{18}$$

The Wigner distribution functions extracted from the filamentation FROG data corresponding to the two distinct filamentation regimes are displayed in Fig. 15.

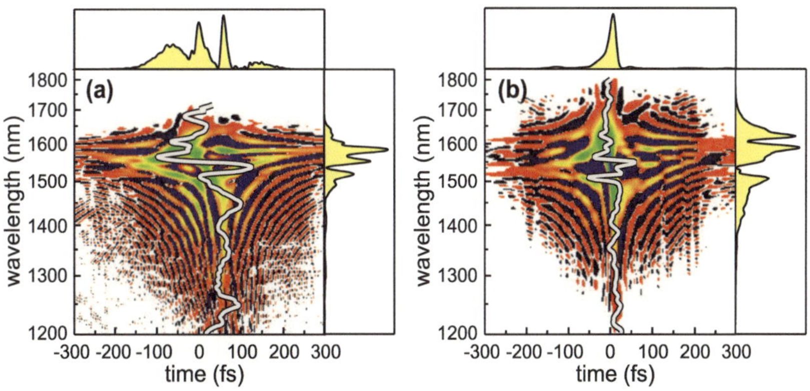

Fig. 15. Wigner distribution functions: (a) filamentation without self-compression extracted from the SHG-FROG data in Fig. 12, (b) self-compression result obtained from Fig. 13. The grey curves indicate the retrieved group delay (GD). The panels above and on the right indicate the temporal intensity profile and pulse spectrum, respectively.

Although the SHG-FROG traces in Figs. 12 and 13 look qualitatively rather different, the Wigner distributions now look more similar. The main difference is that in the self-compression regime [Fig. 15(b)], the group delay is much flatter and exhibits much smaller oscillations than the data obtained in the filamentation regime without self-compression [Fig. 15(a)]. The fact that both cases do not differ qualitatively gives us hope that the filamentation is in principle nicely scalable and that we can control it with further work.

5. Conclusions

In conclusion, we have demonstrated CEP-stable parametric amplification at 1.5 µm signal wavelength with pulse energies up to 12.5 mJ based on a fusion of a DPSS femtosecond Yb:KGW MOPA system and picosecond Nd:YAG solid-state technology. Moreover, we demonstrated multi-mJ IR supercontinuum generation as well as self-compression of CEP-stable 2.2-mJ 74.4-fs 1.57-µm input pulses down to 19.8 fs duration in a single filament in argon with 1.5-mJ output energy and a 66% energy throughput. The output energy was scaled up by 5.6 times over earlier results (Hauri et al., 2007). The output energy and energy throughput can be further increased by systematically optimizing the experimental conditions (input pulse energy and beam diameter, focusing lens and position, gas type and pressure, input polarization, gas cell length etc.). Ultimately, with our 1.6-μm pulses we expect to surpass the present energy limitation [4-5 mJ at 800 nm (Suda et al., 2005; Stibenz et al., 2006)] for gas broadening schemes (Bergé, 2008).

6. Acknowledgments

This work has been supported by the Austrian Science Fund (FWF), grants U33-N16 and F1619-N08, LASERLAB-EUROPE II (JRA HAPPIE), and partly supported by the Lithuanian State Science and Studies Foundation (project No. B-42/2009). O. D. Mücke gratefully acknowledges support from a Lise-Meitner Fellowship by the FWF (project M1094-N14).

Appendix

Within the wavelength interval 0.5 μm < λ < 3.5 μm, the refractive index components of KTP can be evaluated using the Sellmeier equations

$$n_j^2 = A_j + B_j / (\lambda^2 - C_j) - D_j \lambda^2 \tag{A1}$$

with Sellmeier coefficients K_j (K=A,B,C,D and j=x,y,z)

	A	*B*	*C*	*D*
x	3.006700	0.039500	0.042510	0.012470
y	3.031900	0.041520	0.045860	0.013370
z	3.313400	0.056940	0.059410	0.016713

Table 3. Sellmeier coefficients for KTP according to Eq. (A1).

These Sellmeier equations (A1) and coefficients in Table 3 are recommended by C. Bonnin, Cristal Laser, http://www.cristal-laser.fr/prod_pdf/tech1.pdf, and are used for crystal type KTP_F in the SNLO version 41 nonlinear optics code available from A. V. Smith, Sandia National Laboratories, Albuquerque, NM 87185-1423.

7. References

Agrawal, G. P. (2007). *Nonlinear Fiber Optics,* 4th edition, Academic Press, ISBN 978-0-12-369516-1, New York

Agostini, P. & DiMauro, L. F. (2008), Atoms in high intensity mid-infrared pulses, *Contemp. Phys.* 49, 179-197

Baltuška, A., Fuji, T. & Kobayashi, T. (2002), Controlling the Carrier-Envelope Phase of Ultrashort Light Pulses with Optical Parametric Amplifiers, *Phys. Rev. Lett.* 88, 133901-1-4

Bergé, L. (2008), Self-compression of 2 μm laser filaments, *Opt. Express* 16, 21529-21543

Beržanskis, A., Danielius, R., Dubietis, A., Piskarskas, A. & Stabinis, A. (1995), Parametrically induced light diffraction in crystals with second-order susceptibility, *Appl. Phys. B* 60, 421-426

Blaga, C. I., Catoire, F., Colosimo, P., Paulus, G. G., Muller, H. G., Agostini, P. & DiMauro, L. F. (2009), Strong-field photoionization revisited, *Nature Phys.* 5, 335-338

Cavalieri, A. L., Müller, N., Uphues, T., Yakovlev, V. S., Baltuška, A., Horvath, B., Schmidt, B., Blümel, L., Holzwarth, R., Hendel, S., Drescher, M., Kleineberg, U., Echenique, P. M., Kienberger, R., Krausz, F. & Heinzmann, U. (2007), Attosecond spectroscopy in condensed matter, *Nature* 449, 1029-1032

Cerullo, G. & De Silvestri, S. (2003), Ultrafast optical parametric amplifiers, *Rev. Sci. Instrum.* 74, 1-18

Colosimo, P., Doumy, G., Blaga, C. I., Wheeler, J., Hauri, C., Catoire, F., Tate, J., Chirla, R., March, A. M., Paulus, G. G., Muller, H. G., Agostini, P. & DiMauro, L. F. (2008), Scaling strong-field interactions towards the classical limit, *Nature Phys.* 4, 386-389

Dmitriev, V. G., Gurzadyan, G. G. & Nikogosyan, D. N. (1999). *Handbook of Nonlinear Optical Crystals,* 3rd edition, Springer, ISBN 3-540-65394-5, Berlin

Doumy, G., Wheeler, J., Roedig, C., Chirla, R., Agostini, P. & DiMauro, L. F. (2009), Attosecond Synchronization of High-Order Harmonics from Midinfrared Drivers, *Phys. Rev. Lett.* 102, 093002-1-4

Dubietis, A., Jonušauskas, G. & Piskarskas, A. (1992), Powerful femtosecond pulse generation by chirped and stretched pulse parametric amplification in BBO crystal, *Opt. Commun.* 88, 437-440

Dubietis, A., Butkus, R. & Piskarskas, A. P. (2006), Trends in Chirped Pulse Optical Parametric Amplification, *IEEE J. Sel. Top. Quantum Electron.* 12, 163-172

Falcão-Filho, E. L., Gkortsas, V. M., Gordon, A. & Kärtner, F. X. (2009), Analytic scaling analysis of high harmonic generation conversion efficiency, *Opt. Express* 17, 11217-11229

Frolov, M. V., Manakov, N. L., Sarantseva, T. S., Emelin, M. Y., Ryabikin, M. Y. & Starace, A. F. (2009), Analytic Description of the High-Energy Plateau in Harmonic Generation by Atoms: Can the Harmonic Power Increase with Increasing Laser Wavelengths?, *Phys. Rev. Lett.* 102, 243901-1-4

Fuji, T., Ishii, N., Teisset, C. Y., Gu, X., Metzger, T., Baltuška, A., Forget, N., Kaplan, D., Galvanauskas, A. & Krausz, F. (2006), Parametric amplification of few-cycle carrier-envelope phase-stable pulses at 2.1 μm, *Opt. Lett.* 31, 1103-1105

Gordon, A. & Kärtner, F. X. (2005), Scaling of keV HHG photon yield with drive wavelength, *Opt. Express* 13, 2941-2947

Gu, X., Marcus, G., Deng, Y., Metzger, T., Teisset, C., Ishii, N., Fuji, T., Baltuska, A., Butkus, R., Pervak, V., Ishizuki, H., Taira, T., Kobayashi, T., Kienberger, R. & Krausz, F. (2009), Generation of carrier-envelope-phase-stable 2-cycle 740-μJ pulses at 2.1-μm carrier wavelength, *Opt. Express* 17, 62-69

Hauri, C. P., Kornelis, W., Helbing, F. W., Heinrich, A., Couairon, A., Mysyrowicz, A., Biegert, J. & Keller, U. (2004), Generation of intense, carrier-envelope phase-locked few-cycle laser pulses through filamentation, *Appl. Phys. B* 79, 673–677

Hauri, C. P., Lopez-Martens, R. B., Blaga, C. I., Schultz, K. D., Cryan, J., Chirla, R., Colosimo, P., Doumy, G., March, A. M., Roedig, C., Sistrunk, E., Tate, J., Wheeler, J., DiMauro, L. F. & Power, E. P. (2007), Intense self-compressed, self-phase-stabilized few-cycle pulses at 2 μm from an optical filament, *Opt. Lett.* 32, 868-870

Hentschel, M., Kienberger, R., Spielmann, C., Reider, G. A., Milosevic, N., Brabec, T., Corkum, P., Heinzmann, U., Drescher, M. & Krausz, F. (2001), Attosecond metrology, *Nature* 414, 509-513

Itatani, J., Levesque, J., Zeidler, D., Niikura, H., Pépin, H., Kieffer, J. C., Corkum, P. B. & Villeneuve, D. M. (2004), Tomographic imaging of molecular orbitals, *Nature* 432, 867-871

Kalashnikov, M. P., Risse, E., Schönnagel, H. & Sandner, W. (2005), Double chirped-pulse-amplification laser: a way to clean pulses temporally, *Opt. Lett.* 30, 923-925

Kienberger, R., Goulielmakis, E., Uiberacker, M., Baltuska, A., Yakovlev, V., Bammer, F., Scrinzi, A., Westerwalbesloh, T., Kleineberg, U., Heinzmann, U., Drescher, M. & Krausz, F. (2004), Atomic transient recorder, *Nature* 427, 817-821

Koechner, W. (2006). *Solid-State Laser Engineering,* 6th edition, Springer, ISBN 978-0387-29094-2, Berlin

Kraemer, D., Hua, R., Cowan, M. L., Franjic, K. & Miller, R. J. D. (2006), Ultrafast noncollinear optical parametric chirped pulse amplification in KTiOAsO4, *Opt. Lett.* 31, 981-983

Kraemer, D., Cowan, M. L., Hua, R., Franjic, K. & Miller, R. J. D. (2007), High-power femtosecond infrared laser source based on noncollinear optical parametric chirped pulse amplification, *J. Opt. Soc. Am. B* 24, 813-818

Krausz, F. & Ivanov, M. (2009), Attosecond physics, *Rev. Mod. Phys.* 81, 163-234

Liu, H. J., Chen, G. F., Zhao, W., Wang, Y. S., Wang, T. & Zhao, S. H. (2001), Phase matching analysis of noncollinear optical parametric process in nonlinear anisotropic crystals, *Opt. Commun.* 197, 507–514

Meckel, M., Comtois, D., Zeidler, D., Staudte, A., Pavičić, D., Bandulet, H. C., Pépin, H., Kieffer, J. C., Dörner, R., Villeneuve, D. M. & Corkum, P. B. (2008), Laser-Induced Electron Tunneling and Diffraction, *Science* 320, 1478-1482

Miaja-Avila, L., Saathoff, G., Mathias, S., Yin, J., La-o-vorakiat, C., Bauer, M., Aeschlimann, M., Murnane, M. M. & Kapteyn, H. C. (2008), Direct Measurement of Core-Level Relaxation Dynamics on a Surface-Adsorbate System, *Phys. Rev. Lett.* 101, 046101-1-4

Moses, J., Huang, S.-W., Hong, K.-H., Mücke, O. D., Falcão-Filho, E. L., Benedick, A., Ilday, F. Ö., Dergachev, A., Bolger, J. A., Eggleton, B. J. & Kärtner, F. X. (2009), Highly stable ultrabroadband mid-IR optical parametric chirped-pulse amplifier optimized for superfluorescence suppression, *Opt. Lett.* 34, 1639-1641

Mücke, O. D., Sidorov, D., Dombi, P., Pugžlys, A., Baltuška, A., Ališauskas, S., Smilgevičius, V., Pocius, J., Giniūnas, L., Danielius, R. & Forget, N. (2009a), Scalable Yb-MOPA-driven carrier-envelope phase-stable few-cycle parametric amplifier at 1.5 μm, *Opt. Lett.* 34, 118-120

Mücke, O. D., Ališauskas, S., Verhoef, A. J., Pugžlys, A., Baltuška, A., Smilgevičius, V., Pocius, J., Giniūnas, L., Danielius, R. & Forget, N. (2009b), Self-compression of millijoule 1.5 μm pulses, *Opt. Lett.* 34, 2498-2500

Nejadmalayeri, A. H., Herman, P. R., Burghoff, J., Will, M., Nolte, S. & Tünnermann, A. (2005), Inscription of optical waveguides in crystalline silicon by mid-infrared femtosecond laser pulses, *Opt. Lett.* 30, 964-966

Nisoli, M., De Silvestri, S. & Svelto, O. (1996), Generation of high energy 10 fs pulses by a new pulse compression technique, *Appl. Phys. Lett.* 68, 2793-2795

Nisoli, M., De Silvestri, S., Svelto, O., Szipöcs, R., Ferencz, K., Spielmann, C., Sartania, S. & Krausz, F. (1997), Compression of high-energy laser pulses below 5 fs, *Opt. Lett.* 22, 522-524

Paye, J. (1992), The Chronocyclic Representation of Ultrashort Light Pulses, *IEEE J. Quantum Electron.* 28, 2262-2273

Pérez-Hernández, J. A., Roso, L. & Plaja, L. (2009), Harmonic generation beyond the Strong-Field Approximation: the physics behind the short-wave-infrared scaling laws, *Opt. Express* 17, 9891-9903

Popmintchev, T., Chen, M.-C., Cohen, O., Grisham, M. E., Rocca, J. J., Murnane, M. M. & Kapteyn, H. C. (2008), Extended phase matching of high harmonics driven by mid-infrared light, *Opt. Lett.* 33, 2128-2130

Popmintchev, T., Chen, M.-C., Bahabad, A., Gerrity, M., Sidorenko, P., Cohen, O., Christov, I. P., Murnane, M. M. & Kapteyn, H. C. (2009), Phase matching of high harmonic generation in the soft and hard X-ray regions of the spectrum, *Proc. Natl. Acad. Sci. USA* 106, 10516–10521

Sansone, G., Benedetti, E., Calegari, F., Vozzi, C., Avaldi, L., Flammini, R., Poletto, L., Villoresi, P., Altucci, C., Velotta, R., Stagira, S., De Silvestri, S. & Nisoli, M. (2006), Isolated Single-Cycle Attosecond Pulses, *Science* 314, 443-446

Schiessl, K., Ishikawa, K. L., Persson, E. & Burgdörfer, J. (2007), Quantum Path Interference in the Wavelength Dependence of High-Harmonic Generation, *Phys. Rev. Lett.* 99, 253903-1-4

Seres, J., Seres, E., Verhoef, A. J., Tempea, G., Streli, C., Wobrauschek, P., Yakovlev, V., Scrinzi, A., Spielmann, C. & Krausz, F. (2005), Source of coherent kiloelectronvolt X-rays, *Nature* 433, 596

Shan, B. & Chang, Z. (2001), Dramatic extension of the high-order harmonic cutoff by using a long-wavelength driving field, *Phys. Rev. A* 65, 011804(R)-1-4

Sheehy, B., Martin, J. D. D., DiMauro, L. F., Agostini, P., Schafer, K. J., Gaarde, M. B. & Kulander, K. C. (1999), High Harmonic Generation at Long Wavelengths, *Phys. Rev. Lett.* 83, 5270-5273

Shiner, A. D., Trallero-Herrero, C., Kajumba, N., Bandulet, H.-C., Comtois, D., Légaré, F., Giguère, M., Kieffer, J.-C., Corkum, P. B. & Villeneuve, D. M. (2009), Wavelength Scaling of High Harmonic Generation Efficiency, *Phys. Rev. Lett.* 103, 073902-1-4

Stibenz, G., Zhavoronkov, N. & Steinmeyer, G. (2006), Self-compression of millijoule pulses to 7.8 fs duration in a white-light filament, *Opt. Lett.* 31, 274-276

Suda, A., Hatayama, M., Nagasaka, K. & Midorikawa, K. (2005), Generation of sub-10-fs, 5-mJ-optical pulses using a hollow fiber with a pressure gradient, *Appl. Phys. Lett.* 86, 111116-1-3

Takahashi, E. J., Kanai, T., Nabekawa, Y. & Midorikawa, K. (2008a), 10 mJ class femtosecond optical parametric amplifier for generating soft x-ray harmonics, *Appl. Phys. Lett.* 93, 041111-1-3

Takahashi, E. J., Kanai, T., Ishikawa, K. L., Nabekawa, Y. & Midorikawa, K. (2008b), Coherent Water Window X Ray by Phase-Matched High-Order Harmonic Generation in Neutral Media, *Phys. Rev. Lett.* 101, 253901-1-4

Tate, J., Auguste, T., Muller, H. G., Salières, P., Agostini, P. & DiMauro, L. F. (2007), Scaling of Wave-Packet Dynamics in an Intense Midinfrared Field, *Phys. Rev. Lett.* 98, 013901-1-4

Trisorio, A. & Hauri, C. P. (2007), Control and characterization of multiple circularly polarized femtosecond filaments in argon, *Opt. Lett.* 32, 1650-1652

Varanavičius, A., Dubietis, A., Beržanskis, A., Danielius, R. & Piskarskas, A. (1997), Near-degenerate cascaded four-wave mixing in an optical parametric amplifier, *Opt. Lett.* 22, 1603-1605

Varela, O., Zaïr, A., San Román, J., Alonso, B., Sola, I. J., Prieto, C. & Roso, L. (2009), Above-millijoule super-continuum generation using polarisation dependent filamentation in atoms and molecules, *Opt. Express* 17, 3630-3639

Verluise, F., Laude, V., Huignard, J.-P., Tournois, P. & Migus, A. (2000), Arbitrary dispersion control of ultrashort optical pulses with acoustic waves, *J. Opt. Soc. Am. B* 17, 138-145

Vozzi, C., Cirmi, G., Manzoni, C., Benedetti, E., Calegari, F., Sansone, G., Stagira, S., Svelto, O., De Silvestri, S., Nisoli, M. & Cerullo, G. (2006), High-energy, few-optical-cycle pulses at 1.5 μm with passive carrier-envelope phase stabilization, *Opt. Express* 14, 10109-10116

Vozzi, C., Calegari, F., Benedetti, E., Gasilov, S., Sansone, G., Cerullo, G., Nisoli, M., De Silvestri, S. & Stagira, S. (2007), Millijoule-level phase-stabilized few-optical-cycle infrared parametric source, *Opt. Lett.* 32, 2957-2959

Vozzi, C., Calegari, F., Frassetto, F., Poletto, L., Sansone, G., Villoresi, P., Nisoli, M., De Silvestri, S. & Stagira, S. (2009), Coherent continuum generation above 100 eV driven by an ir parametric source in a two-color scheme, *Phys. Rev. A* 79, 033842-1-6

Yakovlev, V. S., Ivanov, M. & Krausz, F. (2007), Enhanced phase-matching for generation of soft X-ray harmonics and attosecond pulses in atomic gases, *Opt. Express* 15, 15351-15364

Zilkie, A. J., Meier, J., Smith, P. W. E., Mojahedi, M., Aitchison, J. S., Poole, P. J., Allen, C. N., Barrios, P. & Poitras, D. (2006), Femtosecond gain and index dynamics in an InAs/InGaAsP quantum dot amplifier operating at 1.55 μm, *Opt. Express* 14, 11453-11459

14

Measurement and Control of Carrier-Envelope Phase in Femtosecond Ti:sapphire Laser

Zhiyi Wei, Hainian Han, Wei Zhang, Yanying Zhao, Jiangfeng Zhu, Hao Teng and Qiang Du
Joint Laboratory of Advanced Measurement Technology, Institute of Physics, Chinese Academy of Sciences, Beijing 100190. PR China

1. Introduction

The emergence of few-cycle laser pulses induced enormous interests on the nonlinear phenomena, such as above-threshold ionization[1~3], high harmonic generation[4, 5] and isolated attosecond generation[6~13] etc. The state-of-the-art few-cycle laser generators, like Kerr-lens mode-locked Ti:sapphire lasers, have delivered directly from oscillator optical pulses with the duration of 5fs [14,15], which corresponds to less than two optical cycles in the near infrared wavelength range. At such short pulse duration, the peak electric field strength of the laser depends strongly on the relative phase between the carrier and the envelope of the pulse. This relative phase is the carrier-envelope phase (CEP). The CEP occurs in the ultrashort pulses, for longer pulses, in general it is not ignored because of the neglectable role. However, the CEP will become significant on the laser matter interaction for few-cycle pulses. It is obvious that the CEP plays an important role in physical processes stated above, especially in attosecond generation schemes, such as interaction between the inert gases and the few-cycle intense pulses [6, 7]and Fourier coherent synthesis among ultrashort lasers with different wavelength[16,17], controlling the CEP is always an important prerequisite.

In fact, the CEP from pulse to pulse in a pulse train that emitted by a mode-locked femtosecond laser is different because of dispersion. The group and phase velocity will differ and cause CEP evolve rapidly when propagating through materials inside the cavity, such as Ti:sapphire crystal, coating mirrors and prisms etc. The pulse-to-pulse CEP change is the carrier-envelope phase offset (CEO). Up to now, several methods have been developed to measure and control the CEO of femtosecond laser pulses train in the frequency domain[18~26]. In this chapter, we will introduce two schemes for measuring the CEO frequency of pulses in femtosecond Ti:sapphire laser. The first and the most widely used technique is the self-reference method, also be termed the *f*-to-2*f* method [19~21]. It is to measure the interference beat frequency between the high frequency and the frequency-doubled low frequency spectral components of an octave spanning spectrum. Normally, a photonic crystal fiber (PCF)[27~31] is used to broaden the laser spectrum to octave-spanning bandwidth in such *f*-to-2*f* CEO frequency measurement scheme. Although the PCF has advantage of high nonlinear coefficient be easy to generate supercontinuum spectrum by self phase modulation (SPM) and four-wave mixing frequency effect, the small core diameter (1~2μm) and high loss (>70%) may cause significant spectrum fluctuation in case of tiny perturbation, and also limit the usable maximum output power. Then we adopted a novel measurement technique based on difference frequency generation (DFG) in a nonlinear crystal,[22~26], which is also named as Monolithic Scheme. Through the interference beat between the fundamental and the DFG spectral components in the overlapped infrared spectrum region, the CEO frequency signal with high signal-to-noise ratio can be achieved, and at the same time the DFG laser in the mid-infrared spectrum region can also be generated. It is worth mentioned that the CEO of the DFG pulses generated in the mid-infrared is zero, in other words, is self-stabilized, which is valuable in attosecond generation. After stabilized by the servo-electronics phase-locked loop, the CEO frequency of laser pulses in Ti:sapphire lasers based on two schemes showed the similar high stability in short terms, but in the long term characteristics, the latter had obvious advantages than the former in hold time and coherence[32~35]. The measurement schemes and control results of CEO frequency of pulses in Ti:sapphire oscillator will be described in detail in the third and fourth segment.

In chirped pulse amplification (CPA) system, although the CEO frequency of seed pulses emitted from femtosecond oscillator stabilized highly, the CEO of amplified laser pulses can change through the amplified crystal, dispersion material and air[36~38]. Lastly we described the measurement and control the CEO frequency of few-cycle high energy laser pulses at 1kHz repetition rate.

2. Carrier-envelope phase offset in femtosecond laser pulse

As already stated above, the CEP is the relative phase between carrier and envelope peak for a single laser pulse and the CEO is the CEP change from pulse to pulse for a pulses train. If we write the electric field of the laser pulse as

$$E(t) = A(t)\cos(\omega_c t + \varphi) \tag{2.1}$$

then φ determines the CEP, often used to be referred "absolute" phase. The CEO is designated as $\Delta\varphi$. This terminology "absolute" is helpful for distinguish between φ and $\Delta\varphi$ and to emphasize the fact that φ is not relative to a second reference beam. In practice, the methods on measurement and control of φ and $\Delta\varphi$ are different completely. In our experiments, we focused on the measurement and control of $\Delta\varphi$ in the frequency domain.

To understand deeply φ and $\Delta\varphi$, the time and frequency domain descriptions will be helpful. In the time domain, the pulses train emitted from the mode-locked femtosecond lasers delivers the same time interval τ, while in the frequency domain, the spectrum can easily be obtained by a Fourier serials expansion, yielding a comb of regularly spaced frequencies, where the comb spacing is inversely proportional to time interval between the adjacent pulses. The corresponding relationship of the pulse train in the time domain and the frequency domain is shown in Fig1 (a) and (b). If $\Delta\varphi$ is zero, the frequency comb distribution of the pulse train from mode-locked Ti:sapphire oscillator is plotted by the dashed lines in Fig.1(b), and if $\Delta\varphi$ isn't zero, there occurs an frequency offset of the whole comb caused by $\Delta\varphi$, defined as f_0.

The CEO in the pulse train emitted by a mode-locked laser occurs because the phase and group velocities inside the cavity are different. Due to dispersion, the successive pulses

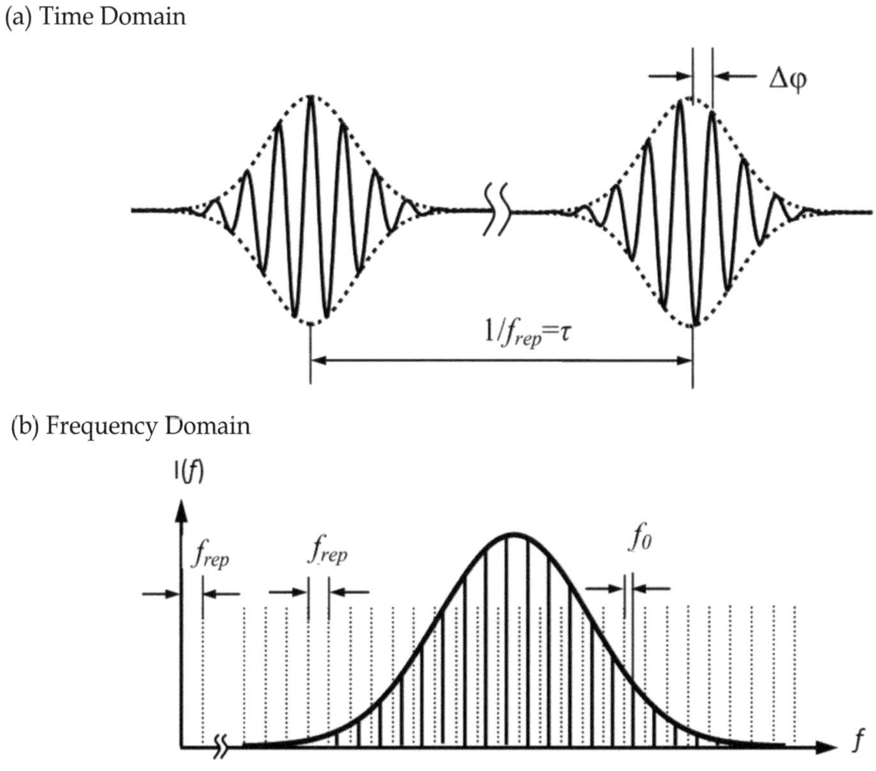

Fig. 1. Pictures of the time-frequency correspondence for a pulse train

output after once per round trip in the cavity will produce the offset phase, it can be expressed as

$$\Delta\varphi = 2\pi\omega_c \cdot l \cdot (1/v_g - 1/v_p) \tag{2.2}$$

Where v_p (v_g) is the mean phase (group) velocity in the laser cavity, l is the length of the laser cavity, and ω_c is the carrier frequency.

In the frequency domain, the optical frequencies of the comb lines can be written as:

$$f_m = mf_{rep} + f_0 \tag{2.3}$$

Where m stands for a large integer of order millions that indexes the comb line, f_{rep} is the repetition rate at less than 1GHz in general, and f_0 is the comb offset due to $\Delta\varphi$.The f_0 has the relationships with $\Delta\varphi$

$$f_0 = \frac{1}{2\pi} f_{rep}\Delta\varphi \tag{2.4}$$

Thus, the task of stabilizing $\Delta\varphi$ is switched to stabilization of f_0.

From Eq. (2.3), we can find that the optical frequency f_m is connected directly to the repetition rate f_{rep} and the CEO frequency f_0. That means a comb system has two degrees of freedom, f_{rep} and f_0, and both need to be controlled for implement a stable frequency comb.

3. Schemes for measuring the CEO frequency

Controlling the CEO is an important prerequisite for many high intensity laser phenomena that are sensitive to the electric field. To control it, we have to measure it firstly. It is easy to measure the repetition rate with a photo diode detector, however, measurement of the CEO is a much more difficult task. From Eqn. (2.4), we see that the pulse-to-pulse phase evolution causes a rigid shift of the frequency comb by f_0. Thus, if we can measure f_0, we can determine the CEO. This gives us an idea that it is possible to measure f_0 with frequency heterodyne coherent techniques. In this segment, we will compare two different frequency measurement schemes for two different home-built mode-locked Ti:sapphire oscillators, that is, *f-to-2f* technique and *0-to-f* technique.

3.1 f-to-2f interferometer method

To measure the CEO frequency with *f-to-2f* technique, it is necessary to broaden the spectrum wider than one octave. However, the spectrum bandwidth generated from an ordinary mode-locked Ti:sapphire oscillator is generally less than 100nm. To produce an-octave spanning spectrum, the photonic crystal fiber (PCF), an ideal material with high effective nonlinearity is used to extend the spectrum by self-phase-modulated the femtosecond laser pulse generated from ordinary Ti:sapphire oscillators.

If the spectrum is sufficiently broad, for example covering an octave, the low frequency component of the spectrum can be written as

$$f_n = nf_{rep} + f_0 \tag{3.1}$$

and the high frequency component of the spectrum is given by

$$f_{2n} = 2nf_{rep} + f_0 \tag{3.2}$$

There occurs spectrum region where high frequency components and the frequency-doubled of the low frequency components overlapped. Measuring the heterodyne beat signal between the f_{2n} and $2f_n$ in the overlapped spectrum, the offset frequency f_0 can be obtained:

$$2f_n - f_{2n} = 2(nf_{rep} + f_0) - (2nf_{rep} + f_0) = f_0 \tag{3.3}$$

Since the high frequency components and the frequency-doubled of low frequency components are from the same spectrum, the f-to-2f measurement technique also is termed "self-referencing". It should be noticed that the octave-spanning spectrum can be produced by the PCF used outside the cavity of the narrow-bandwidth femtosecond Ti:sapphire oscillator stated above, and also can be yielded directly by the ultrabroad-bandwidth femtosecond Ti:sapphire oscillator developed successfully in recent years[39,40].

The typical experimental setup implementing the f-to-2f interferometer is shown in Fig.2. In reality, a group of comb lines contribute phase coherently to the photodiode signal, so even with at most 10 nanoWatts per comb line (a typical amount for a 100MHz spaced comb after broadening in microstructure fiber) there is enough signal to noise (S/N) to properly lock f_0. In our experience the achieved S/N is higher than 30 dB enough for reliable phase lock. Except for implementing such typical apparatus for measuring f_0, there is another f-to-2f interferometer with slight modification in the configuration used in our previous experiments[21]. In our opinion, this new f-to-2f interferometer has some advantages in the structure and operation, let's introduce it curtly. Fig. 3 shows the modificated f-to-2f interferometer configuration. The biggest difference from the standard one is in the position of the PCF and the frequency-doubled crystal. Just this subtle position exchange not only reduces the optical components such as wo lens for focusing on LBO crystal, but also relaxes the need of spectrum abroaden by PCF, it will be enough to extend the spectrum to 400nm in short wavelength And the S/N of CEO frequency by this interferometer is higher than 30 dB, satisfy a tight lock.

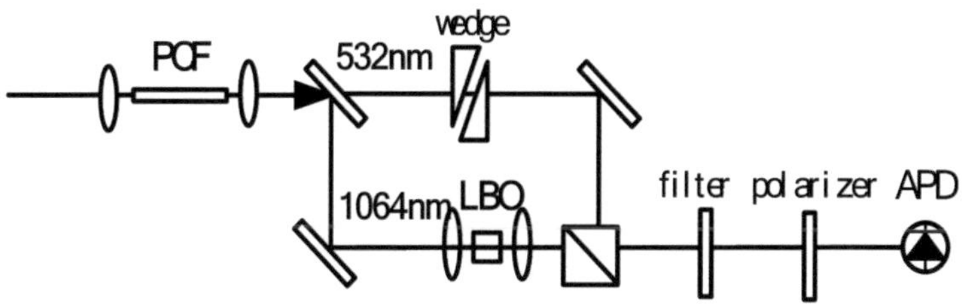

Fig. 2. The standard f-to-2f interferometer configuration

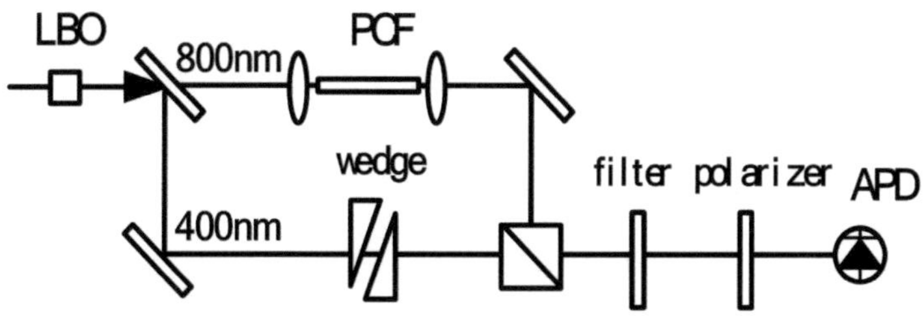

Fig. 3. The configuration of f-to-2f interferometer with modification

3.2 0-to-f method based on difference frequency generation

Although the f-to-2f measurement technique is widely used to measure the CEO frequency of ultrashort pulses emitted from mode-locked Ti:sapphire laser, there have been some disadvantages in the stabilized frequency comb system based on f-to-2f scheme because of its complicated configuration and alignment-sensitive PCF. To generate strong nonlinear effects when ultrashort pulses propagating through it, the core diameter of the PCF has to be made very small, generally at 1um~2um. However, such small core diameter may cause high loss, significant spectrum change in tiny disturb and limit the maximum comb output power. Moreover, excessive dispersion and phase noise of the nonlinear interferometer prevents the broadband spectrum laser from being compressed to few-cycle pulses with low time-jitter. Recently, a novel CEO frequency measurement technique based on the nonlinear difference frequency generation(DFG) and self-phase modulation (SPM) of the few-cycle pulses focused in the monolithic highly nonlinear periodically poled magnesium-oxide-doped lithium niobate (PP-MgO:LN) crystal or other nonlinear crystals is provided[25]. In regions of spectral overlap, a CEO beat frequency between the fundamental wave extended by SPM and the DF wave was observed. This method is termed 0-to-f measurement technique. The schematic of 0-to-f technique by use of DFG is shown in Fig.4. The DF frequency comb (red line) is produced by mixing of the high frequency part and the low frequency part of the spectrum (blue line), given by

$$f_{DF} = (n_h f_{rep} + f_0) - (n_l f_{rep} + f_0) = (n_h - n_l) f_{rep} \quad (3.4)$$

It is independent of f_0, means that the CEO of the DF pulse train is zero and the CEP of every pulse is the same. This is the unique characteristics of 0-to-f technique, which is important for precision frequency metrology and attosecond generation.

If the spectrum of fundamental wave broad enough by SPM, it will overlap with the DF frequency comb, where interference beat signal can be obtained at f_0 by 0-to-f comparison.

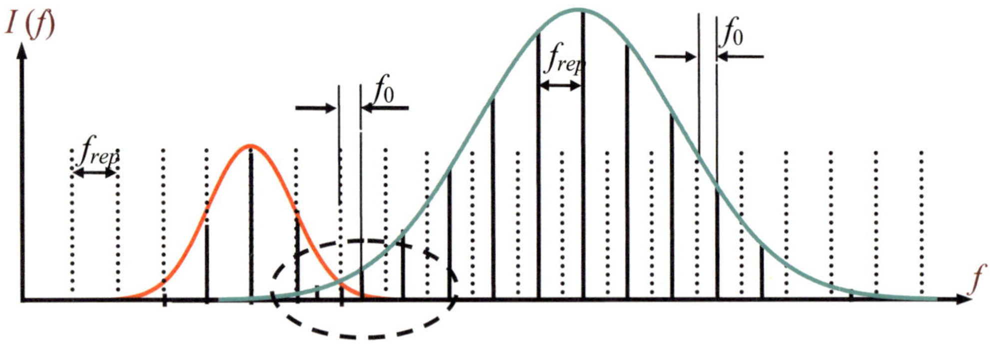

Fig. 4. Schematic of 0-to-f technique for measuring the CEO

The 0-to-f measurement technique is usually implemented in the laser pulse shorter than 10fs with the ultrabroadband spectrum even less than one octave. Such mode-locked Ti:sapphire lasers are available based on the chirped-mirror dispersion compensation technique[41~43]. In our experiment[44,45], the typical laser cavity consists of a pair of concave chirped-mirror, a plane chirped-mirror and an output coupler. The schematic is shown in Fig. 5. The high doped Ti:sapphire crystal (Ti:Sa) was cut at Brewsterd angle with length in 2mm, CM1, CM4 and CM5 are chirped mirrors with high reflectivitity from 580 to 1100nm, CM2 and CM3 is a chirped pair mirror with ROC of 50mm, OC is the output coupler. All these mirrors were optimized choiced based on the calculation for dispersion balance among all optics intracavity. Once the mode-locking was started, ultrabroaden

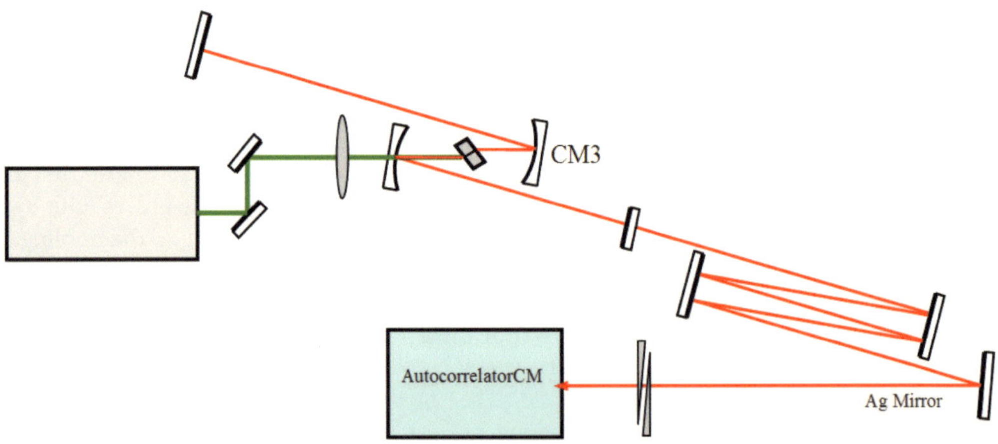

Fig. 5. The mode-locked Ti:sapphire oscillator based on the chirped-mirror dispersion compensation

spectrum can be obtained through the optimized alignment the distance between CM2 and CM3. With the optimized dispersion compensation extracavity with CM 4, CM5 and a pair of silica wedge, we got the interferenmeter resolved autocorrelation trace as figure 6(A) by using a commercial autocorelator (Femtolaser Inc). It infers the pulse duration is shorter as 7.5fs.

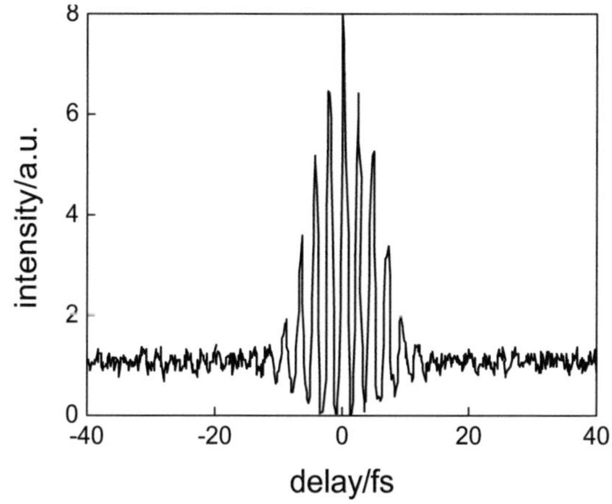

Fig. 6. A. The interferenmeter resolved autocorrelation trace

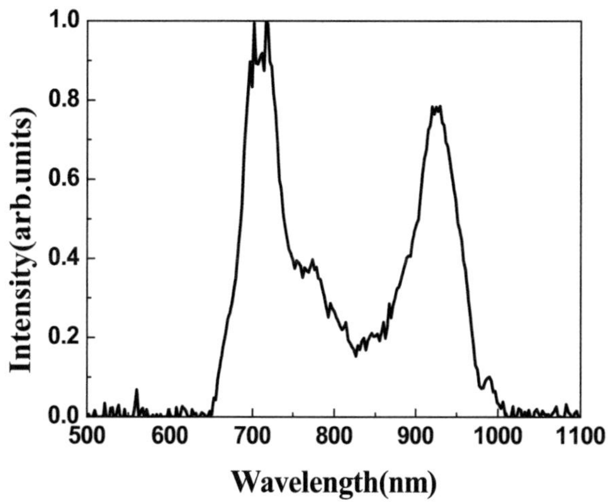

Fig. 6. B Output spectrum of Ti:sapphire oscillator

In 0-to-f measurement technique, the difference frequency is generated between the frequency components in two sides of spectrum, the middle parts of spectrum has little contribution. To generate DFG effectively, we designed and used a special output coupler in the Ti:sapphire laser to enhance the laser intensity in both edges than the central spectrum. With this idea, we generated the artifical spectrum with U-shaped distribution,as Fig.6(B). Two peaks appear around the wavelength of 710nm and 940nm. Such spectrum can greatly optimize the signal-to-noise ratio because of the efficient DFG between two wavelengths at spectrum edges.

Fig. 7. shows the schematic of measuring the CEO frequency of 7.5 fs laser pulses emitted from the Ti: sapphire oscillator by the 0-to-f technique[26]. The few-cycle pulses are reflected

firstly several rounds by two chirped-mirrors outside the cavity to compensate the dispersion caused by the output coupler, and then through a pair of wedges to adjust carefully the residual dispersion. This is a key step for generating the high S/N beat signal. Because both the high and low frequency components for DFG come from the same laser beam, if the dispersion is compensated incompletely, they will not completed overlap in the spatial and tempal domains, which will lead to unable to generate difference frequency.
There is a unique advantage by use of 0-to-f technique to measure the CEO frequency. Except for the high S/N CEO beat signal in the infrared near the fundamental spectrum, the generation of difference frequency with self-stabilized phase happens in the far infrared from the fundamental spectrum at the same time. The difference frequency spectrum is shown in Fig.8[46]. The power of such self-stabilization laser pulses is too low to be measured, however, once through amplification, it will be an ideal light source for some high field applications.

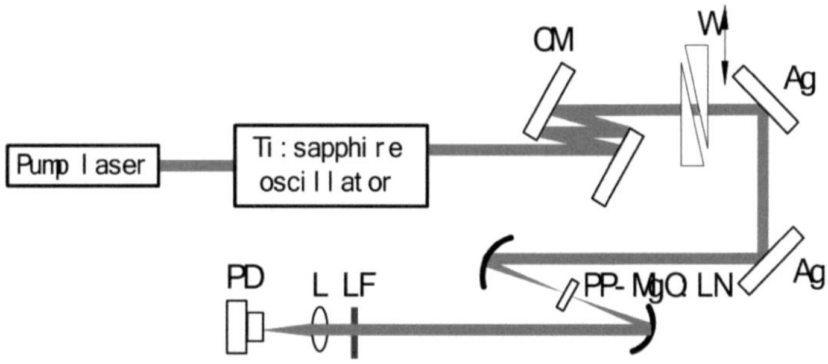

Fig. 7. The Schematic of 0-to-f measurement technique

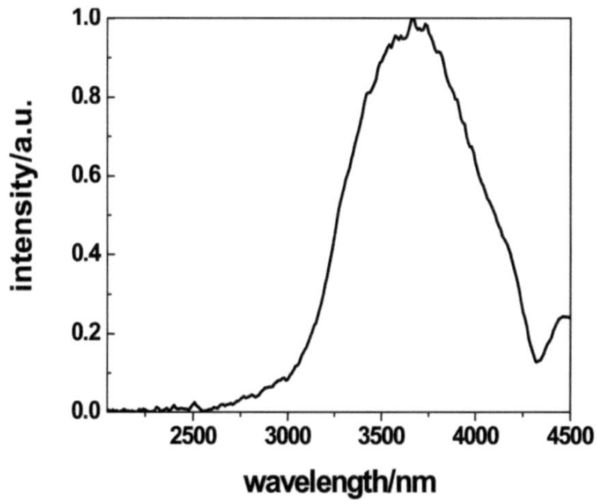

Fig. 8. The spectrum of the DFG with the zero CEO frequency

Comparing with f-to-2f and 0-to-f methods, the latter has obvious advantages, such as higher power, better stability, and no excessive amplitude to phase conversion noise in PCF etc. Additionally, self-stabilization of CE phase by use of DFG, which provides an intrinsically phase-stabilized laser pulses in mid-infrared spectrum, is useful and valuable in the application of the attosecond generation.

4. Robust stabilization of the CEO frequency

The CEO frequency in few-cycle pulses is sensitive to the surrounding environment, such as air flow, temperature shift and even sound noise[47~49]. In practice, anything that can passively reduce the environmental perturbations to the laser relieves the burden placed on servo-loops and leads to a more successful stabilization. From our experiences of stabilizing the CEO frequency in Ti:sapphire laser, cavity design based on the chirped-mirror dispersion compensation is more stable than that based on the prisms dispersion compensation. This is because of beam pointing fluctuations via the prism sequence, which lead to CEP fluctuation, however, CM-based cavities without prisms are free of such fluctuations. In order to realize long term reliable phase lock, some of actions that isolated cavities from the surrounding disturbance include enclosing the cavity in a sealed aluminum box, employing high quality, solid mounts, and using the lowest practical beam height. In addition, it is necessary to use a single frequency, diode-pumped solid state 532nm laser as the pump source, such as Verdi serials made by coherent company, which is helpful to reduce the noise of Ti:sapphire laser.

4.1 Control the repetition rate and the CEO frequency

The repetition rate and the CEO frequency are the two degrees of freedom for few-cycle laser pulses. To stabilize laser frequency corresponds to stabilize these two parameters. In general, the repetition rate and the CEO frequency are locked to the same stable microwave source outside at the same time, such as Rb, Cs or H clocks. Then the stability of the microwave clocks can be transmitted precisely to the optical frequency.

The measurement and control of the repetition rate is relatively easy because of its natural stability, usually shift at Hz level rather than KHz level shift of the CEO frequency. However it is not easy to stabilize with high accuracy. Through glue a mirror to a PZT in the cavity as the controller, then receive the residual transmission light after one mirror by an APD and switch to the electrical signal of the repetition rate, compare with the standard frequency of microwave source and get the control signal after proportional-integral (PI) process to feedback control the PZT, finally the stable cavity length can be realized with experience the long term temperature shift.

Comparatively speaking, control of the CEO frequency is a hard task because it changes fast at KHz level. In this case the responding velocity of PZT is usually unable to follow it, the AOM with high responding velocity used to control the CEO frequency. The AOM can changes lightly the power of the pump laser and lead to the change of the CEO frequency. So in the CEO control feedback loop, the AOM plays the role of controller. In addition, to close the loop, there must be a laser parameter that changes of CEO frequency inside the cavity by hand. A pair of wedges is inserted into the beam path of the cavity to help realization of close loop.

4.2 Stabilization of frequency comb based on DFG

It is well known, under the condition of the same laser average power, each mode power of the comb linearly increases with the repetition rate. Therefore, frequency comb with high repetition rate is more helpful for optical frequency measurements. A mode-locked Ti:sapphire laser with 350MHz repetition rate based on chirped mirrors dispersion compensation technique is developed[35]. Fig.9 is the schematic layout of the stabilized frequency comb system. Chirped-mirror mode-locked Ti:sapphire oscillator consists of five mirrors, of which mirror M4 with the diameter of 6mm, thickness of 2.3mm is glued to a PZT used to control the length of the cavity. A pair of wedges is put between the mirror M3 and M4 to adjust crudely the CEO frequency for loop close. After dispersion pre-compensating outside the cavity by a pair of chirped mirrors, the output laser with average power of 300mW and pulse duration of 7 fs is focused on a 2mm-thick MgO:PPLN crystal in the in-loop 0-to-f unit for measurement of the CEO frequency. And the excess second 0-to-f measurement unit, called out-of-loop, have built for analyzing and comparing the phase noise of the CEO frequency. Figure 9 also shows the electrical feedback system for f_o and f_{rep} locking. The signal analyzer, three microwave frequency synthesizers and two frequency counters are all referenced to a 10MHz standard signal derived from a local TV-Rb clock.

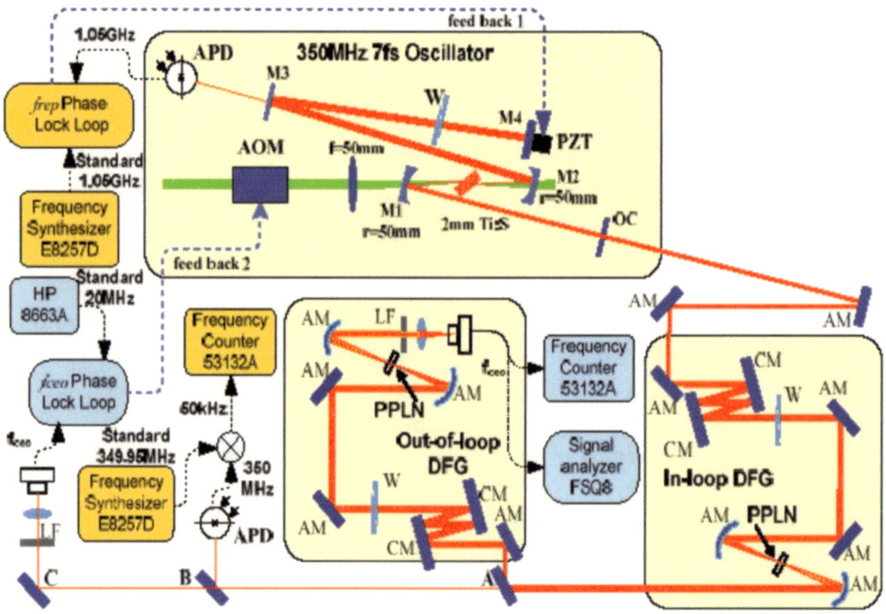

Fig. 9. Setup for stabilizing few-cycle pulses in Ti:sapphire laser

The repetition rate of the femtosecond laser is phase locked to a standard 1.05GHz signal from a frequency synthesizer and stabilized by controlling the cavity length. In the phase locked loop for the stabilization of f_{rep}, the 3rd harmonic of f_{rep} is extracted and mixed with a standard output signal of 1.05GHz from the synthesizer. The phase error signal is filtered by PI processor and fed back to the Ti:sapphire laser oscillator by a PZT mounted on the end mirror M4 to stabilize the cavity length. The locking result of f_{rep} is shown in Fig 10. The continuous locking time is 32673s (Gate time 1s), and the standard deviation is 220μHz, the Allan deviation of locked f_{rep}, starting from 6×10^{-14} (1-s averaging time) and depending on close to τ^{-1} (red line).

Compared with the f-to-2f scheme based on a piece of PCF for octave-spanning spectrum generation and an interferometer for f_0 detection, the monolithic scheme without PCF can make great and outstanding contribution to the long-term stabilization. In the case of no additional servos applied except for two set of phase-locked loop for locking f_0 and f_{rep} simultaneously, the continuous locking time can be exceeded to 32673s (gate time 1-s), i.e., more than 9 hours, which is about twenty-fold longer compared to the locking result of f-to-2f scheme we have reported. The locked result of f_0 over the whole period is shown in Fig. 11 and the standard deviation is about 2.6 mHz. The Allan deviation starts from 2×10^{-11} (1-s gate time) and drops close to τ^{-1} (red line in Fig. 9), which is expected for the phase-locked signal.

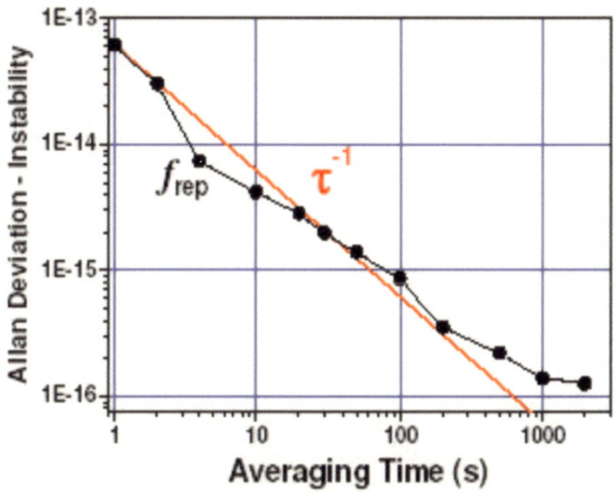

Fig. 10. Allan deviation of f_{rep} at different averaging time

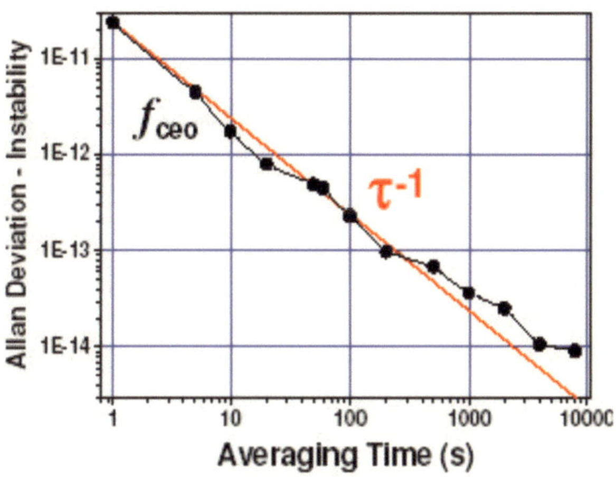

Fig. 11. Allan deviation of f_0 at different averaging time

4.3 In-loop and Out-of-loop phase noise measurement

Long term phase coherence of the CEP became the key parameter for the utility of CEP-stable pulses in many actual nonlinear experiments after control of a pulse train's CEO. As seen from eqn. (2.4), slight offset from f_0 =0 cause an accumulated phase noise in $\Delta\varphi$. Thus the phase noise of $\Delta\varphi$ is displayed by the frequency noise of f_0 and leads to broadening of the linewidth in f_0. In order to investigate the accumulated phase noise of CEP in the above frequency comb based on 0-to-f scheme and compare the noise induced by PCF and PPLN crystal, two same 0-to-f apparatus are built simultaneously, each with its own monolithic PPLN crystal. As shown in Fig. 9, the laser beam is separated into two parts by a half-to-half beam splitter (mirror A). One is focused onto the first MgO:PPLN crystal for f_0 control, the other part is focused onto the second MgO:PPLN crystal for an independent out-of-loop phase noise measurement. Though the SNR of f_0 for each part is about 35dB (100kHz resolution bandwidth), which is lower because the input power of each beam decreases to only one half than before, it is still sufficient for control and phase noise measurement. The in- and out-of-loop single-side-band (SSB) phase-noise power spectral densities (PSD) of f_0 are obtained by injecting these signals into a signal analyzer. The measured SSB phase-noise PSD of f_0 and the noise floor of analyzer are shown in Fig. 12, respectively. The phase error integrated from 1Hz to 100kHz is about 230mrad for in-loop and 242 mrad for out-of-loop. There is a typical 12 mrad increase from in-loop accumulated phase noise to out-of-loop. On the contrary, in the conventional f-to-2f scheme, where a piece of PCF is used for octave-spanning spectrum generation, the phase noise for out-of-loop is usually increased by several times compared with that for in-loop. As reported in Ref. [50], the accumulated phase noise is 0.1 rad for in-loop and more than 0.2 rad for out-of-loop [1Hz, 100kHz], corresponding to an approximate two-fold increase. This significant difference originates from the conversion of amplitude fluctuations to phase fluctuations in PCF. Based on the monolithic scheme, however, there is no additional noise introduced by PCF, and then the difference between in- and out-of-loop is steeply decreased.

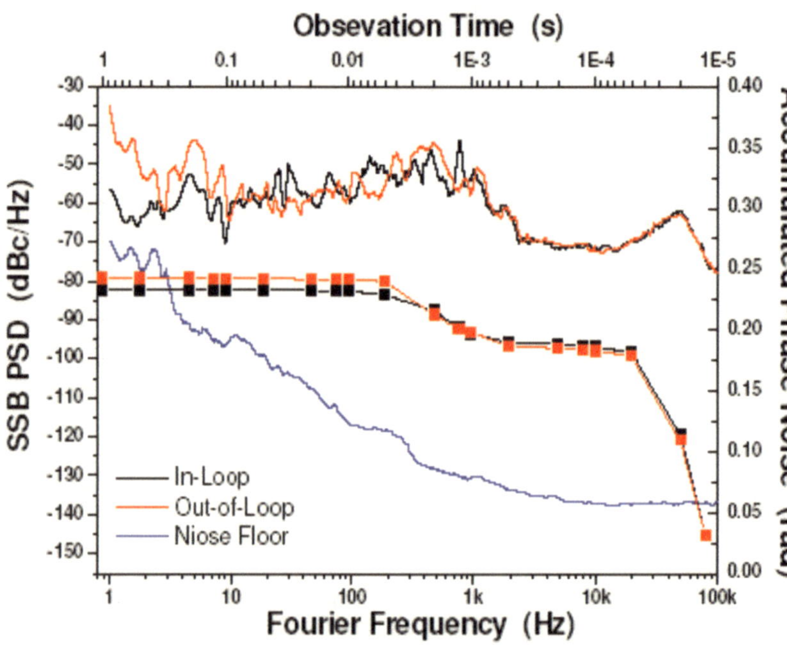

Fig. 12. Bottom-Left: the SSB noise power spectrum of f_0 as the function of Fourier frequency. Black line: in-loop phase error. Red line: out-of -loop phase error. Blue line: the noise floor of the signal analyzer. Top-Right: the accumulated phase noise of in-loop (black closed square) and out-of-loop (red closed square) as the function of observation time.

Here it should be specifically noted that the coherence of f_0 is not a coherence of optical carrier wave with successive pulses, just is the coherence of CEP. In the in-loop and out-of-loop phase measurement, the repetition rate is not locked, so the optical carrier frequency varies with f_{rep} in the case of stable f_0. If f_{rep} is stabilized then optical coherence in pulses train can be realized. Measuring the phase noise of f_0 without locking f_{rep} doesn't affect the coherence result of CEP here. However, considering that the optical frequency has a large multiplication factor of 10^6 at repetition rate, simply locking f_{rep} to a stable microwave clock will be not adequate for realization of the long term coherence of optical frequency. There have been interests of locking f_{rep} to an optical clocks or ultrastable optical cavity, which will most likely be necessary to get much higher stability and longer coherent time.

5. Control the CEP of high energy laser pulse at 1kHz repetition rate

CEP control of the few cycle light pulses enables the generation of optical waveforms with the reproducible electric field profile which has shown its importance in the observed phenomena, such as above-threshold ionization and high-order harmonic generation (HHG). In particular, isolated attosecond pulse generation through HHG is strongly affected by the drift of CEP. Additional CEP drift is occurred in the amplification stage even if the CEP of the seed oscillator is fixed. Several factors including thermal fluctuation of the environment, beam pointing fluctuation and the laser pulse energy fluctuation may result in the drift and fluctuation to the CEP of the amplified pulses. Typically the amplification is operated at a low repetition rate with large pulse energy. To evaluate and further control the CEP slip of the amplified pulses, a single-shot measurement of the spectral interference (SI) is proposed[36,37].

Phase locking is achieved by modulating the pump power of the seed oscillator through the AOM which varies the transmitted power proportional to the phase error signal generated by the phase locking electronics. As a result, for Ti:sapphire CPA system at 1kHz repetition rate seeded by few-cyle pulses at 80MHz repetition rate, given selecting f_0=20MHz and being locked to quarter of f_{rep}, every 4th pulse replicates itself. When a pulse picker is triggered to pick every 80 000th pulse for subsequent amplification, all pulses look alike.

5.1 Spectral interferemetry

Since the repetition rate of the amplified pulses is only 1 kHz, it is difficult to use f-to-2f or DFG method to detect CEP signal. In the year 2001, Kakehata *et al.* [13] proposed a method of SI to measure CEP for single shot pulses. The method uses a dispersive media to delay and spectrally broaden the pulse, so that the interference between the short wavelength portion and its second harmonic could be detected as a periodical curve, from which the CEP frequency could be calculated. The expression of the electric field of a Gaussian profile ultrashort laser pulse is

$$E_{\mathrm{F}}(\omega)=\sqrt{2\pi}\exp\left\{-[(\omega-\omega_{\mathrm{c}})/2]^{2}\right\}\exp(j\varphi_{\mathrm{CE}}), \tag{5.1}$$

Here ω_{c} is the center frequency in its spectrum, and φ_{CE} is the CEP.

Considering the long wavelength portion and its harmonic, when passing through a dispersive media, the short wavelength portion delays τ and appears as a phase delay $\exp(-j\omega\tau)$. Calculating its 1st and 2nd order of electric dipole, the intensity of interference curve is:

$$I(\omega)\propto\left|P_{\tau}^{(1)}(\omega)+jP^{(2)}(\omega)\right|^{2}\propto I_{\mathrm{F}}+I_{\mathrm{SH}}+2\left[I_{\mathrm{F}}I_{\mathrm{SH}}\right]^{1/2}\cos\left[\omega\tau+\varphi_{CE}+\pi/2\right]. \tag{5.2}$$

For a specific dispersive media, the delay time τ is constant. So φ_{CE} is able to extract by Fourier transforming the interference signal. Moreover, the gap of the interference streak is inverse proportional to the delay time, which indicates that only short delay could get large streak gap for interferometers to identify.
The Fourier transform of interference signal (for clearance, the Fourier frequency is k) is:

$$F\left(I(\omega)\right) \propto \exp\left[\frac{k - \tau/2\pi}{2}\right]^2 \exp(j\varphi_{CE}), \tag{5.3}$$

$$\arg\left[F\left(I(\omega)\right)\Big|_{k=\tau/2\pi}\right] \propto \tan^{-1}(\varphi_{CE}). \tag{5.4}$$

In the real experiment, by introducing ~1% of the amplified pulses, about10uJ into a 1 mm thick sapphire plate white light continuum of an octave spanning is produced, then a BBO crystal is used to frequency double the infrared spectrum to generate spectral interference with the blue parts. The SI signal is collected by a spectrometer and is analyzed by the PC software. Fig.13 shows the interference spectrum at the center wavelength of 525nm with the bandwidth of 30nm which covers about 17 interference fringes. It must be mentioned that the prerequisite of the achievement of this interference spectrum is to lock the CEO of the seed pulses train emitted from Ti:sapphire oscillator. The phase-locked loop for feedback control the CEO of multiple-MHz repetition rate seed pulses is called "fast-loop", relatively, "slow-loop" designated to the electrical servo-control CEP for successive kHz repetition rate amplified pulses. It is obvious that there is no interference phenomenon in the spectrum of amplified pulses unless "fast-loop" turned on.

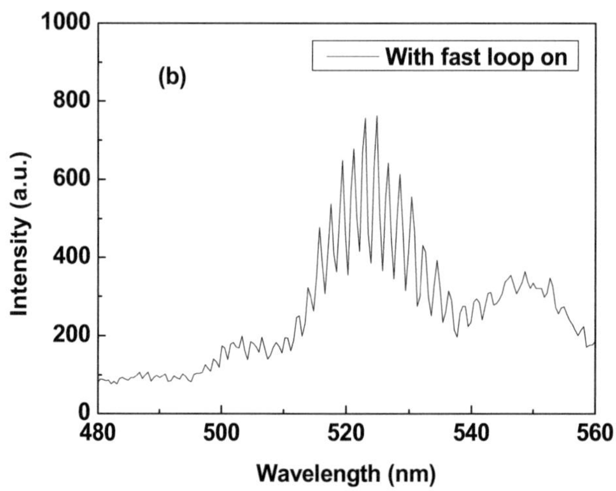

Fig. 13. The interference spectrum of amplified pulses

5.2 Control of CEP in amplified pulses

The CEP control process for amplified pulses is implemented by a real-time software running on a PC, which acquires the output of the interferometer (Ocean Optics, HR2000) and output the control signal by D/A conversion through an DAQ card (National Instrument, PCI-6221). Firstly, the system samples the interference curve and applies FFT to spectrum region of interest to get CEP power and phase spectrum. Secondly, the peak frequency offset of CEP spectrum k is automatically evaluated from the Fourier spectrum so that the corresponding φ_{CE} could be calculated. From equation (9), the transfer function of 'slow loop' phase detection process is:

$$H_2(\varphi_{\text{CE}}(t))\big|_{\varphi_{\text{CE}}<<1} \approx K_2\varphi_{\text{CE}}(t), H_2(s) \approx K_2\frac{1}{s^2}. \tag{5.5}$$

The PI controller, $G_{c2}(s) = K_{p2}\cdot\left(1+\frac{1}{T_{i2}s}\right)$, compensates the phase error and outputs control signal for 'fast loop' to change its reference point so that both PLLs could be closed. As a result, the CEP of the amplified pulses can readily be controlled by both "fast" and "slow" feedback to the seed oscillator through phase locking electronics. To meet equation (4) and tune the controller's parameter for different experiment conditions, the design of controller must has adaptivity. An algorithm of auto-tuning method based on setpoint relay is applied to optimize P/I parameter for shortest rise time and minimum overshot. Experiment shows that the auto-tuning method may have 2-3 time smaller CEP error than manually tune the parameters.

Fig.14 displays the locking results of amplified pulses with the slow loop on and off. In Fig 14, curve A shows CEP drift of the amplified pulses during 5 minutes when only locking the oscillator by 'fast loop', curve B is the result after locking 'slow loop'. It is clear that the CEP is changing greatly in the former state, while the rms error of CEP is only 53 mrad in the later one. The system is able to lock for over 3 hours. For even lower repetition rate lasers, although the SI method is still able to effectively detect CEP, the feedback loop bandwidth is limited by the repetition rate, so the performance of noise suppressing is also limited.

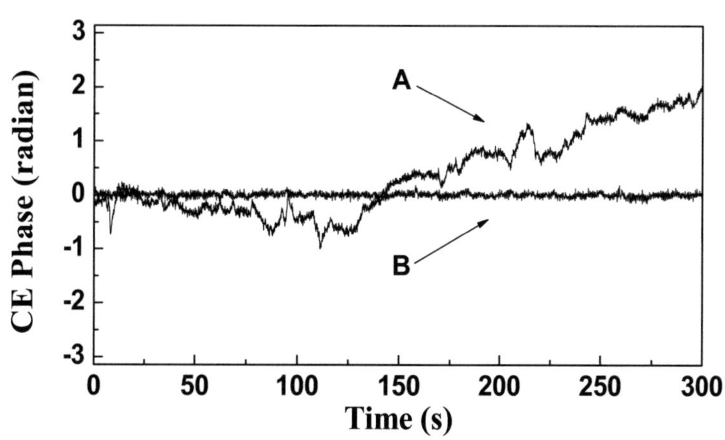

Fig. 14. Amplified pulses CEP varies by time, Curve A is only locking the oscillator; Curve B is locking two PLLs simultaneously.

6. Summary

Frequency comb techniques (stabilization of CEO in oscillator) have brought a truly revolutionary influence to the optical frequency metrology and precision spectroscopy since it emerged about ten years ago. Specially, in recent years a number of nonlinear physical processes dependent on electric filed rather than intensity envelope of the pulse have been investigated, such as coherent control of atomic and molecular system, optimization of high harmonic generation, and dynamic of electron motion on femtosecond and attosecond time scales, where clearly the CEP stabilization in amplified pulses played an important role. We presented a novel monolithic 350MHz frequency comb system based on 0-to-f measurement technique with excellent long term phase lock, and realized stabilization of high energy pulses at 1kHz with the phase fluctuation less than 60 mrad. Accordingly, we believed that CEP stabilized femtosecond laser using advanced techniques will continue to open up new fields of applications in the future.

7. References

[1] T. Brabec, and F. Krausz, ''Intense few-cycle laser fields: Frontiers of nonlinear optics,'' Rev. Mod. Phys. 72, 545 (2000)
[2] P.B. Corkum., "Plasma perspective on strong field multiphoton ionization," Phys. Rev. Lett., 71, 1994-1997 (1993)
[3] G. G. Paulus, F. Grasbon, H. Walther, P. Villoresi, M. Nisoli, S.Stagira, E. Priori, and S. De Silvestri, "Absolute-phase phenomena in photoionization with few-cycle laser pulses.'' Nature (London) 414, 182 (2001)
[4] C. G. Durfee, A. R. Rundquist, S. Backus, C. Herne, M. M. Murnane, and H. C. Kapteyn, ''Phase matching of high order harmonics in hollow waveguides,'' Phys. Rev. Lett. 83, 2187 (1999)
[5] C. Spielmann, N. H. Burnett, S. Sartania, R. Koppitsch, M. Schnurer, C. Kan, M. Lenzner, P. Wobrauschek, and F. Krausz, ''Generation of coherent X-rays in the water window using 5-femtosecond laser pulses,'' Science 278, 661 (1997
[6] I. P. Christov, M. M. Murnane, and H. C. Kapteyn, ''High-harmonic generation of attosecond pulses in the 'single-cycle' regime,'' Phys. Rev. Lett. 78, 1251 (1997)
[7] Paul P M, Toma E S, Breger P, et al. Observation of a train of attosecond pulses from high harmonic generation. Science, 2001, 292(5522): 1689–1692
[8] M. Hentschel, R. Kienberger, Ch. Spielmann, G. A. Reider, N. Milosevic, T. Brabec, P. Corkum, U. Heinzmann, M. Drescher and F. Krausz, "Attosecond metrology," Nature 414, 509–513 (2001).
[9] R. Kienberger, M. Uiberacker, E. Goulielmakis, A. Baltuska, M. Drescher, F. Krausz, "Single sub-fs soft-x-ray pulses: generation and measurement with the atomic transient recorder," J. Morden Optics 20,261 (2005)
[10] R. Kienberger, M. Hentschel, M. Uiberacker, Ch. Spielmann, M. Kitzler, A. Scrinzi, M. Wieland, Th. Westerwalbesloh, U. Kleineberg, U. Heinzmann, M. Drescher, and F. Krausz. "Steering attosecond electron wave packets with light," Science 297, 1144 (2002)
[11] M.Drescher, M. Hentschel, R. Kienberger, M. Uiberacker, V. Yakovlev, A. Scrinzi, Th. Westerwalbesloh, U. Kleineberg, U. Heinzmann and F. Krausz., "Time-resolved atomic inner-shell spectroscopy," Nature 419, 803 (2002)
[12] Baltuška A, Udem Th, Uiberacker M, et al. Attosecond control of electronic processes by intense light fields. Nature, 2003, 421(6293): 611–615
[13] F. Krausz,"Attosecond spectroscopy comes of age," Opt. Photon. News 13, 62–68 (20)

[14] Baltuska A, Wei Z, Pshenichnikov M S, et al. Optical pulse compression to 5 fs at a 1-MHz repetition rate. Opt Lett, 1997, 22(2): 102–104

[15] U. Morgner, F. X. Kärtner, S. H. Cho, Y. Chen, H. A. Haus, J. G. Fujimoto, E. P. Ippen, V. Scheuer, G. Angelow and T. Tschudi, "Sub-two-cycle pulses from a Kerr-lens modelocked Ti:sapphire laser," Opt. Lett. 24, 411–413(1999).

[16] Y. Kobayashi, K.Torizuka, Zhiyi. Wei, Kobayashi Y, Torizuka K, Control of relative carrier-envelope phase slip in femtosecond Ti: sapphire and Cr: forsteritelasers. Opt. Lett. 28, 746 (2002)

[17] Z. Wei, Y. Kobayashi, K. Torizuka, Relative carrier-envelope phase dynamics between passively synchronized Ti: sapphire and Cr:forsterite lasers, Opt. Lett., 27, 2121(2001).

[18] Telle H R, Steinmeyer G, Dunlop A E, et al. Carrier-envelope offset phase control: A novel concept for absolute optical frequency measurement and ultrashort pulse generation. Appl Phys B, 1999, 69(4): 327–332

[19] A. Apolonski, A. Poppe, G. Tempea, C. Spielmann, T. Udem, R. Holzwarth, T. W. Hänsch, and F. Krausz, "Controlling the phase evolution of few-cycle light pulses" Phys. Rev.Lett. 85, 740 (2000)

[20] Jones D J, Diddams S A, Ranka J K, et al. Carrier-envelope phase control of femtosecond mode-locked lasers and direct optical frequency synthesis. Science, 2000, 288(5466): 635–639

[21] Han H N, Wei Z Y, Zhang J et al., Measurement of Carrier-Envelope-Offset in the femtosecond Ti:sapphire laser, Acta Physica Sinica (in Chinese), 2005, 54(1):155~158

[22] Mucke O. D., Kuzucu O., Wong F. N. C., Ippen E. P., Kaertner F. X., Foreman S.M., Jones D. J., Ma L. S., Hall J. L., Ye J., Experimental implementation of optical clockwork without carrier-envelope phase control, 29 2806 Opt. Lett. (2004)

[23] Zinllnellllann M., Gohle Ch, Holzwarth R, Udem Th, Hansch T., Optical clockwork with an offset-free difference-frequency comb: accuracy of sum- and difference-frequency generation Opt. Lett. 29 310(2004)

[24] Fuji T, Apolonski A, Krausz F. Self-stabilization of carrier-envelope offset phase by use of difference-frequency generation. Opt Lett, 2004, 29(6): 632–634

[25] Takao Fuji, Jens Rauschenberger et, Monolithic carrier-envelope phase-stabilization scheme, Opt. Lett. 30 332(2005)

[26] Han H N, Zhao Y Y, Zhang W et al., Measurement of Carrier-Envelope Phase of few cycles Ti:sapphire laser by Difference Frequency Technique, Acta Physica Sinica (in Chinese), 2007, 56(5):2756~2759

[27] J. C. Knight, T. A. Birks, P. St. J. Russell, and D. M. Atkin. "All-silica single-mode fiber with photonic crystal cladding". Opt. Lett. 19, 1547-1549 (1996)

[28] J. K. Ranka, R. S. Windeler, and A. J. Stentz. "Visible continuum generation in air silica microstructure optical fibers with anomalous dispersion at 800nm". Opt. Lett. 1, 25-27 (2000)

[29] Stephane Coen, Alvin Hing Lun Chau, Rainer Leonhardt, and John D. Harvey. "White-light supercontinuum generation with 60-ps pump pulses in a photonic crystal fiber". Opt. Lett. 17, 1356-1358 (2001)

[30] L. Provino, J. M. Dudley, H. Maillotte, N. Grossard, R. S. Windeler, and B. J. Eggleton. "Compact broadband continuum source based on microchip laser pumped microstructured fiber". Electron. Lett. 37, 558-560 (2001)

[31] A. V. Husakou and J. Herrmann. "supercontinuum generation , four-wave mixing, and fission of higher-order solitons in photonic-crystal fibers". J. Opt. Soc. Am. B, 9, 2171-2182 (2002)

[32] H. N. Han, W. Zhang, P. Wang, D. H. Li, and Z.Y. Wei, "Precise control of femtosecond Ti:sapphire laser frequency comb," Acta Physica Sinica(in Chinese) 56, 2760-2764 (2007).
[33] H. N. Han, W. Zhang etc, Control of carrier-envelope phase offset in femtosecond laser with PLL and TV-Rb clock, Acta Physica Sinica(in Chinese) 56, 291 (2007)
[34] W. Zhang, H. N. Han, P. Wang, and Z. Y. Wei, "Stabilization and phase control of femtosecond Ti:sapphire laser with a repetition rate of 90MHz," in 27th International Congress on High-Speed Photography and Photonics, Proc.of SPIE 6279, 62793X1-X8 (2007)
[35] Zhang Wei, Han Hainian, Zhao Yanying, Du Qiang, Wei Zhiyi, "A 350MHz Ti:sapphire laser comb based on monolithic scheme and absolute frequency measurement of 729nm laser" Opt. Exp. 17,6059-6067 (2009)
[36] Kakehata M, Fujihira Y, Takada H, et al. Measurements of carrier-envelope phase changes of 100-Hz amplified laser pulses. Appl Phys B, 74(9): 43 (2002)
[37] Kakehata M, Takada H, Kobayashi Y, et al. Single-shot measurement of carrier-envelope phase changes by spectral interferometry. Opt Lett, 26 1436(2001)
[38] Chang Z. Carrier-envelope phase shift caused by grating-based stretchers and compressors. Appl Opt, 45(32): 8350 (2003)
[39] Sutter D H, Steinmeyer G, Gallmann L, et al. Semiconductor saturable absorber mirror assisted Kerr-lens mode-locked Ti:sapphire laser producing pulses in the two-cycle regime. Opt Lett, 24, 631(1999)
[40] Ell R, Fujimoto J G, Scheuer V, et al. Generation of 5-fs pulses and octave spanning spectra directly from a Ti:sapphire laser. Opt Lett, 26, 373 (2001)
[41] Jung I D, Kartner F X, Matuschek N, et al. Self-starting 6.5-fs pulses from a Ti:sapphire laser. Opt Lett, 22, 1009(1997)
[42] Wei Z Y. Solid state self-mode-locking Ti:Al2O3 laser with chirp mirrors. In: Academic Report Memoir of the 14th National Laser (in Chinese), Beijing, 1999
[43] Liao R, Wen J H, Liu Z G, et al. Generation sub-10-fs pulse from KLM Ti:sapphire laser. Chin Sci Bull (in Chinese), 47, 345(2002)
[44] Zhang Wei, Han Hai-Nian, Teng Hao, and Wei Zhi-Yi, "Generation of 10 fs Ti:sapphire laser at repetition rate of 525MHz and measurement of carrier-envelope phase frequency"Chinese PhysicsB Vol.18,1105(2009)
[45] ZHAO YanYing, WANG Peng, ZHANG Wei et al, Generation of sub-7 fs laser pulse directly from a compact Ti:sapphire laser with chirped mirrors Sci China-Phys Mech Astron 50, 1-6(2007)
[46] Han H N, Zhao Y Y, Zhang W et al. A new monolithic measurement scheme for carrier-envelope offset frequency with enhanced signal-to-noise ratio, The 6th Asia Pacific Laser Symposium(APLS2008), Nagoya, Japan
[47] J. Stenger, H. R. Telle, " Intensity-induced mode shift in a femtosecond laser by a change in the nonlinear index of refraction", Opt. Lett. 25, 1553(2000)
[48] Kevin W. Holman, R. Jason Jones, Adela Marian, Steven T. Cundiff, Jun Ye, "Detailed studies and control of intensity-related dynamics of femtosecond frequency combs from mode-locked Ti:sapphire lasers", IEEE Journal of Selected Topics in Quantum Electronics 9,1018 (2003)
[49] F. W. Helbing, G. Steinmeyer, J. Stenger, H. R. Telle, U. Keller, "Carrier-envelope-offset dynamics and stabilization of femtosecond pulses", Appl. Phys. B74(Suppl.) S35 (2002)
[50] T. M. Fortier, D. J. Jones, J. Ye, and S. T. Cundiff, "Highly Phase Stable Mode-Locked Lasers," IEEE J. Selected Top. Quant. Electron 9, 1002-1010 (2003).

15

Pulse Measurement Techniques Using an Acousto-Optic Programmable Dispersive Filter

Nicolas Forget and Thomas Oksenhendler
FASTLITE, Centre scientifique d'Orsay - Bât.503 - BP 45, 91401 Orsay
France

1. Introduction

Programmable spectral filters, such as acousto-optic programmable dispersive filters (AOPDF, Tournois (1997)) or spatial light modulators inserted in the Fourier plane of zero-dispersion lines (Froehly et al., 1983; Weiner, 2000), have opened up the field of ultrafast pulse shaping and given the ability to manipulate spectral amplitude and phase of broadband ultrashort pulses. These devices have found a great number of applications, among which are phase compensation in chirped-pulse amplification laser chains (Seres et al., 2003; Ohno et al., 2002; Verluise et al., 2000), coherent control experiments within atomic or molecular systems (Tkaczyk et al., 2008; Murphy et al., 2007; Veshapidze et al., 2007; Ogilvie et al., 2006; Yamada et al., 2005) and complex pulse shaping for photo-injectors (Garzella et al., 2006). Surprisingly, pulse shapers have mainly been used to control already well characterized ultrashort pulses but seldom to characterize these pulses themselves. And yet, pulse shapers provide a convenient way to perform quantitative, reliable and versatile pulse measurements.

A few pioneering works have already demonstrated that several of the existing pulse measurement techniques could be implemented with pulse shapers (Galler & Feurer, 2008; Sung et al., 2008; Forget et al., 2007; Oksenhendler et al., 2003; Monmayrant et al., 2003) and that new pulse characterization methods could even be invented (Forget et al., 2007; Grabielle et al., 2009; Lozovoy et al., 2004). The use of pulse shapers for ultrashort pulse metrology has already found practical applications, such as *in situ* pulse compression at the focus of high NA objectives in twophoton microscopy (von Vacano et al., 2006; 2007). Beyond specific applications, pulse shapers are expected to extend the robustness and dynamic range of the existing pulse characterization techniques. Indeed, all techniques suffer from specific weaknesses and cross-check can help to overcome the drawbacks, limitations or ambiguities related to a particular technique and to eliminate spurious results. Spectral phase interferometry for direct electric-field reconstruction (SPIDER) (Iaconis & Walmsley, 1998), for example, requires a precise calibration and determination of the relative delay between the two replicas. Such a requirement is much less stringent for the second-harmonic (SH) frequency resolved optical gating (FROG) technique. Conversely, SH-FROG suffers from time direction ambiguity and phase retrieval is not straightforward. Finally, these two methods are not equivalently robust with respect to complex pulse shapes. Cross-check between SH-FROG and SPIDER results is, however, often made difficult since it involves separate measurement devices, which multiplies the causes of

systematic experimental errors. As we shall demonstrate in this chapter, pulse shapers bring an elegant solution to this issue since both techniques can be implemented with the same device.
Despite the ever growing number of ultrafast pulse characterization techniques that have been proposed and demonstrated for four decades, it is possible to classify the various methods into a small number of categories (Wong & Walmsley, 1995). Since this classification is out of the scope of this paper, we will refer to a recent review paper by Walmsley and Dorrer (Walmsley & Dorrer, 2009). In brief, the characterization techniques relevant to the sub-picosecond pulse range fall within three general classes: spectrography, tomography and interferometry. Besides, all the techniques require at least one time-stationary filter and one time-nonstationary filter followed by an integrative detection. The most popular pulse characterization techniques dedicated to the shortest pulses use a nonlinear effect as a time-nonstationary filter. Most often, the chosen nonlinear effect is second harmonic generation (SHG), for practical reasons related to the efficiency of this process compared to other nonlinear effects. For the time-stationary filter, that is to say for the spectral filter, a set of conventional optics (beam-splitters, delay lines, dispersive materials) is used. Depending on the arrangement and nature of these optics, the pulse retrieval algorithm will differ, giving rise to different pulse characterization techniques. It is of the greatest interest to note that, following the formalism developed by Walmsley and Dorrer, any pulse characterization technique could be implemented with a minimal setup, including no more than a couple of spectral filters and nonlinear stages.
In this chapter we review, both theoretically and experimentally, the practical adaptation of pulse measurement techniques to a setup including one *single* phase and amplitude pulse shaper and one *single* nonlinear stage. Special attention will be drawn on their implementation with AOPDFs, a class of bulk and compact pulse shapers particularly well adapted to this quantitative use. We will focus on pulse durations ranging from a few tens to a few hundreds of fs (20fs-1ps) over the entire visible to near-infrared spectral range (550-1600nm). Section 2 presents the key features of pulse-shaper-based techniques whereas sections 3 to 5 are dedicated to FROG-inspired and SPIDER-inspired techniques. Section 6 will deal with the extension of these techniques to near UV (266-400nm).

2. Formalism and experimental setup

2.1 Representation of pulses fields

Assuming the light field is linearly polarized, the electric field $\varepsilon(t)$ may be described as the real part of an analytic signal $E(t)$:

$$\varepsilon(t) = \mathrm{Re}\{E(t)\} \tag{1}$$

$E(t)$ is a complex function of time with a compact support within some time interval [-T,+T] and can be expressed in terms of carrier and envelope as follows:

$$E(t) = A(t)e^{-i\omega_0 t} \tag{2}$$

where ω_0 is the carrier angular frequency and $A(t)$ is the complex time amplitude. The Fourier transform of $A(t)$, noted $A(\omega)$, is referred to as the complex spectral amplitude of the pulse. Both $A(t)$ and $A(\omega)$ can be described in terms of modulus and phase. The square of the complex time amplitude, $I(t) = |A(t)|^2$, is the time-dependent instantaneous power. $I(\omega) = |A(\omega)|^2$ is referred to as the spectral intensity.

2.2 Spectral filters

All the optically *linear*[1] and *time-stationary*[2] operations on a pulse can be described by a complex spectral filter or, equivalently, by an impulse time response, these two functions being conjugated by Fourier transform. If $A_{in}(\omega)$ is the input spectral amplitude of an optical pulse, then the spectral amplitude at the output of an optical element characterized by the spectral filter $H(\omega)$ is equal to the *product* of $A_{in}(\omega)$ and $H(\omega)$:

$$A_{out}(\omega) = H(\omega)A_{in}(\omega) \tag{3}$$

Conversely, if $A_{in}(t)$ is the time complex amplitude of the input pulse and $H(t)$ is the time impulse response of the optical element, then the complex time amplitude of the output pulse is given by the convolution of $A_{in}(t)$ and $H(t)$:

$$A_{out}(t) = H(t) \otimes A_{in}(t) \tag{4}$$

The spectral filters of some of the most useful operations in optics are given in Table 1.

Description	Time-domain definition	Spectral filter
Attenuation	$A_{out}(t) = kA_{in}(t)$	k
Phase offset	$A_{out}(t) = e^{i\theta}A_{in}(t)$	$e^{i\theta}$
Optical delay	$A_{out}(t) = A_{in}(t-\tau)$	$e^{-i(\omega-\omega_0)\tau}$
Chirp	-	$e^{i\phi_2(\omega-\omega_0)^2}$

Table 1. Spectral filters associated with some common optical operations.

Any combination of linear operations can be described by a single spectral filter. For example, the operation "create two pulse replicas with an optical interpulse delay τ and a relative phase offset of $\pi/4$" is described by the spectral filter:

$$H(\omega) = 1 + e^{i\frac{\pi}{4}}e^{-i(\omega-\omega_0)\tau} \tag{5}$$

or equivalently by the time response function:

$$H(t) = \delta(t) + e^{i\frac{\pi}{4}}e^{i\omega_0\tau}\delta(t-\tau) \tag{6}$$

where $\delta(t)$ stands for the Dirac function.

---

[1] Linear is to be understood in the mathematical sense. An application f is linear if and only if for any values (a, b) of a given vectorial space and for any values of $k \in \mathbb{C}$, $f(a+k.b) = f(a) + k.f(b)$ and $f(0) = 0$.

[2] An operation which effects do not vary with the choice of the origin of time (an operation invariant by translation in time). For example, the operation "delay" is stationary but the operation "multiplication by a sinus" is not stationary.

It must be noted that the *optical delay* used in the former paragraph corresponds mathematically to a pure translation of the electric field $\varepsilon(t)$ along the time dimension. But this is not the case for the time-dependent amplitude (envelope function) $A(t)$ of the pulse. Indeed, the spectral filter associated with a pure translation in time of the amplitude - which we will call a *pure delay* - writes $H(\omega) = e^{-i\omega\tau}$, which is different from the optical delay $H(\omega) = e^{-i(\omega-\omega_0)\tau}$. Physically, an *optical delay* corresponds to a *pure delay* plus a phase shift of the carrier-envelope (Albrecht et al., 1999).
As it has been mentioned already in this chapter, pulse shaper devices offer the possibility to mimic almost any set of conventional optics. However, this statement must be mitigated, since pulse shaping devices require both input and output to be single beam (one beam in - one beam out) and linearly polarized. The equivalent optical setups are thus limited to the ones preserving the space profile and the polarization state. These constraints limit, in turn, the number of pulse measurement techniques that are transposable to pulse shapers, since many techniques require either two shaped beams or, which is almost equivalent, two shaped polarization states. We will hereafter distinguish the measurement techniques requiring a *single* spectral filter from the techniques requiring *two* spectral filters.
To extend the number of adaptable characterization techniques, it is convenient to use in combination with a pulse shaper a polarization multiplexer to enable not only the control of the spectral phase and amplitude but also of the output polarization state as a function of frequency. An example of such a polarization multiplexer is given in subsection 2.4.

2.3 Pulse shaping device

As mentioned before, pulse shaping devices such as the AODPF can be viewed as generic programmable spectral filters since their spectral phase and amplitude can be almost arbitrarily defined. These devices are nevertheless characterized by a finite *bandwidth* and a finite spectral *resolution*. The bandwidth refers to the maximal spectral range over which the device can operate, which might be limited by the absorption edges of the constitutive material, optical coatings or by any other physical mechanism leading to a drop of throughput. Typical bandwidth values for the AOPDF cover about one octave from the visible (>530nm) to the near infrared (<3μm), for example, 550nm-1100nm. As far as pulse measurement is involved, this limits the minimal duration of the measurable ultrashort pulses to a few optical cycles.
The spectral resolution of a pulse shaper refers to the minimal spectral feature, in both amplitude or phase, that can be controlled. Although spectral resolution can be expressed in terms of full width at half maximum (FWHM), it is much more convenient to express the spectral resolution in the conjugate space, that is in the time domain. As a matter of fact, all pulse shaping devices have a finite impulse time response which vanishes outside of a given time window $[0,T_{max}]$. This window will be referred to as the *pulse shaping window*. In application of the sampling theorem, the spectral resolution of a pulse shaper is proportional to the inverse of the pulse shaping window width. It is clear that a finite spectral resolution implies, in the context of pulse characterization, that the largest measurable pulse durations will be limited to a fraction of the pulse shaping window. For an AOPDF, the pulse shaping window depends on the thickness of the acousto-optic crystal and ranges usually from 3ps to 15ps. The experimental data showed in the following sections were recorded by using a ~7ps pulse shaping window, which corresponds to a FWHM spectral resolution of better than 150GHz (about 0.3nm at 800nm).

Other features, however less stringent, distinguish pulse-shaper-generated spectral filters from their conventional counterparts. For example, the spectral filter is not static on long time scales (tens of μs to tens of ms) and the pulse shaping device must be triggered.

2.4 Experimental setup

2.4.1 Optical layout

All the experiments reported in section 3 and section 4 were performed with the same optical layout: an AOPDF, an optional 6mm Calcite retardation plate for polarization multiplexing, a focusing lens, a second harmonic generation stage (thin BBO crystal) and a miniature spectrometer.

The heart of the system is an acousto-optic birefringent crystal of Paratellurite (TeO_2). The action of the AOPDF is based on a quasi-collinear interaction between an optical beam and an acoustic beam (Tournois, 1997). Through the process of acousto-optic diffraction, this interaction gives rise to a diffracted beam when suitable phase matching conditions are met. In the time domain and for low values of acoustic power density, this crystal performs a convolution between the complex amplitude of the input optical pulse and that of the acoustic pulse. In the spectral domain, this device acts as a programmable spectral filter. Without Calcite plate and assuming a perfectly phase-matched SHG, the signal collected by the spectrometer is given by:

$$I(\omega) \propto \left| \int \left[H(t) \otimes A(t) \right]^2 e^{2i\omega_0 t} e^{-i\omega t} dt \right|^2 \tag{7}$$

where $H(t)$ is the time impulse response of the considered spectral filter.

2.4.2 Polarization multiplexing

When polarization control of the output pulses is required (two-filter techniques), a 6mm thick retardation plate made of Calcite is inserted between the AOPDF and the SH crystal. This birefringent plate is cut such that its optical axis is perpendicular to the direction of incidence and orientated at 45 degrees to the polarization plane of the diffracted beam. With this setup, the fraction of the shaped pulse which is extraordinary polarized with respect to the Calcite crystal, is delayed with respect to the ordinary polarized part. This differential delay, noted τ_c, is proportional to the birefringence of the Calcite crystal and to the thickness of the plate. As depicted in Figure 1, if the AOPDF is set to produce two consecutive pulses of adjacent but non-overlapping time supports $A_1(t)$ and $A_2(t)$ and if the birefringent group delay is chosen to match that of the time delay between $A_1(t)$ and $A_2(t)$, then, after propagation through the Calcite plate, four pulses will emerge: two ordinary polarized pulses and two extraordinary pulses. Among them, the only overlapping pulses are the ordinary projection of $A_2(t)$ and the extraordinary projection of $A_1(t)$. If the principal polarization planes of the type II SHG crystal are orientated parallel to those of the Calcite plate, the type II SHG signal will be proportional to the product of these two projections. Mathematically, the spectral filter required to produce the pulse replicas is:

$$H(\omega) = H_1(\omega) + H_2(\omega) e^{-i\omega\tau_c} \tag{8}$$

Assuming a perfect phase matching in a type II second harmonic generation stage, the SH signal collected by the spectrometer will then be equal to:

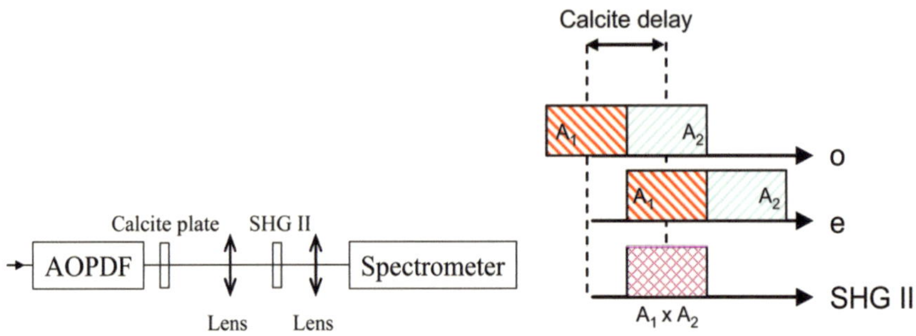

Fig. 1. a) Optical setup. b) Principle of polarization multiplexing.

$$I(\omega) \propto \left| \int \left[H_1(t) \otimes A(t) \right] \left[H_2(t) \otimes A(t) \right] e^{2i\omega_0 t} e^{-i\omega t} dt \right|^2 \quad (9)$$

where $H_1(t)$ and $H_2(t)$ are the time impulse responses of the spectral filters $H_1(\omega)$ and $H_2(\omega)$. The use of this polarization multiplexing technique comes at some cost however: the principal polarization planes in the SHG and Calcite crystals must be aligned with high precision, the effective pulse-shaping window is reduced by a half, and a half of the energy of the shaped pulses does not contribute to the SH generation.

2.4.3 Dispersion precompensation

It is well-known that propagation through thick dispersive media drastically modifies the spectral phase of broadband pulses. Therefore, in order to characterize the input pulses, it is required to compensate for both the dispersion of the AOPDF crystal and focusing lens. When the Calcite plate is used, it is also necessary to compensate for the mean and differential dispersions. As a result, for all measurements, the spectral filter will be defined as the product of the "measurement" filter $H(\omega)$ and a precompensation filter $\Pi(\omega)$. $\Pi(\omega)$ is defined as follows: $|\Pi(\omega)|$ is a supergaussian function of order 6, defining the bandwidth of the measurement, and $\arg(\Pi(\omega))$ is equal to the opposite of the total dispersion of all the optical elements of the device. It must be noted that $\Pi(\omega)$ uses part of the pulse shaping capability. It also limits the minimal measurable pulse duration (~10fs for a gaussian spectrum), since the AOPDF cannot compensate itself over a bandwidth larger than ~300nm. In the following section, $\Pi(\omega)$ will be dropped out for the sake of clarity.

3. Two-filter measurements

A first class of pulse measurement methods relies on a *spectrographic* measurement, that is on the analysis of the spectral content of a time-slice of the pulse to be characterized. The most popular techniques belonging to this class are the multiple flavors of Frequency-Resolved Optical Gating (FROG) (Trebino et al., 1997). In FROG-related techniques, the measured quantity is the spectrally-resolved autocorrelation or cross-correlation function of the pulse. From this bi-dimensional data (spectrum as a function of time delay between pulse replicas), the spectral phase and amplitude can be retrieved by an iterative algorithm. In practice, a SH-FROG setup consists of an interferometer producing two pulse replicas of the pulse to be characterized, a second harmonic crystal and a spectrometer. Since FROG open-space setups use two non collinear beams or two collinear but perpendicularly polarized beams (Fittinghoff et al., 1998; Gallmann et al., 2000), it is clear that FROG is a two-filter measurement method. It is worth noting that FROG does not exhaust the class of spectrographic techniques. The STRUT technique for example, also belongs to this class and has been implemented with a pulse shaping device (Galler & Feurer, 2008).

A second class of measurement techniques relies on an *interferometric* measurement. In this case, the experimental data is unidimensional and the spectral phase is encoded as a spectral or temporal beating signal. The most popular member in this class is SPIDER. As FROG, usual practical implementations of SPIDER make use of two beams.

In this section, we will focus on the implementation of FROG and SPIDER, as two-filter methods, in pulse-shaper-based measurement devices. All the reported data were acquired with the polarization multiplexing setup.

3.1 SH-FROG

For intensimetric FROG, the signal collected by the spectrometer is equal to:

$$I_{\text{FROG}}(\tau,\omega) \propto \left| \int A(t)A(t-\tau)e^{2i\omega_0 t}e^{-i\omega t}dt \right|^2, \tag{10}$$

The corresponding spectral filter is therefore:

$$H_{\text{FROG}} = (\tau,\omega) = 1 + e^{-i\omega(\tau+\tau_c)} \tag{11}$$

where τ_c stands for the differential group delay of the Calcite plate. This function being a two-dimensional function, the measurement cannot be made single-shot with the studied setup. However, by using a discrete number of time delays τ, the FROG trace can sampled in the time direction and used to retrieve the spectral phase and amplitude of the complex envelope $A(t)$.

3.1.1 Trace features

A FROG trace can be divided into three areas: small, large and intermediate delays. In the small delay area, that is to say close to zero delay, the signal is maximal and corresponds to the SHG signal of the pulse:

$$I_{\text{FROG}}(\tau = 0,\omega) \propto I_{\text{SH}}(\omega) \tag{12}$$

At sufficiently large delays, the pulse replicas do not overlap anymore in time and the FROG signal vanishes:

$$I_{\text{FROG}}(\tau \to \infty,\omega) = 0 \tag{13}$$

In the intermediate range, the FROG trace cannot be expressed simply but must remain symmetric with respect to the origin of time. Indeed, changing τ into $-\tau$ in Equation 10 doesn't change the value of the integral:

$$I_{\text{FROG}}(-\tau,\omega) = I_{\text{FROG}}(\tau,\omega) \tag{14}$$

This particular feature implies that FROG method is inherently insensitive to the direction of time, which is known as the "time ambiguity" of FROG traces and retrieved pulse temporal profiles.

3.1.2 Retrieval algorithm

FROG traces cannot be analytically inverted to retrieve the spectral phase (and amplitude) of the corresponding pulse. However, some efficient iterative pulse retrieval algorithms exist, and the Principal Component Generalized Projects algorithm (PCGPA, DeLong et al. (1994); Kane et al. (1999); Reid (1999)) is one of the most powerful among all published algorithms.

Most often the raw experimental data have to be processed and formatted prior to the application of the PCGPA algorithm. In particular, the background signal must be removed with care. Indeed, as the FROG algorithm is based on a time-frequency description, a constant background would be interpreted as continuous (single-frequency) contribution to the signal. Once the background is removed, it is mandatory to re-sample and/or extrapolate the spectra in order to satisfy the Nyquist-Shannon sampling theorem.

When a pulse shaper is used to generate the time delay between the two pulse replicas, the delay step and the number of points along the time direction can be arbitrarily defined. It must be noticed, that the spectral grid on which the spectra must be sampled is only set by the delay span of the measurement. Let the sampled delays be:

$$\tau_k = \tau_{max} \frac{(k-N)}{N} \text{with } k = 1..(2N-1) \tag{15}$$

The SHG spectra will then have to be re-sampled on the frequency grid:

$$\nu_k = 2\nu_0 + \Delta\nu \frac{(k-N)}{N} \text{with } \ k = 1..(2N-1) \tag{16}$$

where $\Delta\nu$ is the Nyquist frequency of the delay sampling:

$$\Delta\nu = \frac{2}{\tau_{max}} \tag{17}$$

and ν^0 is the central frequency of the fundamental pulse (i.e. its carrier frequency). From Equation 17 it follows that the *spectral resolution* of a FROG measurement is inversely proportional to the maximum explored time delay. Since both the delay excursion and the delay step size can be arbitrarily chosen within the pulse shaping window of the AOPDF, the spectral resolution of the measurement can be adapted to the pulse to be measured. Besides, with an AOPDF, the precision on the interpulse delays is estimated to be better than ~100as (10^{-16}s), which is about one magnitude lower than the optical cycle period in the visible and IR spectral ranges.

3.1.3 Experimental example

We used as test pulses a leakage (1μJ) from a kHz amplified system delivering pulses of ~100fs at 800nm. The leakage beam was sent into a 25mm AOPDF and the diffracted beam was focused by a plano-convex lense (f=75mm) in a 100μm-thick BBO crystal cut for a type II SHG at 800nm. The generated SHG beam was refocused at the input of a miniature spectrometer (Avantes, 3648 pixels, 700-900nm). 65 equidistant time delays covering the [-1000fs, 1000fs] interval were used for this measurement. Figure 2 shows the experimental FROG trace, the reconstructed FROG trace, the retrieved spectral phase and amplitude and the retrieved time intensity.

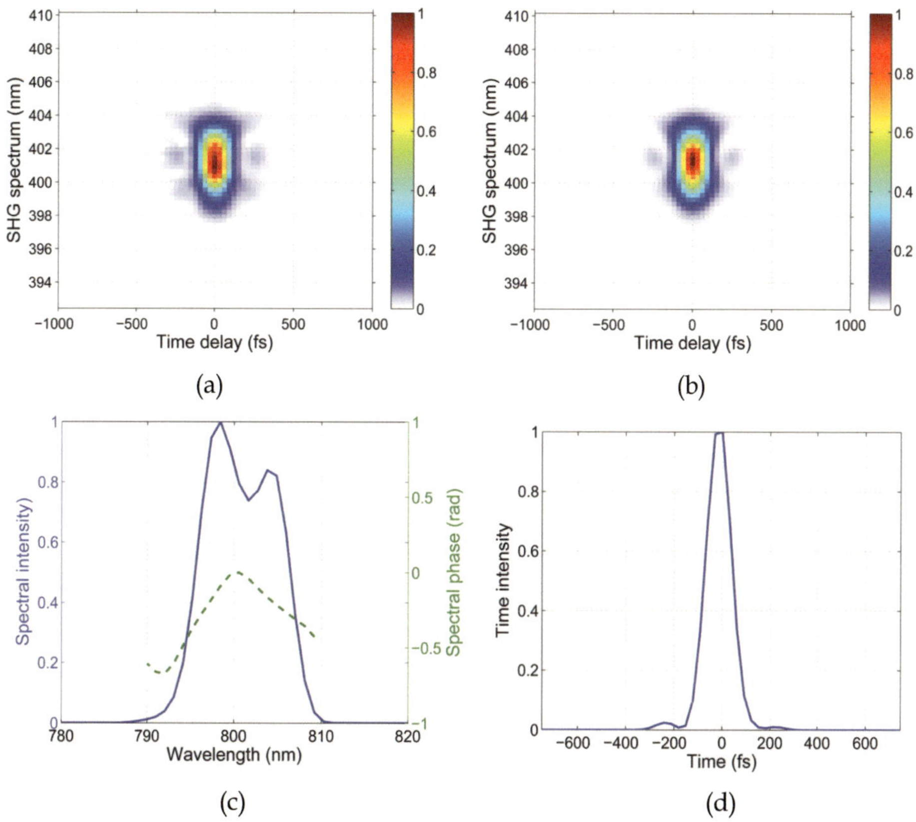

Fig. 2. a) Experimental FROG trace. b) Reconstructed FROG trace. c) Retrieved spectrum and spectral phase. d) Retrieved time intensity (123fs FWHM).

3.2 SPIDER

Contrary to the FROG technique, SPIDER doesn't allow the reconstruction of the spectral intensity (i.e. the spectrum must be measured independently). However, it provides an analytic and unambiguous determination of the spectral phase. SPIDER is based on shearing interferometry in the optical frequency domain, which means, on the interference pattern created by two frequency-shifted replicas of the pulse. Let $A(\omega) = \sqrt{I(\omega)}\exp(i\varphi(\omega)$ be the spectral complex amplitude of a broadband optical pulse, and let $A(\omega + \delta\omega)$ be the spectrally shifted replica of $A(\omega)$. The interference pattern between $A(\omega)$ and $A(\omega + \delta\omega)$, as recorded by a square-law detector, is:

$$S(\omega) = I(\omega) + I(\omega + \delta\omega) + 2\sqrt{I(\omega)}\sqrt{I(\omega + \delta\omega)}\cos\left[\varphi(\omega) - \varphi(\omega + \delta\omega)\right] \tag{18}$$

The phase difference can be easily and algebraically extracted from the interference pattern using standard Fourier processing techniques. The spectral phase of the field, $\varphi(\omega)$, can then be reconstructed by concatenation or integration.

In a conventional SPIDER setup, two replicas of the pulse separated by a delay τ are mixed in a nonlinear crystal with a chirped pulse having a large second-order dispersion. Because of the linear relation between time and instantaneous frequency in the chirped pulse, the instantaneous frequencies are constant during the interaction with each of the short replicas but differ by the spectral shear. The nonlinear interaction shifts the complex amplitudes of the replicas of the initial pulse by two different frequencies separated by a frequency shift $\delta\omega$. Measurement of the phase difference between the two sheared replicas by use of Fourier transform spectral interferometry (FTSI) gives access to the quantity:

$$\phi_{\text{SPIDER}}(\omega) = \omega\tau + \varphi(\omega + \omega_0) - \varphi(\omega + \omega_0 + \delta\omega) \tag{19}$$

The phase of the initial pulse is then obtained in three steps: substraction of the linear term $\omega\tau$, global shift by ω_0 along the frequency axis, and concatenation/integration of the phase by steps of $\delta\omega$. This inversion algorithm is thus totally algebraic and does not rely on iterative procedures.

Despite its numerous advantages (single-shot, analytic, unambiguous...), the SPIDER technique usually requires several precise calibrations. Indeed, the values of the interpulse delay τ, the global shift ω_0 and the spectral shear $\delta\omega$ are key parameters. Although these parameters can be measured by several means, using a bulk pulse shaper such as the AOPDF to produce the pulse replicas and the chirped pulse brings an elegant solution to the calibration issue:

- the interpulse delay is known, without any need for external calibration,
- the global shift is known with high accuracy,
- all the parameters (interpulse delay and spectral shear) can be changed at will.

This measurement is the exact equivalent of the classical SPIDER optical setup: the AOPDF generates two identical pulse replicas of the input pulse and one highly chirped narrowband pulse which is obtained by frequency filtering. The Calcite plate is used to rotate the polarization of the chirped pulse so that the pulse replicas and the chirped pulse can experience a background-free sum-frequency generation in a type II BBO crystal.

The experimental measurements displayed in this section were performed at ~800nm, at the output of an amplified kHz Ti:Sapphire laser system. About 10μJ of energy was sampled from the main beam and used for the pulse measurement. The AOPDF was a 25mm TeO_2 crystal and the SHG crystal was a 20μm-thick BBO crystal cut for a type II phase matching. The interpulse delay was set to 500fs and the spectral shear to 1.2nm at 800nm. Figure 3 shows the raw SPIDER spectrum together with the retrieved spectral phase and time intensity.

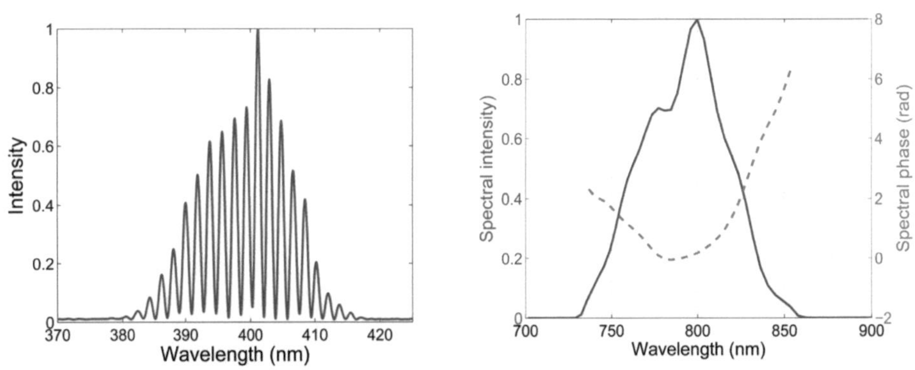

Fig. 3. a) Experimental single-shot SPIDER spectrum. b) Extracted spectral amplitude and phase. Retrieved pulse duration was 28fs FWHM.

4. Single filter measurements

As it was pointed out in section 3, SH-FROG and SPIDER are intrinsically two-filter pulse measurement techniques. Some workarounds, such as polarization multiplexing, can be found to implement these techniques with a single pulse shaper. However, it is much more convenient to use single-filter variants of these techniques: type I phase matching can be used instead of type II and no additional birefringent crystal is required, which removes the drawbacks expressed in 2.4.2. In this section we introduce, for the first time, single-filter variants of SHFROG and SPIDER.

4.1 Interferometric FROG

Interferometric FROG (iFROG) is the most simple variant of the FROG technique to be implemented with a pulse shaping device. The optical setup of iFROG is a Michelson interferometer, followed by a second harmonic generation stage and a spectrometer (Amat-Roldan et al., 2004; 2005). The signal collected by the spectrometer is then equal to:

$$I_{\text{iFROG}}(\tau,\omega) \propto \left| \int \left(A(t) + A(t-\tau)e^{i\omega_0\tau} \right)^2 e^{2i\omega_0 t} e^{-i\omega t} dt \right|^2 \tag{20}$$

where ω_0 is a the optical carrier frequency. Comparing formula 20 and 7, leads to the following spectral filter:

$$H_{\text{iFROG}}(\omega) = 1 + \exp -i\left[(\omega - \omega_0)\tau \right] \tag{21}$$

Given the order of magnitude of ω_0 (several $10^{15}Hz$), the oscillation period is of the order of a few fs. As a result, it is not always possible to sample both correctly the autocorrelation signal over large delay ranges and keep a small number of sampling points. A Gaussian pulse of 200fs FWHM at 800nm, for example, requires a scanning range of ±600fs, which corresponds to a minimal number of sampling points of ~900. By using a pseudo-carrier of angular frequency Ω, this number can be reduced greatly without any loss of information. Such a substitution is impossible when using a mirror-based Michelson interferometer. However, with a pulse shaper, this is possible and for the sake of generality, we will consider the following class of filters:

$$H_{\Omega}(\omega,\tau) = 1 + \exp -i\left[(\omega - \Omega)\tau \right] \tag{22}$$

Compared to FROG traces, iFROG traces exhibit oscillations with respect to delay, with a period equal to the period of the pseudo-carrier (Galler & Feurer, 2008). This feature is equivalent to that of conventional interferometric autocorrelators, for which $\Omega = \omega_0$. As a FROG trace, a iFROG trace can be divided into three areas: small, large and intermediate delays. In the small delay area, that is to say close to delay zero, the signal is maximum and corresponds to the SHG signal of the pulse.

$$I_{\text{iFROG}}(\tau = 0,\omega) \propto 16 I_{\text{SH}}(\omega) \tag{23}$$

At sufficiently large delays, the pulse replica do not overlap any more in time and the iFROG signal is equal to the interference pattern of the time-shifted SHG pulses:

$$I_{\text{iFROG}}(\tau \to \infty,\omega) \propto 4 I_{\text{SH}}(\omega)\cos^2(\delta\omega\tau / 2 - \Omega\tau) \tag{24}$$

where $\delta\omega = \omega - 2\omega_0 + \Omega$ is the frequency offset from the frequency carrier of the SHG signal. As a matter of fact, the iFROG signal does not vanish for large delays as for the FROG signal.
For intermediate delays, the iFROG signal is more complex. It can be expressed as a function of the centered FROG signal:

$$E^{c}_{\mathrm{FROG}}(\tau,\omega) = \int A(t-\tau/2)A(t+\tau/2)e^{i(2\omega_0-\omega)t}dt \tag{25}$$

The iFROG signal is then equal to:

$$I_{\mathrm{iFROG}}(\tau,\omega) \propto I_{\mathrm{SH}}(\omega)\cos^2\left(\frac{\delta\omega\tau}{2}-\Omega\tau\right) + 4I^{c}_{\mathrm{FROG}}(\tau,\omega)$$

$$+8\cos\left(\frac{\delta\omega\tau}{2}-\Omega\tau\right)\mathrm{Re}\,[\,E^{\star}_{\mathrm{SH}}(\omega)E^{c}_{\mathrm{FROG}}(\tau,\omega)\,] \tag{26}$$

This expansion shows that the iFROG signal is the sum of three terms: a SH component, a FROG component and a cross-term. iFROG traces must be, - as all FROG type traces -, symmetric with respect to delay zero. Besides, the autocorrelation trace (i.e. the iFROG trace integrated along the spectral coordinate) exhibits the following features: the signal is positive, reaches a maximum at zero delay and becomes constant at large delays. Besides, the ratio between the signals at zero delay and large delays is equal to 8.
As for FROG, the delay range must be much greater than the pulse duration. More precisely, the maximum delay must be chosen so that the electric field at such delays is truly zero. For example, for a 30fs gaussian pulse, the maximum delay must be set to at least 120-150fs. A rule of the thumb is to choose a value which is equal to 4-5 times the expected pulse duration. The minimum number of points to acquire is a function of the pulse complexity. A good estimation of the pulse complexity is given by the time-bandwidth product of the pulse. Arule of the thumb to choose the right number of points is to multiply the expected time-bandwidth product by 128.

4.1.1 Phase retrieval algorithm

Several phase retrieval algorithms for iFROG have been published already. Most of them, however, were developed for the iFROG with $\Omega = \omega_0$ and did not consider the new degree of freedom given by the ability to choose $\Omega \neq \omega_0$. Mainly, there are two different approaches. From Equation 26 it is clear that if the product $\Omega \times \Delta\tau$, where $\Delta\tau$ is the time support of the measured pulse, is greater than a few units, then the three components of the iFROG signal (SHG, FROG term and cross-term) can be isolated by Fourier filtering. A first method consists in extracting the FROG term (i.e. the low frequency term) and then retrieving the spectral phase and amplitude by using the PCGPA algorithm (Stibenz et al., 2006; Stibenz & Steinmeyer, 2005). A second method consists in using the cross-term (i.e. the term oscillating at $\pm\Omega$), from which the spectral phase can be analytically extracted (MEFISTO technique, Amat-Roldan et al. (2004; 2005)).

4.1.2 Experimental demonstration

The experimental demonstration of iFROG was carried at the SHG output (~600nm) of a collinear optical parametric amplifier (OPA) pumped by a Ti:Sapphire amplified system.

About 1μJ of energy was sampled from the main beam and used for the pulse measurement. The AOPDF was a 25mm TeO_2 crystal and the SHG crystal was a 20μm-thick BBO crystal cut for a type I phase matching. The focusing lens was achromatic with a focal length of 50mm and the spectrometer had a 0.05nm resolution (AvaSpec-2048 from Avantes). Figure 4 displays the raw experimental iFROG trace and iFROG autocorrelation together with the filtered FROG trace and the corresponding reconstructed FROG trace.

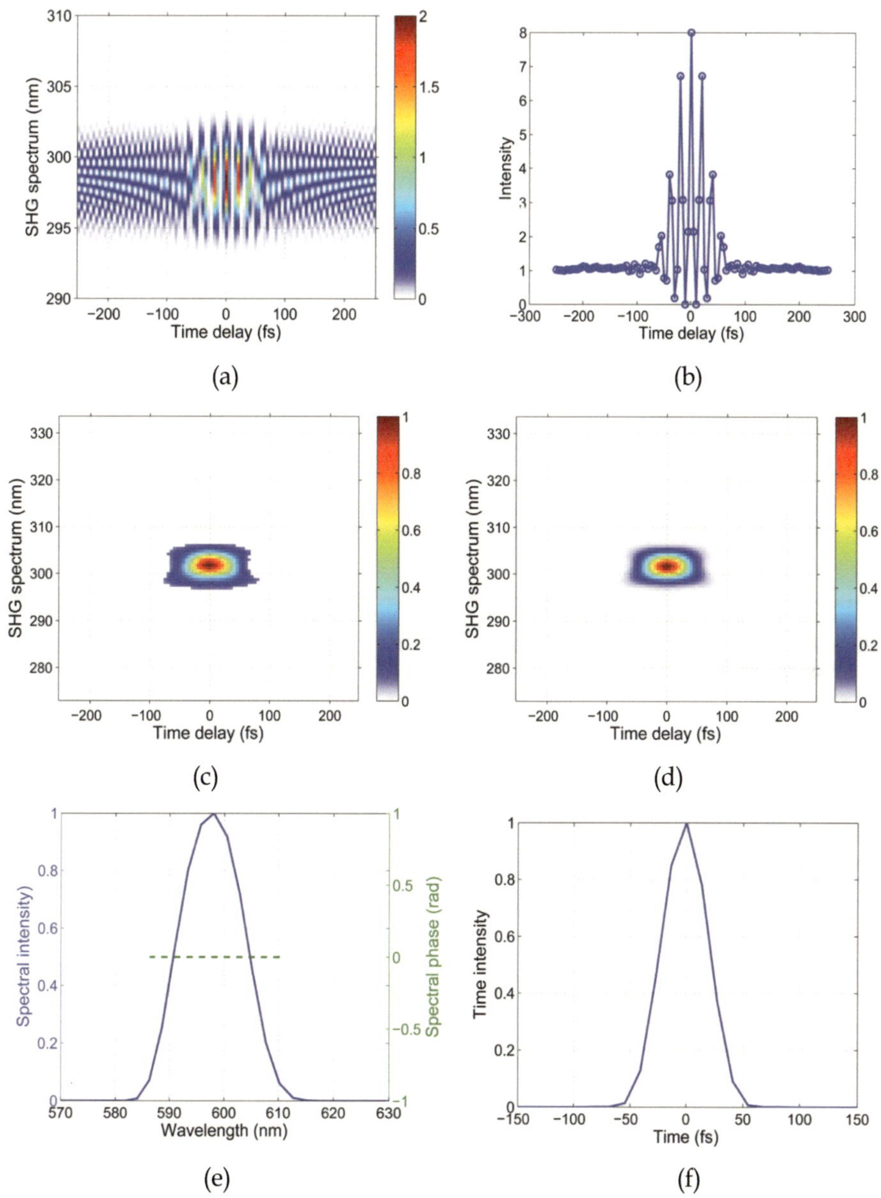

Fig. 4. a) Experimental iFROG trace. b) Experimental iFROG autocorrelation. c) Filtered FROG trace. d) Retrieved FROG trace. e) Retrieved spectral phase and intensity. f) Retrieved time intensity.

4.2 Baseband interferometric FROG

In the limit $\Omega \to 0$, the phase retrieval algorithm detailed in the previous section doesn't work anymore and another algorithm is required. Far from being pathologic, this variant of iFROG is highly interesting from a practical point of view. Indeed, at $\Omega = 0$, the iFROG trace is free from fast oscillations and therefore requires a minimal number of sampling points in the time direction. Given these properties we call this technique *baseband interferometric FROG* or bFROG. What is more, the obtained traces are very close to that of intensimetric SHG-FROG. For bFROG the signal collected by the spectrometer is:

$$I_{\text{bFROG}}(\tau,\omega) \propto \left| \int \left(A(t) + A(t-\tau)e^{i\phi} \right)^2 e^{2i\omega_0 t} e^{-i\omega t} dt \right|^2 \tag{27}$$

The trace features are the same as for iFROG (values at zero delay and at large delays, symmetry) but the analysis of the spectral content of the bFROG map exhibits new features. In particular, the collected signal can be expressed as a function of the centered FROG signal:

$$I_{\text{bFROG}}(\tau,\omega) \propto I_{\text{SH}}(\omega)\cos^2\left(\frac{\delta\omega\tau}{2} - \phi\right) + 4I^c_{\text{FROG}}(\tau,\omega)$$

$$+8\cos\left(\frac{\delta\omega\tau}{2} - \phi\right)\text{Re}\left[E^{\star}_{\text{SH}}(\omega)E^c_{\text{FROG}}(\tau,\omega) \right] \tag{28}$$

As iFROG, bFROG is the sum of three terms: a constant SH component, a FROG component -term. It also shows that if $\phi = \pi/2$ then the bFROG is almost equal to its FROG term for small delays ($\delta\omega\tau/2 \ll 1$).

4.2.1 Phase extraction

Spectral phase and amplitude cannot be retrieved from bFROG traces as from iFROG traces since the SH and FROG components are merged. However, it is still possible to retrieve this information by using a modified version of the PCGPA algorithm used for intensimetric FROG. A detailed description of this algorithm is out of the scope of this paper but, in a few words, the main modification consists in removing the SH complex amplitude prior to the projection step.

4.2.2 Experimental example

The experimental demonstration of bFROG was carried on the same visible source as described in 4.1.2. Figure 5 displays the raw experimental bFROG trace, bFROG autocorrelation and the reconstructed bFROG trace for $\phi = 0$rad. N=65 equidistant time delays covering the [-1000fs, 1000fs] interval were used for this measurement.

4.3 Phase-cycling SPIDER

4.3.1 Principle

In this variant of SPIDER, both the pulse replicas and the chirped pulse are in the same polarization state. In this case, the technique cannot be single-shot anymore. As shown below, the SPIDER signal can nevertheless be extracted by using a phase-cycling technique. For the sake of simplicity, we will consider a combination of four collinear pulses: two time-

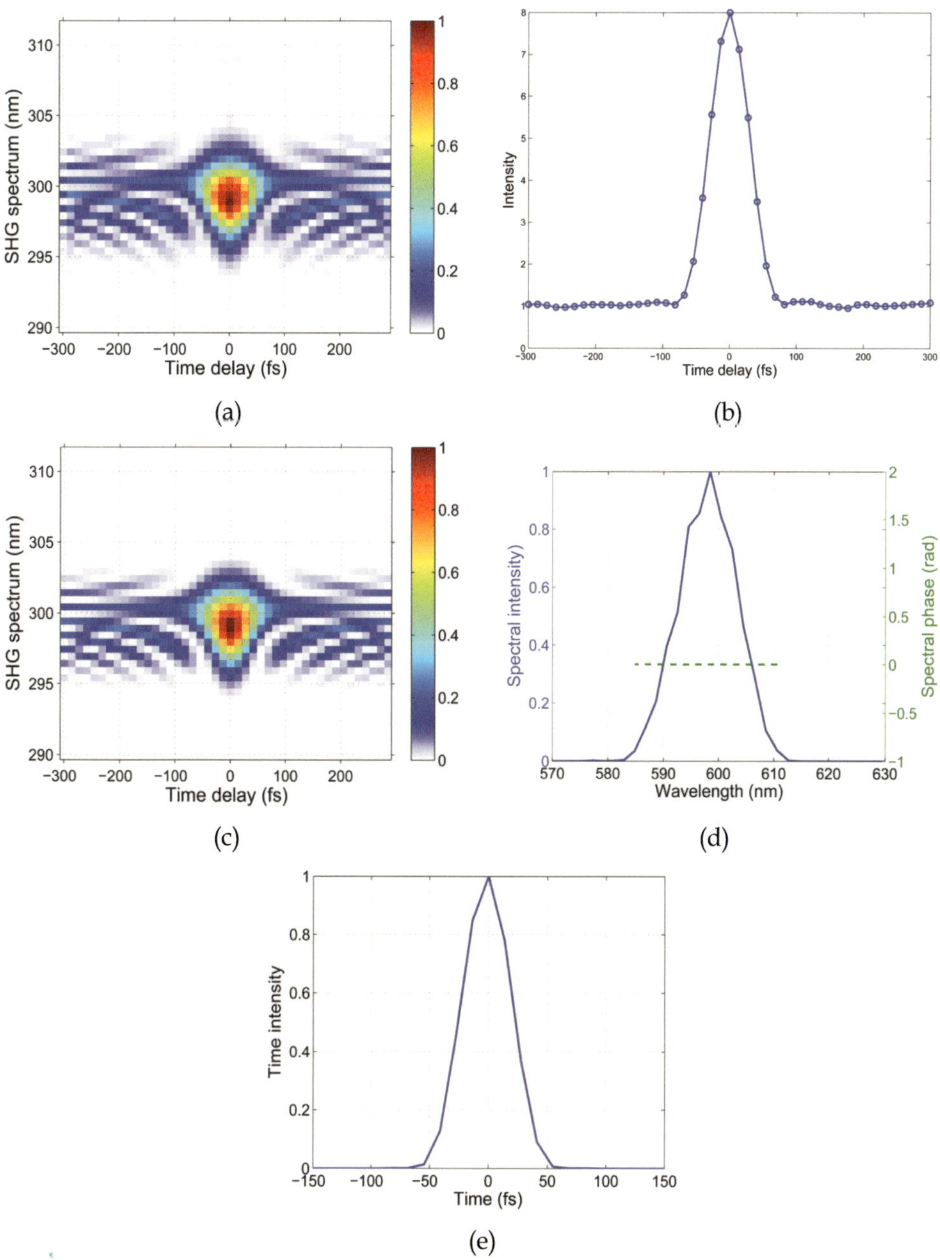

Fig. 5. a) Experimental bFROG trace. b) Experimental bFROG autocorrelation. c) Retrieved FROG trace. d) Retrieved spectral phase and intensity. e) Retrieved time intensity.

delay pulse replicas and two time-delayed quasi-monochromatic pulses (the equivalent of the two parts of the chirped pulse which overlap with the short replicas), the pulses being synchronized by pairs. The complex time-dependent amplitude of the electric fields will be noted with roman capital letters while their Fourier transforms will be noted with a hat. In the time domain, the total electric amplitude of the considered pulse combination is then:

$$E_{\text{tot}}(t) = \left\{ E(t) + M_1(t)e^{i\delta\omega t} + E(t-\tau) + M_2(t-\tau)e^{-i\delta\omega t} \right\} e^{i\omega_0 t} \tag{29}$$

The SHG signal induced by $E_{\text{tot}}(t)$ is proportional to the square of $E_{\text{tot}}(t)$. Since the pulses are much shorter than the time delay τ, cross-products of non-concomitant pulses vanish, which leads to:

$$E_{\text{tot}}^2(t) = \left\{ E^2(t) + M_1^2(t)e^{2i\delta\omega t} + M_2^2(t-\tau)e^{-2i\delta\omega t} + E_2^2(t-\tau) \right.$$

$$\left. +2E(t)M_1(t)e^{i\delta\omega t} + 2E(t-\tau)M_2(t-\tau)e^{-i\delta\omega t} \right\} e^{2i\omega_0 t} \tag{30}$$

For convenience we define:

$$E_{\text{SPIDER}}(t) = E(t)M_1(t)e^{i\delta\omega t} + E(t-\tau)M_2(t-\tau)e^{-i\delta\omega t} \tag{31}$$

This term is indeed the SPIDER signal since it corresponds to two time-delayed and spectrally shifted pulse replicas of the same pulse. The spectral amplitude of $E_{\text{tot}}^2(t)$ is then given by the Fourier transform of Equation 30:

$$\widehat{E_{\text{tot}}^2}(2\omega_0 + \Omega) = \widehat{E^2}(\Omega)\left(1 + e^{-i\Omega\tau}\right)$$

$$+\widehat{M_1^2}(\Omega - 2\delta\omega) + \widehat{M_2^2}(\Omega + 2\delta\omega)e^{-i\Omega\tau - i2\delta\omega\tau} + 2\hat{E}_{\text{SPIDER}}(\Omega) \tag{32}$$

where $\Omega = \omega - \omega_0$ is the frequency offset with respect to the central angular frequency ω_0 and $\hat{E}_{\text{SPIDER}}$ is the Fourier-transform of E_{SPIDER} which is explicitly given by:

$$\hat{E}_{\text{SPIDER}}(2\omega_0 + \Omega) = \hat{E}(\Omega) \otimes \widehat{M_1}(\Omega - \delta\omega) + \hat{E}(\Omega) \otimes \widehat{M_2}(\Omega + \delta\omega)e^{-i\Omega\tau} \tag{33}$$

Without surprise, the SHG spectral amplitude is the sum of two contributions: the individual SHGs of all fours pulses plus all the possible cross-terms. The spectral intensity, which is the only physically accessible quantity is then equal to the square modulus of Equation 32 which is a rather complicated expression. For the sake of compactness, all implicitly clear dependances with the frequency variables will be dropped from now on.

$$I_{\text{tot}}(2\omega_0 + \Omega) = \begin{cases} 4\left|\widehat{E^2}\right|^2 \cos^2(\Omega\tau) + \left|\widehat{M_1^2}\right|^2 + \left|\widehat{M_1^2}\right|^2 & \text{(SHG)} \\ +4\left|\hat{E}_{\text{SPIDER}}\right|^2 & \text{(SPIDER)} \\ +2\text{Re}\left\{\widehat{E^2}\left(1 + e^{-i\Omega\tau}\right)\left(\widehat{M_1^2} + \widehat{M_2^2}e^{-i2\Omega\tau - i2\delta\omega\tau}\right)^\star\right\} & \text{(cross-term)} \\ +4\text{Re}\left\{\hat{E}_{\text{SPIDER}}^\star\left[\widehat{E^2}\left(1 + e^{-i\Omega\tau}\right) + \widehat{M_1^2} + \widehat{M_2^2}e^{-i(\Omega + 2\delta\omega)\tau}\right]\right\} & \text{(cross-term)} \end{cases} \tag{34}$$

Clearly, it is not possible to extract the SPIDER signal from $I(\omega)$, even by using a spectral filter since there are several terms oscillating with respect to $\Omega\tau$. Consequently, it is not possible to extract the SPIDER signal from this pulse combination.

We now consider the two following combinations of pulses:

$$\begin{aligned} E_{++} &= E(t) + M_1(t) + E(t-\tau) + M_2(t-\tau) \\ E_{--} &= -E(t) + M_1(t) - E(t-\tau) + M_2(t-\tau) \end{aligned} \tag{35}$$

Adding π to the phase of $E(t)$ leaves all the terms unchanged except $E_{SPIDER}(\omega)$ which becomes $-E_{SPIDER}(\omega)$. So, by summing $I_{++} = |E_{++}|^2$ and $I_{--} = |E_{--}|^2$, the SPIDER/SHG cross-term can be eliminated:

$$I_{++} + I_{--} = \begin{cases} 8\left|\widehat{E^2}\right|^2 \cos^2(\Omega\tau) + 2\left|\widehat{M_1^2}\right|^2 + 2\left|\widehat{M_1^2}\right|^2 & \text{(SHG)} \\ +8\left|\hat{E}_{\text{SPIDER}}\right|^2 & \text{(SPIDER)} \\ +4\text{Re}\left\{\widehat{E^2}\left(1+e^{-i\Omega\tau}\right)\left(\widehat{M_1^2}+\widehat{M_2^2}e^{-i2\Omega\tau-i2\delta\omega\tau}\right)^\star\right\} & \text{(cross-term)} \end{cases} \quad (36)$$

This signal, however, still contains the parasitic term of the SHG/SHG cross term. This term is narrowband and has a minority weight but still cannot be easily removed. We now consider two additional pulse combinations:

$$\begin{aligned} E_{+-} &= E(t) + M_1(t) - E(t-\tau) + M_2(t-\tau) \\ E_{-+} &= -E(t) + M_1(t) + E(t-\tau) + M_2(t-\tau) \end{aligned} \quad (37)$$

As for I_{++} and I_{--}, summing I_{+-} and I_{-+} eliminates the cross term between the SPIDER and second harmonic terms. Next, the difference between $I_{++} + I_{--}$ and $I_{+-} + I_{-+}$ eliminates all the residual terms but one:

$$I_{++} + I_{--} - I_{+-} - I_{-+} = 32\text{Re}\left\{\hat{E}(\Omega) \otimes \widehat{M_1}(\Omega - \delta\omega) \times \hat{E}^\star(\Omega) \otimes \widehat{M_2}^\star(\Omega + \delta\omega)e^{i\Omega\tau}\right\} \quad (38)$$

Provided that the spectra of $M_1(t)$ and $M_2(t)$ are small with respect to the spectrum of $E(t)$, these spectra can be approximated by two Dirac distributions located at $+\delta\omega$ and $-\delta\omega$ and the combination of the four experimental signals can rewritten as a function of the spectral phase $\varphi(\Omega)$ of the $\hat{E}$:

$$I_{++} + I_{--} - I_{+-} - I_{-+} = 32\hat{E}(\Omega+\delta\omega)\hat{E}(\Omega-\delta\omega)\cos\left(\Omega\tau + \varphi(\Omega+\delta\omega) - \varphi(\Omega-\delta\omega)\right) \quad (39)$$

This combination gives directly the term containing the spectral phase of the pulse $E(t)$.

4.3.2 Spectral filter

The spectral filter corresponding to the pulse replicas are:

$$H_1(\omega) = \exp(-i\Omega\tau / 2) \quad (40)$$

$$H_2(\omega) = \exp(+i\Omega\tau / 2) \quad (41)$$

and the spectral filter of the chirped pulse is given by:

$$H_3(\omega) = \text{Rect}\left(\omega_0, \Delta\omega\right)\exp(i\phi^{(2)}\Omega^2 / 2) \quad (42)$$

The spectral filter is then given by the sum of three filters:

$$H_{\pm\pm}(\omega) = \pm H_1(\omega) \pm H_2(\omega) + H_3(\omega) \quad (43)$$

In these formula, $\Omega = \omega - \omega_0$ stands for the frequency offset from the carrier frequency, $\phi^{(2)}$ is the second order phase (chirp) coefficient and Rect $(\omega_0, \Delta\omega)$ is a passband amplitude function of width $\Delta\omega$ located at frequency ω_0. The mean group delays of the chirped pulse and of the double pulse structure are equal, which is one of the two conditions required to have a non zero time overlap between the chirped pulse and the pulse replicas. The other condition is to choose large enough values for the chirp coefficient $\phi^{(2)}$ and the bandwidth $\Delta\omega$ so that the temporal spread of the chirped pulse is greater than the interpulse delay. In practice however, there is almost no degree of freedom since the optimal ratio $\tau/\phi^{(2)} = \delta\omega$, which corresponds to the spectral shear between the pulse replicas after the SH stage, must be chosen as small as possible but cannot be smaller than the spectral resolution of the AOPDF.

4.3.3 Experimental demonstration

The experimental demonstration of cSPIDER was also carried on the same visible source as described in subsubsection 4.1.2. The interpulse delay was set to 500fs and the spectral shear to 2nm at 600nm. Figure 6 shows the raw acquired spectra together with the extracted signals.

The SPIDER signal, extracted using formula 39, is shown in Figure 6c.

4.4 Homodyme phase-cycling SPIDER

4.4.1 Principle

Homodyne SPIDER (hSPIDER) is a version of SPIDER that uses a self-referencing homodyne detection. This technique has a higher sensitivity than phase-cycling SPIDER and is self-calibrated with regard to the exact value of the interpulse delay τ.

In the world of pulse measurements, the most sensitive method demonstrated to date is spectral interferometry, in which a high-energy well-characterized reference pulse interferes with a weak unknown pulse in a spectrometer. The primary advantage of this technique over self-referencing approaches is the use of homodyne detection: a strong local oscillator provides effective amplification of the weak signal field due to the properties of square-law detectors. On the other hand, the main drawback is the need for detailed knowledge of the local oscillator pulse field (reference pulse), which increases the complexity of the method since an ancillary measurement of the reference pulse is necessary.

The homodyne implementation of SPIDER takes advantage of homodyne detection, while retaining the self-referencing and direct inversion features that are the hallmarks of its simplicity. The method is based on the interference of frequency-sheared replicas of the input pulse with an uncharacterized high-energy reference pulse, which is the SH signal of the pulse to be measured. In hSPIDER, one of the upshifted pulse replicas is replaced by the SH of the input pulse. Let ω_1 be the frequency shift of the upshifted pulse replica with respect to the input pulse and $\varphi_{SHG}(\omega)$ the spectral phase of the SH of the input pulse. The FTSI of the resulting signal is then:

$$\phi_{\text{hSPIDER},1}(\omega) = \omega\tau + \varphi(\omega + \omega_1) - \varphi_{\text{SHG}}(\omega) \tag{44}$$

The spectral phase of the input pulse cannot be directly retrieved since the spectral phase of the SH is unknown. However, with two different frequency shifts, both the linear term $\omega\tau$ and the SHG phase term can be eliminated. hSPIDER measurement hence requires three steps: measurement of the phase difference between the SH and a pulse replica upshifted by ω_1, measurement of the phase difference between the SH and a pulse replica upshifted by ω_2, computation of the difference between these two homodyne phases.

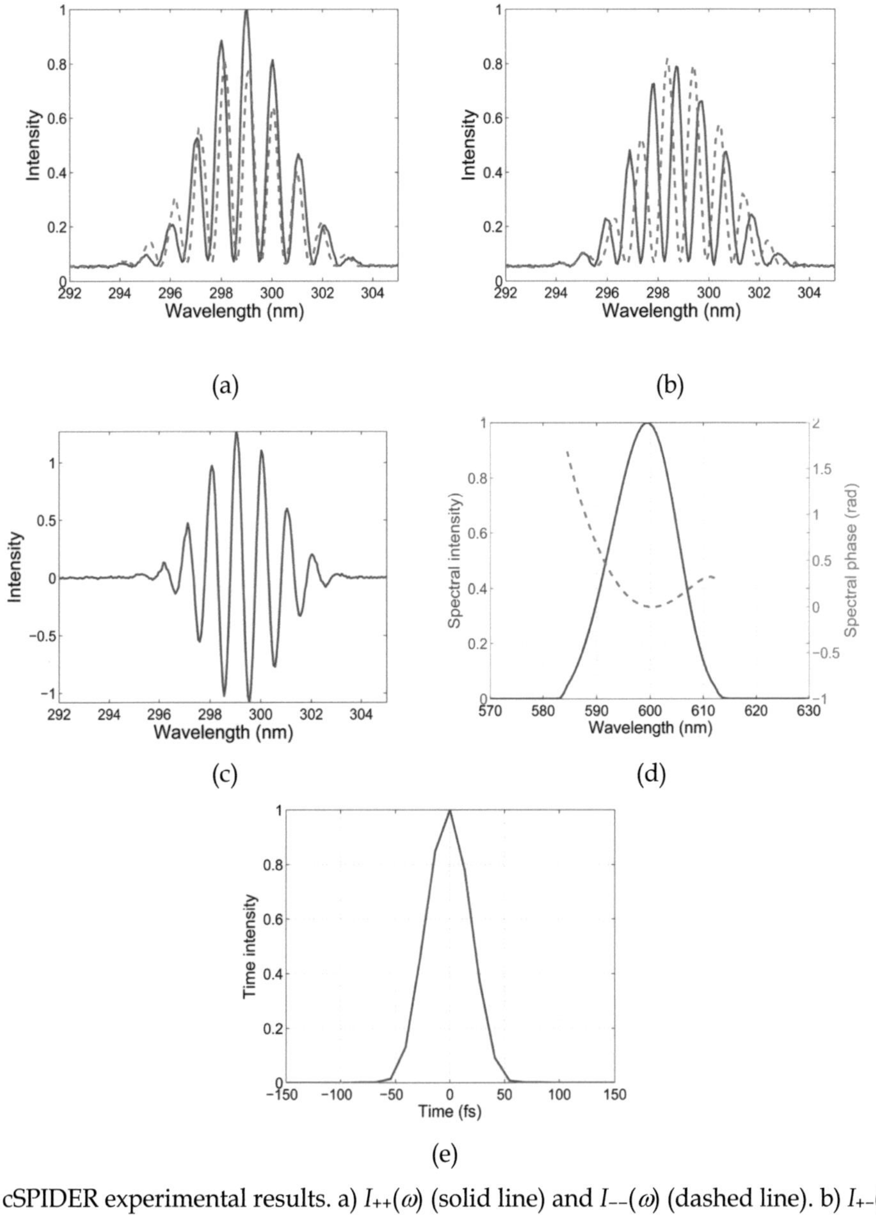

Fig. 6. cSPIDER experimental results. a) $I_{++}(\omega)$ (solid line) and $I_{--}(\omega)$ (dashed line). b) $I_{+-}(\omega)$ (solid line) and $I_{-+}(\omega)$ (dashed line). c) Extracted SPIDER signal. d) Retrieved spectrum and spectral phase. e) Retrieved time intensity.

The resulting phase function is then:

$$\Delta\phi(\omega) = \phi_{\text{hSPIDER},2}(\omega) - \phi_{\text{hSPIDER},1}(\omega) = \varphi(\omega + \omega_2) - \varphi(\omega + \omega_1) \tag{45}$$

which, by concatenation and global shift, yields eventually the spectral phase $\varphi(\omega)$.
As for cSPIDER, it is not possible to generate an isolated upshifted pulse replica by using a single beam and a single polarization state. However, it is possible to extract the relevant

signal by phase-cycling. For hSPIDER, the relevant combination of pulses is the following: two time-delayed pulse replicas and *one* time-delayed quasi-monochromatic pulse. This narrowband pulse is synchronized with one of the pulse replicas and doesn't overlap with the other pulse replica. The total electric field then reads:

$$E_{\pm 1}(t) = \left\{E(t) + E(t-\tau) \pm M_1(t-\tau)e^{-i\delta\omega_1 t}\right\} e^{i\omega_0 t} \tag{46}$$

In the time domain, the SHG signal induced by $E_{\pm,1}(t)$ is proportional to the square of $E_{\pm,1}(t)$. Since the pulses are much shorter than the interpulse delay τ, the cross-product between the first pulse replica and the narrowband pulse vanishes, which leads to the following expression:

$$E^2_{\pm 1}(t) = \left\{E^2(t) + M_1^2(t-\tau)e^{-2i\delta\omega_1 t} + E^2(t-\tau) \pm 2E(t-\tau)M_1(t-\tau)e^{-i\delta\omega_1 t}\right\} e^{i2\omega_0 t} \tag{47}$$

The SH spectral intensity is then equal to the square modulus of Equation 47 which can be written as follows:

$$I_{\pm 1}(2\omega_0 + \Omega) = \begin{cases} \left|\widehat{E^2}(\Omega) \pm 2\widehat{E}(\Omega) \otimes \widehat{M_1}(\Omega - \delta\omega_1)e^{-i\Omega\tau}\right|^2 \\ +\left|\widehat{M_1^2}(\Omega - 2\delta\omega_1)e^{-i\Omega\tau} + \widehat{E^2}(\Omega)e^{-i\Omega\tau}\right|^2 \\ +2\mathrm{Re}\left\{\left(\widehat{E^2}(\Omega) \pm 2\widehat{E}(\Omega) \otimes \widehat{M_1}(\Omega - 2\delta\omega_1)e^{-i\Omega\tau}\right)\right. \\ \left.\times\left(\widehat{M_1^2}(\Omega - \delta\omega_1) + \widehat{E^2}(\Omega)\right)^\star e^{i\Omega\tau}\right\} \end{cases} \tag{48}$$

In Equation 48, the first term is the homodyne SPIDER term, the second term is the SHG parasitic background, and the third term is the cross-product of these two contributions. By summing I_{+1} and I_{-1}, the SHG parasitic background can be eliminated:

$$\begin{aligned} I_{+1} - I_{-1} = {} & 8\mathrm{Re}\left\{\widehat{E^2}^\star(\Omega) \times \left[\widehat{E}(\Omega) \otimes \widehat{M_1}(\Omega - \delta\omega_1)e^{-i\Omega\tau}\right]\right\} \\ & + 8\mathrm{Re}\left\{\widehat{E}(\Omega) \otimes \widehat{M_1}(\Omega - \delta\omega_1)\left(\widehat{M_1^2}(\Omega - 2\delta\omega_1) + \widehat{E^2}(\Omega)\right)^\star\right\} \end{aligned} \tag{49}$$

The combination still contains two terms however: the homodyne SPIDER term, which is an oscillating term with respect to Ω, and a DC term. A Fourier filter around the beat "frequency" τ is, therefore, able to isolate the homodyne term. By using FTSI, it is possible to retrieve, assuming that the narrowband pulse is of negligible width:

$$\phi_{\mathrm{hSPIDER},1}(\omega) = -\Omega\tau + \varphi(\omega + \omega_1) - \varphi_{\mathrm{SHG}}(\omega) \tag{50}$$

By repeating this measurement with another frequency shear, that is with the pulse combinations

$$E_{\pm,2}(t) = \left\{E(t) + E(t-\tau) \pm M_2(t-\tau)e^{-i\delta\omega_2 t}\right\} e^{i\omega_0 t} \tag{51}$$

another homodyne signal can be extracted:

$$\phi_{\mathrm{hSPIDER},2}(\omega) = -\Omega\tau + \varphi(\omega + \omega_2) - \varphi_{\mathrm{SHG}}(\omega) \tag{52}$$

By combining $\phi_{hSPIDER,1}$ and $\phi_{hSPIDER,2}$, the derivative of the spectral phase $\varphi(\omega)$ can be retrieved.

4.4.2 Experimental demonstration

The experimental demonstration of hSPIDER was also carried on the same visible source as described in subsubsection 4.1.2. The interpulse delay was set to 600fs and the spectral shear to 2nm at 600nm. Figure 7 shows the raw acquired spectra together with the extracted signals.

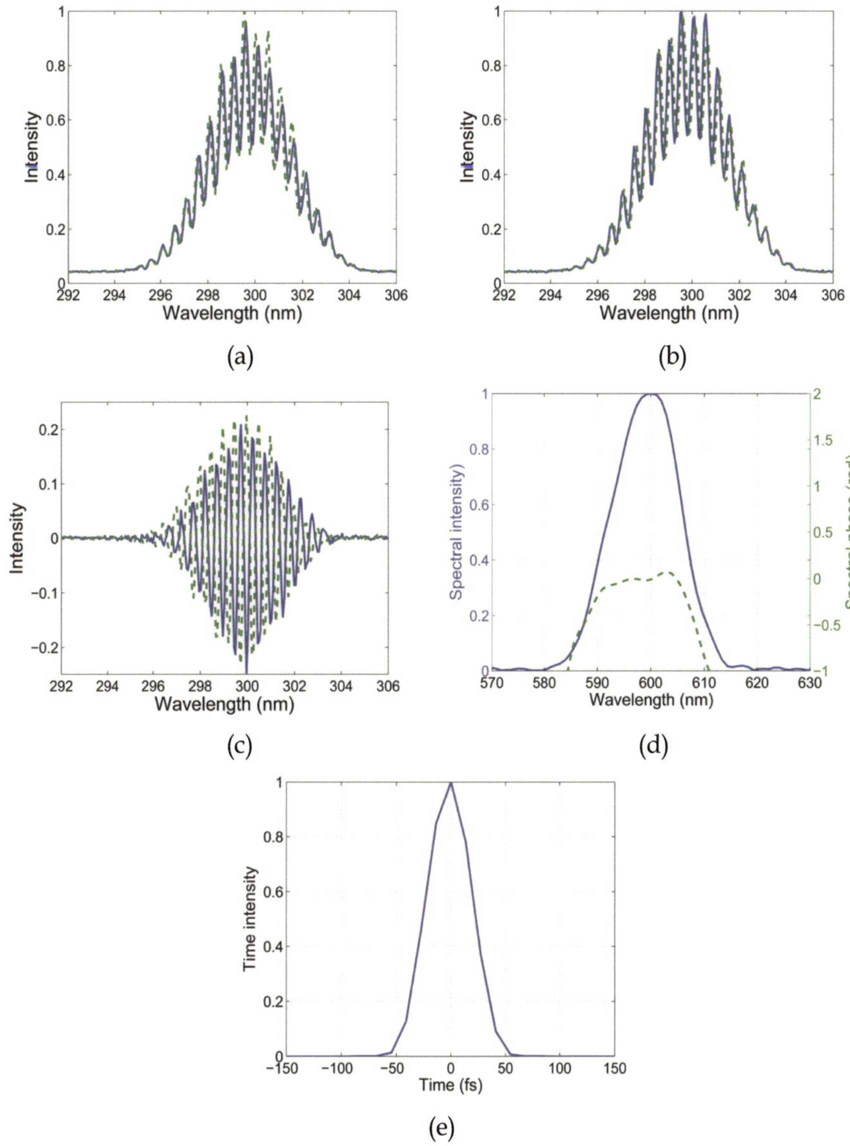

Fig. 7. hSPIDER experimental results. a) $I_{+1}(\omega)$ (solid line) and $I_{-1}(\omega)$ (dashed line). b) $I_{+2}(\omega)$ (solid line) and $I_{-2}(\omega)$ (dashed line). c) Extracted homodyne signals. d) Retrieved spectral phase and intensity. e) Retrieved time intensity.

5. Cross-check

As it was stressed in the introduction, pulse shapers offer the unique possibility to crosscheck between several pulse measurements. All the experiments reported in section 4 have been performed with the same OPA source and can be therefore be compared. The retrieved time intensity from iFROG, bFROG, cSPIDER and hSPIDER measurements are compared in Figure 8. As can be seen, all four pulse measurements agree on the pulse duration as well as on the time shape. On the other hand, Figure 9 is an example of non-consistent measurements. The experiment was performed at ~800nm on a kHz amplified system delivering pulses of ~25fs. A fraction of the output beam was sampled and sent to a thin Calcite plate and then to a Glan polarizer. The Calcite crystal was orientated so that the angle between the input polarization and the principal polarization plane was close to 45 degrees. The group delay between the ordinary and extraordinary polarizations was independently measured to be ~80fs. The polarizer was set to transmit the initial polarization, so that, after propagation through the Calcite plate and the polarizer, a double pulse structure was generated. This double pulse was characterized by both FROG and SPIDER measurements and the result of these measurements is displayed in Figure 9. As can be noted, both measurements retrieve this double structure with the correct interpulse delay but the SPIDER measurement shows a spurious post pulse, whereas the FROG measurement doesn't. This was the result of a bad choice of parameters for the SPIDER measurement: the chosen interpulse delay (300fs) was too small to characterize a 100fs pulse structure.

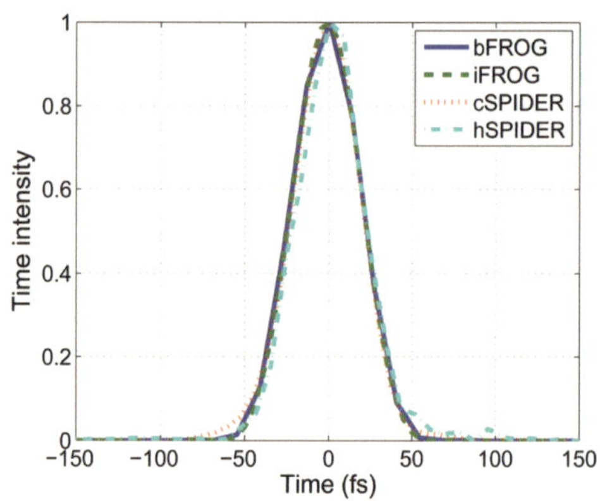

Fig. 8. Compared time intensities retrieved by bFROG, iFROG, cSPIDER and hSPIDER.

6. Extension to the UV

Ultrashort pulse characterization techniques based on second harmonic generation (SHG), such as frequency resolved optical gating (SH-FROG), were developed for the near infrared and cannot be easily transposed to the UV and the visible range because of the lack of suitable and efficient non linear crystals for broadband SHG. Such difficulties can be overcome by the use of a third order non linear effect such as induced birefringence which is achromatic and frequency conserving. However, third order non linear effects usually require non collinear geometry in order to allow the extraction of the useful signal, which greatly increases the complexity of the system. Conversely, a single-beam geometry would greatly simplify the optical setup but comes at the expense of an interferometric control of the delay between pulse replicas and signal discrimination.

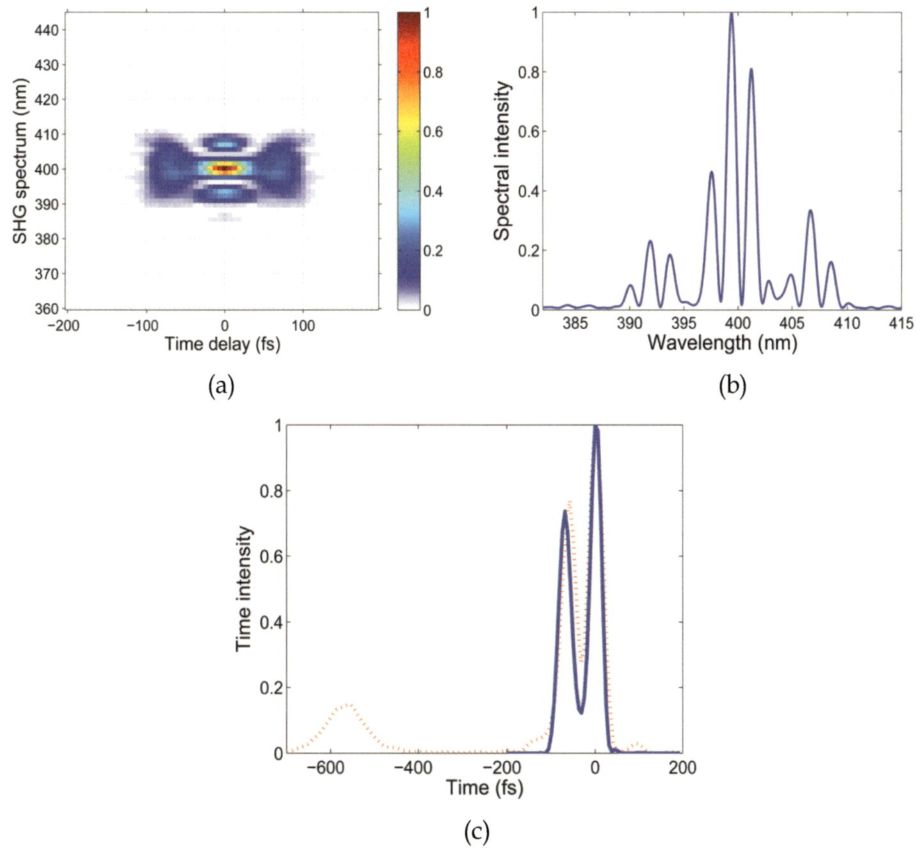

Fig. 9. a) Experimental FROG traces. b) Experimental SPIDER spectrum. c) Compared retrieved time intensities (FROG: solid line. SPIDER: dots).

In this section we demonstrate a single beam, achromatic, programmable and self-compensated spectrally resolved interferometric autocorrelation technique which overcomes such drawbacks. This technique is based on the conjugate use of UV acousto-optic programmable dispersive filter (UV-AOPDF, Coudreau et al. (2006)) and of a recently discovered third order non linear effect, cross-polarized wave (XPW) generation (Canova et al., 2008; Jullien et al., 2005).

6.1 Cross-polarized wave generation

Under a strong illumination by a linearly polarized wave, a nonlinear crystal with an anisotropic cubic susceptibility can generate a cross-polarized wave at the same wavelength as the incident wave. The intensity of the generated XPW signal varies as the cubic power of the intensity of the input wave. Barium fluoride (BaF2) is especially well adapted to XPW generation. First, its high $\chi^{(3)}$ anisotropy allows for good conversion efficiency. Second, its linear optical properties allow for perfect group-velocity matching along the propagating axis and, as a consequence, for a perfect spatial overlap of the cross-polarized waves. Assuming that two replicas of an ultrashort optical pulse $E(t)$ with an interpulse delay τ are incident on the non linear crystal, the XPW intensity writes:

$$I_{XPW}(\tau) \propto \int \left(E(t) + E(t-\tau)\right)^3 dt \tag{53}$$

When spectrally resolved, this signal contains, in theory, all the information required to retrieve the spectral phase and amplitude of $E(t)$.

6.2 Setup and proof of principle

The UV-AOPDF (75mm-long KDP crystal) is set to produce a pair of replicas of the input short pulse $E(t)$ and to vary the interpulse delay while maintaining a constant carrier envelope phase. The diffracted beams are focused into a 3mm thick BaF_2 crystal by a first lens and recollimated by a second lens. A Glan polarizer is used to isolate the XPW signal and a spectrometer records the XPW spectrum (Figure 10). The AOPDF is also used to precompensate the optical dispersion of all optical elements: lenses, KDP and BaF_2 crystals.

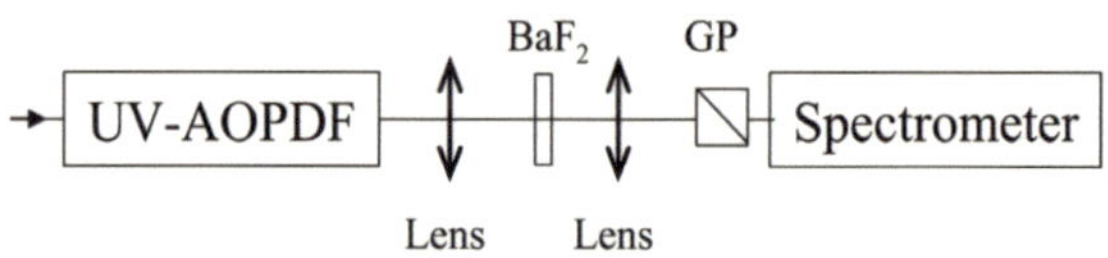

Fig. 10. Experimental setup (GP: Glan polarizer).

First experiments were performed on the SHG output of an amplified Ti:Sapphire system delivering Fourier-transform limited pulses of 40fs at 397nm. The bFROG trace is shown in Figure 11a along with the derived autocorrelation function and shows a good agreement with the expected pulse width. The experimental bFROG trace of a double pulse structure

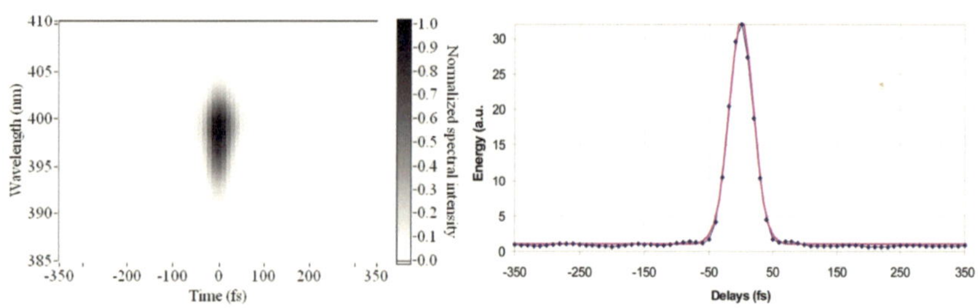

(a) Single pulse.

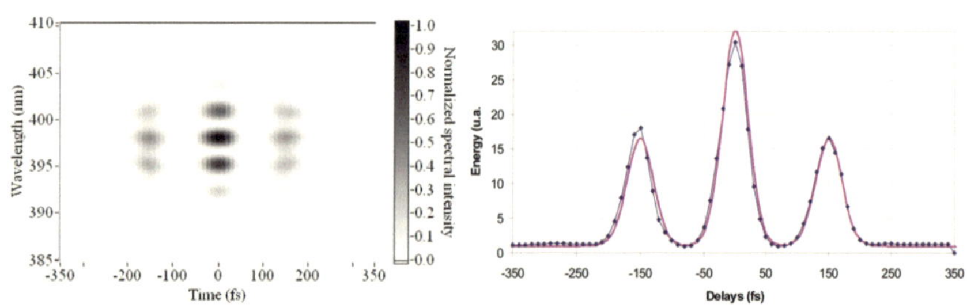

(b) Doublepulse

Fig. 11. a) Experimental bFROG trace (left) and autocorrelation (right) of the 40fs UV pulses. b) Experimental bFROG trace (left) and autocorrelation (right) of a double UV pulse with interpulse delay of 150fs. Dots: experimental. Solid line: theory.

with a pulse separation of 150fs is shown in Figure 11b. The bFROG trace exhibits the double pulse structure as well as the expected spectral modulations with excellent agreement with the theoretical interpulse delay. The XPW conversion efficiency was not measured exactly, but was estimated to be of the order of 1%. Given the extinction ratio of the polarizer (10^4), the signal to background ratio is estimated to better than 100.

7. Conclusion

In this chapter we explored the practical implementation of several pulse characterization techniques on a minimal optical setup comprising an AOPDF, an SHG stage and a spectrometer. Although special interest was brought to SPIDER and SH-FROG related techniques, we demonstrated that two-filter techniques can be made single-filter by slight adaptations. We introduced new pulse measurement techniques (bFROG, cSPIDER, hSPIDER), provided experimental demonstrations of these methods and performed a mutual cross-check. Finally, we showed how some of these techniques could be extended to the near UV spectral range by using a third order nonlinear effect.

8. References

Albrecht, A. W., Hybl, J. D., Faeder, S. M. G. & Jonas, D. M. (1999). Experimental distinction between phase shifts and time delays: Implications for femtosecond spectroscopy and coherent control of chemical reactions, *Journal Of Chemical Physics* 111(24): 10934- 10956.

Amat-Roldan, I., Cormack, I. G., Loza-Alvarez, P. & Artigas, D. (2005). Measurement of electric field by interferometric spectral trace observation, *Optics Letters* 30(9): 1063–1065.

Amat-Roldan, I., Cormack, I., Loza-Alvarez, P., Gualda, E. & Artigas, D. (2004). Ultrashort pulse characterisation with shg collinear-frog, *Opt. Express* 12(6): 1169–1178. URL: *http://www.opticsexpress.org/abstract.cfm?URI=oe-12-6-1169*

Canova, L., Kourtev, S., Minkovski, N., Lopez-Martens, R., Albert, O. & Saltiel, S. M. (2008). Cross-polarized wave generation in the uv region, *Opt. Lett.* 33(20): 2299–2301. URL: *http://ol.osa.org/abstract.cfm?URI=ol-33-20-2299*

Coudreau, S., Kaplan, D. & Tournois, P. (2006). Ultraviolet acousto-optic programmable dispersive filter laser pulse shaping in kdp, *Opt. Lett.* 31(12): 1899–1901. URL: *http://ol.osa.org/abstract.cfm?URI=ol-31-12-1899*

DeLong, K. W., Fittinghoff, D. N., Trebino, R., Kohler, B. & Wilson, K. (1994). Pulse retrieval in frequency-resolved optical gating based on the method of generalized projections, *Opt. Lett.* 19(24): 2152–2154. URL: *http://ol.osa.org/abstract.cfm?URI=ol-19-24-2152*

Fittinghoff, D., Squier, J., Barty, C., Sweetser, J., Trebino, R. & Muller, M. (1998). Collinear type ii second-harmonic-generation frequency-resolved optical gating for use with high-numerical-aperture objectives, *Optics letters* 23(13): 1046–1048.

Forget, N., Coudreau, S., Lepetit, F., Albert, O. & Oksenhendler, T. (2007). Achromatic and single-beam pulse characterization technique for visible-uv pulses based on direct uv pulse shaping and cross-polarized wave generation, *CLEO/Europe and IQEC 2007 Conference Digest*, Optical Society of America, pp. CF4–5.

Forget, N., Joffre, M., Coudreau, S. & Oksenhendler, T. (2007). Toward programmable ultrashort pulse characterization, *CLEO/Europe and IQEC 2007 Conference Digest*, Optical Society of America, pp. CF–16.

Froehly, C., Colombeau, B. & Vampouille, M. (1983). Shaping and analysis of picosecond light pulses, *Progress in optics* 20: 63–153.

Galler, A. & Feurer, T. (2008). Pulse shaper assisted short laser pulse characterization, *Applied Physics B: Lasers and Optics* 90(3): 427–430.

Gallmann, L., Steinmeyer, G., Sutter, D. H., Matuschek, N. & Keller, U. (2000). Collinear type ii second-harmonic-generation frequency-resolved optical gating for the characterization of sub-10-fs optical pulses, *Opt. Lett.* 25(4): 269–271.
URL: *http://ol.osa.org/abstract.cfm?URI=ol-25-4-269*

Garzella, D., Lepetit, F., Gobert, O., Perdrix, M., Hollander, P. & Oksenhendler, T. (2006). Temporal analysis ans shape control of uv high energy laser pulses for photoinjector, *Proc. FEL 2006*, number THPPH001, pp. 552–555.

Grabielle, S., Forget, N., Coudreau, S., Oksenhendler, T., Kaplan, D., Hergott, J.-F. & Gobert, O. (2009). Local spectral compression method for cpa lasers, *CLEO/Europe and EQEC 2009Conference Digest*.

Iaconis, C. & Walmsley, I. (1998). Spectral phase interferometry for direct electric-field reconstruction of ultrashort optical pulses, *Opt. Lett.* 23(10): 792–794.
URL: *http://ol.osa.org/abstract.cfm?URI=ol-23-10-792*

Jullien, A., Albert, O., Ch´eriaux, G., Etchepare, J., Kourtev, S., Minkovski, N. & Saltiel, S. M. (2005). Nonlinear polarization rotation of elliptical light in cubic crystals, with application to cross-polarized wave generation, *J. Opt. Soc. Am. B* 22(12): 2635–2641.
URL: *http://josab.osa.org/abstract.cfm?URI=josab-22-12-2635*

Kane, D., Inc, S. & Santa Fe, N. (1999). Recent progress toward real-time measurement of ultrashort laserpulses, *IEEE Journal of Quantum Electronics* 35(4): 421–431.

Lozovoy, V., Pastirk, I. & Dantus, M. (2004). Multiphoton intrapulse interference. iv. Ultrashort laser pulse spectral phase characterization and compensation, *Optics letters* 29(7): 775–777.

Monmayrant, A., Joffre, M., Oksenhendler, T., Herzog, R., Kaplan, D. & Tournois, P. (2003). Time-domain interferometry for direct electric-field reconstruction by use of an acousto-optic programmable filter and a two-photon detector, *Opt. Lett.* 28(4): 278–280.
URL: *http://ol.osa.org/abstract.cfm?URI=ol-28-4-278*

Murphy, D. S., McKenna, J., Calvert, C. R., Bryan, W. A., English, E. M. L., Wood, J., Turcu, I. C. E., Newell, W. R., Williams, I. D. & McCann, J. F. (2007). Controlling dissociation processes in the d-2(+) molecular ion using high-intensity, ultrashort laser pulses, *JOURNAL OF PHYSICS B-ATOMIC MOLECULAR AND OPTICAL PHYSICS* 40(11): S359–S372.

Ogilvie, J. P., Débarre, D., Solinas, X., Martin, J.-L., Beaurepaire, E. & Joffre, M. (2006). Use of coherent control for selective two-photon fluorescence microscopy in live organisms, *Opt. Express* 14(2): 759–766.
URL: *http://www.opticsexpress.org/abstract.cfm?URI=oe-14-2-759*

Ohno, K., Tanabe, T. & Kannari, F. (2002). Adaptive pulse shaping of phase and amplitude of an amplified femtosecond pulse laser by direct reference to frequency-resolved optical gating traces, *JOSAB* 19(11): 2781–2790.

Oksenhendler, T., Monmayrant, A., Herzog, R., Kaplan, D., Tournois, P. & Joffre, M. (2003). Time-domain interferometry for direct electric-field reconstruction by use of an acousto-optic programmable dispersive filter, *in* D. Miller, M. Murnane, N. Scherer & A. Weiner (eds), *ULTRAFAST PHENOMENA XIII*, Vol. 71 of *SPRINGER SERIES IN CHEMICAL PHYSICS*, pp. 220–222. 13th International Conference on Ultrafast Phenomena, VANCOUVER, CANADA, MAY 12-17, 2002.

Reid, D. (1999). Algorithm for complete and rapid retrieval of ultrashort pulseamplitude and phase from a sonogram, *IEEE Journal of Quantum Electronics* 35(11): 1584–1589.

Seres, E., Herzog, R., Seres, J., Kaplan, D. & Spielmann, C. (2003). Generation of intense 8 fs laser pulses, *Opt. Express* 11(3): 240–247.

Stibenz, G., Ropers, C., Lienau, C., Warmuth, C., Wyatt, A. S., Walmsley, I. A. & Steinmeyer, G. (2006). Advanced methods for the characterization of few-cycle light pulses: a comparison, *Applied Physics B-Lasers And Optics* 83(4): 511–519.

Stibenz, G. & Steinmeyer, G. (2005). Interferometric frequency-resolved optical gating, *Optics Express* 13(7): 2617–2626.

Sung, J., Chen, B.-C. & Lim, S.-H. (2008). Single-beam homodyne spider for multiphoton microscopy, *Opt. Lett.* 33(13): 1404–1406.
URL: *http://ol.osa.org/abstract.cfm?URI=ol-33-13-1404*

Tkaczyk, E. R., Mauring, K., Tkaczyk, A. H., Krasnenko, V., Ye, J. Y., Jr., J. R. B. & Norris, T. B. (2008). Control of the blue fluorescent protein with advanced evolutionary pulse shaping, *Biochemical and Biophysical Research Communications* 376(4): 733 - 737.
URL: *http://www.sciencedirect.com/science/article/B6WBK-4THJ4HB-2/2/478cac9116090743c2a2ea82be5b4970*

Tournois, P. (1997). Acousto-optic programmable dispersive filter for adaptive compensation of group delay time dispersion in laser systems, *Opt. Commun.* 140(4-6): 245–249.

Trebino, R., DeLong, K. W., Fittinghoff, D. N., Sweetser, J. N., Krumbugel, M. A., Richman, B. A. & Kane, D. J. (1997). Measuring ultrashort laser pulses in the timefrequency domain using frequency-resolved optical gating, *Review Of Scientific Instruments* 68(9): 3277–3295.

Verluise, F., Laude, V., Cheng, Z., Spielmann, C. & Tournois, P. (2000). Amplitude and phase control of ultrashort pulses by use of an acousto-optic programmable dispersive filter: pulse compression and shaping, *Opt. Lett.* 25(8): 575–577.

Veshapidze, G., Trachy, M., Jang, H., Fehrenbach, C. & DePaola, B. (2007). Pathway for twocolor photoassociative ionization with ultrafast optical pulses in a rb magneto-optical trap, *Physical Review A* 76(5): 51401.

von Vacano, B., Buckup, T. & Motzkus, M. (2006). In situ broadband pulse compression for multiphoton microscopy using a shaper-assisted collinear spider, *Optics Letters* 31(8): 1154–1156.

von Vacano, B., Buckup, T. & Motzkus, M. (2007). Shaper-assisted collinear spider: fast and simple broadband pulse compression in nonlinear microscopy, *J. Opt. Soc. Am. B* 24(5): 1091–1100.
URL: *http://josab.osa.org/abstract.cfm?URI=josab-24-5-1091*

Walmsley, I. & Dorrer, C. (2009). Characterization of ultrashort electromagnetic pulses, *Advances in Optics and Photonics* 1(2): 308–437.

Weiner, A. (2000). Femtosecond pulse shaping using spatial light modulators, *Review of Scientific Instruments* 71: 1929.

Wong, V. &Walmsley, I. A. (1995). Linear filter analysis of methods for ultrashort-pulse-shape measurements, *Journal Of The Optical Society Of America B-Optical Physics* 12(8): 1491- 1499.

Yamada, H., Yokoyama, K., Teranishi, Y., Sugita, A., Shirai, T., Aoyama, M., Akahane, Y., Inoue, N., Ueda, H., Yamakawa, K., Yokoyama, A., Kawasaki, M. & Nakamura, H. (2005). Selective transition to the closely-lying states cs(7d(3/2)) and cs(7d(5/2)) by femtosecond laser pulses, *Phys.Rev.A* 72(6): 000.

16

Pulse-Shaping Techniques Theory and Experimental Implementations for Femtosecond Pulses

T. Oksenhendler and N. Forget
Fastlite, Centre scientifique d'Orsay - Bât 503
Plateau du Moulon - BP45, 91401 ORSAY
France

1. Introduction

Femtosecond pulses are used in many fields due to their specificities of extreme short duration, ultra high peak power or large spectral bandwidth.

Since the early days of the laser in the 60s, there has been a continous quest to generate shorter and or higher peak power pulses.

Reliable generation of pulses below 100fs occurred the first time in 1981 with the invention of the colliding pulse modelocked (CPM) ring dye laser (Fork R.L. and al., 1981). Despite relative low energy per pulses, the ultrashort pulse duration leads to peak power large enough for non linear pulse compression culminating in pulses as short as 6fs in the visible. Recent advances in laser technology as the use of solid-state gain media, laser diode pumping, fiber laser, have led to simple, reliable, turn key ultrashort laser oscillators with pulse duration ranging form few ps down to 5fs.

Limitation to pulse energy in the range of a microJoule or less in the CPM laser has been overcomed by the Chirped Pulse Amplification (CPA) technique (Strickland D., Mourou G., (1985)). This technique is the optical transposition of a Radar technique developped during the second world war. The basic principle is to spread in time i.e to stretch the ultrashort pulse before amplification. Indeed limitation of the pulse amplification because of the damage threshold of the optics is mainly due to the pulse peak power. A stretch ratio of a million gives the ability to amplify the stretched pulse, without optical damage, by a factor of a million from less than a microJoule to more than a Joule per pulse. After amplification, recompression of the pulse is achieved by an optical set-up that has a very high damage threshold. To obtain the highest peak power, the pulse duration has to be "Fourier transform limited", ie its spectral phase is purely linear. The compensation of the chirp and higher spectral phase order is highly simplified by the ability to introduced an arbitrarly shaped spectral phase.

Application of these ultrashort pulses requires to control of optimize their temporal shape. Dispersion of materials and optical devices has been used to compress, stretch or replicate the pulses. Limitations on the ability to control the pulse temporal shape by classical optical devices have lead to the development of arbitrary pulse shapers. These devices are linear

filters enabling the independent control of the spectral amplitude and phase giving a complete control of the temporal shape of the pulse.
Due to the extreme short duration of the pulses, the temporal control cannot be achieved directly by temporal modulators. The control has to be done in the spectral domain. Two technologies of pulse shapers are widely used: spatial amplitude and phase modulators implemented in a zero-dispersion line or 4-f line, and acousto-optic programmable dispersive filters based on longitudinal Bragg acousto-optic diffraction.
Both techniques will be theoretically reviewed to point out limitations, advantages and drawbacks for femtosecond pulse shaping techniques. Starting from simulations of the two techniques, some specific examples of pulse shaping will be invastigated such as pulse compression, complex square pulse generation, pure linear chirp generation, double pulses with a controlled delay or focal point power density optimization. These examples will then be used to illustrate the limitations, advantages and drawbacks of each technology.
Experimental implementations of these pulse shaping examples will then be briefly presented.
In the last section, we discuss some special topics, Carrier Enveloppe Phase control and indirect pulse shaping.

2. Ultrashort pulses characteristics

The extreme shortness of ultrashort pulses implies a large spectral bandwidth. The temporal and spectral electric fields are dual i.e. Fourier transform. Thus the electric field can be determined either by the temporal phase and amplitude or by the spectral phase and amplitude.
The electric field is a real quantity that can be decomposed as:

$$\mathcal{E}(t)=\frac{1}{2}\left[E(t)+E^{*}(t)\right] \tag{1}$$

with E(t) the complex electric field whose corresponds to positive frequency. This field can be expressed as:

$$E(t) = A(t)\, e^{-i\omega_0 t + i\varphi(t)} \tag{2}$$

where ω_0 is the central pulsation, A(t) its enveloppe and $\varphi(t)$ its temporal phase.
Its Fourier transform corresponds to the spectral components:

$$E(\omega) = A(\omega)\, e^{i\phi(\omega)} = TF\left[E(t)\right]_{\omega} \tag{3}$$

where $A(\omega)$ is the spectral amplitude, and $\phi(\omega)$ the spectral phase.
The spectrum $I(\omega)$, or spectral power density, is the square modulus of the spectral amplitude. Its temporal equivalent, the temporal intensity I(t) equals the square modulus of the temporal amplitude A(t).
One must be aware that the temporal amplitude A(t) depends upon both the spectral amplitude and phase.
We assume a normalized field:

$$\int_{-\infty}^{+\infty} |E(\omega)|^2 \frac{d\omega}{2\pi} = \int_{-\infty}^{+\infty} |E(t)|^2\, dt = 1 \tag{4}$$

The pulse center is then defined by:

$$t_0 = \langle t \rangle = \int_{-\infty}^{+\infty} t\,|E(t)|^2\, dt \tag{5}$$

The pulse duration by:

$$\Delta t = \sqrt{\langle t^2 \rangle - \langle t \rangle^2} = \sqrt{\langle (t - t_0)^2 \rangle} \tag{6}$$

The central frequency by:

$$\omega_0 = \langle \omega \rangle = \int_{-\infty}^{+\infty} \omega\,|E(\omega)|^2 \frac{d\omega}{2\pi} \tag{7}$$

The spectral width or bandwidth by:

$$\Delta\omega = \sqrt{\langle \omega^2 \rangle - \langle \omega \rangle^2} = \sqrt{\langle (\omega - \omega_0)^2 \rangle} \tag{8}$$

Duration and bandwidth are related by:

$$\Delta\omega\Delta t \geq \frac{1}{2} \tag{9}$$

The minimum is obtained for a pure linear spectral phase. As shown by the relations between spectral phase and duration:

$$\langle t \rangle = \int_{-\infty}^{+\infty} t|E(t)|^2\, dt = \left\langle \frac{d\phi}{d\omega} \right\rangle, \tag{10}$$

$$\tau_g = \frac{d\phi}{d\omega}, \tag{11}$$

$$\langle t \rangle^2 = \Delta t_{\varphi=0}{}^2 + \Delta\tau_g{}^2 \text{ with } \Delta\tau_g{}^2 = \left\langle \left(\tau_g - \langle \tau_g \rangle\right)^2 \right\rangle = \left\langle \left(\frac{d\phi}{d\omega} - \left\langle \frac{d\phi}{d\omega} \right\rangle \right)^2 \right\rangle. \tag{12}$$

For $|E(\omega)|$ fixed, the pulse is shortest when τ_g is independent of the frequency ($\Delta\tau_g=0$). The pulse is said to be a Fourier transform pulse or Fourier transform limited.

By opposition when τ_g changes linearly with frequency, the pulse is said to be chirped (as when a bird sings).

To analyse the different effects of the spectral phase, it is used to expand the spectral phase into a Taylor series:

$$\phi(\omega) = \phi(\omega_0) + \phi^{(1)}(\omega_0)(\omega - \omega_0) + \frac{\phi^{(2)}(\omega_0)}{2!}(\omega - \omega_0)^2 + \frac{\phi^{(3)}(\omega_0)}{3!}(\omega - \omega_0)^3 \text{ with } \phi^{(k)}(\omega_0) = \left.\frac{\partial^k \phi(\omega)}{\partial \omega^k}\right|_{\omega_0} \tag{13}$$

The first order spectral phase term corresponds to a delay, the second order (also named chirp) spreads linearly in time the frequency and so stretches the pulse. The third order introduces pre-pulses or post-pulses around the main pulse.
Different set-ups combining gratings, prisms or grisms and optics can modify the spectral phase but not arbitrarily. The different orders cannot be set independently.
The temporal intensity is modified by changing the spectral phase only, but its complete control requires shaping both the spectral phase and amplitude. This kind of control is expressed by a linear filtering.

3. Linear filtering

The femtosecond pulse shaping approach described in this article is based on the linear, time-invariant filter, a concept well-known in electrical engineering. Here we apply linear filtering to generate specially shaped optical waveforms on the picosecond and femtosecond time scale.
Linear filtering can be described in either the time domain or frequency domain, as depicted in figure 1. In the time domain, the filter is characterized by an impulse response function h(t). The output of the filter $E_{out}(t)$ in response to an input pulse $E_{in}(t)$ is given by the convolution of $E_{in}(t)$ and h(t)

$$E_{out}(t) = E_{in}(t) \otimes h(t) = \int E_{in}(t')h(t-t')dt' \tag{14}$$

where ⊗ denotes convolution. If the input is a delta function, the output is simply h(t). Therefore, for a sufficiently short input pulse, the problem of generating a specific output pulse shape is equivalent to the task of fabricating a linear filter with the desired impulse response. Note that instead of the term "impulse response function", which is common in electrical engineering, h(t) may also be called a Green function, which is a common terminology in other fields.

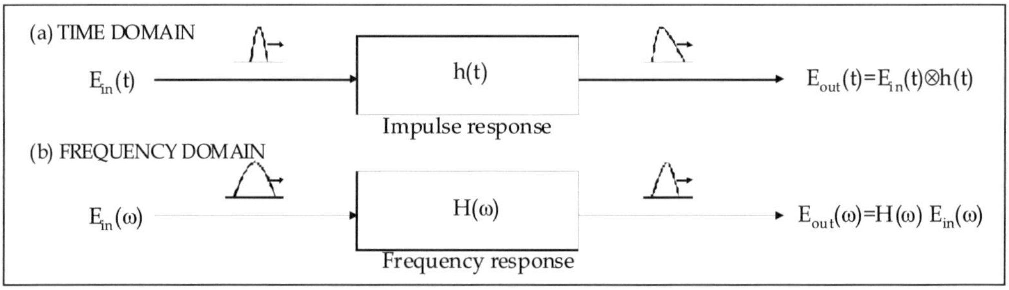

Fig. 1. Pulse shaping by linear filtering. (a) Time domain view. (b) Frequency domain view

In the frequency domain, the filter is characterized by its frequency response H(ω). The output of the linear filter $E_{out}(\omega)$ is the product of the input signal $E_{in}(\omega)$ and the frequency response H(ω) –i.e.,

$$E_{out}(\omega) = E_{in}(\omega)H(\omega) \tag{15}$$

Here $E_{in}(t)$, $E_{out}(t)$, h(t), and $E_{in}(\omega)$, $E_{out}(\omega)$, H(ω), respesctively, are Fourier transform pairs.

A linear filter can emulate any linear combination of pulses such as for examples two, three, four... pulses. Any optical system without frequency conversion or time-dependant component is indeed a linear filter (Michelson or Mach-Zender interferometers, bandpass filters, ...).
For a delta function input pulse $E_{in}(t)$, the input spectrum $E_{in}(\omega)$ is equal to unity, and the output spectrum is equal to the frequency response of the filter. Therefore, due to the Fourier transform relations, generation of a desired ouput waveform can be accomplished by implementing a filter with the required frequency response. As the time scale of the pulses (10fs-100fs) is shorter than any temporal modulator (>10ps), the pulse shaping approach is based on frequency domain and naturally is described by frequency domain point of view.

4. Theory of Pulse shaping techniques

The two main technologies of pulse shaping commonly used are the 4-f pulse shaper, or Fourier transform femtosecond pulse shaping, and the acousto-optic programmable dispersive filter (AOPDF). This part will review theoretically these two technologies and introduces their simulation models in order to determine the frequency response of these filters.

4.1 Femtosecond pulse shaping using spatial light modulators

4.1.1 Analytical analysis

The first use of the pulse shaping apparatus shown in Fig.2 was reported by Froehly, who performed pulse shaping experiments with 30ps input pulses [Froehly (1983)]. Related experiments demonstrating shaping of a few picoseconds pulses by spatial masking within a fiber and grating compressor were performed independently by Heritage and Weiner in 1985. In those experiments a grating pair was used in a dispersive configuration without internal lenses since grating dispersion was needed in order to compress the input pulses which were chirped through non linear propagation in the fiber. The dispersion-free apparatus in Fig.2 was subsequently adopted by Weiner et al. for manipulating pulses on the 100fs time scale, initially using fixed masks and later using programmable Spatial Light Modulators (SLM). The apparatus of Fig. 2 (without the mask) can also be used to introduce dispersion for pulse stretching or compression by changing the grating-lens spacing. This idea was introduced and analyzed by Martinez and is now extensively used for high-power femtosecond chirped pulse amplifier.
The waveform synthesis is achieved by spatial masking of the spatially dispersed optical frequency spectrum. Figure 2 shows the basic pulse shaping apparatus, which consists of a pair of diffraction gratings and lenses, arranged in a configuration known as a "zero dispersion pulse compressor", and a pulse shaping mask. The individual frequency components contained within the incident ultrashort pulse are angularly dispersed by the first diffraction grating, and then focused to small diffraction limited spots at the back focal plane of the first lens, where the frequency components are spatially separated along one dimension. Essentially the first lens performs a Fourier transform which converts the angular dispersion from the the grating to a spatial separation at the back focal plane. Spatially patterned amplitude and phase masks (or a SLM) are placed in this plane in order to manipulate the spatially dispersed optical Fourier components. After a second lens and grating recombine all the frequencies into a single collimated beam, a shaped output pulse is

obtained, with the output pulse shape given by the Fourier transform of the patterned transferred by the masks onto the spectrum.

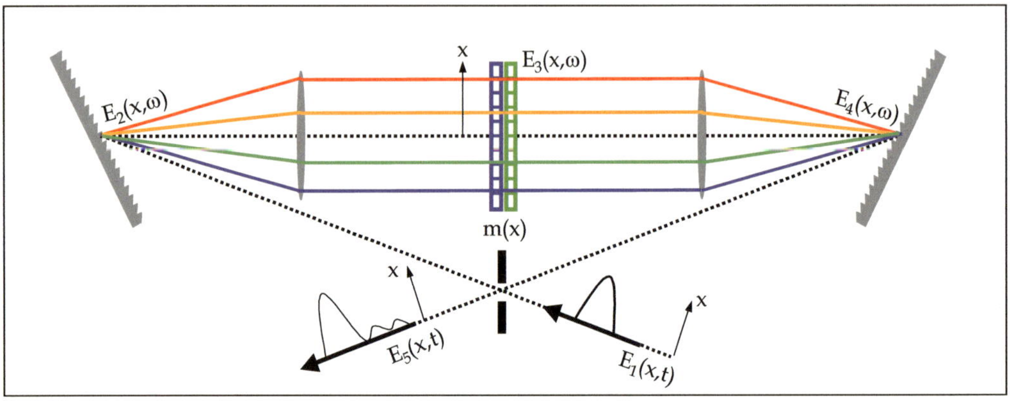

Fig. 2. Basic layout for Fourier transform femtosecond pulse shaping.

In order for this technique to work as desired, one requires that in the absence of a pulse shaping mask, the output pulse should be identical to the input pulse. Therefore, the grating and lens configuration must be truly free of dispersion. This can be guaranteed if the lenses are set up as a unit magnification telescope. In this case the first lens performs a spatial Fourier transform between the plane of the first grating and the masking plane, and the second lens performs a second Fourier transform from the masking plane to the plane of the second grating. The total effect of these two consecutive Fourier transforms is that the input pulse is unchanged in traveling through the system if no pulse shaping mask is present.

Note that this dispersion-free condition also depends on several approximations, e.g., that the lenses are thin and free of aberrations, that chromatic dispersion in passing through the lenses or other elements which may be inserted into the pulse shaper is small, and that the gratings have a flat spectral response. Many optimized designs have been proposed in the litterature to minimize optical aberrations [Monmayrant and Chatel (2003), Weiner(2000),…].

The optimization of the apparatus for a quantitative control requires precise analysis and simulation[Wefers and Nelson (1995), Vaughan and al (2006), Monmayrant (2005)]. In terms of the linear filter formalism, we wish to relate the linear filtering function H(ω) to the actual physical masking function with complex transmittance m(x). To do so, we must determine the relation between the spatial dimension x on the mask and the optical frequency ω. The input grating disperses the optical frequencies angularly:

$$\lambda = p\left(\sin\theta_i + \sin\theta_d\right) \tag{16}$$

where λ is the optical wavelength, p is the spacing between grating lines, and θ_i and θ_d are angles of incidence and diffraction, respectively. The first lens brings the diffracted rays from the first grating parallel. The lateral displacement x of a given frequency component λ from the center frequency component λ_0 immediately after the lens is given by

$$x(\lambda) = f\tan\left[\theta_d(\lambda) - \theta_d(\lambda_0)\right] \tag{17}$$

Expanding x as a power series in angular frequency ω gives

$$x(\omega)=f\left[\left.\frac{\partial\theta_d}{\partial\omega}\right|_{\omega=\omega_0}(\omega-\omega_0)+\frac{1}{2}\left.\frac{\partial^2\theta_d}{\partial\omega^2}\right|_{\omega=\omega_0}(\omega-\omega_0)^2+\ldots\right],\tag{18}$$

where

$$\left.\frac{\partial\theta_d}{\partial\omega}\right|_{\omega=\omega_0}=\frac{2\pi c}{\omega_0{}^2 p\cos\theta_d(\omega_0)}\text{ and }\left.\frac{\partial^2\theta_d}{\partial\omega^2}\right|_{\omega=\omega_0}=\frac{-4\pi c}{\omega_0{}^3 p\cos\theta_d(\omega_0)},\tag{19}$$

c is the speed of light, and ω_0 is the central carrier frequency of the input pulse.
Usually the second order term is neglected [except in Monmayrant thesis and Vaughan and al.] so that the frequency components are laterally dispersed linearly across the mask. However, for very broad bandwidth pulses (pulse with duration <20fs), or precise pulse shaping, this assumption may break down. Subtle second order dispersion effects have been noticed by Weiner and co-workers[Weiner (1988)], and Sauerbrey and co- workers[Vaughan (2006)].
It is assumed that the lateral dispersion of the lenses and gratings is such that the mask can accommodate the entire bandwidth of the input pulse. The "mask bandwidth" depends upon the width of the mask L, the focal length of the lens f, the line spacing of the grating p and the angle of diffraction $\theta_d(\omega_0)$:

$$\Delta\lambda_M=\arctan\left(\frac{L}{f}\right)p\cos\theta_d(\omega_0).\tag{21}$$

To avoid any significant cut, the "mask bandwidth" $\Delta\Omega_M$ has to be larger than the input pulse bandwidth $\Delta\omega$. We shall use as a criteria that $\Delta\Omega_M>3\Delta\omega$.
Considering an ideal mask, without pixelisation and other spurious effect, the space-time coupling used for the temporal or spectral shaping by a spatial mask has some incidence on the shaped pulse [Danailov (1989), Wefers (1995), Wefers (1996), Sussman (2008)]. The principal issue is that the spectral content - and hence time evolution - at each point within the output beam is not the same. Following the notations introduced on Fig.2 and by considering the input field without space-time coupling, the electric field incident upon the pulse shaping apparatus (immediately prior to the grating) is defined in the slowly varying envelope approximation as

$$E_1(x,t)=E_{in}(x)A(t)e^{-i\omega_0 t+i\varphi(t)}.\tag{22}$$

Following the results of Martinez [Martinez (1986)], the electric field immediately after the grating in frequency and position space is given by

$$E_2(x,\Omega)=\sqrt{\beta}E_{in}(\beta x)A(\Omega)e^{i\gamma\Omega x+i\phi(\Omega)}\tag{23}$$

with $\beta=\cos\theta_i/\cos\theta_d$, $\gamma=2\pi/\omega_0 p\cos\theta_d$, and $\Omega=\omega-\omega_0$, where $\phi(\Omega)=\phi(\omega)$ and θ_i and θ_d are the angles of incidence and diffraction respectively, and p the grating line spacing.
The electric field profile in the focal plane of the lens is given by the spatial Fourier transform of (23) with the substitution $k=2\pi x/\lambda_0 f$, where f is the focal length of the lens and λ_0 is the center wavelength of the input field. The electric field is then multiplied by the mask filter m(x) to give

$$E_3(x,\Omega)=\sqrt{2\pi/\beta\lambda_0 f}\tilde{E}_{in}(2\pi x/\beta\lambda_0 f+\gamma\Omega/\beta)A(\Omega)e^{i\phi(\Omega)}m(x) \qquad (24)$$

where $\tilde{E}_{in}(k)$ is the spatial fourier transform of $E_{in}(x)$.
To determine the electric field profile immediately before the second grating, a spatial Fourier transform of Eq.(24) is taken again with the substitution $k=2\pi x/\lambda_0 f$, giving

$$E_4(x,\Omega)=\left(\sqrt{2\pi\beta}/\lambda_0 f\right)\left[E_{in}(-\beta x)A(\Omega)e^{i\phi(\Omega)-i\gamma\Omega x}\otimes M(2\pi x/\lambda_0 f)\right] \qquad (25)$$

where $M(k)$ is the spatial Fourier transform of the mask pattern m(x) and $\otimes$ denotes a convolution.
Again following Martinez, the inverse transfer function of the second grating (which is anti-parallel to the first) gives the electric field profile after the grating as

$$E_5(x,\Omega)=\left(\sqrt{2\pi}/\beta\lambda_0 f\right)\left[E_{in}(-x)A(\Omega)e^{i\phi(\Omega)-i\gamma\Omega x/\beta}\otimes M(2\pi x/\beta\lambda_0 f)\right] \qquad (26)$$

Taking the spatial Fourier transforms of (26) yields the electric field profile of the output waveform in the spatial frequency domain

$$\tilde{E}_{out}(k,\Omega)=\tilde{E}_5(k,\Omega)=\tilde{E}_{in}(-k,\Omega)m(\lambda_0 f(\gamma\Omega+\beta k))=\tilde{E}_{in}(-k)A(\Omega)e^{i\phi(\Omega)}m(\lambda_0 f(\gamma\Omega+\beta k)). \qquad (27)$$

In space and time it is expressed as a convolution

$$E_{out}(x,t)=\left(\sqrt{2\pi}/\gamma\lambda f\right)e^{i\omega_0 t}\int E_{in}(-(x+t\beta/\gamma),t-t')M(-2\pi t'/\gamma\lambda f)dt'. \qquad (28)$$

The space-time coupling appears as a coupling between the spatial and spectral frequencies onto the mask. If the mask does not modify the beam, it cancels out. But if the mask introduces a modulation then the output pulse will be modified both on its spectral and spatial dimensions. Due to this coupling, no simple expression of the pulse shaper response function H(ω) can be given without the strong hypothesis that this effect is negligeable.
To illustrate this effect, we will consider a pure delay, and a quadratic phase sweep to compensate for an initial chirp of the input pulse.
For a pure delay, the spectral phase is linear and the mask is given by

$$m(\omega)=e^{-i\omega\tau}. \qquad (29)$$

Applying eq. (27) with this mask and an inverse spatial Fourier transform yields the output electric field

$$E_{out}(x,\Omega)=E_5(x,\Omega)=E_{in}(x+(\beta/\gamma)\tau)A(\Omega)e^{i\phi(\Omega)-i\omega\tau}. \qquad (30)$$

The output beam is spatially shifted and this shift is proportionnal to the applied delay. Quantitavely, the slope of this time-dependent lateral shift is given by

$$v=\partial x/\partial t=-\beta/\gamma=\frac{-cp\cos\theta_i}{\lambda}, \qquad (31)$$

Which for typical parameters (p=1000-line/mm gratings, λ=800nm) is ≈0.2mm/ps. Equation (31) shows that this slope depends only on the angular dispersion produced by the grating.

However, the effect of this lateral shift is measured relative to the spot size of the unshaped incident pulse. Spatially large input pulses reduce the effect of space time coupling but also reduce the spot size on the mask.
We now consider a mask pattern consisting of a quadratic phase sweep

$$m(\omega) = e^{-\frac{i}{2}\phi^{(2)}\Omega^2}. \tag{32}$$

This quadratic spectral phase sweep produces a "chirped" pulse with a temporally broadened envelope and an instantaneous carrier frequency that varies linearly with time under that envelope. The delay associated with each spectral components varies linearly ($\tau(\Omega)=\phi^{(2)}\Omega$). So from Eq.(30), by replacing τ by $\tau(\Omega)$, the spatial dependance becomes coupled with the optical frequency. Exact calculations have been done by Wefers[1996] and Monmayrant [2005]. These analyses point out a complex spatio-temporal coupling modifying the beam divergence and even the compression of the initial pulse. Supposing that the initial pulse has gaussian shapes in space and spectral amplitude, and is "chirped" as

$$E_{in}(x,\Omega) = E_{in}(x)A(\Omega)e^{i\frac{\phi_{in}^{(2)}}{2}\Omega^2} = e^{-x^2/\Delta x^2} e^{-\Omega^2/\Delta\Omega^2} e^{-i\frac{\phi_{in}^{(2)}}{2}\Omega^2}. \tag{33}$$

$$E_{in}(x,t) \propto e^{-x^2/\Delta x^2} \exp\left(-\left(\left(1/\Delta\Omega^2\right) - i\left(\phi_{in}^{(2)}/2\right)/4\left(\left(1/\Delta\Omega^2\right)^2 + \left(\phi_{in}^{(2)}/2\right)^2\right)\right)t^2\right). \tag{34}$$

Then the effect of the pulse shaper should be to recompress this pulse to its best compressed pulse

$$E_{out,best\ compressed}(x,t) \propto e^{-x^2/\Delta x^2} e^{-\Delta\Omega^2 t^2/4}. \tag{35}$$

The exact calculation with the spatio-temporal coupling yields to

$$E_{out}(x,t) \propto e^{-\Phi_x x^2} e^{-\Phi_t t^2} e^{\Phi_{xt} xt} e^{i\mathrm{X}_x x^2} e^{i\mathrm{X}_t t^2} e^{-i\mathrm{X}_{xt} xt}, \tag{36}$$

where $\Phi_x = \left(1/\Delta x_p{}^2\right)$, $\Phi_t = \left(\left(1/\Delta\Omega^2\right) + \left(v^2\phi^{(2)2}/\Delta x_p{}^2\right)\right)/4\Delta$,
$\Phi_{xt} = \left(\left(A\phi^{(2)}/v\right)\left(\left(1/\Delta\Omega^2\right) + \left(v^2\phi^{(2)2}/\Delta x_p{}^2\right)\right) + \left(v\phi^{(2)}/\Delta x_p{}^2\right)\left(\phi^{(2)} + \phi_{in}{}^{(2)} + 2A\phi^{(2)2}/2\right)\right)/4\Delta$, $\mathrm{X}_x = A/v^2$,
$\mathrm{X}_{xt} = \left(\left(v\phi^{(2)}/\Delta x_p{}^2\right)\left(\left(1/\Delta\Omega^2\right) + \left(v^2\phi^{(2)2}/\Delta x_p{}^2\right)\right) - \left(\left(\alpha\phi^{(2)}/v\right)\right)\left(\phi^{(2)} + \phi_{in}{}^{(2)} + 2\alpha\phi^{(2)2}/2\right)\right)/4\Delta$, and
$\Delta = \left(\left(1/\Delta\Omega^2\right) + \left(v^2\phi^{(2)2}/\Delta x_p{}^2\right)\right)^2 + \left(\phi^{(2)} + \phi_{in}{}^{(2)} + 2A\phi^{(2)2}/2\right)^2$, $v = -\beta/\gamma = -pc\cos\theta_i/\lambda_0$,
$A = \left(2\phi^{(2)}\right)/\left(\left(2\phi^{(2)}\right)^2 + \left(\Delta x^2/v^2\right)^2\right)$, $\Delta x_p = \Delta x\sqrt{\left(1 + \left(2v^2\phi^{(2)}/\Delta x^2\right)\right)}$, $\theta = -\arctan\left(2\phi^{(2)}v^2/\Delta x^2\right)$.
This equation illustrates the degree of complexity of the spatio-temporal coupling. The pulse temporal ans spatial characteristics are modified by the pulse shaping. The temporal amplitude and phase are altered through respectively Φ_t and X_t. The spatial properties are affected through the dependance of Φ_x (amplitude) and X_x (phase) on $\phi^{(2)}$. The pure space-time coupling is expressed by Φ_{xt} and X_{xt}.

Consider that the chirp introduced by the pulse shaper optimally compresses the pulse. With Δx=2mm (half-width at 1/e), v=0.15mm/ps, $\Delta\Omega$=25ps^{-1} (half-width at 1/e), $\phi_{in}^{(2)}$=160000fs^2, the pulse is stretched to 1ps with a Fourier limit of 20fs (half-width at 1/e). The optimal chirp compensation is $\phi^{(2)}$=-160000fs^2. The optimally compressed pulse half-width at 1/e is then given by $\Delta t=1/4\sqrt{\Phi_t}$=22.6fs. The 10% error is due to the decrease of Φ_t when $\phi^{(2)}$ increases. These values are extreme and in most of the cases, the introduced chirp is small enough not to impact the recompression. On the spatial characteristics the modifications are small compared to the beam size, the output beam size is Δx_p=1.998mm compared to Δx=2mm at the input.
To decrease the effect of this coupling, the ratio v/Δx has to be kept small compare to the value of $\phi^{(2)}$, i.e. large input beams and highly dispersive gratings (p>600lines/mm).
As shown by Wefers [1996], it cannot by removed by a double pass configuration except for pure amplitude shaping. Despite its relatively small incidence on the output beam, this coupling can be very important when focusing the shaped pulse as shown by Sussman [2008] and Tanabe [2005].
To further analyze this pulse shaping technology, the mask has to be defined. The different technologies of spatial modulators are acousto-optic modulators (AOM) [Warren (1997)], Liquid Crystals Spatial Light Modulator diffraction-based approach [Vaughan (2005)], and Liquid Crystals Spatial Light Modulator. In the following, the mask used is a double Liquid Crystal Spatial Light Modulators (LC SLM) as described in Wefers (1995). The arbitrary filter is the combination of two LC SLM's whose LC's differ in alignment by 90 deg. This would produce independent retardances for orthogonal polarizations. The LC's for the two masks are respectiveley aligned at -45 and +45deg from the x axis, the incident light were polarized along the x axis, and the two LC SLM's are followed by a polarizer aligned along the x axis, the filter in this case for pixel n is given by

$$B_n = \exp\left\{i\left[\Delta\phi^{(1)} + \Delta\phi^{(2)}\right]/2\right\}\cos\left\{\left[\Delta\phi^{(1)} - \Delta\phi^{(2)}\right]/2\right\} = A_n e^{i\phi_n}, \tag{37}$$

where the dependence on the voltage for pixel n $\Delta\phi^{(i)}$ $[V_n^{(i)}]$ is implicitly included. In this case neither mask acts alone as a phase or amplitude mask, but the two in combination are capable of independent attenuation and retardance. Furthermore, as the respective LC SLM's act on orthogonal polarizations, light filtered by one mask is unaffected by the second mask. As shown by Wefers and Nelson, this eliminates multiple-diffraction effects of the two masks.
As discussed previously, spatially large input pulses reduce the space-time coupling effect. Each dispersed frequency component incident upon the mask has a finite spot size associated with it. However, this blurs the discrete features of the mask, the incident frequency components should be focused to a spot size comparable with or less than the pixel width. If the spot size is too small, replica waverforms that arise from discrete Fourier sampling will be unavoidable. On the other hand, if the spot size is too big, the blurring of the mask will give rise to substantial diffraction effects. As the spatial profile of a wavelength on the mask is the Fourier transform of the spatial profile on the grating. Minimizing the space-time coupling by using spatially large input pulses, discrete Fourier sampling and pulse replica cannot be avoid as the following analysis (suggested by Vaughan [2005] and Monmayrant[2005]) will show.

The modulating function m(x) is simply the convolution of the spatial profile S(x) of a given spectral component with the phase and amplitude modulation applied by the LC SLM,

$$m(x) = S(x) \otimes \sum_{n=-N/2}^{N/2} squ\left(\frac{x - x_n}{\delta x}\right) A_n \exp(i\phi_n), \tag{38}$$

where x_n is the position of the nth pixel, A_n and ϕ_n are the amplitude and phase modulation applied by the nth pixel ($A_n\exp(i\phi_n)=B_n$), δx is the separation of adjacent pixels, and the top-hat function squ(x) is defined as

$$squ(x) = \begin{cases} 1 & |x| \le \frac{1}{2} \\ 0 & |x| > \frac{1}{2} \end{cases}. \tag{39}$$

The spatial profile S(x) of a given spectral component is directly the Fourier transform of the input spatial profile as

$$S(x) = TF\left[E_{in}(x_{in})\right]_{x = 2\pi x_{in}/\lambda f \beta}, \tag{40}$$

where f is the focal length,
Here, the grating dispersion is assumed to be linear by

$$x(\omega) = \alpha(\omega - \omega_0), \text{where } \alpha = \frac{2\pi c f}{\omega_0^2 p \cos\theta_d(\omega_0)}. \tag{41}$$

Thus the position of the nth pixel x_n corresponds to a frequency $\Omega_n = n\delta\Omega$, where the frequency Ω_n of the nth pixel is defined relative to the center frequency ω_0 by $\Omega_n = \omega_n - \omega_0$, and where $\delta\Omega$ is the frequency separation of adjacent pixels corresponding to δx:

$$\delta\Omega = \frac{\delta x \omega_0^2 p \cos\theta_d(\omega_0)}{2\pi c f}. \tag{42}$$

Assuming also that the spatial field profile of a given spectral component is a Gaussian function $S(x)=\exp(-x^2/\Delta x^2)$, the modulation function may be written as

$$m(\Omega) = \exp\left(\frac{-\Omega^2}{\Delta\Omega_x^2}\right) \otimes \sum_{n=-N/2}^{N/2} squ\left(\frac{\Omega - \Omega_n}{\delta\Omega}\right) A_n \exp(i\phi_n). \tag{43}$$

Here the width of the spatial Gaussian function has been expressed in terms of $\Delta\Omega_x$, the spectral resolution of the grating-lens pair, where $\Delta\Omega_x = \Delta x \delta\Omega / \delta x$. The spot size Δx (measured as half-width at 1/e of the intensity maximum, assuming a Gaussian input beam profile) is dependent upon the input beam diameter D (half-width at 1/e), the focal length f and the angles of incidence and diffraction of the grating according to

$$\Delta x = \left(\left(\lambda_0 f \cos(\theta_i)\right) / \left(\pi D \cos(\theta_d)\right)\right). \tag{44}$$

The width of the Gaussian function expressed in frequency is

$$\Delta\Omega_x = \left(\omega_0 p\cos(\theta_i)\right)/(\pi D). \tag{45}$$

If we assume that the input pulse is a temporal delta function, $E_{in}(\Omega)=1$. The output field corresponds to the response function of the filter and its Fourier transform yields an expression of the impulse response function:

$$E_{out}(t) = h(t) \propto \exp\left(-\Delta\Omega_x{}^2 t^2/4\right)\mathrm{sinc}\left(\delta\Omega t/2\right)\sum_{n=-N/2}^{N/2} A_n \exp\left(i(\Omega_n t + \phi_n)\right). \tag{46}$$

The summation term describes the basic properties of the output pulse, such as would be obtained by modulating amplitude and/or phase of the input pulse at the point Ω_n with a grating-lens apparatus that has perfect spectral resolution. The sinc term is the Fourier transformation of the top-hat pixel shape, where the width of the sinc function is inversely proportional to the pixel separation δx, or equivalently, $\delta\Omega$. The Gaussian term results from the finite spectral resolution of the grating lens-pair, where the width of the Gaussian function is inversely proportional to the spectral resolution $\Delta\Omega_x$. Collectively, the product of the Gaussian and sinc terms is known as the time window. Therefore to increase the time window, both the frequency separation of adjacent pixel $\delta\Omega$ and the spectral resolution $\Delta\Omega_x$ have to be increased.

The expression of the impulse response function (eq.46) contains a summed term that is a complex Fourier series. A property of Fourier series (with evenly-spaced frequency samples) is that they repeat themselves with a period given by the reciprocal of the frequency increment $T_0=1/\delta\Omega$. These pulses repetitions, refered as sampling replica, are a cause of concern since they can degrade the quality of the desired output waveform.

While eq. 46 provides a compact and useful analytical result, it considers only the LC SLM with perfect pixels and spatial spot size. It neglects some important limitations of these devices. First, the pixels of the LC SLM are not perfectly sharp, and there are gap regions between the pixels whose properties are somewhat intermediate between those of the adjacent pixels. Second, LC SLMs typically have a phase range that is only slightly in excess of 2π. Fortunately since phases that differ by 2π are mathematically equivalent, the phase modulation may be applied modulo 2π. Thus, whenever the phase would otherwise exceed integer multiples of 2π, it is "wrapped" back to be within the range of 0-2π. Although smoothing of the pixelated phase and/or amplitude pattern might in general sound desirable, when it is combined with the phase-wraps, distortions in the spectral phase and/or amplitude modulation are introduced at phase-wrap points. Third, while the pixels are evenly distributed in space, the frequency components of the dispersed spectrum are not. This nonlinear mapping of pixel number to frequency makes difficult the determination of an exact analytical expression for $m(\Omega)$.

The contribution of the gaps has been taken into account in the litterature (Wefers [1995], Montmayrant[2005]) as a constant complex amplitude. This analysis supposes that the gap region does not depend upon the neighbour pixels. As the filter in each gap is assumed to be the same, the gaps simply reproduce the single input pulse at time zero with a reduced complex amplitude given by $(1-r)B_g$ where r is the ratio of the pixel width ($r\delta x$) by the pixel pitch δx and B_g its complex response. The expression for $m(x)$ including the gaps is

$$m(x) = S(x) \otimes \sum_{n=-N/2}^{N/2}\left[\left(squ\left((x-x_n)/r\delta x\right)A_n \exp(i\phi_n)\right) + \left(squ\left(\left(x-x_n+{\delta x}/{2}\right)/\left((1-r)\delta x\right)\right)B_g\right)\right]. \tag{47}$$

With the approximation of linear spectral dispersion, the filter response function can be expressed as:

$$h(t) \propto E_{in}\left(\left(p\omega_0 \cos\theta_i/2\pi\right)t\right)\left\{\left[r\,\mathrm{sinc}\left(r\delta\Omega t/2\right)\sum_{n=-\infty}^{\infty} A_n e^{i(\Omega_n t+\phi_n)}\right]+\left[(1-r)\mathrm{sinc}\left((1-r)\delta\Omega t/2\right)\sum_{n=-\infty}^{\infty} B_g e^{i\Omega_n t}\right]\right\}. \quad (48)$$

The time extent of the contribution of the gap is a lot longer than the pixel one. The theoretical ratio in intensity is $(r/(1\text{-}r))^2$ in the order of thousand for up-to-date LC SLM. But the experimental ratio is about 40 to 100. This order of magnitude is due to the hypthesis that the gap region is the same and that the pixel edges are perfectly sharp. The smoothing of the phase between pixels has to be considered.

The smoothing function has been first introduced by Vaughan and al. but without explicit expression, and on a phase mask only. In fact no simple analytical model can reproduce this effect. It will be introduce in the simulation part.

The phase wraps used to extend the phase modulation of the LC SLM above its limited excursion of 2π by applying a phase that is "wrapped"back into 0-2π as

$$\phi_{applied,n} = \mathrm{mod}_{2\pi}\left[\phi_{desired,n}\right]. \quad (49)$$

Due to the mathematical equivalence of phase values that differ by integer multiples of 2π, there are an infinite number of ways to "unwrap"the applied phase. Sampling replica pulses constitute an important class of these equivalent phase functions, and their phase as a function of pixel,$\phi_{replica,n}$, may be described by

$$\phi_{replica,n} = \phi_{applied,n} + 2\pi Rn, \quad (50)$$

where R is the sampling replica order and may be any non zero integer (0 corresponds to the desired pulse). In the case of linear spectral dispersion, $\phi_{replica,n}$ for different values of R differ by a linear spectral phase $2\pi R\omega/\delta\Omega$, which corresponds to a temporal shift of $R/\delta\Omega$. This is another explanation of the sampling replica that are temporally separated by $1/\delta\Omega$. In the case of a non linear spectral dispersion, the different replica phases do not differ by a linear spectral phase but rather by a non linear one. The quadratic term will introduce a second order spectral phase (chirp) linearly depending on the replica number R. A very explicit illustration is given by Vaughan and al.(2006), but no analytical expression could be given for the non linear dispersion.

Finally, the modulation function can be expressed analytically as

$$m(\Omega) = squ\left[\Omega/N\delta\Omega\right]\left\{S(\Omega)\otimes comb\left[\Omega/\delta\Omega\right]\left[squ\left(\Omega/r\delta\Omega\right)H(\Omega)+squ\left(\left(\Omega+\delta\Omega/2\right)/\left((1-r)\delta\Omega\right)\right)B_g\right]\right\}, \quad (51)$$

where $comb\left[\Omega\right]=\sum_{n=-\infty}^{\infty}\delta(\Omega-n)$, N is the number of pixels, $H(\Omega)$ is the desired transfer function. This function combines the pixelization, the gap effect, the input beam spatial dimension, the limited number of pixel. The impulse response function is then given by

$$M(t) \propto sinc\left[N\delta\Omega t/2\right]\otimes\left[E_{in}\left(\left(\frac{p\omega_0\cos\theta_i}{2\pi}\right)t\right)\left\{\begin{matrix}\left(r\,\mathrm{sinc}\left(r\delta\Omega t/2\right)\left(comb\left[\delta\Omega t\right]\otimes h(t)\right)\right)+\\ r\,\mathrm{sinc}\left(\pi(1-r)\delta\Omega t/2\right)\left(comb\left[\delta\Omega t\right]\otimes B_g\right)\end{matrix}\right\}\right]. \quad (52)$$

where N is the pixels number, $\delta\Omega$ is the frequency extent of a the pixel pitch, S(x) is the spatial profile of the input pulse, r the ratio between the pixel size and the pixel pitch, h(t) the ideal impulse response function and B_g the gap complex transmission.
The figure 3 illustrates the different contributions of this model on the output temporal intensity.

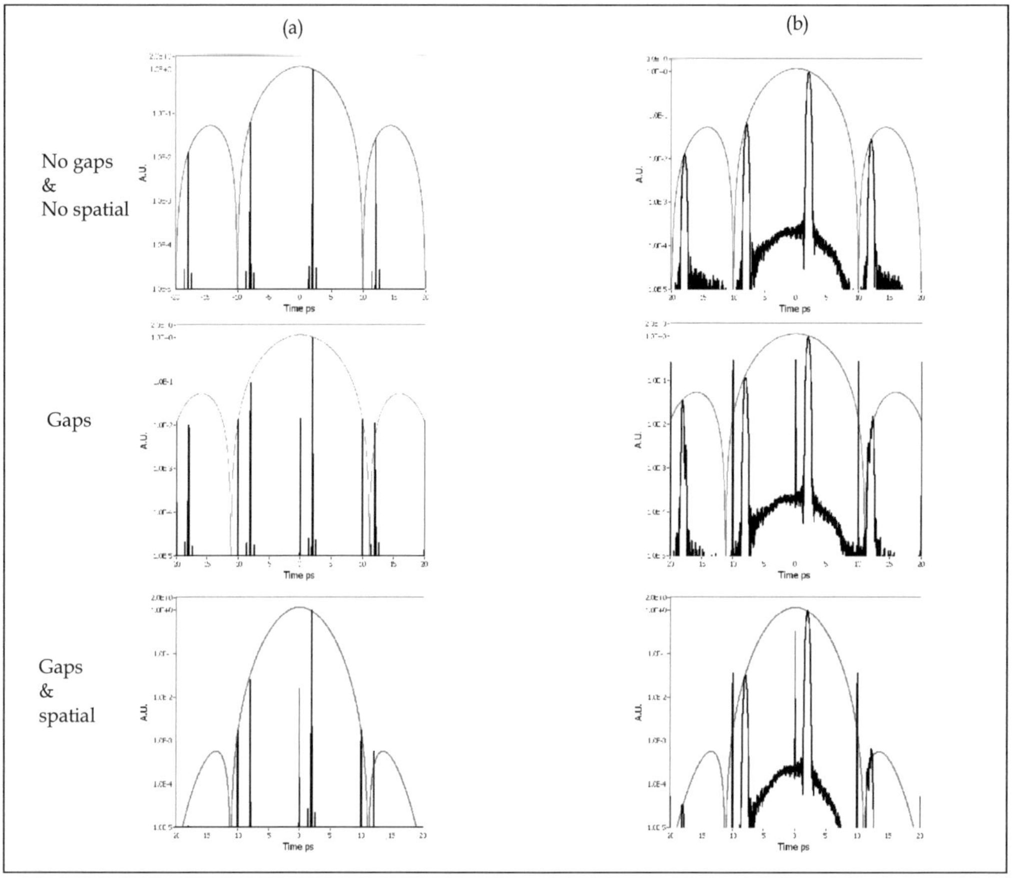

Fig. 3. output temporal intensity examples in logarithmic scale for a 4-f pulse shaper (f=220mm,2000lines/mm, δx=100µm, r=0.9, D=1.7mm half-width at 1/e, B_g=1) with (a) a delay 2000fs, (b) adding a chirp 4000fs^2 to the delay. The first row does not include contribution of gaps and spatial filtering, second row includes gaps contribution, third row gaps and spatial input beam profile contribution. The black line is the output waveform, the grey line the envelope of the filter response pulse shaper pixels.

Other contributions can only be numerically simulated as the non linear dispersion, the smoothing effect, the spatio-temporal coupling.
The pulse replicas can be filtered out as the spatio-temporal coupling by using a spatial filter at the output (cf Fig.5). This filtering effect is only efficient if the filter select the lowest Hermite-Gaussian mode as shown by Thurston and al. (1986). Regenerative amplifiers or monomode optical fibers are good fundamental Hermite-Gaussian mode filters. A simple iris cannot be considered as such a filter as shown by Wefers (1995). With perfect filtering, the filter modulation becomes

$$m_{filtered}(t) \propto Filter(t) \cdot m(t) \propto sinc[N\delta\Omega t/2] \otimes \{rh(t) + (1-r)B_g\}. \quad (53)$$

The filter function Filter(t) introduced by the spatial filtering decreases the overall efficiency and does not filter out the contribution of the gaps. It can be estimated as applying another enveloppe on the time profile with a restricted area limiting the time window. The contribution of the filters response has to be taken into account for exact pulse shaping.

4.1.2 4-f pulse shapers numerical simulations

4-f pulse shapers are commonly used with a simple iris aperture filtering directly at the output before the experiment. As seen in the previous part, the filter response can be affected by limitations of the 4-f apparatus (spatio-temporal coupling, non linear dispersion) and of the LC SLMs (smoothing) that cannot be expressed analytically. Complex input pulse and pulse shaping as multiple pulses or square pulses can only be simulted numerically. This part gives an adavnced numerical models combining models used in the litterature (Wefers [1995], Vaughan [2005], Monmayrant [2005], Sussman [2008], Tanabe [2002], Tanabe [2005]).

The effects of pulse propagation through a pulse shaper have been carefully detailed by Danailov [1989] and Wefers [1995]. As Tanabe (2005) and Sussman (2008), the propagation is simulated by a Fresnel propagation as:

$$\tilde{E}(k_x,\omega,z) = e^{-i\pi k_x^2(z-z_0)c/\omega}\tilde{E}(k_x,\omega,z_0), \quad (54)$$

where $\tilde{E}(k_x,\omega,z)$ is the spatial Fourier transform of the electric field, $U_{Fresnel}(k_x) = e^{-i\pi k_x^2(z-z_0)c/\omega}$ is the Fresnel propagator. The field will be simulated at a focal plane as oftenly used experimentally.

For the shaper in Fig.4, there are 17 different steps from input beam to focal field, as enumerated below.

1. An input beam E(x,t,0) is propagated from its origin to the diaphragm aperture of the pulse shaper by Fresnel propagation.
2. An iris aperture spatially of diameter D_{iris} filters the beam: $E(x,t,z) \to \mathrm{Rect}(x/D_{iris})E(x,t,z)$.
3. The beam is propagated form the iris to the input grating by Fresnel propagation.
4. The beam is dispersed by the input grating by applying Martinez: $E(x,\Omega,z) \to \sqrt{\beta}e^{2i\pi\gamma\Omega x}E(\beta x,\Omega,)$
5. The beam is Fresnel propagated a distance f.
6. A perfect thin lens of focal length f introduces a quadratic spatial phase: $E(x,\Omega,z) \to E(x,\Omega,z)e^{-i\pi\Omega f x^2/2c}$.
7. The beam is Fresnel propagated a distance f.
8. The spatial mask is applied via multiplication: $E(x,\Omega,z) \to E(x,\Omega,z)m(x)$
9. The beam is Fresnel propagated a distance f.
10. A perfect thin lens of focal length f introduces a quadratic spatial phase : $E(x,\Omega,z) \to E(x,\Omega,z)e^{-i\pi\Omega f x^2/2c}$.

11. The beam is Fresnel propagated a distance f.
12. The second grating is applied in the inverted geometry by applying Martinez: $E(x,\Omega,z)\rightarrow\left(1/\sqrt{\beta}\right)e^{2i\pi\gamma\Omega x}E(-x/\beta,\Omega,z)$.
13. The beam is Fresnel propagated from the grating to the output iris.
14. The beam is spatially filtered by the iris: $E(x,t,z)\rightarrow \mathrm{Rect}(x/D_{iris})E(x,t,z)$.
15. The beam is Fresnel propagated a distance L
16. A thin lens of focal length f_L is applied: $E(x,\Omega,z)\rightarrow E(x,\Omega,z)e^{-i\pi\Omega f_L x^2/2c}$
17. The beam is propagated to the focal plane.

The spatio-temporal coupling is directly include in these steps. All the other effects can be introduced directly on the mask and grating functions.
The non linear dispersion is estimated through a modification of the mask by introducing:

$$\Omega(x)=ax+bx^2+O(x^2), \tag{55}$$

where $a=p\omega_0^{\,2}\cos\theta_d/2\pi cf$, $b=p^2\omega_0^{\,3}\left(\cos\theta_d\right)^2/8\left(\pi cf\right)^2$. This contribution has to be corrected for the main pulse but still remains for the replica.
The pixelization is introduced on the mask by

$$m(x)=\sum_{n=-N/2}^{N/2-1} rect\left[\frac{x-x_n}{\delta x}\right]A_n e^{i\phi_n}. \tag{56}$$

The smoothed-out pixel regions may cause an entirely different class of output waveform distortions from the pixel gap as mentionned by Vaughan and al. (2006). Although the exact nature of the smooth pixel boundaries is expected to be highly dependent upon the specific device that is being considered, it has been approximated by convolving a spatial response function L(x) with an idealized phase modulation function that would result in the case of sharply defined pixel and gap regions (Vaughan [2005]). But no explicit smoothing function has been given in the litterature. Moreover this approximation stands only for a phase only pulse shaper. The exact analysis of a phase step between two adjacent pixels is very complex. A simple model can consider that the phase introduced by a LC SLM is given by

$$\phi(\lambda,V)=\frac{2\pi\Delta n(\lambda,V)e_{LC}}{\lambda}=e^{P(V)}+C. \tag{57}$$

Despite the sharp edges of the pixel, a relaxation process occurs in the Liquid Crystal material whose anisotropy is very strong ($\varepsilon_{//}\approx 5\,\varepsilon_{\perp}$) [Khoo (1993)]. For an up-to-date LC SLM, the pixel pitch is 100µm and the gap 2µm, the thickness is about 10µm. Without taking into account the anisotropy, the smoothing is about 1/20 of the pixel pitch independantly of the gap size. With the anisotropy, the smoothing covers more than half the pixel. A rather good smoothing function is a Lorentzian:

$$L(x)=\frac{\Gamma/2\pi}{x^2+\left(\Gamma/2\right)^2}, \tag{58}$$

where Γ is the width.

With the relaxation, the small gaps completely disappear. This smoothing has to be done on the potential of the LC SLM directly. So from the desired phase modulation on both LC SLMs, the potential is calculated, smoothed by the Lorentzian, and discretized according to to the voltage resolution of the device.

So the estimation of the mask modulation can include the non-linear dispersion, the pixelization and pixels smoothing by applying the following algorithm:

1. From a regular array of points in the space domain of the mask x_n, estimation of the corresponding frequencies with the non linear dispersion : Ω_n.
2. Determination of the amplitude and phase of the ideal mask on these frequencies: $A_n(\Omega_n)$ and $\phi_n(\Omega_n)$.
3. Determination of the frequencies relative to each pixel: Ω_k^{pixel}.
4. Pixelization of the phase and amplitude by applying the same phase and amplitude over a pixel i.e. for $\Omega_n \in [\Omega_k^{pixel}, \Omega_{k+1}^{pixel}]$.
5. Pixels smoothing by:
   a. Estimation of the phases on the two LC SLMs: $\Delta\phi^{(1)} = \left(a\cos\left(A_n\right) + \phi_n\right)/2, \Delta\phi^{(2)} = \left(a\cos\left(A_n\right) - \phi_n\right)/2$.
   b. Determination of the voltage on the pixels by inverting eq.57: $V_1 = f^{-1}\left(\Delta\phi^{(1)}\right), V_2 = f^{-1}\left(\Delta\phi^{(2)}\right)$.
   c. Smoothing of this voltages by convolving with the Lorentzian function (eq.58): $V_{i,smoothed}\left(\Omega\right) = L\left(\Omega\right) \otimes V_i\left(\Omega\right)$.
   d. Calculation of the two LC SLMs phases: $\Delta\phi^{(i)}{}_{smoothed} = f\left(V_{i,smoothed}\right)$.
   e. Calculation of the mask modulation from eq.37.

The numeric propagation of pulses is efficiently achieved using the fast Fourier transform (FFT) and its inverse (IFFT), for transforming between space to frequency and time to frequency. Care should be taken to assure that the sampling is done correctly. Propagating through large distances or studying the intensity close to the focal point requires resampling the spatial grid. The spatio-temporal complete simulation requires a bidimensionnal grid in space and time restricting the resolution in time. Specific study of sampling replica, pixels smoothing effects and gaps should be done with a simplified model without the space-time coupling. For example, for a pulse shaper with 640 pixels and pixel gaps about 3% of the pixel pitch, the number of sampling points (>10000) is too high for this bidimensionnal simulation. The simplification consists in directly multiplying the input pulse by the mask function in the frequency domain as

$$E_{out}\left(\Omega\right) = E_{in}\left(\Omega\right)M(\Omega) \propto E_{in}\left(\Omega\right)TF\left[E_{xin}\left(t/v\right)TF^{-1}\left[M\left(\Omega\right)\right]\right], \quad (59)$$

where M(Ω) is calculated by the algorithm described just beneath.

These models are in quantitative agreement with experimental published results. The different contributions (pixelization, non-linear dispersion, pixel gaps and pixels smoothing) are illustrated on the figure 4 below on a 100fs Fourier transform pulse at 800nm delayed by –2ps, or stretched by a $7.10^5 fs^2$ chirp, with a pulse shaper using two LC SLMs 0f 640 pixels (pixel pitch=100µm, pixel size=0.97), a focal length of 200mm and a 2000lines/mm grating with equal input and output angles. The input beam diameter is 2.3mm gaussian shape.

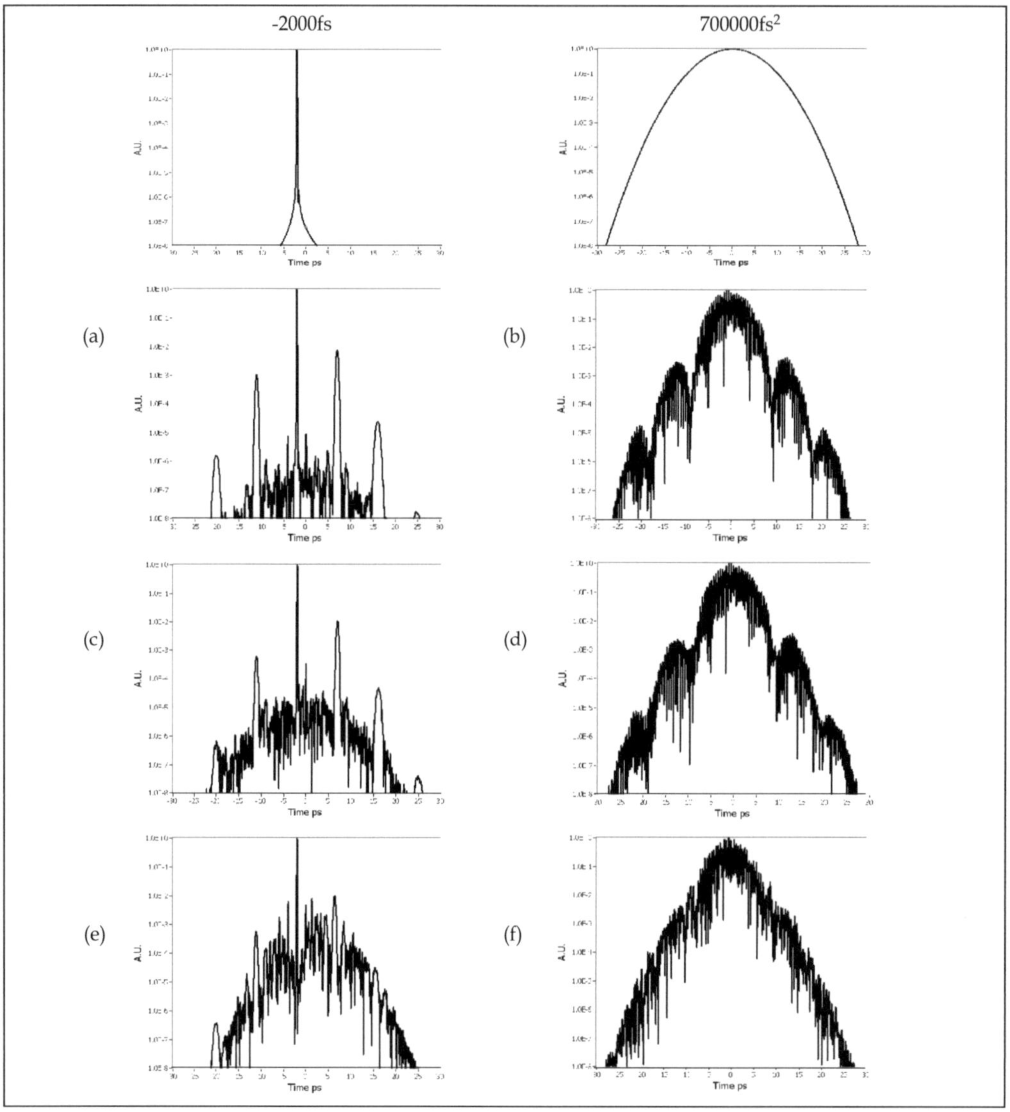

Fig. 4. Contributions on pulses with a –2ps delay or a 0.7ps² chirp of (a),(b) non-linear dispersion, (c),(d) pixel gaps, (e),(f) pixels smoothing.

4.1.3 Conclusions on 4-f pulse shapers

This pulse shaper technology based on the coupling between space and time in a 4f-zero dispersion line apparatus allows complex pulse shaping over a large range of pulse characteristics. Its optical set-up allows to adapt the performances of the pulse shaper. Despite its relative simple concept, its optimization requires a trade-off between parameters and side effects.

The parameters are: p the grating pitch, f the focal length, θ_i the incidence angle, θ_d the diffracted angle, δx the pixel pitch, N the number of pixels and D the input beam diameter.

The relevant characteristics are:

- The spectral bandwidth: $\Delta\Omega_M = \left(\omega_0^{\,2} p\cos\theta_d / 2\pi c\right)\arctan\left(N\,\delta x/f\right)$,
- spectral resolution or initial time window : $\delta\Omega = 1/\delta T = \delta x \omega_0^{\,2} p\cos\theta_d \,/\, 2\pi c f$,
- spatio-temporal slope : $v = -\gamma/\beta = 2\pi \,/\, \omega_0 p\cos\theta_i$,
- real time window (spatial filtering) : $\Delta T = D \,/\, v$,
- Rayleigh length at the mask: $z_R = \lambda f^2 \,/\, \pi D^2$.

The Rayleigh length has to be larger than the two LC SLMs mask thickness which is typically about 1mm.
To decrease the spatio-temporal coupling, v/D has to be minimized, but this also reduces the time window. Thus a trade-off between the side-effects of the spatio-temporal coupling and the required time window and the pulse replica has to be done.
As mentionned by Wefers (1995), Monmayrant (2005) and Tanabe (2002), the pixel gaps and some other effects can be compensated for by iterative algorithms. As the models are not precise enough, this compensation has to be done experimentally (Tanabe).
The effects of misalignement and tolerances of the optical set-up is beyond the scope of this chapter but can be very significant on the output waveform as shown by Wefers (1995), Tanabe (2002).

4.2 Acousto-optic programmable dispersive filter

The second pulse shaping technology has been invented by Pierre Tournois in 1997 (Tournois (1997)). The basic idea is to make a programmable Bragg grating or chirped mirror. Through an acousto-optic longitudinal Bragg cell, the acousto-optic diffraction directly transfers the amplitude and phase modulation of the acoustic wave onto the optical diffracted beam.
A schematic of the AOPDF is shown on fig.5. An acoustic wave is launched in an acousto-optic birefringent crystal by a transducer excited by a temporal RF signal. The acoustic wave propagates with a velocity V along the z-axis of the crystal and hence reproduces spatially the temporal shape of the RF signal. Two optical modes can be coupled efficiently by acousto-optic interaction in the case of phase matching. If there is locally only one spatial frequency in the acoustic grating , then only one optical frequency can be diffracted at a position z. The incident optical short pulse is initially polarized onto the fast axis polarization of the birefringent crystal. Every optical frequency ω travels a certain distance before it encounters a phase matched spatial frequency in the acoustic grating. At this position z(ω), part of the energy is diffracted onto the slow axis polarization. The pulse leaving the device onto the extraordinary polarization will be made up all the spectral components that have been diffracted at various positions. Since the velocities of the two polarizations are different, each optical frequencies will see a different time delay τ(ω) given by:

$$\tau(\omega) = \frac{z(\omega)}{v_{g1}} + \frac{L - z(\omega)}{v_{g2}}, \qquad (60)$$

where L is the crystal length, v_{g1} and v_{g2} are the group velocities of ordinary and extraordianry modes respectively.

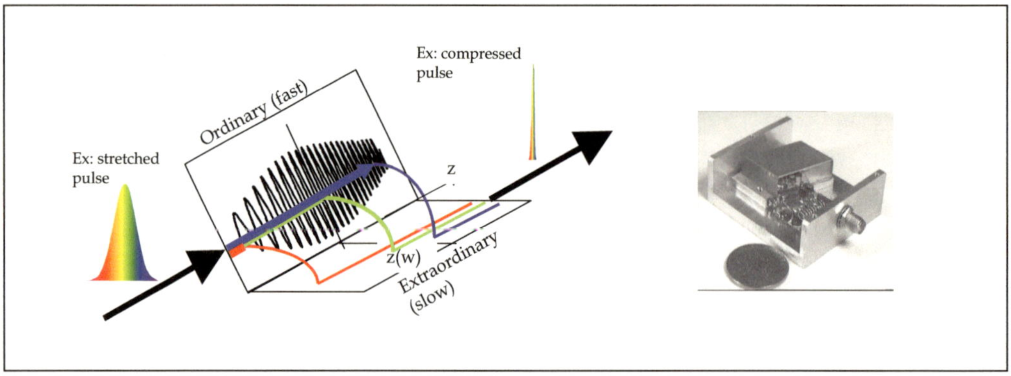

Fig. 5. Schematic of the AOPDF.

The amplitude of the output pulse, or diffraction efficiency, is controlled by the acoustic power at position z(ω). The optical output $E_{out}(t)$ of the AOPDF is a function of the optical input $E_{in}(t)$ and of the acoustic signal S(t). More precisely, it has been shown (Tournois (1997)), for low value of acoustic power density, to be proportionnal to the convolution of the optical input and of the scaled acoustic signal:

$$E_{out}(t) \propto E_{in}(t) \otimes S(t/\alpha) \Leftrightarrow E_{out}(\omega) \propto E_{in}(\omega) S(\alpha\omega), \tag{61}$$

where the scaling factor α is the ratio of the acoustic frequency to the optic frequency.
In this formulation, the AOPDF is exactly a linear filter whose filter response is S(αω). Thus by generating the proper function, one can achieve any arbitrary convolution with a temporal resolution given by the inverse of the available filter bandwidth.
This physical discussion qualitatively explains the principle of the AOPDF. A more detailed analysis is given in the following part based on a first order theory of operation, and second order influence will then be estimated.

4.2.1 First order theory of the AOPDF

The acousto-optic crystal considered in this part is Paratellurite TeO_2. The propagation directions of the optical and acoustical waves are in the P-plane which contains the [110] and [001] axis of the crystal. The acoustic wave vector K makes an angle θ_a with the [110] axis. The polarization of the acoustic wave is transverse, perpendicular to the P-plane, along the [/110] axis. Because of the strong elastic anisotropy of the crystal, the K vector direction and the direction of the Poynting vector are not collinear. The acoustic Poynting vector makes an angle β_a with the [110] axis. When one sends an incident ordinary optical wave polarized along the [/110] direction with a vector k_0 which makes an angle θ_0 with the [110] axis, it interacts with the acoustic wave. An extraordinary optical wave polarized in the P-plane with a wave vector k_d is diffracted with an angle θ_d relative to the [110] axis. To maximize the interaction length for a given crystal length, and hence to decrease the necessary acoustic power, the incident ordinary beam is aligned with the Poynting vector of the acoustic beam, i.e. $\beta_a=\theta_0$. Figure 10 shows the k-vector geometry related to the acoustical and optical slowness curves. V_{110} and V_{001} are the phase velocities of the acoustic shear waves along the [110] axis and along the [001] axis respectively. n_o and n_e are the ordinary and extraordinary indices on the [110] axis and n_d is the extraordinary index associated with the diffracted beam direction at angle θ_d.

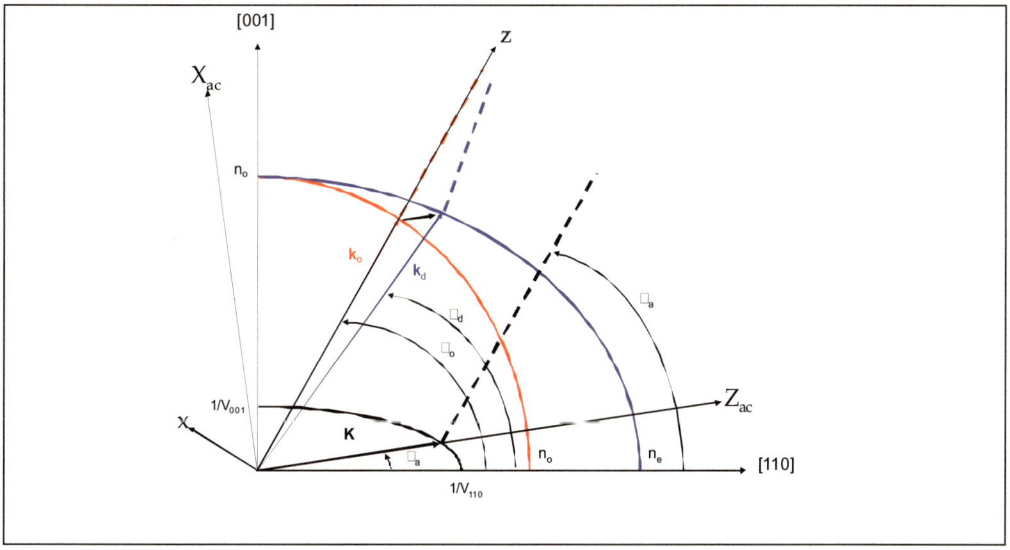

Fig. 6. Acoustic and optic slowness curves and k-vector diagram

The optical anisotropy $\Delta n=(n_e-n_o)$ being generally small as compared to n_o, the following relations can be obtained to first order in $\Delta n/n_o$:

$$\delta n = n_d - n_o = \Delta n.\cos^2\theta_0 , \tag{62}$$

$$V(\theta_a) = \sqrt{V_{001}{}^2 \sin^2\theta_a + V_{110}{}^2\cos^2\theta_a} , \tag{63}$$

$$\theta_d - \theta_0 = -(\Delta n/n_0).\cos^2\theta_0.\tan(\theta_0-\theta_a), \tag{64}$$

$$K = k_0(\Delta n/n_0).(\cos^2\theta_0/\cos(\theta_0-\theta_a)), \tag{65}$$

$$\alpha = (V(\theta_a)\Delta n/c).(\cos^2\theta_0/\cos(\theta_0-\theta_a)), \tag{66}$$

where c is the speed of light.
The single frequency solution of the coupled mode theory for plane waves (Yariv and Yeh) allows to relate the diffracted light intensity to the incident light intensity and to the acoustic power density $P(\alpha\omega)$ present in the interaction area by the formula:

$$I_d(\omega) = I_0(\omega)\left(\frac{P(\alpha\omega)}{P_0}\right)\frac{\pi^2}{4}\mathrm{sinc}^2\left[\frac{\pi}{2}\sqrt{(P(\alpha\omega)/P_0)+\left(\frac{\delta\varphi}{\pi}\right)^2}\right] \text{ with } P_0 = \frac{1}{2M_2}\left[\frac{\lambda}{L\cos(\theta_0-\theta_a)}\right]^2, \tag{67}$$

with $\delta\varphi$ is an asynchronous factor proportional to the product of the departure δk from the phase matching condition and of the interaction length along the acoustic wave vector K:

$$\delta\varphi = \frac{\delta k}{\pi}L\cos(\theta_0-\theta_a) \approx \frac{\Delta n}{n_o}k_0 L\cos\theta_0\left[\frac{-\delta\omega}{\omega} + \delta\theta_0\left(2\tan\theta_0 - \tan(\theta_0-\theta_a)\right)\right], \tag{68}$$

L being the interaction length along the optical wave vector k_0, λ the wavelength of the light in vacuum, ρ the density of TeO_2 crystal, p an elasto-optic coefficient, and M_2 the merit factor given by:

$$M_2 = \frac{n_o^3 \left[n_d(\theta_d)\right]^3 \left[p(\theta_0,\theta_a)\right]^2}{\rho\left[V(\theta_a)\right]^3} \text{ with } p(\theta_0,\theta_a) = -0.17\sin\theta_a\cos\theta_0 + 0.09\sin\theta_0\cos\theta_a. \quad (69)$$

From eq.67, with a perfect matching condition (δφ=0), complete diffraction of an optical frequency ω corresponds to an acoustic power density $P(\alpha\omega)=P_0$. As the interaction is longitudinal or quasi-collinear the efficiency of diffraction is excellent. P_0 is in the order of few mW/mm².

The spectral resolution and angular aperture are defined by the phase matching condition through the condition that the efficiency $\eta=I_d/I_0=0.5$ for $\delta\varphi=\pm0.8$ when $P(\alpha\omega)=P_0$ as:

$$\left(\frac{\delta\lambda}{\lambda}\right)_{1/2} = \left(\frac{\delta\omega}{\omega}\right)_{1/2} = \frac{0.8}{\Delta n\cos^2\theta_0}\frac{\lambda}{L}, \quad (70)$$

$$(\delta\theta_0)_{1/2} = \left(\frac{\delta\lambda}{\lambda}\right)_{1/2}\frac{1}{\left[2\tan\theta_0 - \tan(\theta_0-\theta_a)\right]}. \quad (71)$$

By using conventionnal acousto-optic technology, diffraction efficiencies can be up to 50% over 100nm. If Δλ is the incident optical bandwidth, the number of programming points N and the estimation of the acoustic power density to maximally diffract the whole bandwidth will be:

$$N = \frac{\Delta\lambda}{(\delta\lambda)_{1/2}} = \frac{\Delta nL\cos^2\theta_0}{0.8}\frac{\Delta\lambda}{\lambda^2}, \quad (72)$$

$$P_N = NP_0 = \frac{1.25\Delta n\cos^2\theta_0}{2M_2\cos^2(\theta_0-\theta_a)}\frac{\Delta\lambda}{L}. \quad (73)$$

The different applications of the AOPDFs call for two different cut optimizations of the TeO_2 crystal. When the goal is to control the spectral phase and amplitude in the largest possible bandwidth, to obtain the shortest possible pulse, the diffraction efficiency has to be maximized and hence P_0 minimized (Wide Bandc cut). When the goal is to shape the input pulse width with the higher resolution, the optimization is a trade-off between the spectral resolution and the diffraction efficiency (High Resolution cut). The parameters for the Wide Band and High Resolution AOPDFs for λ=800nm are given in table 1.

Since Paratellurite crystals are dispersive, the acoustic to optic frequency ratio α depends on the wavelength through the spectral dispersion of optical anisotropy.

The dispersion becomes very large below λ=480nm. For limited bandwidth Δλ, the dispersion of the crystal can be compensated b y programming an acoustic wave inducing an inverse phase variation in the diffracted beam. This self-compensation is, however, limited by the maximum group delay variation given by:

AOPDFs name	L mm	θ_a deg	θ_0 deg	θ_d-θ_0 deg	α 10^{-7}	$n_o(\delta\theta_0)_{1/2}$ deg	P_0 MW/ mm²	$(\delta\lambda)_{1/2}$ nm	T ps	$\Delta\lambda$ (η=0.5 for 0.6W/mm²) nm	N
Wide Band 25 (WB25)	25	8	58.5	1.25	1.42	0.04	4.5	0.6	3.7	100	170
High Resolution 25 (HR25)	25	3.9	38.5	1.60	2.3	0.045	3.8	0.25	8.0	50	200
Wide Band 45 (WB45)	45	8	58.5	1.25	1.42	0.022	1.4	0.33	6.7	180	540
High Resolution 45 (HR45)	45	3.9	38.5	1.60	2.3	0.025	1.2	0.14	14.4	90	640

Table 1. Standard AOPDFs parameters.

$$\delta\tau_g = \tau_{gd} - \tau_{g0} = \left(n_{gd} - n_{g0}\right)\frac{L}{c} = \left(n_{gd} - n_{g0}\right)\cos^2\theta_0 \frac{L}{c}. \tag{74}$$

More precisely, when the dispersion of the crystal is compensated by an adapted acoustic waveform, all the wavelength in the optical bandwidth $\Delta\lambda=\lambda_2-\lambda_1$ have to experience the same group delay time, i.e. the same group index $n_{g0}(\lambda_1)=n_{gd}(\lambda_2)$. The maximum bandwidth of self compensation depends upon the central wavelength and the crystal type (cf table 2). If the bandwidth of operation is larger than this maximum bandwidth $\Delta\lambda$, it is necessary to use an outside compressor. The major component of the dispersion in TeO_2 is the second order. If this second order is externally compensated this leads to a new limit bandwidth $\Delta\lambda_1>\Delta\lambda$ associated to higher orders compensation.

Central lambda	650 nm	800 nm	1064 nm	1550 nm
$\Delta\lambda$ nm/ $\phi^{(2)}$ fs² / $\Delta\lambda_1$ nm WB 25	45 / 17300 / 300	70 / 12800 / 560	150/ 8900 / >600	500/ 5370 />1000
$\Delta\lambda$ nm/ $\phi^{(2)}$ fs² / $\Delta\lambda_1$ nm HR 25	130 / 17300 / >400	200 / 12800 / >800	420/ 8900 / >800	800/ 5370 />1000
$\Delta\lambda$ nm/ $\phi^{(2)}$ fs² / $\Delta\lambda_1$ nm WB 45	80 / 17300 / >400	125 / 23000 / 800	270/ 1600 / >800	900/ 9660 />1000
$\Delta\lambda$ nm/ $\phi^{(2)}$ fs² / $\Delta\lambda_1$ nm HR 45	230 / 17300 / >400	360 / 23000 / >800	725/ 1600 / >800	>1000/ 9660 />1000

Table 2. Self-compensation bandwidth $\Delta\lambda$, second order dispersion and higher order limited bandwidth $\Delta\lambda_1$.

4.2.2 Rigourous theory of the AOPDF

The first order theory is a good approximation despite strong hypothesis of acoustic and optic plane waves, acoustic and optic single frequencies. The validity of these two hypothesis is studied in the following parts.

4.2.2.1 From the single frequency to the multiple frequencies

The multi-frequencies general approach (Laude (2003)) is complex and not actually required for the simulation of the AOPDF (Oksenhendler (2004)). In the AOPDF crystal geometry, as only one diffracted mode can exist, the coupled-wave equation can be simplified and expressed in a matrix notation such as:

$$\frac{\partial \hat{D}}{\partial z} = jM(z)\hat{D} \text{ where } M(z) = \begin{pmatrix} 0 & \kappa(z) \\ \kappa^*(z) & 0 \end{pmatrix},\ \hat{D}(z) = \begin{pmatrix} D_0 \\ D_1 \end{pmatrix} \tag{75}$$

with $\kappa(z) = -\frac{j}{4}k_0\sqrt{n_0^3}\sqrt{n_d^3}L\int A(\omega_{ac})e^{-j\psi(\omega_{ac})}\left(e^{j(k_{0z}-k_{1z}+K(\omega_{ac}))z}\right)d\omega_{ac}$,

where the index 0,1 corresponds respectively to the incident and diffracted beam, D is the electric displacement vector, A the acoustic complex amplitude.
This equation can be solved independently of the number of acoustic frequencies considered. The solutions are:

$$\begin{pmatrix} D_0(L) \\ D_1(L) \end{pmatrix} = \exp\left(j\int_0^L M(z)dz\right)\begin{pmatrix} 1 \\ 0 \end{pmatrix}. \tag{76}$$

The difference with the first order theory is within 1% on the spectral amplitude. The spectral phase is conserved even in the saturated or over saturated regime because it comes directly from the phase matching condition (fig.6).

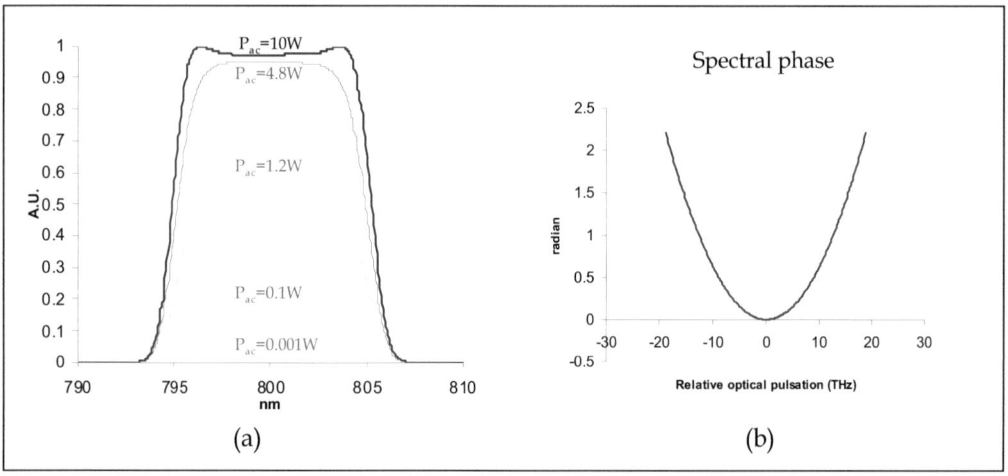

Fig. 7. Simulation of acousto-optic diffraction for (a) spectral amplitude, (b) spectral phase.

The first order can then be used to precompensate the saturation within few percents but exact pulse shaping requires to monitor and loop on the spectral amplitude. The spectral phase is automatically conserved through the Bragg phase matching condition.

4.2.2.2 Acoustic beam limitation

This coupled-wave analysis considers plane waves. Due to the size of the beam relatively to the wavelength, the acoustic wave cannot be considered as a single plane wave. The acoustic beam finite dimension D_a results in the limitation in spatial aperture of each wave that allows to represent the acoustic field in the components of angular spectrum as:

$$\begin{aligned} e_{ac}(t,X_{ac},Z_{ac},\xi,\omega_{ac}) &= A(\xi,\omega_{ac})\exp\left[i\left(\omega_{ac,0}t + K_{\varphi z_{ac}}Z_{ac} + K_{\varphi x_{ac}}X_{ac}\right)\right] \\ &= \frac{A_0 L D_a}{2}\operatorname{sin}c\left(\frac{(\omega_{ac}-\omega_0)L}{2V_0}\right)\operatorname{sin}c\left(\frac{\omega_{ac}\sin(\xi)D_a}{2V(\xi)}\right) \\ &\times\exp\left[i\left(\omega_0 t + \frac{\omega_{ac}}{V(\xi)}\cos(\xi)Z_{ac} + \frac{\omega_{ac}}{V(\xi)}\sin(\xi)X_{ac}\right)\right]. \end{aligned} \tag{77}$$

where X_{ac} and Z_{ac} are the coordinate along the acoustic central wavevector (cf. fig.10), ξ is the relative angle between the wavevector and the Z_{ac} direction, ω_{ac} the acoustic pulsation ($\omega_{ac}=2\pi f_{ac}$), $A(\xi,\omega_{ac})$ the amplitude of the acoustic plane wave of direction ξ and frequency ω_{ac}, $\omega_{ac,0}$ the central acoustic pulsation ($\omega_{ac,0}=2\pi f_{ac,0}$).
Due to the strong anisotropy of the crystal (Zaitsev (2003)), the phase matching condition or Bragg synchronism condition can be rewritten as:

$$\delta k(\xi) = 2\pi\left[\frac{\Delta n}{\lambda}\cos^2\theta_0 - \frac{\omega_{ac}(\xi)}{2\pi V(\xi)}\cos(\theta_0-\theta_a+\xi)\right] = 0 \Leftrightarrow \frac{\lambda\omega_{ac}(\xi)}{2\pi V(\xi)\Delta n} = \frac{\cos^2(\theta_0)}{\cos(\theta_0-\theta_a+\xi)} \tag{78}$$

The acoustic matched frequency can be expressed from the other parameters as:

$$\omega_{ac}(\xi) = \frac{\omega_{ac}(0)V(\xi)\cos(\theta_0-\theta_a)}{V(0)\cos(\theta_0-\theta_a+\xi)}. \tag{79}$$

The expression for the diffracted light field can be written as a superposition of plane waves:

$$E_d(t,X_{ac},Z_{ac}) \propto \int E_{in}(t,X_{ac},Z_{ac})e_{ac}\left(t,X_{ac},Z_{ac},\xi,\omega_{ac}(\xi)\right)d\xi. \tag{80}$$

The intensity of the diffrated field can be expressed as the superposition of the plane wave contribution with propagation angle ξ:

$$I_d(\omega_0) \propto \int_{-\pi/2}^{\pi/2}\operatorname{sin}c^2\left[\frac{L}{2V_0}\left(\omega_{ac}(\xi)-\omega_0\right)\right]\operatorname{sin}c^2\left[\omega_{ac}(\xi)\sin\xi\frac{D_a}{2V_0}\right]d\xi. \tag{81}$$

The acoustic wave velocity can be developped under a small deviation of the angle ξ as:

$$\begin{aligned} &V(\xi) = V_0\left(1 + a\xi + b\xi^2\right), \\ &\text{where } a = -\cot(\theta_0-\theta_a), \\ &b = \frac{\left(V_{001}{}^2 - V_{110}{}^2\right)}{2V_0^2}\left(\cos 2\theta_a - \frac{\left(V_{001}{}^2 - V_{110}{}^2\right)}{4V_0^2}\sin 2\theta_a\right), \end{aligned} \tag{82}$$

a characterizes the acoustic "walk-off" angle and b is the acoustic field spread.
The acoustic pulsation changes on the angle according to the parabolic law:

$$\omega_{ac}(\xi) = \omega_{ac,0}\left[1 + \left(b + \tfrac{1}{2}\right)\xi^2\right]. \tag{83}$$

The diffracted field intensity has the form

$$I_d(\omega_0) \propto \int_{-\pi/2}^{\pi/2} \sin c^2\left[\frac{L\omega_{ac,0}}{2V_0}\left[1+\left(b+\tfrac{1}{2}\right)\xi^2\right]-\omega_0\right]\sin c^2\left[\frac{\omega_{ac,0}\xi}{V_0}\frac{D_a}{2}\right]d\xi . \quad (84)$$

As expected from the first order theory, any divergence of the beams decreases the resolution of the device. While optical beam direction modifies mostly linearly the peak diffraction position in frequency, the acoustic direction has a quadratic dependance in the AOPDF configuration which modifies the symetry of the diffracted field intensity profile versus the frequency (or wavelength) as shown on figure 8.

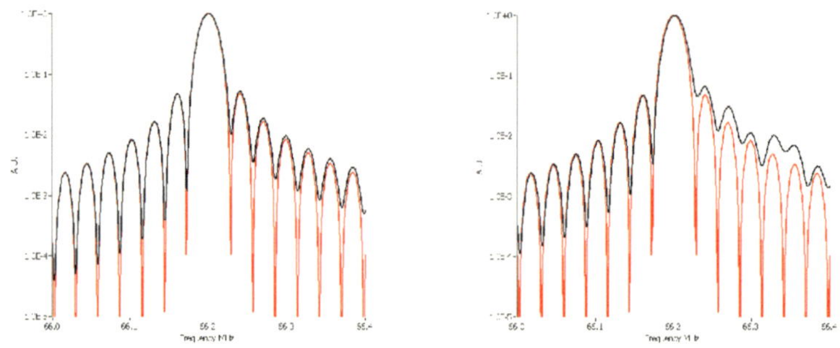

(a) (b)

Fig. 8. Simulation of acousto-optic diffraction for (a) D_a=8mm, L=25mm, (b) D_a=2mm, L=25mm, the black and red curves are respectively with and without acoustic beam limitation.

Considering a gaussian optical beam of 2mm at 800nm, its rayleigh length is z_R=15m and its divergence about $\delta\theta\approx 250\mu rad$. Without initial divergence, the resolution of the device is not affected by such a divergence.
This effect can be combined with the multi-frequencies through the momentum mismatch δk of optical and acoustic wave vectors:

$$\delta k(\xi) = k_{0z} - k_{1z} + K_z\left(\omega_{ac}(\xi)\right) = 2\pi\left[\frac{\Delta n}{\lambda}\cos^2\theta_0 - \frac{\omega_{ac}(\xi)}{2\pi V(\xi)}\cos(\theta_0 - \theta_a + \xi)\right] \quad (85)$$

and summation over the acoustic spectral and spatial frequencies.
These effects can be neglected in standard configuration.

4.2.2.3 Walk-off contribution

The main physical effect not already considered is the walk-off of the diffracted beam and of the acoustic beam. These walk-offs are due to the anisotropy of the crystal. The figure 13 illustrates the two walk-offs and their consequences on the output diffracted beam.
These effects combine each other also with the diffracted beam direction dispersion and finally result in a diffracted beam whose angular chirp is compensated by an adequate output face orientation but the spatial chirp illustrated on fig.9.a still remains. The effect is only a variation of the position of the different frequencies spatially. The maximum value corresponds to the walk-off over the complete crystal length and is given in table 1 for the

different crystals. By opposition to the 4f-line, this effect is not a coupling between optical frequencies and beam direction but rather a coupling between optical frequencies and beam position. The consequence of this coupling on a focal spot is very small.

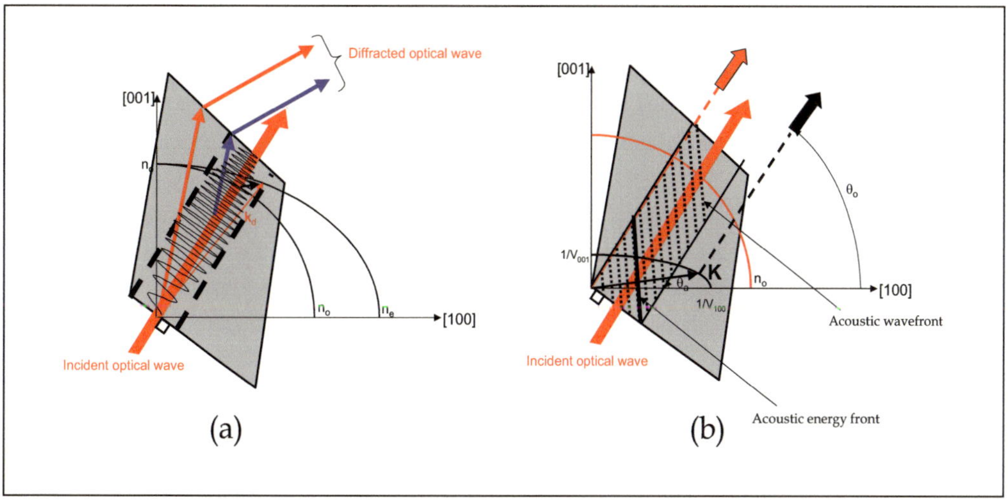

Fig. 9. Illustration of optical walk-off (a), acoustic walk-off (b).

The simulation of the device considering the walk-offs requires the following steps:

1. An input beam E(x,t,0) is propagated from its origin to the diaphragm aperture of the AOPDF by Fresnel propagation.
2. An iris aperture of diameter D_{iris} spatially filters the beam: $E(x,t,z) \to \mathrm{Rect}(x/D_{iris})E(x,t,z)$
3. The spatio-temporal response fonction of the AOPDF H(x,ω) is applied to the pulse
4. The beam is propagated a distance L to the lens
5. A thin lens of focal length f_L is applied
6. The beam is propagated to the focal plane.

The spatio-temporal caracteristics of the pulse shaping are directly include in the filter response function H(x,ω). This function is estimated from the desired ideal filter function H(ω)=A(ω)exp(iϕ(ω)) by applying the following algorithm:

1. Estimation of the dispersion of the crystal
2. Adding the compensation of the spectral phase introduced by the crystal to the output phase
3. Determination of the acoustic wave in the crystal
4. Determination of the crystal length L(x) integrating the prismatic output face
5. Estimation of the acoustic temporal window corresponding to L(x)
6. Introduction of the acoustic walk-off
7. Estimation of the acoustic filter function $H_{ac}(x,\omega)$
8. For each x, calculus of the acoustic delay $\tau(\omega_{ac})$ and determination of its longitudinal position in the crystal : $Z(x,\omega_{ac})$
9. Estimation of the walk-off for each pulsation: $W(\omega_{ac})$
10. For each ω_{ac}, an effective walk-off displacement X is estimated : $X(\omega_{ac})$
11. The filter optical function is calculated from $H_{ac}(x,\omega)$ including saturation of the diffraction and its correction: H(x,ω)

12. The ouput pulse is estimated by multiplying the input pulse $E(x,\omega)$ by the filter function $H(x,\omega)$ and applying the walk-off as $E_{out}(x,\omega)=E_{out}(x-X(\omega),\omega)$.

This model strongest hypothesis is the localization of the diffraction at a specific position $Z(x,\omega_{ac})$. As long as the bandwidth is large enough, this hypothesis is valid.

In the extreme case, a monochromatic acoustic wave fills completely the crystal. There are no specific position but in the same time there are no chromatic displacement at the output because of the monochromaticity!

Figure 10 shows simulation of a chirp, a delay and two pulses for the temporal intensity.

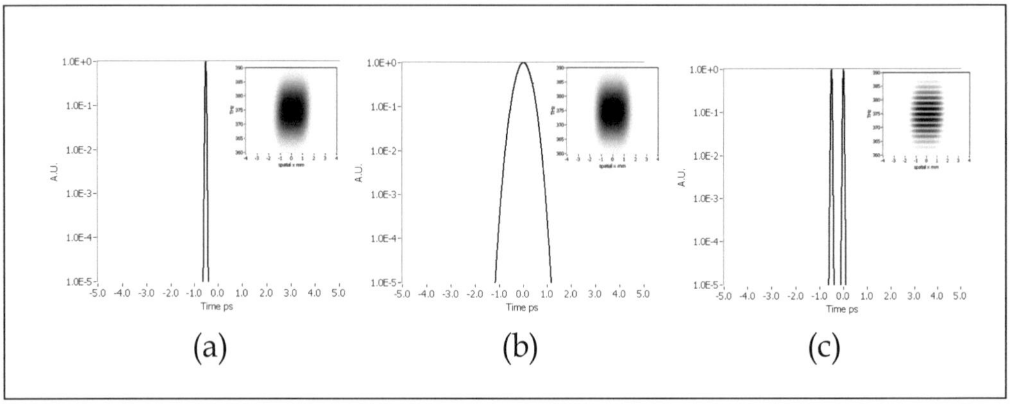

Fig. 10. AOPDF simulation including walk-off, temporal intensity in logarithmic scale and spatio-temporal vizualisation (inset) for (a) a 500fs delay, (b) a chirp 10000fs^2, (c) two pulses delayed by 500fs.

4.2.3 Conclusions on AOPDFs

The AOPDF devices are bulk and all the characteristics depend upon the parameters of the crystal. Different orientations and crystal lengths give either higher diffraction efficiency or higher resolution from 3ps, 0.6nm to 14ps,0.15nm at 800nm. The principal limitations are :

- the limited length of the crystal limiting the temporal window,
- the propagation of the acoustic wave (about 30µs for the crystal length),
- the efficiency of diffraction.

The crystal cuts optimization is a trade-off between the temporal window maximization and the efficiency of diffraction. The efficiency is determined by the acoustic power for each wavelength. This power depends upon the acoustic pulse shape and the maximal RF power acceptable by the transducer for the acoustic generation. This power is in the range of 10W. With an optimal chirp acoustic pulse high efficiency of diffraction can be achieved over large bandwidth. But if the acoustic pulse is compressed, then all the wavelength "share" the 10W peak power. This effect can decrease the efficiency of diffraction by an important factor. For example compensation of third order spectral phase influences the efficiency of diffraction.

The propagation of the acoustic wave through the crystal at about 720m/s implies to synchronize the acoustic wave generation with the optical pulse. The jitter on the synchronization has a direct incidence on the absolute phase of the diffracted pulse. The delay in acoustic generation can be directly linked to the optical delay by the acoustic and optical temporal windows lengths. This effect will be review on the CEP control off the

pulse. As the crystal length equivalent acoustic duration is about 30µs, a single acoustic wave can be synchronized perfectly with a single optical pulse only for laser repetition rate below 30kHz. For higher repetition rate, laser pulses will be diffracted by a same acoustic pulse at different position in the crystal leading to distortions.
As seen on the Fig.10, the walk-off in the crystal modifies spatially the output beam. The main effect is to spread the different wavelength at different positions. Depending upon the crystal caracteristics, the maximal displacement is in the range of 0.5mm or 1.5mm. This effect can be completely nullify by a double pass configuration as shown in the experimental implementations.

5. Pulse shaping examples

In this section, we compare the results obtain with the two technologies simulated with the models described in the previous part on identical pulse shaping examples. The ultrashort pulse considered is 20fs gaussian shape with a 2.3mm spatial gaussian shape.
This pulse will be pulse shaped by the four devices (table 3) to obtain first the best compressed pulse at the focal point of a perfect lens with an initial dispersion of $2000fs^2$ and $50000fs^3$ at 800nm, secondly a square pulse with a time/bandwidth product of 100, thirdly double pulses.
The table 3 sums up the parameters and caracteristics of the different pulse shapers used in this part at 800nm.

Parameters	4f-SLM 128 pixel pitch=100µm gap=3µm f=200mm $\theta_I=\theta_d$=20deg p=600lines/mm	4f-SLM 640 pixel pitch=100µm gap=3µm f=500mm $\theta_I=\theta_d$=20deg p=1200lines/mm	AOPDF WB25	AOPDF HR45
Spectral resolution $\delta\lambda$ nm	0.8	0.16	0.6	0.15
Bandwidth $\Delta\lambda$ nm	100	100	>200	>200
Temporal window @ 3dB (10%)	2.5ps	12 ps	3ps	15ps
Spatio-temporal slope or maximal walk-off	0.783mm/ps	0.293mm/ps	0.2mm/ps	0.1mm/ps
Number of points on 100nm	128	640	167	667

Table 3. Pulse shapers parameters.

5.1 Compressed pulses and focal spots

In this part, we simulate the compression obtained for (1) a pulse but stretched by $2000fs^2$ and $50000fs^3$ spectral phase with a gaussian shape of 50nm full-width at half-maximum, 2.3mm diameter. The compressed pulse for each pulse shaper will be caracterized on for its ps contrast, and the comparison between ideal compressed pulse energy distribution around the focus and the simulated one.
The ps contrast is shown directly by the intensity profile on a logarithmic scale.
The energy distribution is represented by plotting the energy distribution in the focal spot area and its difference with the ideally compressed pulse.
The initial pulse is stretched in time over about 100fs by the chirp and with a trailing edge due to the third order spectral phase on one ps at 10^{-6}. The compression of this pulse by the

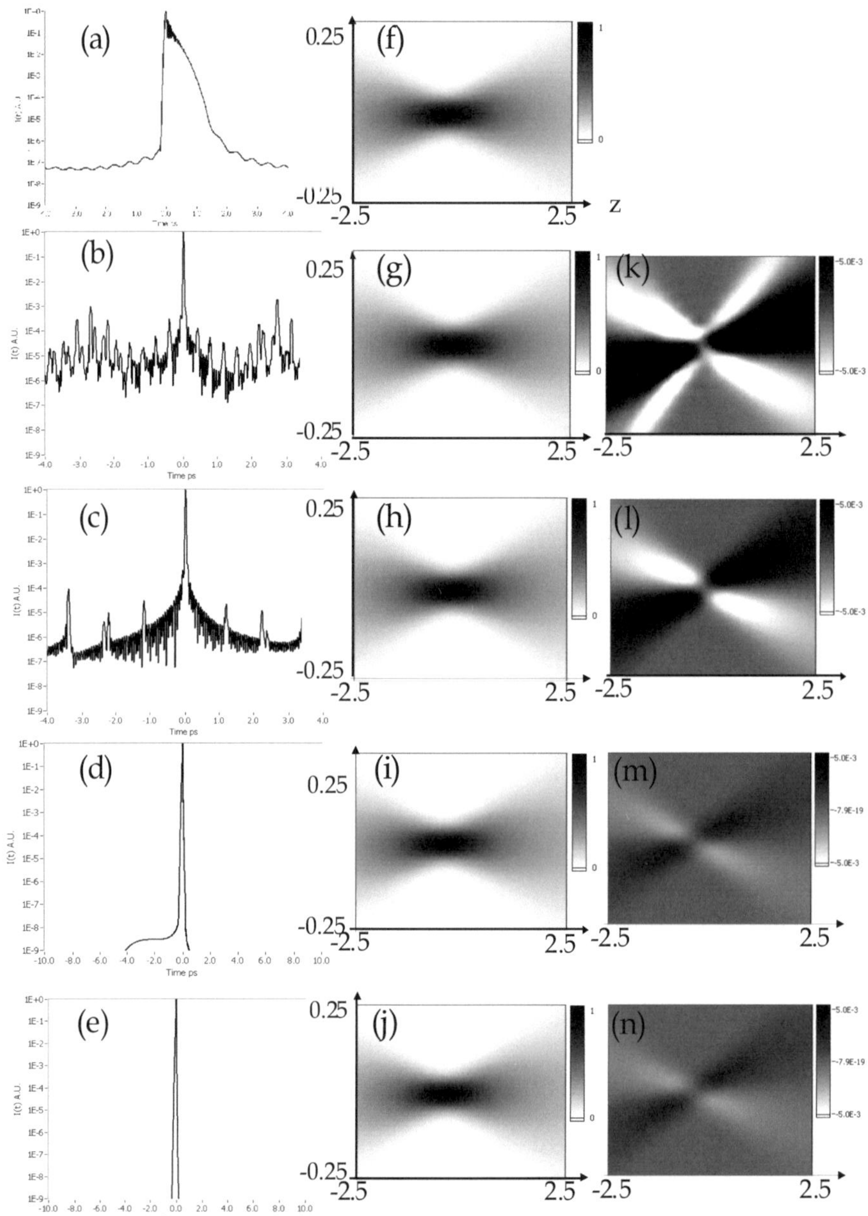

Fig. 11. Compression of 50nm gaussian shape pulse with $\phi^{(2)}$=2000fs^2 and $\phi^{(3)}$=50000fs^3. The first column shows the temporal intensities in logarithmic scale of (a) initial stretched pulse, (b) SLM128 compression, (c) SLM640 compression, (d) AOPDF WB25 compression, (e) AOPDF HR45. The second and third columns contain respectively peak power and energy distribution maps for respectively (f) initial stretched pulse, (g,k) SLM128, (h,l) SLM640, (i,m) AOPDF WB25, (j,n) AOPDF HR45.

pulse shaper points out specificities on the contrast. Due to pixelization, phase wraps and smoothing, the SLM128 and SLM640 4f-pulse shapers create pulse replicas at a level 10^{-3} and 10^{-4} respictevely. These pulses mainly due to gaps, pixelization and phase wraps are on top-off a background at 10^{-6} due to pixel smoothing. On the focal spots the deviation of the energy distribution is about 1% for these two devices. The AOPDFs recompression leads to a pulse without any temporal alteration for the High Resolution 45 device and a small pedestal at 10^{-8} for the Wide Band 25. This pedestal is due to a cut of the acoustic wave which is slightly longer than the crystal itself. On the focal spot the effect of the walk-off is more than 10 times smaller than for 4-f pulse shaper.

High contrast compression of pulses clearly requires no pixelization, gaps, phase wraps or smoothing and thus is better achieved by AOPDFs.

5.2 Square pulse with Time/Bandwidth product >100

Temporal square pulses are of interest both for Free Electron Laser electron bunch seeding and for optical parametric amplification pumping. The temporal intensity required has fast rise time (<500fs) and a flat top pulse without ripples. The very first idea to obtain such a pulse is to apply a flat temporal phase and Fourier transform the square root of this temporal intensity profile. This leads to a spectral amplitude with a sinus cardinal shape. This shape has a large pedestal bandwidth and a relatively narrow central peak. Considering that the initial pulse to shape has a gaussian shape with 50nm fwhm, obtaining a 2.3ps square pulse with a sinus cardinal spectral shape will loose most of the energy and required a very precise shaping on the edge of the spectrum.

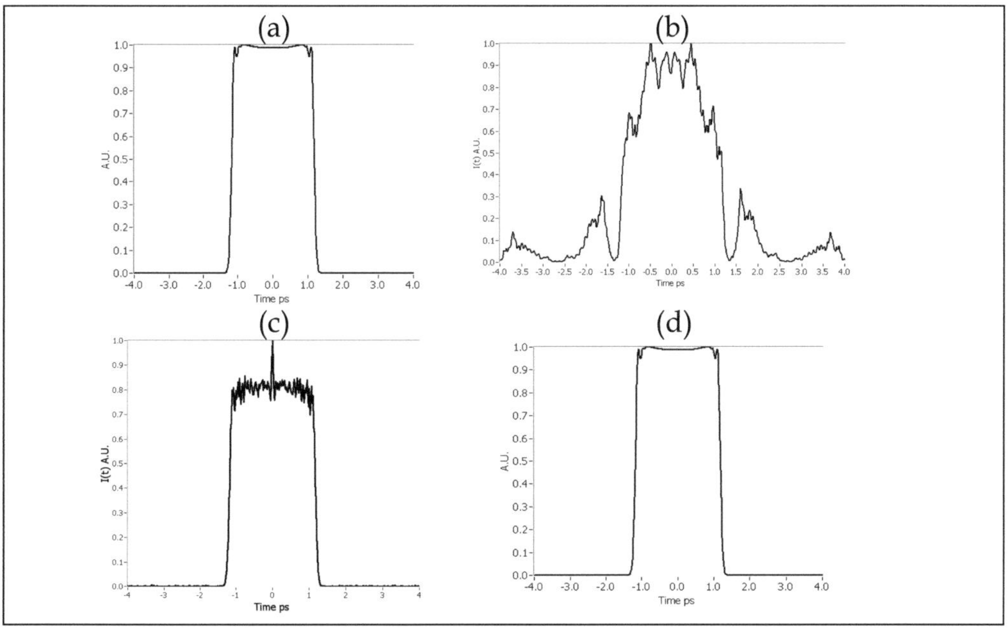

Fig. 12. Creation of a square temporal intensity pulse from a 50nm gaussian shape Fourier transform pulse by applying an amplitude filter super gaussian of order 6 with a fwhm of 52nm and a spectral phase with $\phi^{(2)}$= 20 10^3 fs2,$\phi^{(4)}$=-5.2 10^6 fs^4, $\phi^{(6)}$= 600 10^6 fs^6, with (a) ideal pulse shaper, (b) SLM128, (c) SLM640, (d) AOPDF WB25 or AOPDF HR45.

If no temporal phase shape is required, the optimal shape is a mix of amplitude and phase shaping. The fast rise time will be obtain by linear phase on the sides, the flat top by a chirp. The optimal shaping is obtained by applying a super gaussian amplitude filter of order 6 and fwhm 52nm on top of the 50nm fwhm gaussian spectrum, and a spectral phase composed by $\phi^{(2)}$= 20 10^3 fs2,$\phi^{(4)}$=-5.2 10^6 fs^4, $\phi^{(6)}$= 600 10^6 fs^6. Such pulse shaping is shown on the fig.16 below for the different pulse shapers and for an initial Fourier transform pulse.

The effects of pixelization and smoothing clearly modify the temporal shape for the 4-f pulse shapers. The effect of the beam size is not taken into account in this simulation. The central peak for the SLM640 pulse shaper can be compensated as proposed by Wefers and Nelson but only by an experimental feedback loop. This kind of feedback loop depends upon the measurement technique and accuracy. For the AOPDF, no default appears as long as the acoustic wave fits in the crystal.

5.3 Double pulses

To evaluate the quality of the pulse shaping for compressed pulse at the focal spot, three parameters are considered: the temporal intensity integrated over the focal volume considered; the spatial overlap integrals, within the focal volume, between frequencies; the relative strength of multiphoton transition probabilities, integrated across the focal volume. The last parameters have been introduced by Sussman [2008] to be relevant criteria in multiphoton experiments. The normalized overlap integrals measure the spatial variation of different colors through the focus:

$$O_{\lambda_1\lambda_2} = \frac{\int I_{\lambda_1} I_{\lambda_2} dxdz}{\sqrt{\left(\int I_{\lambda_1}{}^2 dxdz \int I_{\lambda_2}{}^2 dxdz\right)}}, \tag{86}$$

where $I_{\lambda 1}(x,z)$ represents the field intensity for a particular wavelength. The values of the overlap integrals $O_{\lambda 1\lambda 2}$ may vary considerably. A value of 1 indicates that two colors have a complete overlap throughout the focal volume. Variations below 1 are significant for any multiphoton experiments.

The probability for a vertical n-photon transition having no intermediate resonances is given by the nth-order power spectrum:

$$P_n \propto \left| \int (E(t,x,z))^n e^{-i\omega t} dt \right|^2. \tag{87}$$

Without space-time coupling, this transition probability is proportionnal to the peak intensity I at a point: $P_n \propto I^n$. Therefore, the ratio between the nth-order and the mth-order multiphoton transitions $P_n/P_m{}^{n/m}$ should be constant if only the peak intensity - and not the spectral content- at a spatial point is altered. In order to quantify this effect, this ratio is integrated over the focal volume, normalized and compare with the non space-time coupled one as

$$r_{mn} = \frac{\int \left| \left(P_n / P_m{}^{n/m} \right) - 1 \right| dxdz}{\int dxdz}. \tag{88}$$

In the absence of space-time coupling, these values would be zero. However, since the spectral content at each point is modified, the ratios are not constant and their deviation from 1 can be considerable. The global value is integrated over the volume of interest, for example the focal volume.

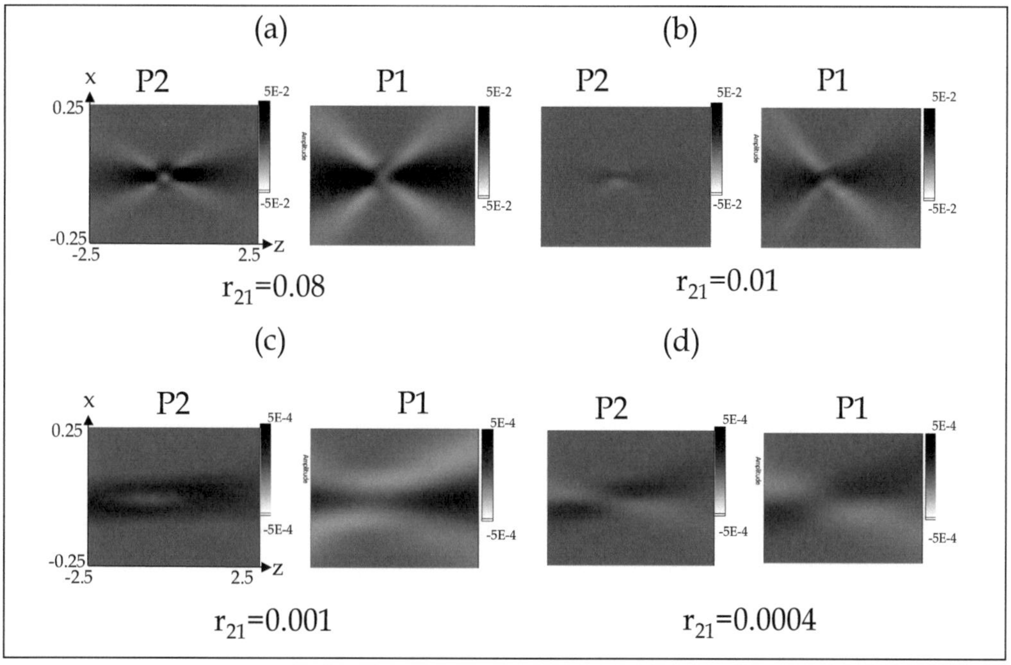

Fig. 13. Comparison of two pulses delayed by +250fs and -250fs and maps of the differences of energy P1 and power P2 at the focal plane of a 100m focal length lens for (a) SLM128, (b) SLM640, (c) AOPDF WB25 and (d) AOPDF HR45.

The influence of the spatio-temporal coupling is caracterized both by the r_{21} ratio and the energy and power difference maps. As pointed out by the spatio-temporal parameter, the SLM128 has the highest coupling, and the AOPDF HR45 the lowest. These effects can be very important when the pulse shaper is used to optimize a non linear effect at the focus on a non strictely homogeneous media. Indeed, modification of the spatial profile instead of the temporal one can be the predominant effect in the optimization.

5.4 Conclusions

Depending upon the relevant parameters of a pulse shaping experiment, the pulse shaper has to be adapted. An adequate simulation of the pulse shaper should estimate the minimum requirements. Also experimental implementation and alignement tolerances that are beyond the scope of this chapter should be taken into account. Feedback loop can be used to optimize the temporal shape. But some inherent defaults of the pulse shaping technology can also compromise the experimental results and cannot be compensated by any feedback loop. Moreover the accuracy and dynamic of the measurement that should be used for the loop is already an experimental challenging part.

6. Experimental implementations

This part gives examples of experimental implementations of pulse shaping with their advantages and limitations. As pulse shapers can be applied to a wide range of applications, experimental implementations are reviewed in the scope of the laser source from oscillators to multi-TeraWatt laser systems. Example of feedback loop with measurement devices will also be given.

6.1 Oscillators

Applications of pulse shaping with oscillators are direct pulse shaping of the output train of pulses. Among them are for examples, multiphoton microscope imaging, white light optimization for spectroscopy. The pulse shaper is used directly before the experiment.
For 4-f pulse shaper there are no modification of the pulse shaping response function due to the high repetition rate of the laser. The mask can be considered as fix. An imaging relay optics should be used to avoid magnification of the space-time coupling effects [Tanabe (2005)]. Many different kind of experiments have benefit from optimization by feedback loop such as multiphoton microscopy or coherent control (fig.14). This feedback can be either used to optimize the pulse shape [Coello 2008], or to directly optimize an experimental result by blind algorithms [Assion 1998, Brixner 2000]. Refreshing rate of standard LC SLM is about few tens of millisecond. Higher modulation refreshing rate can be obtain by using two lines LC SLM and switching from one line to the other.
For AOPDF, the acoustic wave is moving at 720m/s in the crystal. Therefore, the acoustic pulse is moving in the crystal from pulse to pulse. Synchronization of the measurement system with the acoustic wave is then needed to eliminate the measurement with a partial acoustic wave in the crystal. In standard 25mm crystals, the complete acoustic time window is about 30µs. Thus depending upon the duration of the acoustic wave (Δt_a) used for the shaping, this acoustic wave is totally include in the crystal during 30µs-Δt_a. The measurement has to be gated on during this time and off in between to consecutive acoustic pulse (Δt_a) (fig.14). This drawback of AOPDF can be overcomed by its higher refreshing rate (up to 30kHz) that can be used for differential measurements between two pulses shapes [Ogilivie 2006] eventually with heterodyne detection.

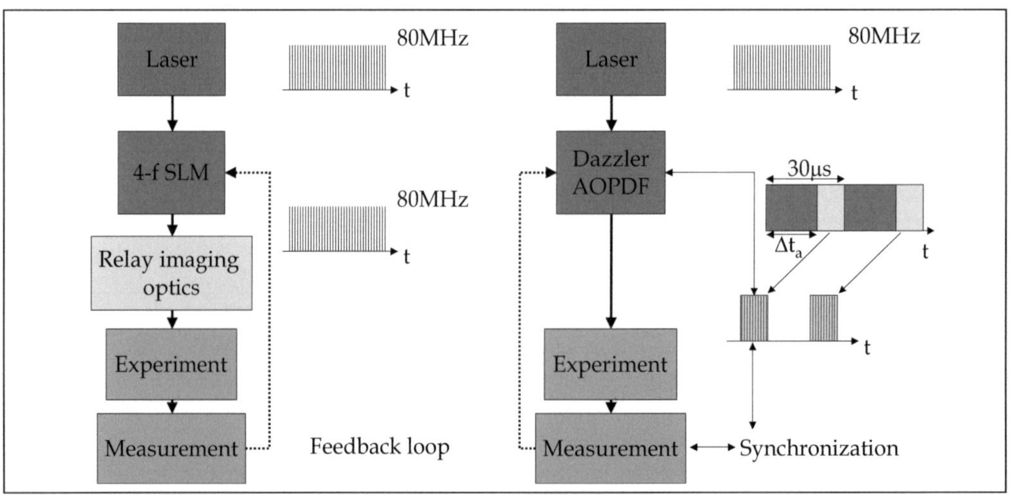

Fig. 14. Implementation of 4-f pulse shaper and AOPDF at the output of an oscillator.

New implementations of these two techniques for multiphoton or CARS microscopy, coherent control are currently published demonstrating higher efficiency or sensitivity.

6.2 Amplified systems

In amplified system, depending upon the damage threshold energy and non linearities in the pulse shaper, the device is inserted in the laser chain or at the output.
The irradiance limit ($W.m^{-2}$) of pulse shaper is different for AOPDF and for 4f-pulse shaper. For AOPDF, this limit is defined upon the non-linearities in the massive TeO2 crystal. To avoid any significative distorsion, this limit is defined as an upper value of cumulative effect of self-phase modulation (B-integral):

$$B = \max\left(\varphi_{nonlinear}\right) = \max\left(\int \frac{2\pi}{\lambda} n_2 I(t,l)\, dl\right), \tag{89}$$

where n_2 is the nonlinear index ($n_2 \approx 23 \times 10^{-13}$esu $\approx 67 \times 10^{-16}$ cm^2/W for TeO_2), λ is the optical wavelength, I(t,l) the irradiance or intensity ($W.m^{-2}$). Depending upon the tolerances on spatial and temporal distorsions due to this slef-phase modulation, the B-integral limit is set to 0.1, 1 or π [Perry (1994)]. Considering a medium value of 1, the input irradiance can be up to 200μJ/cm^2 for 30fs pulses [Monmayrant (2005)]. This value includes the dispersion of the pulse by the crystal itself. A 200μJ/cm^2 pulse of 25fs and a 220μJ/cm^2 pulse of 35fs at 800nm lead to the same B-intergal of 1. Thus depending upon the pulse duration and size, pulses up to mJ level can be used with AOPDF before compression.
For 4f-pulse shaper, the gratings can have high damage threshold and the limitation is about 1GW/cm^2. This implies that a 4f-pulse shaper can be used at the output of an amplified system at the milliJoule level.
The figure 15 details the different position where the pulse shaper can be inserted. The stretcher and compressor are linear elements and do not affect the pulse shaping. The amplifier can be either a CPA amplifier or an optical parametric amplifier. Amplifiers are in general non linear because of saturation, red shift, four wave mixing [Liu (1995)].

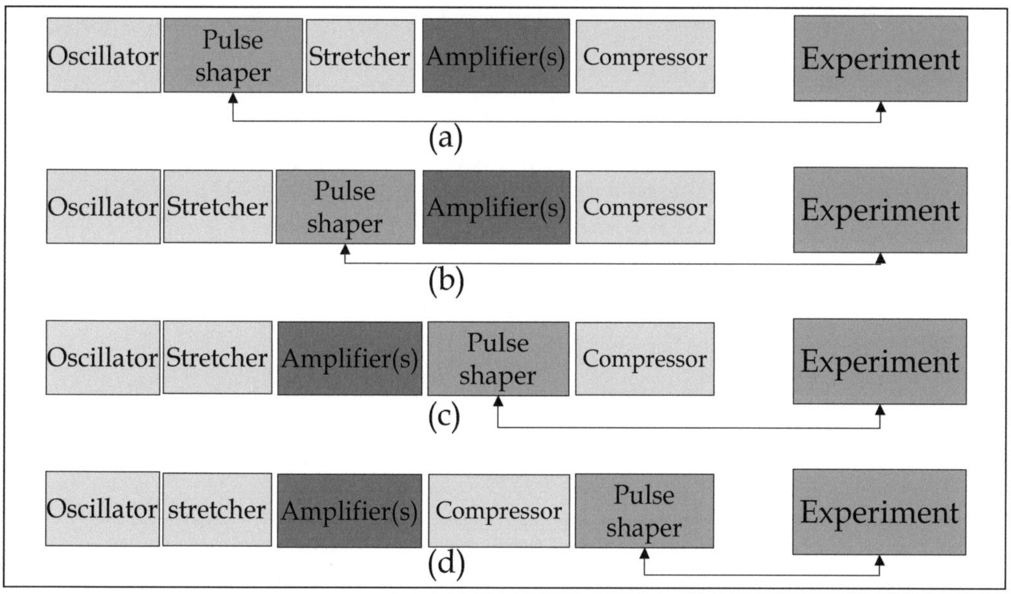

Fig. 15. Implementations of a pulse shaper amplified systems: (a), (b) indirect pulse shaping, (c), (d) direct pulse shaping.

Depending upon the position of the pulse shaper, before or after the "non linear" element, the amplifier(s), the shaping will be linear or not. The linear case is "named direct pulse shaping" because the shaping will be transmitted from the pulse shaper to the experiment directly. In the non linear case, or "indirect pulse shaping", the shaping introduced is altered by the non linear element. Indeed, as the amplification distords the spectral amplitude, when the pulse shaper is inserted before the amplifier, the shaping is modified by the amplifier. Indirect pulse shaping restricts the possibilities of pulse shaping as mentionned in the Special Topics hereafter. As an example, a double pulse injected in the amplifier will result in multiple pulses at the output [Liu (1995), Boyle (2001)]. Limitations are due the amplifier caracteristics. In CPA systems, the phase shaping can be considered as a small perturbation of the Chirped introduced and is generally conserved linearly through the amplification. Main limitation comes from spectral amplitude distorsions. Part of the distortions can be pre-compensated by a feedback loop but not all of them.

6.3 Feedback loops

As mentionned previously, there is a strong interest of feeding back the pulse shaper with measurements of the shaped pulses. This loop can overcome some defaults of the pulse shaper or optimize experimental results such as compression via non linear generation optimization or specific excitation in coeherent control.

To illustrate such a loop, optimization of the pulse duration by flattening the spectral phase will be described for the two techniques.

Many publications describes feedback loop for optimization with 4f-pulse shaper after amplification but very few with the shaper before the amplifier. When the pulse shaper is directly before the amplifier, the feedback loop is used to overcome inaccuracy of the pulse shaper. Few publications present loops with advanced measurement devices as Frequency Resolved Optical Gating (FROG) [Kane (1993)] or Spectral Phase Interferometry for Direct Electric-field Reconstruction (SPIDER) [Iaconis (1998)] with 4f-pulse shaper. Direct shaping optimization with a SPIDER (pulse shaper just in front of the SPIDER) on a ultrashort pulse (3.8fs) requires 5 iterations [Schenkel (2003)].

For AOPDF, optimization in indirect pulse shaping configuration with a SPIDER is regularly used in CPA laser system with a single correction step.

This feedback loop can be sketched as:

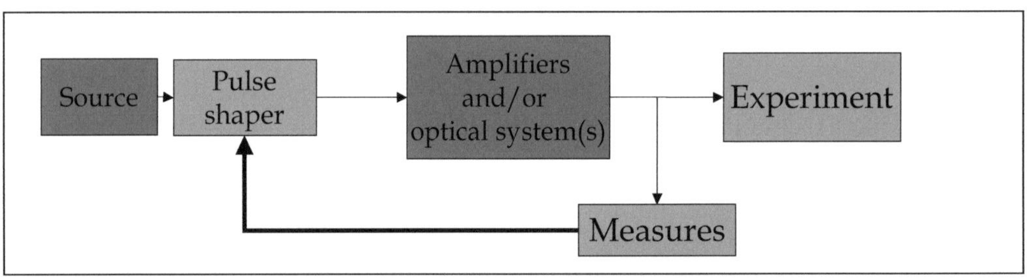

Fig. 16. Implementations of a pulse shaper feedback loop for pulse compression optimization.

The feeding parameters are the spectral amplitude and phase, but they can be obtained through different type of mesurements. With SPIDER, care should be taken on the spectral phase on the sides of the spectrum where the measurement is inaccurate. Polynomial prolongation from valid area of measurement shouold be used to avoid significant cut due to the pulse shaping [Oksenhendler (2003)]. With FROG, the feedback is more efficient if done by direct reference to FROG traces [Ohno (2002)].

In conclusion, implementing a feedback loop with a pulse shaper can overcome indirect shaping or inaccuracy of the pulse shaper. Special attention should be taken onto convergence and limitation problem. As an example if the compression of the pulse requires a shaping that is out of the scope of the pulse shaper used, the optimal solution will not be the compressed pulse (best optimum) but a second and local optimum that can be achevied by the pulse shaper. User should check that the solution seeked is in the scope of the loop.

7. Special topics

As described previously, a precise determination of the key parameters of the experiment should be done to select and adapt the proper pulse shaping technology. This part deals with some examples that are experimentally important and point out the relevant parameters to consider and the technologies that can be applied.

7.1 Carrier Envelope Phase control

In attosecond pulse generation, the Carrier Envelope Phase has to be stable to get a stable generation. The time dependence of the electric field associated with an optical pulse can be described as a fast sinusoidal oscillation, called the carrier, multiplied by a more slowly varying envelope function. When the pulse propagates through a medium, the relative position between the carrier wave and envelope will in general change due to chromatic dispersion causing a difference between phase velocity and group veocity, and possibly also due to optical nonlinearities. The carrier envelope offset phase (or absolute phase) of a pulse is defined as the difference between the phase of the carrier wave and the envelope position, the latter being converted to a phase value.

As the pulse shaper has the ability to control directly the realtive absolute phase, it can control the carrier envelope phase.

For a 4-f pulse shaper, this phase is directly introduced by the mask and can be changed at the refreshing time of the mask (about 10ms). Two effects can introduce fluctuations on the carrier envelope phase: beam pointing fluctuations and mechnanical vibration of the optical components of the pulse shaper.

For an AOPDF pulse shaper, the absolute phase value is directly linked to the phase value of the acoustic wave and its relative position in the crystal. The synchronization of the acoustic wave generation with the optical pulse fixes this position. A delay of $\delta\tau$ on this acoustic synchronization will result in an optical delay $\delta\tau_{opt}\approx\delta\tau(3ps/30\mu s)$. This delay introduces a modification of the absolute phase $\delta\phi=2\pi(\delta\tau_{opt}.f_{opt})$. So for a phase modification lower than $\pi/20$, the jitter on the acoustic wave generation has to be lower than 500ps. This low jitter operation requires a synchronization of the RF clock with the oscillator laser repetition rate. With this synchronization, the AOPDF can control the phase with an accuracy determined by the residual jitter, at refreshing time down to 30kHz. As for 4-f pulse shaper, the beam pointing will affect the carrier envelope phase stability. But mechanical vibration as it is a bulk crystal are less important.

7.2 Indirect pulse shaping

In some experiments, pulse shapers cannot be used directly on the pulse to shape. For example on high peak power TeraWatt laser system, the pulse shaper can't stand the output energy of the laser. In these cases, the pulse shaper is used before non linear systems as amplifiers or non-linear elements. The shaping of the wanted pulse is then obtained through an indirect process. This indirect pulse shaping modifies the possibilities of shaping. For example, with a Ti:Saphire CPA amplifier, the spectral amplitude is modified by redshift as shown on the figure 17. The spectral phase introduced by the pulse shaper is only a perturbation compared with the chirp of the stretcher. The amplifier is linear for the spectral phase and non linear for the amplitude. The limitations on the pulse shapes are due to the distorsion in the amplifier. A feedback loop on the amplitude can overcome some of the distorsions.

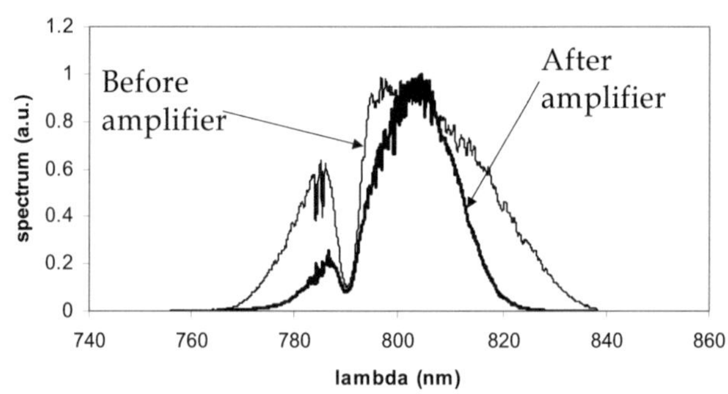

Fig. 17. Indirect pulse shaping.

8. Conclusion

Pulse shaping techniques introduces the concept of linear filter in the field of ultrashort pulses. Optimization of pulse duration, simulation of linear optical set-up (interferometers, pass band filters ...) can be achieved, modified or adapted directly by programming different filter function. This capability opens new possibilities such as illustrated by the pulse measurement techniques using an acousto-optic programmable dispersive filter.

An careful analysis of the pulse shaper limitations and defaults such as pulse replica, space-time coupling, should be done, prior to the choice of the pulse shaper technology, to fit the experimental needs.

9. References

Assion A., Baumert T., Bergt M., Brixner T., Kiefer B., Seyfried V., Strehle, Gerber G., "Control of chemical reactions by feedback-optimized phase-shaped femtosecond laser pulses", Science Vol.282, pp. 919-922 (1998)

Boyle M., Thoss A., Zhavaronko N., Korn G., Kaplan D., Oksenhendler T., "Amplitude and phase shaping in a high average power 1kHz CPA-laser", CLEO 2001

Brixner T., Oehrlein A., Strehle M., Gerber G., "Feedback-controlled femtosecond pulse shaping", Appl.Phys.B 70, pp. S119-S124 (2000)

Coello Y., Lozovoy V.V, Gunaratne T.C., Xu B., Borukhovich I., Tseng C.-H. Weinacht T., Dantus M., "Interference without an interferometer: a different approach to measuring, compressing, and shaping ultrashort pulses", JOSA B, Vol.25, pp. A140-A150 (2008)

Danailov M.B., Christov I.P, (1989)."Time-space shaping of light pulses by Fourier optical processing", J.Mod.Opt. Vol 36, 725 (1989).

Dugan M.A., Tull J.X., Warren W.S., (1997). J.Opt.Soc.Am.B Vol.14, 2348 (1997).

Fork R.L., (1981). R.L.Fork, B.I. Greene, C.V.Shank, Appl.Phys.Lett. 38 671 (1981).

Froehly C., Colombeau B., Vampouille M., (1983). *Progressinf Optics*, edited by E.Wolf (north-Holland, Amsterdam, 1983), Vol.20, pp.65-153.

Iaconis C., Walmsley I.A.,"Spectral phase interferometry for direct electric-field reconstruction of ultrashort pulses", Opt.Lett., Vol. 23, pp.792-794 (1998)

Kane D.J., Trebino R., "Single-shot measurement of the intensity and phase of an arbitrary ultrashort pulse by using frequency-resolved optical gating", Opt.Lett., Vol.18, pp. 823-825 (1993)

Khoo I.-C., Wu S.-T., "Optics and nonlinear optics of liquid crystals", World Scientific Publishing, Singapore (1993)

Laude V., "General solution of the coupled-wave equations of acousto-optics", J.Opt.Soc.Am.B, Vol.20, No.12, 2307-2314 (2003)

Liu X., Wagner R., Maksimchuk A., Goodman E., Workman J., Umstadter D., Migus A., "Nonlinear temporal diffraction and frequency shifts resulting from pulse shaping in chirped-pulse amplification systems", Opt.Lett. Vol.20, pp. 1163-1165 (1995)

Martinez O.E., (1986), "Grating and prism compressors in the case of finite beam size", J.Opt.Soc.Am.B vol.3, pp.929-934 (1986)

Monmayrant A., Chatel B., (2003). "Newphase and amplitude high resolution pulse shaper", Rev.Sci.Instrum., Vol.75, No.8, 2668-2671 (2003).

Monmayrant A., "Façonnage et caractérisation d'impulsions ultracourtes. Contrôle cohérent de systèmes simples.", PhD Thesis, Université Paul Sabatier Toulouse III, 2005.

Monmayrant A., Arbouet A., Girard B., Chatel B., Barman A., Whitaker B.J., Kaplan D., "AOPDF-shaped optical parametric amplifier output in the visible", App.Phys.B, 81, pp. 177-180 (2005)

Ogilvie J.P., Débarre D., Solinas X., Martin J.-L., Beaurepaire E., Joffre M. « Use of coherent control for selective two-photon fluorescence microscopy in live organisms", Opt.Expr., Vol.14, pp. 759-766 (2006)

Ohno K., Tanabe T., Kannari F., "Adaptative pulse shaping of phase and amplitude of amplified femtosecond pulse laser by direct reference to frequency-resolved optical gating traces", JOSA B, Vol.19, pp.2781-2790, (2002).

Oksenhendler T., Rousseau P., Herzog R., Gobert O., Perdrix M., Meynadier P., "20Hz Femtosecond laser amplifier optimization using an AOPDF pulse shaper and a SPIDER", CLEO 2003

Oksenhendler T., "Mesures et contrôles temporels dans le domaine des lasers ultrabrefs", PhD thesis, Ecole Polytechnique 2004.

Perry M.D., Ditmire T., Stuart B.C., "Self-phase modulation in chirped-pulse amplification", Opt.Lett, Vol. 19, pp. 2149-2151 (1994)

Schenkel B., Biegert J., Keller U., Vozzi C., Nisoli M., Sansone G., Stagira S., De Silvestri S., Svelto O., "Generation of 3.8fs pulses from adaptative compression of a cascaded hollow fiber supercontinuum", Opt.Lett., Vol.28, pp.1987-1989 (2003)

Strickland D., Mourou G., (1985). "Compression of amplified chirped optical pulses, Opt.Comm. 55, 219-221, (1985)

Sussman B.J., Lausten R., Stolow A., (2008). "Focusing of light following a 4-f pulse shaper: Considerations for quantum control", Phys.Rev.A 77, 043416-1043416-11 (2008).

Tanabe T., Tanabe H., Teramura Y., Kannari F., "Spatiotemporal measurements based on spatial spectral interferometry for ultrashort optical pulses shaped by a Fourier pulse shaper", J.Opt.Soc.Am.B Vol.19,2795-2802 (2002)

Tanabe T., Kannari F., Korte F., Koch J., Chichkov B., "Influence of spatiotemporal coupling induced by an ultrashort laser pulse shaper on a focused beam profile", Appl.Opt. Vol.44, 1092-1098 (2005)

Thurston R.N., Heritage J.P., Weiner A.M., Tomlinson W.J., "Analysis of Picosecond Pulse Shape Synthesis by Spectral Masking in a Grating Compressor", IEEE J.Quantum Electron.22, 682-696 (1986).

Tournois P., "Acousto-optic programmable dispersive filter for adaptative compensation of group delay time dispersion in laser systems", Opt.Comm, Vol.140, 245-249 (1997)

Vaughan J.C., Hornung T., Feurer T., Nelson K.A, (2005). "Diffraction-based femtosecond pulse shaping with a 2D SLM", Opt.Lett. 30, 323-325 (2005).

Vaughan J.C., Feurer T., Stone K.W., Nelson K.A., (2006), "Analysis of replica pulses in femtosecond pulse shaping with pixelated devices", Opt.Exp. Vol.14, No.3,1314-1328 (2006).

Verluise F., Laude V., Cheng Z., Spielmann C., Tournois P., "Amplitude and phase control of ultrashort pulses by use of an acousto-optic programmable dispersive filter: pulse compression and shaping", Opt.Lett. Vol.25, 575-577 (2000)

Wefers M.M., Nelson K.A., (1995). "Analysis of programmable ultrashort waveform generation using liquid-crystal spatial light modulators", J.Opt.Soc.Am.B, Vol.12, No.7, 1343-1362 (1995).

Wefers M.M., Nelson K.A., (1996). "Space-time profiles of shaped ultrafast optical waveforms", IEEE J.Quant.Elec Vol.32, No.1, 161-171 (1996).

Weiner A.M., (2000). "Femtosecond pulse shaping using spatial light modulators", Rev.Sci.Instrum., Vol.71, No.5, 1929-1960 (2000).

Yariv A., Yeh P., "Optical waves in crystals", Wiley, New York, 1984.

17

Femtosecond Filamentation in Temperature Controlled Noble Gas

Zhenming Song[2], Yun Wei[1], Shiying Cao[1], Weipeng Kong[2], Dongqing Pang[2], Ruxin Li[3], Qingyue Wang[2] and Zhigang Zhang[1]
[1]*Institute of Quantum Electronics, State Key Laboratory of Advanced Optical Communication Systems and Networks, School of Electronics Engineering and Computer Science, University of Beijing, Beijing 100871*
[2]*Ultrafast Laser Laboratory, School of Precision Instrument and Optoelectronics Engineering, Key Laboratory of Optoelectronic Information Technical Science, EMC, University of Tianjin, Tianjin 300072*
[3]*State Key Laboratory of High Field Laser Physics, Shanghai Institute of Optics and Fine Mechanics, Chinese Academy of sciences, P.O. Box 800-211 Shanghai 201800*
China

1. Introduction

Intense few to monocycle pulses are the mJ or sub-mJ ultrashort pulses which have one to two optical periods (800 nm corresponding to 2.67 fs). They play more and more important roles in today's scientific and technological research fields, such as time-resolved measurements of electron dynamics in atoms and molecules, high-order harmonics and isolate attosecond pulse generation. To date, the major way of achieving intense few to monocycle pulses is to compress the spectra broadened through hollow fiber filled with noble gases or filamentation.

Combined with the technique of noble gas-filled hollow fiber and feedback phase compression, Yamashita group has achieved the shortest pulse duration 2.6 fs in the visible light range (Yamashita et al., 2006; Yamane et al., 2003; Matsubara et al., 2007). However, due to the damage at the entrance of the fiber, self-focusing and gas breakdown, the pulse energy in a gas-filled hollow fiber is limited to μJ level. In addition, the loss in the hollow fiber also contributes to the low efficiency. Spectrum broadening through filamentation was presented as one of the extremely simple technique to generate intense few to monocycle pulses (Hauri et al. 2004; Hauri et al., 2005; Zaïr et al., 2007). In addition, the pulse self-compression can be realized in the filamentation (Couairon et al., 2006; Couairon et al., 2005). However, the competition of the multi-filament (Shen, 1984; Bergé et al., 2004; Vidal & Johnston, 1996; Braun et al., 1995; Skupin et al., 2004; Hosseini et al., 2004) when the peak power is several times above the critical power introduces instability between the filaments and restricts the achievement of high quality ultrashort intense pulses. The spectrum broadening through hollow fiber with a gradient pressure seems to be an efficient way to overcome the breakdown at the entrance of the fiber, and preserve the spatial and spectral qualities of the pulses with final energy of multi-mJ (Suda et al., 2005; Sung et al., 2006), but it is difficult to realize in the filamentation, because differential pumping does not only result in a large consumption of expensive noble gases, but also in a gas flow that disturbs the spatial beam stability and the spectral characteristics.

In this Chapter, we present a novel technique that could have the same effect as gradient pressure scheme to produce controllable nonlinear refraction index: the gradient temperature in a sealed tube filled with noble gas (Song et al., 2008a; Song et al., 2008b; Cao et al., 2009). That is, in practice, to heat the entrance and cool the exit end of the tube, so a temperature gradient can be formed along the tube. The gas pressure in this sealed tube reaches equilibrium but the gas density is gradually increased along the tube. This gradient temperature should have the similar effect as that of gradient pressure, because these two schemes both have increasing gas density distribution along the propagation of the pulse. Therefore, the incident pulse will be allowed to have a higher energy because the lower gas density at the entrance. The spectrum broadening will go on along the tube because the increasing gas density will compensate the decreasing peak power due to loss and dispersion. The ideas of gradient pressure and gradient temperature schemes are shown in Fig. 1. In the gradient pressure case the gas flow has disturbance to the system which will affect the stability and quality of the output beam while in the gradient temperature case the gas reaches equilibrium to avoid the disadvantages of the gradient pressure. Theoretical and experimental results will be shown in section 3 and 4 respectively to illustrate and prove the feasibility and effectiveness of the gradient temperature scheme and a final conclusion will be drawn in section 5 of this Chapter. Before we show the results we shall simply make some efforts on the filamentation and dynamics in filamentation in the section below.

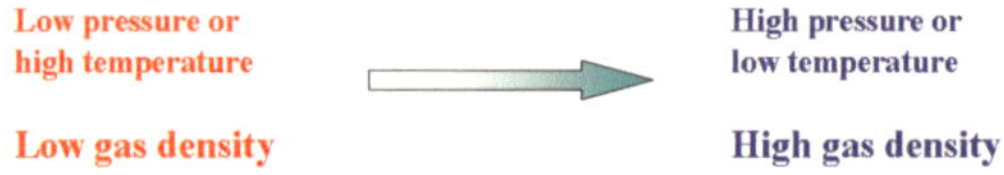

Fig. 1. Illustration of the gradient pressure and gradient temperature technique.

2. Theoretical model of dynamics in filamentation

Filamentation is a well-known phenomenon of the propagation of intense laser pulses in any kinds of Kerr media (gases, liquids, and transparent solid materials) under certain conditions. It is the result of the balance between the self-focusing due to the nonlinear refractive index of the medium dependence on the intensity and the defocusing due to the formation of the plasma and beam diffraction when a sufficient intense ultrashort laser pulse propagates in the transparent medium. A single filament always occurs when the incident pulse with a peak power between one and a few times P_c where P_c is the critical power values of self-focusing. For a CW Gaussian beam, the self-focusing critical power is expressed by (Marburger, 1975):

$$P_c = \lambda_0^2 / 2\pi n_0 n_2 \tag{1}$$

where λ_0 is the wavelength in vacuum, n_0 and n_2 are the linear and nonlinear refraction index of the medium. However, when the laser peak power is much higher r than the self-focusing critical power, the incident beam typically breaks into several long and narrow filaments, a phenomenon known as multiple filamentation (MF). Multiple filamentation is an unstable state because there is the interaction among the filaments (Shen, 1984; Bergé et al., 2004; Vidal & Johnston, 1996; Braun et al., 1995; Skupin et al., 2004; Hosseini et al., 2004), which limits the energy that each filament can carry.

2.1 Model Equations

In order to describe the dynamic of filamentation, a 3D extended nonlinear Schrödinger equation coupled with the electron density equation are usually employed. The extended nonlinear Schrödinger equation for the electric field envelop $\varepsilon(r, t, z)$ in a reference frame ($t = t_{lab}$-z/v_g) moving at the group velocity v_g can be written as,

$$\begin{aligned} \partial\varepsilon / \partial z &= i(\partial^2\varepsilon / \partial r^2 + \partial\varepsilon / \partial r / r) / 2k_0 n_0 - ik''(\partial^2\varepsilon / \partial t^2) / 2 \\ &+ik_0 n_2 |\varepsilon|^2 \varepsilon - ik_0 \rho\varepsilon / 2n_0\rho_c - \sigma\rho\varepsilon / 2 - U_i W(I)(\rho_{nt} - \rho)\varepsilon / 2I \end{aligned} \tag{2}$$

where the evolution of the electron density ρ follows the equation,

$$\partial\rho / \partial t = W(I)(\rho_{nt} - \rho) + \sigma\rho|\varepsilon|^2 / U_i \tag{3}$$

where the terms on the right-hand side of Eq. (2) describe transverse beam diffraction, GVD, nonlinear self-focusing, absorption and defocusing due to the electron density ρ, and absorption by photo-ionization; the terms on the right-hand side of Eq. (3) are the generation of the plasma by photon ionization (involving multiphoton ionization and tunnel ionization) and avalanche (multiplication of the electrons in the laser field). In Eqs. (2) and (3), k_0 is the wave number in vacuum, k'' the second order dispersive coefficient, ρ_c the critical plasma density above which the plasma becomes opaque, ρ_{nt} the neutral atom density, σ the cross-section for inverse bremsstrahlung, U_i the ionization potential, $I = \varepsilon\varepsilon^*$ the laser intensity and $W(I)$ the photoionization rate. Equations (2) and (3) can be solved simultaneously by the split-step method (Agrawal, 2007). The photoionization rate from Keldysh theory (Keldysh, 1965) can be calculated by the PPT (Perelemov-Popov-Terent'ev) model (Perelemov et al., 1966) which reduced to tunnelling (Ammosov et al., 1966) or multiphoton ionization models under certain conditions. An initial collimated Gaussian beam following a lens read as:

$$\varepsilon(r,t,0) = \varepsilon_g \exp(-r^2 / w_0^2 - t^2 / t_0^2 - ikr^2 / 2f) \tag{4}$$

where ε_g is the peak amplitude of the incident electric field, $k = n_0 k_0$ the wave number in the medium, f the focal length of the lens, t_0 and w_0 are the half temporal width and the spatial radius of the $1/e^2$intensity (for), respectively.

2.2 Description of ionization

Plasma generation and plasma defocusing are important dynamics in the filamentation formation. It is very important to construct the ionization model to describe the propagation of the intense pulse in filamentation. That is, how to calculate the parameter $W(I)$ in Eq. (3). Optical-field-ionization (OFI) covers both the multiphoton and tunnel regimes which we usually use the adiabaticity parameter γ to distinguish with. For an atom with ionization potential U_i in a linearly polarized laser field with frequency ω_0 and peak electric field E, γ is defined as:

$$\gamma = \omega_0 \sqrt{2mU_i} / eE \tag{5}$$

where e is the electron charge and m the electron mass. OFI occurs in the tunnel regime when $\gamma \ll 1$ and in the multiphoton regime when $\gamma \gg 1$.

Keldysh's theory (Keldysh, 1965) and its subsequent developments (PPT model) (Perelemov et al., 1966) led to a general formula for OFI rates valid for any atom or ion with the orbital quantum number l and its projection m onto the quantization axis and charge state Z:

$$\begin{aligned} W(\omega_0, E) &= \omega_{a.u.} \sqrt{6/\pi} \left| C_{n^*,l^*} \right|^2 f(l,m)(U_i / 2U_H) \\ &\times A_m(\omega_0, \gamma)(2E_0 / E\sqrt{1+\gamma^2})^{2n^* - |m| - 3/2} \exp(-2E_0 g(\gamma) / 3E) \end{aligned} \tag{6}$$

where U_H is the ionization potential of hydrogen, $\omega_{a.u.}$ is defined by $\gamma = 1$ from Eq. (5) for hydrogen or by $\hbar\,\omega_{a.u.} = 2U_H$:

$$\omega_{a.u.} = eE_H / \sqrt{2mU_H} \approx 4.1 \times 10^{16} s^{-1} \tag{7}$$

$$E_H = e^5 m^2 / \hbar^4 (4\pi\varepsilon_0)^3 \approx 5.14 \times 10^{11} V/m \tag{8}$$

$$E_0 = E_H \times (U_i / U_H)^{3/2} \tag{9}$$

For a given atom, the dimensionless constant C_{n^*,l^*}

$$\left| C_{n^* l^*} \right|^2 = 2^{2n^*} / n^* \Gamma(n^* + l^* + 1) \Gamma(n^* - l^*) \tag{10}$$

is adapted from the formula well known for the hydrogen atom (Perelemov et al., 1966) by replacing the principal n and orbital l quantum numbers by their effective counterparts

$$n = Z(U_i / U_H)^{-1/2} \tag{11}$$

$$n^* = n - \delta l \tag{12}$$

$$l^* \approx n^* - 1 \tag{13}$$

respectively, and

$$\delta l = n - (U_i / U_H)^{-1/2} \tag{14}$$

is the quantum defect. Here Γ in Eq. (10) is the gamma function (Abramowitz & Stegun, 1972).

Other functions in Eq. (6) read as:

$$f(l,m) = (2l+1)!(l+|m|)!/2^{|m|}(|m|)!(l-|m|)! \\ f(0,0) = 1 \tag{15}$$

$$A_m(\omega_0,\gamma) = 4\gamma^2/\sqrt{3\pi}|m|!(1+\gamma^2) \\ \times\exp[-\alpha(k-v)]\Phi_m(\sqrt{\beta(k-v)}) \tag{16}$$

$$\Phi_m(x) = x^{2|m|+1}/2\times\int_0^1 e^{-x^2t}t^{|m|}/\sqrt{1-t}dt \tag{17}$$

$$\beta(\gamma) = 2\gamma/\sqrt{1+\gamma^2} \tag{18}$$

$$\alpha(\gamma) = 2[\sinh^{-1}\gamma - \gamma/\sqrt{1+\gamma^2}] = \begin{cases} 2\gamma^3/3 & when\,\gamma \ll 1 \\ 2\log(2\gamma-1) & when\,\gamma \gg 1 \end{cases} \tag{19}$$

$$\alpha(\gamma) = 3/2\gamma\times[(1+1/2\gamma^2)\sinh^{-1}\gamma - \sqrt{1+\gamma^2}/2\gamma] \\ = \begin{cases} 1-\gamma^2/10+9\gamma^4/280 & when\,\gamma \ll 1 \\ 3/2\gamma\times(\log 2\gamma - 1/2) & when\,\gamma \gg 1 \end{cases} \tag{20}$$

$$v = U_i(1+1/2\gamma^2)/\hbar\omega_0 \\ v_0 = U_i/\hbar\omega_0 \tag{21}$$

Single ionization of atoms in the fundamental state with $l = m = 0$, $n^* = (U_i / U_H)^{-1/2}$ and $|C_{n^*,l^*}|^2 = 2^{2n^*}/n^*\Gamma(2n^*)$ has explicit description for tunnel (ADK model) (Ammosov et al., 1986) and multiphoton limits.

When $\gamma \ll 1$, it is the tunnel ionization case where $A_m(\omega_0,\gamma) \to 1$, and we have

$$W(E) = \omega_{a.u.}\sqrt{6/\pi}\left|C_{n^*,l^*}\right|^2(U_i/2U_H)(2E_0/E)^{2n^*-3/2}\exp(-2E_0/3E) \tag{22}$$

We can see from Eq. (22) that the probability of tunnel ionization is independent of the frequency of the wave.

When $\gamma \gg 1$, it is the multiphoton case, by taking into account only the term $\kappa = K \equiv \mathrm{mod}(v_0+1)$ in the series for $A_m(\omega_0,\gamma)$, and by using the relation $v_0 = U_i/\hbar\omega_0 = E_0/2E\gamma$, we have

$$W(\omega_0,E) = \omega_{a.u.}(4^{2n^*}/\pi\sqrt{2})\left|C_{n^*,l^*}\right|^2(U_i/2U_H) \\ \times v_0^{2n^*+2K-3/2}e^{2K-v_0}\Phi_0(\sqrt{2(K-v_0)})(E/E_0)^{2K} \tag{23}$$

We can rewrite Eq. (23) as

$$W(\omega_0,I) = \sigma_K I^K \tag{24}$$

where

$$W(\omega_0, E) = \omega_{a.u.}(4^{2n^*} / \pi\sqrt{2})\left|C_{n^*,l^*}\right|^2 (U_i / 2U_H) \times v_0^{2n^*+2K-3/2} e^{2K-v_0} \Phi_0(\sqrt{2(K-v_0)}) / I_0^{\,K} \tag{25}$$

$$I_0 = \varepsilon_0 n_0 c E_0^2 / 2 \tag{26}$$

where ε_0 is the electric permittivity of free space, c is the speed of light in vacuum.

3. Simulation on gradient temperature (Song et al., 2008a; Song et al., 2008b)

3.1 Model of simulation

In our simulation model, to simplify the calculation and hold the essential physical dynamic characteristics, we just only consider the fundamental mode (the spatial profile of which is not changing along propagation) of the coupled leaky modes propagating in the hollow fiber. We also neglect the interaction and energy transfer between the fundamental and high-order modes because the attenuation length of high-order modes is much smaller than that of the fundamental.

We use the standard nonlinear (1+1) dimension Schrödinger equation to simulate and analyze the evolution dynamic of the pulse propagation both in temporal and spectra domain. The nonlinear Schrödinger equation for the electric field envelope $u(z, t)$ in a reference frame moving at the group velocity v_g takes the following form (assuming propagation along the z axis) (Agrawal, 2007):

$$\frac{\partial u}{\partial z} = -\frac{\alpha}{2}u + \frac{i\beta_2}{2}\frac{\partial^2 u}{\partial T^2} + i\gamma[|u|^2 u + \frac{i}{\omega_0}\frac{\partial}{\partial T}(|u|^2 u) - T_R u\frac{\partial |u|^2}{\partial T}] \tag{27}$$

The terms on the right hand side of the equation are the loss, second order dispersion, self-phase modulation, self-steepening and Raman scattering, respectively. Here c is the speed of light in vacuum, ω_0 the central angle frequency, a the loss, β_2 the GVD (group velocity dispersion) and T_R is related to the slope of the Raman gain spectrum. The nonlinear coefficient $\gamma = n_2\omega_0/cA_{eff}$ where n_2 is the nonlinear refractive index and A_{eff} the effective cross section area of the hollow fiber. Equation (27) and the parameters in the equation characterize propagation of the fundamental mode.

The initial envelop of the pulse is in the following form (Tempea & Brabec, 1998; Courtois et al., 2001), which is a simplification expression of Eq. (4):

$$u(0,t) = \sqrt{\frac{2P_{in}}{\pi w_0^2}} \exp(-\frac{t^2}{t_0^2}) \tag{28}$$

here P_{in} is the peak power of the incident pulse, w_0 the spot size (for $1/e^2$ intensity) of the fundamental beam (we assume that the beam focused on the entrance section of the hollow fiber matches the radius of the fundamental mode in our calculation mode), t_0 the half temporal width at the $1/e^2$ points of the pulse intensity distribution.

Equations (27) and (28) can be solved by the split-step Fourier method (Agrawal, 2007) in which the propagation is broken into consecutive steps of linear and nonlinear parts. The linear part including loss and dispersion can be calculated in the spectrum domain by Fourier transform, while the nonlinear part which includes other terms on the right hand of Eq. (27) was solved in the time domain by Runge-Kutta method. The convergence of the solution can be easily checked by halving the step size to see if the calculation results are nearly unchanged.

Although the studies of filamentation in many gases have been focused by scientists and technologists (Akturk et al., 2007; Fuji et al., 2007; Dreiskemper & Botticher, 1995), argon (Ar) is the most frequency used gas for generation of ultrashort intense femtosecond pulses. In simulation in this chapter, we employ Ar as the medium to reveal the essence of gradient temperature technology.

The loss and waveguide dispersion relations of the hollow fiber can be expressed as (Marcatili & Schmeltzer, 1964):

$$\frac{\alpha}{2}=\left(\frac{2.405}{2\pi}\right)^2\frac{\lambda^2}{2a^3}\frac{v^2+1}{(v^2-1)^{1/2}} \tag{29}$$

$$\beta_{waveguide}=\frac{2\pi}{\lambda}\left[1-\frac{1}{2}\left(\frac{2.405\lambda}{2\pi a}\right)^2\right] \tag{30}$$

where v is the refractive index ratio between the material of the hollow fiber (glass in our case) and the inner gas (argon in our case), $\lambda = \lambda_0/n$, the wavelength in the medium (λ_0 the wavelength in vacuum), a the bore radius of the hollow fiber. The propagation constant β icluding contributions from both waveguide part (Eq. (30)) and material part:

$$\beta_{material}=\frac{\omega}{c}n \tag{31}$$

The relation between propagation constant β and m order dispersion coefficient β_m is:

$$\beta_m=\left(\frac{d^m\beta}{d\omega_m}\right)_{\omega=\omega_0} \tag{32}$$

For gases, the refractive index is a function of both temperature and pressure (Lehmeier, 1985):

$$n=\left(2\frac{n_0^2-1}{n_0^2+2}\frac{pT_0}{p_0T}+1\right)^{1/2}\left(1-\frac{n_0^2-1}{n_0^2+2}\frac{pT_0}{p_0T}\right)^{-1/2} \tag{33}$$

For argon, at the standard condition (T=273.15K, p=1atm), we have (Dalgarno & Kingston, 1960):

$$n_0^2-1=5.547\times10^{-4}\left(1+\frac{5.15\times10^5}{\lambda_0^2}+\frac{4.19\times10^{11}}{\lambda_0^4}+\frac{4.09\times10^{17}}{\lambda_0^6}+\frac{4.32\times10^{23}}{\lambda_0^8}\right) \tag{34}$$

In the above equations, p is the pressure, p_0 the pressure at normal conditions (1 atm), T the temperature, T_0 the temperature at normal conditions (273.15 K), n the refractive index of the medium, and n_0 the refractive index of the medium at normal conditions (T=273.15 K, p=1 atm). In Eq. (34), the unit of λ_0 is Å (10^{-10} m).

Before we do the simulations of the evolution of the pulse under gradient temperature, we first check the effect of the temperature on the hollow fiber and the medium (Ar as in our case) qualities such as loss, refractive index, etc. Figures 2 and 3 show the loss and refractive index as a function of the temperature. They all keep nearly constant during the interval from 300 K to 600 K. We can conclude that compared with room temperature, higher temperature does not introduce extra attenuation during the pulse propagation.

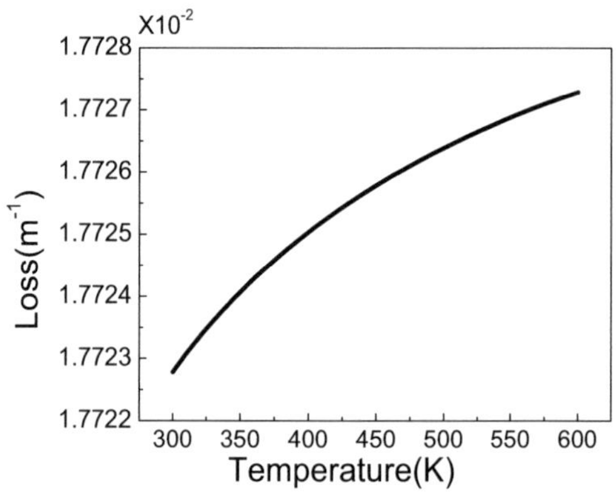

Fig. 2. Loss as a function of temperature for Ar in hollow fiber (bore diameter 500 μm, pressure1 atm).

To simplify and catch the essence physics process, we define a factor *TF* which represents the gas gradient temperature factor through the ideal gas equation: $TF=pT_0/p_0T$. It is obvious to see that the factor *TF* is proportional to the gas density (proportional to the gas pressure while inversely proportional to the gas temperature). When the gas pressure is 1 atm, the gradient factor *TF* is 1 for 300 K, and 0.5 for 600 K.

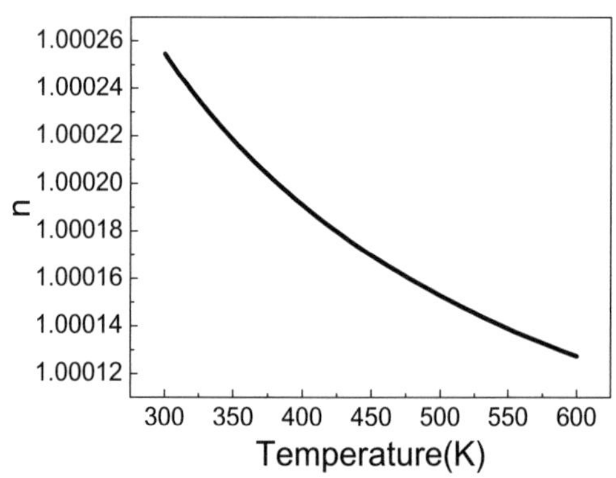

Fig. 3. Refractive index as a function of temperature for Ar at 1 atm.

The nonlinear refractive index and GVD are both proportional to the factor *TF* (Mlejnek et al., 1998):

$$n_2 = 4.9 \times 10^{-23} \times TF(\mathrm{m}^2/\mathrm{W}) \tag{35}$$

$$\beta_2 = 2.6 \times 10^{-29} \times TF(\mathrm{s}^2/\mathrm{m}) \tag{36}$$

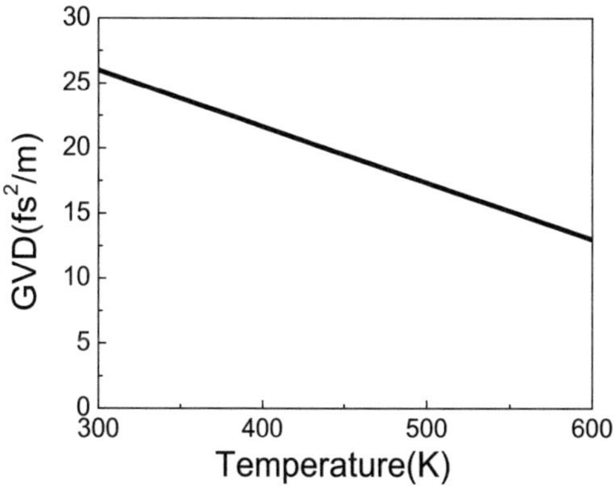

Fig. 4. GVD as a function of temperature (bore diameter 500 μm, pressure1 atm).

Now we can calculate the GVD and nonlinear refractive index by Eqs. (30)-(36). The results are shown in Figs. 4 and 5. As we can see from these figures, a higher temperature at 600 K decreases both the GVD and nonlinear refractive index n_2 by a factor of 2 for the room temperature 300 K. The decreasing GVD gives the pulse a chance to slow down the pulse broadening in time domain; while the decreasing nonlinear refractive index increases the critical power for self-focusing P_c (see Eq. (1)). Therefore, at a higher temperature, the pulse broadening in time domain slows down and P_c is higher. If the tube is sealed and is locally heated at the entrance, and cooled at the exit end, the gas temperature gradient will be formed along the tube and so will the nonlinear refractive index. P_c at the hot side of the tube (entrance) will be higher than the cold end (exit end), like in the case of gradient pressure (see Fig. 1).

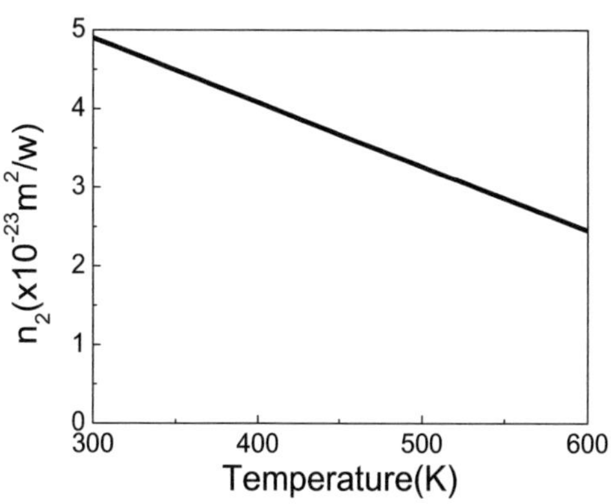

Fig. 5. Nonlinear refractive index as a function of temperature (pressure 1 atm).

3.2 Spectrum broadening

As the incident pulse propagating along the hollow fiber filled with argon, the peak power of the pulse is continuously decreasing due to the dispersion and loss. However, the decreasing temperature along the fiber provides a gradually increasing nonlinear coefficient which partly compensates the decreasing peak power, the spectrum broadening can go on till the end of the tube. For the argon gas at atmospheric pressure and temperature of 600 K, P_c is 1.2 GW; while for the room temperature, 300 K, it is 2.1 GW, i.e. the critical power for 600 K is twice of that for 300 K. This means that the energy of the incident pulse will be allowed twice higher as that of the pulse under room temperature for the same pulse width.

We did the simulation on the spectrum broadening for the uniform and gradient temperature cases in the hollow fiber. The bore diameter of the hollow fiber was 500 μm and the length of the fiber was 60 cm. The temperature conditions are: condition 1: uniform room temperature (T = 300 K); condition 2: temperature linearly decreasing from 600 K to 300 K along the hollow fiber; condition 3: temperature linearly decreases from 600 K to 300 K in the first half and increases from 300 K to 600 K in the second half of the fiber, i.e., the triangle temperature. The incident peak power of the pulses was set to be $2P_c$ and the pressure was 0.2 atm, thus, for a 30 fs pulse, the incident pulse energy should be 0.6 mJ and 1.2 mJ, for room temperature 300 K (uniform case) and 600 K (gradient temperature case), respectively.

By solving Eq. (27) coupled with the initial condition in Eq. (28), we obtained the spectra and phases of the output pulses under the above three conditions, which are shown in Figs. 6 and 7. It is obvious that the output spectrum bandwidth of the pulse increases from 250 nm (about 675 nm to 925 nm, uniform temperature case) to 350 nm (about 625 nm to 975 nm, linear and triangle gradient temperature case). However, the triangle shaped gradient temperature does not seem to make visible difference from the linear gradient temperature case. We also plot the spectrum evolution of the triangle shaped temperature in Fig. 7(b). We can see that the spectrum starts to expand at about 20 cm and the profile collapses along the fiber. We will discuss on the spectrum broadening quantatively in the following subsection.

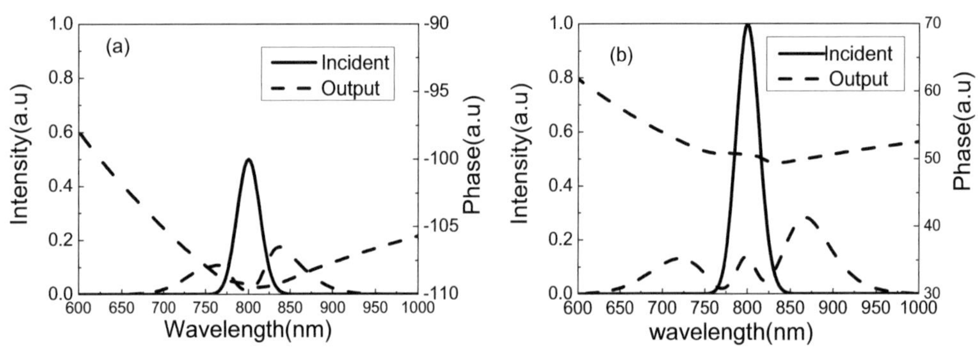

Fig. 6. Spectrum & phase for (a) uniform temperature (300 K), (b) linear gradient temperature (600 K to 300 K) and. Other conditions: bore diameter of the hollow fiber: 500 μm, fiber length: 60 cm, filled argon gas pressure: 0.2 atm, incident pulse width: 30 fs, pulse energy: 0.63 mJ for the uniform case and 1.26 mJ for the gradient case.

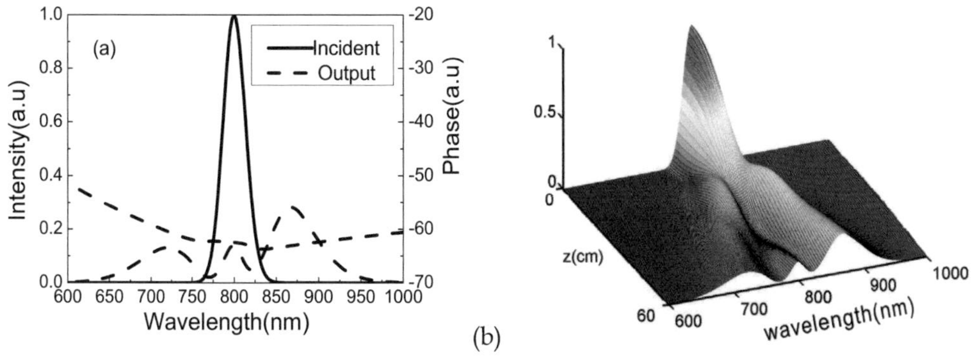

Fig. 7. (a) Spectrum & phase (b) Spectra evolution for triangle gradient temperature (600 K to 300 K to 600 K). Other conditions: bore diameter of the hollow fiber: 500 μm, fiber length: 60 cm, argon gas pressure: 0.2 atm, incident pulse width: 30 fs, pulse energy: 1.26 mJ.

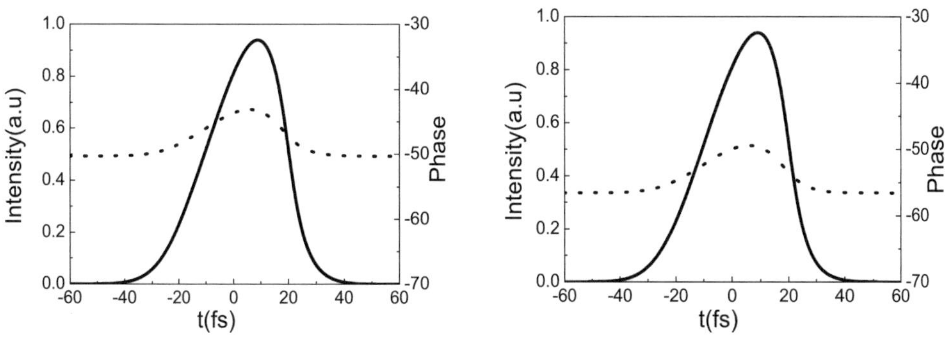

Fig. 8. Pulse profiles & phases for the (a) linear gradient and (b) triangle gradient temperature in Fig. 6 (b) and 7 (a).

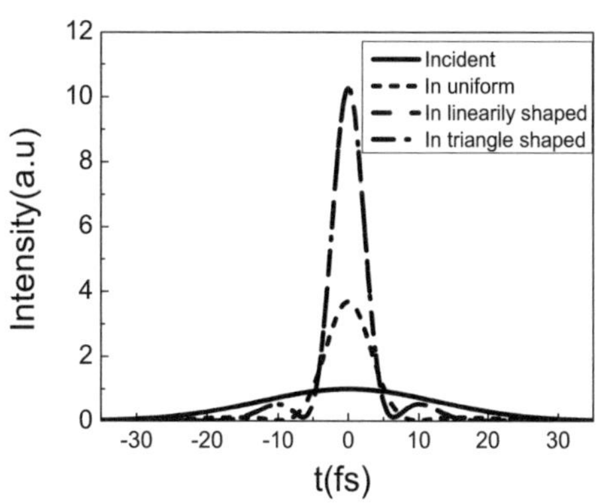

Fig. 9. Pulses profiles after ideal compression (spectra are shown in Figs. 6 and 7(a) respectively).

The output pulse profiles and phases of the linear and triangle gradient temperature are shown in Figs. 8(a) and 8(b) respectively. Still, we cannot see much difference between the linear and triangle cases for the output pulse. The transform limited pulse after ideal compression for the three conditions are shown in Fig. 9. The pulse width after ideal compression is 5 fs in the gradient temperature case (both linear and triangle gradient cases), which is 2/3 of pulse width in the uniform temperature case (7.5 fs). In addition, the pulse energy we can obtain in the gradient temperature scheme is twice higher as that in the uniform temperature scheme.

3.3 Discussions on spectrum broadening

When a pulse propagates through a Kerr medium whose length is L, the spectrum broadening S_p of the pulse is approximately determined by the integral below (Agrawal, 2007):

$$S_p = \int_0^L n_2(z)P(z)dz \tag{37}$$

where $n_2(z)$ is the nonlinear refractive index at position z, $P(z)$ the peak power of the pulse at position z. We use Eq. (37) to discuss the spectrum broadening comparing with the simulation we did in the above subsection.

First, this integral can approximately determinate the spectrum broadening quantatively. If we take $n_2(z)P(z)$ as a variable and set it equal everywhere along the medium, the nonlinear Schrödinger equation (Eq. (27)) is actually the same in every z of the medium. The result is that the final pulse temporal and spectral profiles (normalized with themselves) are the same, which means that they are only different with intensity.

Second, from the integral we can see that the spectrum broadening will not be much broader in the gradient temperature case than that in the uniform case. But from the energy point, we can see that the incident energy will be allowed twice higher than uniform temperature. This is a big priority of gradient temperature. Our intention is to achieve not only ultrashort but also intense pulses. The energy is also a main final object which we focus on.

Third, from the integral in Eq. (32) we can deduce that the spectrum broadening in triangle gradient will be almost the same as that in the linear gradient case. This is true and can be verified by our simulation results (see Figs. 6 (b) and 7 (a)). In fact, the difference of linear and triangle gradient scheme excluding real experimental conditions in simulation is small. Their different effects can be seen from experiments more obviously. Triangle gradient scheme's priority is that this design gives even better pulse compression, avoids cyclic compression stages, and therefore limits the energy loss as shown in Ref (Couairon et al., 2005). From ideal theoretical point, these two schemes have almost the same ability of spectrum broadening. From the experimental point, triangle scheme has priority to linear project and it is a little more complex. Although this experimental conclusion is obtained from gradient pressure scheme, we can expect the same results in gradient temperature case.

3.4 Ideal gradient line shape

In the above simulation, we set the input pulse peak power related to the critical self-focusing power P_c. Inversely, we can derive an ideal gradient shape for a giving pulse, which means that at every step of evolution, we change the temperature so as to make the pulse's peak power equals to the critical power of self-focusing. Figure 10 shows the ideal gradient shape for a 30 fs, 0.1 mJ incident pulse. The *TF* differential increases along the fiber, which implies that the peak power of the pulse along the tube drops faster and faster during the evolution. If we can realize such a gradient temperature, we can avoid multi-filament formation everywhere along the tube. In fact, in the linear gradient shape case, a moderately increasing the length of the tube (corresponding to decreasing the slope of the linear line) or decreasing the peak power of the input pulse will avoid the self-focusing or filament formation everywhere along the tube and the result of the spectrum broadening is still much broader than the uniform temperature case.

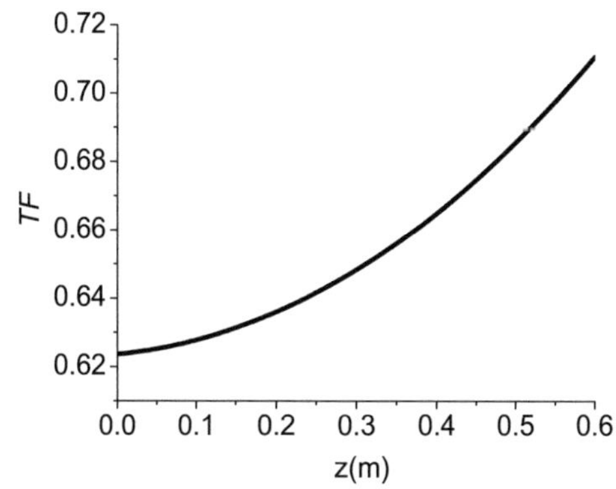

Fig. 10. Ideal gradient shape for 0.1mJ.

4. Experimental results (Cao et al., 2009)

As we mentioned in the introduction, spectrum broadening through filamentation was one of the most extremely simple and robust techniques to generate intense few to monocycle pulses with less sensitivity to the experiment conditions. In real experiments, an aperture (Cook et al., 2005), rotating lens, anamorphic prisms, circular spatial phase mask (Pfeifer et al., 2006), periodic amplitude modulation of the transverse beam profile (Kandidov et al., 2005), introducing beam astigmatism (Fibich et al., 2004) or incident beam ellipticity (Dubietis et al., 2004) in the laser beam prior to focusing have been used to stabilize the pointing fluctuations of a single filament. In the previous section, we show the priority of the gradient temperature scheme by theoretical simulation. In this section, we will verify the robustness of this scheme by showing the experimental results.

We show our experimental setup in Fig. 11. The laser pulse was produced from a set of conventional chirped pulse amplification (CPA) Ti: sapphire laser system. This laser system produced linearly polarized pulses of 37 fs pulse at the central wavelength of 805 nm. The energy of the pulses was 2 mJ and the repetition rate was 1 kHz. The beam diameter of the pulses was 10 mm (at $1/e^2$ of the peak intensity). In this experiment, four silver mirrors were used to couple the amplified pulses into the sealed silica tube, where M1, M2 and M3 were the plane mirrors and FM1 was a concave mirror with a 1.7 m radius of curvature. A hard aperture A1 as an attenuator and a beam profile shaper was inserted in front of the concave mirror FM1. The output pulse was focused by a concave mirror, FM2, into a pulse compression system consisting of two negative dispersion mirrors, CM1 and CM2. The negative dispersion mirrors were rectangles of size 10 × 30 mm^2. Each reflection contributed a GDD of 50 fs^2within wavelength area of 680~1100 nm. The pulse after compression was reflected by plane mirrors, M4 and M5, then through a beam split mirror, BS1, into SPIDER. The Ar gas filled in the tube was controlled and monitored to be below the maximum pressure of 3 atm, because a higher gas pressure may blow up the windows of the tube. The focal point in the tube was measured as 47 cm from the input window. The spot size of the focused pulse was 100 μm. To make a temperature gradient along the propagation of the pulse, a 20 cm heating length furnace was used to heat the tube. The 100 cm long high-temperature and high-gas-pressure resistance silica tube with the inner diameter of 25 mm was sealed off with two 1-mm thick fused silica Brewster windows. The tube was inserted into the transverse center of the furnace. Two ends of the tube were cooled by air. To avoid the expansion of the tube and make the furnace easy to move along the tube, between the external side of the tube and the internal side of the furnace, there was a 2 mm wide gap. The temperature of the furnace was controlled by a temperature controller between 25 °C and 500 °C with a ± 5 °C precision. It should be noted that the temperature we mention in the following text in this section is the temperature at the longitude center of the furnace. With this configuration, the temperature at the heating point could be increased from 25°C to 500°C within 35 minutes. The above experimental setup is the same as that was used in broadening the spectrum through filamentation, expecting for the additional furnace. Therefore, an additional furnace and temperature controller are sufficiently easy to modify the traditional filamentation setup to our experimental setup.

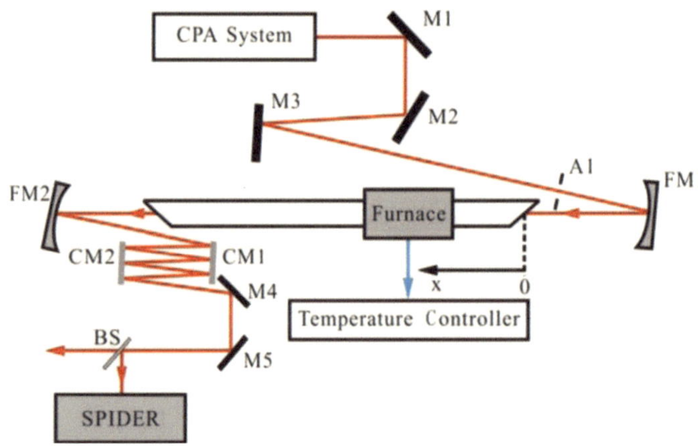

Fig. 11. The schematic of the experimental setup

To know the actual temperature distribution inside the tube, we inserted a thermistor and moved it along the tube to measure the temperature. The measured temperature distribution at a maximum central temperature of 500°C is shown in Fig. 12. The temperature rapidly drops down to the room temperature outside the furnace, so that the temperature distribution is of a triangular shape, with the temperature gradient of about 2403 °C/m. According to our simulation results and discussions in the former section, the priority of the triangle gradient is that it gives an even better pulse compression, avoids cyclic compression stages, and limits the energy loss. As the temperature is distributed along the tube, there should be a gas flowing from the hot to the cool position. However, in the experiment, the temperature variation was a very slow process. We did not observe the instability caused by the gas turbulence. In general, the radial thermal distribution could also cause thermal lensing effect. In our case, because the inner tube diameter was only 25 mm, the radial temperature difference between the wall and the center was measured to be only 2–3 °C, so that the thermal lensing effect could be neglected.

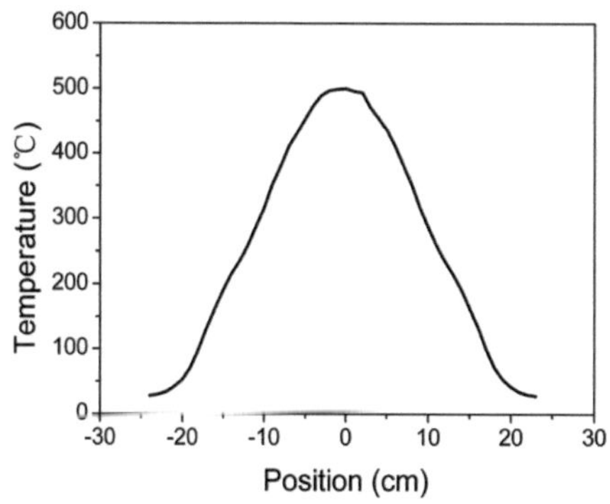

Fig. 12. Temperature distribution along the tube when the temperature at the furnace central (zero point at x-axis) is 500 °C.

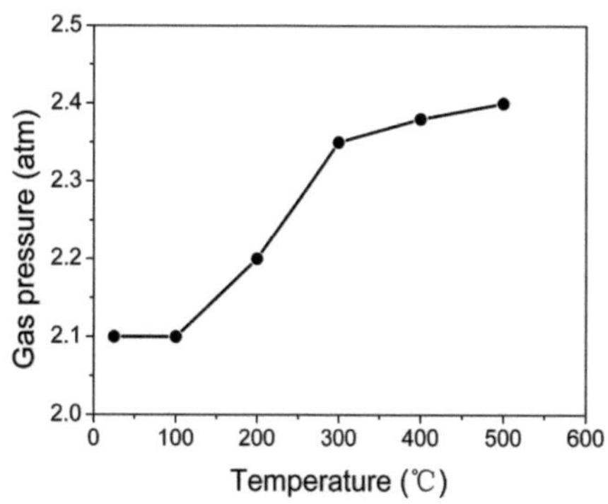

Fig. 13. Measured gas pressure as a function of the heated temperatures when the initial gas pressure is 2.1 atm

As for the sealed tube, the gas pressure in total should be uniform and increase with the temperature. The measured gas pressure as a function of the heated temperatures when the initial gas pressure is 2.1 atm is shown in Fig. 13. Generally speaking, the influence of the pressure and temperature should be separately examined. However, since we just wanted to investigate the filamentation process in a sealed tube with the change of the temperature, we did not attempt to separate the temperature and pressure effects in our experiment. Moreover, the pressure change within 100 °C was only a few percent. This small change does not introduce noticeable difference in the material parameters for argon gas such as GVD, n_2, and the average electron collision time.

4.1 Filament controll and spectrum broadening by gradient temperature

To check the influence of the temperature, we changed the local temperature in the tube and measured the beam pattern and the broadened spectrum. The beam pattern was taken by an ordinary digital camera looking at the surface of a white paper positioned at the plane orthogonal to the beam path and 3 m away from the exit window of the tube and the broadened spectrum was measured by a spectrometer (Ocean Optics, SD2000). When the furnace was turned off, the temperature was kept at room temperature 25 °C inside the tube, and the gas density was uniform along the tube. Pulses with energy of 1.2 mJ (32.4 GW peak power, about 6.5 times higher than the critical power at 2.1 atm) after the aperture A1 were coupled into the tube, and output pulse energy of the pulses was 1.1 mJ. A single filament began to appear at 3 cm before the focal point and the filament was about 40 cm long at 1.7 atm. By increasing the gas pressure to above 2.1 atm, the single filament broke into multiple filaments, as is shown in Fig. 14(a). The inserted picture is the output beam profile of the multi-filament in the far field, where three filament spots can be identified. The interactions among multi-filament result in shot-to-shot fluctuations in the filamentation pattern. As the temperature was increased to 200 °C, the gas pressure in the tube was increased to 2.2 atm, a little higher than that at 25 °C (see Fig. 13). Although the heated gas will flow to the cool end and be kept at the same temperature in a long term, the gas temperature of the exit end of the tube was found still 25 °C. When the furnace was increased to 200 °C, the mult-filament turned to become a single filament, as shown in Fig. 14(b). There was only one single filament that has a good beam profile. Further increasing the temperature to 300 °C or higher, the single filament collapsed and disappeared, as shown in Fig. 14(c). Although the gas pressure was also increased at the same time, the higher gas pressure was caused by the accelerated activity of the gas atoms, but not because of the increase of the number of the gas atoms. Gas atoms moved from the position of higher temperature to that of lower temperature, which resulted in that the gas density was lower at the entrance and higher at the end of the filament. Higher self-focusing critical power P_c induced by the higher gas

Fig. 14. Filament pattern at temperature of (a) 25 °C; (b) 200 °C; (c) 300 °C; (d) 25 °C. The inserted pattern in every picture is the output beam profile.

density was effective to avoid the occurrence of the multi or even single filamentation. Inversely, by decreasing the temperature from 300 °C to the initial room temperature 25 °C, multi-filament appeared gradually, which is shown in Fig. 14(d). It was almost the same as in the initial state (Fig. 14(a)) of our experiment.

At the temperate of 25 °C and input pulse energy of 1.2 mJ, we measured the output spectra at different gas pressure. The results are shown in Fig. 15. The output spectra toward the short wavelength became wider with the increase of gas pressure, which resulted from the increase of the number of the filled gas atoms. At the gas pressure of 2.1 atm, multi-filament was formed in the gas-filled tube.

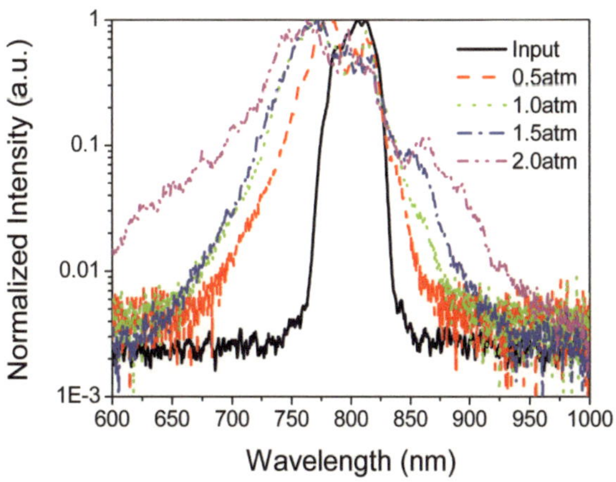

Fig. 15. Spectra at different gas pressures with the input pulse energy of 1.2 mJ and the heated temperature of 25 °C

Fig. 16 shows the evolution of output spectra at different temperatures of the entrance of the filament, when the gas pressure was 2.1 atm and incident pulse energy was 1.2 mJ. It can be seen that the spectral width was broadened to about twice as that of the incident spectrum. For a single filament at 200 °C, the spectrum broadening is due to an increasing phase contribution from ionization-induced spectrum broadening and interaction with the plasma. Whereas, in the case of non-filament at above 300 °C, the spectrum broadening is due to the dominant self-phase modulation (SPM) rooted from n_2, which becomes weak with the increase of the temperature. The further increasing of the temperature results only in a narrower broadened spectrum. When the filament disappears at high temperature, it means that the self-focusing critical power is high.

Therefore, we can increase the input pulse energy up to the new self-focusing critical power. The final results are shown in Fig. 17. At the temperature of 25 °C and incident pulse energy of 1.2 mJ, filament was formed at 2.1 atm, shown as the point A in Fig. 17. Then, when the temperature at the entrance of the filament was increased to 300 °C, the filament disappeared, shown as the point B in Fig. 17. After increasing the pulse energy from 1.2 mJ to 1.54 mJ at 300 °C, the filament appeared again, shown as the point C in Fig. 17. After increasing the temperature from 300 °C to 400 °C at 1.54 mJ, the filament disappeared again, shown as the point D in Fig. 17. It indicates that the filament can appear or disappear by increasing the temperature and input pulse energy in turn. Meanwhile, if the temperature

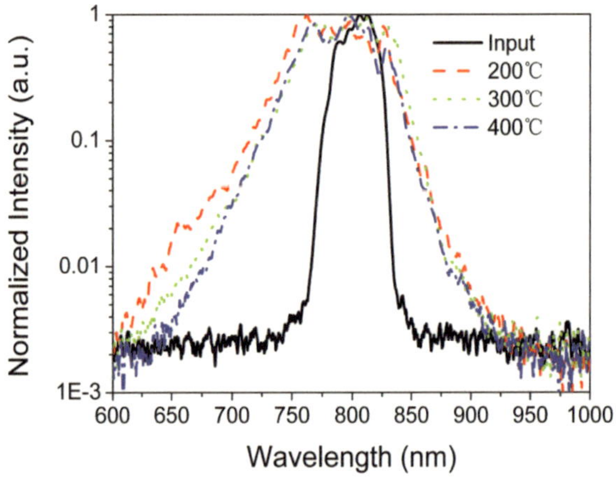

Fig. 16. Spectra at different temperatures with the input pulse energy of 1.2 mJ and the initial gas pressure of 2.1 atm

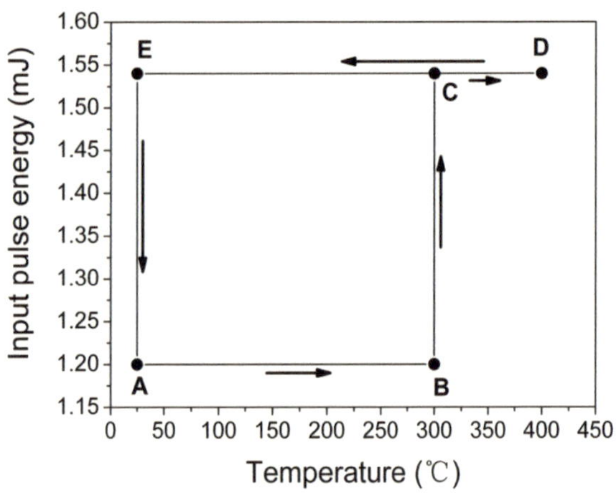

Fig. 17. The cycle between filament and no-filament by changing the temperature and input pulse energy in turn. Filament appears at points A, C, and E and disappears at points B and D.

was decreased to 25 °C at the incident pulse energy of 1.54 mJ, the filament also appeared, shown as the point E in Fig. 17, and further when the pulse energy was decreased to 1.20 mJ, the filament was the same as in the initial state. The above experimental results indicate that the filament can be controlled by adjusting the local self-focusing critical power by the temperature, although the broadened spectrum narrows with the increase of the temperature. More incident pulse energy can be allowed in the tube at the higher local temperature. The presented method is simple and feasible to operate with only a heating furnace, without continuing consumption of expensive gases comparing with the gradient pressure scheme.

4.2 Self compression in temperature controlled filamentation

The filamentation of intense femtosecond laser pulses will lead to a remarkable pulse self-compression. The filamentation of ultrashort laser pulses is a banlance between the beam self-focusing by the optical Kerr effect, beam defocusing due to the plasma, pulse self-steepening, and beam diffraction. The propagating pulse suffers significant reshaping in both time and space domain. This reshaping process will lead to the self compression of the intense femtosecond pulse. In this part, we investigate the self compression of the femtosecond pulse propagation in temperature controlled filamentation. We start fromed 0.7 mJ incident pulse with 25 °C and 2 atm. In this condition, it could only form a very short filament at the focal point. The output pulse reduced from 35 fs to 23.5 fs due to self compression, with a Fourier transform limit of 16 fs (see Fig. 18).

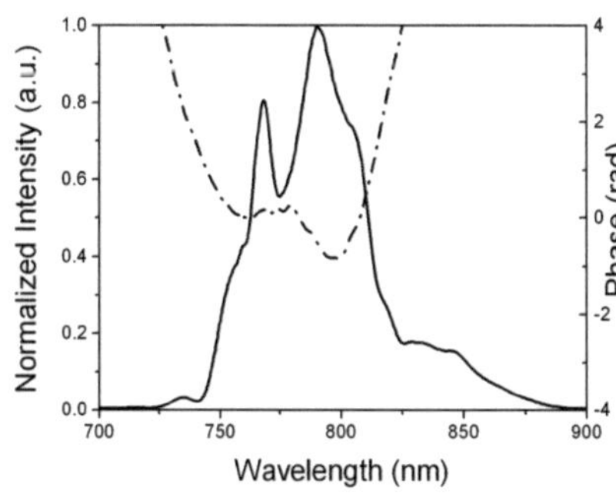

(a)

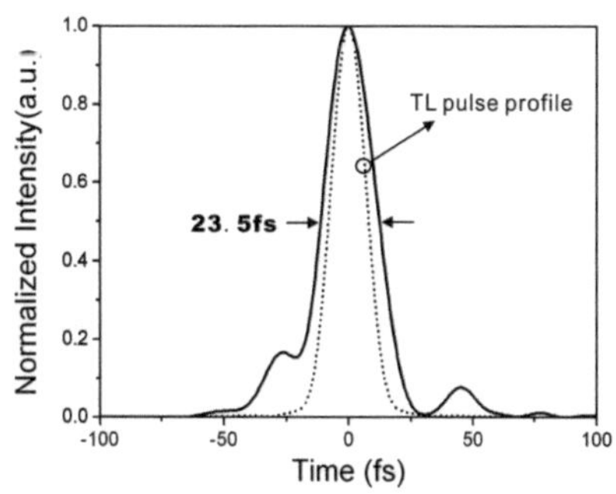

(b)

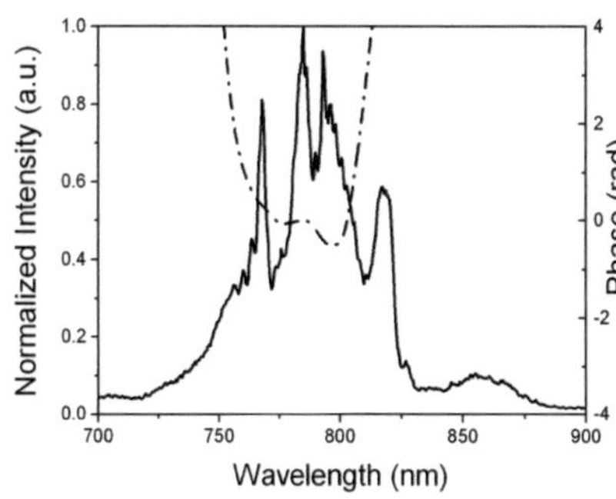

(c)

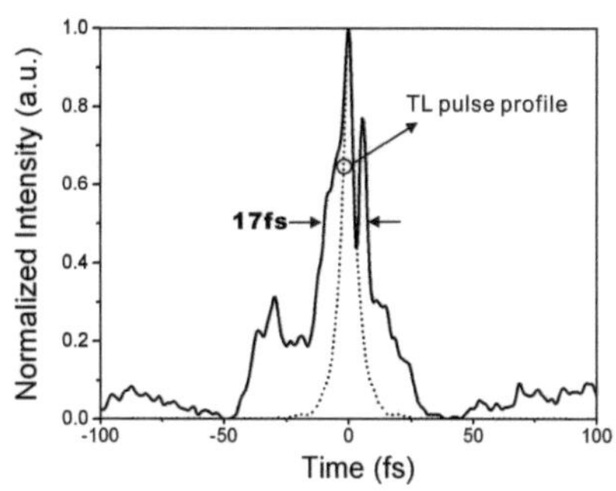

(d)

Fig. 18. The spectra and phases of (a) 0.7 mJ, (c) 0.8mJ incident pulse at 25 °C. Pulse profiles after self compression and Fourier transform limit corresponding to (a) and (c) are shown in (b) and (d) respectively.

When the energy of incident pulse was 0.8 mJ, the pulse further reduced to 17 fs, with a Fourier transform limit of 5.5 fs. As the energy of input pulse was increased to 1.3 mJ, filament in the tube split to multifilaments at 5 cm after focal point, and the spots split and converged rapidly evidenced by far field observation. The transform of multifilaments to single filament can be controlled by temperature. We could observe remarkable multifilaments with pulse energy up to 1.7 mJ. When the temperature of heat center reached

170°C, multifilaments converged to single filament and at this temperature, multifilaments will reoccur if pulse energy increases to 2.7 mJ. When the temperature of heat center was 400°C, multifilaments shrunk to single filament again. We increased the temperature to 450°C, filamentation was not obvious and the pulse is self compressed to 19 fs with a Fourier transform limit of 14.5 fs (see Fig. 19). Filament disappeared as the temperature increases to 500 °C. The width of output pulse reduced to 24.5 fs and its transform limited pulse is 15 fs (see Fig. 19). The energy of the self compressed pusle increased by nearly 2 mJ compared to the case of 0.8 mJ (Fig. 19).

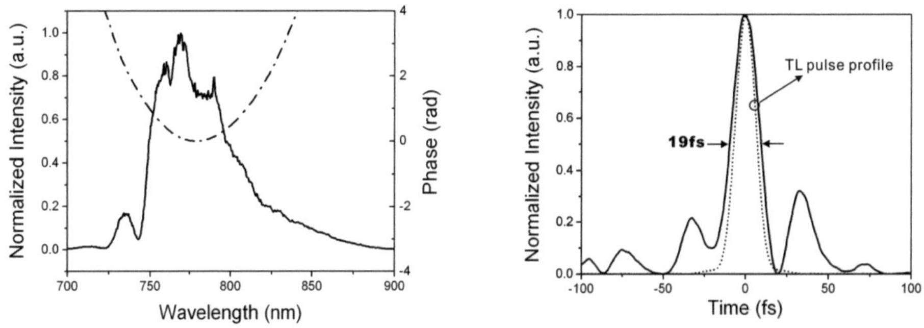

(a) (b)

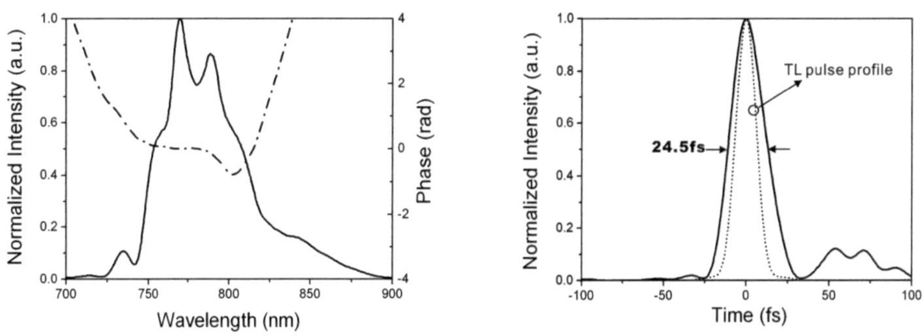

(c) (d)

Fig. 19. The spectra and phases of (a) 450 °C, (c) 500°C with 2.7 mJ incident pulse. Pulse profiles after self compression and Fourier transform limit corresponding to (a) and (c) are shown in (b) and (d) respectively.

Single filament reoccured when the energy of incident pulse increased to 5.7 mJ, and the width of output pulse was 69 fs. Compared with multifilaments before heating, the pulse energy of self compression was increased nearly 4 mJ by heating the gas to 500 °C with overcoming the emergence of multifilaments.

We can define the self-compression ratio, S, as the ratio between the width of Fourier transform limit and that of self compression pulse which exhibits the condition of self compression without any dispersion compensation. S = 1 is the ideal value (see table 1). Preliminary analysis shows that at the same temperature, high energy promotes self compression, but the self compression rate for high energy is low, which can approach the ideal value after dispersion compensation. For the same energy, self compression rate differs slightly at different gradient temperature, which is higher at lower temperature.

Temperature (°C)	Pulse Energy (mJ)	Measured pulse width (fs)	Transform limited pulse width (fs)	S
25 (No Gradient)	0.7	23.5	16	0.6809
25 (No Gradient)	0.8	17	5.5	0.3235
450 (Gradient)	2.7	19	14.5	0.7632
500 (Gradient)	2.7	24.5	15	0.6122

Table 1. Self compression rate S at different energy and temperatre schemes.

4.3 Pulse compression with dispersion compensation by chirp mirror after temperature controlled filamentation

To obtain intense ultrashort pulse, dispersion compensation is need after the filamentation. We started from a 2.4 mJ incident pulse under 25 °C and 2 atm condition, and we observed multifilaments. The width of spectrum was broaden 3 times of that of the incident pulse, and the transform limited pulse was 6 fs. The pulse compression was difficult under this condition because of the strong fluctuation caused by multifilaments. When we increased the temperature to 380 °C, the pressure in the tube was about 2.3 atm, and multifilaments gradually shrunk to a 35 cm long single filament, starting at 3 cm before the focal point. Figure 20 shows the narrowing of the pulse width due to high temperature.

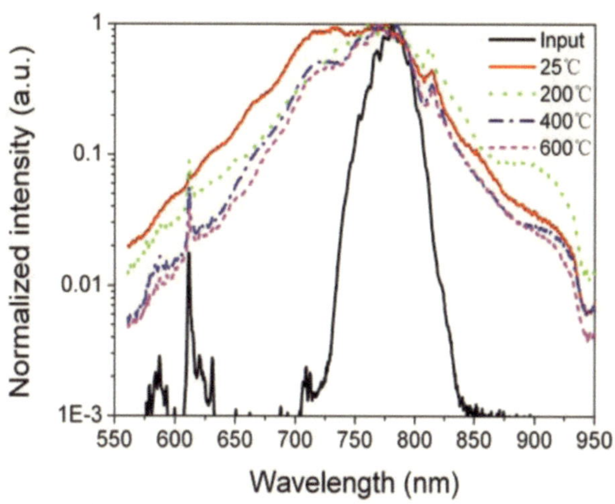

Fig. 20. Spectra of a 2.4 mJ incident pulse at 2 atm at different temperature.

In Fig. 21, we show the phases and spectrum of the pulse reflected by the chirp mirror for 6 and 8 bounces, measured by a SPIDER. We can see that after 3 bounces between the chirp mirror, the spectrum is not flat. While after 4 bounces, the spectrum becomes flat and the GDD turns from positive to negative. We can get a 1.6 mJ, 15 fs output pulse compared with 8 fs of transform limited,. Certainly the dispersion compensation is not complete and finely compensation is needed.

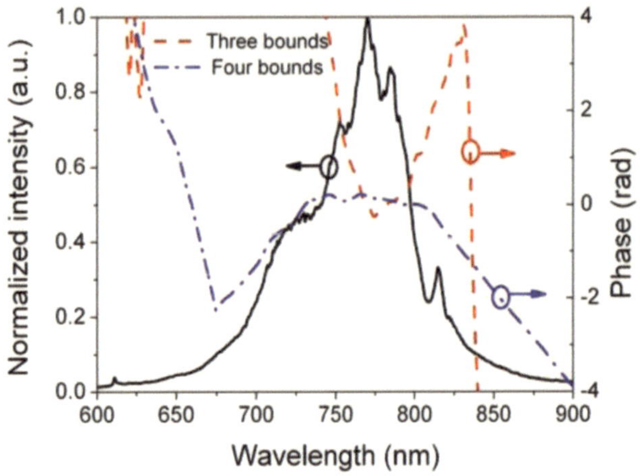

Fig. 21. Spectrum and phases for 3 and 4 bounces between chirp mirror.

5. Summary

In this chapter, a novel technology for generating intense few to monocycle light pulse was proposed and demonstrated. This technology has similar effect as the gradient pressure scheme while avoid the disadvantage of gas flow and consumption of expensive noble gases.

A model for simulation of the pulse evolution in a gradient temperature hollow fiber filled with argon gas has been established. The simulation results show that in the gradient temperature scheme, the incident pulse energy can be much higher than that of the uniform case, which is similar to the gradient pressure. In the gas of gradient temperature, the pulse spectra can be broadened more than that in the case of uniform temperature. Shorter pulses can be obtained after a further compression.

We also verified the effectiveness and feasibility of the scheme of gradient temperature. The entrance of the filament was heated by a furnace and the two ends of the tube were cooled with air, which resulted in the temperature gradient distribution along the tube. The presented method is easily done with only a furnace, without the large consumption of noble gas and turbulence. Although the temperature gradient is not linear, we observed that multiple filaments were shrunken into a single filament and then filament disappears by increasing the temperature to some degree, which indicates that the critical power increases with temperature due to the gas atoms squeezed to the other end of the tube where the temperature is lower. Also, the filament can appear and disappear by controlling the local temperature and incident pulse energy in turn. The spectrum of the exit pulses is not expanded so much in comparison with the case of the same pressure and the same pulse energy, because the total gas atoms number is unchanged in the sealed tube. However, higher pulse energy is allowed to incident into the tube and a round trip pass of the tube is expected to expand the spectrum further with self-compression.

The gradient temperature technique has a great advantage that the temperature is easier to control than gradient pressure by differential pumping. Another merit is that the gas in the tube is relatively steady without flow, which is very important for keeping the output spectra stable. Not long after heating the gas to a high temperature at part of the sealed tube, the inner gas pressure will reach an equilibrium and the gas density in the tube will be gradient while the pressure in the tube will be equal everywhere. Because the pressure in the sealed tube is uniform, the convection and instabilities does not appear in our experiments. In contrast, in our experiment, the spectra and the light spot are very stable. For the pulse of same incident peak power, the spectra expansion in the gradient temperature is not as large as in the uniform temperature case. This is because the high temperature reduces the nonlinearity. However, because of this, a higher input energy can be sent through the tube, such that at the end of the tube, the peak power of the pulse is still high enough to expand the spectrum. This is the main reason that the transform limited pulse is shorter in gradient temperature tube than in the uniform temperature one. The drawback of the scheme is that the gas density difference cannot be as large as in the scheme using differential pumping. In addition, a big temperature difference may break the glass tube.

This technique offers one more degree of freedom to control the filamentation in a gas-filled tube for the intense monocycle pulse generation without gas consumption and turbulence and opens a new way for multi mJ level monocycle pulse generation through filamentation in the noble gas.

6. References

Abramowitz, M. & Stegun, I.A. (1972). *Handbook of Mathematical Functions*, Dover, ISBN : 0-486-61272-4, New York

Agrawal, G.P. (2007). *Nonlinear Fiber Optics* (4th Edition), Academic, ISBN : 0-123-69516-3, San Diego

Akturk, S.; D'Amico, C. ; Franco, M. ; Couairon, A. & Mysyrowicz, A. (2007). Pulse shortening, spatial mode cleaning, and intense terahertz generation by filamentation in xenon. *Phys. Rev. A*, Vol. 76, No. 6, (Dec. 2007) page numbers (3819-3825), ISSN: 1050-2947

Ammosov, M. V. ; Delone, N. B. & Krainov, V. P. (1986). Tunnel ionization of complex atoms and of atomic ions in an alternating electromagnetic field. *Sov. Phys. JETP*, Vol. 64, No. 6, (Dec. 1986) page numbers (1191-1194), ISSN : 0038-5646

Bergé, L.; Skupin, S. ; Lederer, F. ; Méjean, G. ; Yu, J. ; Kasparian, J. ; Salmon, E. ; Wolf, J.P. ; Rodriguez, M. ; Wöste, L. ; Bourayou, R. & Sauerbrey, R. (2004). Multiple filamentation of terawatt laser pulses in air. *Phys. Rev. Lett.*, Vol. 92, No. 22, (Jun. 2004) page numbers (225002 :1-4), ISSN: 0031-9007

Braun, A.; Korn, G. ; Liu, X. ; Du, D. ; Squier J. & Mourou, G. (1995). Self-channeling of high-peak-power femtosecond laser pulses in air. *Opt. Lett.*, Vol. 20, No. 1, (Jan. 1995) page numbers (73-75), ISSN: 0146-9592

Cao, S.; Kong, W. ; Wang, Z. ; Song, Z. ; Qin, Y. ; Li, R. ; Wang, Q. & Zhang Z. (2009). Filamentation control in the temperature gradient argon gas. Appl. Phys. B, Vol. 94, No. 2, (Feb. 2009) page numbers (265-271), ISSN: 0946-2171

Cook, K.; Kar, A. & Lamb, R.A. (2005). White-light filaments induced by diffraction effects. *Opt. Express*, Vol. 13, No. 6, (Mar. 2005) page numbers (2025-2031), ISSN: 1094-4087

Couairon, A.; Biegert, J. ; Hauri, C.P. ; Kornelis, W. ; Helbing, F.W. ; Keller, U. & Mysyrowicz, A. (2006). Self-compression of ultra-short laser pulses down to one optical cycle by filamentation. *J. Mod. Opt.*, Vol. 53, No. 1, (Oct. 2006) page numbers (75-85), ISSN: 0950-0340

Couairon, A.; Franco, M.; Mysyrowicz, A.; Biegert, J. & Keller, U. (2005). Pulse self-compression to the single-cycle limit by filamentation in a gas with a pressure gradient. *Opt. Lett.*, Vol. 30, No. 19, (Oct. 2005) page numbers (2657–2659), ISSN: 0146-9592

Courtois, C.; Couairon, A. ; Cros, B. ; Marquès, J.R. & Matthieussent, G. (2001). Propagation of intense ultrashort laser pulses in a plasma filled capillary tube: Simulations and experiments. Phys. Plasmas, Vol. 8, No. 7, (Jul. 2001) page numbers (3445-3456), ISSN: 1070-664X

Dalgarno, A & Kingston, A.E. (1960). The refractive indices and Verdet constants of the inert gases. *Proc. R. Soc. London, Ser. A,* Vol. 259, No. 1298 (Dec. 29, 1960), page numbers (424-431), ISSN : 1471-2946

Dreiskemper, R. & Botticher, W. (1995). Current filamentation of strongly preionized high pressure glowdischarges in Ne/Xe/HCl mixtures. *IEEE Transactions on Plasma Science*, Vol. 23, No. 6, (Dec. 1995) page numbers (987-995), ISSN: 0093-3813

Dubietis, A.; Tamosauskas, G. ; Fibich, G. & Ilan, B. (2004). Multiple filamentation induced by input-beam ellipticity. Opt. Lett., Vol. 29, No. 10, (May. 2004) page numbers (1126-1128), ISSN: 0146-9592

Fibich, G.; Eisenmann, S. ; Ilan, B. & Zigler, A. (2004). Control of multiple filamentation in air. *Opt. Lett.*, Vol. 29, No. 15, (Aug. 2004) page numbers (1772–1774), ISSN: 0146-9592

Fuji, T.; Horio, T. & Suzuki, T. (2007). Generation of 12 fs deep-ultraviolet pulses by four-wave mixing through filamentation in neon gas. *Opt. Lett.*, Vol. 32, No. 17, (Sep. 2007) page numbers (2481-2483), ISSN: 0146-9592

Hauri, C.P.; Guandalini, A.; Eckle, P.; Kornelis, W.; Biegert, J. & Keller, U. (2005). Generation of intense few-cycle laser pulses through filamentation – parameter dependence. *Opt. Express*, Vol. 13, No. 19, (Sep. 2005) page numbers (7541-7547), ISSN: 1094-4087

Hauri, C. P.; Kornelis, W.; Helbing, F.W.; Heinrich, A.; Couairon, A.; Mysyrowicz, A.; Biegert, J. & Keller, U. (2004). Generation of intense, carrier-envelope phase-locked few-cycle laser pulses through filamentation. *Appl. Phys. B*, Vol. 79, No. 6, (Oct. 2004) page numbers (673-677), ISSN: 0946-2171

Hosseini, S.A.; Luo, Q. ; Ferland, B. ; Liu, W. ; Chin, S.L. ; Kosareva, O.G. ; Panov, N.A. ; Aközbek, N. & Kandidov, V.P. (2004). Competition of multiple filaments during the propagation of intense femtosecond laser pulses. *Phys. Rev. A*, Vol. 70, No. 03, (Sep. 2004) page numbers (3802-3813), ISSN: 1050-2947

Kandidov, V.P.; Akozbek, N. ; Scalora, M. ; Kosareva, O.G. ; Nyakk, A.V. ; Luo, Q. ; Hosseini, S.A. & Chin, S.L. (2005). Towards a control of multiple filamentation by spatial regularization of a high-power femtosecond laser pulse. *Appl. Phys. B*, Vol. 80, No. 2, (Feb. 2005) page numbers (267-275), ISSN: 0946-2171

Keldysh, L.V. (1965). Ionization in the field of a strong electromagnetic wave. *Sov. Phys. JETP,* Vol. 20, No. 5, (May 1965) page numbers (1307-1314), ISSN : 0038-5646

Lehmeier, H.J.; Leupacher, W. & Penzkofer, A. (1985). Nonresonant third order hyperpolarizability of rare gases and N2 determined by third harmonic generation. *Opt. Commun.*, Vol. 56, No. 1, (Nov. 1985) page numbers (67-72), ISSN: 0030-4018

Marburger, J.H. (1975). Self-focusing: Theory. *Prog. Quant. Electr.* Vol. 4, page numbers (35–110), ISSN: 0079-6727

Marcatili, E.A.J. & Schmeltzer, R.A. (1964). Hollow metallic and dielectric waveguides for long distance optical transmission and lasers (Long distance optical transmission in hollow dielectric and metal circular waveguides, examining normal mode propagation). *Bell Syst. Tech. J*, Vol. 43, (Jul. 1964) page numbers (1783-1809), ISSN: 0005-8580

Matsubara, E.; Yamane, K.; Sekikawa, T. & Yamashita, M. (2007). Generation of 2.6 fs optical pulses using induced-phase modulation in a gas-filled hollow fiber. *J. Opt. Soc. Am. B*, Vol. 24, No. 4, (Apr. 2007) page numbers (985–989), ISSN: 0740-3224

Mlejnek, M.; Wright, E.M. & Moloney, J.V. (1998). Femtosecond pulse propagation in argon: A pressure dependence study. *Phys. Rev.* E, Vol. 58, No. 4, (Oct. 1998) page numbers (4903-4910), ISSN: 1539-3755

Perelemov, A.M. ; Popov, V.S. &. Terent'ev, M.V. (1966). Ionization of atoms in an alternating electric field. *Sov. Phys. JETP,* Vol. 23, No. 5, (May 1966) page numbers (924-933), ISSN : 0038-5646

Pfeifer, T.; Gallmann, L. ; Abel, M.J. ; Neumark, D.M. & Leone, S.R. (2006). Circular phase mask for control and stabilization of single optical filaments. *Opt. Lett.*, Vol. 31, No. 15, (Aug. 2006) page numbers (2326–2328), ISSN: 0146-9592

Shen Y.R. (1984). *The Principles of Nonlinear Optics*, Wiley-Interscience, ISBN : 0-471-43080-3, NewYork

Skupin, S.; Bergé, L. ; Peschel, U. ; Lederer, F. ; Méjean, G. ; Yu, J. ; Kasparian, J. ; Salmon, E. ; Wolf, J.P. ; Rodriguez, M. ; Wöste, L. ; Bourayou, R. & Sauerbrey, R. (2004). Filamentation of femtosecond light pulses in the air: Turbulent cells versus long-range clusters. *Phys. Rev. E*, Vol. 70, No. 4, (Oct. 2004) page numbers (6602-6616), ISSN: 1063-651X

Song, Z.; Qin, Y. ; Zhang, G. ; Cao, S. ; Pang, D. ; Chai, L. ; Wang, Q. ; Wang, Z. & Zhang, Z. (2008). Femtosecond pulse propagation in temperature controlled gas-filled hollow fiber. *Opt. Commun.*, Vol. 281, No. 15-16, (Aug. 2008a) page numbers (4109-4113), ISSN: 0030-4018

Song, Z.; Zhang, G. ; Cao, S. ; Pang, D. ; Chai, L. ; Wang, Q. & Zhang Z. (2008b). Simulation of Femtosecond Pulse Propagation through Hollow Fibre Filled with Noble Gases of Gradient Temperature. *Chin. Phys. Lett.*, Vol. 25, No. 1, (Jan. 2008) page numbers (129-132), ISSN: 0256-307X

Suda, A.; Hatayama, M. ; Nagasaka, K. & Midorikawa, K. (2005). Generation of sub-10-fs, 5-mJ-optical pulses using a hollow fiber with a pressure gradient. *Appl. Phys. Lett.*, Vol. 86, No. 11, (Mar. 2005) page numbers (111116 :1-3), ISSN: 0003-6951

Sung, J.H.; Park, J.Y. ; Imran, T. ; Lee, Y.S. & Nam, C.H. (2006). Generation of 0.2-TW 5.5-fs optical pulses at 1 kHz using a differentially pumped hollow-fiber chirped-mirror compressor. *Appl. Phys. B*, Vol. 82, No. 1, (Jan. 2006) page numbers (5-8), ISSN: 0946-2171

Tempea, G. & Brabec, T. (1998). Theory of self-focusing in a hollow waveguide. *Opt. Lett.*, Vol. 23, No. 10, (May. 1998) page numbers (762-764), ISSN: 0146-9592

Vidal F. & Johnston, T.W. (1996). Electromagnetic beam breakup: multiple filaments, single beam equilibria, and radiation. *Phys. Rev. Lett.*, Vol. 77, No. 7, (Aug. 1996) page numbers (1282-1285), ISSN: 0031-9007

Yamane, K.; Zhang, Z.; Oka, K.; Morita, R.; Yamashita, M. & Suguro, A. (2003). Optical pulse compression to 3.4fs in the monocycle region by feedback phase compensation. *Opt. Lett.*, Vol. 28, No. 22, (Nov. 2003) page numbers (2258-2260), ISSN: 0146-9592

Yamashita, M.; Yamane, K. & Morita, R. (2006). Quasi-automatic phase-control technique for chirp compensation of pulses with over-one-octave bandwidth-generation of few- to mono-cycle optical pulses. *IEEE J. Sel. Top. Quantum Electron*, Vol. 12, Issue. 2, (March-April 2006) page numbers (213-222), ISSN: 1077-260X

Zaïr, A.; Guandalini, A.; Schapper, F.; Holler, M.; Biegert, J.; Gallmann, L.; Couairon, A., Franco, M.; Mysyrowicz, A. & Keller, U. (2007). Spatio-temporal characterization of few-cycle pulses obtained by filamentation. *Opt. Express*, Vol. 15, No. 9, (Apr. 2007) page numbers (5394-5404), ISSN: 1094-4087

18

Diffraction Gratings for the Selection of Ultrashort Pulses in the Extreme-Ultraviolet

Luca Poletto, Paolo Villoresi and Fabio Frassetto
CNR-National Institute for the Physics of Matter & Dep. of Information Engineering
LUXOR - Laboratory for UV and X-Ray Optical Research
Padova
Italy

1. Introduction

We discuss the use of diffraction gratings to perform the spectral selection of ultrashort pulses in the extreme-ultraviolet and soft X-ray spectral regions, ranging in the 3-100 nm wavelength range. The main application of such a technique is the spectral selection of high-order laser harmonics and free-electron-laser pulses in the femtosecond time scale. We present the design and realization of both single- and double-grating monochromators using an innovative grating geometry, namely the off-plane mount. The performances of existing instruments are shown. The use of diffraction gratings to change the phase properties of the pulse, e.g. to compress it to shorter temporal duration close to the Fourier limit, is also discussed.

Extreme-ultraviolet (XUV) and X-ray photons have been used for many fundamental discoveries and outstanding applications in natural sciences (Wiedermann, 2005). They have played a crucial role in basic research and medical diagnostics, as well as in industrial research and development. The main reason for this success is that the wavelength, which determines the smallest distance one can study with such a probe, is comparable to the molecular and atomic dimension. On the other side, the advent of femtosecond (1 fs = 10^{-15} s) lasers has revolutionized many areas of science from solid-state physics to biology (Diels & Rudolph, 2006). The significance of the femtosecond time regime is that atomic motion which governs structural dynamics, such as phase transitions and chemical reactions, occurs on the vibrational timescale of ~100 fs. While femtosecond optical lasers have offered unique insights into ultra-fast dynamics, they are limited by the fact that the structural arrangement and motion of nuclei are not directly accessible from measured optical properties.

The availability of coherent and tunable sources in the XUV and X-rays with characteristics similar to those of ultrashort lasers in the visible and near-infrared opens the way to a completely new class of experiments both in fundamental and applied research (Patel, 2002). It requires joining the competences in the ultrafast techniques with those on instrumentation and experiments in the XUV and X-rays. The handling of the photons emitted by such sources requires particular attention to the management of high intensity pulses, to the preservation of the ultrashort pulse duration and to the effects of the optical components on the phase of the pulse.

Here, we deal with the problem of making the spectral selection of a XUV and soft X-ray pulse while preserving its duration in the femtosecond, or even shorter, time scale. The technique is useful for high-order laser harmonics and free-electron-laser pulses.
High-order harmonics (HHs) generated by the interaction between an ultra-short laser pulse and a gas jet are currently considered as a relevant source of coherent XUV and soft X-ray radiation of very short time duration and high peak brilliance, with important applications in several areas both in fundamental research and in advanced technology (Jaegle, 2006). Owing to the strong peak power of a femtosecond laser pulse, a nonlinear interaction with the gas jet takes place and produces odd laser harmonics (i.e. of order $2n+1$ with n integer), well above the order of 100. When the laser beam and its second harmonic are used together, a full spectrum of even ($2n$) and odd ($2n+1$) harmonics is obtained. In this way a conversion from near-infrared or visible light into XUV and soft-x-ray radiation takes place, giving rise to a XUV source with the same properties of the generating laser in terms of coherence and short pulse duration. The HH spectrum is described as a sequence of peaks corresponding to the harmonics of the fundamental laser wavelength and having an intensity distribution characterized by a vast plateau, whose extension is related to the laser pulse intensity. The radiation generated with the scheme of the HHs generated by laser pulses of a few optical cycles recently become the tool for the investigation of matter with sub-femtosecond, or attosecond, resolution (1 as = 10^{-18} s) (Kienberger & Krausz, 2004; Corkum & Krausz, 2007). The access to this unexplored time domain opens new frontiers in atomic, molecular and solid-state science (Marciak-Kozlowska, 2009), as it becomes possible to do experiments with an unprecedented time resolution and intensity.
Another way to obtain very intense ultrashort and tunable pulses in the XUV is the use of free-electron-laser (FEL) generation. FELs share the same optical properties as conventional lasers but they use different operating principles to form the beam, i.e. a relativistic electron beam as the lasing medium which moves freely through a magnetic structure that induces radiation, the so-called undulator (Saldin et al., 2000). The lack of suitable mirrors in the XUV and X-rays regimes prevents the operation of a FEL oscillator; consequently, FEL emission in the XUV and X-ray has to be obtained in a single pass through the undulator. In this case it is possible to feed an electron beam into the undulators with a much smaller emittance than achievable in storage rings. Some XUV and X-ray FEL facilities are now running worldwide: we can cite FLASH (see http://hasylab.desy.de/facilities/flash/index_eng.html) in Hamburg (Germany) and SLAC (see http://lcls.slac.stanford.edu) in Stanford (USA).
Let us consider an ultrashort pulse of XUV radiation that has a wavelength in the 4-100 nm range and that is mixed with the radiation of different spectral ranges. The spectral selection of such a pulse requires the use of a monochromator. As examples of the experimental problems to be addressed by such an instrument in HH generation, we can cite the extraction of a single harmonic (or a group of harmonics) within a broad HH spectrum to obtain an ultrafast pulse at a suitable XUV wavelength, later to be scanned in a given range. Monochromators can be useful also for FEL radiation, both to increase the spectral purity of the fundamental FEL emission and to select the FEL high-order harmonics at shorter wavelengths while rejecting the most intense fundamental. The monochromator called for this purpose is also called to preserve the pulse temporal duration of the XUV pulse as short as in the generation process. This is crucial in order to have both high temporal resolution and high peak power.

The study and design of such compensated monochromator extends the usual domain of the geometrical optics and XUV diffraction grating mountings to include the analysis of the XUV pulse transformation in both spectrum and spectral phase. The monochromator can be modelled as a filter with a complex frequency response $K(\omega)$, that includes both the spectral transmission and the distorsion in the spectral phase as a function of the frequency (Akhmanov et al. 1992). Since the XUV pulse at the generation may be produced to be close to its transform limit, any modification of its complex spectrum results in a severe time broadening as described by its Fourier transform. For a Gaussian profile with no modulation of either phase or frequency, the product of the spectral width at half-height $\Delta\omega_{1/2}$ times the duration at half-height $\Delta\tau_{1/2}$ has a lower limit expressed by the relation

$$\Delta\omega_{1/2}\Delta\tau_{1/2} = 4\ln 2 = 2.77\,. \tag{1}$$

Two are the conditions that have to be verified by the monochromator to maintain the time duration expressed by Eq. 1 after the monochromatization: 1) the band-pass $\Delta\omega_m$ transmitted by the monochromator has to be greater than the bandwidth of the pulse $\Delta\omega_{1/2}$ and 2) the complex transfer function $K(\omega)$ has to be almost constant within the bandwidth. In case of HH selection, since harmonic peaks are well separated, the first condition is verified if the monochromator selects the whole spectral band of a single harmonic (or a group of them), so no modifications in the Fourier spectrum are induced. The case of FEL radiation is also similar: the bandwidth of the monochromator has to be larger than the intrinsic FEL bandwidth. The second condition is almost always verified if the monochromator is realized by reflecting optics: the variations of reflectivity of the coating within the bandwidth of the pulse are usually negligible, so $K(\omega)$ can be considered almost constant, although lower than unity.

2. Grating monochromators for spectral selection of ultrashort pulses

The simplest way to obtain the spectral selection of ultrashort pulses with very modest time broadening is the use of a multilayer mirror in normal incidence, which does not alter the pulse time duration up to fractions of femtosecond and is moreover very efficient: in fact, the functions of selecting a single spectral pulse and focusing it can be demanded to a single concave optics, maximizing then the flux. The choice of the type of multilayer can be made among many couple of materials (i.e. the spacer and the absorber) to optimize the response in a given spectral region (e.g. see http://www-cxro.lbl.gov/multilayer/survey.html). Monochromators with one (Wieland et al. 2001) or two (Poletto & Tondello, 2001) multilayer mirrors have been proposed and realized. The main drawback of the use of multilayer optics is the necessity of many different mirrors to have the tunability on a broad spectral region.

The spectral selection of XUV ultrashort pulses can also be accomplished by an ordinary diffraction grating used in reflection mode. In this case, the major mechanism that alters the time duration of the pulse is the difference in the lengths of the optical paths of the rays diffracted by different grating grooves. In fact, a single grating gives inevitably a time broadening of the ultrafast pulse because of the diffraction: the total difference in the optical paths of the rays diffracted by N grooves illuminated by radiation at wavelength λ is $\Delta OP = Nm\lambda$, where m is the diffracted order. The effect is schematically illustrated in Fig. 1. It follows that the longer the exposed area of the grating and the higher the groove density, the longer is the time difference between the arrival of the ray diffracted from the first illuminated groove and that diffracted from the last one. As an order-of-magnitude estimate, let us consider a 300 gr/mm grating illuminated by radiation at 40 nm over a length of 10 mm at half intensity; the total number of grooves involved in the diffraction is 3000, corresponding to a maximum delay in the first diffracted order of 120 mm, i.e. 400 fs full-width-at-half-maximum (FWHM). This delay is usually irrelevant for picosecond pulses, but, in case of femtosecond pulses, it reduces dramatically both the time resolution capability and the peak intensity at the output. Nevertheless, single-grating monochromators can be adopted for the spectral selection of ultrashort pulses because of the simplicity of the design and the high efficiency due to the use of a single optical element. In fact, in the XUV and soft X-ray spectral region the reflectivity is usually poor and conventional schemes generally use a single grating. Furthermore, a time resolution in the 100-300 fs time scale can be tolerated on many pump-probe experiments, that is in the range of the temporal response of low-density gratings.

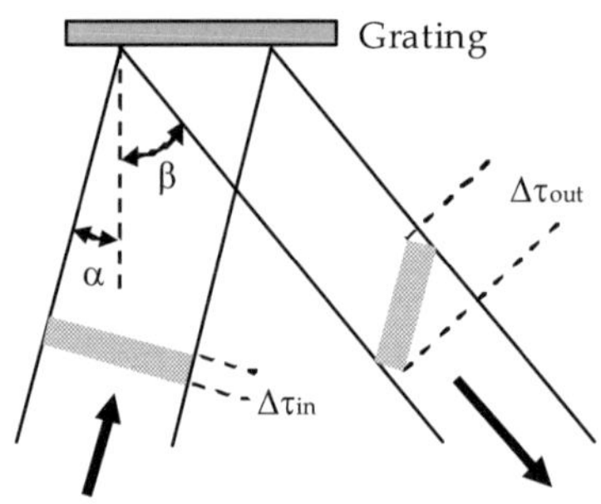

Fig. 1. Schematic illustration of the tilt of the intensity wavefront of an ultrashort pulse diffracted by a grating in the classical diffraction geometry. The incidence angle is indicated as α, the diffracted angle as β, the pulse width at the input as $\Delta\tau_{in}$, the pulse width at the output as $\Delta\tau_{out}$. The broadening is $\Delta OP = Nm\lambda = N\sigma^{-1}(\sin\alpha - \sin\beta)$, where σ is the grating groove density.

The temporal broadening given by a grating poses a problem that cannot be solved with conventional monochromator designs. To select effectively a portion of an extended spectrum without introducing a difference in the optical path length between different rays in the beam, one must use a so called time-delay compensated optical mounting, which involves the use of two gratings in a subtractive configuration to compensate for the dispersion. In the following, such a configuration will be defined as *Time-Delay Compensated (TDC) monochromator*, in the sense that the second grating compensates for the time and spectral spread introduced by the first one. These concepts are well known in the fields of chirped pulse laser amplification and, more generally, propagation of ultrafast light pulses (Walmsley, 2001) and are here discussed in the XUV spectral domain.

From the point of view of the ray paths, two are the conditions that the design must comply: 1) the differences in the path lengths of rays with the same wavelength but with different entrance directions within the beam aperture that are caused by the first grating must be compensated by the second grating, and 2) two rays at different wavelengths within the spectrum of the pulse to be selected have to be focused on the same point, i.e. the global spectral dispersion has to be zero. Both these conditions are satisfied by a scheme with two equal concave gratings mounted with opposite diffraction orders (Villoresi, 1999): the incidence angle on the second grating is equal to the diffraction angle of the first grating. The spectral selection is performed by a slit placed in an intermediate position between the gratings, where the radiation is focused by the first grating. In the design, only two optical elements are used, namely the concave gratings, since the grating itself provides both the spectral dispersion through diffraction and the focusing due to the curvature of the surface. The schematic of the TDC configuration in case of using two toroidal gratings at normal incidence in the classical geometry is shown in Fig. 2.

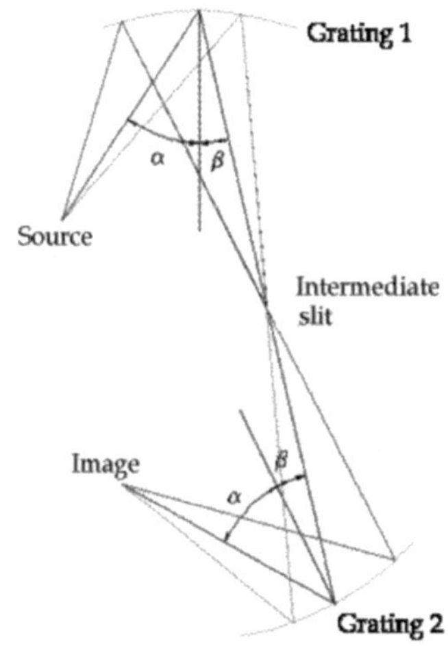

Fig. 2. Schematic of the TDC configuration with two toroidal gratings used at normal incidence.

This design has been proved to be very effective in time compensation of few femtoseconds or even lower for wavelengths longer than ≈40 nm, where the normal-incidence reflectivity of conventional coatings is high so that normal-incidence configurations can be adopted. It can be shown by ray-tracing simulations that a TDC monochromator in the Seya-Namioka mounting can be operated in the 40-100 nm region with residual broadening at the output in the 0.5-1 μm range, i.e. 1-3 fs (Villoresi, 1999). The main advantages of this design are the simplicity, due to the use of only two optical elements, the tunability in a broad spectral band by a simple rotation of both gratings and the possibility of being operated also in spectral regions where multilayer-coated optics are not available (e.g. for wavelengths higher than 60 nm). Unfortunately, the main drawback of using two normal-incidence gratings is the low efficiency in the XUV. By choosing a suitable XUV coating (e.g. gold or platinum), the efficiency of a single grating can be estimated in the 0.1-0.2 range, so the monochromator efficiency results 0.01-0.04.

For wavelengths below ≈35 nm, the normal-incidence configuration cannot be adopted because of the low reflectivity of conventional coatings in normal incidence. Therefore, the gratings have to be operated in grazing incidence. The compensation in this case is harder, due to the intrinsic difficulties arising from grazing-incidence mountings, which are very sensitive to aberrations and misalignments. A configuration with two toroidal gratings has already been presented and discussed (Villoresi, 1999). The time compensation is again very effective, but once the grating radii and the subtended angle are chosen, the compensation is optimum only in a very narrow spectral region, so the tunability is impaired unless changing the geometry for different spectral regions. Also in this case, the global efficiency of the monochromator is expected to be rather poor. The efficiency is obviously the major factor discriminating among different monochromators: an instrument with low output flux could be not useful for scientific experiments.

We can summarize the main differences between single- and double-grating monochromators as follows: 1) the single-grating design is simpler and more efficient, but its temporal response is longer; 2) the double-grating design is more complex and less efficient, but the temporal response can be as short as fractions of femtosecond. The choice between the two configurations has to be done as a trade-off between photon flux and time resolution.

3. The off-plane grating geometry: characteristics and performance

The main limitation to the use of TDC configurations with gratings in the classical mount is the poor efficiency of the design and the difficulty in achieve a broad tunability in grazing incidence. Both these drawback are overtaken by adopting a different geometry, the so called off-plane mount for gratings at grazing incidence. This mount differs from the classical one in that the incident and diffracted wave vectors are almost parallel to the grooves (Cash, 1982). The geometry is shown in Fig. 3.

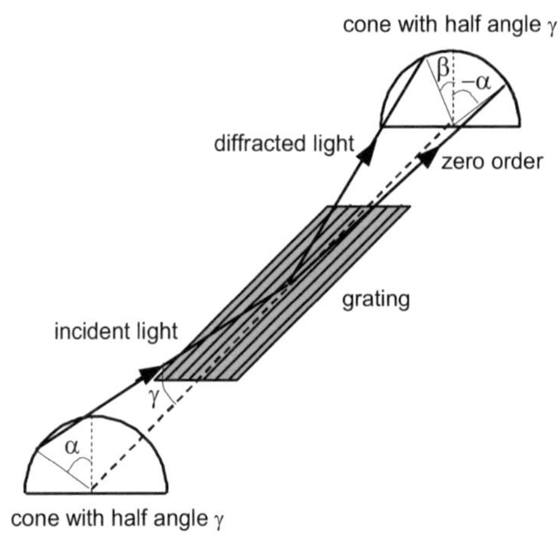

Fig. 3. The off-plane geometry.

The direction of the incoming rays is described by two parameters, the altitude and the azimuth. The altitude γ is the angle between the direction of the incoming rays and the direction of the grooves. It defines the half-angle of the cone into which the light is diffracted: all the rays leave the grating at the same altitude angle at which they approach. The azimuth α of the incoming rays is defined to be zero if they lie in the plane perpendicular to the grating surface and parallel to the rulings, so $-\alpha$ is the azimuth of the zero order light. Let β define the azimuth of the diffracted light at wavelength λ and order m. The grating equation is written as

$$\sin\gamma\,(\sin\alpha + \sin\beta) = m\lambda\sigma \qquad (2)$$

where σ is the groove density.
The blaze condition of maximum efficiency is described as in the classical diffraction geometry: the light has to leave the grating in such a way that it performs a specular reflection on the groove surface, that is $\alpha + \beta = 2\delta$, where δ is the grating blaze angle. In addition, shadowing effects from adjacent grooves must be avoided, that is $\alpha = \delta$. It follows that the highest efficiency of a blazed grating in the off-plane mount is achieved when $\alpha = \beta = \delta$, that is when each groove of the grating is seen by the incident ray as a portion of a plane mirror. The grating equation in the blaze condition becomes $2 \sin\gamma \sin\delta = m\lambda\sigma$, indicating that both incident and diffracted rays at the blaze wavelength lie in a plane that is parallel to the direction of the grooves and also perpendicular to their surface.
It has been theoretically shown and experimentally measured that the diffraction efficiency of the off-plane mount is close to the reflectivity of the coating, so much higher efficiencies than in the classical diffraction mount can be obtained in the XUV (Petit, 1980). The off-plane mount is then an appropriate candidate to realize XUV and soft X-ray monochromators for ultrashort pulses with high efficiency (Poletto, 2004a; Pascolini et al., 2006).
Gratings in the off-plane mount are usually operated in parallel light. Therefore, a monochromator requires at least three optical elements: 1) a first concave mirror at grazing incidence that collimates the light coming from the entrance slit, 2) the plane grating and 3) a second concave mirror that focuses the light on the exit slit. In order to keep the grating on-blaze for each wavelength (i.e. maximum efficiency over the whole bandwidth of operation), the grating equation in blaze condition $2 \sin\gamma \sin\delta = m\lambda\sigma$ should be satisfied in the entire spectral bandwidth: this requires the variation of the altitude γ, since the blaze angle δ is fixed. In this case, the wavelength scanning is rather complex: the grating is translated along an axis perpendicular to its surface, so the altitude is changed. In order to keep a constant output direction toward the focusing mirror, two additional plane mirrors are inserted in the design, and are rotated with the wavelength (Werner & Wisser, 1984). A schematic of the monochromator is shown in Fig. 4(a).

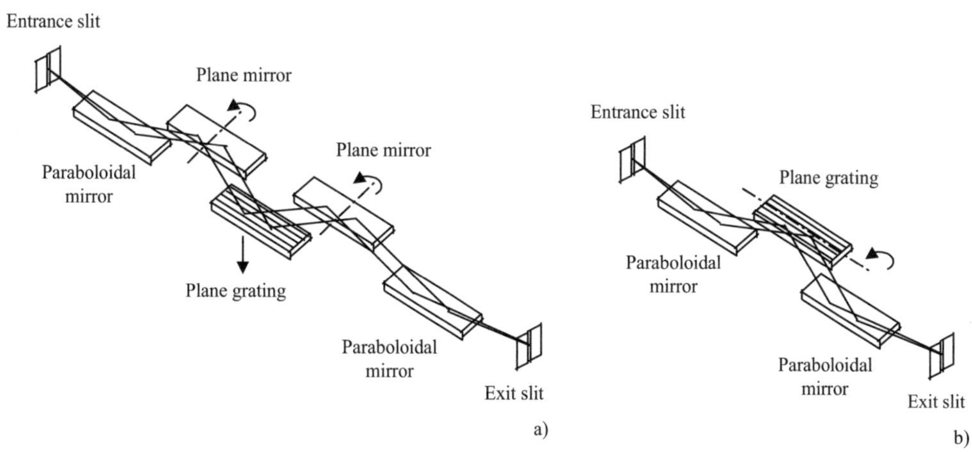

Fig. 4. Monochromator with plane gratings in the off-plane mount. a) Grating on-blaze at each wavelength: the wavelength scanning is performed by changing the altitude and keeping constant the azimuth. b) Grating on-blaze at one wavelength: the wavelength scanning is performed by changing the azimuth and keeping constant the altitude.

A simpler layout consists in performing the wavelength scanning by rotating the grating around an axis passing through the grating center and parallel to the groove direction, so the azimuth is changed while the altitude γ is kept constant (Werner & Wisser, 1984). For each wavelength, the relation $\alpha = \beta$ is satisfied, so the grating equation is expressed as

$$2 \sin\gamma \sin\alpha = m\lambda\sigma. \quad (3)$$

The blaze condition of maximum efficiency is satisfied only at the wavelength λ_B for which $\alpha(\lambda_B) = \beta(\lambda_B) = \delta$: $\lambda_B = 2 \sin\gamma \sin\delta / m\sigma$. At different wavelengths, the efficiency decreases because the grating is operated off-blaze. No additional plane mirrors are required in this configuration. A schematic of the monochromator is shown in Fig. 4(b).
The main advantage the off-plane mount is the high efficiency in the XUV and soft X-ray domains. We summarize here some experimental results obtained on gratings to be used for the spectral selection of ultrashort pulses. The efficiency of three plane gold-coated gratings manufactured by Richardson Gratings (Newport) has been measured in the 10-130 nm spectral region at the beamline BEAR of the ELETTRA Synchrotron (Trieste, Italy). The gratings were operated in the same way as in geometry shown in Fig. 4(b): the altitude angle γ is kept constant and, for each desired wavelength, the azimuth angle α is selected from Eq. 3 to have $\alpha = \beta$. The condition of maximum efficiency is satisfied only at the blaze wavelength. Far from it, the efficiency decreases. The efficiency is here intended as the absolute efficiency, i.e. the ratio between the flux measured in the diffracted order and the total flux on the grating. The parameters of the measurements are resumed in Tab. 1.

Grating	*Grooves/mm*	*Blaze angle* δ	*Altitude angle* γ	*Blaze wavelength (1st order)*	*Wavelength*
#1	600	7°	11.4°	80 nm	50-130 nm
#1	600	7°	3°	21.5 nm	10-45 nm
#2	400	9.7°	3°	45 nm	20-90 nm
#3	400	4.5°	6.6°	45 nm	20-90 nm

Table 1. Gratings under test in the off-plane geometry.

The efficiency curves of grating #1 are shown in Fig. 5. The peak efficiency at the blaze wavelength is about 0.5. The grating is very selective when operated on-blaze: at 1st order blaze wavelength, both 0th and 2nd orders are negligible. As expected, 1st order efficiency decreases when the grating is rotated off-blaze to select other portions of the spectrum. At wavelengths longer than 1st order blaze wavelength $\lambda > \lambda_{B\text{-}1st}$, where the grating blaze angle is lower than the azimuth angle $\delta < \alpha$, the radiation is split between 1st and 0th orders. 2nd order efficiency is null because 2nd order diffraction is far from the blaze condition. At wavelengths shorter than 1st order blaze wavelength $\lambda < \lambda_{B\text{-}1st}$, where the grating blaze angle is higher than the azimuth angle $\delta > \alpha$, the radiation is split between 1st and 2nd orders, for which the blaze condition would be $\delta = (\alpha + \beta_{2nd})/2$. 2nd order efficiency increases and reaches its maximum at $\lambda_{B\text{-}1st}/2$, that is the 2nd order blaze wavelength. Then, the same grating can be operated in a large spectral region with high efficiency using 1st order diffraction around the 1st order blaze wavelength, and 2nd order diffraction around 2nd order blaze wavelength.
The efficiency curves of grating #2 and #3 operated in the 20-90 nm region are shown in Fig. 6. The peak efficiency of grating #2 is exceptionally high, even more than 0.7. Again, the gratings are very selective when operated on-blaze.

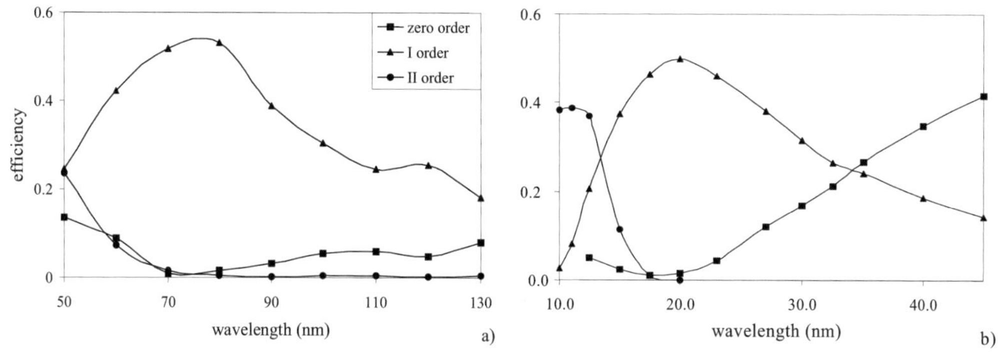

Fig. 5. Efficiency curves of grating #1: a) grating operated at $\gamma = 11.4°$ in the 50-130 nm region; b) grating operated at $\gamma = 3°$ in the 10-45 nm region.

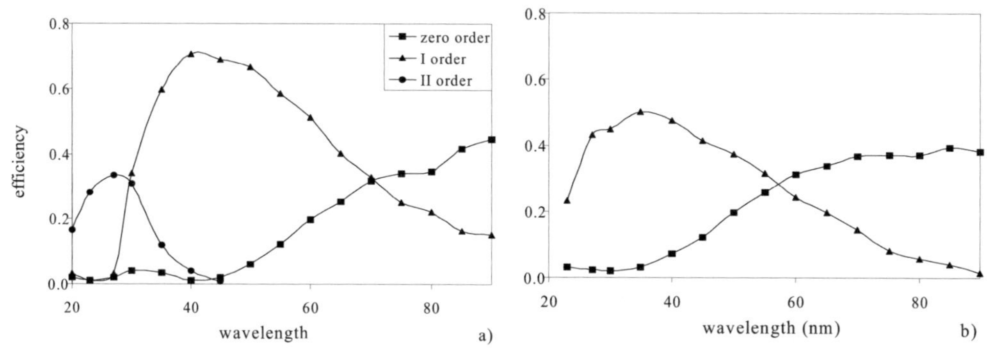

Fig. 6. Efficiency curves of grating #2 and #3 operated at $\gamma = 3°$ in the 20-90 nm region: a) grating #2; b) grating #3.

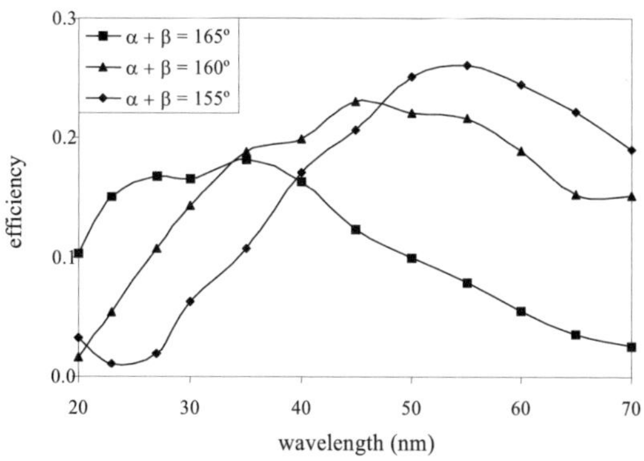

Fig. 7. Efficiency curves of grating #3 operated in classical configuration at constant inclusion angle, i.e. $\alpha + \beta = K$.

The diffraction efficiency of grating #3 was also measured in the classical grazing-incidence configuration, when the incoming rays are perpendicular to the grooves. The measurements were performed at constant inclusion angle, i.e. $\alpha + \beta = K$, where α is the incidence angle and β the diffraction angle. The curves in the 20-70 nm region are shown in Fig. 7. The maximum efficiency depends on the inclusion angle and ranges in the 0.18-0.25 interval. With respect to the off-plane mount, the efficiency of the same grating is lower by a factor two to three. This confirms the advantage of the off-plane mount in terms of high throughput from the monochromator.

4. Design of a single-grating monochromator in the off-plane geometry

The single-grating configuration in the off-plane geometry adopts a plane grating operated in parallel beam and two concave mirrors, as shown in Fig. 4(b). The first mirror collimates the light coming from the point source; the grating is operated in the condition $\alpha = \beta$; the second mirror focuses the diffracted light on the exit slit. All the optical elements are operated in grazing incidence. The wavelength scan is provided by rotating the grating around an axis passing through the grating center and parallel to the direction of the grooves, as shown in Eq. 3. In this design, the azimuth changes with the wavelength while the altitude γ is kept constant, so that the maximum efficiency condition is fulfilled only at the blaze wavelength $\lambda_B = 2 \sin\gamma \sin\delta / m\sigma$ which depends on the blaze angle of the grating profile. At different wavelengths, the efficiency decreases because the grating is operated off-blaze. Such a mount is very efficient in a broad spectral region, as shown from the experimental curves of Fig. 5 and Fig. 6.
The driving parameter is the FWHM bandwidth at the output $\Delta\lambda_{FWHM}$, calculated at first diffraction order as

$$\Delta\lambda_{FWHM} = \frac{W}{\sigma\, q_{out}} \tag{4}$$

where W is the width of the exit slit, that is typically in the range 50-200 μm, and q_{out} is the length of the output arm of the focusing mirror, i.e. the distance between the center of the mirror and the exit slit, that is typically in the range 300-500 mm. The bandwidth has to be as broad as the bandwidth of the ultrashort pulse to be selected, to avoid the broadening of the pulse duration as expressed in Eq. 1.
In case of selection of the single harmonic Hn (n odd) within the HH spectrum, i.e. to have the adjacent harmonics completely filtered out, the slit aperture is

$$W = \sigma\, q_{out} \frac{\Delta\lambda_{H(n-2)-H(n+2)}}{2} \cong \sigma\, q_{out}\, \lambda_0 \frac{2}{n^2} \tag{5}$$

where $\Delta\lambda_{H(n-2)-H(n+2)}$ indicates the difference in wavelength between the two harmonics H(n-2) and H(n+2), that is expressed by $\Delta\lambda_{H(n-2)-H(n+2)} = 4\lambda_0/n^2$ where λ_0 is the laser fundamental wavelength, that is 800 nm for the Ti:Sa laser.
Once the geometric parameters have been chosen (i.e. output arm and slit width), the spectral resolution $\mathcal{R} = \lambda/\Delta\lambda$ given by the off-plane mount is proportional only to the grating groove density. On the contrary, the resolution given by the classical mount is proportional also to the inverse of the cosine of the diffraction angle, so at extreme grazing conditions the resolution becomes higher for a given groove density. Since the resolution of the off-plane mount is independent from the grazing angle, low-resolution configurations in the off-plane mount can be designed even at extreme grazing incidence for XUV and soft X-rays.

The instrumental temporal response of the off-plane grating $\Delta\tau_{FWHM}$ depends on the groove density and on the illuminated area

$$\Delta\tau_{FWHM} = \frac{\sigma\lambda q_{in}\mathcal{D}_{FWHM}}{c} \tag{6}$$

where $\mathcal{D}_{FWHM}$ is the FWHM divergence of the source, c is the speed of light in vacuum and q_{in} is the length of the input arm of the collimating mirror, i.e. the distance between the source and the center of the mirror. The mirrors have typically a toroidal shape, then the aberrations at the output are minimized if they have equal arms, i.e. $q_{in} = q_{out} = q$.
Since the grooves are almost parallel to the input direction, the number of grooves that are illuminated in the off-plane mount is independent from the altitude angle, $N = \sigma\, q\, \mathcal{D}_{FWHM}$. Therefore, once the groove density has been defined, the temporal broadening is univocally determined. This allows to operate the grating even at extreme grazing conditions without changing its temporal response. On the contrary, the number of grooves illuminated in the classical mount is proportional to the inverse of the cosine of the incident angle, so the temporal response at grazing incidence changes at each wavelength depending on the working conditions. Then, the operation in the off-plane mount gives more flexibility than the classical mount in the choice of the incidence angle at very short wavelengths and a temporal response that varies linearly with the wavelength.

Spectral region	10-80 nm
Concave mirrors	Toroidal surfaces
Grazing angle	3°
Arms	300 mm
Gratings	4 plane gratings
Altitude	3.5°
Grating 1	Low resolution, long wavelengths
Groove density	60 gr/mm
Blaze angle and blaze wavelength	1.3°, 45 nm
Plate factor	55.6 nm/mm
Grating 2	Low resolution, short wavelengths
Groove density	120 gr/mm
Blaze angle and blaze wavelength	1.1°, 20 nm
Plate factor	27.8 nm/mm
Grating 3	Medium resolution, long wavelengths
Groove density	300 gr/mm
Blaze angle and blaze wavelength	6.5°, 46 nm
Plate factor	11.1 nm/mm
Grating 4	Medium resolution, short wavelengths
Groove density	500 gr/mm
Blaze angle and blaze wavelength	4.7°, 20 nm
Plate factor	6.7 nm/mm
Exit slit	50-300 μm

Table 2. Parameters of the monochromator of the ASTRA-ARTEMIS beamline at Rutherford Appleton Laboratory (UK).

It is clear from Eq.s 4-5 that the higher the spectral resolution, the longer the temporal response. The aim of the optical design is to find the best trade-off between spectral resolution and temporal broadening.

As an example, let us report the characteristics of a low-resolution monochromator for the spectral selection of HHs in the 80-10 nm region (Frassetto et al., 2008). The monochromator has been realized in the LUXOR laboratory in Padova (Italy) and is installed in the ASTRA-ARTEMIS beamline of the Central Laser Facility of the Rutherford Appleton Laboratory (UK). The design parameters are resumed in Tab. 2. The monochromator is operated without an entrance slit: the HH generation point is the source point for the instrument. It has four different gratings to cover the 10-80 nm region with different spectral resolutions. The gratings are mounted on a swivel motorized stage for the rotation and a linear motorized stage for the selection of the grating to be used. The radiation exiting from the slit is finally focused to the experimental chamber by a suitable toroidal mirror at grazing incidence. A schematic of the monochromator and a picture of the instrument are shown in Fig. 8.

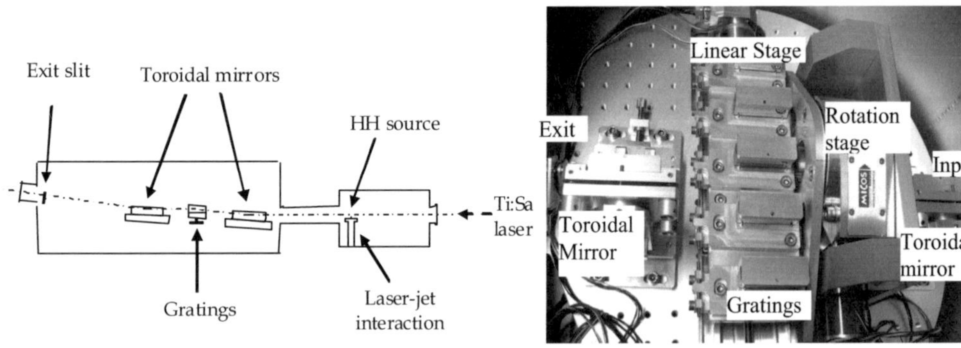

Fig. 8. The monochromator for HHs of the beamline ASTRA-ARTEMIS at Rutherford Appl. Lab. (UK). a) Schematic. b) Internal view of the monochromator chamber.

The pulse duration of the ultrashort pulse is altered by the use of the grating. The FWHM spread of the optical paths at the output of the monochromator is calculated from Eq. 6 as 15-25 fs for the two low-resolution gratings and 45-70 fs for the two medium-resolution gratings. In case of broad bandwidth, as that needed to select the whole single harmonic, time-resolved experiments with resolution of few tens of femtoseconds can be supported by the monochromator in the off-plane mount.

One of the calibration spectra taken at LUXOR with a hollow-cathode XUV lamp and a HH spectrum measured at ASTRA-ARTEMIS are shown in Fig. 9. The spectra have been acquired by a channeltron (CEM) detector mounted after the exit slit through the grating rotation. The spectral resolution is very close to the values predicted by the ray-tracing calculations.

The efficiency of the instrument is here defined as the ratio between the output and the entrance flux: it includes the reflectivity of the two toroidal mirrors and the grating diffraction efficiency. The peak values measured at the blaze wavelength of each grating are in the 0.20-0.28 range. These are extremely good values in the XUV, confirming the benefits of the off-plane mount in terms of expected throughput.

Finally, the output photon flux has been measured by an absolutely calibrated XUV photodiode to be as high as $2.5{\cdot}10^6$ photons/shot on H25 (32 nm) focused on a spot of 150 μm in diameter. Taking 50 fs as a conservative estimate of the duration, we have about $1.4{\cdot}10^6$ W/cm^2 at 32 nm.

With respect to monochromators adopting the classical grating geometry, the off-plane mount gives higher efficiency and shorter temporal response. This innovative configuration can be used for the spectral selection of ultrashort pulses in a single-grating configuration when a monochromatic beam in the time scale of few tens of femtoseconds is required at the output.

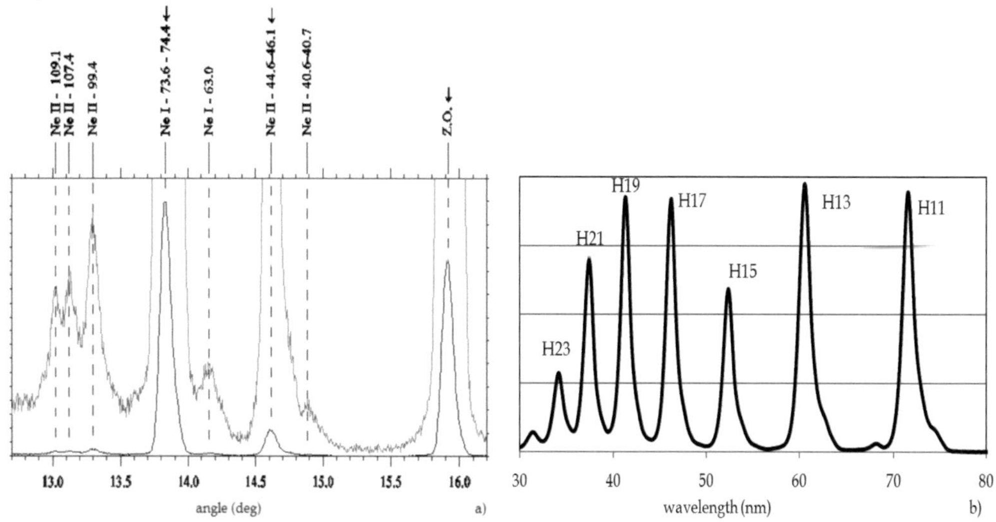

Fig. 9. Spectra taken with the single-grating monochromator in the off-plane mount. a) Spectrum of the emission lines of the hollow-cathode lamp with Ne gas; the over-imposed plots refer to spectral lines with lower intensities. b) HH spectrum generated in Argon with 1-mJ 25-fs Ti:Sa laser operated at 1 kHz repetition rate.

5. Design of a double-grating monochromator in the off-plane geometry

The single-grating monochromator gives temporal resolution in the 20-100 fs range. If the duration of the output pulse has to be preserved as short as in the generation process, a double-grating configuration has to be adopted. Here we will discuss the design and realization of a TDC monochromator using plane gratings in the off-plane mount.

The off-plane geometry can be applied to the design of TDC monochromators using two equal gratings operated at the same altitude and azimuth (Poletto, 2004b; Poletto & Villoresi, 2006). A schematic of the configuration with plane gratings, requiring six optical elements, is shown in Fig. 10. The monochromator is divided in two equal sections, each of them with two toroidal mirrors and a plane grating. The first section gives a spectrally dispersed image of the source on the intermediate plane, where a slit carries out the spectral selection. Only a selected portion of the spectrum is propagating through the slit toward the second section that compensates both for the temporal spread and spectral dispersion and gives a spectrally selected stigmatic image on its focal plane.

The first mirror of each section acts as the collimator, the second mirror as the condenser. The four mirrors are operated at equal grazing angle and unity magnification to minimize the aberrations, i.e. the input arm of each of the two collimators is equal to the output arm of each of the two condensers. With reference to Fig. 10, the term "*input arm*" refers to the two collimators and indicates the distance between the source and the vertex of mirror M1 and the distance between the slit and the vertex of mirror M3; the term "*output arm*" refers to the two condensers and indicates the distance between the vertex of mirror M2 and the slit and the distance between the vertex of mirror M4 and the output focal point. The four mirrors are then equal, i.e. they have the same tangential and sagittal radii.

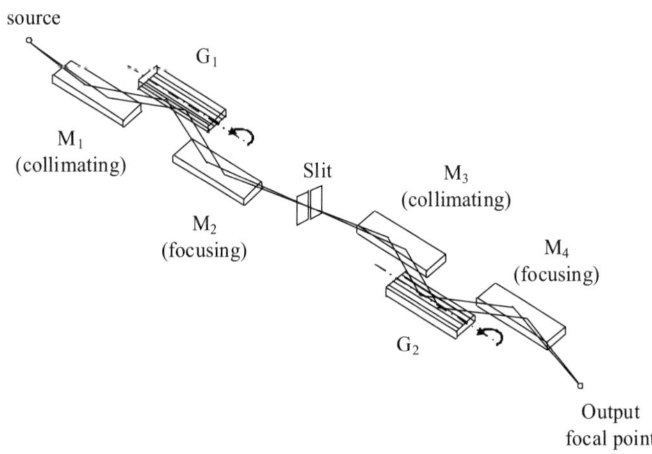

Fig. 10. Schematic of a time-delay compensated monochromator with plane gratings in the off-plane mount.

The wavelength scanning is performed by rotating the gratings around an axis tangent to their vertex and parallel to the grooves, as in the single-grating design already discussed. The axis of rotation is parallel to the plane which defines the direction of propagation of light. The rotation is defined to be zero when the two gratings are operated as mirrors, i.e. $\alpha=\beta=0$, and positive if the grating is rotated clockwise when looking to it from the direction of propagation of the light. The rotation $\Delta\alpha$ to select the wavelength λ is calculated from the grating equation as

$$\Delta\alpha = \pm\arcsin\left(\frac{\lambda\sigma}{2\sin\gamma}\right) \tag{7}$$

where the two gratings have opposite rotations to realize the TDC configuration. The bandwidth transmitted by the intermediate slit is given by Eq. 4 and Eq. 5.

It can be demonstrated (Poletto, 2009) that the off-plane mount is more robust than the classical one in terms of the tolerances in the alignment to achieve time-delay compensation in the femtosecond time scale.

A TDC monochromator for the selection of HHs in the 20-70 nm region has been designed and realized at LUXOR (Padova) and is now installed in the Department of Physics of Politecnico Milano (Italy) (Poletto & Villoresi, 2006; Poletto et al., 2007). Its characteristics are resumed in Tab. 3. The toroidal mirrors are operated with 250 mm arms at 3° grazing angle to have high reflectivity in the whole spectral region of operation. Two 400 gr/mm plane gratings at 3° altitude angle are used. The accepted angular aperture is 10 mrad.

The optical performance and the analysis of the compensation of the optical paths have been carried out by a ray-tracing program suitable to simulate the temporal response of monochromators. The results of the simulations are reported in Tab. 4 for some of the harmonics. The FWHM spreads of the path lengths in a single harmonic are shown both at the slit plane and at the output plane. The differences in the path lengths are almost completely cancelled by the compensated configuration. The residual spread at a fixed harmonic is less than one femtosecond at the shortest wavelengths of the spectral interval of operation.

Spectral region	15-70 nm
Concave mirrors	Toroidal surfaces
Input/output arms	350 mm
Grazing angle	3°
Gratings	**Plane**
Groove density	400 grooves/mm
Altitude angle	3°
Blaze angle and blaze wavelength	6.5°, 30 nm (1st order)

Table 3. Parameters of the TDC monochromator.

Harmonic order	FWHM ΔOP_{SLIT}	FWHM ΔOP
H15 (53.3 nm)	32 μm (105 fs)	0.6 μm (2.0 fs)
H19 (42.1 nm)	25 μm (85 fs)	0.4 μm (1.2 fs)
H31 (25.8 nm)	15 μm (50 fs)	0.15 μm (0.5 fs)

Table 4. Ray-tracing simulations of the TDC monochromator: ΔOP_{SLIT} and ΔOP are the spreads of the optical paths respectively in the intermediate slit plane and at the output.

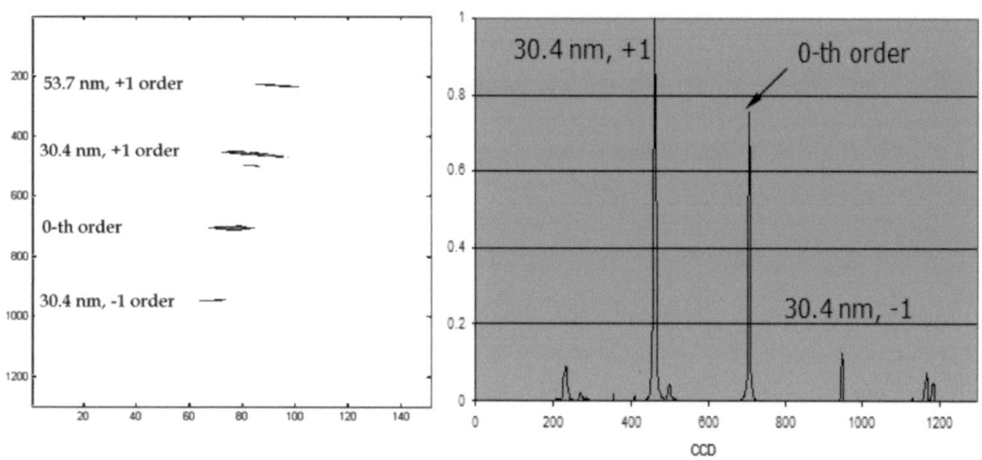

Fig. 11. Spectrum of He emissions from an hollow-cathode lamp, taken by a XUV-enhanced CCD detector placed in the output plane. Both gratings are operated in the condition $\alpha = 0$. A cross section along the CCD column num. 70 is shown on the right.

One of the spectra of He emissions taken during the calibration phase by a CCD camera placed in the output plane is shown in Fig. 11. The spectrum has been acquired by using both gratings in the condition $\alpha = 0$. The first section acts simply as a relay section that focuses the 0th order radiation on the intermediate slit. The grating of the second section disperses the spectrum on an arc, as expected from the off-plane geometry. Furthermore, it

is clear that the most intense spectral line is the HeII 30.4 nm emission, indicating that the blaze wavelength of the gratings is close to 30 nm, according to the optical design.
A HH spectrum is shown in Fig. 12. The spectrum has been acquired by a CEM detector placed at the output. The different wavelengths have been scanned by the rotation of both gratings. HHs from H13 to H25 are clearly distinguished.
The absolute efficiency of the monochromator was measured using the radiation from a hollow-cathode spectral lamp, emitting lines with a narrow spectral profile. The region of operation has been characterized by using Neon and Helium gases. The efficiency has been obtained from the comparison of the input and output power at a given wavelength. The total transmission efficiency in the spectral region of operation is shown in Fig. 13. The peak efficiency at 30 nm is about 0.17, whereas for an extended part of the curve it is above 0.10. This efficiency is exceptionally high for a six-optics setup in grazing incidence and is mostly the result of the off-plane mount of the gratings.

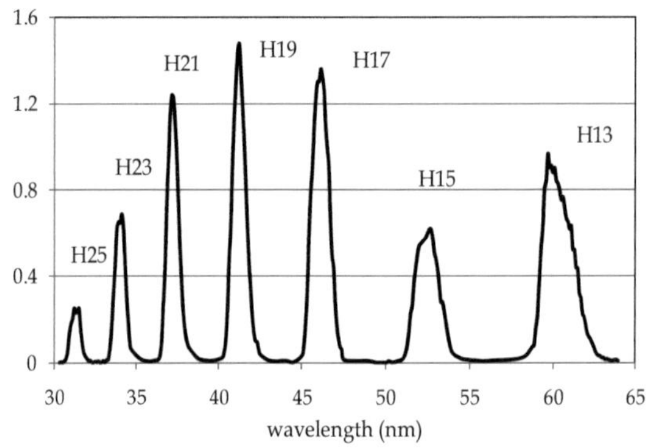

Fig. 12. HH spectrum generated in Argon with 0.3-mJ 25-fs Ti:Sa laser operated at 1 kHz repetition rate.

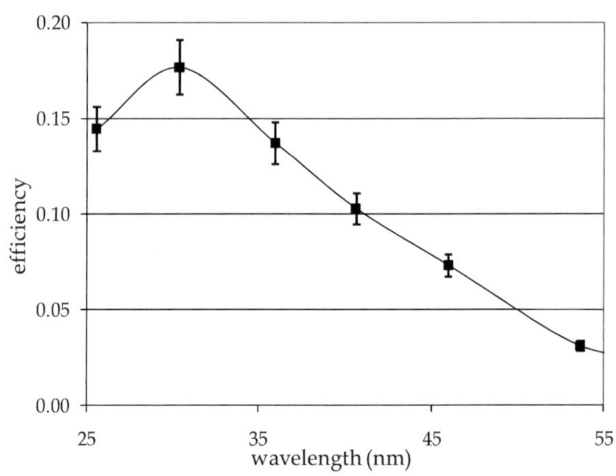

Fig. 13. Absolute efficiency of the TDC monochromator. Both gratings are operated in 1st diffraction order.

The absolute flux at the output has been measured by an absolutely calibrated XUV photodiode. The flux at the output of the monochromator ranges from $5 \cdot 10^5$ photons/shot at

H17 to 1.3·10[6] photons/shot at H23, generated in Argon with 0.3-mJ 25-fs Ti:Sa laser operated at 1 kHz repetition rate.

The XUV spot size at the output has been measured to be around 150 μm FWHM in diameter. Despite the presence of six optical elements, the TDC monochromator gives high-quality imaging properties.

Finally, the temporal profile of the XUV pulses at the output of the monochromator has been measured by a cross-correlation technique (Poletto et al., 2008a). The experimental setup is shown in Fig. 14. The Ti:Sa laser pulse (25-fs duration and 1-kHz repetition rate) is split in two parts using a drilled mirror with a central hole. The inner part is focused into the Argon cell for HH generation. The XUV radiation propagates inside the monochromator that realizes the spectral selection of one of the HHs. Both the XUV beam and the outer annular part of the infrared beam, after a variable-delay stage, are focused onto the same Argon jet that is located in the output focal point of the monochromator for the cross-correlation measurement. The photoelectrons generated by single-photon absorption of the XUV pulses are collected by a time-of-flight spectrometer. The duration of the XUV harmonic pulses is obtained by measuring the cross-correlation between the XUV and the IR pulses.

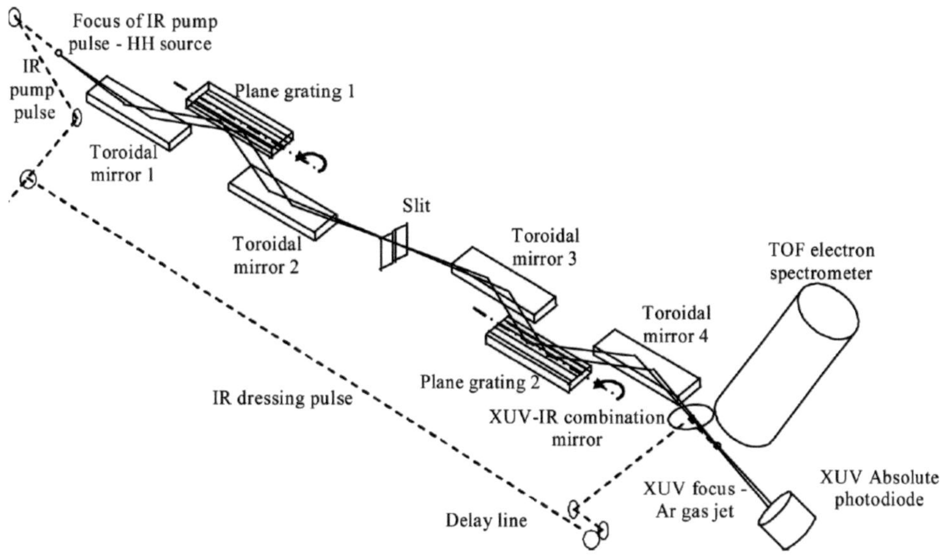

Fig. 14. Experimental setup to emasure the duration of the XUV pulse at the output of the TDC monochromator.

The XUV pulse ionizes Argon in the presence of the infrared field; when the two pulses overlap in time and space on the gas jet, sidebands appear in the photoelectron spectrum, spectrally shifted by the infrared photon energy, determined by the absorption of one harmonic photon plus either the absorption or the emission of one infrared photon. The sideband amplitude as a function of the delay τ between the XUV and infrared pulses provides the cross-correlation signal. Fig. 15 show (dots) the temporal evolution of the amplitude of the first sideband vs delay for H19 and H23. The measured cross-correlation traces can be well fitted assuming an XUV pulse duration T_X =13±0.5 fs FWHM for H19 and

T_X =8±1 fs FWHM for H23. The relative durations of the XUV and generating pulses turn out to be in good agreement with what expected from numerical simulations, which show that for plateau harmonics, as the case of H19, $T_X \approx T_0/2$, where T_0 is the duration of the generating pulse (25 fs in our case), while the XUV duration decreases in the cut-off region (Sansone et al., 2004). To our knowledge, this was the first measurement to confirm the ability of the double-grating configuration to make the spectral selection of ultrashort pulses and preserve their duration.

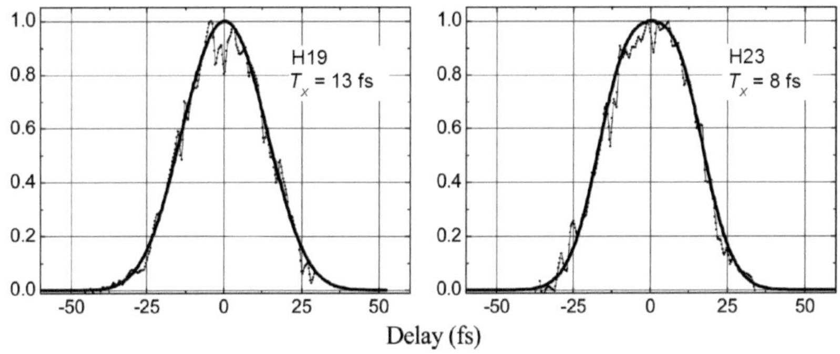

Fig. 15. Amplitude of the first sideband for H19 and H23 as a function of the delay between the XUV and IR pulses. The dots are the experimental results. T_X is the retrieved XUV pulse duration

The off-plane mount of the gratings is the core of the design of the TDC monochromator: it assures high efficiency, tunability in a broad spectral region, good focusing properties and compensation of the optical paths in the femtoseconds time scale.

6. Design of a double-grating attosecond compressor for ultrashort pulses

Radiation generated with the scheme of HHs by the use of laser pulses of few optical cycles is at present the tool for the investigation of matter with attosecond time resolution (Marciak-Kozlowska, 2009). The generation of trains of pulses (Paul et al., 2001; Lopez-Martens et al., 2005) as well as of isolated bursts (Kienberger et al., 2004; Sansone et al., 2006) of radiation in XUV region has been experimentally demonstrated. In both cases, the pulses are positively chirped as the result of the different duration of the quantum paths that contribute to the different portions of the emitted spectrum, so they can be compressed by introducing a suitable device with a negative group-delay-dispersion (GDD). This has been performer successfully by using a thin metallic aluminum filter. The main drawbacks of this compression method are the filter absorption, which may exceed one order of magnitude, and, even more, the rigidity of the compressor properties, as they derive from the elemental properties of the foil. For example, by using the polarization grating technique with neon as a target atom, the generation of pulses at 45 eV with a large spectrum that supports sub-100-as duration has occurred (Sola et al., 2006b), but not compressed so far for lack of suitable filter elements.

We present here a design of an instrument that aims to solve the problem of temporal compression of attosecond pulses by exploiting the influence on the pulse phase of a grating compressor. Its design develops from the scheme of the double-grating TDC monochromator already discussed by a careful analysis of the effect on the phase of the ultrashort pulses.

The optical layout of the compressor is shown in Fig. 16. The design is similar to the TDC monochromator, but the toroidal mirrors are substituted by parabolic mirrors. The four mirrors are operated at equal grazing angle and at unity magnification, i.e., the input arm p of the collimators is equal to the output arm p of the focusers. The choice of parabolic mirrors is resulting from the analysis of residual aberrations in the optical path lengths, which shows that the use of the more simple toroidal mirrors is unacceptable for the brevity of the compressed pulses under study. After the reflection on P1, the beam is collimated and propagates toward G1. The azimuth α_1 of the incident rays on G1 is the same for all the wavelengths. The radiation is diffracted at the azimuth angle $\beta_1(\lambda)$ and collected by P2, which realizes a spectrally dispersed image of the source on its focal plane. This is the source plane for the second section of the compressor. Due to the symmetry of the configuration, the azimuth angle $\alpha_2(\lambda)$ of incident rays on G2 is equal to the azimuth angle $\beta_1(\lambda)$. Since the gratings are equal and operated at the same altitude, the azimuth angle β_2 of the rays diffracted from G2 is constant with the wavelength and equal to α_1. Finally, the rays are focused on the output focal point by P4. As in the case of the TDC monochromator, the design satisfies the condition of time-delay compensation, i.e., the differences in the path lengths of rays with the same wavelength but with different entrance directions within the beam aperture that are caused by the first grating are compensated by the second grating.

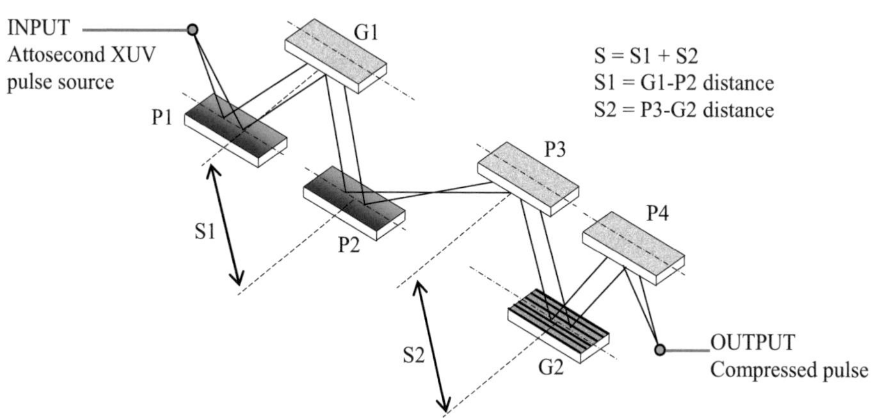

Fig. 16. Layout of the XUV attosecond compressor.

Since rays at different wavelengths are diffracted by G1 at different azimuth angles, they do not make the same optical path. The group delay, intended as the propagation time of rays from the source point to the output focal point, is dependent on the wavelength. The condition for zero GDD is to have G1 imaged on G2, which is realized when

$$S = S1 + S2 = 2p \quad (8)$$

where S1 and S2 are, respectively, the distances G1–P2 and P3–G2. For $S < 2p$, G1 is imaged behind G2 and the resulting GDD is positive. For $S > 2p$, G1 is imaged before G2 and the resulting GDD is negative. The three cases are illustrated in Fig. 17.

The chirp of the pulse can be varied by changing S = S1 + S2. It is simpler to keep S1 constant and to finely adjust S2 for changing the GDD. The variation of S2 is performed by mounting G2 on a linear translator and moving it along an axis coincident with the straight line that connects G2 to P3. Since the beam diffracted from G2 is collimated in a constant direction, the radiation reflected from P4 is focused always on the same output point. Therefore, the compressor has no moving parts except the translation of G2 for the fine tuning of the GDD.

The modeling of the compressor is done by ray-tracing simulations. The group delay is calculated for different values of the distance S. The GDD is then defined as the derivative of the group delay with respect to the frequency. The higher effects on the phase of the pulse can be analogously calculated by successive derivatives.

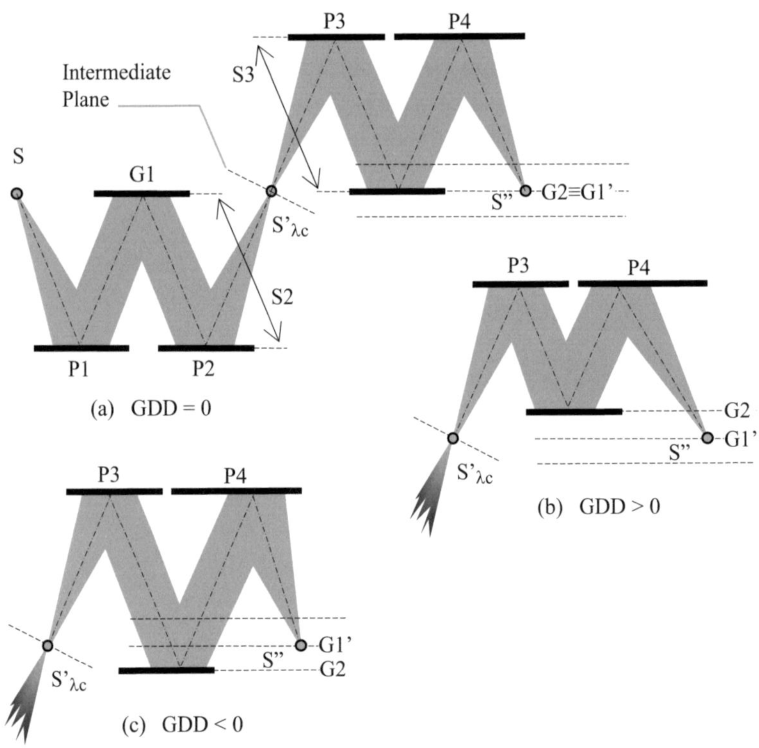

Fig. 17. Operation of the XUV attosecond compressor: a) GDD equal to zero; b) positive GDD; c) negative GDD.

As a test case of the optical configuration, the design of a compressor in the 12-24 nm region is presented. The grazing angle on the mirrors is chosen to be 3° for high reflectivity. The acceptance angle is 6 mrad, which matches the divergence of XUV ultrashort pulses. The size of the illuminated portion of the paraboloidal mirrors results 23 mm × 1.2 mm. On such a small area, manufacturers routinely can produce paraboloidal mirrors with very-high quality finishing, both in terms of figuring errors ($\lambda/30$ at 500 nm) and of slope errors (less than 2 μrad rms). Since the gratings are ruled on plane substrates, the surface finishing is even better than on curved surfaces. Such precision on the optical surfaces is essential to have time-delay compensation in the range of tens of attoseconds. The altitude and blaze angles of the gratings have been selected to optimize the time-delay compensation.

Particular attention must be given to isolate the compressor optics from environmental vibrations and to precisely align the components in order to realize correct implementation of the optimal geometry.

The global efficiency of such a compressor can be predicted to be in the 0.10–0.20 range, on the basis of the efficiency measurements made on the existing off-plane monochromator and already discussed.

The time-delay compensation of the compressor has been analyzed with ray-tracing simulations. Once the grating groove density is selected, the time-delay compensation depends on the choice of altitude and azimuth angles. Both these parameters have to be optimized in order to minimize the residual spreads of the optical paths. In the case presented here, the choice of 200 gr/mm groove density, 1.5° altitude, and 4.3° azimuth gives a residual spread of less than 10 as FWHM in the whole 12-24 nm spectral region. The characteristics of the compressor are resumed in Tab. 5.

Pulse spectral interval	12-24 nm
Mirrors	**Off-axis paraboloidal**
Input/output arms	200 mm
Acceptance angle	6 mrad × 6 mrad
Grazing angle	3°
Size of illuminated area	23 mm × 1.2 mm
Gratings	**Plane**
Groove density	200 gr/mm
Altitude angle	1.5°
Blaze angle	4.5°
Distances	S > 400 mm for negative GDD

Table 5. Parameters of the compressor for the 12-24 nm region.

The calculated results of the compressor's phase properties are shown in Fig. 18. For $S > 400$ mm, the GDD is always negative as predicted and the values reported are in agreement with what is required to compress the pulse as resulting from the HH generation modeling.

An example of compression of a pulse with a positive GDD, modeled according to the results obtained using the polarization gating technique (Sola et al., 2006; Sansone et al., 2006), is presented in Fig. 19. Note the clear time compression to a nearly single-cycle pulse.

The scheme of the compressor is very versatile: it can be designed with high throughput in any spectral interval within the 4-60 nm XUV region. By a simple linear translation of a single grating, the instrument introduces a variable group delay in the range of few hundreds attoseconds with constant throughput and either negative or positive group-delay dispersion. The extended spectral range of operation and the versatility in the control of the group delay allows the compression of XUV attosecond pulses beyond the limitations of the schemes based on metallic filters.

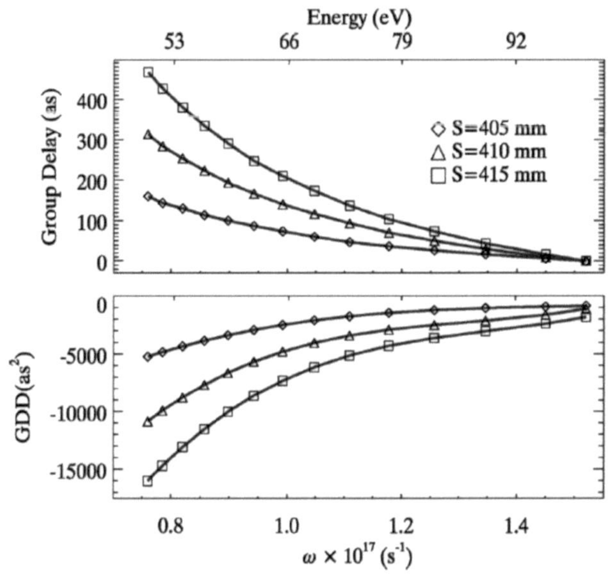

Fig. 18. Phase properties of the compressor with the parameters listed in Tab. 5: group delay (top) and group-delay dispersion (GDD).

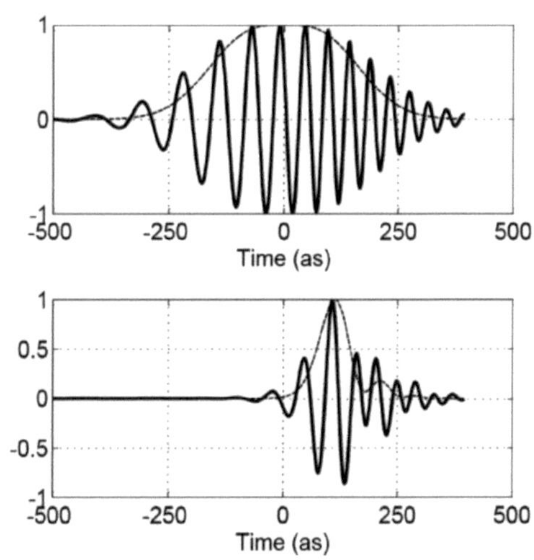

Fig. 19. Simulation of the compression of a chirped XUV pulse at the output of the compressor with the parameters listed in Tab. 5. Input pulse parameters: central energy 73 eV (17 nm), bandwidth 25 eV (6 nm), positive chirp with GDD = 5100 as^2. Compressor: S = 410 mm. FWHM durations: input 350 as, output 75 as.

7. Conclusions

The use of diffraction gratings to perform the spectral selection of ultrashort pulses in the XUV spectral region has been discussed. The main applications of such technique are the spectral selection of high-order laser harmonics and free-electron-laser pulses in the femtosecond time scale.

The realization of monochromators tunable in a broad spectral band in the XUV requires the use of gratings at grazing incidence. Obviously, the preservation of the time duration of the pulse at the output of the monochromator is crucial to have both high temporal resolution and high peak power. A single grating gives a temporal broadening of the ultrafast pulse because of the diffraction. This effect is negligible for picosecond or longer pulses, but is dramatic in the femtosecond time scale. Nevertheless, it is possible to design grating monochromators that do not alter the temporal duration of the pulse in the femtosecond time scale by using two gratings in a time-delay compensated configuration. In such a configuration, the second grating compensates for the time and spectral spread introduced by the first one.

Therefore, the grating monochromators for ultrafast pulses are divided in two main families:

1. the single-grating configuration, that gives intrinsically a temporal broadening of the ultrafast pulse, but is simple and has high efficiency since it requires the use of one grating only;
2. the time-delay compensated configuration with two gratings, that has a much shorter temporal response, in the femtosecond or even shorter time scale, but is more complex and has a lower efficiency.

Once the experimental requirements are given, aim of the optical design is to select the configuration that gives the best trade-off between time response and efficiency.

The efficiency is obviously the major factor discriminating among different designs: an instrument with low efficiency could be not useful for scientific experiments. An innovative configuration to realize monochromators with high efficiency and broad tunability has been discussed. It adopts gratings in the off-plane mount, in which the incident light direction belongs to a plane parallel to the direction of the grooves. The off-plane mount has efficiency higher than the classical mount and, once the grating groove density has been selected, it gives minimum temporal response at grazing incidence.

Both single- and double-grating monochromators in the off-plane mount can be designed. In particular, we have presented in details two applications to the selection of high-order harmonics, one using a single-grating design and one in a time-delay compensated configuration. In the latter case, the XUV temporal response at the output of the monochromator has been measured to be as short as few femtoseconds, confirming the temporal compensation given by the double-grating design.

Finally, the problem of temporal compression of broadband XUV attosecond pulses by means of a double-grating compressor has been addressed. The time-delay compensated design in the off-plane configuration has been modified to realize an XUV attosecond compressor that can introduce a variable group-delay dispersion to compensate for the intrinsic chirp of the attosecond pulse.

This class of instruments plays an important role for the photon handling and conditioning of future ultrashort sources.

8. Acknowledgment

The authors would like to remember the essential contribution of Mr. Paolo Zambolin (1965-2005) to the mechanical design of the time-delay compensated monochromator at LUXOR (Padova, Italy). The experiments on the beamline ARTEMIS at Rutherford Appleton Laboratory (UK) are carried on under the management of Dr. Emma Springate and Dr. Edmund Turcu. The experiments with high-order harmonics at Politecnico Milano (Italy) are carried on under the management of Prof. Mauro Nisoli and Dr. Giuseppe Sansone.

9. References

Akhmanov, S. A.; Vysloukh, V. A. & Chirkin, A. S. (1992) *Optics of Femtosecond Laser Pulses,* American IOP, New York

Cash, W. (1982). Echelle spectrographs at grazing incidence, *Appl. Opt.* Vol. 21, pp. 710–717

Corkum, P.B. & Krausz, F. (2007). Attosecond science. *Nat. Phys.* Vol. 3, pp. 381-387

Chiudi Diels, J.-C.; Rudolph, W. (2006). *Ultrashort Laser Pulse Phenomena,* Academic Press Inc., Oxford

Frassetto, F.; Bonora, S.; Villoresi, P.; Poletto, L.; Springate, E.; Froud, C.A.; Turcu, I.C.E.; Langley, A.J.; Wolff, D.S.; Collier, J.L.; Dhesi, S.S. & Cavalleri, A. (2008). Design and characterization of the XUV monochromator for ultrashort pulses at the ARTEMIS facility, *Proc. SPIE* Vol. 7077, Advances in X-Ray/EUV Optics and Components III, no. 707713, S. Diego (USA), August 2008, SPIE Publ., Bellingham

Frassetto, F.; Villoresi, P. & Poletto, L. (2008). Optical concept of a compressor for XUV pulses in the attosecond domain, *Opt. Exp.* Vol. 16, pp. 6652-6667

Kienberger, R. & Krausz, F. (2004). Subfemtosecond XUV Pulses: Attosecond Metrology and Spectroscopy, in *Few-cycle laser pulse generation and its applications,* F.X. Kärtner ed., pp. 143-179, Springer, Berlin

Kienberger, R.; Goulielmakis, E.; Uberacker, M.; Baltuska, A.; Yakovlev, V.; Bammer, F.; Scrinzi, A.; Westerwalbesloh, T.; Kleineberg, U.; Heinzmann, U.; Drescher, M. & Krausz, F. (2004). Atomic transient recorder, *Nature* Vol. 427, pp. 817-822

Jaegle', P. (2006). *Coherent Sources of XUV Radiation,* Springer, Berlin

Lopez-Martens, R.; Varju, K.; Johnsson, P.; Mauritsson, J.; Mairesse, Y.; Salières, P.; Gaarde, M.B.; Schafer, K.J.; Persson, A.; Svanberg, S.; Wahlström, C.-G. & L'Huillier, A. (2005). Amplitude and phase control of attosecond light pulses, *Phys. Rev. Lett.* Vol. 94, no. 033001

Marciak-Kozlowska, J. (2009). *From Femto-to Attoscience and Beyond,* Nova Science Publishers Inc., New York

Patel, N. (2002). Shorter, Brighter, Better. *Nature* Vol. 415, pp. 110-11

Petit, R. (1980). *Electromagnetic theory of gratings,* Springer, Berlin

Pascolini, M.; Bonora, S.; Giglia, A.; Mahne, N.; Nannarone, S. & Poletto, L. (2006) Gratings in the conical diffraction mounting for an EUV time-delay compensated monochromator, *Appl. Opt.* Vol. 45, pp. 3253-3562

Paul, P.M.; Toma, E.S.; Breger, P.; Mullot, G.; Auge, F.; Balcou, P.; Muller, H.G. & Agostini, P. (2001). Observation of a train of attosecond pulses from high harmonic generation, *Science* Vol. 292, pp. 1689–1692

Poletto, L. & Tondello, G. (2001). Time-compensated EUV and soft X-ray monochromator for ultrashort high-order harmonic pulses, *Pure Appl. Opt.* Vol. 3, pp. 374-379

Poletto, L.; Bonora, S.; Pascolini, M.; Borgatti, F.; Doyle, B.; Giglia, A.; Mahne, N.; Pedio, M. & Nannarone, S. (2004). Efficiency of gratings in the conical diffraction mounting for an EUV time-compensated monochromator, *Proc. SPIE* Vol. 5534, Fourth Generation X-Ray Sources and Optics II, pp. 144–153, Denver (USA), August 2004, SPIE Publ., Bellingham

Poletto, L. (2004). Time-compensated grazing-incidence monochromator for extreme-ultraviolet and soft X-ray high-order harmonics, *Appl. Phys. B* Vol. 78, pp. 1013-1016

Poletto, L. & Villoresi, P. (2006). Time-compensated monochromator in the off-plane mount for extreme-ultraviolet ultrashort pulses, *Appl. Opt.* Vol. 45, pp. 8577-8585

Poletto, L.; Villoresi, P.; Benedetti, E.; Ferrari, F.; Stagira, S.; Sansone, G.; Nisoli, M. (2007). Intense femtosecond extreme ultraviolet pulses by using a time-delay compensated monochromator, *Opt. Lett.* Vol. 32, pp. 2897-2899

Poletto, L.; Villoresi, P.; Benedetti, E.; Ferrari, F.; Stagira, S.; Sansone, G.; Nisoli, M. (2008). Temporal characterization of a time-compensated monochromator for high-efficiency selection of XUV pulses generated by high-order harmonics, *J. Opt. Soc. Am. B* Vol. 25, pp. B44-B49

Poletto, L.; Frassetto, F. & Villoresi, P. (2008). Design of an Extreme-Ultraviolet Attosecond Compressor, *J. Opt. Soc. Am. B* Vol. 25, pp. B133-B136

Poletto, L. (2009). Tolerances of time-delay compensated monochromators for extreme-ultraviolet ultrashort pulses, *Appl. Opt.* Vol. 48, pp. 4526-4535

Saldin, E.L.; Schneidmiller, E.A. & Yurkov, M.V. (2000). *The Physics of Free Electron Lasers*, Springer, Berlin

Sansone, G.; Vozzi, C.; Stagira, S. & Nisoli, M. (2004). Nonadiabatic quantum path analysis of high-order harmonic generation: role of the carrier-envelope phase on short and long paths, *Phys. Rev. A* Vol. 70, no. 013411

Sansone, G.; Benedetti, E.; Calegari, F.; Vozzi, C.; Avaldi, L.; Flammini, R.; Poletto, L.; Villoresi, P.; Altucci, C.; Velotta, R.; Stagira, S.; De Silvestri, S. & Nisoli, M. (2006). Isolated single-cycle attosecond pulses, *Science* Vol. 314, pp. 443–446

Sola, J.; Mevel, E.; Elouga, L.; Constant, E.; Strelkov, V.; Poletto, L.; Villoresi, P.; Benedetti, E.; Caumes, J.-P.; Stagira, S.; Vozzi, C.; Sansone, G. & Nisoli, M. (2006). Controlling attosecond electron dynamics by phase-stabilized polarization gating, *Nat. Phys.* Vol. 2, pp. 319–322

Villoresi, P. (1999). Compensation of optical path lengths in extreme-ultraviolet and soft-x-ray monochromators for ultrafast pulses, *Appl. Opt.* Vol. 38, pp. 6040-6049

Walmsley, I.; Waxer, L. & Dorrer, C. (2001). The role of dispersion in ultrafast optics, *Rev. Sci. Instr.* Vol. 72, pp. 1–28

Werner, W. (1977). X-ray efficiencies of blazed gratings in extreme off-plane mountings, *Appl. Opt.* Vol. 16, pp. 2078–2080

Werner, W. & Visser, H. (1981). X-ray monochromator designs based on extreme off-plane grating mountings, *Appl. Opt.* Vol. 20, pp. 487-492

Wiedemann, H. (2005). *Synchrotron Radiation*, Springer, Berlin

Wieland, M.; Frueke, R.; Wilhein, T.; Spielmann, C.; Pohl, M. & Kleinenberg, U. (2002) Submicron extreme ultraviolet imaging using high-harmonic radiation, *Appl. Phys. Lett.* Vol. 81, pp. 2520-2522

19

High-Harmonic Generation

Kenichi L. Ishikawa
Photon Science Center, Graduate School of Engineering, University of Tokyo
Japan

1. Introduction

We present theoretical aspects of high-harmonic generation (HHG) in this chapter. Harmonic generation is a nonlinear optical process in which the frequency of laser light is converted into its integer multiples. Harmonics of very high orders are generated from atoms and molecules exposed to intense (usually near-infrared) laser fields. Surprisingly, the spectrum from this process, high-harmonic generation, consists of a plateau where the harmonic intensity is nearly constant over many orders and a sharp cutoff (see Fig. 5).
The maximal harmonic photon energy E_c is given by the cutoff law (Krause et al., 1992),

$$E_c = I_p + 3.17U_p, \tag{1}$$

where I_p is the ionization potential of the target atom, and $U_p[\text{eV}] = E_0^2 / 4\omega_0^2 = 9.337 \times 10^{-14} I$ [W/cm^2] (λ [μm])^2 the ponderomotive energy, with E_0, I and λ being the strength, intensity and wavelength of the driving field, respectively. HHG has now been established as one of the best methods to produce ultrashort coherent light covering a wavelength range from the vacuum ultraviolet to the soft x-ray region. The development of HHG has opened new research areas such as attosecond science and nonlinear optics in the extreme ultraviolet (xuv) region.
Rather than by the perturbation theory found in standard textbooks of quantum mechanics, many features of HHG can be intuitively and even quantitatively explained in terms of electron rescattering trajectories which represent the semiclassical three-step model and the quantum-mechanical Lewenstein model. Remarkably, various predictions of the three-step model are supported by more elaborate direct solution of the time-dependent Schrödinger equation (TDSE). In this chapter, we describe these models of HHG (the three-step model, the Lewenstein model, and the TDSE).
Subsequently, we present the control of the intensity and emission timing of high harmonics by the addition of xuv pulses and its application for isolated attosecond pulse generation.

2. Model of high-harmonic generation

2.1 Three Step Model (TSM)

Many features of HHG can be intuitively and even quantitatively explained by the semiclassical three-step model (Fig. 1)(Krause et al., 1992; Schafer et al., 1993; Corkum, 1993). According to this model, in the first step, an electron is lifted to the continuum at the nuclear position with no kinetic energy through tunneling ionization (*ionization*). In the second step, the subsequent motion is governed classically by an oscillating electric field (*propagation*). In the third step, when the electron comes back to the nuclear position, occasionally, a harmonic, whose photon energy is equal to the sum of the electron kinetic energy and the ionization potential I_p, is emitted upon *recombination*. In this model, although the quantum mechanics is inherent in the ionization and recombination, the propagation is treated classically.

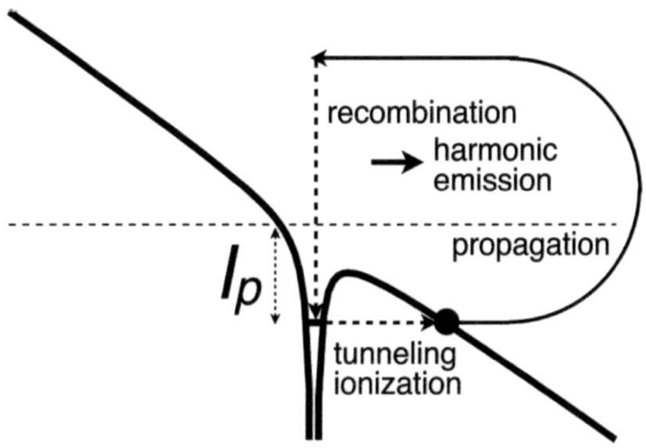

Fig. 1. Three step model of high-harmonic generation.

Let us consider that the laser electric field $E(t)$, linearly polarized in the z direction, is given by

$$E(t) = E_0 \cos \omega_0 t, \tag{2}$$

where E_0 and ω_0 denotes the field amplitude and frequency, respectively. If the electron is ejected at $t = t_i$, by solving the equation of motion for the electron position $z(t)$ with the initial conditions

$$z(t_i) = 0, \tag{3}$$

$$\dot{z}(t_i) = 0, \tag{4}$$

we obtain,

$$z(t) = \frac{E_0}{\omega_0^2}\left[\left(\cos\omega_0 t - \cos\omega_0 t_i\right) + \left(\omega_0 t - \omega_0 t_i\right)\sin\omega_0 t_i\right]. \tag{5}$$

It is convenient to introduce the phase $\theta \equiv \omega_0 t$. Then Equation 5 is rewritten as,

$$z(\theta) = \frac{E_0}{\omega_0^2}\left[\left(\cos\theta - \cos\theta_i\right) + \left(\theta - \theta_i\right)\sin\theta_i\right], \tag{6}$$

and we also obtain, for the kinetic energy E_{kin},

$$E_{kin}(\theta) = 2U_p\left(\sin\theta - \sin\theta_i\right)^2. \tag{7}$$

One obtains the time (phase) of recombination t_r (θ_r) as the roots of the equation $z(t) = 0$ ($z(\theta) = 0$). Then the energy of the photon emitted upon recombination is given by $E_{kin}(\theta_r) + I_p$.

Figure 2 shows $E_{kin}(\theta_r)/U_p$ as a function of phase of ionization θ_i and recombination θ_r for $0 < \theta_i < \pi$. The electron can be recombined only if $0 < \theta_i < \pi/2$; it flies away and never returns to the nuclear position if $\pi/2 < \theta_i < \pi$. $E_{kin}(\theta_r)$ takes the maximum value $3.17U_p$ at $\theta_i = 17°$ and $\theta_r = 255°$. This beautifully explains why the highest harmonic energy (cutoff) is given by $3.17U_p + I_p$. It should be noted that at the time of ionization the laser field, plotted in thin solid line, is close to its maximum, for which the tunneling ionization probability is high. Thus, harmonic generation is efficient even near the cutoff.

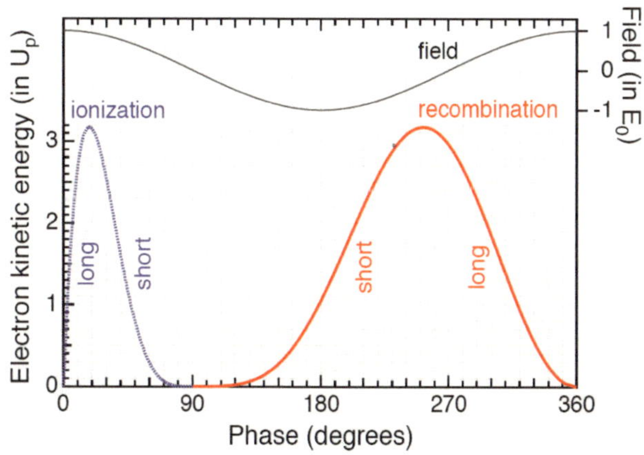

Fig. 2. Electron kinetic energy just before recombination normalized to the ponderomotive energy $E_{kin}(\theta_r)/U_p$ as a function of phase of ionization θ_i and recombination θ_r. The laser field normalized to the field amplitude $E(t)/E_0$ is also plotted in thin solid line (right axis).

For a given value of E_{kin}, we can view θ_i and θ_r as the solutions of the following coupled equations:

$$(\cos\theta_r - \cos\theta_i) + (\theta_r - \theta_i)\sin\theta_i = 0, \tag{8}$$

$$(\sin\theta_r - \sin\theta_i)^2 = \frac{E_{kin}}{2U_p} \tag{9}$$

The path $z(\theta)$ that the electron takes from $\theta = \theta_i$ to θ_r is called *trajectory*. We notice that there are two trajectories for a given kinetic energy below $3.17U_p$. $17° < \theta_i < 90°$, $90° < \theta_r < 255°$ for the one trajectory, and $0° < \theta_i < 17°$, $255° < \theta_r < 360°$ for the other. The former is called *short* trajectory, and the latter *long* trajectory.

If (θ_i, θ_r) is a pair of solutions of Equations 8 and 9, $(\theta_i + m\pi, \theta_r + m\pi)$ are also solutions, where m is an integer. If we denote $z(\theta)$ associated with m as $z_m(\theta)$, we find that $z_m(\theta) = (-1)^m z_{m=0}(\theta - m\pi)$. This implies that the harmonics are emitted each half cycle with an alternating phase, i.e., field direction in such a way that the harmonic field $E_h(t)$ can be expressed in the following form:

$$E_h(t) = \cdots + F_h(t + 2\pi/\omega_0) - F_h(t + \pi/\omega_0) + F_h(t) - F_h(t - \pi/\omega_0) + F_h(t - 2\pi/\omega_0) - \cdots. \tag{10}$$

One can show that the Fourier transform of Equation 10 takes nonzero values only at odd multiples of ω_0. This observation explains why the harmonic spectrum is composed of odd-order components.

In Fig. 3 we show an example of the harmonic field made up of the 9th, 11th, 13th, 15th, and 17th harmonic components. It indeed takes the form of Equation 10. In a similar manner, high harmonics are usually emitted as a train of bursts (*pulse train*) repeated each half cycle of the fundamental laser field. The harmonic field as in this figure was experimentally observed (Nabekawa et al., 2006).

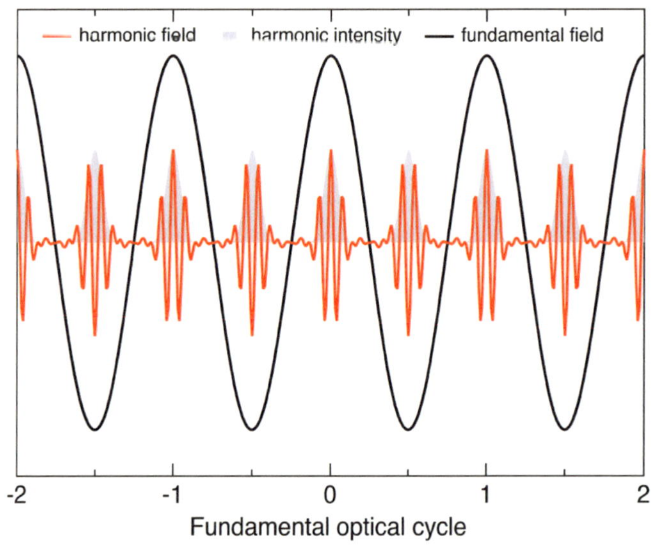

Fig. 3. Example of the harmonic field composed of harmonic orders 9, 11, 13, 15, and 17. The corresponding harmonic intensity and the fundamental field are also plotted.

2.2 Lewenstein model

The discussion of the propagation in the preceding subsection is entirely classical. Lewenstein et al. (Lewenstein et al., 1994) developed an analytical, quantum theory of HHG, called *Lewenstein model*. The interaction of an atom with a laser field $E(t)$, linearly polarized in the z direction, is described by the time-dependent Schrödinger equation (TDSE) in the length gauge,

$$i\frac{\partial\psi(r,t)}{\partial t}=\left[-\frac{1}{2}\nabla^2+V(r)+zE(t)\right]\psi(r,t), \tag{11}$$

where $V(\mathbf{r})$ denotes the atomic potential. In order to enable analytical discussion, they introduced the following widely used assumptions (*strong-field approximation, SFA*):

- The contribution of all the excited bound states can be neglected.
- The effect of the atomic potential on the motion of the continuum electron can be neglected.
- The depletion of the ground state can be neglected.

Within this approximation, it can be shown (Lewenstein et al., 1994) that the time-dependent dipole moment $x(t) \equiv \langle \psi(\mathbf{r},t) \mid z \mid \psi(\mathbf{r},t)\rangle$ is given by,

$$x(t)=i\int_{-\infty}^{t} dt'\int d^3\mathbf{p}\, d^*(\mathbf{p}+\mathbf{A}(t))\cdot\exp[-iS(\mathbf{p},t,t')]\cdot E(t')d(\mathbf{p}+\mathbf{A}(t'))+\text{c.c.}, \tag{12}$$

where $\mathbf{p}$ and $d(\mathbf{p})$ are the canonical momentum and the dipole transition matrix element, respectively, $\mathbf{A}(t) = -\int \mathbf{E}(t)dt$ denotes the vector potential, and $S(\mathbf{p}, t, t')$ the semiclassical action defined as,

$$S(\mathbf{p},t,t') = \int_{t'}^{t} dt'' \left(\frac{[\mathbf{p} + \mathbf{A}(t'')]^2}{2} + I_p \right). \tag{13}$$

If we approximate the ground state by that of the hydrogenic atom,

$$\mathbf{d}(\mathbf{p}) = -\frac{i8\sqrt{2}(2I_p)^{5/4}}{\pi} \frac{\mathbf{p}}{(p^2 + 2I_p)^3}. \tag{14}$$

Alternatively, if we assume that the ground-state wave function has the form,

$$\psi(\mathbf{r}) = (\pi\Delta^2)^{-3/4} e^{-\mathbf{r}^2/(2\Delta^2)}, \tag{15}$$

with $\Delta\,(\sim I_p^{-1})$ being the spatial width,

$$\mathbf{d}(\mathbf{p}) = i\left(\frac{\Delta^2}{\pi}\right)^{3/4} \Delta^2 \mathbf{p} e^{-\Delta^2 \mathbf{p}^2/2}. \tag{16}$$

In the spectral domain, Equation 12 is Fourier-transformed to,

$$\hat{x}(\omega_h) = i\int_{-\infty}^{\infty} dt \int_{-\infty}^{t} dt' \int d^3\mathbf{p}\, d^*(\mathbf{p} + \mathbf{A}(t)) \cdot \exp[i\omega_h t - iS(\mathbf{p},t,t')] \cdot E(t') d(\mathbf{p} + \mathbf{A}(t')) + \text{c.c.}. \tag{17}$$

Equation 12 has a physical interpretation pertinent to the three-step model: $E(t')d(\mathbf{p}+\mathbf{A}(t'))$, $\exp[-iS(\mathbf{p}, t, t')]$, and $d^*(\mathbf{p} + \mathbf{A}(t))$ correspond to ionization at time t', propagation from t' to t, and recombination at time t, respectively.

The evaluation of Equation 17 involves a five-dimensional integral over $\mathbf{p}$, t, and t', i.e., the sum of the contributions from all the paths of the electron that is ejected and recombined at arbitrary time and position, which reminds us of Feynman's path-integral approach (Salières et al., 2001). Indeed, application of the saddle-point analysis (SPA) to the integral yields a simpler expression. The stationary conditions that the first derivatives of the exponent $\omega_h t - S(\mathbf{p}, t, t')$ with respect to $\mathbf{p}$, t, and t' are equal to zero lead to the saddle-point equations:

$$p(t - t') + \int_{t'}^{t} A(t'')dt'' = 0, \tag{18}$$

$$\frac{[p + A(t')]^2}{2} = -I_p, \tag{19}$$

$$\frac{[p + A(t)]^2}{2} = \omega_h - I_p, \tag{20}$$

Using the solutions (p_s, t_s, t_s'), $\hat{x}(\omega_h)$ can be rewritten as a coherent superposition of quantum trajectories s:

$$\hat{x}(\omega_h) = \sum_s \left(\frac{\pi}{\varepsilon + \frac{i}{2}(t_s - t_s^{'})} \right)^{3/2} \frac{i2\pi}{\sqrt{\det S^{''}(t,t')|_s}} d^*(p_s + A(t_s))$$

$$\times \exp[i\omega_h t_s - iS(p_s, t_s, t_s^{'})] E(t_s^{'}) d(p_s + A(t_s^{'})), \tag{21}$$

where ε is an infinitesimal parameter, and

$$\det S^{''}(t,t')|_s = \left(\frac{\partial^2 S}{\partial t \partial t'}\bigg|_s \right)^2 - \frac{\partial^2 S}{\partial t^2}\bigg|_s \frac{\partial^2 S}{\partial t'^2}\bigg|_s, \tag{22}$$

$$\frac{\partial^2 S}{\partial t \partial t'} = \frac{(p + A(t))(p + A(t'))}{t - t'}, \tag{23}$$

$$\frac{\partial^2 S}{\partial t^2} = -\frac{2(\omega_h - I_p)}{t - t'} - E(t)(p + A(t)), \tag{24}$$

$$\frac{\partial^2 S}{\partial t'^2} = \frac{2I_p}{t - t'} + E(t')(p + A(t')), \tag{25}$$

The physical meaning of Equations 18-20 becomes clearer if we note that $p + A(t)$ is nothing but the kinetic momentum $v(t)$. Equation 18, rewritten as $\int_{t'}^{t} v(t^{''})dt^{''} = 0$, indicates that the electron appears in the continuum and is recombined at the same position (nuclear position). Equation 20, rewritten together with Equation 19 as $v(t)^2/2 - v(t')^2/2 = \omega_h$, means the energy conservation. The interpretation of Equation 19 is more complicated, since its right-hand side is negative, which implies that the solutions of the saddle-point equations are complex in general. The imaginary part of t' is usually interpreted as tunneling time (Lewenstein et al., 1994).
Let us consider again that the laser electric field is given by Equation 2 and introduce $\theta = \omega_0 t$ and $k = p\omega_0/E_0$. Then Equations 18-20 read as,

$$k = -\frac{\cos\theta - \cos\theta^{'}}{\theta - \theta^{'}}, \tag{26}$$

$$(k - \sin\theta^{'})^2 = -\frac{I_p}{2U_p} = -\gamma^2, \tag{27}$$

$$(k - \sin\theta)^2 = \frac{\omega_h - I_p}{2U_p}, \tag{28}$$

where γ is called the *Keldysh parameter*. If we replace I_p and $\omega_h - I_p$ in these equations by zero and E_{kin}, respectively, we recover Equations 8 and 9 for the three-step model. Figure 4 displays the solutions (θ, θ') of these equations as a function of harmonic order. To make our discussion concrete, we consider harmonics from an Ar atom (I_p = 15.7596eV) irradiated by a laser with a wavelength of 800 nm and an intensity of 1.6 × 10^{14}W/cm^2. The imaginary part of θ' (Fig. 4 (b)) corresponds to the tunneling time, as already mentioned. On the other hand, the imaginary part of θ is much smaller; that for the long trajectory, in particular, is nearly vanishing below the cutoff (≈ 32nd order), which implies little contribution of tunneling to the recombination process. In Fig. 4 (a) are also plotted in thin dashed lines the trajectories from Fig. 2, obtained with the three-step model. We immediately notice that the Lewenstein model predicts a cutoff energy E_c,

$$E_c = 3.17U_p + gI_p \quad (g \approx 1.3), \tag{29}$$

slightly higher than the three-step model (Lewenstein et al., 1994). This can be understood qualitatively by the fact that there is a finite distance between the nucleus and the tunnel exit (Fig. 1); the electron which has returned to the position of the tunnel exit is further accelerated till it reaches the nuclear position. Except for the difference in E_c, the trajectories from the TSM and the SPA (real part) are close to each other, though we see some discrepancy in the ionization time of the short trajectory. This suggests that the semi-classical three-step model is useful to predict and interpret the temporal structure of harmonic pulses, primarily determined by the recombination time, as we will see later.

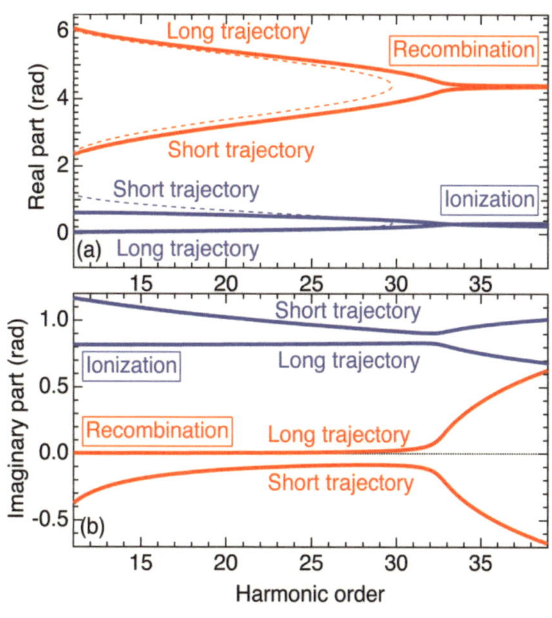

Fig. 4. (a) Real and (b) imaginary parts (radian) of the solutions θ (for recombination) and θ' (for ionization) of Equations 26-28 as a function of harmonic order ω_h/ω_0. The value of I_p = 15.7596eV is for Ar. The wavelength and intensity of the driving laser are 800 nm and 1.6 × 10^{14}W/cm^2. Thin dashed lines in panel (a) correspond to the three-step model.

2.3 Gaussian model

In the *Gaussian model*, we assume that the ground-state wave function has a form given by Equation 15. An appealing point of this model is that the dipole transition matrix element also takes a Gaussian form (Equation 16) and that one can evaluate the integral with respect to momentum in Equation 12 analytically, without explicitly invoking the notion of quantum paths. Thus, we obtain the formula for the dipole moment $x(t)$ as,

$$x(t) = i\Delta^{-7} \int_{-\infty}^{t} (2C(t,t'))^{3/2} E(t')$$
$$\times\{A(t)A(t') + C(t,t')[1 - D(t,t')(A(t) + A(t'))] + C^2(t,t')D^2(t,t')\}$$
$$\times\exp\left(-i[I_p(t-t') + B(t,t')] - \frac{[A^2(t) + A^2(t')]\Delta^2 - C(t,t')D^2(t,t')}{2}\right)dt', \quad (30)$$

where $B(t, t')$, $C(t, t')$, and $D(t, t')$ are given by,

$$B(t,t') = \frac{1}{2}\int_{t'}^{t} dt'' A^2(t'') \quad , \quad (31)$$

$$C(t,t') = \frac{1}{2\Delta^2 + i(t-t')} \quad , \quad (32)$$

$$D(t,t') = [A(t) + A(t')]\Delta^2 + i\int_{t'}^{t} dt'' A(t'') \quad . \quad (33)$$

The Gaussian model is also useful when one wants to account for the effect of the initial spatial width of the wave function within the framework of the Lewenstein model (Ishikawa et al., 2009b).

2.4 Direct simulation of the time-dependent Schrödinger equation (TDSE)

The most straightforward way to investigate HHG based on the time-dependent Schrödinger equation 11 is to solve it numerically. Although such an idea might sound prohibitive at first, the TDSE simulations are indeed frequently used, with the rapid progress in computer technology. This approach provides us with exact numerical solutions, which are powerful especially when we face new phenomena for which we do not know a priori what kind of approximation is valid. We can also analyze the effects of the atomic Coulomb potential, which is not accounted for by the models in the preceding subsections. Here we briefly present the method developed by Kulander et al. (Kulander et al., 1992) for an atom initially in an s state. There are also other methods, such as the pseudo-spectral method (Tong & Chu, 1997) and those using the velocity gauge (Muller, 1999; Bauer & Koval, 2006).

Since we assume linear polarization in the z direction, the angular momentum selection rule tells us that the magnetic angular momentum remains $m = 0$. Then we can expand the wave function $\psi(\mathbf{r}, t)$ in spherical harmonics with $m = 0$,

$$\psi(\mathbf{r},t) = \sum_l R_l(r,t) Y_l^0(\theta,\phi). \quad (34)$$

At this stage, the problem of three dimensions in space physically has been reduced to two dimensions. By discretizing the radial wave function $R_l(r, t)$ as $g_l^j = r_j R_l(r_j,t)$ with

$r_j = (j - \frac{1}{2})\Delta r$, where Δr is the grid spacing, we can derive the following equations for the temporal evolution (Kulander et al., 1992):

$$i\frac{\partial}{\partial t}g_l^j = -\frac{c_j g_l^{j+1} - 2d_j g_l^j + c_{j-1} g_l^{j-1}}{2(\Delta r)^2} + \left(\frac{l(l+1)}{2r_j^2} + V(r_j)\right) g_l^j \tag{35}$$

$$+r_j E(t)\left(a_l g_{l+1}^j + a_{l-1} g_{l-1}^j\right)$$

$$= (H_0 g)_l^j + (H_I g)_l^j, \tag{36}$$

where the coefficients are given by,

$$c_j = \frac{j^2}{j^2 - 1/4}, \quad d_j = \frac{j^2 - j + 1/2}{j^2 - j + 1/4}, \quad a_l = \frac{l+1}{\sqrt{(2l+1)(2l+3)}}. \tag{37}$$

Here, in order to account for the boundary condition at the origin properly, the Euler-Lagrange equations with a Lagrange-type functional (Kulander et al., 1992; Koonin et al., 1977),

$$\mathcal{L} = \langle \psi \,|\, i\partial / \partial t - (H_0 + H_I(t)) \,|\, \psi \rangle, \tag{38}$$

has been discretized, instead of Equation 11 itself. c_j and d_j tend to unity for a large value of j, i.e., a large distance from the nucleus, with which the first term of the right-hand side of Equation 35 becomes an ordinary finite-difference expression. The operator H_0 corresponds to the atomic Hamiltonian and is diagonal in l, while H_I corresponds to the interaction Hamiltonian and couples the angular momentum l to the neighboring values $l \pm 1$.
Equations 35 and 36 can be integrated with respect to t by the *alternating direction implicit (Peaceman-Rachford)* scheme,

$$g_l^j(t + \Delta t) = [I + iH_0 t / 2]^{-1}[I + iH_I t / 2]^{-1}[I - iH_I t / 2][I - iH_0 t / 2], \tag{39}$$

with Δt being the time step. This algorithm is accurate to the order of $\mathcal{O}(\Delta t^3)$, and approximately unitary. One can reduce the difference between the discretized and analytical wave function, by scaling the Coulomb potential by a few percent at the first grid point (Krause et al., 1992). We can obtain the harmonic spectrum by Fourier-transforming the dipole acceleration $\ddot{x}(t) = -\partial_t^2 \langle z(t) \rangle$, which in turn we calculate, employing the Ehrenfest theorem, through the relation $\ddot{x}(t) = \langle \psi(\mathbf{r},t) | \cos\theta / r^2 - E(t) | \psi(\mathbf{r},t) \rangle$ (Tong & Chu, 1997), where the second term can be dropped as it does not contribute to the HHG spectrum.
$V(r)$ is the bare Coulomb potential for a hydrogenic atom. Otherwise, we can employ a model potential (Muller & Kooiman, 1998) *within the single-active electron approximation (SAE)*,

$$V(r) = -[1 + Ae^{-r} + (Z - 1 - A)e^{-Br}] / r, \tag{40}$$

where Z denotes the atomic number. Parameters A, and B are chosen in such a way that they faithfully reproduce the eigenenergies of the ground and the first excited states. One can account for nonzero azimuthal quantum numbers by replacing a_l by (Kulander et al., 1992),

$$a_l^m = \sqrt{\frac{(l+1)^2 - m^2}{(2l+1)(2l+3)}}. \tag{41}$$

In Fig. 5 we show an example of the calculated harmonic spectrum for a hydrogen atom irradiated by a Ti:Sapphire laser pulse with a wavelength of 800 nm ($\hbar\omega_0$ = 1.55eV) and a peak intensity of 1.6×10^{14}W/cm^2. The laser field $E(t)$ has a form of $E(t) = f(t) \sin\omega_0 t$, where the field envelope $f(t)$ corresponds to a 8-cycle flat-top pulse with a half-cycle turn-on and turn-off. We can see that the spectrum has peaks at odd harmonic orders, as is experimentally observed, and the cutoff energy predicted by the cutoff law.

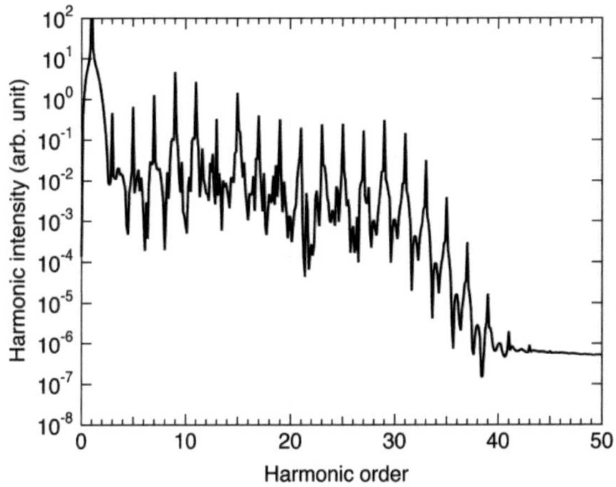

Fig. 5. HHG spectrum from a hydrogen atom, calculated with the Peaceman-Rachford method. See text for the laser parameters.

3. High-harmonic generation by an ultrashort laser pulse

Whereas in the previous section we considered the situation in which the laser has a constant intensity in time, virtually all the HHG experiments are performed with an ultrashort (a few to a few tens of fs) pulse. The state-of-the-art laser technology is approaching a single-cycle limit. The models in the preceding section can be applied to such situations without modification.

For completeness, the equations for the recombination time t and ionization time t' in the three-step model is obtained by replacing I_p in the right-hand side of Equation 19 by zero as follows:

$$-A(t')(t-t') + \int_{t'}^{t} A(t'')dt'' = 0, \tag{42}$$

$$\frac{\left[A(t) - A(t')\right]^2}{2} = \omega_h - I_p. \tag{43}$$

The canonical momentum is given by $p = -A(t')$.

In the Lewenstein model, any form of electric field $E(t)$ can be, through Fourier transform, expanded with sine waves, defined in the complex plane. Thus the saddle-point equations 18-20 can be solved at least numerically.

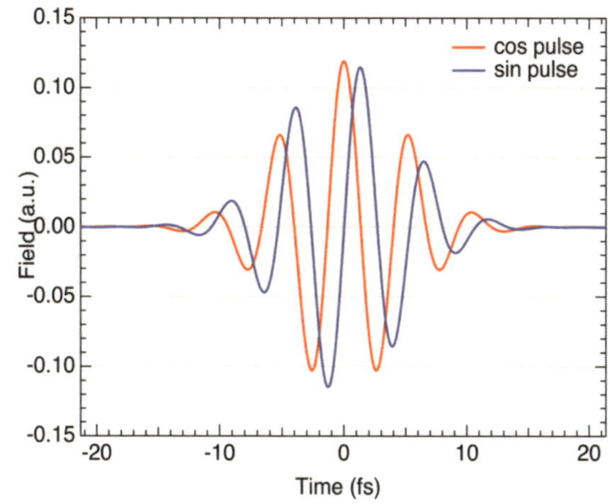

Fig. 6. Electric fields of cos and sin pulses.

In this subsection, let us consider HHG from a helium atom irradiated by au ultrashort laser pulse whose central wavelength is 800 nm, temporal profile is Gaussian with a full-width-at-half-maximum (FWHM) pulse duration $T_{1/2}$ of 8 fs (1.5 cycles), and peak intensity of 5 × 10^{14}W/cm^2. There are two particular forms of electric field, as shown in Fig. 6,

$$E(t) = f(t)\cos\omega_0 t \quad (cos\ pulse), \tag{44}$$

and,

$$E(t) = f(t)\sin\omega_0 t \quad (sin\ pulse), \tag{45}$$

where the field envelope $f(t)$ is given by,

$$f(t) = E_0 e^{-(2\ln 2)t^2/T_{1/2}^2}. \tag{46}$$

In general, when the field takes a form of,

$$E(t) = f(t)\cos(\omega_0 t + \phi_0), \tag{47}$$

ϕ_0 is call *carrier-envelope phase (CEP)*. The CEP is zero and $-\pi/2$ for cos and sin pulses, respectively.

3.1 Cos pulse

Figure 7 (a) displays the real part of the recombination (t) and ionization (t') times calculated with the saddle-point equations for the 1.5-cycle cos pulse. The recombination time from the three-step model, also shown in this figure, is close to the real part of the saddle-point solutions. By comparing this figure with the harmonic spectrum calculated with direct simulation of the TDSE (Fig. 7 (b)), we realize that the steps around 400 and 300 eV in the spectrum correspond to the cutoff of trajectory pairs C and D. Why does not a step (cutoff) for pair B appear? This is related to the field strength at time of ionization, indicated with vertical arrows in Fig. 7 (a). That for pair B (~ -5 fs) is smaller than those of pairs C (~ -2.5 fs) and D (~ 0 fs). Since the tunneling ionization rate (the first step of the three-step model) depends exponentially on intensity, the contribution from pair B is hidden by those from

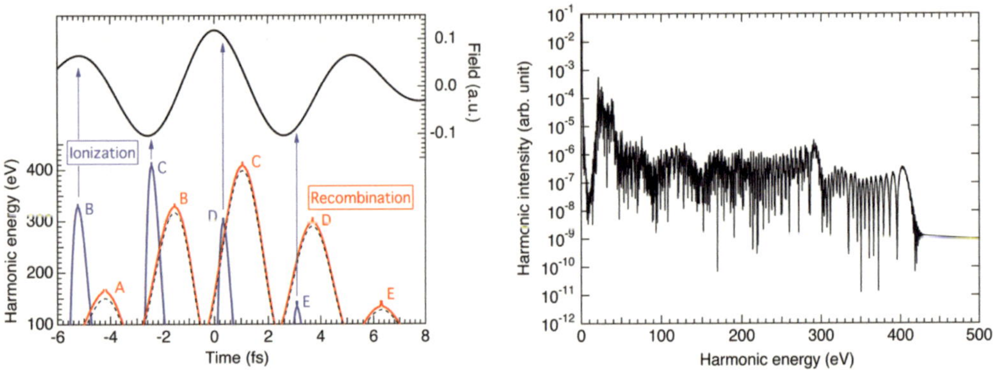

(a) (b)

Fig. 7. (a) Real part of the recombination (red) and ionization times (blue) calculated from the saddle-point equations for the cos pulse. Each trajectory pair is labeled from A to E. The black dashed line is the recombination time from the three-step model. The electric field is also shown in black solid line. (b) Harmonic spectrum calculated with direct simulation of the TDSE.

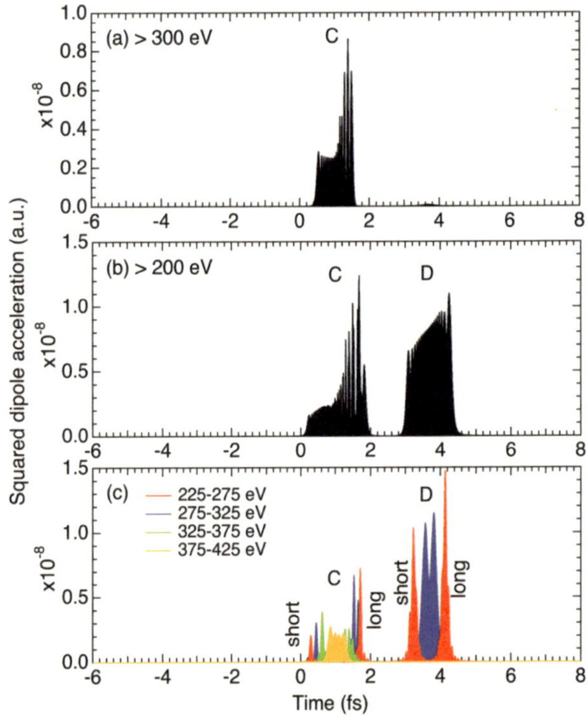

Fig. 8. Temporal profile of the TDSE-calculated squared dipole acceleration (SDA), proportional to the harmonic pulse intensity generated by the cos pulse, (a) at $\hbar\omega_h > 200$ eV, (b) at $\hbar\omega_h > 300$ eV, (c) for different energy ranges indicated in the panel. Labels *C* and *D* indicate corresponding trajectory pairs in Fig. 7 (a). Labels "short" and "long" indicate short and long trajectories, respectively.

pairs C and D. It is noteworthy that the trajectory pair C for the cutoff energy (~ 400 eV) is ionized *not* at the pulse peak but half cycle before it. Then the electron is accelerated efficiently by the subsequent pulse peak.

From the above consideration, and also remembering that harmonic emission occurs upon recombination, we can speculate the following:

- The harmonics above 200 eV consists of a train of two pulses at $t \approx 1$ (C) and 3.5 fs (D).
- By extracting the spectral component above 300 eV, one obtains an isolated attosecond pulse at $t \approx 1$ fs (C).
- The emission from the short (long) trajectories are positively (negatively) chirped, i.e., the higher the harmonic order, the later (the earlier) the emission time. The chirp leads to temporal broadening of the pulse.

These are indeed confirmed by the TDSE simulation results shown in Fig. 8, for which the calculated dipole acceleration is Fourier transformed, then filtered in energy, and transformed back into the time domain to yield the temporal structure of the pulse radiated from the atom.

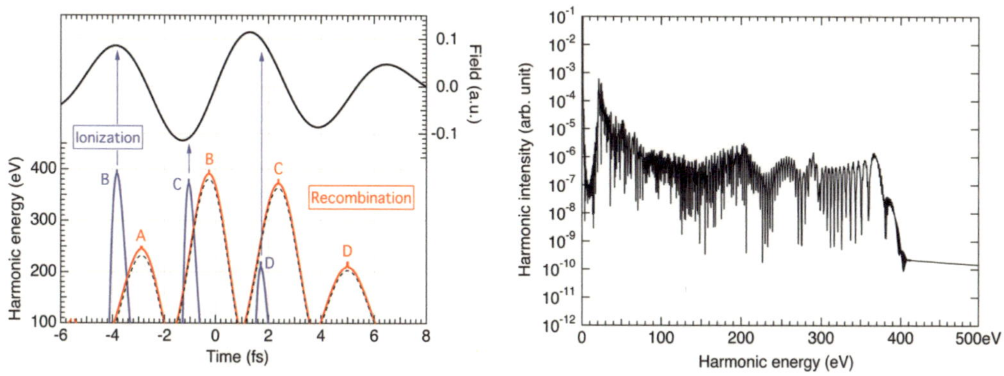

(a) (b)

Fig. 9. (a) Real part of the recombination (red) and ionization times (blue) calculated from the saddle-point equations for the sin pulse. Each trajectory pair is labeled from A to D. The black dashed line is the recombination time from the three-step model. The electric field is also shown in black solid line. (b) Harmonic spectrum calculated with direct simulation of the TDSE.

3.2 Sin pulse

Let us now turn to the sin pulse. The harmonic spectrum (Fig. 9 (b)) has two-step cutoffs at 370 and 390 eV associated with trajectory pairs C and B (Fig. 9 (a)), respectively. The latter (B) is much less intense, since the laser field is weaker at the ionization time for pair B than for pair C as can be seen from Fig. 9 (a), leading to smaller tunneling ionization rate. The inspection of Fig. 9 (a), similar to what we did in the previous subsection, suggests that, noting field strength at the time of ionization,

- The harmonics above 100 eV consists of a train of two pulses at $t \approx 2.5$ (C) and 5 fs (D). The contribution from pairs A and B are negligible due to small ionization rate.
- By extracting the spectral component above 220 eV, one obtains an isolated attosecond pulse at $t \approx 2.5$ fs (C), *not* double pulse (B and C).

- As in the case of the cos pulse, the emission from the short (long) trajectories are positively (negatively) chirped. The chirp leads to temporal broadening of the pulse.

The second point indicates that broader harmonic spectrum is available for isolated attosecond pulse generation with the sin pulse than with the cos pulse, implying potentially shorter soft x-ray pulse generation if the intrinsic chirp can be compensated (Sansone et al., 2006). It also follows from the comparison between Figs. 7 and 9 that for the case of few-cycle lasers high-harmonic generation is sensitive to the CEP (Baltuška et al., 2003).

The examples discussed in this section stresses that the Lewenstein model and even the semiclassical three-step model, which is a good approximation to the former, are powerful tools to understand features in harmonic spectra and predict the temporal structure of generated pulse trains; the emission times in Figs. 8 and 10 could be predicted quantitatively well without TDSE simulations. One can calculate approximate harmonic spectra using the saddle-point analysis from Equation 21 or using the Gaussian model within the framework of SFA. On the other hand, the direct TDSE simulation is also a powerful tool to investigate quantitative details, especially the effects of excited levels and the atomic Coulomb potential (Schiessl et al., 2007; 2008; Ishikawa et al., 2009a).

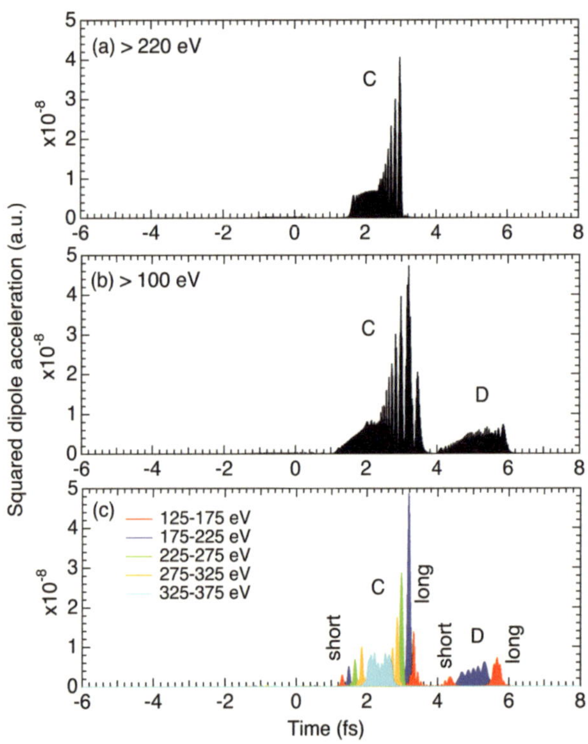

Fig. 10. Temporal profile of the TDSE-calculated squared dipole acceleration (SDA), proportional to the harmonic pulse intensity generated by the sin pulse, (a) at $\hbar\omega_h > 100$ eV, (b) at $\hbar\omega_h > 220$ eV, (c) for different energy ranges indicated in the panel. Labels *C* and *D* indicate corresponding trajectory pairs in Fig. 9 (a). Labels "short" and "long" indicate short and long trajectories, respectively.

4. High-harmonic generation controlled by an extreme ultraviolet pulse

In this section, we discuss how the addition of an intense extreme ultraviolet (xuv) pulse affects HHG (Ishikawa, 2003; 2004). Whereas the xuv pulse is not necessarily harmonics of the fundamental laser, let us first consider how a He^+ ion behaves when subject to a fundamental laser pulse and an intense 27th or 13th harmonic pulse at the same time. Due to its high ionization potential (54.4 eV) He^+ is not ionized by a single 27th and 13th harmonic photon. Here, we are interested especially in the effects of the simultaneous irradiation on harmonic photoemission and ionization. The fundamental laser pulse can hardly ionize He^+ as we will see later. Although thanks to high ionization potential the harmonic spectrum from this ion would have higher cut-off energy than in the case of commonly used rare-gas atoms, the conversion efficiency is extremely low due to the small ionization probability. It is expected, however, that the addition of a Ti:Sapphire 27th or 13th harmonic facilitates ionization and photoemission, either through two-color frequency mixing or by assisting transition to the $2p$ or $2s$ levels. The direct numerical solution of the time-dependent Schrödinger equation shows in fact that the combination of fundamental laser and its 27th or 13th harmonic pulses dramatically enhance both high-order harmonic generation and ionization by many orders of magnitude.
To study the interaction of a He^+ ion with a combined laser and xuv pulse, we solve the time-dependent Schrödinger equation in the length gauge,

$$i\frac{\partial\Phi(\mathbf{r},t)}{\partial t}=\left[-\frac{1}{2}\nabla^2-\frac{2}{r}-zE(t)\right]\Phi(\mathbf{r},t), \tag{48}$$

where $E(t)$ is the electric field of the pulse. Here we have assumed that the field is linearly polarized in the z-direction. To prevent reflection of the wave function from the grid boundary, after each time step the wave function is multiplied by a $\cos^{1/8}$ mask function (Krause et al., 1992) that varies from 1 to 0 over a width of 2/9 of the maximum radius at the outer radial boundary. The ionization yield is evaluated as the decrease of the norm of the wave function on the grid. The electric field $E(t)$ is assumed to be given by,

$$E(t)=F_F(t)\sin(\omega t)+F_H(t)\sin(n\omega t+\phi), \tag{49}$$

with $F_F(t)$ and $F_H(t)$ being the pulse envelope of the fundamental and harmonic pulse, respectively, chosen to be Gaussian with a duration (full width at half maximum) of 10 fs, ω the angular frequency of the fundamental pulse, n the harmonic order, and ϕ the relative phase. The fundamental wavelength is 800 nm unless otherwise stated. Since we have found that the results are not sensitive to ϕ, we set $\phi = 0$ hereafter.

4.1 HHG enhancement

In Fig. 11 we show the harmonic photoemission spectrum from He^+ for the case of simultaneous fundamental and 27th harmonic (H27) irradiation. The peak fundamental intensity I_F is 3×10^{14}W/cm^2, and the peak H27 intensity I_{H27} is 10^{13}W/cm^2. For comparison we also show the spectra obtained when only the fundamental pulse of the same intensity is applied to a He^+ ion and a hydrogen atom. For the case of Fig. 11, the cutoff energy is calculated from Equation 1 to be 70 eV (H45) for a hydrogen atom and 111 eV (H73) for He^+. The cutoff positions in Fig. 11 agree with these values, and harmonics of much higher orders are generated from He^+ than from H. We note, however, that the harmonic intensity from He^+ when only the laser is applied is extremely low compared with the case of H. This is because the large ionization potential, though advantageous in terms of the cut-off energy, hinders ionization, the first step of the three-step model. The situation changes completely if the H27 pulse is simultaneously applied to He^+. From Fig. 11 we can see that the conversion efficiency is enhanced by about seventeen orders of magnitude (Ishikawa, 2003; 2004). Moreover, the advantage of high cutoff is preserved. Figure 12 shows the harmonic

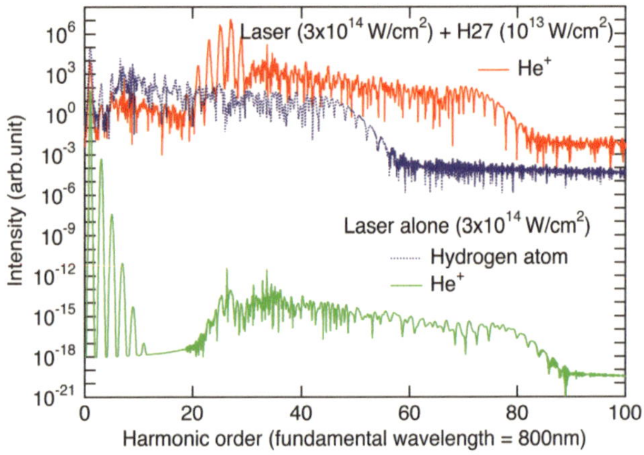

Fig. 11. Upper solid curve: harmonic spectrum from He^+ exposed to a Gaussian combined fundamental and its 27th harmonic pulse with a duration (FWHM) of 10 fs, the former with a peak intensity of $3 \times 10^{14}W/cm^2$ and the latter $10^{13}W/cm^2$. The fundamental wavelength is 800 nm. Lower solid and dotted curves: harmonic spectrum from He^+ and a hydrogen atom, respectively, exposed to the fundamental pulse alone. Nearly straight lines beyond the cut-off energy are due to numerical noise.

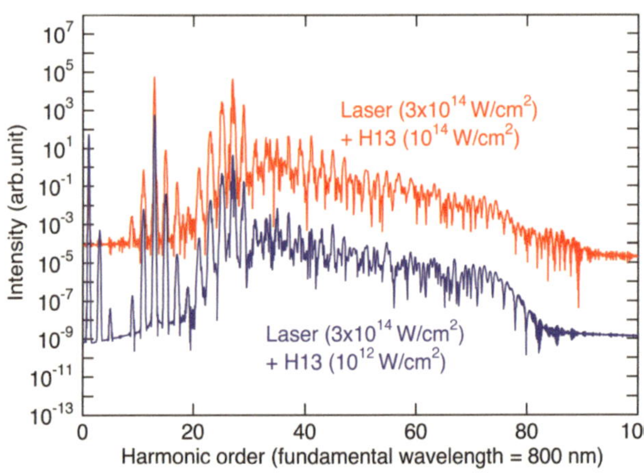

Fig. 12. Harmonic spectrum from He^+ exposed to a Gaussian combined fundamental and its 13th harmonic pulse with a duration (FWHM) of 10 fs, the former with a peak intensity of $3\times10^{14}W/cm^2$, and the latter $10^{14}W/cm^2$ (upper curve) and $10^{12}W/cm^2$ (lower curve). The fundamental wavelength is 800 nm. Note that the horizontal axis is of the same scale as in Fig. 11.

photoemission spectrum from He^+ for the case of simultaneous fundamental and H13 irradiation. Again the harmonic intensity is enhanced by more than ten orders of magnitude compared to the case of the laser pulse alone. We have varied the fundamental wavelength between 750 and 850 nm and found similar enhancement over the entire range.

4.2 Enhancement mechanism

The effects found in Figs. 11 and 12 can be qualitatively understood as follows. The H27 photon energy (41.85 eV) is close to the 1*s*-2*p* transition energy of 40.8 eV, and the H13 photon (20.15 eV) is nearly two-photon resonant with the 1*s*-2*s* transition. Moreover, the 2*p* and 2*s* levels are broadened due to laser-induced dynamic Stark effect. As a consequence, the H27 and H13 promote transition to a virtual state near these levels. Depending on the laser wavelength, resonant excitation of 2*s* or 2*p* levels, in which fundamental photons may be involved in addition to harmonic photons, also takes place. In fact, the 2*s* level is excited through two-color two-photon transition for the case of Fig. 11 at $\lambda_F = 800$ nm, as we will see below in Fig. 15, and about 8% of electron population is left in the 2*s* level after the pulse. Since this level lies only 13.6 eV below the ionization threshold, the electron can now be lifted to the continuum by the fundamental laser pulse much more easily and subsequently emit a harmonic photon upon recombination. Thus the HHG efficiency is largely increased. We have found that the harmonic spectrum from the superposition of the 1*s* (92%) and 2*s* (8%) states subject to the laser pulse alone is strikingly similar to the one in Fig. 11. Hence the effect may also be interpreted as harmonic generation from a coherent superposition of states (Watson et al., 1996). On the other hand, at a different fundamental wavelength, e.g., at $\lambda_F = 785$ nm, there is practically no real excitation. Nevertheless the photoemission enhancement (not shown) is still dramatic. This indicates that fine tuning of the xuv pulse to the resonance with an excited state is *not* necessary for the HHG enhancement. In this case, the H27 pulse promotes transition to a virtual state near the 2*p* level. Again, the electron can easily be lifted from this state to the continuum by the fundamental pulse, and the HHG efficiency is largely augmented. This may also be viewed as two-color frequency mixing enhanced by the presence of a near-resonant intermediate level. In general, both mechanisms of harmonic generation from a coherent superposition of states and two-color frequency mixing coexist, and their relative importance depends on fundamental wavelength. A similar discussion applies to the case of the H13 addition. The comparison of the two curves in Fig. 12 reveals that the harmonic spectrum is proportional to I_{H13}^2, where I_{H13} denotes the H13 peak intensity. We have also confirmed that the photoemission intensity is proportional to I_{H27} for the case of H27. These observations are compatible with the discussion above.

4.3 Ionization enhancement

Let us now examine ionization probability. Table 1 summarizes the He^{2+} yield for each case of the fundamental pulse alone, the harmonic pulse alone, and the combined pulse. As can be expected from the discussion in the preceding subsection, the ionization probability by the combined pulse is by many orders of magnitude higher than that by the fundamental laser pulse alone. Especially dramatic enhancement is found in the case of the combined fundamental and H27 pulses: the He^{2+} yield is increased by orders of magnitude also with respect to the case of the H27 irradiation alone. This reflects the fact that field ionization from the 2*s* state by the fundamental pulse is much more efficient than two-photon ionization from the ground state by the H27 pulse. In Fig. 13, we plot the He^{2+} yield as a function of I_{H27} for a fixed fundamental intensity of 3×10^{14}W/cm^2. The ionization probability is linear in I_{H27} except for the saturation at $I_{H27} > 10^{13}$W/cm^2. This is compatible with our view that an H27 photon promotes 1*s*-2*s* two-color two-photon transition, followed by field ionization.

I_F (W/cm^2)	I_{H27} (W/cm^2)	I_{H13} (W/cm^2)	He^{2+} yield
3×10^{14}	–	–	2.29×10^{-15}
–	10^{13}	–	4.79×10^{-6}
3×10^{14}	10^{13}	–	0.173
–	–	10^{14}	1.25×10^{-4}
3×10^{14}	–	10^{14}	2.04×10^{-4}

Table 1. The He^{2+} yield for various combinations of a Gaussian fundamental and its 27th or 13th harmonic pulses with a duration (FWHM) of 10 fs and a peak intensity listed in the table. λ_F = 800 nm.

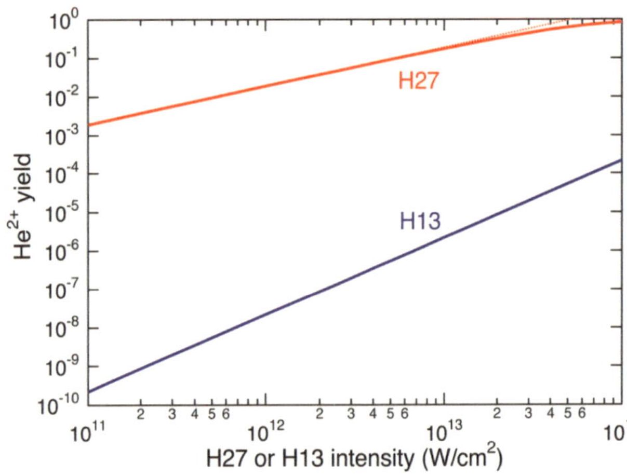

Fig. 13. The He^{2+} yield as a function of peak intensity of the 27th (upper curve) and 13th (lower curve) harmonic pulse when He^+ is exposed to a Gaussian combined fundamental and its 27th harmonic pulse with a duration (FWHM) of 10 fs. The wavelength λ_F and the peak intensity I_F of the fundamental pulse are 800 nm and 3×10^{14}W/cm^2, respectively.

Figure 14 shows the dependence of the He^{2+} yield on the peak intensity I_F of the fundamental pulse. Surprisingly, the yield is not monotonically increasing in I_F: ionization is decreased with an increasing laser intensity at $I_F > 3 \times 10^{14}$W/cm^2 in the presence of the H27 pulse and at $2 \times 10^{14} < I_F < 4 \times 10^{14}$W/cm^2 in the presence of the H13 pulse. In order to clarify the origin of this counter-intuitive behavior, we have calculated the dependence of the He^{2+} yield on the fundamental wavelength λ_F for different values of fundamental intensity. The results are shown in Fig. 15 for the cases of $I_{H27} = 10^{13}$W/cm^2 and $I_F = 5 \times 10^{13}$, 8×10^{13}, 3×10^{14}, 5×10^{14}, 10^{15}W/cm^2. The curve for $I_F = 5 \times 10^{13}$W/cm^2 peaks around λ_F = 820 nm, for which the H27 photon energy is exactly resonant with the 1*s*-2*p* transition. At $I_F = 8 \times 10^{13}$W/cm^2 we can see a second peak around λ_F = 793 nm, for which $26\hbar\omega$ (= $27\hbar\omega - \hbar\omega$) coincides with the 1*s*-2*s* transition energy. This indicates that at this wavelength the 2*s* level is resonantly populated through two-color two-photon excitation, as we alluded earlier.

With increasing laser intensity, the peaks are shifted to longer wavelengths due to the laser-induced dynamic Stark effect, and at $I_F = 3 \times 10^{14}$W/cm^2 the second peak, now higher than the first one, is located around 800 nm, which leads to high ionization yield at this wavelength. At $I_F = 3 \times 10^{14}$W/cm^2, we can also see a third peak around $\lambda_F = 770$ nm corresponding to two-color $2p$ excitation involving one H27 and two fundamental photons. The dynamic Stark effect induces not only the peak shift but also the peak broadening, which results in the decrease of peak heights at higher fundamental intensity, leading to the decrease of the ionization yield at $I_F > 3 \times 10^{14}$W/cm^2 in Fig. 14. Thus, the role of the

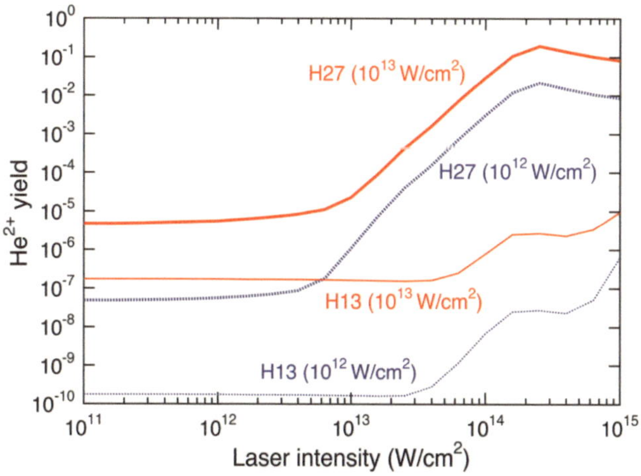

Fig. 14. The He^{2+} yield as a function of peak intensity I_F of the fundamental laser pulse when He^+ is exposed to a Gaussian combined fundamental and its 27th or 13th harmonic pulse with a duration (FWHM) of 10 fs. The fundamental wavelength λ_F is 800 nm. The peak intensity of each harmonic pulse is indicated in the figure along with its order.

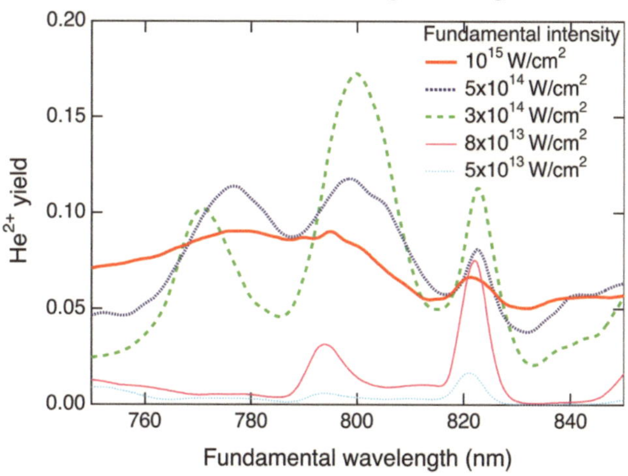

Fig. 15. The He^{2+} yield as a function of fundamental wavelength when He^+ is exposed to a Gaussian combined fundamental and its 27th harmonic pulse with a duration (FWHM) of 10 fs. The peak intensity I_{H27} of the 27th harmonic pulse is fixed at 10^{13}W/cm^2, and we plot the results for five different values of fundamental peak intensity I_F indicated in the figure. Note that the wavelength of the 27th harmonic varies with that of the fundamental pulse.

fundamental laser pulse is three-fold: to lift the electron in an excited (real or virtual) level to the continuum through optical-field ionization, to assist 1*s*-2*s*, 2*p* transitions through two-color excitation, and to induce dynamic Stark shift and broadening. The interplay of these three leads to a complicated behavior seen in Fig. 14.

4.4 Remarks

The mechanism of the HHG enhancement discussed above is that the xuv addition increases the field ionization rate by promoting transition to (real or virtual) excited states, from which ionization is much easier than from the ground state. Hence, the enhancement is effective *only when the ionization rate by the fundamental pulse alone is not high enough.* If HHG is already optimized by a sufficiently intense laser pulse, the xuv addition does not increase the HHG yield significantly.

Although the cases involving resonant transitions are highlighted above, the resonance with an excited state is *not* necessary for the enhancement. Figure 16 shows the harmonic spectra from He for $\lambda = 1600$nm with and without the XUV field ($\hbar\omega_X = 17.05$eV). For the case of the driving laser alone with a peak intensity I of 1.6×10^{14}W/cm^2 (blue curve), the ionization yield Y_I is very low (1.7×10^{-5}%). The addition of the xuv pulse with an intensity of 2.3×10^{11}W/cm^2 increases Y_I to 0.31% and enhances the harmonic yield accordingly (red curve). In this case, the xuv photon energy is not resonant with transition to excited states. Nevertheless, the HHG yield is enhanced by orders of magnitude. The dramatic enhancement effect has been experimentally demonstrated by the use of mixed gases (Takahashi et al., 2007); also in this experiment, the booster xuv pulse (harmonics from Xe) was non-resonant with the target atom (He).

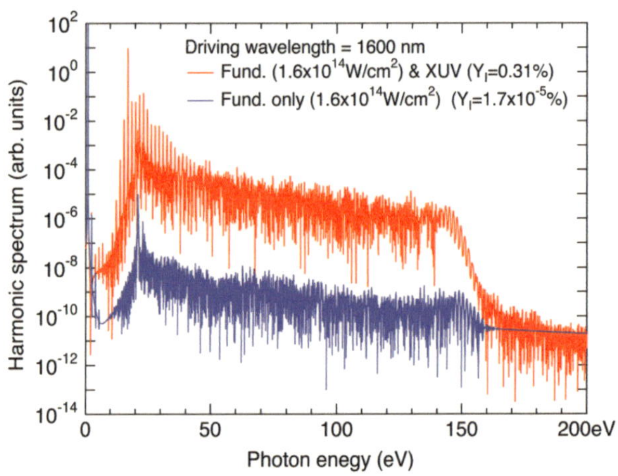

Fig. 16. Upper curve: harmonic spectrum from He exposed to a 35 fs Gaussian combined driving and XUV pulse ($\hbar\omega_X = 17.05$eV), the former ($\lambda = 1600$ nm) with a peak intensity of 1.6×10^{14}W/cm^2 and the latter 2.3×10^{11}W/cm^2. Lower curve: harmonic spectra for the cases of the driving pulse alone, with an intensity of 1.6×10^{14}W/cm^2.

4.5 Single attosecond pulse generation using the enhancement effect

The progress in the high-harmonic generation (HHG) technique has raised significant interest in the generation of attosecond pulses. As we have seen in Subsec. 2.1, the

photoemission process is repeated every half-cycle of the laser optical field and produces an attosecond pulse train (Fig. 3). On the other hand, a single attosecond pulse (SAP), in particular, is critical for the study of the electron dynamics inside atoms (Krausz & Ivanov, 2009). If the driving laser pulse is sufficiently short that the effective HHG takes place only within one half-cycle, the cut-off region of the spectrum may become a continuum, corresponding to a single recollision (Figs. 8 and 10). The first SAPs (Baltuška et al., 2003) were obtained on this basis. Isolated attosecond pulses have also been realized by other methods such as *ionization shutter* (Sekikawa et al., 2004) and *polarization gate* (Corkum et al., 1994; Sansone et al., 2006), and *two-color scheme* (Pfeifer et al., 2006) has also been proposed.
The enhancement effect discussed in the preceding subsections provides a means to control HHG and new physical insights (Schafer et al., 2004; Ishikawa et al., 2009b). As an example, in this subsection, we present an alternative method of SAP generation using multi-cycle laser pulses, called *attosecond enhancement gate for isolated pulse generation (AEGIS)* (Ishikawa et al., 2007). Let us first describe AEGIS qualitatively (Figs. 17(a) and 18(a)). The underlying mechanism of the HHG enhancement by the addition of xuv pulses is that the seed pulse induces transition to real or virtual excited levels, facilitating optical-field ionization, the

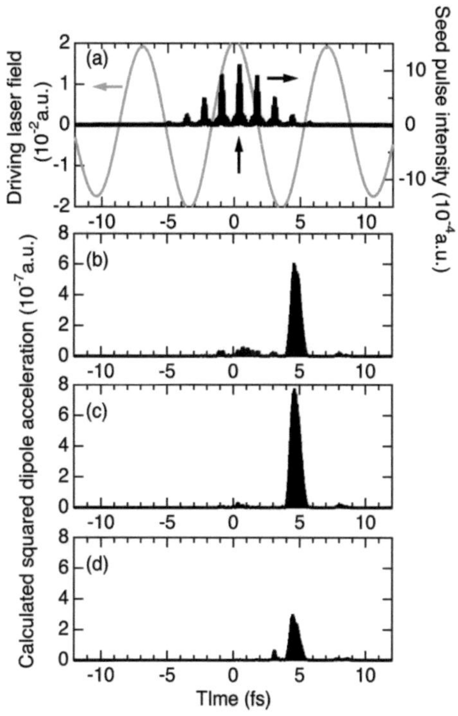

Fig. 17. Soft-x-ray pulse generation by the combination of $\lambda_d = 2.1\mu$m and $\lambda_{sf} = 800$nm. (a) temporal profile of the seed harmonic pulse intensity (black, right axis) with a global pulse width of 5 fs and the driving laser field (gray, left axis) with a pulse duration of 30 fs. The vertical arrow indicates the pulse expected to act as a gate to enhance HHG. (b) temporal profile of the calculated squared dipole acceleration (SDA), proportional to the generated pulse intensity, around 30 nm wavelength. (c) SDA when the seed pulse train has a global pulse width of 3 fs. (d) SDA when, in addition, the delay of the seed pulse train is a quarter cycle of the driving field.

first step of the three-step model. When the seed is composed of a train of attosecond pulses, this can be viewed as repeated attosecond enhancement gates. Let us consider that the fundamental pulse generating the seed harmonic pulse, referred to as the *seed fundamental* pulse hereafter, and the driving laser pulse which will be combined with the seed pulse have different wavelengths. For example, when the seed fundamental wavelength λ_{sf} and the driving wavelength λ_d are 800 nm and 2.1μm, respectively, a seed harmonic pulse train even composed of several attosecond pulses repeated every 1.33 fs is confined within one cycle (7 fs) of the driving laser, as is schematically shown in Fig. 17 (a). Hence, we would expect that the enhanced harmonic emission forms a SAP. When λ_{sf}= 2.1μm and λ_d= 800nm, conversely, seed harmonic pulses are separated by 3.5 fs, which is longer than the driving laser cycle (2.67 fs). If the seed harmonic pulse train and the driving pulse are superposed as shown in Fig. 18(a), only the central pulse would contribute to HHG in the cutoff region, resulting in SAP generation.

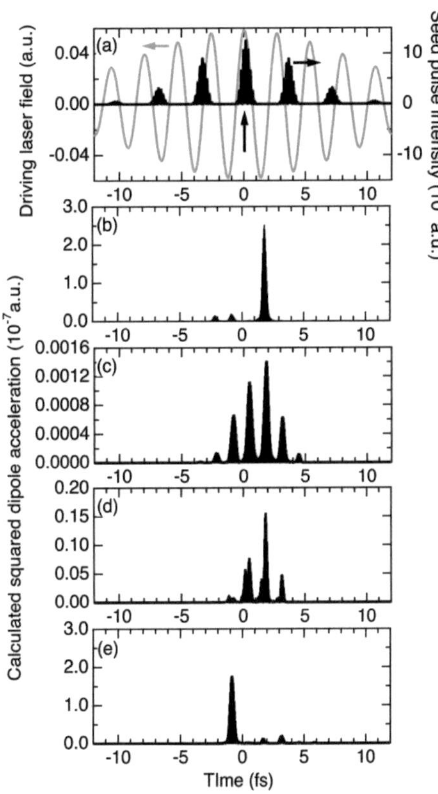

Fig. 18. Soft-x-ray pulse generation by the combination of λ_d= 800nm and λ_{sf}= 2.1μm. (a) temporal profile of the seed harmonic pulse intensity (black, right axis) with a global pulse width of 10 fs and the driving laser field (gray, left axis) with a pulse duration of 15 fs. The vertical arrow indicates the pulse expected to act as a gate to enhance HHG. (b) temporal profile of the calculated SDA, proportional to the generated pulse intensity, around 24 nm wavelength. (c) SDA in the absence of the seed pulse (d) SDA when λ_{sf}= 800nm (e) SDA when the delay of the seed pulse train (λ_{sf}= 2.1μm) is a quarter cycle of the driving field.

Furthermore, the kinetic energy of the recombining electron and, thus, the emitted photon energy in the three-step model depends on the electron's time of release, and that, in particular, an electron ionized by tunneling at $\omega t = \phi_c$, $\phi_c + 180°$, . . . ($\phi_c \approx 17°$), around the pulse peak contributes to cut-off emission. Therefore, when only one pulse of the seed pulse train is adjusted to satisfy this relation, marked by vertical arrows in Figs. 17(a) and 18(a), only that pulse could contribute to harmonic emission in the cut-off region, which would further favor SAP generation, even though the driving pulse is relatively long and the seed contains multiple pulses.

We now confirm this qualitative idea, using direct numerical solution of the time-dependent Schrödinger equation. The harmonic spectrum is calculated by Fourier transforming the dipole acceleration, then high-pass filtered as is done in experiments with multilayer mirrors, and transformed back into the time domain to yield the temporal structure of the pulse radiated from the atom.

Let us first consider the case where λ_{sf} = 800nm and λ_d = 2.1μm. Figure 17(a) displays the seed harmonic pulse train used in the present simulation. The pulse train

$$E_s(t) = E_{s0}(t - \phi_c / \omega_d) \sum_{q(odd)=11}^{19} f_q \cos[q\omega_{sf}(t - \phi_c / \omega_d)], \quad (50)$$

is composed of the 11th to 19th harmonics. We use experimentally observed values (Takahashi et al., 2002) for the harmonic mixing ratio $(f_{11}^2, f_{13}^2, f_{15}^2, f_{17}^2, f_{19}^2)$ = (0.50, 0.34, 0.07, 0.04, 0.05), and the common amplitude envelope $E_{s0}(t-\phi_c/\omega_d)$ is assumed to be of a Gaussian temporal profile centered at $t = \phi_c/\omega_d$ with a full width at half maximum (FWHM) of 5 fs, referred to as a global pulse width hereafter. The sum of peak intensity of each component is 10^{13}W/cm^2. Such a train of ca. 7 pulses would typically be generated by applying a Ti:Sapphire laser pulse of a duration ≈ 15 fs to a Xe gas, and even higher intensity has experimentally been realized using 20 mJ laser pulses (Takahashi et al., 2002). The intensity of the resulting harmonic pulse is proportional to that of the seed harmonic pulse, but its relative temporal structure is not affected by the latter. The driving pulse $E_d(t)$, also shown in Fig. 17(a), is assumed to have a Gaussian temporal intensity profile centered at t = 0 with a FWHM of 30 fs. The peak intensity I_d is 1.5 × 10^{13}W/cm^2. Figure 17(b) presents the calculated squared dipole acceleration (SDA), which is proportional to the intensity of the pulse radiated from a Ne atom subject to the seed harmonic and the driving pulses simultaneously, as would be obtained after reflected by a multilayer X-ray mirror whose reflectivity peaks in the cut-off region around 30 nm (H67-H75). As we have expected, we can see that only the central pulse in the seed (Fig. 17(a)) acts as a gate for dramatic enhancement of HHG in the cut-off region and that a practically single pulse with a FWHM of 800 as is obtained, even though small satellite pulses are present. If we use a seed pulse train with a 3 fs global pulse width, composed of ca. 5 pulses, we can suppress the satellite pulses, as is shown in Fig. 17(c). It should be noted that the driving pulse alone would generate virtually no harmonics for this driving intensity. This indicates that even if the driving laser is not sufficiently intense for HHG, the combination with a seed pulse can serve as efficient means to generate a harmonic single pulse. If we use a higher driving intensity, on the other hand, we obtain a single pulse of shorter wavelength and duration; for the case of I_d = 10^{14}W/cm^2, e.g., a 580 as single pulse would be obtained around 10 nm wavelength. Although the seed contains harmonic components (H15-H19) which may induce direct ionization, the obtained pulses are mainly due to H11 and H13 with a photon energy below the ionization threshold. We have confirmed this by simulations excluding H15-H19.

Let us next turn to the case where λ_{sf}= 2.1μm and λ_d = 800nm. Figure 18(a) displays the seed harmonic pulse train,

$$E_s(t) = E_{s0}(t-\phi_c/\omega_d) \sum_{q(odd)=15}^{23} f_q \cos[q\omega_{sf}(t-\phi_c/\omega_d)], \tag{51}$$

containing ca. 5 pulses, composed of the 15th to 23rd harmonics. The harmonic mixing ratio is $(f_{15}^2, f_{17}^2, f_{19}^2, f_{21}^2, f_{23}^2)$ = (0.0625, 0.25, 0.375, 0.25,0.0625), and the common amplitude envelope with a FWHM of 10 fs peaks at $t = \phi_c/\omega_d$. The sum of peak intensity of each component is 10^{13}W/cm^2. The driving pulse $E_d(t)$, also shown in Fig. 18(a), is assumed to have a FWHM of 15 fs and a peak intensity I_d of 1.2 × 10^{14}W/cm^2. Figure 18(b) presents the calculated SDA, as would be obtained after reflected by a multilayer X-ray mirror whose reflectivity peaks in the cut-off region around 24 nm (H31-H35). The result, containing ca. 4 pulses, for the case of the driving pulse alone is shown in Fig. 18(c). From Fig. 18(b), we can see that only the middle pulse in the seed significantly boosts HHG in the cut-off region, thus leading to a single pulse with a duration of 350 as. This is shorter than in Fig. 17, probably because the phase range of the driving field relevant with the cut-off region translates to a shorter time interval due to a shorter driving wavelength. The small satellite pulses are further suppressed if the seed pulse is composed of a larger number of orders, for which each pulse in the train becomes shorter.

5. References

Baltuška, A., Udem, Th., Uiberacker, M., Hentschel, M., Goulielmakis, E., Gohle, Ch., Holzwarth, R., Yakovlev, V. S., Scrinzi, A., Hänsch, T.W. & Krausz, F. (2003). Attosecond control of electronic processes by intense light fields. *Nature* 421(6923): 611–615.

Bauer, D. & Koval, P. (2006). Qprop: A Schrödinger-solver for intense laser-atom interaction. *Comp. Phys. Comm.* 174(5): 396–421.

Corkum, P. B. (1993). Plasma perspective on strong-field multiphoton ionization. *Phys. Rev. Lett.* 71(13): 1994–1997.

Corkum, P. B., Burnett, N. H. & Ivanov, M. Y. (1994). Subfemtosecond pulses. *Opt. Lett.* 19(22): 1870–1872.

Ishikawa, K. (2003). Photoemission and ionization of He+ under simultaneous irradiation of fundamental laser and high-order harmonic pulses. *Phys. Rev. Lett.* 91(4): 043002.

Ishikawa, K. L. (2004). Efficient photoemission and ionization of He+ by a combined fundamental laser and high-order harmonic pulse *Phys. Rev. A* 70(1): 013412.

Ishikawa, K. L., Takahashi, E. J. & Midorikawa, K. (2007). Single-attosecond pulse generation using a seed harmonic pulse train. *Phys. Rev. A* 75(2): 021801(R).

Ishikawa, K. L., Schiessl, K., Persson, E. & Burgdörfer, J. (2009). Fine-scale oscillations in the wavelength and intensity dependence of high-order harmonic generation: Connection with channel closings. *Phys. Rev. A* 79(3): 033411.

Ishikawa, K. L., Takahashi, E. J. & Midorikawa, K. (2009). Wavelength dependence of highorder harmonic generation with independently controlled ionization and ponderomotive energy. *Phys. Rev. A* 80(1): 011807.

Koonin, S. E., Davies, K. T. R., Maruhn-Rezwani, V., Feldmeier, H., Krieger, S. J. & Negele, J. W. (1977). Time-dependent Hartree-Fock calculations for ^{16}O + ^{16}O and ^{40}Ca + ^{40}Ca reactions. *Phys. Rev. C* 15(4): 1359–1374.

Krause, J. L., Schafer, K. J. & Kulander, K. C. (1992). High-order harmonic generation from atoms and ions in the high intensity regime. *Phys. Rev. Lett.* 68(24): 3535–3538.

Krausz, F. & Krausz, F. (2009). Attosecond physics. *Rev. Mod. Phys.* 81(1): 163–234.

Kulander, K. C., Schafer, K. J. & Krause, J. L. (1992). Time-dependent studies of multiphoton processes, In: *Atoms in intense laser fields*, Gavrila, M. (Ed.), 247-300, Academic Press, ISBN 0-12-003901-X, New York.

Lewenstein, M., Balcou, Ph., Ivanov, M. Yu., L'Huillier, A., & Corkum, P. B. (1994). Theory of high-harmonic generation by low-frequency laser fields. *Phys. Rev. A* 49(3): 2117–2132.

Muller, H. G. & Kooiman, F. C (1998). Bunching and focusing of tunneling wave packets in enhancement of high-order above-threshold ionization. *Phys. Rev. Lett.* 81(6): 1207–1210.

Muller, H. G. (1999). An efficient propagation scheme for the time-dependent Schrödinger equation in the velocity gauge. *Laser Phys.* 9(1): 138–148.

Nabekawa, Y., Shimizu, T., Okino, T., Furusawa, K., Hasegawa, H., Yamanouchi, K., & Midorikawa (2006). Conclusive evidence of an attosecond pulse train observed with the mode-resolved autocorrelation technique. *Phys. Rev. Lett.* 96(8): 083901.

Pfeifer, T., Gallmann, L., Abel, M. J., Neumark, D. M. & Leone, S. R. (2006). Single attosecond pulse generation in the multicycle-driver regime by adding a weak second-harmonic field. *Opt. Lett.* 31(7): 975–977.

Salières, P., Carré, B., Le Déroff, L., Grasbon, F., Paulus, G. G.,Walther, H., Kopold, R., Becker, W., Milošević, D. B., Sanpera, A. & Lewenstein, M. (2001). Feynman's path-integral approach for intense-laser-atom interactions. *Science* 292(5518): 902–905.

Sansone, G., Benedetti, E., Calegari, F., Vozzi, C., Avaldi, L., Flammini, R., Poletto, L., Villoresi, P., Altucci, C., Velotta, R., Stagira, S., De Silvestri, S. & Nisoli, M. (2006). Isolated single-cycle attosecond pulses. *Science* 314(5798): 443–446.

Schafer, K. J., Yang, B., DiMauro, L. F. & Kulander, K. C. (1993). Above threshold ionization beyond the high harmonic cutoff. *Phys. Rev. Lett.* 70(11): 1599–1602.

Schafer, K. J., Gaarde, M. B., Heinrich, A., Biegert, J. & Keller, U. (2004). Strong field quantum path control using attosecond pulse trains. *Phys. Rev. Lett.* 92(2): 023003.

Schiessl, K., Ishikawa, K. L., Persson, E. & Burgdörfer, J. (2007). Quantum path interference in the wavelength dependence of high-harmonic generation. *Phys. Rev. Lett.* 99(25): 253903.

Schiessl, K., Ishikawa, K. L., Persson, E. & Burgdörfer, J. (2008). Wavelength dependence of high-harmonic generation from ultrashort pulses. *J. Mod. Opt.* 55(16): 2617–2630.

Sekikawa, T., Kosuge, A., Kanai, T. & Watanabe, S. (2004). Nonlinear optics in the extreme ultraviolet. *Nature* 432(7017): 605–608.

Takahashi, E. J., Nabekawa, Y., Otsuka, T., Obara, M. & Midorikawa, K. (2002) Generation of highly coherent submicrojoule soft x rays by high-order harmonics. *Phys. Rev. A* 66(2): 021802(R).

Takahashi, E. J., Kanai, T., Ishikawa, K. L., Nabekawa, Y. & Midorikawa, K. (2007). Dramatic enhancement of high-order harmonic generation. *Phys. Rev. Lett.* 99(5): 053904.

Tong, X.-M. & Chu, S.-I (1997). Theoretical study of multiple high-order harmonic generation by intense ultrashort pulsed laser fields: A new generalized pseudospectral timedependent method. *Chem. Phys.* 217(2–3): 119–130.

Watson, J. B., Sanpera, A., Chen, X. & Burnett, K. (1996). Harmonic generation from a coherent superposition of states. *Phys. Rev. A* 53(4): R1962–R1965.

20

High-Order Harmonic Generation from Low-Density Plasma

Tsuneyuki Ozaki[1], Rashid Ganeev[1], Masayuki Suzuki[2] and Hiroto Kuroda[2]
[1]*Institut national de la recherche scientifique*
[2]*Institute for Solid-State Physics, University of Tokyo*
[1]*Canada*
[2]*Japan*

1. Introduction

High-order harmonic generation (HHG) of ultra-short laser pulses is an effective method to generate extreme ultraviolet (XUV) radiation and attosecond light pulses. Currently, various techniques are used for HHG, such as the interaction of intense laser with gases (Kazamias et al., 2003), with low-density laser-produced plasmas (Ganeev, 2007), and by reflecting relativistic intensity laser pulses from solid surfaces (Norreys et al., 1996). Such harmonics are promising sources of coherent soft x-ray radiation with various applications, such as XUV nonlinear optics and spectroscopic studies. In particular, the major advantages of HHG are very good spatial quality and femtosecond to attosecond temporal resolution. Single attosecond XUV pulses (with pulse durations of several hundred attoseconds) from HHG has been generated, by irradiating neon gas atoms with intense, few-cycle laser pulse (Drescher et al., 2001), and the current record for the shortest pulse ever generated is 80 attoseconds (Goulielmakis et al., 2008). Furthermore the maximum cut-off energy of harmonics has been extended up to 1.3 keV (corresponding to a wavelength of 0.95 nm) (Seres et al., 2005). This ultra-short nature of harmonics has lead to several applications of pivotal importance, such as controlling the motion of bound electrons in molecular dissociation (Kling et al., 2006), mapping of attosecond electron wave packet motion (Niikura et al., 2005), and nonlinear phenomena in the XUV region (Nabekawa et al., 2005).

To extend the harmonic cut-off to shorter wavelengths, one needs to use nonlinear medium with higher ionization potential. This is because generally speaking, the harmonic cut-off energy $E_{cut\text{-}off}$ is determined by the cut-off rule $E_{cut\text{-}off} = I_p + 3.17\ U_p$, where I_p is the ionization potential and U_p the ponderomotive energy (Corkum, 1993). Thus, we should observe increased cut-off energy with increased laser intensity. However, the highest photon energies demonstrated to date by HHG have generally been limited not by the cut-off rule, but by effects that come into play as the gas is ionized by the laser. For example, if the gas medium is ionized before the peak of the laser pulse arrives, the neutral atom can only interact with a lower "saturation intensity", which is the intensity where 98% of the gas is singly ionized. The saturation intensity can be increased using shorter-duration laser pulses or using atoms with a large ionization potential, thus allowing the atoms to survive higher laser intensity before ionizing. For example, instead of rare-gas atoms, rare-gas like ions in laser-produced plasmas of alkali metal has been used in the past for HHG (Akiyama et al., 1992; Kubodera et al., 1993; Wahlstrom et al., 1995). The basic idea was to capitalize on the high ionization potential of alkali metal, which is higher than those of any other gas atom. The cut-off energies of the HHG using KrF excimer lasers (248 nm wavelength) were 64.92 eV and 105 eV, for potassium (Akiyama et al., 1992) and lead ions (Kubodera et al., 1993) produced in laser-plasmas. On the other hand, experiments using Ti:sapphire lasers as the pump have only demonstrated cut off up to the 27th harmonic (29.4 nm wavelength) using sodium and potassium ions (Wahlstrom et al., 1995). In all of the above experiments with rare-gas-like ions, the plateau was never observed. Recently, we have demonstrated the 63rd harmonic (12.63 nm wavelength, 98 eV photon energy) using laser ablation boron ions irradiated by 150 fs Ti:sapphire laser pulse (Ganeev et al., 2005d). Experimental evidence show that these harmonics result from the interaction of femtosecond laser pulse with singly charged ions and neutral atoms. In this experiment, the plateau has been observed.

Another important goal of HHG research is to increase its conversion efficiency. In the past, such conversion efficiency enhancement has been pursued by controlling the phase matching condition in gas filled capillary (Rundquist, 1998) or gas cell (Tamaki et al., 1999). Using xenon gas-filled hollow-core fiber, strong 15th harmonic at a wavelength of 53.3 nm was obtained with a conversion efficiency of 4×10^{-5} (Constant et al., 1999). Conversion efficiency of 3×10^{-5} was obtained for the 23rd harmonic at a wavelength of 35 nm, by using argon gas cell (Kazamias et al., 2003). The strongest output energy of harmonics were 7 μJ for the 11th harmonic at a wavelength of 72.7 nm, 4.7 μJ for 13th harmonic at a wavelength of 62.3 nm, and 1 μJ for 15th harmonic at a wavelength of 54 nm using xenon gas cell (Takahashi et al., 2002). More recent results have reported an enhancement of high-order harmonic conversion efficiency when a two-color (fundamental and second harmonic field) orthogonally polarized driving field is used (Kim et al., 2005). In this scheme, the conversion efficiency of the 38th harmonic (21.6 nm) was 5×10^{-5}, using helium gas jet. An alternative approach is to use resonance enhancement in HHG, which has been proposed by theoretical calculations (Gaarde & Schafer, 2001; Faria et al., 2002; Taieb et al., 2003b). Experimentally, resonance enhancement of the 13th and the 15th harmonics using argon gas medium has been observed (Toma et al., 1999). However the enhancement in these works were a factor of two to three, and the harmonic spectrum consisted of multiple harmonics.

In this chapter, we will review recent developments in the generation of intense high-order harmonics using lowly ionized plasma as the nonlinear medium. While early experiments have shown high-order harmonic generation using similar technique (Akiyama et al., 1992; Kubodera et al., 1993; Wahlstrom et al., 1995), the harmonic spectra lacked a plateau, and consequently the cut-off was limited. In contrast, recent results have shown clear plateau (Ganeev et al., 2005d) with a cut-off as high as the 101st order (Ganeev et al., 2007). A unique intensity enhancement of a single high-order harmonic that dominates the spectrum has been observed in the XUV region (Ganeev et al., 2006). Such enhancement of a single harmonic is an advantage for several important applications of coherent XUV radiation. One example is photoelectron spectroscopy for understanding excited-state electron dynamics, in which there is a need to select one harmonic while eliminating neighboring harmonics. This single enhancement technique will allow the generation of a quasi-monochromatic coherent x-ray source, without complexities of using monochromators, multilayer mirrors and filters. Another characteristic of this plasma harmonic method is that one could use any target that could be fabricated into solids. Profiting from this feature, there has been several works on pioneering high-order harmonic generation from nanostructured material, such as metallic nanoparticles and fullerenes. This chapter is organized as follows. In Section 2, I will review the basic experimental setup for plasma harmonic generation. In Section 3, I will review the various cut-off of plasma harmonics and their relation to the ionization potential of the target material. In Section 4, intensity enhancement of a single harmonic order is discussed, and in Section 5, on high-order harmonics from nanostructured material.

2. Experimental setup

A schematic diagram of the typical experimental setup is shown in Fig. 1. The pump laser is usually a high-intensity femtosecond laser system, such as the Ti:sapphire laser. A pre-pulse is split from the amplified laser beam using a beam splitter before the pulse compressor. The typical pre-pulse energy is 5 - 17 mJ with pulse duration of a few hundred ps. The main femtosecond pump pulse typically has energy of 5 - 17 mJ. A cylindrical or spherical lens focuses the pre-pulse onto a solid target placed within a vacuum chamber, which generates a laser ablation plume that contains low-charged ions. The size of the focus on the target surface is typically 1 mm, and the intensity of the pre-pulse was varied between 10^{10} - 10^{11} Wcm^{-2}. The main pulse is focused onto the ablation plasma by a spherical lens, with a delay between 20 to 100 ns after the pre-pulse irradiation. The typical intensity of the main pulse at the plasma plume is 10^{14} - 10^{15} Wcm^{-2}. The spectrum of the generated high-order harmonics was measured by grazing incidence spectrometer with a gold-coated Hitachi 1200 grooves/mm flat-field grating (such as the Hitachi 001-437). A gold-coated grazing-incidence cylindrical mirror was used to image the target harmonics at the ablation plume onto detector plane. The XUV spectrum is detected using a two-dimensional x-ray detector

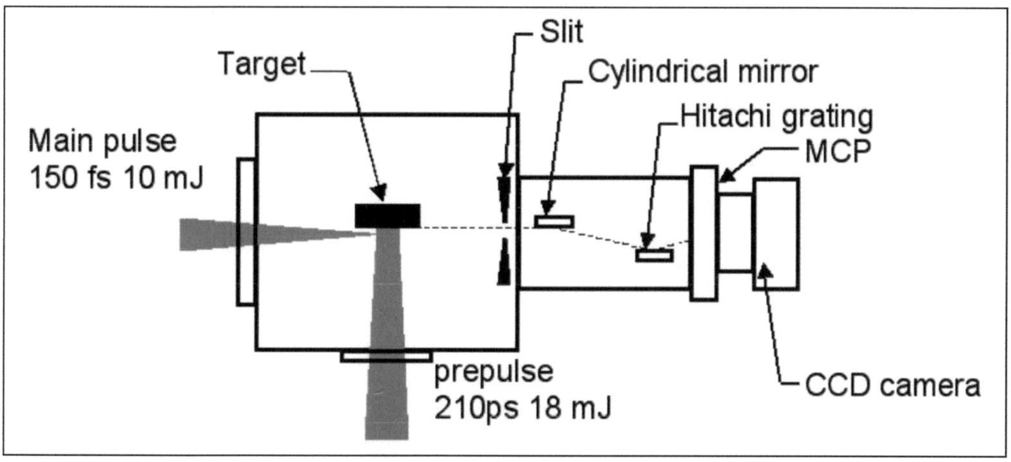

Fig. 1. Schematic of experimental setup for HHG measurement from the laser ablation plume pumped by femtosecond laser pulse. First the prepulse was focused on the target surface. Then the main pulse was focused into the preplasma that was produced by the prepulse. Delay between prepulse and main pulse is 100 ns. The generated high-order harmonics were measured by grazing incidence spectrometer and detected by microchannel plate with CCD camera.

(such as a micro-channel plate with a phosphor screen read-out), and the spectrum is recorded using a CCD camera. To obtain the energy of the harmonics generated, the absolute calibration of the spectrometer system is performed using the following method (Ganeev et al., 2005a)[27]. First, the intensity of the 3rd harmonic generated from a nonlinear crystal is measured, and its readout intensity is compared with that of the 3rd harmonic generated from the laser ablation plasma by using a commercial spectrograph. The intensity of the 3rd harmonic was measured with a power meter. Then, one compares the intensity of the 3rd harmonic with that of the 9th harmonic from the laser ablation plumes using the commercial spectrograph. Finally, one compares the intensity of the 9th harmonic with that of the high-order harmonics in the XUV region using the flat-field spectrometer.

3. Cut-off of plasma harmonics

Figure 2 and 3 shows the typical HHG spectra from silver and vanadium laser ablation plumes. There are two characteristics of these spectra that show them to be harmonics. First, only odd-order harmonics are observed, and second, these harmonics disappear by simply rotating the polarization of the femtosecond laser pulse from linear to circular. The spectrum from vanadium ablation (Fig. 3) shows high-order harmonics up to the 71st order (11.20 nm wavelength). The conversion efficiency of the 71st harmonic was measured to be 1.6×10^{-7}. Using laser plasma of silver (Fig. 2), the cut-off is at the 59th harmonic (13.47 nm) with a conversion efficiency of 6×10^{-6}. Typical plateau and cut-off were observed in both spectra. The appearance of the plateau in the spectrum indicates that these harmonics are generated in the non-perturbative regime. Therefore the cut-off energy of the harmonics from the laser ablation plume could be explained by the three-step model (Corkum, 1993). In the three-step model, the cut-off energy rule is given by $E_{cut\text{-}off} = I_p + 3.17U_p$, [eV]. Here, I_p is the ionization potential of the nonlinear medium, and U_p is the ponderomotive potential, where $U_p = e^2E^2/4m\omega^2 = 9.33\times10^{-14}\, I\lambda^2$, where e and m are electron charge and mass, and E, ω, I, λ, are the laser field amplitude, laser frequency, laser intensity (in units of Wcm^{-2}) and laser wavelength (in units of μm), respectively. In previous works, a plot of the harmonic cut-off energy as a function of the ionization potential showed that the cut-off depended strongly on the second ionization potential of the ablation atom (Ganeev et al., 2005b). For example, the harmonic cut-off energy for boron was 98 eV (12.6 nm wavelength), which has a second ionization potential of 24.12 eV. For silver, which has a second ionization potential of 21.04 eV, the cut-off was at 95 eV (13.03 nm wavelength), showing high correlation between the two. However, our experimental results with vanadium seem to deviate from this rule. The second ionization potential of vanadium is 14.65 eV, but the cut-off is at the 71st harmonic (11.20 nm wavelength), with photon energy of 110 eV. That is, the second ionization potential of vanadium is lower than that of silver, but the cut-off energy of vanadium is higher than that of silver. The estimated maximum cut-off energy of vanadium was 85.6 eV by using the cut-off rule equation and assuming that the U_p is the same as that of silver. However, experiments show that the cut-off energy is higher than this estimated value. We attribute the observation of the 71st harmonic to the contribution from doubly charged ions. The third ionization potential of vanadium is 29.31 eV, which is higher than the second ionization potential value of boron. Therefore the observation of cut-off energy by using vanadium is higher than that of boron.

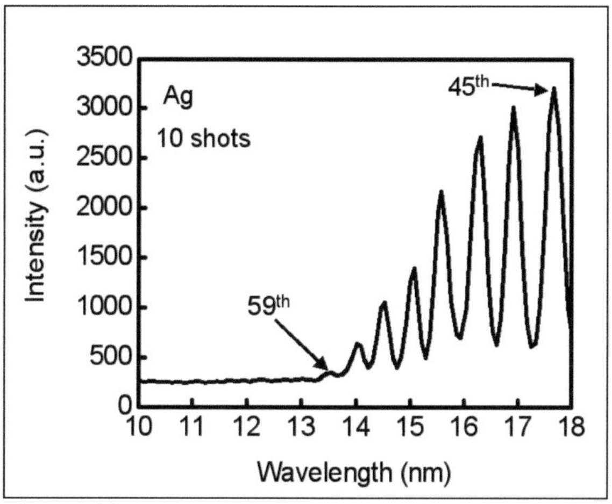

Fig. 2. HHG from the laser ablation silver plume, with a cut-off at the 59th harmonic (13.47 nm wavelength).

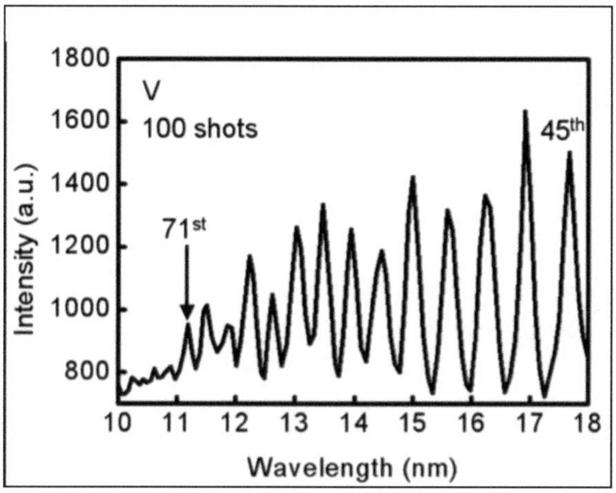

Fig. 3. HHG from the laser ablation vanadium plume, with a cut-off at the 71st harmonic (11.20 nm wavelength).

We attribute the observation of the vanadium cut-off at the 71st harmonic to contribution from doubly charged ions to the harmonic spectrum. The third ionization potential of vanadium is 29.31 eV, which is higher than the second ionization potential of boron. Therefore, harmonic generation from doubly- ionized vanadium will allow a cut-off higher than those from boron plasma. To investigate the nature of HHG from doubly charged ions in other ablation plumes, we have also measured the harmonic spectra generated from niobium, titanium, lead, and germanium. Table 1 shows the experimentally observed cut-off energy of HHG from laser plasma of niobium, titanium, vanadium, lead, and germanium plumes, generated by femtosecond laser pulses. The second ionization potentials of niobium, titanium, vanadium, lead, and germanium are 14.32, 13.58, 14.65, 15.03, and 15.93 eV, respectively. Furthermore, the third ionization potentials of these materials are 25.04, 27.49, 29.31, 31.94, and 34.22 eV, respectively. In the experiments, niobium, titanium, and vanadium showed relatively high cut-off energies of 76.42, 95.14, and 110.7 eV, respectively.

	Nb	Ti	V	Pb	Ge
Maximum cut-off energy (eV)	76.42	95.14	110.7	54.59	38.99
Second ionization (eV)	14.32	13.58	14.65	15.03	15.93
Third ionization (eV)	25.04	27.49	29.31	31.94	34.22

Table 1. The harmonic cut-off energy and ionization potentials for Nb, Ti, V, Pb, and Ge atoms.

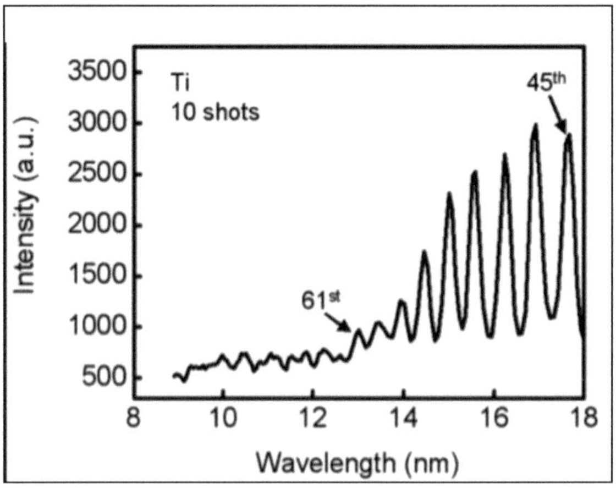

Fig. 4. HHG spectrum from laser ablation titanium plume, showing a cut-off at the 61st harmonic (13.03 nm wavelength).

Interestingly, however, the maximum cut-off energy of lead and germanium were limited to 54.59 and 38.99 eV, respectively. For example, the 61st harmonic (13.03 nm wavelength, 95eV photon energy) from the laser ablation titanium plume have been obtained as can be seen in Fig. 4. This is contrary to our hypothesis that higher ionization potentials generally yield higher cut-off harmonics, since, for example, the second and third ionization potentials of germanium are higher than vanadium, but the cut-off energy of germanium is only one-third of vanadium. The above results show that simple comparison of ionization potentials alone cannot definitively predict the relative cut-off harmonics among different atomic species. From our comparison with boron ablation, we can safely say that the HHG from the laser ablation niobium, titanium and vanadium were generated from doubly charged ions. Therefore, the relatively low cut-off from lead and germanium ablation can be attributed to the smaller contribution of harmonics from doubly charged ions. In these experimental conditions, the laser ablation plumes were not optimized for the lead and germanium target. As a result, self-defocusing of the femtosecond pump laser occurred from the electron by the leading edge of femtosecond laser pulse and the focusing intensity is decreased. Fig. 5 shows the maximum cut-off energy as a function of the ionization potential of the laser ablation plume used for nonlinear medium. Lead, zinc, indium, gold, silver, boron, titanium, and vanadium were used as the nonlinear medium in this experiment. The maximum cut-off energy has been increased when the ionization potential of the nonlinear medium increased. From Fig. 5, we can clearly see a good correlation between the cut-off energy and the ionization potential of the target material. This relatively clear correlation is also due to the fact that there is a maximum pump laser intensity that the ion can interact with, before it further ionizes. Usually, this maximum intensity can be approximated by the barrier suppression intensity I_{BSI}, where I_{BSI} [eV] ~ 4×10^9 $(I_P)^4/ Z_i^2$. Therefore, the maximum pump intensity (and as such U_p) is also a function of the ionization potential, and thus resulting in the strong correlation between the cut-off energy and I_p.

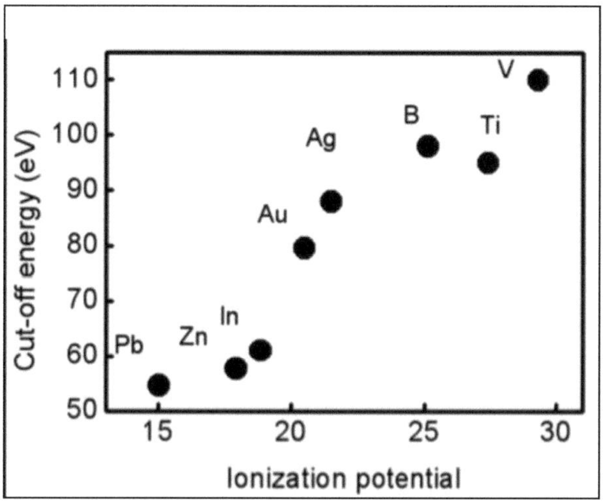

Fig. 5. The maximum observed cut-off energy as a function of the ionization potential of the nonlinear medium.

4. Intensity enhancement of single harmonic

Interesting results have been observed when the laser ablation of tin and indium plasma is used for harmonic generation. Fig. 6 shows the spectra of HHG from tin plasma pumped by femtosecond laser pulse. Plateau and cut-off of the harmonics have both been observed in these experiments. Although the signal is weak, high-order harmonics up to the 23rd order (34.56 nm wavelength) was observed with tin. However, the 17th order harmonic (at a wavelength of 46.76 nm) was unusually strong using tin laser ablation plume, as shown in Fig. 6 (Suzuki et al., 2006). The intensity of the 17th harmonics was 20 times higher than those of its neighbors. The conversion efficiency of the 17th harmonic was measured to be about 1.1×10^{-4}, and this output energy of 1.1 μJ was obtained from the pump laser energy of 10 mJ. To confirm the nature of this strong emission at the wavelength of 46.76 nm, we investigated the effect of the pump laser polarization on HHG. Fig. 7 shows the intensity of the 46.76 nm harmonic as a function of the laser polarization ellipticity. A quarter-wave plate was installed after the focusing lens to change the pump laser polarization from the linear to circular. The line intensities shown in Fig. 7 are normalized to that generated using linearly polarized pump laser, corresponding to the zero position. The intensity of emission at the wavelength of 46.76 nm decreased after 10-degree rotation of the quarter-wave plate, which completely disappeared after 50-degree rotation of the quarter-wave plate. This tendency is consistent with that of gas harmonics, which leads us to conclude that the strong emission at 46.76 nm should be generated by HHG. To investigate the mechanism of enhancement for the 17th harmonic, the central wavelength of the pump laser pulse was changed from 795 nm to 778 nm. Fig. 8 shows the HHG spectra from the laser-ablated tin with three different laser wavelengths of 795 nm, 782nm, and 778 nm (Suzuki et al., 2006).

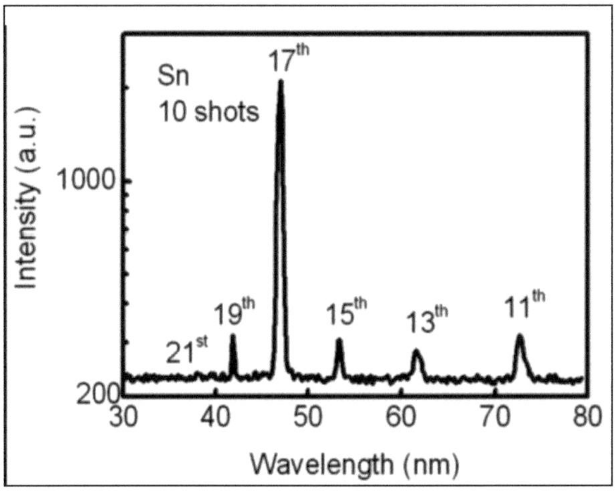

Fig. 6. HHG spectra from tin laser ablation irradiated by femtosecond laser pulse. Strong 17th harmonic (46.76 nm wavelength) is observed. (Suzuki et al., 2006)

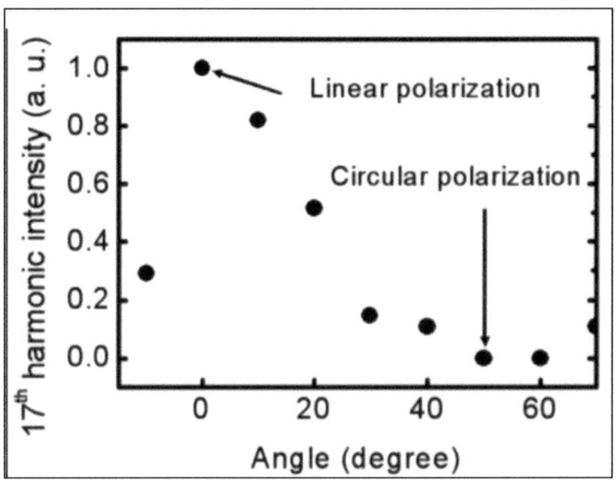

Fig. 7. Intensity of the 17th harmonic (46.76 nm wavelength) as a function of the quarter waveplate angle. The laser polarization is varied from linear (0 degree) to circular (45 degree). (Suzuki et al., 2006)

In Fig. 8(a), one sees that the intensity of the 17th harmonic using 795 nm wavelength pump dominates the harmonic spectrum. The intensity of the 17th harmonic is 20 times higher than that of other harmonics. However, in Fig 8(b), the intensity of the 17th harmonic using 782 nm wavelength pump is decreased, and is almost the same as that of other harmonics. In Fig. 8(c), the intensity of the 17th harmonic with 778 nm wavelength pump is further decreased. In this case, the 17th harmonic intensity is weaker than that of the 13th and 11th harmonics. The above results show that the intensity of the 17th harmonic gradually decreased as the wavelength of the pump laser become shorter. In past work, the Sn II ion has been shown to posses a strong transition of the $4d^{10}5s^25p$ $^2P_{3/2}$ - $4d^95s^25p^2$ (^1D) $^2D_{5/2}$ at the wavelength of 47.20 nm (Duffy & Dunne, 2001). The gf-value of this transition has been calculated to be 1.52 and this value is 5 times larger than other transition from ground state of Sn II. Therefore, the enhancement of the 17th harmonic with 795 nm wavelength laser pulse can be explained be due to resonance with this transition. By changing the pump laser wavelength from 795 nm to 778 nm, the wavelength of the 17th harmonic is changed from 46.76 nm to 45.76 nm. Therefore, the wavelength of the 17th harmonic pumped by laser wavelength of 778 nm is farther away from the $4d^{10}5s^25p\ ^2P_{3/2}$ - $4d^95s^25p^2\ (^1D)\ ^2D_{5/2}$ transition, at the wavelength of 47.20 nm. As a result, the resonance condition of the 17th order harmonic is weaker when pumped by a 778 nm, compared with the case for 795 nm pump.

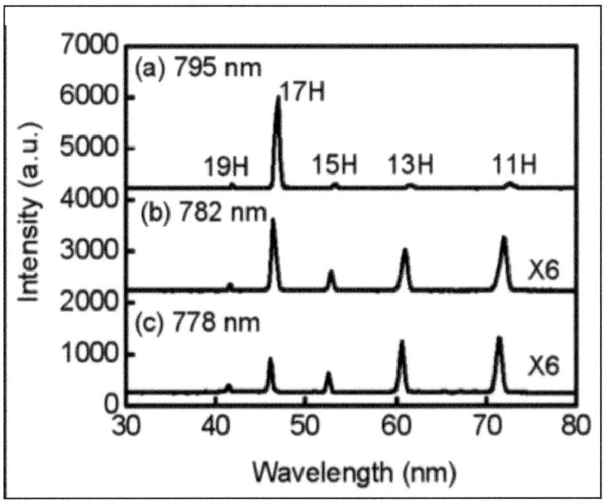

Fig. 8. HHG spectra from tin laser ablation for pump laser with central wavelength of (a) 795 nm, (b) 782 nm, and (c) 778 nm. The intensity of the spectra (b) and (c) are multiplied by 6 times. (Suzuki et al., 2006)

Fig. 9 shows the typical spectra of HHG from the laser ablation indium plume. In this experiment, the indium plasma was produced by a low-energy laser pulse, instead of the conventional gas medium. Exceptionally strong 13th harmonic at a wavelength of 61.26 nm have obtained as can be seen in Fig. 9. Using a 10 mJ energy Ti:sapphire laser pulse at a wavelength of 796.5 nm, the conversion efficiency of the 13th harmonic at a wavelength of 61 nm was about 8×10^{-5}, which was two orders of magnitude higher than its neighboring harmonics. The output energy of the 13th harmonic was measured to be 0.8 μJ. A cut-off of the 31st order at a wavelength of 25.69 nm has observed in this experiment.
For indium, the $4^d10^5s^2\ ^1S_0$ - $4d^95s^25p\ (^2D)\ ^1P_1$ transition of In II, which has an absorption oscillator strength (gf–value) of 1.11 (Duffy et al., 2001)[30], can be driven in to resonance with the 13th order harmonic. Fig. 10 shows the HHG spectra at the wavelength of 796 and 782 nm. The intensity of the 13th harmonic for indium is attributed to such resonance of the harmonic wavelength with that of a strong radiative transition. By changing the laser wavelength from 796 nm to 782 nm, the 15th harmonic at the wavelength of 52.13 nm increased, and the intensity of the 13th harmonic decreased at the same time. The reason of the 15th harmonic enhancement is due to resonance with the $4d^{10}5s5p\ ^3P_2$ - $4d^95s5p^2\ (^2P)\ ^3F_3$ transition of In II, which has a gf-value of 0.30. The enhancement of the 15th order harmonic intensity is lower than that of the 13th harmonic because the gf-value of $4d^{10}5s5p\ ^3P_2$ - $4d^95s5p^2\ (^2P)\ ^3F_3$ transition is lower than that of the $4d^{10}5s^2\ ^1S_0$ - $4d^95s^25p\ (^2D)\ ^1P_1$ transition.

Furthermore the central wavelength of the 13th harmonic was driven away from resonance with the $4d^{10}5s^2$ $^1S_0$ - $4d^95s^25p$ (^2D) $^1P_1$ transition when using 782 nm wavelength laser, thereby decreasing the 13th order harmonics.

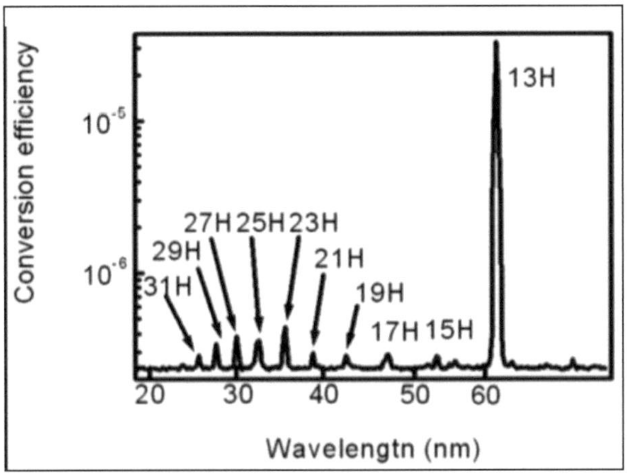

Fig. 9. Spectrum of the HHG from the laser ablation indium plume. The conversion efficiency is 8×10^{-5}.

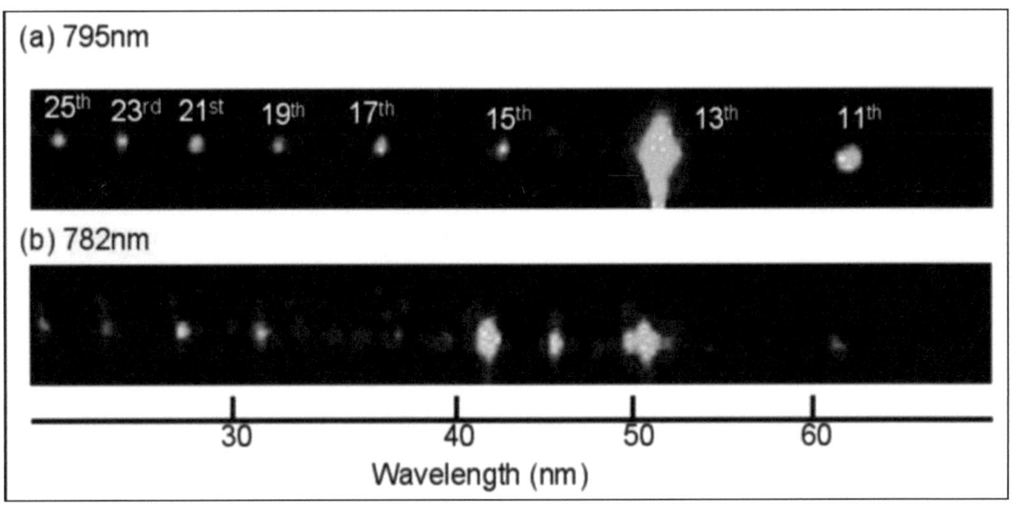

Fig. 10. HHG spectra from indium laser ablation for pump laser with central wavelength of (a) 795 nm, (b) 782 nm. The intensity of the 13th harmonic is two orders of magnitude higher than its neighboring harmonics.

There have been several discussions on the reason for this intensity enhancement of a single harmonic order. Taïeb et al. (Taieb et al., 2003a) have shown theoretically that if there is resonance between a specific harmonic order and a radiative transition, then considerable population could result on the upper state of the transition. Since electrons that ionizes from this excited state has a non-zero initial velocity, the electron driven by the laser electric field could recollide multiple times with the parent atom, thus increasing the harmonic emission. There is also work that explains the phenomenon to a better phase matching condition under the presence of a strong radiative transition (Elouga-Bom et al., 2008). A strong absorption line will greatly modify the index of refraction near its wavelength. Under appropriate conditions, this could greatly improve the coherence length of a harmonic order close to this absorption, thus greatly increasing the intensity of this harmonic. Simulations using the actual parameters for indium plasma have shown that this theory explains well the intense 13th harmonic of indium.

5. High-order harmonics from nanostructured material

5.1 Silver nanoparticles

First, we performed harmonic generation experiments using silver nanoparticles glued on various substrates. We observed the nanoparticles used in this experiment with a scanning tunneling microscope, and we confirmed that their size varied between 90 to 110 nm. We initially verified that harmonics generated from the substrates themselves (glue, tape and glass) without the nanoparticles, was negligible compared with those from silver plasma. We fabricated the target so that a slab silver target was next to the nanoparticle target, with the two target surfaces at the same height. This target was placed on to the target holder, so that they interacted with both the prepulse and main pump laser at the same intensities. First, the prepulse and main pulse was aligned using a solid silver target, to search for conditions for maximum harmonic intensity within the plateau. Next, the target was translated so that the prepulse beam now irradiates the Ag nanoparticle target.

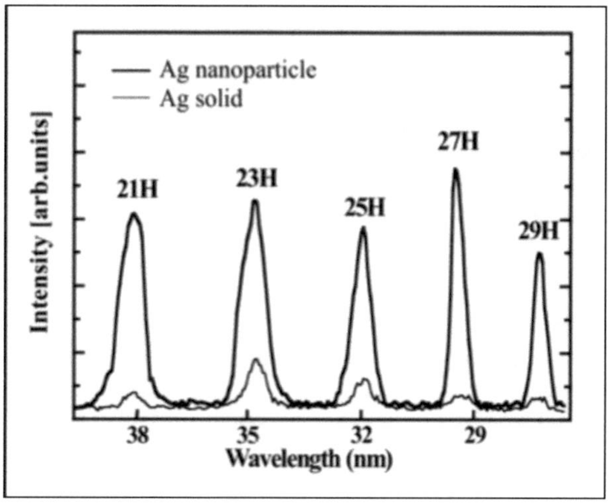

Fig. 11. Harmonic distribution in mid-plateau region for produced from bulk Ag target (thin lineout) and Ag nanoparticle plasma (thick lineout).

We compared the harmonic yield for silver nanoparticle targets with those from bulk silver targets, under the same prepulse and main pulse conditions. Fig. 11 shows the lineout of the harmonic spectra between the 21st and the 29th harmonics within the plateau. One clearly sees that the HHG intensity from the nanoparticle target was more than six times higher compared with that from bulk silver target. We can estimate the energy of these harmonics based on calibrations we have performed using longer (130 fs) pulses (Ganeev et al., 2005a). For 130 fs pump lasers, we have measured a conversion efficiency of 8×10^{-6} for bulk silver target. This would be a conservative estimate of the conversion efficiency for bulk silver targets in the present work, which uses shorter 35 fs pulses. We therefore estimate a minimum harmonic conversion efficiency of 4×10^{-5} from silver nanoparticles within the plateau region. For the maximum main pump laser energy of 25 mJ used, the energy of the 21st to the 29th harmonics is evaluated to be more than 1 μJ.

When we compare the cut-off observed for harmonics from nanoparticle and slab silver targets, we also noted a slight extension of the harmonic cut-off for nanoparticles (Fig. 12). Harmonics up to the 67th order (103 eV photon energy) was observed in these studies with silver nanoparticles, while, for bulk silver target, the cut-off was at the 61st order (94 eV photon energy) under the same conditions. This slight extension of harmonic cut-off agrees with past observations, which noted similar extension in the cut-off for argon clusters, compared with isolated atoms (Donnelly et al., 1996)[19]. This difference has been explained by the increase in the effective binding energy of electrons in the cluster. The higher binding energy will allow the cluster to interact with laser intensities that are much higher than for isolated atoms, resulting in the extended cut-off for the former. In past works with Ar (Donnelly et al., 1996), the cut-off for clusters was at the 33rd order, compared with the 29th order cut-off for monomer harmonics.

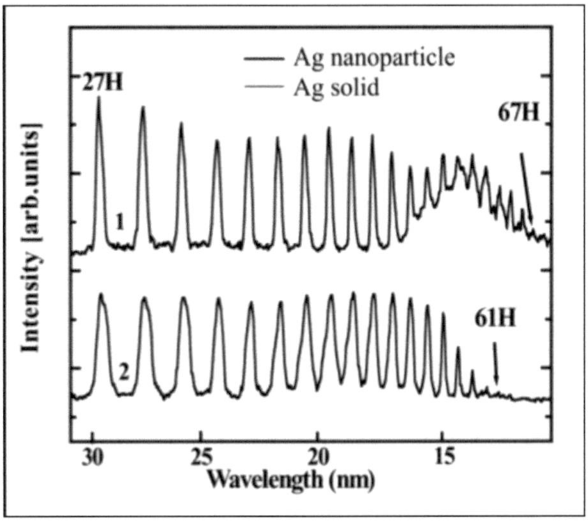

Fig. 12. High-order harmonic spectra generated from (1) silver nanoparticle plasma, and (2) plasma produced from bulk silver target.

Next, we studied the dependence of the harmonic yield on the pump intensity. However, the measurement was made difficult by the rapid shot-to-shot change in the harmonic intensity from Ag nanoparticle target. For experiments with solid slab targets, stable harmonic generation can be obtained for about ten minutes at 10 Hz repetition rates, without translating for a new target surface. However, for nanoparticle targets, the harmonics were strong for the first few shots, which were followed by a rapid decrease in harmonic yield when the plasma was created at the same target position. We attribute this effect to evaporation of the thin layer of nanoparticles. The first shot results in a strong harmonic spectrum, with the typical plateau-like structure starting from the 17th order. Then, for the second and third shots, the intensity of the harmonics decreased drastically, and, for the fourth shot and after, the harmonics almost disappeared. We repeated the experiments with nanoparticles many times, revealing the same feature. We also observed that, when we used different material as the substrate, there was a different behavior of the shot-to-shot decrease in harmonic yield. Another interesting feature found in the experiments with nanoparticle targets was that the prepulse intensity necessary for HHG was lower than that used for bulk targets.

These observations give us a rough picture of the ablation for nanoparticle targets. The material directly surrounding the nanoparticles is polymer (epoxy glue), which has a lower ablation threshold than metallic materials. Therefore, the polymer starts to ablate at relatively low intensities, carrying the nanoparticle with it, resulting in the lower prepulse intensity. Polymer also has a lower melting temperature than metals. Therefore, repetitive irradiation of the target leads to melting and change in the properties of the target. This results in the change in conditions of the plasma plume, resulting in a rapid decrease in the harmonic intensity with increased shots. The different shot-to-shot harmonic intensities for different substrates can be explained by the different adhesion properties of nanoparticles to the substrate.

Due to such rapid change in the conditions for harmonic generation with nanoparticle targets, it was difficult to define precisely the dependence of harmonic yield on prepulse and main pulse intensities. Nevertheless, approximate measurements of the dependence of harmonic yield on the main pulse intensity for Ag nanoparticles have shown a saturation of this process at relatively moderate intensities ($I_{fp} \approx 8\times10^{14}$ W cm^{-2}).

Harmonics from plasma nanoparticles also displayed several characteristics similar to gas harmonics. First, the harmonic intensity decreased exponentially for the lower orders, followed by a plateau, and finally a cut-off. Next, the harmonic intensity was strongly influenced by the focus position of the main pump laser, along the direction parallel to the harmonic emission. The strongest harmonic yield was obtained when the main pump laser was focused 4 to 5 mm after the nonlinear medium. We observed the same tendency of the harmonics using bulk silver target. The typical intensity of the pump laser for maximum harmonic yield was between 5×10^{14} to 2×10^{15} W cm^{-2}. These results agree with those of gas harmonics (Lindner et al., 2003), and are due to the selective short-trajectory-generated harmonics when the pump laser is focused after the medium. Harmonics from short-trajectories have a flat and large area on-axis, with excellent phase matching conditions, resulting in the higher harmonic yield. In our case, we needed to focus the pump laser away from the medium, since the total intensity that would be produced at focus would exceed the barrier suppression intensity for multiply charged ions. This would result in over-ionization of the plasma, leading to the decrease in the harmonic yield.

To study the size effect of nanoparticles, we performed harmonic generation experiments using colloidal silver targets, which are contains blocks of silver with sizes between 100 to 1000 nm. We confirmed the size of the silver blocks by viewing with a scanning tunneling microscope. The results showed that the harmonic yield for these sub-μm-sized silver blocks was much lower than that from nanoparticles, and was comparable to those from bulk silver targets. We also noted a tendency of slightly extended harmonic cut-off for smaller particle sizes. The cut-offs for the harmonics were at the 61st, 63rd and 67th order, for bulk silver, sub-μm silver colloid and silver nanoparticle targets, respectively.

These studies have shown that the increasing the particle size over some limit is undesirable due to the disappearance of enhancement-inducing processes. The observed enhancement of harmonic yield for plasma plume with 90 to 110 nm size nanoparticles can probably be further improved by using smaller nanoparticles.

These experiments show that the size of the nanoparticles is of essential importance for harmonic generation. To gain maximum HHG conversion efficiency, it is essential to know the maximum tolerable particle size for increased harmonic yield. On one hand, increasing the size of the particles increases its polarizability, and large polarizability of a medium is critical for efficient harmonic generation (Liang et al., 1994). On the other hand, the increase in particle size leads to phenomena that reduces harmonic yield (such as HHG only from surface atoms (Toma et al., 1999), and reabsorption of harmonics).

The increased HHG efficiency for silver nanoparticles might also be an important factor for explaining the high conversion efficiency of HHG from plasma produced from bulk silver targets. Silver has been known to be a highly efficient material for plasma HHG, but up to now the reason was not clear (Taieb et al., 2003a). However, it is known that nanoclusters (such as Ag_2 and Ag_8) and nanoparticles are abundantly produced by laser ablation. Since our laser plume expanded adiabatically for 100 ns before irradiation by the main pulse, one can expect that the silver plume from bulk silver target also contained many nanomaterials, which would contribute to increasing the HHG efficiency.

5.2 Other nanoparticles

To study what parameters affect the strong harmonics from nanoparticles, we next performed experiments using nanoparticles of different materials. An example of the harmonic spectrum from chromium oxide (Cr_2O_3) nanoparticle target is shown in Fig. 13(a). The spectrum from nanoparticle targets showed a featureless plateau with a cut-off at the 31st harmonic, with harmonic yield that is much stronger than those from bulk Cr_2O_3 targets (Fig. 13(b)). Another important observation is that the relative intensities between harmonic orders differ for different targets. For nanoparticle targets, the harmonic spectrum resembles those observed from gas, with a plateau followed by a cut-off. However, harmonics from bulk Cr_2O_3 target has a characteristic enhancement of the 29th order, and a cut-off at the 35th harmonic, which has also been observed in previous studies of HHG in chromium plasma (Ganeev et al., 2005c)[24]. For bulk chromium oxide targets, the 29th harmonic is about 10 times stronger than the lower 27th harmonic. Such enhancement was not observed with Cr_2O_3 nanoparticle targets at moderate prepulse intensities (5×10^9 W cm-2). We should note that by further increasing the prepulse intensity to 9×10^9 W cm^{-2}, we could generate intense 29th harmonic from Cr_2O_3 nanoparticle targets. This is a sign of ionization of the nanoparticles in the plasma, since enhanced single harmonic in chromium has previously been attributed to the proximity of the 29th harmonic with the giant 3p - 3d ionic transitions of singly ionized chromium ions (Ganeev et al., 2005c). The delay between the prepulse and main pulse in these experiments was kept at 25 ns.

High-order harmonics from other nanoparticles also showed similar features, with a notable enhancement of low-order harmonics at the plateau and a decrease in the harmonic cut-off compared with harmonics using bulk targets. For example, Fig. 13(c) and (d) show the harmonic spectrum for manganese titanium oxide ($MnTiO_3$) nanoparticles and bulk targets, respectively. The $MnTiO_3$ nanoparticles show relatively strong 19th and 21st harmonics, with a cut-off at the 25th order, whereas the bulk MnTiO3 targets show only weak harmonics that are comparable to noise. Increasing the femtosecond pump intensity did not lead to extension of the harmonic cut-off for nanoparticle targets, which is a sign of saturation of the

HHG in these media. Also, at relatively high femtosecond pump intensities, we noted a decrease in the harmonic conversion efficiency due to the onset of negative effects (such as increase in the free electron density, self-defocusing and phase mismatch). Similar effects were also observed when we increased the prepulse intensity, which is attributed to the increase in the free electron density of the plasma, resulting in phase mismatch.

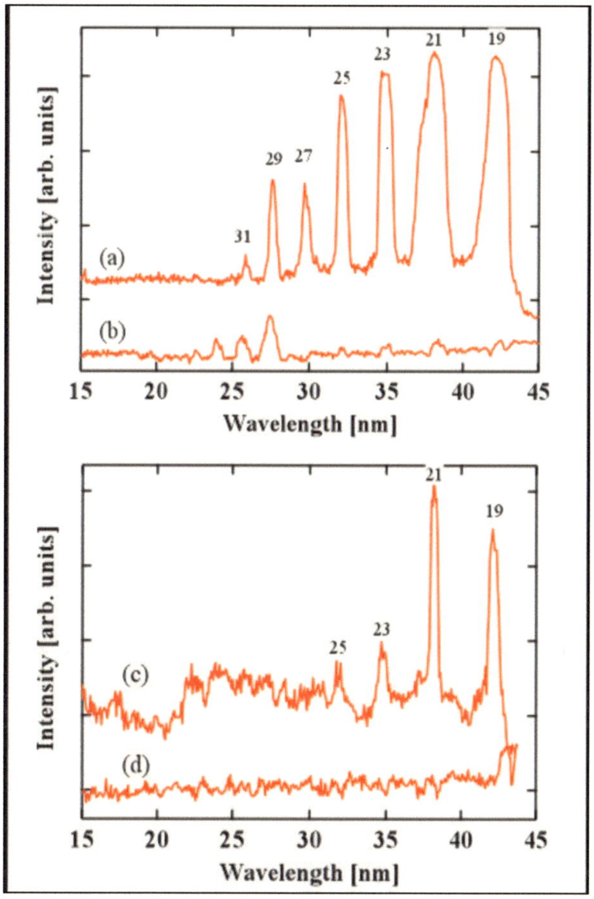

Fig. 13. Harmonic spectrum for (a) chromium oxide nanoparticles, (b) chromium oxide bulk, (c) manganese titanium oxide nanoparticles, and (d) manganese titanium oxide bulk targets.

A comparison of the low-order harmonic generation using the single atoms and multiparticle aggregates has previously been reported for Ar atoms and clusters (Donnelly et al., 1996)[7]. It was demonstrated that a medium of intermediate-sized clusters with a few thousand atoms of an inert gas has a higher efficiency for generating the harmonics, compared with a medium of isolated gas atoms of the same density. The reported enhancement factor for the 3rd to 9th harmonics from gas jets was about 5. In our HHG experiments with the laser-ablated nanoparticles, these observations were extended toward the higher-order harmonics and stronger enhancement for the harmonics up to the 25th order was achieved. These results have also shown that the dependence of the HHG efficiency on the prepulse and main pulse intensity is much more prominent for nanoparticles than for monatomic particles.

Since nanoparticles are smaller than the laser wavelength, they contain many equivalent, optically active electrons at effectively the same point in the laser field. This leads to the possibility that each of these electron oscillators may contribute coherently to a global nanoparticle dipole. However, this statement is true only for low-order harmonics. For high-order harmonic generation (such as those considered in this paper), the dipole approximation is inapplicable, because the harmonic radiation wavelength is shorter than the size of nanoparticles (about 100 nm).

We would like to point out that in our experiments with nanoparticles, the intense harmonics were observed (i) only for lower orders, (ii) when the intensity of the picosecond prepulse (which generates the plasma plume) was moderate. When the prepulse intensity was increased, phenomena that are explained by the presence of ions appeared. For example, enhancement of the 29th harmonic in chromium is related to the giant 3p - 3d ionic transitions of Cr^+, which started to appear for Cr_2O_3 nanoparticles when the prepulse intensity was increased to 9×109 W cm^{-2}. These results suggest that one major reason for the intense harmonics from nanoparticles is the contribution from neutral atoms. Since neutral atoms are larger compared with its ions, the recombination probability of the electron wave packet that was liberated by the laser electric field is also larger for neutral atoms. As a result, the neutral atoms emit stronger harmonics than ions, but with a lower cut-off due to its lower ionization potential.

5.3 C_{60} fullerenes

5.3.1 Harmonic generation from C_{60} fullerenes

A problem with experiments using nanoparticles is that there is always a distribution in their size and shape. Since phenomena such as ionization and nonlinear response to intense laser fields should vary with nanoparticle dimensions, it becomes difficult to determine how the various characteristics of the nanostructured material affect harmonic generation. To study HHG from a more uniform nano-material, we decided to next explore C_{60} fullerenes. In our previous experiments, we demonstrated HHG from laser-produced plasma of fullerene targets (Ganeev et al., 2009). In that work, we showed that (i) the harmonics lying within the spectral range of SPR in C_{60} (20 - 22 eV) are enhanced, (ii) the harmonic efficiency from C_{60} targets are 20 to 25 times larger for the 13th harmonic compared with those generated from carbon monomer rich plasma, and (iii) the harmonic cut-off in C_{60} is lower (19th order) than carbon but extends beyond the value (11th order) predicted by the three-step model. Here, we present a more detailed account of HHG from C_{60} fullerenes.

Fig. 6 shows the harmonic spectra from C_{60} for different delays between the pump pulse and the femtosecond driving pulse. HHG by ablation of bulk materials is influenced by the temporal delay between the pump pulse and driving pulse, as it results in a change in the atomic density and plasma length of the nonlinear medium. To study their effects on the harmonic intensity, we varied the delay from 18 ns to 100 ns. Our measurements showed no significant changes in the harmonic intensities in C_{60} (see Fig. 13(a) and (b)) for delays of 22 ns and 63 ns, with some two-fold increase of harmonic efficiency for the shorter delay. By comparing with calibrated harmonics from silver plasma (Ganeev, 2007), we estimate the efficiency of the 13th harmonic from fullerene plasma to be near 10-4.

However, for bulk targets such as C, Cr and Mn, no harmonics were observed from plasmas when we used the shorter delays, which is contrary to the case of C_{60}. This can be attributed to the non-optimal plasma conditions, since it requires time for the plasma to ablate on the bulk surface and expand into the area where the femtosecond beam interacts with the plasma. This can also be inferred from the lower pump pulse intensity ($I_{pp} \sim 2\times10^{9}$ W cm^{-2}) needed for HHG from C_{60}-rich target, compared with that needed for bulk targets [$I_{pp} > 10^{10}$ W cm^{-2}]. We believe that short delays lead to more favorable evaporation conditions and higher particle density for the cluster-rich medium compared with the monatomic medium, thus resulting in a higher harmonic yield. Usually for heavy bulk targets, the strong harmonics were observed using longer delays (40-70 ns). The use of light targets (B, Be, Li) showed an opposite tendency, where one can obtain effective HHG for shorter delays. The optimization is related to the presence of a proper density of particles within the volume where harmonics are generated, which depends on the propagation velocity of the plasma front. For C_{60}, one can expect to optimize HHG at longer delays due to the larger weight of the fullerene particles. However, one also needs to take into account the possibility of the presence of the fragments of C_{60} in the plume, in which case, the density of the medium within the laser-interaction region becomes sufficient even for shorter delays.

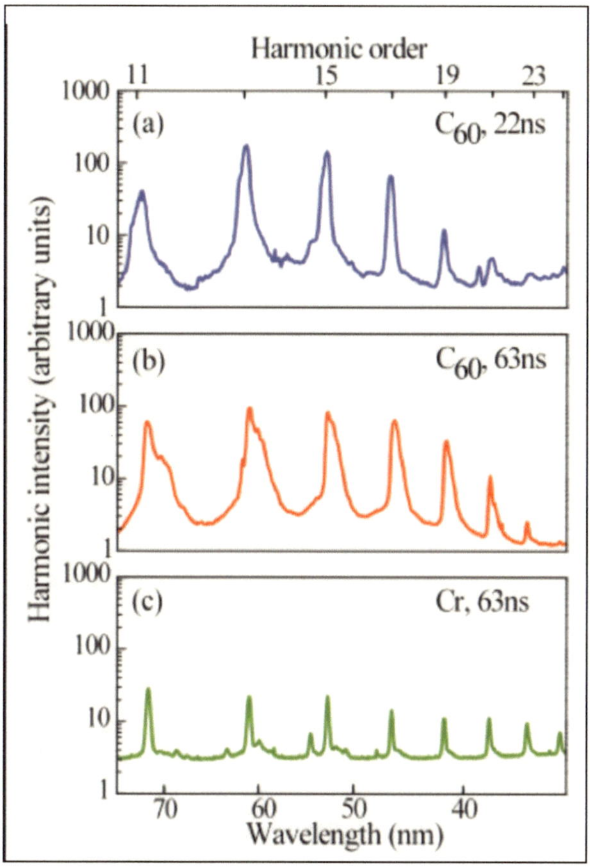

Fig. 13. Harmonic generation observed in C_{60} plasma at (a) 22 ns and (b) 63 ns delays between the prepulse and main pulse and (c) in chromium plasma

An interesting feature of the fullerene harmonic spectra is that the spectral width is about three to four times broader compared with those generated in plasma rich with monatomic particles (1.2 nm and 0.3 nm FWHM, respectively). For comparison, Fig. 13(c) shows the harmonic spectra for Cr bulk targets. Broader width of the harmonics can be explained by self-phase modulation and chirping of the fundamental radiation propagating through the fullerene plasma. Broadening of the main beam bandwidth causes the broadening of the harmonic's bandwidth. Increase in the harmonic bandwidth with delay can be explained by the longer length of the fullerene plasma for the longer delay, and thus stronger self-phase modulation of the femtosecond pump laser.

The intensities of the pump pulse and driving pulse are crucial for optimizing the HHG from C_{60}. Increasing the intensity of the driving pulse did not lead to an extension of the cut-off for the fullerene plasma, which is a sign of HHG saturation in the medium. Moreover, at relatively high femtosecond laser intensities, we observed a decrease in the harmonic output, which can be ascribed to phase mismatch resulting from higher free electron density. We observe a similar phenomenon when the pump pulse intensity on the surface of fullerene-rich targets is increased above the optimal value for harmonic generation. This reduction in harmonic intensity can be attributed to phenomena such as the fragmentation of fullerenes, an increase in free electron density, and self-defocusing. At relatively strong ablation intensity for fullerene film ($I_{pp} > 1\times10^{10}$ W cm^{-2}), we observed only the plasma spectrum, without any sign of harmonics.

The stability of C_{60} molecules to ionization and fragmentation is of particular interest, especially for their application as a medium for HHG. The structural integrity of the fullerenes ablated off the surface should be intact until the driving pulse arrives. Therefore, the pump pulse intensity is a sensitive parameter. At lower intensities the density of clusters in the ablation plume would be low, while at higher intensities one can expect fragmentation. C60 has demonstrated both direct and delayed ionization and fragmentation processes and is known to survive even in intense laser fields. This can be attributed to the large number of internal degrees of freedom that leads to the fast diffusion of the excitation energy (Bhardwaj et al., 2003). At 796 nm, multiphoton ionization is the dominant mechanism leading to the ionization of C_{60} in a strong laser field. The collective motion of π electrons of C_{60} can be excited by multiphoton process. Since the laser frequency is much smaller than the resonance frequency of π electrons, barrier suppression and multiphoton ionization are the dominant mechanisms leading to the ionization in a strong laser field.

Another important parameter that affects the stability of HHG process is the thickness of the fullerene target. We obtained stable harmonic generation with low shot-to-shot variation in harmonic intensity by moving the fullerene film deposited on the glass substrate after several laser shots. This avoids decrease in the fullerene density due to ablation of the thin film. The number of laser shots at the same target position that resulted in stable harmonic emission decreased drastically with the film thickness. For example, in a 10- μm film, the harmonic emission disappeared after 70 to 90 shots, whereas in a 2-μm film, the harmonics disappeared after 5 to 7 shots.

To understand the origin of the harmonic emission in C_{60}, we studied its dependence on the polarization of the main pulse. This also allows one to distinguish the plasma emission from the HHG. HHG is highly sensitive to laser polarization, since the trajectories of the recolliding electrons are altered significantly for elliptically polarized pump lasers, thus inhibiting the recombination process. We noted that the harmonic signal drop rapidly and disappear with ellipticity of the laser polarization. For circular polarization, as expected, the harmonic emission disappears and the resulting background spectrum corresponds to the plasma emission.

Does the influence of plasmon resonance on the HHG in fullerene plasma depend on the wavelength of the driving field? To address this question, we also studied HHG using the second harmonic (396 nm, 4 mJ, 35 fs) of the main pulse (793 nm, 30 mJ). The low second harmonic conversion efficiency did not allow us to achieve the laser intensities reached with the 793 nm fundamental laser. As a result, we were able to generate harmonics up to the 9th order of the 396 nm driving pulse, while simultaneously generating harmonics using the 793 nm laser. Harmonic generation using two main pulses (793 nm and 396 nm) did not interfere with each other, due to different focal positions of these two beams (~2 mm in the Z-axis and ~0.2 mm in the X-axis). Therefore, the two HHG processes occurred in different regions of the laser plasma. Here, the Z-axis is the axis of propagation of the driving beam, and the X-axis is the axis vertical to the Z-axis. This axis is defined by the walk-off direction of the second harmonic with respect to the fundamental driving pulse.

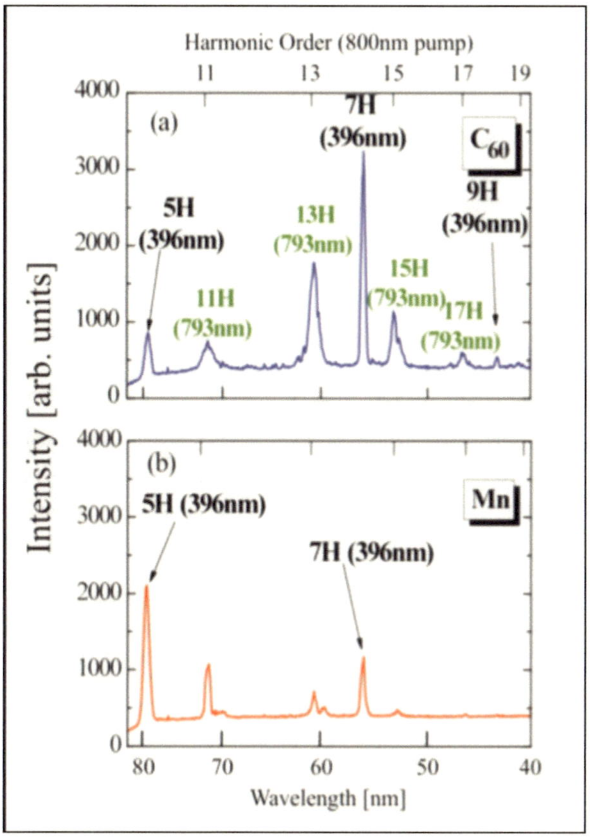

Fig. 14. Harmonic spectra from (a) C_{60} and (b) Mn plasma, when both the 793 nm and 396 nm laser were simultaneously focused on the laser-produced plasma.

Fig. 14(a) shows the HHG spectrum from C_{60} fullerene optimized for the second harmonic driving pulse. The energy of the second harmonic is ~1/7th of the fundamental. One can see the enhancement of the 7th harmonic (which is within the spectral range of the SPR of C_{60}) compared with the 5th harmonic. This behavior is similar to that observed for the 793 nm driving pulse. For comparison, we present in Fig. 14(b) the optimized harmonics generated using the 396 nm pump and the weak harmonics from the 793 nm radiation in manganese plasma. One can see a decrease in harmonic intensity from the Mn plasma for each subsequent order, which is a common case, when one uses a nonlinear optical medium containing atomic or ionic particles. These studies confirmed that, independent of the driving pulse wavelength, the harmonics near SPR in C_{60} are enhanced.

5.3.2 Simulations of C_{60} harmonic spectra

To understand the influence of the absorptive properties of surface plasmon resonance on the harmonic emission spectrum in C_{60}, we simulated the emission spectrum using parameters that are roughly identical to those used in experiments. The HHG efficiency can be understood by three length parameters. For optimum HHG, the length of the nonlinear medium L_{med} should be (a) larger than the coherence length $L_{coh} = \pi/\Delta k$, which is defined by the phase mismatch between the fundamental and harmonic fields ($\Delta k = k_q - qk_0$ where k_q and k_0 are the harmonic and fundamental wave vectors, respectively) and depends on the density and ionization conditions, and (b) smaller than the absorption length of the medium $L_{abs} = 1/\rho\sigma$, where ρ is the atomic density and σ is the ionization cross-section.

The photoionization cross-section of C_{60} is well known, both experimentally and theoretically. It displays a giant and broad plasmon resonance at ~ 20 eV (around the 11th, 13th and 15th harmonics, with a bandwidth of 10 eV FWHM). We calculated the absorption length using the estimated fullerene density in the interaction region (5×10^{16} cm^{-3}) and the known photoionization cross-sections. The absorption length varies from 0.8 mm (for the 7th and 17th harmonic) to 0.3 mm (for the 11th, 13th and 15th harmonic), suggesting that harmonics near the plasmon resonance should be more strongly absorbed in the medium (whose length is estimated to be about 0.8 - 1 mm). Due to this increased absorption in C_{60}, we expect a dip in the harmonic spectrum for the 11th - 15th harmonics. Our calculations also point out that harmonics produced in bulk carbon target are not absorbed by the nonlinear medium. With an assumed medium length of 1 mm, theoretical spectra are obtained by using the proper wavelength-dependent index of refraction and dispersion data.

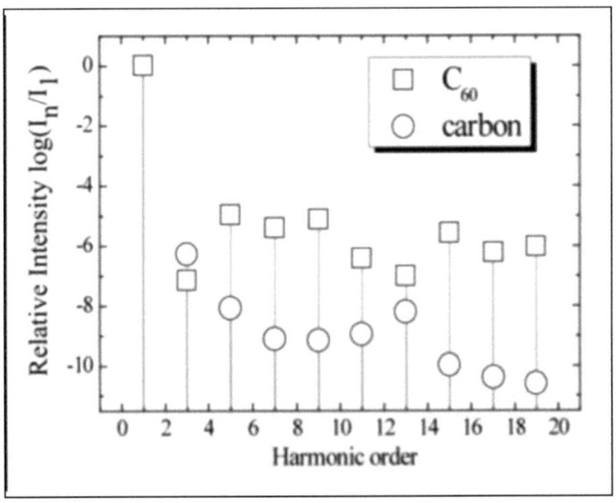

Fig. 15. Calculated relative intensities of harmonics generated from neutral carbon mono-atom and C_{60} fullerene molecule.

From our calculations, we find that for bulk carbon, the influence of absorption on the harmonic yield is negligible and as a result the overall harmonic spectrum is determined by dispersion. The harmonic yield decreases with increasing order as it becomes difficult to phase match higher orders. In C_{60}, absorption of harmonics by the nonlinear medium is dominant due to large photoabsorption cross-sections. The effect of dispersion only lowers the HHG efficiency but does not affect the overall shape of the spectrum. As a result, one expects the harmonic yield to decrease considerably near the surface plasmon resonance, if one does not consider the nonlinear optical influence of this resonance on the harmonic efficiency in this medium. On the contrary, in our experiment, we observed a notable enhancement of these harmonics in the fullerene-rich plume (Figs. 13 and 14). This is a signature of multi-electron dynamics in a complex molecule such as C_{60} and has no atomic analogue.

To understand the origin of enhancement of harmonic yield near SPR, we theoretically studied the interaction of monatomic carbon and fullerene C_{60} molecule with a strong laser pulse by the time-dependent density functional theory (TDDFT) (Runge & Gross, 1984). In the TDDFT approach, the many-body time-dependent wave-function is replaced by the time-dependent density, which is a simple function of the three-dimensional vector r. n(r,t) is obtained with the help of a fictitious system of non-interacting electrons by solving the time-dependent Kohn-Sham equations. These are one-particle equations, so it is possible to treat large systems such as fullerenes. For all calculations we used the OCTOPUS code (Marques et al., 2003) with norm-conserving non-local Troullier-Martins pseudopotentials (Troullier & Martins, 1991), Slater exchange, Perdew and Zunger correlation functionals (Perdew & Zunger, 1981) and grid spacing of 0.6 Å for parallelepiped box of 8×8×60 Å.

We analyzed the relative harmonic intensities calculated for C_{60} and bulk carbon (Fig. 8). A significant increase in HHG efficiency for C_{60} molecule can be attributed to additional oscillation of the time-dependent dipole in the C_{60} molecule. This can be a sign of an induced collective plasmon-like response of the molecule to external field. At the same time the cut-off for the carbon atom is higher than that for a fullerene molecule. Treating relatively high-order harmonics with our simulation codes can become inaccurate, due to an exponential cut-off of the exchange and correlation potential. The effects of correlation for lower harmonics are nevertheless conserved, so a collective oscillation can be responsible for the relative increase of the time-dependent dipole and, respectively, HHG conversion efficiency observed in plasma of fullerene molecules.

6. Conclusion

In this chapter, we have reviewed recent developments in the generation of intense high-order harmonics using lowly ionized plasma as the nonlinear medium. We have shown recent results that demonstrate clear plateau with a cut-off as high as the 101st order. A unique intensity enhancement of a single high-order harmonic that dominates the spectrum has been observed in the XUV region. Such enhancement of a single harmonic is an advantage for several important applications of coherent XUV radiation. One example is photoelectron spectroscopy for understanding excited-state electron dynamics, in which there is a need to select one harmonic while eliminating neighboring harmonics. This single enhancement technique will allow the generation of a quasi-monochromatic coherent x-ray source, without complexities of using monochromators, multilayer mirrors and filters. Another characteristic of this plasma harmonic method is that one could use any target that could be fabricated into solids. Profiting from this feature, there has been several works on pioneering high-order harmonic generation from nanostructured material, such as metallic nanoparticles and fullerenes.

7. References

Akiyama, Y., Midorikawa, K., Matsunawa, Y., Nagata, Y., Obara, M., Tashiro, H. & Toyoda, K. (1992). Generation of high-order harmonics using laser-produced rare-gas-like ions *Phys. Rev. Lett.* 69, 2176-2179.

Bhardwaj, V., Corkum, P. & Rayner, D. (2003). Internal laser-induced dipole force at work in c-60 molecule *Phys Rev Lett* 91, 203004.

Constant, E., Garzella, D., Breger, P., Mevel, E., Dorrer, C., Le Blanc, C., Salin, F. & Agostini, P. (1999). Optimizing high harmonic generation in absorbing gases: Model and experiment *Physical Review Letters* 82, 1668-1671.

Corkum, P. (1993). Plasma perspective on strong-field multiphoton ionization *Phys. Rev. Lett.* 71, 1994-1997.

Donnelly, T., Ditmire, T., Neuman, K., Perry, M. & Falcone, R. (1996). High-order harmonic generation in atom clusters *Phys. Rev. Lett.* 76, 2472-2475.

Drescher, M., Hentschel, M., Kienberger, R., Tempea, G., Spielmann, C., Reider, G., Corkum, P. & Krausz, F. (2001). X-ray pulses approaching the attosecond frontier 291, 1923-1927.

Duffy, G. & Dunne, P. (2001). The photoabsorption spectrum of an indium laser produced plasma *J Phys B-At Mol Opt* 34, L173-L178.

Duffy, G., Van Kampen, P. & Dunne, P. (2001). 4d -> 5p transitions in the extreme ultraviolet photoabsorption spectra of snii and sniii *J Phys B-At Mol Opt* 34, 3171-3178.

Elouga-Bom, L.-B., Bouzid, F., Vidal, F., Kieffer, J.-C. & Ozaki, T. (2008). Correlation of plasma ion densities and phase matching with the intensities of strong single high-order harmonics *J Phys B-At Mol Opt* 41, 215401.

Faria, C. F. D., Kopold, R., Becker, W. & Rost, J. M. (2002). Resonant enhancements of high-order harmonic generation *Physical Review A* 65, -.

Gaarde, M. B. & Schafer, K. J. (2001). Enhancement of many high-order harmonics via a single multiphoton resonance *Physical Review A* 6401, -.

Ganeev, R., Baba, M., Suzuki, M. & Kuroda, H. (2005a). High-order harmonic generation from silver plasma *Physics Letters A* 339, 103-109.

Ganeev, R., Suzuki, M., Baba, M. & Kuroda, H. (2005b). Generation of strong coherent extreme ultraviolet radiation from the laser plasma produced on the surface of solid targets *Appl Phys B-Lasers O* 81, 1081-1089.

Ganeev, R., Suzuki, M., Baba, M. & Kuroda, H. (2005c). Harmonic generation from chromium plasma *Appl Phys Lett* 86, 131116.

Ganeev, R., Suzuki, M., Baba, M., Kuroda, H. & Ozaki, T. (2005d). High-order harmonic generation from boron plasma in the extreme-ultraviolet range *Optics Letters*.

Ganeev, R., Suzuki, M., Baba, M., Kuroda, H. & Ozaki, T. (2006). Strong resonance enhancement of a single harmonic generated in the extreme ultraviolet range *Optics Letters* 31, 1699-1701.

Ganeev, R. A. (2007). High-order harmonic generation in a laser plasma: A review of recent achievements *J Phys B-At Mol Opt* 40, R213-R253.

Ganeev, R. A., Bom, L. B. E., Abdul-Hadi, J., Wong, M. C. H., Brichta, J. P., Bhardwaj, V. R. & Ozaki, T. (2009). Higher-order harmonic generation from fullerene by means of the plasma harmonic method *Physical Review Letters* 102, 013903.

Ganeev, R. A., Bom, L. B. E., Kieffer, J.-C., Suzuki, M., Kuroda, H. & Ozaki, T. (2007). Demonstration of the 101st harmonic generated from a laser-produced manganese plasma *Phys Rev A* 76, 023831.

Goulielmakis, E., Schultze, M., Hofstetter, M., Yakovlev, V. S., Gagnon, J., Uiberacker, M., Aquila, A. L., Gullikson, E. M., Attwood, D. T., Kienberger, R., Krausz, F. & Kleineberg, U. (2008). Single-cycle nonlinear optics *Science* 320, 1614-1617.

Kazamias, S., Douillet, D., Weihe, F., Valentin, C., Rousse, A., Sebban, S., Grillon, G., Auge, F., Hulin, D. & Balcou, P. (2003). Global optimization of high harmonic generation *Phys. Rev. Lett.* 90, 193901.

Kim, I. J., Kim, C. M., Kim, H. T., Lee, G. H., Lee, Y. S., Park, J. Y., Cho, D. J. & Nam, C. H. (2005). Highly efficient high-harmonic generation in an orthogonally polarized two-color laser field *Physical Review Letters* 94, -.

Kling, M. F., Siedschlag, C., Verhoef, A. J., Khan, J. I., Schultze, M., Uphues, T., Ni, Y., Uiberacker, M., Drescher, M., Krausz, F. & Vrakking, M. J. J. (2006). Control of electron localization in molecular dissociation *Science* 312, 246-248.

Kubodera, S., Nagata, Y., Akiyama, Y., Midorikawa, K., Obara, M., Tashiro, H. & Toyoda, K. (1993). High-order harmonic-generation in laser-produced ions *Phys Rev A* 48, 4576-4582.

Liang, Y., Augst, S., Chin, S. L., Beaudoin, Y. & Chaker, M. (1994). High harmonic-generation in atomic and diatomic molecular gases using intense picosecond laser-pulses - a comparison *Journal of Physics B-Atomic Molecular and Optical Physics* 27, 5119-5130.

Lindner, F., Stremme, W., Schatzel, M., Grasbon, F., Paulus, G., Walther, H., Hartmann, R. & Struder, L. (2003). High-order harmonic generation at a repetition rate of 100 khz *Phys Rev A* 68, 013814.

Marques, M. A. L., Castro, A., Bertsch, G. F. & Rubio, A. (2003). Octopus: A first-principles tool for excited electron-ion dynamics *Computer Physics Communications* 151, 60-78.

Nabekawa, Y., Hasegawa, H., Takahashi, E. J. & Midorikawa, K. (2005). Production of doubly charged helium ions by two-photon absorption of an intense sub-10-fs soft x-ray pulse at 42 ev photon energy *Physical Review Letters* 94, -.

Niikura, H., Villeneuve, D. & Corkum, P. (2005). Mapping attosecond electron wave packet motion *Phys Rev Lett* 94, 083003.

Norreys, P., Zepf, M., Moustaizis, S., Fews, A., Zhang, J., Lee, P., Bakarezos, M., Danson, C., Dyson, A., Gibbon, P., Loukakos, P., Neely, D., Walsh, F., Wark, J. & Dangor, A. (1996). Efficient extreme uv harmonics generated from picosecond laser pulse interactions with solid targets *Phys. Rev. Lett.* 76, 1832-1835.

Perdew, J. P. & Zunger, A. (1981). Self-interaction correction to density-functional approximations for many-electron systems *Physical Review B (Condensed Matter)* 23, 5048.

Rundquist, A. (1998). Phase-matched generation of coherent soft x-rays *Science* 280, 1412-1415.

Runge, E. & Gross, E. K. U. (1984). Density-functional theory for time-dependent systems *Physical Review Letters* 52, 997.

Seres, J., Seres, E., Verhoef, A. J., Tempea, G., Strelill, C., Wobrauschek, P., Yakovlev, V., Scrinzi, A., Spielmann, C. & Krausz, F. (2005). Source of coherent kiloelectronvolt x-rays *Nature* 433, 596-596.

Suzuki, M., Baba, M., Ganeev, R., Kuroda, H. & Ozaki, T. (2006). Anomalous enhancement of a single high-order harmonic by using a laser-ablation tin plume at 47 nm *Optics Letters* 31, 3306-3308.

Taieb, R., Veniard, V., Wassaf, J. & Maquet, A. (2003a). Roles of resonances and recollisions in strong-field atomic phenomena. Ii. High-order harmonic generation *Phys Rev A* 68, 033403.

Taieb, R., Veniard, V., Wassaf, J. & Maquet, A. (2003b). Roles of resonances and recollisions in strong-field atomic phenomena. Ii. High-order harmonic generation *Physical Review A* 68, -.

Takahashi, E., Nabekawa, Y. & Midorikawa, K. (2002). Generation of 10-μj coherent extreme-ultraviolet light by use of high-order harmonics *Optics Letters*.

Tamaki, Y., Itatani, J., Nagata, Y., Obara, M. & Midorikawa, K. (1999). Highly efficient, phase-matched high-harmonic generation by a self-guided laser beam *Physical Review Letters* 82, 1422-1425.

Toma, E., Antoine, P., De Bohan, A. & Muller, H. (1999). Resonance-enhanced high-harmonic generation *J Phys B-At Mol Opt* 32, 5843-5852.

Troullier, N. & Martins, J. L. (1991). Efficient pseudopotentials for plane-wave calculations *Phys Rev B Condens Matter* 43, 1993-2006.

Wahlstrom, C., Borgstrom, S., Larsson, J. & Pettersson, S. (1995). High-order harmonic-generation in laser-produced ions using a near-infrared laser *Phys Rev A* 51, 585-591.

21

An Attosecond Soft x-ray Nanoprobe: New Technology for Molecular Imaging

Sarah L Stebbings, Jeremy G Frey and William S Brocklesby
University of Southampton
United Kingdom of Great Britain and Northern Ireland

1. Introduction

The ability to image matter on the microscopic scale is of fundamental importance to many areas of research and development including pharmacology, materials science and nanotechnology. Owing to its generality, x-ray scattering is one of the most powerful tools available for structural studies. The major limitation however is the necessity of producing suitable crystalline structures - this technique relies upon many x-ray photons being scattered from a large number of molecules with identical orientations. As it is neither possible nor desirable to crystallise every molecule of interest, this has provided a huge drawback for most biotechnologies. Although improvements in both sources and detectors have had a strong impact in this area, driving down the required sample size, the need for macroscopic crystalline samples remains a fundamental bottleneck. Fortunately recent technological developments in the generation and sub-micron focusing of soft x-rays (SXRs) have provided a route for bypassing the need for a regular, crystalline structure.

For the purposes of this chapter, SXRs are defined as electromagnetic radiation with wavelengths from 1 - 50 nm, which correspond to photon energies of 1.2 keV - 25 eV respectively. As their wavelengths are on a comparable scale to objects such as proteins, cells and quantum dots, SXRs are ideally suited for imaging these targets with a high spatial resolution. Furthermore water is transparent and carbon opaque to SXRs whose wavelengths lie between 2 - 4 nm, the so-called water window. This offers clear potential for the imaging of biological molecules within their native, aqueous environment, something that would be impossible using traditional x-ray crystallography experiments.

Unsurprisingly there has been great interest in the production and application of SXRs across a wide range of scientific endeavours including, but not limited to, resolving electron motion (Drescher et al. 2002), production of isolated attosecond pulses (Goulielmakis et al., 2008) and x-ray diffraction microscopy (Sandberg et al., 2008). To date there are three major approaches employed to generate SXRs.

The free electron laser (FEL) such as the one located at DESY, Hamburg in Germany, exploits the interactions of electrons within an alternating magnetic field to produce SXR radiation. Electrons are accelerated up to relativistic speeds before being passed into undulator. The undulator consists of a series of magnets that produce an alternating field that causes the electrons to oscillate and emit SXR radiation. These electrons are then able to interact with the radiation to form micro bunches leading to a significant increase in the SXR intensity. Using this source, researchers have produced some impressive images via

holographic (Rosenhahn et al., 2009) and diffraction techniques (Bogan et al., 2008). Due to the properties of the SXR source, the objects that were being imaged were destroyed after only one laser "shot". This is unfortunate as it places a major limitation on the potential quality and reproducibility of the data.
A second approach is to employ a synchrotron source such as the Diamond light source at the Rutherford Appleton Laboratory in Oxfordshire, UK. Here electrons are accelerated up to relativistic speeds in a linear accelerator, booster synchrotron and storage ring. There are a series of bending magnets within the storage ring that control the electron trajectories and cause them to emit synchrotron radiation. This radiation typically ranges from the infrared (wavelength, λ = 700 nm) to gamma rays ($\lambda = 10^{-3}$ nm), easily encompassing the SXR range of the electromagnetic spectrum. Further arrays of magnets within the storage ring cause the electrons to wiggle in a similar manner to the undulator in a FEL, resulting in a more tuneable and intense light beam. The generality of this source has been demonstrated in recent work investigating the structure of metallic nanowires (Humphrey et al., 2008) and the characterisation of 3D molecular orbitals (Beale et al., 2009). In common with FELs, synchrotrons are multi-user, large-scale facilities whose cost and beam time can be restrictive to many researchers. Fortunately there is a third approach to producing SXRs that is a fraction of the cost and can fit in a standard size laboratory.
This chapter describes the development and implementation of such a source of sub-femtosecond (1 fs = 10^{-15} seconds) SXR duration pulses that can be focused down to the nanometre (1 nm = 10^{-9} metres) scale. Consequently this source has the potential to reach down in scale in both time and space that are of enormous benefit to a wide range of fields such as engineering, physical and biological sciences, significantly extending upon the generality of traditional x-ray scattering experiments. In contrast to the synchrotron and FEL sources, this source exploits the highly nonlinear interaction between an intense, femtosecond laser field with a gas medium such as argon in order to produce SXR radiation via a process known as laser-driven high harmonic generation.

2. Semi-classical and quantum mechanical approaches to high harmonic generation

Laser-driven high harmonic generation is an effective and relatively cheap way in which to produce SXRs using an intense laser field and a gas medium such as argon. It is a highly nonlinear process which can most easily be understood in terms of the so-called three step model (Corkum, 1993; Schafer et al., 1993; Lewenstein et al., 1994). In this model the combination of the intense laser field with the atomic potential increases the tunnelling probability of the valence electron through the modified potential barrier as shown in figure 1(a). This electron is then accelerated by the laser field while in the continuum, figure 1(b). In the case of a linearly polarised laser field, the electron will subsequently be driven back to its parent ion when the field reverses direction as shown in figure 1(c). This process occurs every half cycle of the laser pulse - multi cycle driving laser pulses will result in a series of SXR photon bursts which will coherently interfere. Fourier transforming this interference yields the characteristic high harmonic spectrum (Stebbings et al., 2008). The spectrum is a comb of frequencies up to a maximum energy known as the cut-off, E_{max}, is given by equation (1).

$$E_{\max} = I_p + 3.17U_p \approx \lambda^2 I \tag{1}$$

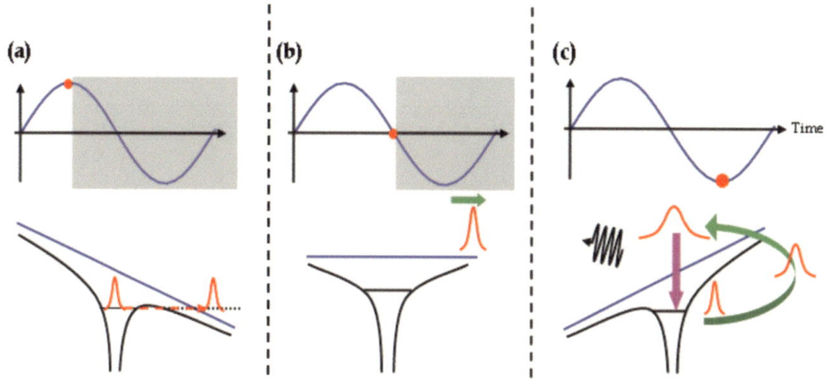

Fig. 1. Semi classical 3 step model of laser-driven high harmonic generation: (a) tunnel ionisation followed by (b) acceleration within the continuum and (c) recombination and emission of an SXR photon. The valence electron is shown in red and the time-dependent evolution of the laser field along with its effect on the atomic potential is shown in the top and bottom panels respectively.

Where I_p is the ionisation potential of the valence electron, U_p is the pondermotive potential, λ and I are the laser wavelength and intensity respectively.

Although the semi-classical model allows much of the underlying physics of high harmonic generation to be understood, it is necessary to solve the time-dependent Schrödinger equation (Jordan & Scrinzi, 2008) in order to fully describe the behaviour of a single atom within an intense, ultrafast laser field. Starting with a simple one dimensional time-dependent Schrödinger equation, a "soft" Coulomb potential is used as described by equation 2.

$$V \propto \frac{1}{\sqrt{x^2 + a_0{}^2}} \tag{2}$$

The next step is to solve the time-independent Schrödinger equation

$$H\psi = E\psi \tag{3}$$

in order to create a ground state wavefunction. This wavefunction is then allowed to propagate in time in the presence of the applied laser field. The final step is to carry out a numerical integration using the Crank-Nicholson scheme in order to solve equation 4

$$i\frac{h}{2\pi}\frac{d\psi}{dt} = \frac{h^2}{2m}\frac{d^2\psi}{dt^2} + \left(V_{static} + V_{laser}\right)\psi \tag{4}$$

and yield the electron acceleration as a function of time which is then Fourier transformed to give the theoretical high harmonic SXR spectrum.

This model was applied to an atom within a laser field of intensity, $I = 3 \times 10^{14}$ W/cm^2 for pulse durations of 30 fs and 7 fs, the results of which can be seen in figures 2 (a) and (b) along with the corresponding high harmonic spectra in figures 2 (c) and (d) respectively.

As can be seen in figures 2 (a) and (b), part of the valence electron wavefunction is ionised into the continuum at specific times during the optical cycle of the laser pulse, while a

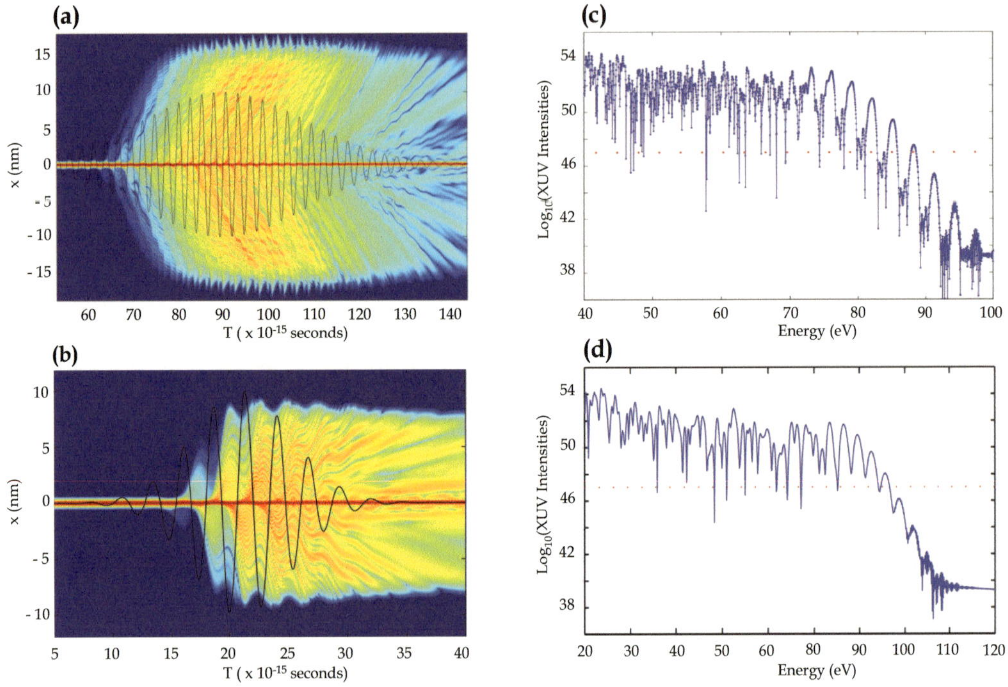

Fig. 2. Solutions to the 1D time-dependent Schrödinger equation. (a) and (b) are $\log_{10}$ (electron density) plotted as a function of x and time, t for 30 fs and 7 fs laser pulses which are also shown. Performing a Fourier transformation on (a) and (b) yields the corresponding high harmonic spectra shown in (c) and (d).

significant portion remains bound to the parent ion. Once in the continuum the ionised portion of the wavefunction undergoes acceleration. Upon returning to the core, it subsequently interferes with the bound part which sets up an oscillating dipole. The result of this oscillation is the emission of the SXR photons. As with the semi classical model a multi cycle laser pulse will result in a series of bursts of SXR photons which will typically result in the characteristic high harmonic spectrum and a train of sub-femtosecond XUV pulses. Comparing figures 2(c) and (d), the individual harmonics are much better defined in the case of a 35 fs, as compared to a 7 fs, duration laser pulse which is consistent with this argument. Reducing the pulse duration still further to the few-cycle regime (corresponding to a pulse duration to 3 fs) a single XUV burst would be expected with the resulting spectrum shown in figures 3 (a) and (b) respectively. In contrast to the calculated harmonic spectra in figures 2 (c) and (d), there is no clear harmonic structure shown to arise from a few-cycle 3 fs pulse, figure 3 (b). Unlike multi-cycle laser pulses, few-cycle laser pulses are used in conjunction with spectral filtering to produce isolated SXR pulses. Currently the record for the shortest duration single SXR pulse produced via laser-driven high harmonic generation stands at 80 as (Goulielmakis et al., 2008). These isolated, sub-femtosecond duration pulses are essential for the investigations into electron dynamics (Kling et al., 2008), time-resolved inner shell spectroscopy (Drescher et al., 2002) and electron correlations (Hu & Collins, 2006) in real time.

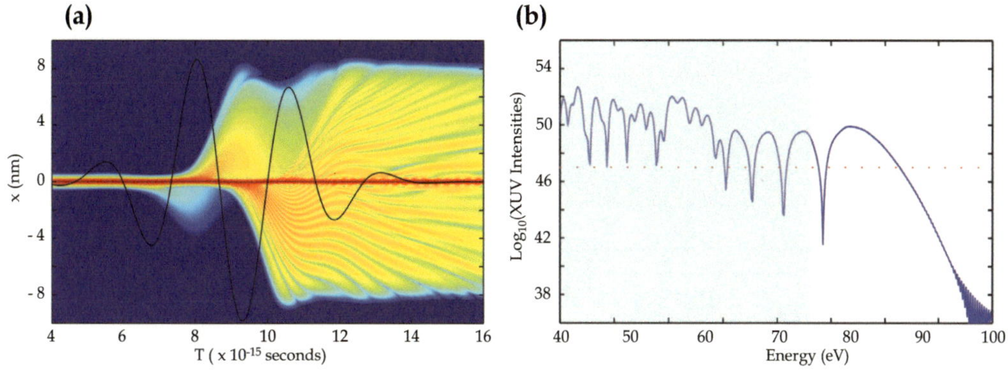

Fig. 3. (a) Few-cycle laser pulse leading to a single SXR photon burst and (b) the resulting spectrum. The shading indicates the part of the spectrum that is filtered in order to produce an isolated attosecond SXR pulse.

For the purposes of imaging sub-micron objects a train of sub-femtosecond XUV pulses is sufficient provided the train is no longer than 10 fs. As long as this is the case then the object that is being imaged will not have sufficient time to undergo Coulomb explosion and fragmentation while in the intense XUV field (Neutze et al., 2000). The configuration and operating parameters of the attosecond SXR nanoprobe is discussed in the following section.

3. Experimental realisation

3.1 Laser source

The experimental system comprises of two parts: the source and generation. As these are independent of one another any improvements in the technology of one can be easily incorporated into the full system.

A commercial Ti: Sapphire chirped pulse amplication (Spitfire Pro, Spectra Physics) laser system is used to produce 3 mJ pulses of duration between 35 – 45 fs at a central wavelength of 790 and 1 kHz repetition rate, a schematic of which is shown in figure 4. The diagnostics used to characterise this output consists of a FROG (*f*requency *r*esolved *o*ptical *g*ating) and a home-built M^2 meter (Praeger 2008). Following characterisation, the beam can either be used to drive the two high harmonic generation lines (denoted 1 and 2) or undergo further compression via filamentation as shown in figure 4.

Filamentation is a straightforward technique that is used to compress intense femtosecond laser pulses. The laser is focused into a 1 m long cell filled with approximately 1 bar of argon gas. As it propagates through the gas it undergoes competing effects in the spectral regime: self focusing from the $\chi^{(3)}$ Kerr effect and self-defocusing due to the plasma index,

$$n_{plasma} = \sqrt{1 - \frac{\omega_p^2}{\omega^2}} \qquad (4)$$

where n_{plasma} = plasma refractive index, ω and ω_p are the radiation and plasma frequecies respectively. Likewise, in the time domain the laser pulses undergo broadening and blue shifting due to the ionisation of the surrounding gas which competes with the simultaneous recompression from the gas dispersion effects. Provided the balance between these

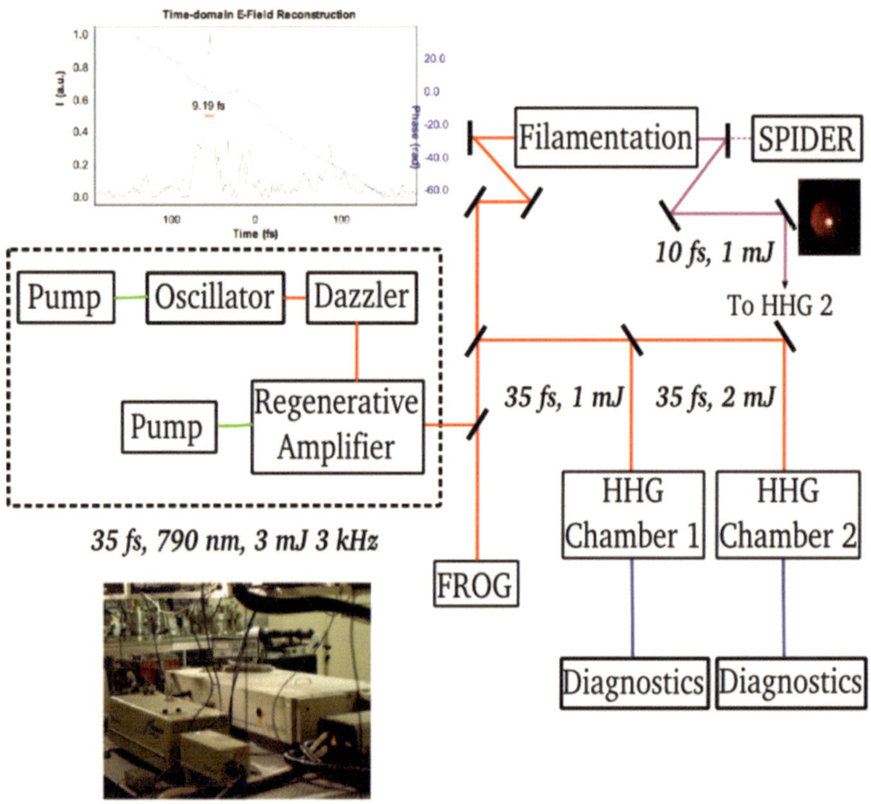

Fig. 4. Laser source used to generate a 10 fs train of sub-femtosecond duration pulses in the XUV for scattering and diffraction imaging. Also shown is the laser mode after filamentation and the corresponding pulse duration measurement in the time domain.

competing processes in both the spectral and temporal domains is correct it is possible to produce spectrally broad pulses of only a few femtosecond duration (Hauri et al., 2005). The filamentation set-up shown in figure 4 is able to produce 10 fs pulses over a spectral range of 450 - 900 nm which are characterised using SPIDER (*s*pectral *p*hase *i*nterferometry for *d*irect *e*lectric-field *r*econstruction). Work is on-going to stabilise this output and use it in a number of high harmonic generation experiments. Ultimately the aim is to use these compressed pulses in conjunction with quasi phase matching techniques to access the water window (Gibson et al., 2003). However, for the work presented in the remainder of this chapter, the uncompressed output from the regenerative amplifier i.e. 35 - 45 fs duration pulses were used to drive the high harmonic generation process.

3.2 Capillary high harmonic generation

While laser-driven high harmonic generation has been investigated in a number of different experimental configurations including gas cells (Sutherland et al., 2004; Kazamias et al., 2003; Tamaki et al., 1999), gas jets (Levesque et al., 2007; Paul et al., 2001) and combined capillary and jet geometries (Heinrich et al., 2006), it is SXR production from a single capillary waveguide that is discussed here.

35 fs, 1 mJ laser pulses are focused into a hollow capillary waveguide of 7 cm length and inner diameter of 150 μm to give a peak intensity in excess of 10^{14} W/cm^2. A pair of two 300 μm diameter holes were drilled 2 cm from either end of the waveguide that allow gas, typically argon, to be fed into the capillary and defining a central region of constant pressure.

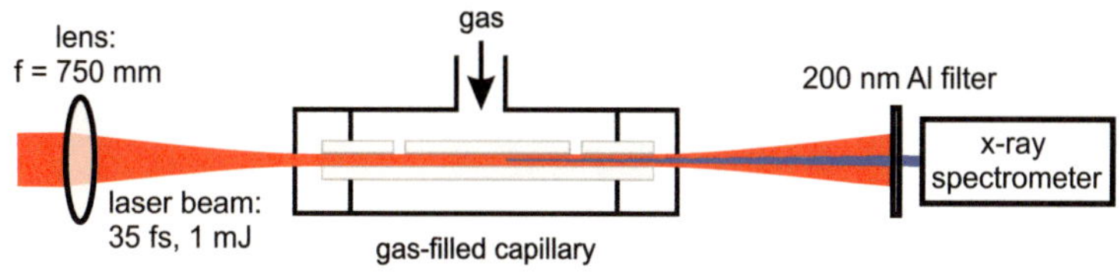

Fig. 5. Schematic of laser-driven high harmonic generation within a capillary waveguide. Reproduced with permission from the Institute of Physics Publishing Ltd (Stebbings et al., 2008).

As both the laser and SXR beams are collinear, a 200 nm aluminium filter is used to block the laser while allowing approximately 10% of the SXRs to be transmitted. This can subsequently be characterised using a grazing incidence spectrometer, Andor x-ray CCD camera or focused and used for scattering and diffraction experiments.

In this configuration the generation process produces SXRs in a coherent and well collimated beam which has a 1 mrad divergence. This output is loosely tunable over a wavelength range of 18 - 40 nm and has approximately 10^7 to 10^8 SXR photons per pulse per harmonic at a repetition rate of 1 kHz under optimal conditions. This equates to an 1μW and 100 kW average and peak powers respectively.

In order to obtain these optimal conditions, and therefore maximise the SXR flux, the phase mismatch, Δk, between the collinear laser and SXR beams must be minimised according to equation 5.

$$\Delta k = qk_{\lambda} - k_q \quad (5)$$

Where q = harmonic order, k_{λ} and k_q are the wavevectors of the laser and SXR beams respectively. Phase matching in a capillary is achieved via the relative balance between the refractive indices of the waveguide, neutral gas and plasma. Consequently the amount of ionisation within the capillary plays a critical role due to its influence on the amount of plasma generated as well as the propagating laser beam (Froud et al., 2006). Too little ionisation and no tunnelling, while too much result in a large phase mismatch, Δk - both cases will result in a significant depletion in the observed SXR flux.

In order to improve understanding of SXR production and therefore optimise the source for maximum flux via phase matching, it is essential to characterise the beam and predict the effects such as gas pressure, species and laser propagation have on the flux produced. These aspects were investigated in some detail and are described in following sections.

4. SXR characterisation: spatio-spectral beam imaging

Although physical processes such as phase matching and beam propagation play a fundamental role in laser-driven high harmonic generation, they are complex effects that must be understood before the full potential of capillary waveguides as SXR sources can be realised. With this in mind a novel technique that permits the full characterisation of the SXR in the spectral and spatial domains from a single SXR diffraction image has been developed.

The experimental configuration was based on figure 5 but with two significant differences:

1. A grid supported 200 nm thick aluminium filter. The grid consisted of 38 μm nickel wires that defined a series of 335 μm square aluminium apertures that provided the diffracting object.

And

2. An Andor CCD camera (instead of a grazing incidence spectrometer) that was used to record the image of the entire SXR beam.

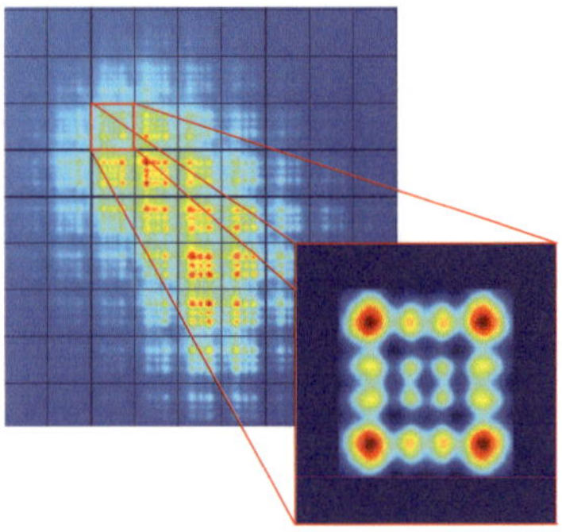

Fig. 6. Fresnel diffraction image of the entire SXR beam generated from 60 mbar of argon gas within a capillary waveguide. The inset shows the diffraction from one aluminium square aperture. This varies as a function of aperture position and is a convolution of a number of different SXR wavelengths. See also (Praeger et al., 2007).

The distances between the SXR source (exit of the capillary waveguide) and the filter and the filter and the camera (defined as the image plane) were 50 cm and 100 cm respectively - well within the Fresnel diffraction regime.

As before the laser (30 fs, 1 mJ, 780 nm) was focused into the capillary waveguide to give peak intensities in excess of 10^{14} W/cm^2. Argon gas was then fed into the system to give a constant pressure equal to 60 mbar within the central region of the capillary. The filter again blocked the laser while allowing approximately 10% of the SXRs to be transmitted and subsequently imaged by the camera over 100 laser shots as shown in figure 6. The image shown in figure 6 was recorded over 100 laser shots corresponding to a total time of 100 ms. Although there is an overall intensity distribution in the SXR beam it is the diffraction from the individual apertures created by the nickel mesh that is of most interest. By examining the image in figure 6 it becomes clear that there is a strong spatial dependence in the diffraction from the aluminium square apertures. Furthermore the source is known to produce a series of harmonic orders (Froud et al., 2006). This means each of these patterns is a convolution of number of different SXR wavelengths. In order to extract any useful information from this image the contributions of each wavelength must be deconvolved from the diffraction pattern from each aperture. Fortunately a number of reasonable assumptions can be made to make these calculations easier.

1. Treat the diffraction from each aperture separately as any contributions from neighbouring apertures are negligible. This is reasonable as photons arriving at any particular point on the image plane are more likely to have taken a direct rather than an indirect path by more than three orders of magnitude.
2. Neglect the spectral width of the harmonics, as any interference between the harmonic will be averaged out over the exposure time of the camera.

The diffraction pattern from each aperture is then calculated using Fresnel theory (Born & Wolf, 1999) for each SXR wavelength up to the 51 st harmonic order that corresponds to a wavelength of approximately 15 nm. This upper limit is defined by the calculated cut-off energy, E_{max}, given in equation (1). The contributions to the overall diffraction pattern from each aperture are then added as an incoherent sum according to equation 6. (Praeger et al., 2007)

$$P_m(x,y,i,j) = \sum_{h=1}^{\infty} P_c(x,y,i,j,h) I(i,j,h) \tag{6}$$

Where (x, y) are the spatial co-ordinates of the aperture (i, j), h is an odd integer representing the high harmonic orders, P_c and P_m are the calculated and measured diffraction patterns respectively. $I(i, j, h)$ is the weighting factor which is directly proportional to the spectral intensities of each harmonic order. Knowing the measured spectrum, P_m, from one aperture allows linear algebra to be applied in order to determine the weighting factors to give the best fit to the experimental data. Ultimately this allows the reconstruction of the high harmonic spectra that are compared to previously measured data using the grazing incidence spectrometer under similar experimental conditions as shown in figure 7.

In contrast to the camera, the spectrometer is only able to image a small section of the overall SXR beam at any one time. It is possible to scan the spectrometer over the entire beam but maintaining accurate alignment is extremely challenging. Another issue is the integration times required to obtain the data sets - using the spectrometer this typically takes tens of seconds as compared to the 100 ms needed to record the image in figure 6. Due to the highly nonlinear high harmonic process an extended integration time can encompass significant changes in the SXR output from the capillary waveguide. The slight offset between harmonic peaks and the harmonics of fundamental laser is due to blue shifting of the propagating laser (Froud et al., 2006) though it is not accounted for here.

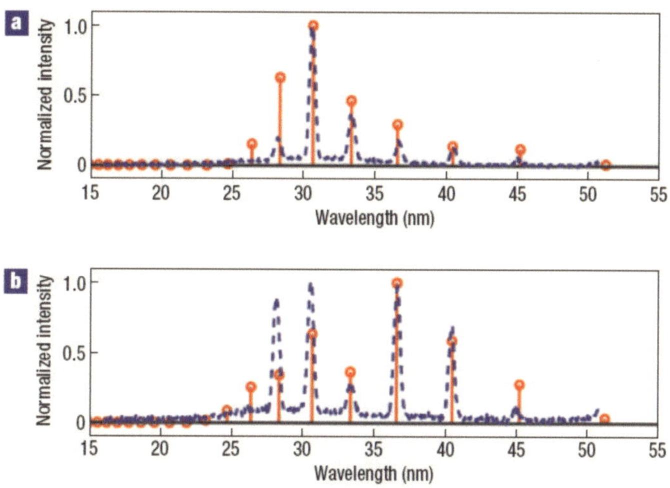

Fig. 7. High harmonic spectra recorded for argon gas pressures of (a) 60 and (b) 80 mbar. Blue dashed and red solid lines denote experimental and reconstructed data respectively. The integration time for the experimental data was 20 seconds. Compare this to the 100 ms required to record the diffraction image in figure 6. Reprinted by permission from Macmillan Publishers Ltd: Nature Physics (Nature Physics 3, 176–179 (1 March 2007), doi:10.1038/nphys516) copyright (2007). (Praeger et al., 2007) http://www.nature.com/nphys/index.html

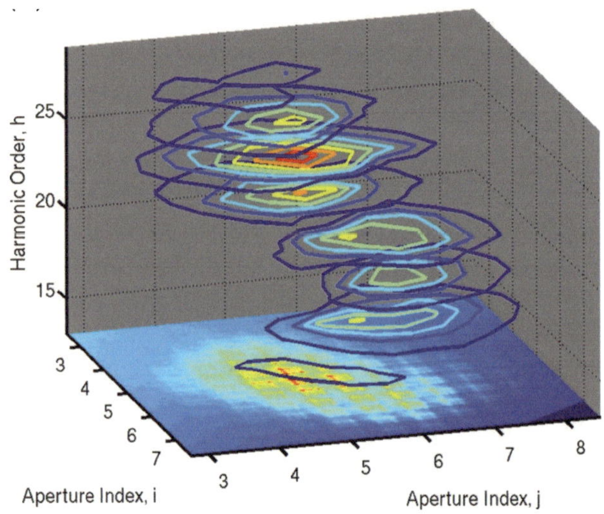

Fig. 8. Spatially resolved reconstructed spectrum for 60 mbar of argon within the capillary waveguide. Coloured lines denote the relative spectral intensities of the harmonic orders. Reprinted by permission from Macmillan Publishers Ltd: Nature Physics (Nature Physics 3, 176–179 (1 March 2007), doi:10.1038/nphys516) copyright (2007). (Praeger et al., 2007) http://www.nature.com/nphys/index.html

Nevertheless there is still good agreement between the reconstructed and the experimental data recorded by the spectrometer in figure 7. While this is a useful comparison and a confirmation of the robustness of the spatio-spectral imaging technique, there is plenty more information that can be extracted from the diffraction image such as the one shown in figure 6. Figure 8 shows the spectral intensity of all of the different SXR wavelengths (as harmonic order of the fundamental laser) as a function of two spatial dimensions corresponding to the aperture indices (*i*, *j*). Essentially this is a plot of the spectral intensity of each wavelength spatially across the entire SXR beam generated from the capillary waveguide.

Further information can be extracted from the diffraction image in order to compare the transverse spatial intensity profiles as a function of harmonic order and gas pressure recorded at 60 and 160 mbar respectively. The spatial profiles are all broadly cylindrical symmetric with the high order harmonics (shortest SXR wavelengths) as a spot located close to the centre of the capillary waveguide. The lower orders are located further out and typically show an annular shaped profile. Possible reasons for these observations, including noncollinear and changes in the radial phase matching conditions are discussed in an earlier publication and the reader is referred therein (Praeger et al., 2007).

One of the major benefits of the technique presented here is that a significant amount of spatial and spectral information of the SXR beam can be obtained from one diffraction image. Although the image in figure 6 was recorded over 100 ms, it would be possible to obtain the same information from a single laser shot provided the signal to noise ratio was good enough. Operation in single shot mode opens up applications such as the study of transient phenomena in the high harmonic generation process and the elimination of shot-to-shot noise arising from the laser system.

For the purposes of this chapter however, these results have provided valuable insight into the radial distribution of capillary high harmonic generation and prompted further investigations into the radially dependent SXR production.

5. Molecular variation of capillary-produced SXRs

There are a number of factors that have a significant effect on the SXR production from the laser-driven high harmonic generation process. Not least the propagation geometry of the driving laser field is of critical importance in obtaining ideal phase matching conditions. In gas jets this is achieved via the Guoy phase shift and is limited by the confocal parameter of the focused laser. In contrast capillaries provide an extended interaction length between the laser and the gas media implying the amount of ionisation within the waveguide will profoundly affect the SXR output. This is due to its effects on the propagation of the laser mode and the phase matching via the changes in the refractive index of the plasma generated.

Building on the previous experiments such as those described in section 4, the SXR production was investigated from a number of media of increasing complexity - argon, nitrogen and nitrous oxide - as a function of gas pressure using the experimental configuration shown in figure 5.

In addition a simple spatially and temporally dependent phase matching model was developed in order to predict the overall spectral envelopes of the harmonic spectra without the need for invoking complex quantum mechanical calculations. This is an extension to the phase matching theory originally proposed by Durfee and co-workers (Durfee et al., 1999). As shown in the previous section SXR production in a capillary displays a strong spatial dependence. Although previous work (Durfee et al., 1999) allowed for the radial dependence of ionisation across the capillary in the laser propagation via modal averaging, the variation in radial phase matching conditions was not included. However as the Rayleigh range of the SXRs is long compared to the interaction then SXRs produced at different radii within the capillary will not necessarily interfere. This means that there will be different phase matching criteria as a function of time ***and*** position across the waveguide that are allowed for in the extended phase matching model described here.

The driving laser pulse is assumed to have a Gaussian temporal profile and propagate through the capillary waveguide in the EH_{11} mode (Marcatili & Schmeltzer, 1964), which allows the calculation of the spatially and dependent ionisation using the so-called ADK tunneling theory (Ammosov et al., 1986; Popov, 2004). This is then used to calculate the phase mismatch, Δk, according to equation 7. (Stebbings et al., 2008)

$$\Delta k \approx \frac{q u_{ij} \lambda_0}{4\pi a^2} + N_e r_e \left(q\lambda_0 - \lambda_q\right) - \frac{2\pi N_{atm}}{\lambda_q}\left(\delta(\lambda_0) - \delta(\lambda_q)\right) \tag{7}$$

Where q is the harmonic order, u_{ij} is the ith root of the Bessel function, J_{i-1}, with i and j both equal to 1 denoting the EH_{11} mode, λ_0 and λ_q the wavelengths of the laser and harmonic (order q) respectively, a is the capillary radius, N_e is the number of electrons, r_e the classical electron radius, a is the radius of the capillary waveguide, N_{atm} the number of neutral gas atoms (or molecules), P is the gas pressure, $\delta(\lambda_0)$ and $\delta(\lambda_q)$ are the dispersion of the neutral gas at the laser and harmonic wavelengths respectively.

The phase mismatch is then substituted into equation 8 in order to determine the intensity as a function of position, r, time, t, and harmonic order, q. (Stebbings et al., 2008)

$$I(r,t,q) \propto dN\left(\frac{1 + e^{-\alpha L} - 2e^{-\alpha L}\cos(\Delta k L)}{\alpha^2 + \Delta k^2}\right) \tag{8}$$

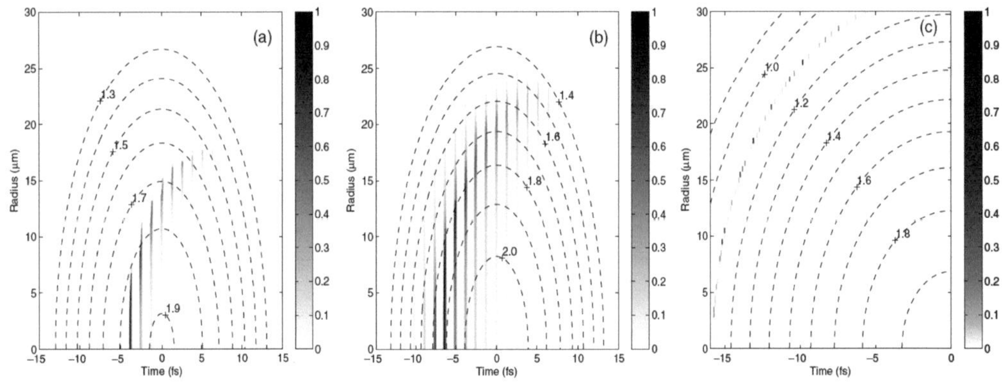

Fig. 9. Theoretical intensity distribution, $I(r, t, q)$, summed over all harmonic intensities for (a) argon, (b) nitrogen and (c) nitrous oxide. The dashed lines denote the iso-intensity contours which are multiplied by 10^{14} W/cm^2. The maximum intensity has also been normalised to 1. Reproduced with permission from Institute of Physics Publishing Ltd (Stebbings et al., 2008).

Where dN is the number of ions produced in a given time interval, α is the field attenuation coefficient and L is the length of the capillary waveguide. $I(r,t,q)$ can then be plotted as a function of time and capillary radius as shown in figure 9.

Figure 9 (a) shows the theoretical phase matched SXR build up from argon gas. Significant intensity is predicted over intensities from 1.6 to 1.8 x 10^{14} W/cm^2 and out to a capillary radius, r = 20 μm. Initially the intensity starts to build up on the rising edge of the laser pulse at t = -5 fs, close to the centre of the waveguide. As the laser intensity increases with time, the phase matched SXR production is pushed out to larger radii. This effect is due to the increasing ionisation fraction along the central core of the waveguide. There is no SXRs observed beyond r = 20 μm and -5 fs $\leq t \leq$ 10 fs as the laser intensity is too low.

Similar trends are also observed for nitrogen shown in figure 9 (b). In contrast to argon however, phase matched SXRs are seen to laser intensity, I = 1.4 x 10^{14} W/cm^2 due to higher ionisation fractions being produced earlier in the pulse, and out to a radius, r = 25 μm. These observations are consistent with the lower ionisation potential of nitrogen as compared to argon.

The results from nitrous oxide, figure 9 (c), are significantly different from either argon or nitrogen. The SXRs are highly localized over laser intensities from 1 to 1.2 x 10^{14} W/cm^2 implying only a small fraction of the driving laser pulse is exploited. These observations are in part explained by the differences in the ionisation properties of nitrous oxide as compared to argon and nitrogen, though other properties such as the quantity of neutral gas and plasma will also play a significant role.

The SXR spectra for argon, nitrogen and nitrous oxide were measured over a pressure range of 5 to 200 mbar. By calculating $I(r, t)$ for each harmonic order, q, a direct comparison between the measured data and theory can be achieved. A selection of these results is shown in figure 10. For further data sets please refer to (Stebbings et al., 2008). Also shown for completeness are the predicted curves using the original phase matching theory (Durfee et al., 1999).

The spatial and temporal theory predicts the overall shape of the spectral envelope and cut-off well for argon and nitrogen for a variety of pressures up to 70 mbar showing a marked

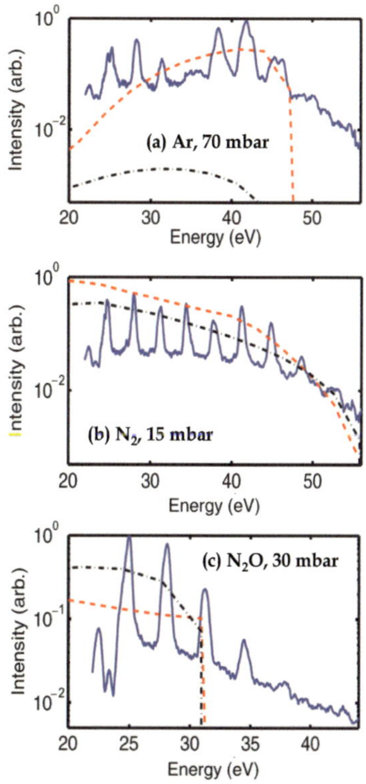

Fig. 10. Selection of experimental SXR spectra for (a) argon, (b) nitrogen and (c) nitrous oxide for different gas pressures as shown by the solid blue lines. The dashed red and blue dashed-dotted curves denote the extended and original phase matching theories. Reproduced with permission from Institute of Physics Publishing Ltd. (Stebbings et al., 2008)

improvement over the previous theory. This agreement decreases for higher pressures as there are a number of complex effects including broadening and splitting of the harmonics. It is also proposed that high pressures can cause temporal distortions in the laser pulse. As these effects lie outside the scope of the simple phase matching theory here it is not surprising that the agreement with the experimental data decreases with increasing pressure.

The predicted cut-off according to equation 1 for argon and nitrogen is 50 eV which is significantly lower than that observed experimentally for most experimental pressures investigated. The reasons for these differences are attributed to absorption of the SXRs by the gas medium as well as phase matching effects, neither of which is accounted for in equation 1. Only by including phase matching can a match between the theoretical and experimental cut-off be achieved highlighting the importance of phase matching effects in capillary high harmonic generation.

In contrast both theories are unable to predict the spectral envelopes arising from high harmonic generation in nitrous oxide an example of which is shown in figure 10 (c). The phase matching calculations are extremely sensitive to the amount of ionisation. Unfortunately this fraction was determined using ADK theory which is an atomic tunneling theory and therefore not ideally suited to a triatomic molecule.

The improvement in agreement between the extended phase matching theory and the experimental data highlights the importance of including spatially and temporally dependent ionisation within the capillary waveguide. However further developments including the incorporation of pressure dependence in the laser pulse profile and a rigorous tunneling theory for triatomic molecules would yield further improvements. Furthermore it should help researchers achieve optimal conditions for SXR production from a capillary waveguide.

6. Determination of the complex refractive index of sub-micron nanospheres in the SXR region

Table-top sources of SXR radiation offer the potential for unprecedented insight into attosecond dynamics with sub-micron spatial resolution. The high flux (ranging from 10^8 - 10^{11} photons per harmonic per second depending upon generation conditions) and excellent coherence of the SXR beam has opened up many new areas for the analysis of nanoscale objects and systems. In this section the source was used to provide a direct measurement of the complex refractive index of polystyrene spheres within the SXR region.
The nanospheres (average diameter of 194 nm with an uncertainty of < 5 %) were deposited on a 50 nm thick silicon nitride membrane (Duke Scientifics, 1.05 g/cm^3) to form a well ordered, single layer hexagonal close packed array as shown in figure 11. Although the ordering was good it was not perfect and a number of defects can be seen in the array in figure 11.

Fig. 11. Scanning electron microscope image of self assembled nanospheres.

In contrast to earlier experiments, laser pulses of 40 fs duration, 1.5 mJ energy at a central wavelength of 800 nm and 1 kHz repetition rate were focused with a dielectric mirror (f = 500 mm) into a gas cell containing 10 mbar of argon. As before the laser and SXR beams were collinear with the laser being blocked by two 100 nm thick aluminium filters. These allowed the transmission of the SXRs to the interaction chamber as shown in figure 12. The SXR beam was subsequently focused by a MoSi multilayer mirror to a spot of 22 μm diameter. The 2D array of nanospheres was placed at the focus and the scattered SXRs detected in transmission by an x-ray CCD camera (Andor, 1024 x 1024 pixels where 1 pixel = 13.5 μm) in the far field regime.
The recorded diffraction pattern in figure 13 shows a number of interesting features. The first of these is in the radial direction where there are multiple spots. These spots were thought to arise from the diffraction of the multiple harmonic wavelengths making up the SXR beam. This was confirmed by removing the MoSi mirror and measuring the spectrum of the SXR beam, which was found to have a wavelength range of 35 - 24 nm

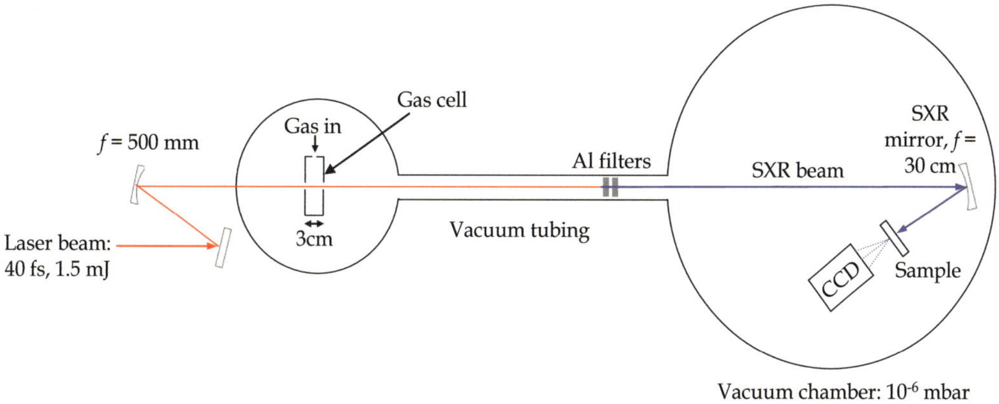

Fig. 12. Schematic of the experimental configuration employed to determine the complex refractive index of a 2D array of 194 nm diameter polystyrene spheres. This set-up is described in detail in (Mills et al., 2008).

(corresponding to photon energies of 36 - 51 eV respectively). The separation of these wavelengths matched the separation of the radial diffraction spots observed in figure 13. Other features include the lack of rotational symmetry and the broadening in the axial direction of the diffraction image. In the experimental configuration (figure 12), the spherical SXR mirror was used in an off-axis geometry. This resulted in a curved wavefront at the focus which led to the lack of rotational symmetry observed in the diffraction image. The broadening in the axial direction was attributed primarily to the defects in the 2D nanosphere array (some of these can be seen in figure 11).

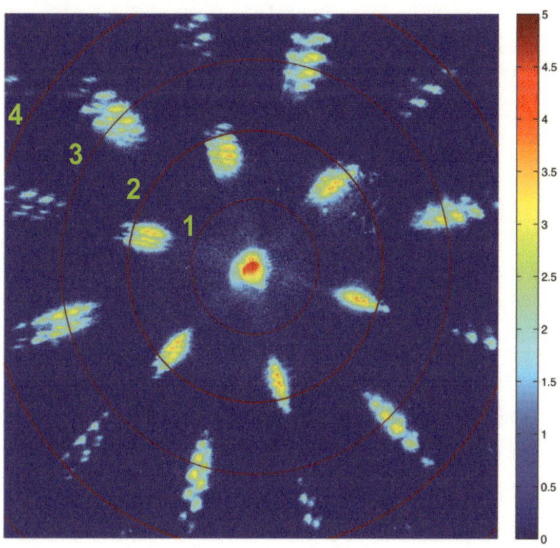

Fig. 13. Far field diffraction pattern from the 2D hexagonal close packed array of 194 nm polystyrene spheres. The integration time was 240 seconds and the intensity is on a logarithmic scale. The red circles denote the angle of 100 mrad x n where n is the contour number. Reprinted and adapted (original in black and white) from (Mills et al., 2008).

The diffraction image shown in figure 13 is the scattering function from a single sphere multiplied by an array of delta functions. Therefore the image can be thought of as sampling the angular distribution of SXR light scattered from a single sphere. This can be determined analytically using Mie theory and is dependent upon the size (for a spherical object this is the radius) of the scatterer, its complex refractive index and the wavelengths of the incident SXR light. As the wavelengths and radius are known, it is possible to use Mie theory to determine the complex refractive index (equation 9), n_c, of the polystyrene spheres within the SXR region of the electromagnetic spectrum.

$$n_c = \alpha + i\beta \tag{9}$$

Where α and β are parameters to be determined. These parameters were found from the diffraction image, recorded in figure 13, by measuring the intensity of the SXR wavelength as a function of angle. A linear least square fitting routine was then used to fit the Mie and measured diffraction peak intensities for the first three scattering orders. The results are shown for a SXR wavelength of 27.6 nm in figure 14.

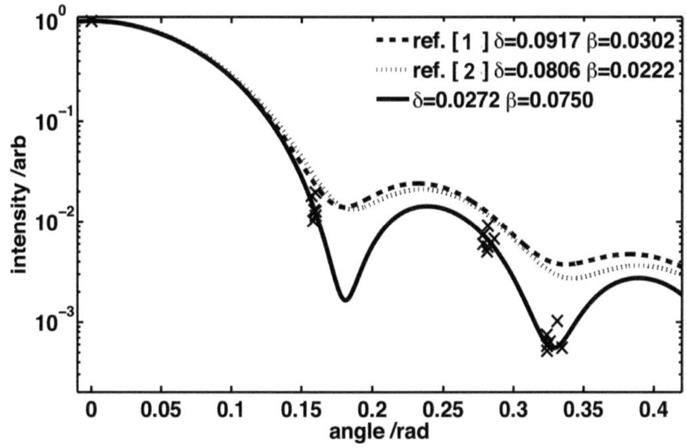

Fig. 14. Comparison between the Mie scattering theory (solid line) for light scattered from a single sphere and the experimental diffraction peaks for a SXR wavelength of 27.6 nm and with complex refractive indices from (dashed line, ref [1] in figure) NIST, Chantler et al., and (dotted line, ref [2] in figure) the Centre for X-ray Optics and determined from the work presented here (solid black line). Reprinted with permission from (Mills et al., 2008).

The complex refractive index parameters, α and β, of this and other SXR wavelengths (26.9 and 29.6 nm) were determined and compared to extrapolated data from CXRO (Centre for X-ray Optics) and NIST (National Institute of Standards and Technology, Chantler, et al.) as shown in figure 15.

Although the general trend of the extrapolated curves follows the experiment, there is a large discrepancy between the magnitudes thus highlighting the importance of obtaining experimental data within this region (Mills et al., 2008).

Using a single diffraction image, the technique described here can be extended to other SXR wavelengths and many other nanostructures provided they are periodic and will therefore diffract light in a manner that is dependent upon their complex refractive indices.

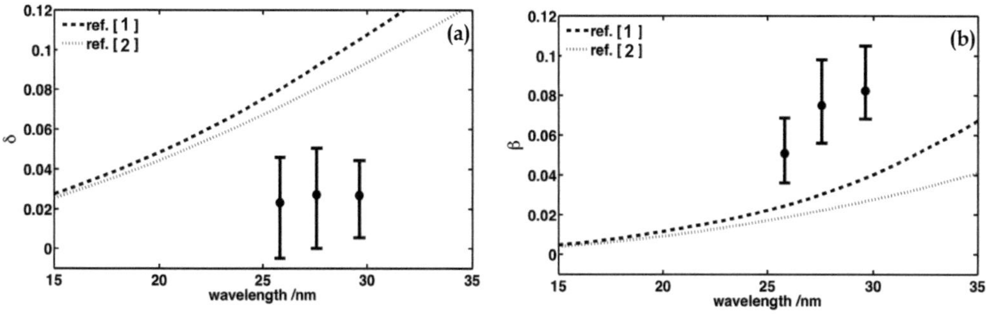

Fig. 15. Comparison of the experimentally determined and extrapolated [1] CXRO and [2] NIST (Chantler, et al., 1995) data for (a) α and (b) β. Reprinted with permission from (Mills et al., 2008).

7. Future work & summary

The table-top, laser-driven SXR source described here routinely produces a total flux of approximately 10^8 photons per harmonic per second at 1 kHz repetition rate to give approximately 1 μW average and 100 kW peak powers respectively. The SXR beam itself has a high spatial coherence and is well-collimated with a divergence of < 1 milli radian. Typically the SXRs have wavelengths from 18 - 50 nm which are loosely tunable depending upon the generation conditions within the capillary waveguide.

The investigation of SXR production from molecular gases e.g. N_2 and N_2O were of particular interest as they provided different combinations of gas absorption, index and ionisability with which to change the spectrum of the emitted SXR radiation. Several interesting effects were observed in the SXR spectra (figure 10 and references therein) that combine the effects of phase matching and the propagation of the driving laser beam. Only by developing a model that incorporates the change in ionisation within the waveguide as a function of time within the laser pulse and radial distance across the capillary could the SXR spectral envelope be predicted.

In order to understand the effects of capillary propagation of the generated SXRs, several diagnostic techniques have been developed and described in this chapter. Using the Ni grids which act as a support for the 200 nm thick free-standing aluminium filter (used to separate the SXR and laser beams) significant spectral and spatial detail of the SXR beam has been observed. Fresnel diffraction from each grid square was used to reconstruct the spectrum within each square leading to a 2D spectral profile across the beam. An understanding of the beam is critical to understand the effect of the radial distribution of ionisation on the SXR production and for further optimisation of the overall SXR flux.

Initial scattering investigations from an array of sub-micron latex spheres have proved extremely successful. The results from these experiments have demonstrated the direct measurement of the complex refractive index of polystyrene nanosphers by analysis of the resulting SXR diffraction intensity distribution. This technique further demonstrates the applicability of a high harmonic SXR source for nanoscale analysis and can be extended to determine the complex refractive index of a material for a range of wavelengths from one diffraction image. These investigations will be extended to other periodic nanostructures in order to gain new insights as well as test the applicability of the technique. The technique of coherent diffraction imaging, CDI, is currently being tested and applied to the SXR diffraction image (figure 13) in order to reconstruct the 2D hexagonal close packed array of polystyrene spheres (figure 11). Future work will combine the technique of coherent diffraction imaging (Gerchberg & Saxton, 1972; Fienup, 1978) with experimental SXR diffraction in order to reconstruct the overall conformation and shape of a wide variety of nanoscale objects.

In the long-term, the plan is to combine quasi phase matching techniques (Gibson et al., 2003) with compression of the driving laser pulse from 40 to 10 fs duration in order to access the water window. Once this regime is reached, further SXR scattering experiments will be carried out. The analysis of the resulting SXR diffraction patterns (via coherent diffraction imaging) will permit the shape of sub-micron objects such as biological molecules within their native, aqueous environments to be determined.

8. References

Ammosov, M. V.; Delone, N. B. & Krainov, V. P. (1986). Tunnel ionization of complex atoms and of atomic ions in an alternating electromagnetic field. *Soviet Physics - JETP*, 64, 1191 - 1194

Beale, T. A. W.; Bland, S. R.; Johnson, R. D.; Hatton, P. D.; Cezar, J. C.; Dhesi, S. S.; Zimmermann, M. V.; Prabhakaran, D. & Boothroyd, A. T. (2009). Thermally induced rotation of 3d orbital stripes in $Pr(Sr_{0.1}Ca_{0.9})_2Mn_2O_7$. *Physical Review B, Condensed Matter and Materials Physics*, 79, 5, (February 2009) 054433-1 - 054433-5, ISSN 1550-235x (online), 1098-0121 (print)

Bogan, M. J.; W. Benner, H.; Boutet, S.; Rohner, U.; Frank, M.; Barty, A.; Seibert, M. M.; Maia, F.; Marchesini, S.; Bajt, S.; Woods, B.; Riot, V.; Hau-Riege, S. P.; Svenda, M.; Marklund,E.; Spiller,E.; Hajdu, J. & Chapman, H. N. (2008). Single Particle X-ray Diffractive Imaging. *Nano Letters*, 8, 1, (January 2008) 310 - 316, ISSN 1530-6992 (online), 1530-6984 (print)

Born, M. & Wolf, E. (1999). *Principles of Optics, Theory of Propagation, Interference and Diffraction of Light*, Cambridge University Press, 0521642221, Cambridge UK.

Centre for X-ray Optics, http://henke.lbl.gov/optical_constants/getdb2.html.

Chantler, C. T.; Olsen, K.; Dragoset, R. A.; Chang, J.; Kishore, A. R.; Kotochigova, S. A. & Zucker, D. S. X-ray form factor, attenuation and scattering tables (version 2.1). http://physics.nist.gov/ffast, National Institute of Standards and Technology, Gaithersburg, MD.

Corkum, P.B. (1993). Plasma perspective on strong field multiphoton ionization. *Physical Review Letters*, 71, 13, (September 1993) 1994 - 1997, ISSN 1079-7114 (online), 0031-9007 (print)

Drescher, M.; Hentschel, M.; Kienberger, R.; Uiberacker, M.; Yakovlev, V.; Scrinzi, A.; Westerwalbesloh, Th.; Kleineberg, U.; Heinzmann U. & Krausz, F. (2002). Time-resolved atomic inner-shell spectroscopy. *Nature*, 320, 6909, (October 2002) 803 - 807, ISSN 0028-0836

Durfee, C. G.; Rundquist, A. R.; Backus, S.; Herne, C.; Murnane, M. M. & Kapteyn, H. C. (1999). Phase matching of high-order harmonics in hollow waveguides. *Physical Review Letters*, 83, 11, (September 1999) 2187 – 2190, ISSN 1079-7114 (online), 0031-9007 (print)

Froud, C. A.; Rogers, E. T. F.; Hanna, D. C.; Brocklesby, W. S.; Praeger, M.; de Paula, A. M.; Baumberg, J. J & Frey, J. G. (2006). Soft-x-ray wavelength shift induced by ionization effects in a capillary. *Optics Letters*, 31, 3, (February 2006) 374 - 376, ISSN 0146-9592

Fienup, J.; (1978). Reconstruction of an object from the modulus of its Fourier transform. *Optics Letters,* 3, 1, (July 1978) 27 - 29, ISSN 0146-9592

Gerchberg, W. & Saxton, W. O. (1972). A practical algorithm for the determination of the phase from image and diffraction plane pictures. *Optik,* 35, 237 – 246

Gibson, E. A.; Paul, A.; Wagner, N.; Tobey, R.; Gaudiosi, D.; Backus, S.; Christov, I. P.; Aquila, A.; Gullikson, E. M.; Attwood, D. T.; Murnane M. M. & Kapteyn, H. C. (2008). Coherent Soft X-ray Generation in the Water Window with Quasi-Phase Matching. *Science,* 302, 5642, (October 2003) 95 - 98, ISSN 1095-9203 (online), 0036-8075 (print)

Goulielmakis, E.; Schultze, M.; Hofstetter, M.; Yakovlev, V. S.; Gagnon, J.; Uiberacker, M.; Aquila, A. L.; Gullikson, E. M.; Attwood, D.T.; Kienberger, R.; Krausz F. & Kleineberg, U. (2008). Single-Cycle Nonlinear Optics. *Science,* 320, 1614, (June 2008) 1614 - 1617, ISSN 1095-9203 (online), 0036-8075 (print)

Hauri, C. P.; Guandalini, A.; Eckle, P.; Kornelis, W.; Biegert, J. & Keller, U. (2005). Generation of intense few-cycle laser pulses through filamentation parameter dependence. *Optics Express,* 13, 19, (September 2005) 7541 - 7547, ISSN 1094-4087

Heinrich, A.; Kornelis, W.; Anscombe, M. P.; Hauri, C. P.; Schlup, P.; Biegert, J. & Keller, U. (2006). Enhanced VUV-assisted high harmonic generation. *J. Phys. B: At. Mol. Opt. Phys.,* 39, 13, (June 2006) S275 – S281, ISSN 0953-4075 (print) 1361-6455 (web)

Hu, S. X. & Collins, L. A. (2006). Attosecond pump probe: exploring ultrafast electron motion inside an atom. *Physical Review Letters,* 96, (February 2006) 073004-1 - 073004-4, ISSN 1079-7114 (online), 0031-9007 (print)

Humphrey, D. S.; Cabailh, G.; Pang, C. L.; Muryn, C. A.; Cavill, S. A.; Marchetto, H.; Potenza, A.; Sarnjeet S. Dhesi, S. S. & Thornton, G. (2009). Self assembled metallic nanowires on a dielectric support: Pd on rutile TiO_2 (110). *Nano Letters,* 9, 1, (January 2009) 155-159, ISSN 1530-6992 (online), 1530-6984 (print)

Jordan, G. & Scrinzi, A. (2008). Core polarization effects in molecular high harmonic generation. *New Journal of Physics,* 10, (February 2008) 025035-1 - 025035-18, ISSN 1367-2630 (online)

Kazamias, S.; Douillet, D.; Weihe, F.; Rousse, A.; Sebban, S.; Grillon, G.; Augé, F.; Hulin, D.& Balcou, Ph. (2003). Global optimization of high harmonic generation. *Physical Review Letters,* 90, (May 2003) 193901-1 - 193901-4, ISSN 1079-7114 (online), 0031-9007 (print)

Kling, M. F. and Vrakking, M. J. J.; (2008). Attosecond electron dynamics. *Annual Review in Physical Chemistry,* 59, (May 2008) 463-492, ISSN

Levesque, J.; Zeidler, D.; Marangos, J. P.; Corkum, P. B. & Villeneuve, D. M. (2007). High Harmonic Generation and the Role of Atomic Orbital Wave Functions. *Physical Review Letters,* 98, 18, (April 2007) 183903-1 - 183903-4, ISSN 1079-7114 (online), 0031-9007 (print)

Lewenstein, M.; Balcou, Ph.; Ivanov, M. Yu.; C. L.; L'Huillier, A. & Corkum, P. B. (1994). Theory of high harmonic generation by low-frequency laser fields. *Physical Review A,* 49, 3, (March 1994) 2117-12132, ISSN 1094-1622 (online), 1050-2947 (print), 1538-4446 (CD-ROM)

Marcatili, E. A. J. & Schmeltzer, R. A. (1964). Hollow metallic and dielectric waveguides for long distance optical transmission and lasers (long distance optical transmission in hollow dielectric and metal circular waveguides, examining normal mode propagation). *Bell Systems Technology Journal,* 43, (July 1964) 1783-1809, ISSN 00058580

Reprinted with permission from Mills, B.; Chau, C. F.; Rogers, E. T. F.; Grant-Jacob, J.; Stebbings, S. L.; Praeger, M.; de Paula, A. M.; Froud, C. A.; Chapman, R. T.;

Butcher, T. J.; Baumberg, J. J.; Brocklesby, W. S. & Frey, J. G. Direct measurement of the complex refractive index in the extreme ultraviolet spectral region using diffraction from a nanosphere array. 93 (23) 231103-1 – 231103-3 (December 2008). Copyright (2008) American Institute of Physics. ISSN Print + Online: 0003-6951 Online only: 1077-3118

Neutze, R.; Wouts, R.; van der Spoel, D.; Weckert, E. & Hajdu, J. (2000). Potential for biomolecular imaging with femtosecond x-ray pulses. *Nature*, 406, 803, (August 2000) 752-757, ISSN 0028-0836

Paul, P. M.; Toma, E. S.; Berger, P.; Mullot, G.; Augé, F.; Balcou, Ph.; Muller, H. G. & Agostini, P. (2001). Observation of a train of attosecond pulses from high harmonic generation. *Science*, 292, 5522, (June 2001) 1689 - 1692, ISSN 1095-9203 (online), 0036-8075 (print)

Praeger, M.; de Paula, A. M.; Froud, C. A.; Rogers, E. T. F.; Stebbings, S. L.; Brocklesby, W. S.; Baumberg, J. J. & Frey, J. G. (2007). Spatially resolved soft X-ray spectrometry from single-image diffraction. *Nature Physics*, 3, (February 2007) 176-179, ISSN 1745-2473

Praeger, M. (2008). Development and spatio-spectral mapping of a capillary high harmonic source. *Faculty of Engineering, Science and Mathematics, School of Physics and Astronomy*, University of Southampton, PhD thesis (March 2008)

Popov, V. S. (2004). Tunnel and multiphoton ionization of atoms and ions in a strong laser field (Keldysh theory). *Physics -Usp*, 3, (September 2004) 855-885

Rosenhahn, A.; Staier, F.; Nisius, T.; Schäfer, D.; Barth, R.; Christophis, C.; . Stadler, L-M.; Streit-Nierobisch, S.; Gutt, C.; Mancuso, A.; Schropp, A.; Gulden, J.; Reime, B.; Feldhaus, J.; Weckert, E.; Pfau, B.; Günther, C. M.; Könnecke, R.; Eisebitt, S.; Martins, M.; Faatz, B.; Guerassimova, N.; Honkavaara, K.; Treusch, R.; Saldin, E.; Schreiber, S.; Schneidmiller, E. A.; Yurkov, M. V.; Vartanyants, I.; Grübel, G.; Grunze, M. & Wilhein, T. (2009). Digital In-line Holography with femtosecond VUV radiation provided by the free-electron laser FLASH. *Optics Express*, 17, 10, (May 2009) 8220 - 8228, ISSN 1094-4087

Sandberg, R. L.; Song, C.; Wachulak, P. W. W.; Raymondson, D. A.; Paul, A.; B Amirbekian, B.; Lee, E.; Sakdinawat, A. E.; Voraki, C. La- O-; Marconi, M. C.; Menoni, C. S.; Murnane, M. M.; Rocca, J. J.; Kapteyn, H. C. & Miao, J. (2008). High numerical aperture tabletop soft x-ray diffraction microscopy with 70-nm resolution. *Proceedings of the National Academy of Sciences of the United States of America*, 105, 1, (January 2008) 24 - 27, ISSN 1091-6490 (online)

Schafer, K. J.; Yang, B.; DiMauro, L. F. & Kulander, K. C. (1993). Above threshold ionization beyond the high harmonic cutoff. *Physical Review Letters*, 70, 11, (March 1993) 1599 - 1602, ISSN 1079-7114 (online), 0031-9007 (print)

Stebbings, S.L.; Rogers, E. T. F.; de Paula, A. M.; Praeger, M.; Froud, C. A.; Mills, B.; Hanna, D. C.; Baumberg, J. J.; Brocklesby, W. S. & Frey, J. G. (2008). Molecular variation of capillary-produced soft x-ray high harmonics. *J. Phys. B: At. Mol. Opt. Phys.*, 41, 14, (July 2008) 145602-1 - 145602-8, ISSN 0953-4075 (print) 1361-6455 (web) Editor Adrian Corrigan.

Sutherland, J.; Christensen, E.; Powers, N.; Rhynard, S.; Painter, J. & Peatross, J. (2004). High harmonic generation in a semi-infinite gas cell. *Optics Express*, 12, 19, (September 2004) 4430 - 4436, ISSN 1094-4087

Tamaki, Y.; Itatani, J.; Nagata, Y.; Obara, M. & Midorikawa, K. (1999). Highly efficient, phase-matched high-harmonic generation by a self-guided laser beam, 82, 7, (February 1999) 1422 - 1425, ISSN 1079-7114 (online), 0031-9007 (print)

22

Relativistic Nonlinear Thomson Scattering: Toward Intense Attosecond Pulse

Kitae Lee[1], Sang-Young Chung[2], and Dong-Eon Kim[3]
[1]*Korea Atomic Energy Research Institute,*
[2]*Pusan national University,*
[3]*Pohang University of Science and Technology*
Republic of Korea

1. Introduction

Over many millennia of human history, mankind has been interested in how events change in time, namely their dynamics. However, the time resolution of recording individual steps has been limited to direct sensory perception such as the eye's ability (0.1 sec. or so) to recognize the motion, until 1800 AD when the technical revolution occurred following the industrial revolution. A famous motion picture of a galloping horse by E. Muybridge in 1878 is a good example of the technological development in time-resolved measurement. By this time, the nanosecond time resolution has been achieved; however, it took another century to break the nanosecond barrier as shown in Fig.1. The Advent of a laser has paved ways to ever shorter time resolution: in the 1980's, the picosecond barrier was broken and the femtosecond science and technology has rapidly progressed in the 1990's; at the turn of the 21st century, the femtosecond barrier has been broken (Hantschel et al., 2001), opening up the era of attosecond science and technology. The current shortest duration of a pulse achieved is 80 attoseconds around 100 eV of photon energy (Goulielmakis et al., 2008).

Femtosecond science and technology have allowed us to explore various ultrafast phenomena in physical (Siders et al., 1999), chemical (Zewail, 2000) and biological (Vos et al., 1999) systems. A great number of ultrafast atomic motions in biology, chemistry, and physics have been investigated with optical probes. In physics, the nature of atomic re-arrangements during phase transitions and the relation between amorphous, liquid and crystalline states has been interest (Afonso et al., 1996; Huang et al., 1998). Along with much interest in spintronics during the last decade, efforts have been made to understand spin dynamics in various pure and complex magnetic systems. In chemistry, the real time observation of atomic motions in chemical reactions has been long thought for. Femtosecond optical and IR technology has served this purpose in excellent ways. Femtosecond pulses have pumped molecules to create wavepackets. The observation of the motion of the wavepackets using femtosecond pulse probe or other methods has provided rich information on chemical reactions (Zewail, 2000). The various chemical bonds such as covalent, ionic, dative, metallic, hydrogen and van der Waals bonds have been studied in the varying complexity of molecular systems from diatomics to proteins and DNA. All of

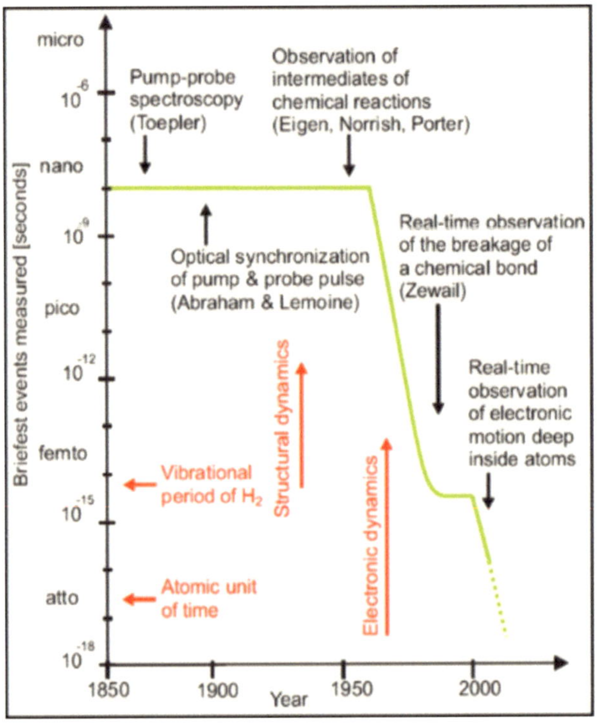

Fig. 1. Evolution of techniques for real-time observation of microscopic processes (Krausz et al., 2009).

these successful applications of femtosecond optical technology to follow reaction dynamics in chemistry have led to the award of Nobel prize to Prof. Zewail in 1999 who has initiated femtochemistry in the 1990's. There are also many fundamental biological processes taking place in femtosecond time scale. Good examples are photo-induced isomerizations (Gai et al., 1998) and ligand dissociations (Perutz et al., 1998). Femtosecond optical pulses have been used for the investigation of these processes.

During the last several years, isolated attosecond pulses have been successfully exploited to control the localization of electron under chemical reaction (Kling et al., 2006), observe inner shell transitions in atoms (Drescher et al., 2002), and electron tunneling across Coulomb barrier (Uiberacker et al., 2007), electron transport in condensed matter (Cavalieri et al., 2007).

The demand on and the emphasis to the understanding of the ultrafast phenomena and the control of them is ever increasing. The Department of Energy of the United States of America asked the Basic Energy Science Advisory Committee (BESAC) to identify grand challenges in science to be pursued in the 21st century. The BESAC has identified 5 grand challenges in their report, the summary of which has been published in the July issue of Physics Today in 2008 (Fleming & Ratner, 2008). The first of the grand challenges is to control material properties at the level of electronic motion. The report recommended the development of attosecond and femtosecond metrology to measure and control electron dynamics.

Optical or ultraviolet light allows one to probe the dynamical changes of excited electronic states in a sample. This is the core of interest in many research topics such as electron-hole dynamics in solids, or excitation transfer in photosynthesis, bond breaking in chemical reactions. However, no or partial information on structure can be obtained if the process under investigation involves the structural changes.

X-ray imaging or diffraction has the potential to provide a global picture of structural changes in the fields of atomic and molecular physics, plasma physics, material science, chemistry and life science (Bloembergen, 1999; http://tesla.desy.de/new_pages/TDR_CD/PartV/fel.html). Since x-ray photons are scattered from all the electrons in a sample, the intensity of diffraction is directly related to the electronic density. Since the structural changes occur in the time scale of 100 fs or so, following the charge re-arrangement, the electronic density measured in this time scale closely reflects the atomic structure. On the other hand, attosecond x-ray pulse will enable us to follow even faster motions of electrons in intra-atomic or intra-molecular processes, which has not been achieved yet because of the lack of such sources.

Femtosecond time-resolved x-ray diffraction experiments have been used to study structural processes such as the observation of atomic structure and dynamics [18], the investigation of ultrafast phase transition in solid (Gaffney & Chapman, 2008) and time-resolved biomolecular imaging (Sokolowski-Tinten et al., 2003). These results are very impressive and of landmark: however, there are still many phenomena yet to be explored and a variety of attosecond and femtosecond X-ray sources yet to be further developed. They comprise a great challenge to the scientific community.

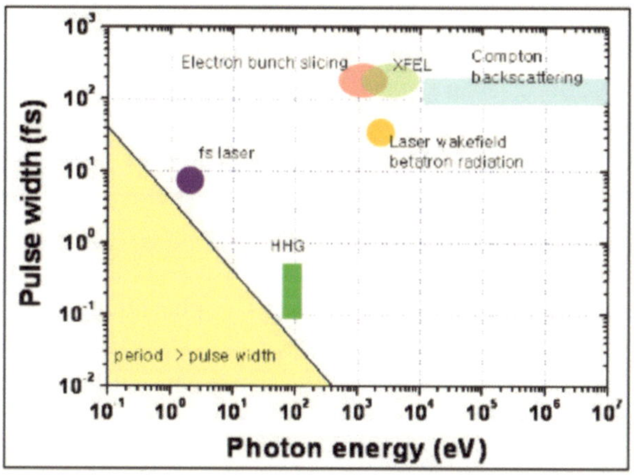

Fig. 2. Currently-available ultrafast light sources are plotted in terms of photon energy and pulse duration.

Figure 2 shows ultrafast sources available currently or in the immediate future. It is conspicuous that there is no source available in keV or higher energies faster than 10 fs. In fact, for wider exploration and manipulation of electron dynamics in a vast spectrum of natural phenomena, attosecond or a few fs keV pulses are demanded.

Several schemes have been proposed and/or demonstrated to generate an ultrashort keV x-ray pulse: the relativistic Doppler shift of a backscattered laser pulse by a relativistic electron beam (Sprangle et al., 1992; Hartemann, 1998; Esarey et al., 1993a; Chung et al., 2009), the harmonic frequency upshift of a laser pulse by relativistic nonlinear motion of electrons (Vachaspati, 1962; Brown & Kibble, 1964; Esarey et al,, 1992; Chen et al., 1998, 2000; Ueshima et al, 1999; Kaplan & Shkolnikov, 2002; Banerjee et al., 2002; Lee et al., 2003a, 2003b, 2005, 2008; Phuoc et al., 2003; Kim et al., 2009), high order harmonic generation in the interaction of intense laser pulse with solids (Linde et al., 1995, 1996, 1999; Norreys et al., 1996; Lichters et al., 1996; Tarasevitch et al., 2000) and x-ray laser using inner shell atomic transitions (Kim et al., 1999, 2001).

Ultrafast high-intensity X-rays can be generated from the interaction of high intensity femtosecond laser via Compton backscattering (Hartemann et al., 2005), relativistic nonlinear Thomson scattering (Ueshima et al., 1999; Kaplan & Shkolnikov 2002; Banerjee et al., 2002) and laser-produced betatron radiation (Phuoc et al., 2007). In synchrotron facilities, electron bunch slicing method has been adopted for experiments (Schoenlein, 2000; Beaud et al., 2007). Moreover, X-ray free electron lasers (Normile, 2006) were proposed and have been under construction. The pulse duration of these radiation sources are in the order of a few tens to hundred fs. There are growing demands for new shorter pulses than 10 fs.

The generation of intense attosecond or femtosecond keV lights via Thomson scattering (Lee et al., 2008; Kim et al., 2009) is attractive, because the radiation is intense and quasi-monochromatic. This radiation may be also utilized in medical (Girolami et al., 1996) and nuclear physics (Weller & Ahmed, 2003) area of science and technology.

When a low-intensity laser pulse is irradiated on an electron, the electron undergoes a harmonic oscillatory motion and generates a dipole radiation with the same frequency as the incident laser pulse, which is called Thomson scattering. As the laser intensity increases, the oscillatory motion of the electron becomes relativistically nonlinear, which leads to the generation of harmonic radiations. This is referred to as **relativistic nonlinear Thomson scattered (RNTS) radiation**. The RNTS radiation has been investigated in analytical ways (Esarey et al., 1993a; Chung et al., 2009; Vachaspati, 1962; Brown et al., 1964; Esarey & Sprangle, 1992; Chen et al., 1998; Ueshima et al., 1999; Chen et al., 2000; Kaplan & Shkolnikov, 2002; Banerjee et al., 2002). Recently, such a prediction has been experimentally verified by observing the angular patterns of the harmonics for a relatively low laser intensity of 4.4×10^{18} W/cm^2 (Lee et al., 2003a, 2003b). Esarey *et al.* (Esarey et al., 1993a) has investigated the plasma effect on RNTS and presented a set of the parameters for generating a 9.4-ps x-ray pulse with a high peak flux of 6.5×10^{21} photons/s at 310 eV photon energy using a laser intensity of 10^{20} W/cm^2. Ueshima *et al.* (Ueshima et al., 1999) has suggested several methods to enhance the radiation power, using particle-incell simulations for even a higher intensity. Kaplan and Shkolnikov *et al.* (Kaplan & Shkolnikov, 2002) proposed a scheme for the generation of zeptosecond (10^{-21} sec) radiation using two counter-propagating circularly polarized lasers, named as lasertron.

Recently, indebted to the development of the intense laser pulse, experiments on RNTS radiation have been carried out by irradiating a laser pulse of 10^{18}–10^{20} W/cm^2 on gas jet targets (Kien et al., 1999; Paul et al., 2001; Hertz et al., 2001). A numerical study in the case of single electron has been attempted to characterize the RNTS radiation (Kawano et al., 1998) and a subsequent study has shown that it has a potential to generate a few attosecond x-ray pulse (Harris & Sokolov, 1998). Even a scheme for the generation of a zeptosecond x-ray pulse using two counter propagating circularly polarized laser pulses has been proposed (Kaplan & Shkolnikov, 1996).

In this chapter, we concern RNTS in terms of the generation of ultrafast X-ray pulses. The topics such as fundamental characteristics of RNTS radiations, coherent RNTS radiations, effects of the high-order fields (HOFs) under a tight-focusing condition, and generation of an intense attosecond x-ray pulse will be discussed in the following sections.

2. Fundamental characteristics of RNTS radiations

In this section, the dynamics of an electron under an ultra-intense laser pulse and some fundamental characteristics of the RNTS radiations will be discussed (Lee et al., 2003a, 2003b).

2.1 Electron dynamics under a laser pulse

The dynamics of an electron irradiated by a laser field is obtained from the relativistic Lorentz force equation:

$$\frac{d}{dt}\left(\gamma\vec{\beta}\right) = -\frac{e}{m_e c}\left(\vec{E}_L + \vec{\beta}\times\vec{B}_L\right), \tag{1}$$

The symbols used are: electron charge (e), electron mass (m_e), speed of light (c), electric field ($\vec{E}_L$), magnetic field ($\vec{B}_L$), velocity of the electron divided by the speed of light ($\vec{\beta}$), and relativistic gamma factor ($\gamma = 1/\sqrt{1-\beta^2}$). It is more convenient to express the laser fields with the normalized vector potential, $\vec{a} = e\vec{E}_L / m_e\omega_L c$, where ω_L is the angular frequency of the laser pulse. It can be expressed with the laser intensity I_L in W/cm^2 and the laser wavelength λ_L in micrometer as below:

$$a = 8.5\times 10^{-10}\lambda_L\sqrt{I_L}\ . \tag{2}$$

Eq. (1) can be analytically solved under a planewave approximation and a slowly-varying envelope approximation, which lead to the following solution (Esarey et al., 1993a):

$$\gamma\vec{\beta} = \gamma_o\vec{\beta}_o + \vec{a} + \frac{a^2 + 2\gamma_o\vec{a}\cdot\vec{\beta}_o}{2q_o}\hat{z}, \tag{3}$$

$$\gamma = \frac{\left|\vec{a} + \gamma_o\vec{\beta}_{o\perp}\right|^2 + 1 + q_o^2}{2q_o}, \tag{4}$$

where $q_o = \gamma_o\left(1-\beta_{oz}\right)$ and the subscript $\perp$ denotes the direction perpendicular to the direction of laser propagation (+z). The subscript, 'o' denotes initial values. When the laser

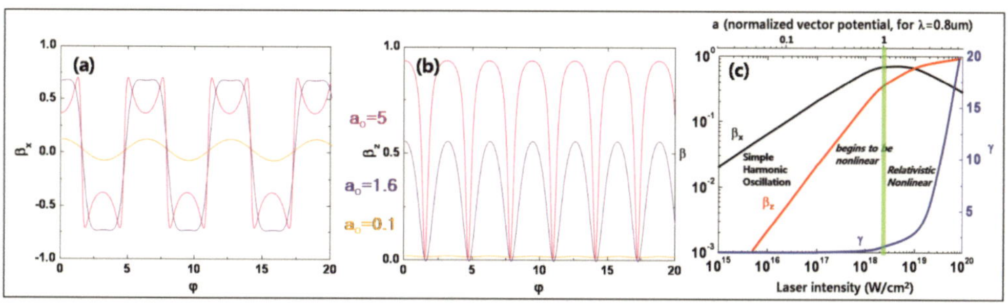

Fig. 3. Dynamics of an electron under a laser pulse: Evolution of (a) transverse and (b) longitudinal velocities, and (c) peak values on laser intensities. The initial velocity was set to zero for this calculation. Different colors correspond to different a_o's in (a) and (b).

intensity is low or $|a| << 1$, the electron conducts a simple harmonic oscillation but as the intensity becomes relativistic or $|a| \geq 1$, the electron motion becomes relativistically nonlinear. Figure 3 (a) and (b) show how the electron's oscillation becomes nonlinear due to relativistic motion as the laser intensity exceeds the relativistic intensity. One can also see that the drift velocity along the +z direction gets larger than the transverse velocity as $|a| \geq 1$ [Fig. 3 (c)].

2.2 Harmonic spectrum by a relativistic nonlinear oscillation

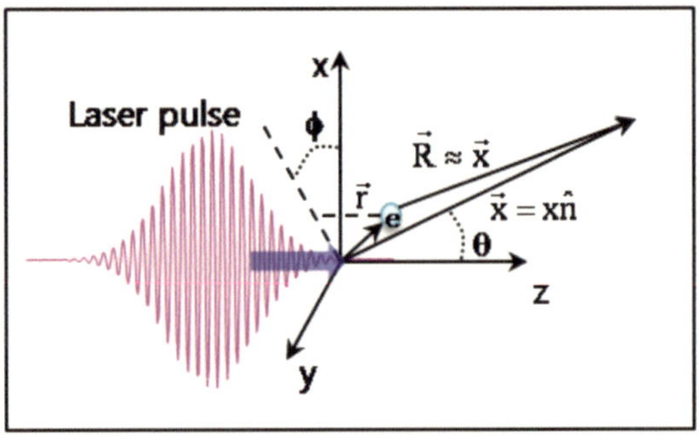

Fig. 4. Schematic diagram for the analysis of the RNTS radiations

Once the dynamics of an electron is obtained, the angular radiation power far away from the electron toward the direction, $\hat{n}$ [Fig. 4] can be obtained through the Lienard-Wiechert potential (Jackson, 1975)

$$\frac{dP(t)}{d\Omega} = \left|\vec{A}(t)\right|^2 \tag{5}$$

$$\vec{A}(t) = \sqrt{\frac{e^2}{4\pi c}} \left[\frac{\hat{n} \times \left\{ \left(\hat{n} - \vec{\beta}\right) \times \dot{\vec{\beta}} \right\}}{\left(1 - \hat{n} \cdot \vec{\beta}\right)^3} \right]_{t'} \tag{6}$$

where t′ is the retarded time and is related to t by

$$t = t' + \frac{x - \hat{n} \cdot \vec{r}(t')}{c}. \tag{7}$$

Then the angular spectrum is obtained by

$$\frac{d^2 I}{d\omega d\Omega} = 2\left|\vec{A}(\omega)\right|^2, \tag{8}$$

where $\vec{A}(\omega)$ is the Fourier transform of $\vec{A}(t)$. These formulae together with Eq. (1) are used to evaluate the scattered radiations. Under a planewave approximation, the RNTS spectrum can be analytically obtained (Esarey et al., 1993a). Instead of reviewing the analytical process, important characteristics will be discussed along with results obtained in numerical simulations.

Figure 5 shows how the spectrum is changed, as the laser intensity gets relativistic. The spectra were obtained by irradiating a linearly-polarized laser pulse on a counter-propagating relativistic electron with energy of 10 MeV, which is sometimes called as nonlinear Compton backscattering. One can see that higher order harmonics are generated as the laser intensity increase. It is also interesting that the spacing between harmonic lines gets narrower, which is caused by Doppler effect (See below). The cut-off harmonic number has been numerically estimated to be scaled on the laser intensity as $\sim a^3$ (Lee et al., 2003b).

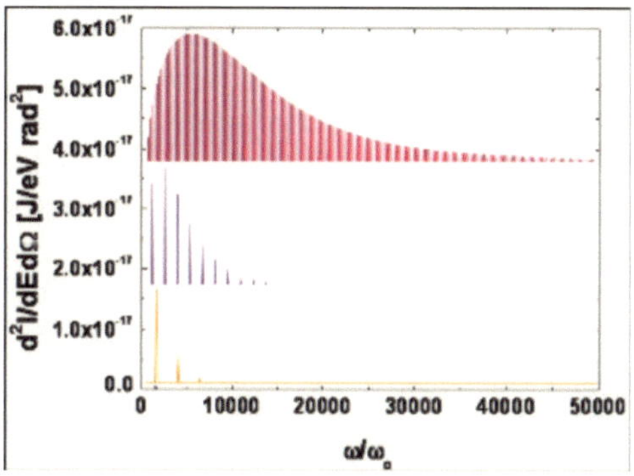

Fig. 5. Spectra of RNTS in a counter-propagating geometry for different laser intensities, a_o=0.1, 0.8, 1.6, and 5 from bottom. (The spectrum for a_o=0.1 is hardly seen due to its lower intensity.)

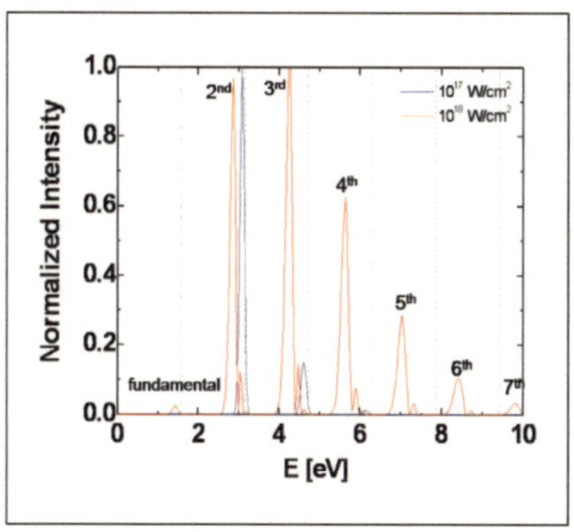

Fig. 6. Red-shift of harmonic frequencies on laser intensity. The spectra were obtained at the direction of $\theta = 90^o$ and $\phi = 0^o$ from an electron initially at rest. The vertical dotted lines indicate un-shifted harmonic lines. For this calculation, a linearly polarized laser pulse with a pulse width in full-with-half-maximum (FWHM) of 20 fs was used.

As shown in Fig. 6, the fundamental frequency, ω_1^s shifts to the red side as the laser intensity increases. This is caused by the relativistic drift velocity of the electron driven by $\vec{v} \times \vec{B}_L$ force. Considering Doppler shift, it can be obtained as (Lee et al., 2006)

$$\frac{\omega_1^s}{\omega_L} = \frac{4\gamma_o^2(1-\beta_{zo})}{4\gamma_o^2\left(1-\vec{\beta}_o \cdot \hat{n}\right) + a_o^2 \dfrac{(1-\cos\theta)}{(1-\beta_{zo})}}. \tag{9}$$

In the case of an electron initially at rest ($\gamma_o = 1$, $\beta_o = 0$), this leads to the following formula

$$\frac{\omega_1^s}{\omega_L} = \frac{1}{1 + \frac{a_o^2}{4}(1 - \cos\theta)}. \tag{10}$$

Note that the amount of the red shift is different at different angles. The dependence on the laser intensity can be stated as follows. As the laser intensity increases, the electron's speed approaches the speed of light more closely, which makes the frequency of the laser more red-shifted in the electron's frame. No shift occurs in the direction of the laser propagation. The parasitic lines in the blue side of the harmonic lines are caused by the different amount of the red-shift due to rapid variation of laser intensity.

The angular distributions of the RNTS radiations show interesting patterns depending on harmonic orders [Fig. 7]. The distribution in the forward direction is rather simple, a dipole radiation pattern for the fundamental line and a two-lobe shape for higher order harmonics. There is no higher order harmonic radiation in the direction of the laser propagation. In the backward direction, the distributions show an oscillatory pattern on θ and the number of peaks is equal to the number of harmonic order. Thus there is no even order harmonics to the direction of $\theta = 180^o$.

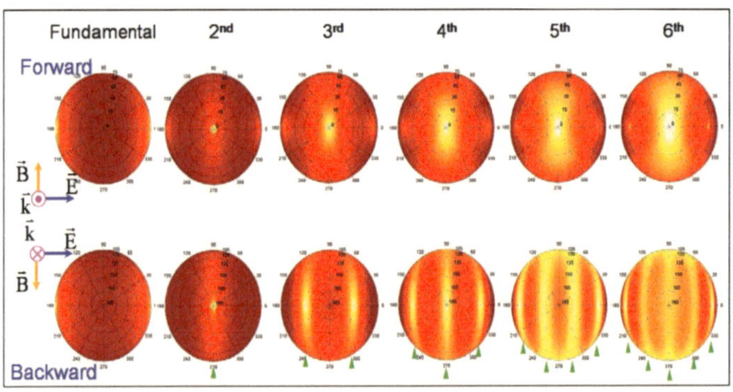

Fig. 7. Angular distributions of the RNTS harmonic radiations from an electron initially at rest. This was obtained with a linearly polarized laser pulse of 10^{18} W/cm^2 in intensity, 20 fs in FWHM pulse width. The green arrows in the backward direction indicate nodes.

For a laser intensity of 10^{20} W/cm^2 (a_o=6.4), the harmonic spectra from an electron initially at rest are plotted in Fig. 8 for different laser polarizations. In the case of a linearly polarized laser, the electron undergoes a zig-zag motion in a laser cycle. Thus the electron experiences severer instantaneous acceleration than in the case of a circularly polarized laser, in which case the electron undergoes a helical motion. This makes RNTS radiation stronger in intensity and higher in photon energy in the case of a linearly polarized laser. The most different characteristics are the appearance of a large-interval modulation in the case of a linear polarization denoted as '1' in Fig. 8 (a). This is also related with the zig-zag motion of the electron during a single laser cycle. During a single cycle, the electron's velocity becomes zero instantly, which does not happen in the case of the circular polarization. Thus a double peak radiation appears in a single laser cycle as shown in Fig. 9 (a). Such a double peak structure in the time domain makes the large-interval modulation in the energy spectrum. In both cases, there are modulations with small-interval denoted by '2' in the Fig. 8 (a) and (b). This is caused by the variation of the laser intensity due to ultra-short laser pulse width. Such an intensity variation makes the drift velocity different for each cycle then the time interval between radiation peaks becomes different in time domain, which leads to a small-interval modulation in the energy spectrum.

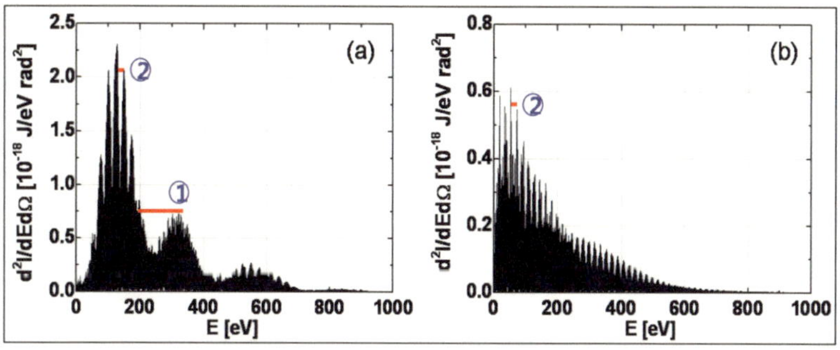

Fig. 8. RNTS spectra from an electron initially at rest on laser polarizations: (a) linear and (b) circular. The laser intensity of 10^{20} W/cm^2 (a_o=6.4) and the FWHM pulse width of 20 fs were used. Note that harmonic spectra are deeply modulated. See the text for the explanation.

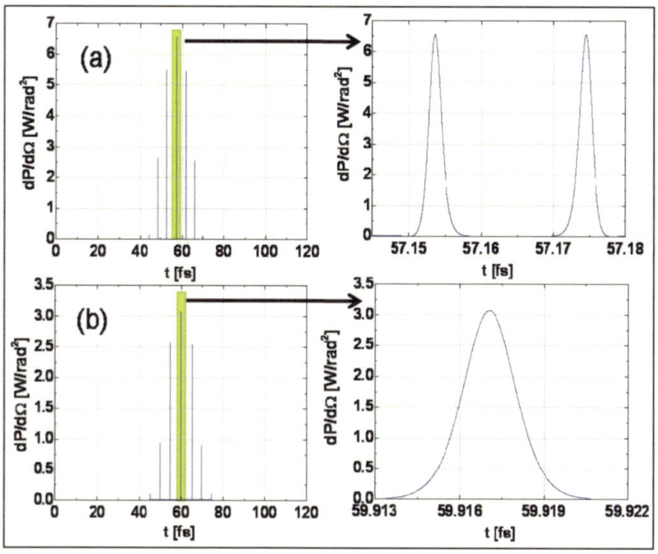

Fig. 9. Temporal shape of the RNTS radiations on polarizations with the same conditions as in Fig. 8: (a) linear and (b) circular polarization. The figures on the right hand side are the zoom-in of the marked regions in green color.

The temporal structure or the angular power can be seen in Fig. 9. As commented above, in the case of the linear-polarization, it shows a double-peak structure. One can also see that the pulse width of each peak is in the range of attosecond. This ultra-short nature of the RNTS radiation makes RNTS deserve a candidate for as an ultra-short intense high-energy photon source. The pulse width is proportional to the inverse of the band width of the harmonic spectrum, and thus scales on the laser intensity as $\sim a^{-3}$ (Lee et al., 2003b). The peak power is analytically estimated to scale $\sim a^{5}$ (Lee et al., 2003b).

The zig-zag motion of an electron under a linearly polarized laser pulse makes the radiation appears as a pin-like pattern in the forward direction as shown in Fig. 10 (a). However the radiation with a circularly polarized laser pulse shows a cone shape [Fig. 10 (b)] due to the helical motion of the electron. The direction of the peak radiation, θ_p was estimated to be $\theta_p \approx 2\sqrt{2} / a_o$ (Lee et al., 2006).

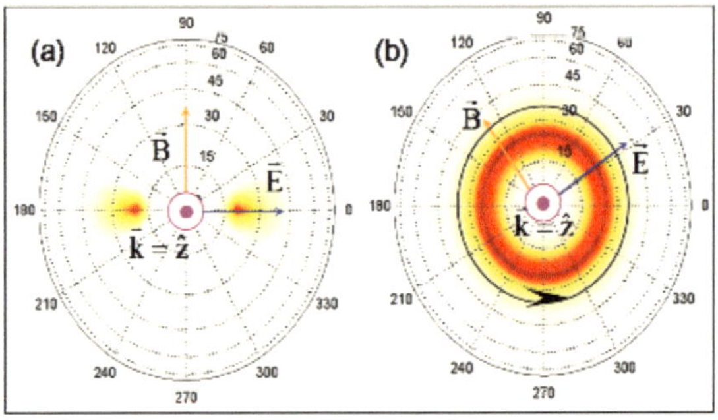

Fig. 10. Angular distributions of the RNTS radiations for different polarizations (a) linear and (b) circular polarization. The laser intensity of 10^{20} W/cm^2 (a_o=6.4) and the FWHM pulse width of 20 fs are used as in Fig. 8.

3. Coherent RNTS radiations

In the previous section, fundamental characteristics of the RNTS radiation are investigated in the case of single electron. It was also shown that the RNTS radiation can be an ultra-short radiation source in the range of attosecond. To maintain this ultra-short pulse width or wide harmonic spectrum even with a group of electrons, it is then required that the radiations from different electrons should be coherently added at a detector. In the case of RNTS radiation, which contains wide spectral width, such a requirement can be satisfied only if all the differences in the optical paths of the radiations from distributed electrons to a detector be almost the same. This condition can be practically restated: all the time intervals that scattered radiations from different electrons take to a detector, Δt_{int} should be comparable with or less than the pulse width of single electron radiation, Δt_{rad} as shown in Fig. 11. In the following subsections, two cases of distributed electrons, solid target and elelctron beam will be investigated for the coherent RNTS radiations.

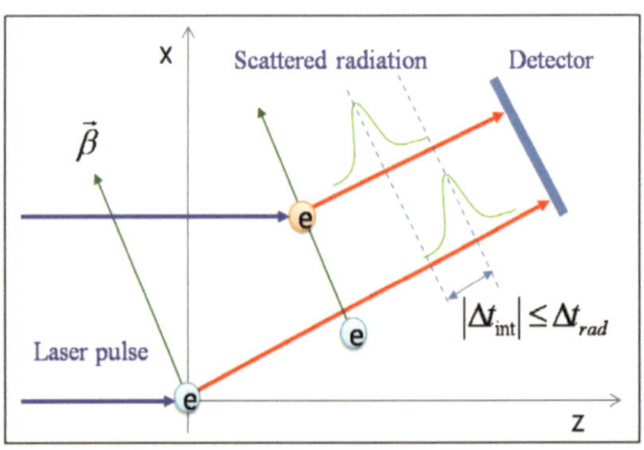

Fig. 11. Schematic diagram for the condition of coherent RNTS radiation.

3.1 Solid target

In the case of a solid target for distributed electrons (Lee et al, 2005), the time intervals that radiations take to a detector can be readily obtained with the following assumptions as the first order approximation: (1) plane wave of a laser field, (2) no Coulomb interaction between charged particles, thus neglecting ions, and (3) neglect of initial thermal velocity distribution of electrons during the laser pulse. With these assumptions, the radiation field $\vec{f}_i(t)$ by an electron initially at a position, $\vec{r}_i$, due to irradiation of an ultra-intense laser pulse propagating in the +z direction can be calculated from that of an electron initially at origin, $\vec{f}_o(t)$ by considering the time intervals between radiations from the electron at $\vec{r}_i$ and one at origin, Δt_i,

$$\Delta t_i = \Delta t'_i - \frac{\hat{n}\cdot\vec{r}}{c}, \tag{11}$$

where $\Delta t'_i = z_i / c$ is the time which the laser pulse takes to arrive at the i-th electron from origin: $\vec{f}_i(t) = \vec{f}_o(t - \Delta t_i)$. Then all the radiation fields from different electrons are summed on a detector to obtain a total radiation field, $\vec{F}(t)$ as

$$\vec{F}(t) = \sum_i \vec{f}_o(t - \Delta t_i). \tag{12}$$

The condition for a coherent superposition in the z-x plane can now be formulated by setting Eq. (11) to be less than or equal to the pulse width of single electron radiation, Δt_{rad}. This leads to the following condition [See Fig. 12]:

$$|z\tan\xi - x| \le \frac{c\Delta t_{rad}}{\sin(2\xi)}. \tag{13}$$

Equation (13) manifests that RNTS radiations are coherently added to the specular direction of an incident laser pulse off the target, if the target thickness, T_{hk} is restricted to

$$T_{hk} \le \frac{c\Delta t_{rad}}{\sin\xi}. \tag{14}$$

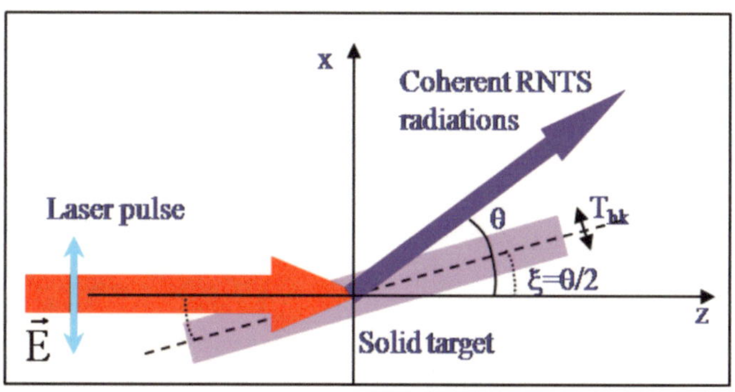

Fig. 12. Schematic diagram for a coherent RNTS condition with an ultrathin solid target.

Since the incident angle of the laser pulse can be set arbitrarily, one can set θ to the direction of the radiation peak of single electron, θ_p. For a linearly polarized laser with an intensity of 4x10^{19} W/cm^2, and a pulse duration of 20 fs FWHM, $\theta_p = 27^o$ and $\Delta t_{rad} = 5$ attosecond for a single electron. Equation (14) then indicates that the target thickness should be less than 7 nm. With these laser conditions, harmonic spectra were numerically obtained to demonstrates the derived coherent condition [Fig. 13].

The spectra in Fig. 13 (a) were obtained for a thick cylindrical target of 1 μm in thickness and radius, and 10^{18} cm^{-3} in electron density under the normal incidence of a laser on its base. The spectrum in Fig. 13 (b) is for the case of oblique incidence on an ultra-thin target of 7 nm in thickness, 5 μm in width, 20 μm in length, 10^{16} cm^{-3} in electron density, and $\xi = 13.5^o$, which were obtained with Eqs. (13) and (14). From Fig. 13 (b), which corresponds to the condition for coherent RNTS radiation, one can find that the spectrum from thin film (a group of electrons) has almost the same structure as that from a single electron radiation [Inset in Fig. 13 (b)] in terms of high-energy photon and a modulation. On the other hand, in the case of Fig. 13 (a), the harmonic spectra show much higher intensity at low energy part, which is caused by an incoherent summation of radiations.

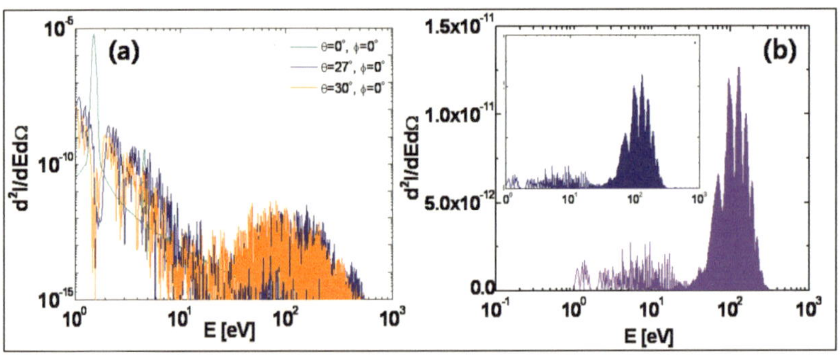

Fig. 13. RNTS spectra obtained under (a) incoherent and (b) coherent conditions. In (a), the spectra obtained in three different directions are plotted, while (b) were obtained in the specular direction. One can see that the spectrum in the coherent condition is very similar with that obtained from single electron calculation (inset of (b)).

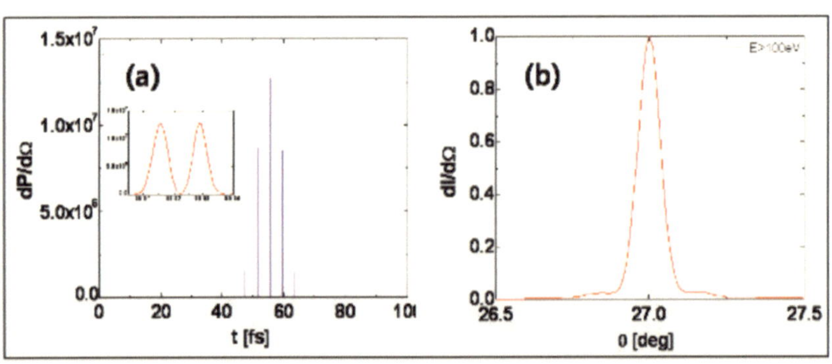

Fig. 14. (a) Temporal shape and (b) angular distribution in the case of the coherent condition [Fig. 13 (b)].

The temporal shape at the specular direction for the case of coherent condition [Fig. 13 (b)] is plotted in Fig. 14 (a), which shows an attosecond pulse. The direction-matched coherent condition also leads to a very narrow angular divergence as shown in Fig. 14 (b). It should be mentioned that with a thick cylinder target, the radiation peak appears at $\theta = 0^{o}$, because the dipole or fundamental radiation becomes dominant in that direction.

3.2 Electron beam

Exploiting a solid target for a coherent RNTS radiation may involve a complicated plasma dynamics due to an electrostatic field produced by a charge separation between electrons and ions. Instead, an idea using an electron beam has been proposed (Lee et al., 2006).

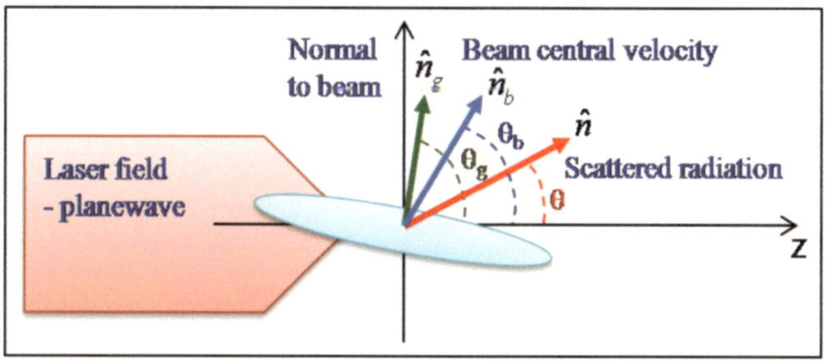

Fig. 15. Schematic diagram for the analysis of a coherent RNTS radiation with an electron beam.

Following similar procedure in the previous section, the RNTS harmonic spectrum can be obtained with that from an electron at center and its integration over initial electron distributions with phase relationships as

$$\vec{A}(\omega) \approx \vec{A}_c(\omega) \int dV_o f\left(\vec{r}_o, \vec{\beta}_o\right) \exp\left(i\delta \frac{\omega}{\omega_o} \right), \tag{15}$$

where $\vec{A}_c(\omega)$ is the angular spectral field from the central electron. The $\delta / k_o = \left(1 - \hat{n}\cdot\vec{\beta}_o\right) / \left(1 - \beta_{oz}\right) z - \hat{n}\cdot\vec{r}_o$ represents the phase relations between scattered radiations due to different initial conditions of the electrons. The distribution function can be assumed to have a Gaussian profile with cylindrical symmetry:

$$f\left(\vec{r}_o, \vec{\beta}_o\right) = \frac{N}{(2\pi)^3 R^2 L \sigma_\Gamma \sigma_{\beta'}^2} \exp\left[-\frac{x_o^{\,2} + y_o^{\,2}}{2R^2} - \frac{z_o^{\,2}}{2L^2} \right] \times \exp\left[-\frac{\beta_{ox}^2 + \beta_{oy}^2}{2\beta_b^2 \sigma_{\beta'}^2} - \frac{\left(\gamma_o - \gamma_b\right)^2}{2\gamma_b^2 \sigma_\Gamma^2} \right], \tag{16}$$

where the following parameters are used: the number of electrons (N), radius (R), length (L), fractional energy spread (σ_Γ), and divergence ($\sigma_{\beta'}$). γ_b is the relativistic gamma factor of the beam, and β_b its corresponding velocity divided by the speed of light. In the above formula, the beam velocity and the axis of the spatial distribution of the beam have the same directions and directed to +z, but below, the direction of the beam velocity ($\hat{n}_b$) and the axis of the beam ($\hat{n}_g$) are allowed to have different directions, as shown in Fig. 15. The integration of Eq. (15) by taking the first order of ($\vec{\beta}_b - \vec{\beta}_o$) in δ leads to the following formula for the coherent spectrum:

$$\vec{A}(\omega) \approx NF(\omega)\vec{A}_c(\omega), \tag{17}$$

$$F(\omega) \approx \frac{1}{\sqrt{1+k^2 l^2}} \exp\left[-\frac{k^2}{2(1+k^2 l^2)}\left(Q_r^2 + Q_\beta^2\right)\right], \tag{18}$$

$$Q_r^2 = L^2 N_{gz}^2 + R^2\left(N_{gx}^2 + N_{gy}^2\right), \tag{19}$$

$$Q_\beta^2 = k^2 R^2 T^2\left[L^2\left(N_{gx}n_{gz} - N_{gz}n_{g\theta z}\right)^2 + N_{gy}^2\left(R^2 n_{g\theta z}^2 + L^2 n_{gz}^2\right)\right], \tag{20}$$

with $k = \omega / c$ and the other parameters being

$$l^2 = T^2\left(L^2 n_{gz}^2 + R^2 n_{g\theta z}^2\right), \tag{21}$$

$$T^2 = \left(\frac{\beta_b}{1-\beta_{bz}}\right)^2\left(\frac{\sigma_\Gamma^2}{\left(\gamma_b^2 - 1\right)^2} N_{bz}^2 + \sigma_{\beta'}^2\left(N_{bx}^2 + N_{by}^2\right)\right), \tag{22}$$

$$\vec{N}_s = \vec{M}_s \cdot (\hat{n} - p_b \hat{z}), \tag{23}$$

$$\vec{M}_s^T = \left[\hat{n}_{s\theta} \quad \hat{n}_{s\varphi} \quad \hat{n}_s\right], \tag{24}$$

$$p_b = \frac{1 - \vec{\beta}_b \cdot \hat{n}_r}{1 - \beta_{bz}}. \tag{25}$$

In Eqs. (23) and (24), the subscript, 's' represents either 'g' or 'b'. $\hat{n}_{s\theta}$ and $\hat{n}_{s\varphi}$ are two unit vectors perpendicular to $\hat{n}_s$. Equation (18) or the coherent factor, $F(\omega)$ shows that, as the beam parameters get larger, the coherent spectrum disappears from high frequency. This manifests that the phase matching condition among electrons is severer for high frequencies.

For the radiation scattered from an electron beam to be coherent up to a frequency ω_c, the above coherent factor $F(\omega)$, should be almost 1, or the exponent should be much smaller than 1 in the desired range of frequency. In the z-x plane, $N_{gy} = 0$; then, this leads to the following relations, one for the angular relation:

$$N_{gx} = \sin\left(\theta - \theta_g\right) + \sin\theta_g \frac{1 - \beta_b \cos(\theta_b - \theta)}{1 - \beta_b \cos\theta_b} = 0, \tag{26}$$

and the other for the restriction on the electron beam parameters:

$$k_c L N_{gz} \sqrt{\frac{1 + k_c^2 R^2 T^2 \sin^2\theta_g}{1 + k_c^2 l^2}} < 1. \tag{27}$$

Eq. (26) also shows why the direction of the beam velocity (θ_b) is set to be different from the axis of the beam distribution (θ_g); otherwise, N_{gx} cannot be zero. The physical meaning of Eqs. (26) and (27) is that time delays between electrons should be less than the pulse width generated by a single electron as commented in the previous section. This equation can be used to find θ_g for given θ_b and θ which can be set to the optimal condition obtained from the single electron calculation. For the realization of the coherent condition, the most important things are the length of the electron beam (L) and the condition to minimize N_{gz}. To minimize N_{gz}, θ should be near 0° but not 0° at which only dipole radiation appears. From single electron calculation, it has been found that when $\theta_b \approx 0^o$ or in the case of a co-propagation (laser and electron beam propagate near the same direction), such a condition can be fulfilled.

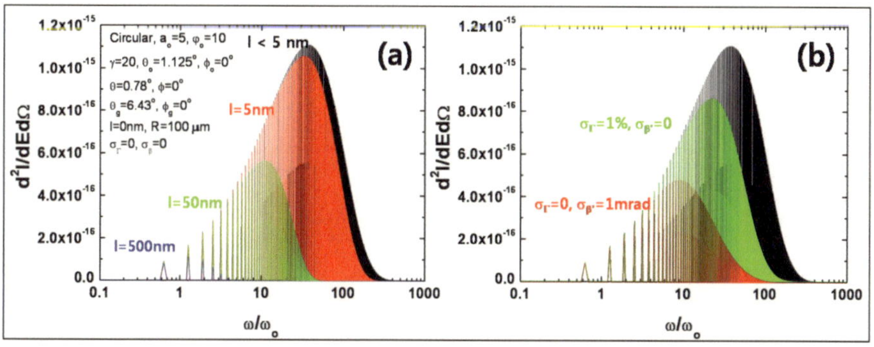

Fig. 16. Coherent RNTS radiation spectra for different beam parameters: (a) beam length and (b) other beam parameters. For better view, only envelops are plotted.

From the single electron calculation (the radiation from an electron of γ_0 = 20 under irradiation of a circularly polarized laser of a_0 = 5), it has been found that the peak radiation appears at $\theta = 0.78^o$ when $\theta_b = 1.125^o$. The insertion of these data into Eq. (26) and (27) leads to $\theta_g = 6.43^o$ and the beam length being restricted to a few nanometers. Coherent RNTS spectra for different electron beam parameters are plotted in Fig. 16. As expected, one can see that the coherent spectral intensity decreases at high frequencies as the beam length increases. These calculations show that the coherent conditions for the beam length and beam divergence are most stringent. However, with a moderate condition, the broadening of the coherent spectrum is still enough to generate about a 100-attosecond pulse.

4. Effects of the high-order laser fields under tight-focusing condition

Paraxial approximation is usually used to describe a laser beam. However, when the focal spot size gets comparable to the laser wavelength, it cannot be applied any more. This is the situation where the RNTS actually takes place. A tightly-focused laser field and its effects on the electron dynamics and the RNTS radiation will be discussed in this section.

4.1 Tightly focused laser field

The laser fields propagating in a vacuum are described by a wave equation. The wave equation can be evaluated in a series expansion with a diffraction angle, $\varepsilon = w_0 / z_r$, where w_0 is beam waist and z_r Rayleigh length. It leads to the following formulas for the laser fields having linear polarization in the x-direction (zeroth-order) and propagating in the +z direction (Davis, 1979; Salamin, 2007),

$$E_x = E\left\{S_0 + \epsilon^2\left[\xi^2 S_2 - \frac{\rho^4 S_3}{4}\right] + \epsilon^4\left[\frac{S_2}{8} - \frac{\rho^2 S_3}{4} - \frac{\rho^2(\rho^2 - 16\xi^2)S_4}{16} - \frac{\rho^4(\rho^2 + 2\xi^2)S_5}{8} + \frac{\rho^8 S_6}{32}\right]\right\}, \quad (28)$$

$$E_y = E\xi\upsilon\left\{\epsilon^2[S_2] + \epsilon^4\left[\rho^2 S_4 - \frac{\rho^4 S_5}{4}\right]\right\}, \quad (29)$$

$$E_z = E\xi\left\{\epsilon[C_1] + \epsilon^3\left[-\frac{C_2}{2} + \rho^2 C_3 - \frac{\rho^4 C_4}{4}\right] + \epsilon^5\left[-\frac{3C_3}{8} - \frac{3\rho^2 C_4}{8} + \frac{17\rho^4 C_5}{16} - \frac{3\rho^6 C_6}{8} + \frac{\rho^8 C_7}{32}\right]\right\}. \quad (30)$$

$$B_x = 0, \quad (31)$$

$$B_y = E\left\{S_0 + \epsilon^2\left[\frac{\rho^2 S_2}{2} - \frac{\rho^4 S_3}{4}\right] + \epsilon^4\left[-\frac{S_2}{8} + \frac{\rho^2 S_3}{4} + \frac{5\rho^4 S_4}{16} - \frac{\rho^6 S_5}{4} + \frac{\rho^8 S_6}{32}\right]\right\}, \quad (32)$$

$$B_z = E\upsilon\left\{\epsilon[C_1] + \epsilon^3\left[\frac{C_2}{2} + \frac{\rho^2 C_3}{2} - \frac{\rho^4 C_4}{4}\right] + \epsilon^5\left[\frac{3C_3}{8} + \frac{3\rho^2 C_4}{8} + \frac{3\rho^4 C_5}{16} - \frac{\rho^6 C_6}{4} + \frac{\rho^8 C_7}{32}\right]\right\}. \quad (33)$$

The laser fields are written up to the 5th order in ε. In above equations, $E = E_o(w/w_o)g(t - z/c)\exp(-r^2/w^2)$, $w = w_0\sqrt{1+(z/z_r)^2}$, $z_r = \pi w_0^2/\lambda$, $\xi = x/w_0$, $\upsilon = y/w_0$, $\zeta = z/z_r$, and $\rho^2 = \xi^2 + \upsilon^2$. $g(t - z/c)$ is a laser envelop function. C_n and S_n are defined as

$$\begin{aligned} C_n &= \left(\frac{w_0}{w}\right)^n \cos(\psi + n\psi_G); \qquad n = 0,1,2,3..., \\ S_n &= \left(\frac{w_0}{w}\right)^n \sin(\psi + n\psi_G), \end{aligned} \quad . \quad (34)$$

where $\psi = \psi_0 + \omega t - kz - kr^2/2R + \psi_G$ and $R = z + z_r^2/z$. ψ_0 is a constant initial phase and k is the laser wave number, $2\pi/\lambda$. ψ_G is the Gouy phase expressed as

$$\psi_G = \tan^{-1}\frac{z}{z_r}. \quad (35)$$

The zeroth order term in ε is a well known Gaussian field. One can see when ε cannot be neglected: when the focal size gets comparable to the laser wavelength, a field longitudinal to the propagation direction appears and the symmetry between the electric and the magnetic fields is broken.

Because ε is proportional to $1/w_0$, the high order fields (HOFs) become larger for smaller beam waist. Figure 17 shows that E_y and E_z get stronger as w_o decreases. The peak field strengths of E_y and E_z amount to 2.6% and 15% of E_x at w_o = 1 μm, respectively. In the case of a counter-interaction between an electron and a laser pulse, HOFs much weaker than the zeroth-order field does not affect the electron dynamics. However, when the relativitic electron is driven by a co-propagating laser pulse, weak HOFs significantly affect the electron dynamics and consequently the RNTS radiation.

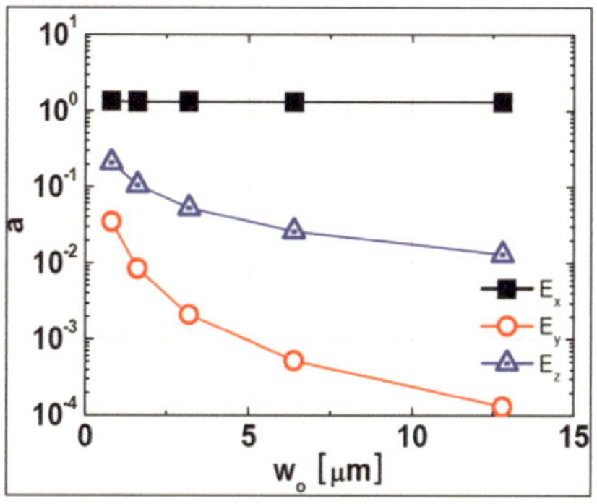

Fig. 17. The strength of laser electric fields against the beam waist size are plotted in unit of the normalized vector potential. The laser field is evaluated at (w_o/2, w_o/2, 0) with the zeroth-order laser intensity of a_0 = 2.2.

4.2 Dynamics of an electron electron with a tightly focused laser

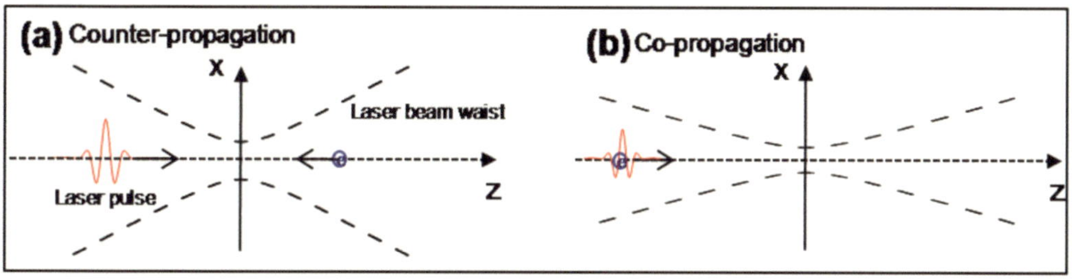

Fig. 18. Two interaction schemes between a relativistic electron and a laser pulse:
(a) counter-propagation and (b) co-propagation.

The dynamics of a relativistic electron under a tightly-focused laser beam is investigated by the Lorentz force equation [Eq. (1)]. One can consider two extreme cases of interaction geometry as shown in Fig. 18. The counter-propagation scheme, or Compton back-scattering scheme is usually adopted to generate monochromatic x-rays. It has been shown in the previous section that the co-propagation scheme is more appropriate to generate the coherent RNTS radiation. For such schemes, the effect of HOFs will be investigated.

In the z-x plane, $E_y = B_z = 0$, then the Lorentz force equation for γ and β_x (transverse velocity) in the case of the counter-propagation scheme ($\vec{\beta} \approx -1\hat{z}$), can be approximated as,

$$\frac{d\gamma}{d\tau} \approx a_0\beta_x - a_{odd}^H, \tag{36}$$

$$\frac{d(\gamma\beta_x)}{d\tau} \approx 2a_0 + \left(a_{even}^H + b_{even}^H\right), \tag{37}$$

where $\tau = \omega_L t$, a_0 is the zeroth-order laser field in unit of the normalized potential. a_{odd}^H is odd HOFs of electric field (or longitudinal electric fields). a_{even}^H and b_{even}^H are even HOFs of electric and magnetic fields, respectively [see Eqs. (28)-(33)]. From above equations, γ can be analytically obtained considering only the first HOF as

$$\frac{\gamma}{\gamma_o} \approx 1 + \eta_o^2 - \frac{\alpha_1}{2\gamma_o^2}\left(1 - \eta_o^2\right), \tag{38}$$

where $\eta_o^2 = \left(1 + \alpha_0^2\right) / 2\gamma_0^2$ and α_1 is the integration of the first-order electric field over phase. For a highly relativistic electron or $\gamma_o >> 1$, the above equations show that the HOFs contribute to the electron dynamics just as a small correction to the zeroth-order field. Figure 19 (a) shows the time derivatives of the gamma factor and the transverse velocity. It is hardly to notice any change due to HOFs.

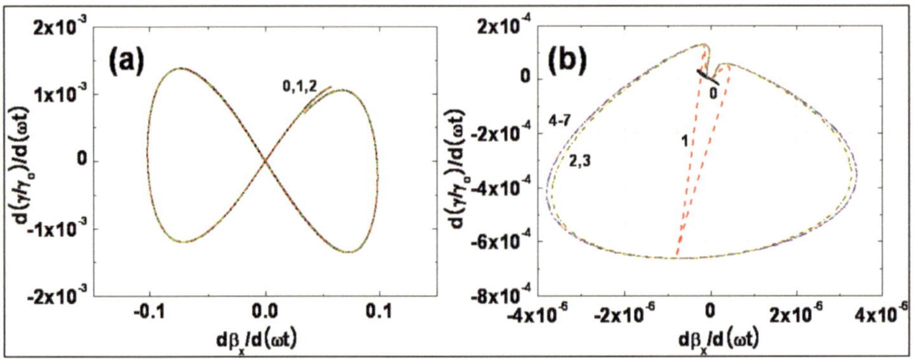

Fig. 19. Variation of the time derivatives of the relativsitic gamma factor and the transverse velocity for (a) counter-propagation and (b) co-propagation schemes. In this calculation, the laser pulse with λ = 0.8 μm, w_o = 4 μm, a_o = 10, and Δt_{FWHM} = 5 fs interacts with an electron with E_o = 100 MeV. The numbers in the figures indicate up to which order HOFs are included.

However, when the electron co-propagates with the laser pulse ($\vec{\beta} \approx +1\hat{z}$), the situation dramatically changes. The Lorentz force equations are

$$\frac{d\gamma}{d\tau} \approx a_0\beta_x + a_{odd}^H, \tag{39}$$

$$\frac{d(\gamma\beta_x)}{d\tau} \approx \frac{a_0}{2\gamma^2} + \left(a_{even}^H - b_{even}^H\right). \tag{40}$$

The relativistic gamma factor is given by

$$\frac{\gamma}{\gamma_o} \approx \frac{1 + \alpha_0^2}{1 - 2\gamma_o\alpha_1}. \tag{41}$$

Now the change of the gamma factor becomes significant and γ gets even smaller than its initial value. This cannot happen in the counter-propagation scheme [Eq. (38)]. The acceleration in the transverse direction can be dominated by the HOFs when $a_0 / \gamma^2 << 1$. This section deals with this kind of case, where $a_0 \leq 10$ and $\gamma_0 >> 10$. As expected, Figure 19 (b) shows that the time derivative of the gamma factor increases with inclusion of the first order HOF and that of the transverse velocity is significantly enhanced with the inclusion of the second-order HOF. Even though the zeroth order field is much stronger than the HOFs, the high gamma factor makes it negligible compared with the HOFs. It is the difference between the high order electric and the magnetic fields that contributes to such a dramatic change in the dynamics. Higher order fields than the second order just contributes to the dynamics as a small correction in this case (In some special cases where the spatial distribution of HOFs gets important near axis, they can be more considerable.). It is also interesting to note that the scaling of the transverse acceleration on the gamma factor changes from $\propto \gamma_0^{-3}$ to $\propto \gamma_0^{-1}$ due to the inclusion of the second-order field.

4.3 Radiation from a co-propagating electron with a tightly focused laser

As shown in Sec. 2, nonlinear motion of an electron contributes to the harmonic spectra, or ultra-short pulse radiation. Thus it can be inferred that the enhancement of nonlinear dynamics with inclusion of HOFs, increase of gamma factor (or electron energy variation) by the first order field and the transverse acceleration by the second-order field, might enhance the RNTS radiations.

Figure 20 shows the effect on RNTS as HOFs are included. Figure 20 is the RNTS radiation obtained from the dynamics in Fig. 19 (b). The number on the plot is the number of order of HOF, up to which HOFs are included. Note that as the HOFs are included, the pulse duration gets shorter and the spectrum gets broader according. The radiation intensity is also greatly enhanced. It should be notices that it is the transverse acceleration that significantly enhances the RNTS radiations. Figure 20 shows the shorter pulse width or wider spectral width by a factor of more than 5 and the higher intensity by an order of magnitude.

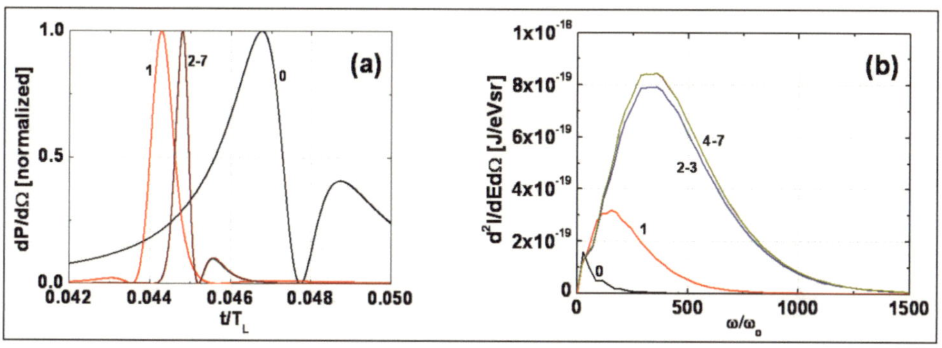

Fig. 20. (a) The normalized temporal structure of the radiation from the dynamics presented in Fig. 19 (b). Each radiation is plotted in the direction of the maximum radiation, which is -0.03° and -0.16° for zeroth-order and first-order, respectively and converges to -0.12° for higher orders. The peak powers for each plot are 2.1, 77, and 580 W/rad2. (b) The harmonic spectrum for the same case.

In the tight-focusing scheme, the strong radiation can be assumed to be generated within the focal region. That is, the electron radiates when it passes through Rayleigh range approximately in length scale, or $\Delta t' = 2z_r / \beta c$ in the electron's own time. Then for $\beta \approx 1$, the period in the detector's own time, Δt can be approximately obtained as

$$\Delta t \approx \frac{2z_r(1-\beta)}{c\beta} \approx \frac{z_r}{c\gamma^2} = \frac{\pi}{c\lambda}\frac{w_0^2}{\gamma^2}. \tag{42}$$

The average photon energy can be approximated to the mean photon energy and then it can be estimated from the inverse of the pulse width according to the Fourier transform as

$$E_{av} \approx \frac{E_{\max} + E_{\min}}{2} \approx \frac{E_{\max}}{2} \approx \frac{\hbar\Delta\omega}{2} \propto \frac{1}{\Delta t} \propto \frac{\gamma^2}{w_o^2} . \tag{43}$$

This shows that the average photon energy scales inversely with the square of the beam waist, as also shown later. The total radiation power, radiated power integrated over whole angle, from an accelerated electron is described as (Jackson, 1999)

$$P(t') = \frac{2}{3}\frac{e^2}{c}\gamma^6[(\dot{\vec{\beta}})^2 - (\vec{\beta}\times\dot{\vec{\beta}})^2] . \tag{44}$$

The radiated powers in the electron's own time (retarded time) $P(t')$ and in the detector's own time $P(t)$ are related to each other by the relation of $P(t) = P(t')(dt'/dt) = P(t')/(1-\hat{n}\cdot\vec{\beta})$ [Eq. (11)]. When $|\vec{\beta}| \approx 1$ and $\hat{n}$ is almost parallel to $\vec{\beta}$, $(1-\hat{n}\cdot\vec{\beta})$ can be approximated to $1/(2\gamma^2)$, which leads to $P(t) \approx 2\gamma^2 P(t')$. When $\dot{\vec{\beta}}$ is parallel to $\vec{\beta}$, the radiated power integrated over all the angles is given by

$$P_{||}(t) = \frac{4}{3}\frac{e^2}{c}\gamma^8\dot{\beta}_{||}^2 . \tag{45}$$

On the other hand, when $\dot{\vec{\beta}}$ is perpendicular to $\vec{\beta}$ the radiation power is expressed as

$$P_{\perp}(t) = \frac{4}{3}\frac{e^2}{c}\gamma^6\dot{\beta}_{\perp}^2 . \tag{46}$$

$\dot{\vec{\beta}}$ can be obtained from the Lorentz force equation [Eq. (1)]. The acceleration term $d(\gamma\vec{\beta})/dt$ can be expanded as $\gamma\dot{\vec{\beta}} + \dot{\gamma}\vec{\beta}$. The first term contributes to $P_{\perp}$, while the second term to $P_{||}$. From this relation, the $\dot{\beta}$ can be expressed with an effective field $\vec{F} = \vec{E} + \vec{\beta}\times\vec{B}$ and γ. Then Eqs. (45) and (46) can be re-written as

$$P_{||}(t) \cong \frac{4}{3}\frac{e^4}{mc^3}\gamma^8\left|\frac{\vec{F}_{||}}{\beta^2\gamma^3}\right|^2 \cong \frac{4}{3}\frac{e^4}{mc^3}\gamma^2|\vec{F}_{||}|^2 \tag{47}$$

and

$$P_{\perp}(t) \cong \frac{4}{3}\frac{e^4}{mc^3}\gamma^6\left|\frac{\vec{F}_{\perp}}{\gamma}\right|^2 = \frac{4}{3}\frac{e^4}{mc^3}\gamma^4|\vec{F}_{\perp}|^2 , \tag{48}$$

respectively. The effective fields for different harmonic orders can be approximated as

$$\vec{F}_{(0)} = ES_1(1-\beta)\hat{x} \cong \frac{ES_1}{2\gamma^2}\hat{x} . \tag{49}$$

$$\vec{F}_{(1)} = E\xi\varepsilon\left[C_2\right]\hat{z}\,. \tag{50}$$

$$\vec{F}_{(2)} = E\varepsilon^2\left\{-(1-\beta)\frac{\rho^4 S_4}{4} + \left[\xi^2 - \frac{\rho^2}{2}\beta\right]S_3\right\}\hat{x} \cong E\varepsilon^2 S_3\left[\xi^2 - \frac{\rho^2}{2}\beta\right]\hat{x}\,. \tag{51}$$

Then the total radiation energy can be calculated by the multiplication of the pulse width and the power. Using the estimated Δt of Eq. (42), the total radiation energy of single electron, I, can be estimated from Eqs. (47) and (48) as follows:

$$I_{||} \propto w_0^2\,|\,\vec{F}_{||}\,|^2\,. \tag{52}$$

$$I_{\perp} \propto w_0^2\gamma^2\,|\,\vec{F}_{\perp}\,|^2\,. \tag{53}$$

With the effective fields [Eqs. (49)-(51)] and the radiation energy relations of Eqs. (47) and (48), the radiation energy for the field of certain order is evaluated as follows:

$$I_{(0)} \propto \varepsilon^0\gamma^{-2}\,. \tag{54}$$

$$I_{(1)} \propto \varepsilon^1\gamma^0\,. \tag{55}$$

$$I_{(2)} \propto \varepsilon^2\gamma^2\,. \tag{56}$$

The effective strength of the zeroth-order field is much smaller than those of HOFs because $1/\varepsilon$ is only ~20 and much smaller than γ^2 ($\gamma^2 > 10000$ in the current study). From this, one can see that the magnitudes of the radiation energies can be ordered as $I_{(0)} << I_{(1)} << I_{(2)}$. Since the transverse field is more effective than the longitudinal field for scattering radiation, $I_{(1)}$ is smaller than $I_{(2)}$. The HOFs higher than the second-order do not make significant contribution to radiation; they can be just considered as a small correction.

5. Generation of an intense attosecond x-ray pulse

In the case of the interaction of an electron bunch with a laser, there are three major interaction geometries: counter-propagation (Compton backscattering), 90°-scattering, and co-propagation (0°-scattering) geometry. To estimate the pulse width of the radiation in these interaction geometries, we need to specify the length (L_{el}) and diameter (L_T) of the electron bunch, the pulse length (L_{laser}) of the driving laser and the interaction length (confocal parameter, L_{conf}). At current technology, the diameter of an electron bunch is typically 30 μm. The typical pulse widths of an relativistic electron bunch (v~c) and femtosecond high power laser are about 20 ps and 30 fs, respectively, which corresponds to L_{el}~ 6 mm and L_{laser} ~ 9 μm in length scale. For the beam waist of 5 μm at focus, the confocal parameter is L_{conf} ~ 180 μm for 800 nm laser wavelength. In other words, $L_e >> L_{conf} >> L_{laser}$ are rather easily satisfied. In the following, we confine our simulation to this case. In this situation, the radiation pulse width is then roughly estimated to be $\Delta t_{counter}$~$2L_{conf}/c$=600 fs, Δt_{90}~$(L_T+L_{laser})/c$=130 fs, and Δt_{co}~ L_{laser}/c=5 fs, for counter-propagation, 90°-scattering, and co-propagation geometry, respectively. When we consider the aspect of the x-ray pulse duration, the co-propagation geometry is considered to be adequate as an interaction geometry.

The co-propagating interactions between a femtosecond laser pulse and an electron bunch were demonstrated as series of simulations. The simulations are similar to those in Section 4. The difference is that the electron bunch and the pulsed laser, co-propagating along the +z direction, meet each other in the center of the tightly focused region (z = 0). The interaction between electrons is ignored because it is much weaker than the interaction between the laser field and electrons.

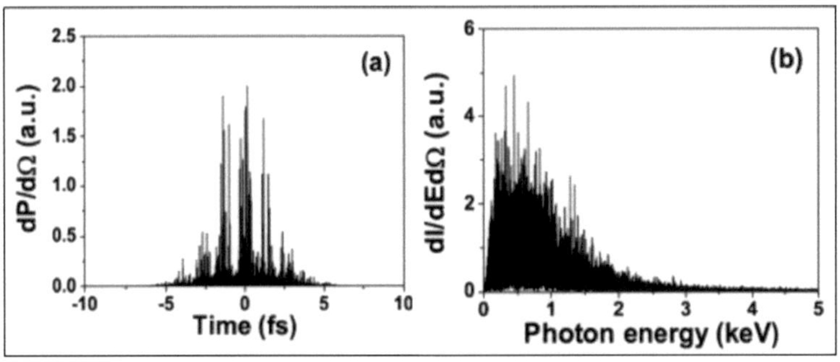

Fig. 21. (a) Temporal structure and (b) spectrum of the radiation from the co-propagation interaction between a 5 fs FWHM laser with 5 μm beam waist and an electron bunch of 200 MeV energy. The laser intensity at focus is 2.1 × 10^{20} W/cm^2 (a_0 = 10).

Figure 21 shows the temporal structure and the spectrum of the radiation from the interaction between an electron bunch and a co-propagating tightly focused fs laser. The pulse width and the wavelength of the laser is 5 fs FWHM and 800 nm (1.55 eV), respectively. The laser is focused to a beam waist of 5 μm at z=0 with an intensity of 2.1x10^{20} W/cm^2 ($a_0 = 10$). The electron bunch has a radius of 30 μm and a length of 30 μm (or a pulse of 100 fs) and a normalized emittance of 2 mm mrad. The energy of the electron bunch is 200 MeV and the energy spread is 0.1 %. The electron bunch consists of 3.0×10^4 electrons which are randomly sampled with the Gaussian distribution [Eq. 16)] throughout the bunch. The both centers of the electron bunch and the laser meet at the center of the focus (z=0). The radiation is detected at the angle of $\theta = 0$. Figure 21 (a) shows that the width of the X-ray pulse radiated from the electron bunch is about 5 fs, which is the same as that of the laser pulse, as mentioned in the above. The spectrum [Fig.21(b)] shows that very high-energy photons are produced as mentioned in previous section.

Figure 22 shows the total radiated energy and the averaged photon energy. The calculations have been done for various electron energies. To check the effect of the HOFs, the simulation have been carried out for various combinations of high order fields: (1) the zeroth-order only, up to (2) the first-, (3) the second-, and (4) the seventh-order fields. The electron bunch consists of 7.5×10^3 electrons. The total radiated energy has been obtained by the integration of the angular radiation energy over the angle $\theta = 1/\gamma$. The conditions are the same as those of Fig. 21, unless otherwise mentioned. Each point of data is the average of the four sets of simulations. The standard deviations are always smaller than 5 % and the error bars are omitted because they are not visible in the log scale.

Figure 22 (a) also shows the γ dependence of the total radiated energies $I_{(0)}$, $I_{(0,1)}$, $I_{(0-2)}$ and $I_{(0-7)}$. The fitting to the simulation data shows $I_{(0)} \propto \gamma^{-2.03}$ which is a good agreement with Eq. (53). In case of $I_{(0,1)}$ which includes both the zeroth- and the first-order field, the first-order field mainly contributes to the radiation and the total radiated energy $I_{(0,1)}$ is independent of γ as Eq. (54). $I_{(0-2)}$, which contains the fields up to the second-order, is proportional to $\gamma^{+1.94}$, indicating that its behaviour is dominated by the second-order field, as Eq. (55). $I_{(0-7)}$ is almost the same as $I_{(0-2)}$, implying that the fields higher than the second-order make a negligible contribution.

The study about an electron acceleration using a tightly-focused laser field by Salamin et al. showed that at least the fields up to the fifth-order are required to describe correctly the dynamics of an electron (Salamin & Keitel, 2002). In that calculation, a low energy ($\gamma < 24$) electron was injected to a focused region at a finite angle ($5^\circ < \theta < 12^\circ$) with respect to the laser propagation. Even though the fields higher than the third-order are small, HOFs can still deflect the path of such a low energy electron. Because the strength of HOfs is sensitive to the position in the case of a tight focus, the interaction dynamics can be different with a low energy electron. However, for a high energy electron ($\gamma > 100$), which is effectively heavy, the deflection of the electron's path by HOFs can be neglected.

As expected in Eq. (43), the average photon energy of the radiation is also proportional to γ^2 as manifested in Fig. 22 (b). Figure 22 (c) shows the dependence of the radiated photon energy on beam waist when the energy of electron is 1 GeV and the radius of the electron bunch 30 μm (the laser parameters are the same as before). HOFs up to the seventh-order were included. It shows that the average photon energy is proportional to w_0^{-2} as expected in Eq. (43). Hence the photon energy is tunable by changing the beam waist as well as the electron energy. The average photon energy is 38 keV for an electron energy of 2 GeV, but higher photon energy is expected for higher electron energy. As mentioned in the introduction it can fill the region shorter than 5 fs and higher than 10 keV of Fig. 2.

Because both of the photon energy E and the total radiated energy I is proportional to γ^2, the number of the radiated photons I / E is independent of γ. For an electron bunch of 1 nC charge, $a_0 = 10$, $w_0 = 5$ μm and the pulse width of 5 fs FWHM, the number of radiated photon is 4.2×10^6. For the counter-propagation interaction (180° collision) between the same electron bunch and laser pulse, the number of the radiated photons is 1.2×10^8 (Hartemann et al., 2005). The radiated photon number in the co-propagation interaction is only 30 times less than the 180° collision when HOfs are taken into account.

This represents a remarkable enhancement if we notice that when only the zeroth-order field is considered, the photon numbers in the co-propagation interaction are 3×10^3 and 3×10^7 times less than those in the counter-propagation interaction, for an electron energy of 50 MeV and 1 GeV, respectively. In terms of the photon number per unit time, both interaction geometry are comparable because the radiation pulse width from co-propagation interaction (< 5 fs) is ~20 times smaller than that from 180° collision (> 100 fs).

The study also shows that the radiation efficiency of an electron increases as the focal spot gets smaller. The radiation efficiency can be defined by the radiated energy divided by both the number of the electron effectively participating in the radiation N_{eff} and the energy of the driving laser I_L:($I/N_{eff} \cdot I_L$). Figure 23 shows the radiation efficiency of an electron with respect to the change of the beam waist. For fair comparison, the laser intensity is kept constant for different beam waists (a_0 = 10 for all the data in Fig. 23). It is shown that the efficiency scales inversely to the 4th power of the beam waist. This scaling can be understood

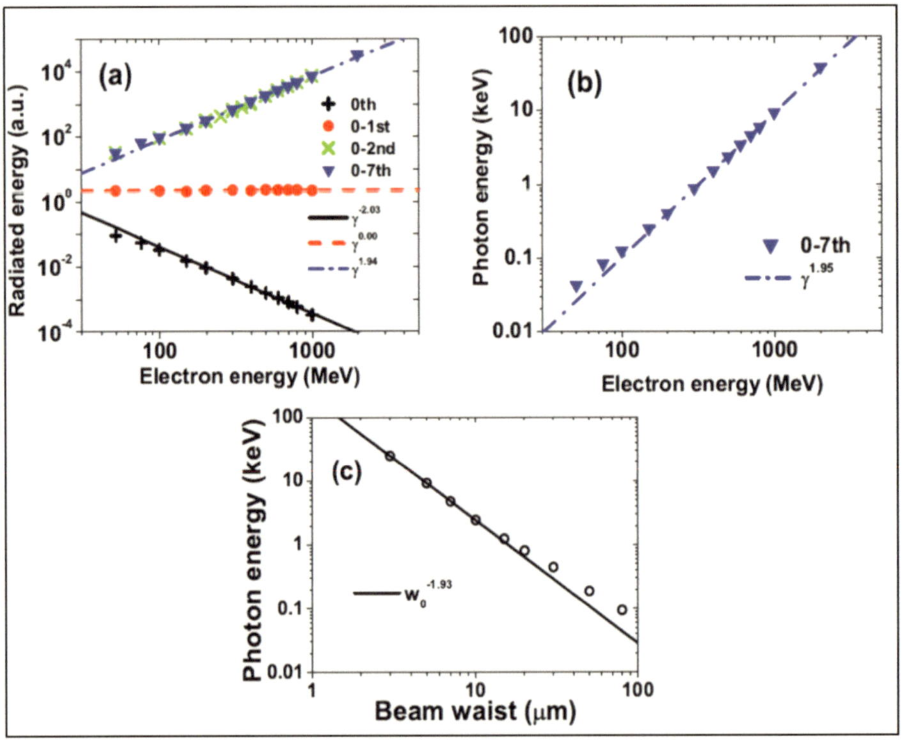

Fig. 22. Dependence of (a) the total radiated energy I and (b) the average photon energy E on γ. The effect of high-order fields in ε is also shown in (a) for different combination of high-order fields. (c) The dependence of the average photon energy on w_0. The normalized vector potential of the laser a_0 is 10, beam waist of laser w_0=5 μm and radius of the electron bunch 30 μm.

as follows: The radius of the electron bunch is assumed to be much larger and the electrons are uniformly distributed so that the electron number participating in the radiation N_{eff} is proportional to the square of the laser beam waist. As explained previously, the total radiated energy from a single electron, $I_{(0-7)} \cong I_{(2)} \propto \varepsilon^2\gamma^2 \propto w_0^{-2}$. Because the laser intensity is kept constant, the energy of the driving laser is then proportional to w_0^2. Hence, $I/(I_L N_{eff})$ is proportional to the w_0^{-4} as shown in Fig. 23. If the density of the electron is constant and the radius of the electron bunch is much larger than the beam waist, the number of electrons participating in the radiation is proportional to w_0^2 and the radiated energy per laser energy, I/I_L is proportional to w_0^{-2}. In a real electron bunch the electron density is not uniform but larger at the center than in the outer part, thus I/I_L decreases more rapidly than w_0^{-2}. In other words, the smaller focal spot size is preferred to the larger one for higher flux of photons for the same laser intensity because the radiation efficiency of electron is higher and the laser energy needed is less. Note in Fig. 23 that the slope changes near the beam waist of 20 μm. It is because the contribution from HOFs (mainly the 2nd order) decreases as the beam waist increases.

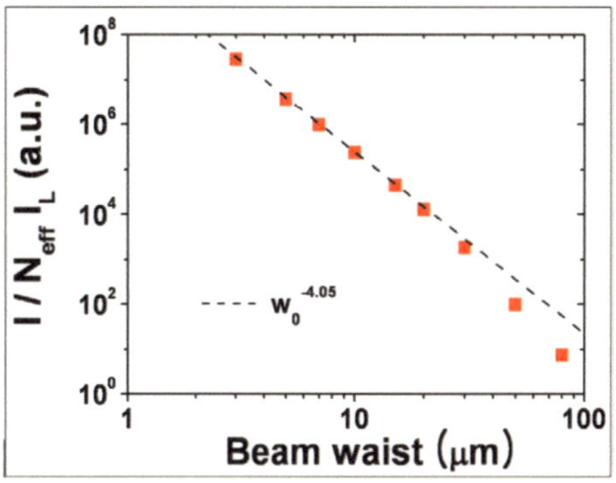

Fig. 23. Change of the radiation efficiency with respect to the beam waist. The radiation efficiency is here defined by the radiation energy I divided by the number of the electrons effectively participating in the radiation N_{eff} and the energy of the driving laser I_L : $I/(I_L N_{eff})$ the laser intensity is kept constant ($a_0 = 10$) for different beam waists.

6. References

Afonso, C. N.; Solis, J.; Catalina, F.; & Kalpouzos, C. (1996). Existence of Electronic Excitation Enhanced Crystallization in GeSb Amorphous Thin Films upon Ultrashort Laser Pulse Irradiation, *Appl. Phys. Lett.*, Vol. 76, 2519-2522, ISSN 0003-6951

Banerjee, S.; Valenzuela, A. R.; Shah, R. C.; Maksimchuk, A. & Umstadter, D. (2002). High harmonic generation in relativistic laser–plasma interaction, *Phys. Plasmas*, Vol. 9, 2393-2398, ISSN 1070-664X

Beaud, P.; Johnson, S. L.; Streun, A.; Abela, R.; Abramsohn, D.; Grolimund, D.; Krasniqi, F.; Schmidt, T.; Schlott, V. & Ingold, G. (2007). Spatiotemporal Stability of a Femtosecond Hard-X-Ray Undulator Source Studied by Control of Coherent Optical Phonons, *Phys. Rev. Lett.*, Vol. 99, 174801-1-4, ISSN 0031-9007

Bloembergen, N. (1999). From nanosecond to femtosecond, science, *Rev. of Mod. Phys.*, Vol. 71, S283-S287, ISSN 0034-6861

Brown, L. S. & Kibble, T. W. B. (1964). Interaction of Intense Laser Beams with Electrons, *Phys. Rev.*, Vol. 133, A705-A719, ISSN

Cavalieri, A.; Müller, L. N.; Uphues, Th.; Yakovlev, V. S.; Baltuška, A.; Horvath, B.; Schmidt, B.; Blümel, L.; Holzwarth, R.; Hendel, S.; Drescher, M.; Kleineberg, U.; Echenique, P. M.; Kienberger, R.; Krausz , F. & Heinzmann, U. (2007). Attosecond spectroscopy in condensed matter, *Nature*, Vol. 449, 1029-1032, ISSN 0028-0836

Chen, S.-Y.; Maksimchuk, A. & Umstadter, D. (1998). Experimental observation of relativistic nonlinear Thomson scattering, *Nature*, Vol. 396, 653-655, ISSN 0028-0836

Chen, S.-Y.; Maksimchuk, A.; Esarey, E. & Umstadter, D. (2000). Observation of Phase-Matched Relativistic Harmonic Generation, *Phys. Rev. Lett.*, Vol. 84, 5528-5531, ISSN 0031-9007

Chung, S-Y.; Yoon, M-H. & Kim, D-E. (2009). Generation of Attosecond X-ray and gamma-ray via Compton backscattering, *Optics Express*, Vol. 17, 7853-7861, ISSN 1094-4087

Davis, L. W. (1979). Theory of electromagnetic beams, *Phys. Rev. A*, Vol. 19, 1177-1179, ISSN 1050-2947

Drescher, M.; Hentschel, M.; Kienberger, R.; Uiberacker, M.; Yakovlev, V.; Scrinzi, A.; Westerwalbesloh, Th.; Kleineberg, U.; Heinzmann, U. & Krausz, F. (2002). Time-resolved atomic inner-shell spectroscopy, *Nature*, Vol. 419, 803-807, ISSN 0028-0836

Esarey, E. & Sprangle, P. (1992). Generation of stimulated backscattered harmonic radiation from intense-laser interactions with beams and plasmas, *Phys. Rev. A*, Vol. 45, 5872-5882, ISSN 1050-2947

Esarey, E.; Ride, S. K. & Sprangle, P. (1993a). Nonlinear Thomson scattering of intense laser pulses from beams and plasmas, *Phys. Rev. E*, Vol. 48, 3003-3021, ISSN 1539-3755

Esarey, E.; Ting, A.; Sprangle, P.; Umstadter, D. & Liu, X. (1993b). Nonlinear analysis of relativistic harmonic generation by intense lasers in plasmas, *IEEE Trans. Plasma Sci.*, Vol. 21, 95-104, ISSN 0093-3813

Fleming, G. & Ratner, M. (2008). Grand challenges in basic energy, sciences, *Physics Today*, July, Vol. 28, 8-80, ISSN

Gaffney, K. J. & Chapman, H. N. (2008). Imaging Atomic Structure and Dynamics with Ultrafast X-ray Scattering, *Science*, Vol. 316, 1444-1448, ISSN 0036-8075

Gai, F.; Hasson, K.; McDonald, C. J. C. & Anfinrud, P. A. (1998). Chemical Dynamics in Proteins: The Photoisomerization of Retinal in Bacteriorhodopsin, *Science*, Vol. 279, 1886-1891, ISSN 0036-8075

Girolami, B.; Larsson, B.; Preger, M.; Schaerf, C. & Stepanek, J. (1996). Photon beams for radiosurgery produced by laser Compton backscattering from relativistic electrons, *Phys. Med. Biol.*, Vol. 41, 1581-1596, ISSN 0031-9155

Goulielmakis, E.; Schultze, M.; Hofstetter, M.; Yakovlev, V. S.; Gagnon, J.; Uiberacker, M.; Aquila, A. L.; Gullikson, E. M.; Attwood, D. T.; Kienberger, R.; Krausz, F. & Kleineberg, U. (2008). Single-cycle nonlinear optics, *Science*, Vol. 320, 1614-1617, ISSN 0036-8075

Harris, S. E. & Sokolov, A. V. (1998). Subfemtosecond Pulse Generation by Molecular Modulation, *Phys. Rev. Lett.*, Vol. 81, 2894-2897, ISSN 0031-9007

Hartemann, F. V. (1998). High-intensity scattering processes of relativistic electrons in vacuum, *Phys. Plasmas*, Vol. 5, 2037-2047, ISSN 1070-664X

Hartemann, F. V.; Brown,W. J.; Gibson, D. J.; Anderson, S. G.; Tremaine, A. M.; Springer, P. T.; Wootton, A. J.; Hartouni, E. P. & Barty, C. P. J. (2005). High-energy scaling of Compton scattering light sources, *Phys. Rev. ST AB*, Vol. 8, 100702-1-17, ISSN 1098-4402

Hentschel, M.; Kienberger, R.; Spielmann, Ch. G.; Reider, A.; Milosevic, N.; Brabec, T.; Corkum, P.; Heinzmann, U.; Drescher, M.; & Krausz, F. (2001). Attosecond metrology, *Nature*, Vol. 414, 509-513, ISSN 0028-0836

Hertz, E., Papadogiannis, N.A., Nersisyan, G., Kalpouzos, C., Halfmann, T., Charalambidis, D., and Tsakiris, G.D. (2001), Probing attosecond pulse trains using "phase-control" techniques, Phys. Rev. A, 64, (051801-1-4) 1050-2947

Huang, L.; Callan, J. P.; Glezer, E. N.; & Mazur, E. (1998). GaAs under Intense Ultrafast Excitation: Response of the Dielectric Function, *Phys. Rev. Lett.*, Vol. 80, 185-188, ISSN 0031-9007

Mourou, G. A.; Barty, C. P. J. & Perry, M. D. (1998). Ultrahigh-Intensity Lasers: Physics of the Extreme on a Tabletop, *Phys. Today*, Vol. 51, Iss. 1, 22-28, ISSN 0031-9228

Jackson, J. D. (1975). *Classical Electrodynamics*, 2nd ed., Wiley, ISBN 047143132X, New York

Kaplan, A. E. & Shkolnikov, P. L. (2002). Lasetron: A Proposed Source of Powerful Nuclear-Time-Scale Electromagnetic Bursts, *Phys. Rev. Lett.*, Vol. 88, 074801-1-4, ISSN 0031-9007

Kaplan, A.E. & Shkolnikov, P.L.(1996). Subfemtosecond pulses in the multicascade stimulated Raman scattering, *J. Opt. Soc. Am. B*, Vol. 13, 347-354, ISSN 0740-3224

Kawano, H.; Hirakawa, Y. & Imasaka, T. (1998). Generation of high-order rotational lines in hydrogen by four-wave Raman mixing in the femtosecond regime, *IEEE J. Quantum Electron.*, Vol. 34, 260-268, ISSN 0018-9197

Kien,F.L.; Liang, J.Q.; Katsuragawa, M.; Ohtsuki, K.; Hakuta, K. & Sokolov, A.V. (1999). Subfemtosecond pulse generation with molecular coherence control in stimulated Raman scattering, *Phys. Rev. A*, Vol. 60, 1562-1571, ISSN 1050-2947

Kim, D.; Lee, H.; Chung, S. & Lee, K. (2009). Attosecond keV x-ray pulses driven by Thomson scattering in a tight focus regime, *New Journal of Physics*, Vol. 11, 063050-1-12, ISSN 1367-2630

Kim, D.-E.; Toth, C. & Barty, C. P. J. (1999). Population inversion between atomic inner-shell vacancy states created by electron-impact ionization and Coster-Kronig decay, *Phys. Rev. A: Rap. Comm.*, Vol. 59, R4129-R4132, ISSN 1050-2947

Kim, D.; Son, S. H.; Kim, J. H.; Toth, C. & Barty, C. P. J. (2001). Gain characteristics of inner-shell photoionization-pumped L23M1 transition in Ca, *Phys. Rev. A*, Vol. 63, 023806-1-6, ISSN 1050-2947

Kling, M. F.; Siedschlag, Ch.; Verhoef, A. J.; Khan, J. I.; Schultze, M.; Uphues, Th.; Ni, Y.; Uiberacker, M.; Drescher, M.; Krausz, F. & Vrakking, M. J. J. (2006). Control of Electron Localization in Molecular Dissociation, *Science*, Vol. 312, 246-248, ISSN 0036-8075

Krausz, F. & Ivanov, M. (2009). Attosecond physics, *Rev. of Mod. Phys*, Vol. 81, 163-234, ISSN 0034-6861

Lee, K.; Cha, Y. H.; Shin, M. S.; Kim, B. H. & Kim, D. (2003a). Temporal and spatial characterization of harmonics structures of relativistic nonlinear Thomson scattering, *Opt. Express*, Vol. 11, 309-316, ISSN 1094-4087

Lee, K.; Cha, Y. H.; Shin, M. S.; Kim, B. H. & Kim, D. (2003b). Relativistic nonlinear Thomson scattering as attosecond x-ray source, *Phys. Rev. E*, Vol. 67, 026502-1-7, ISSN 1539-3755

Lee, K.; Kim, B. H. & Kim, D. (2005). Coherent radiation of relativistic nonlinear Thomson scattering, *Phys. Plasmas*, Vol. 12, 043107-1-8, ISSN 1070-664X

Lee, K.; Park, S. H.; Jeong, Y. U. (2006). Nonlinear Thomson Scattering for Attosecond X-ray Pulse Generation, *J. Korean Phys. Soc.*, Vol. 48, 546-553, ISSN 0374-4884

Lee, H.; Chung, S.; Lee, K. & Kim, D. (2008). A study of the Thomson scattering of radiation by a relativistic electron of a tightly-focused co-propagating femtosecond laser beam, *New Journal of physics*, Vol. 10, 093024-1-11, ISSN 1367-2630

Lichters, R.; Meyer-ter-Vehn, J. & Pukhov, A. (1996). Short-pulse laser harmonics from oscillating plasma surfaces driven at relativistic intensity, *Phys. Plasmas*, Vol. 3, 3425-3437, ISSN 1070-664X

von der Linde, D.; Engers, T.; Jenke, G.; Agostini, P.; Grillon, G.; Nibbering, E.; Mysyrowicz, A. & Antonetti, A. (1995). Generation of high-order harmonics from solid surfaces by intense femtosecond laser pulses, *Phys. Rev. A*, Vol. 52, R25-R27, ISSN 1050-2947

von der Linde, D. & Rzazewski, K. (1996). High-order optical harmonic generation from solid surfaces, *Appl. Phys. B: Lasers Opt.*, Vol. 63, 499-506, ISSN 0946-2171

von der Linde, D. (1999). Generation of high order optical harmonics from solid surfaces, *Appl. Phys. B: Lasers Opt.*, Vol. 68, 315-319, ISSN 0946-2171

Neutze, R.; Wouts, R.; van der Spoel, D.; Weckert,E. & Hajdu, J. (2000). Potential for biomolecular imaging with femtosecond X-ray pulses, *Nature*, Vol. 406, 752-757, ISSN 0028-0836

Norreys, P. A.; Zepf, M.; Moustaizis, S.; Fews, A. P.; Zhang, J.; Lee, P.; Bakarezos, M.; Danson, C. N.; Dyson, A.; Gibbon, P.; Loukakos, P.; Neely, D.; Walsh, F. N.; Wark, J. S. & Dangor, A. E. (1996). Efficient Extreme UV Harmonics Generated from Picosecond Laser Pulse Interactions with Solid Targets, *Phys. Rev. Lett.*, Vol. 76, 1832-1835, ISSN 0031-9007

Normile, D. (2006). MATERIALS SCIENCE: Japanese Latecomer Joins Race To Build a Hard X-ray Laser, *Science*, Vol. 314, 751-752, ISSN 0036-8075

Paul, P. M.; Toma, E. S.; Breger, P.; Mullot, G.; Auge, F.; Balcou, Ph.; Muller, H. G. & Agostini, P. (2001). Observation of a Train of Attosecond Pulses from High Harmonic Generation, *Science*, Vol. 292, 1689-1692, ISSN 0036-8075

Perry, M. D. & Mourou, G. (1994). Terawatt to Petawatt Subpicosecond Lasers, *Science*, Vol. 264, 917-924, ISSN 0036-8075

Perutz, M.; Wilkinson, F. A. J.; Paoli, M. & Dodson, G. G. (1998). THE STEREOCHEMICAL MECHANISM OF THE COOPERATIVE EFFECTS IN HEMOGLOBIN REVISITED, *Annu. Rev. Biophys. Biomol. Struct.*, Vol. 27, 1-34, ISSN 1056-8700

Salamin, Yousef I. & Keitel, Christoph H. (2002). Electron Acceleration by a Tightly Focused Laser Beam, *Physical Review Letters* Vol. 88, 095005, ISSN 1079-7114

Salamin, Y. I. (2007). Fields of a Gaussian beam beyond the paraxial approximation, *Applied Physics B* Vol. 86, 319-326, ISSN 1432-0649

Sarachik, E. S. & Schappert, G. T. (1970). Classical Theory of the Scattering of Intense Laser Radiation by Free Electrons, *Phys. Rev. D*, Vol. 1, 2738-2753, ISSN

Schoenlein, R. W. (2000). Generation of Femtosecond Pulses of Synchrotron Radiation, *Science*, Vol. 287, 2237-2240, ISSN 0036-8075

Sheffield, J. (1975). *Plasma Scattering of Electromagnetic Radiation*, Academic Press, ISBN 0126387508, New York

Siders, C. W.; Cavalleri, A.; Sokolowski-Tinten, K.; Toth, C.; Guo, T.; Kammler, M. H.; v. Hoegen, M. R.; Wilson, K.; v. d. Linde, D. & Barty, C. P. J. (1999). Detection of Nonthermal Melting by Ultrafast X-ray Diffraction, *Science*, Vol. 286, 1340-1342, ISSN 0036-8075

Sokolowski-Tinten, K.; Blome, C.; Blums, J.; Cavalleri, A.; Dietrich, C.; Tarasevitch, A.; Uschmann, I.; Förster, E.; Kammler, M.; Horn-von-Hoegen, M. & von der Linde, D. (2003). Femtosecond X-ray measurement of coherent lattice vibrations near the Lindemann stability limit , *Nature*, Vol. 422, 287-289, ISSN 0028-0836

Sprangle, P.; Ting, A.; Esarey, E & Fisher, A. (1992). Tunable, short pulse hard x-rays from a compact laser synchrotron source, *J. Appl. Phys.*, Vol. 72, 5032-5038, ISSN 0021-8979

Ta Phuoc, K. ; Rousse, A.; Pittman, M.; Rousseau, J. P.; Malka, V.; Fritzler, S.; Umstadter, D. & Hulin, D. (2003). X-Ray Radiation from Nonlinear Thomson Scattering of an Intense Femtosecond Laser on Relativistic Electrons in a Helium Plasma, *Phys. Rev. Lett.*, Vol. 91, 195001-1-4, ISSN 0031-9007

Ta Phuoc, K.; Fitour, R.; Tafzi, A.; Garl, T.; Artemiev, N.; Shah, R.; Albert, F.; Boschetto, D.; Rousse, A.; Kim, D-E.; Pukhov, A. & Seredov, V. (2007). Demonstration of the ultrafast nature of laser produced betatron radiation, *Phys. Plasmas*, Vol. 14, 080701.1-4, ISSN 1070-664X

Tarasevitch, A.; Orisch, A.; von der Linde, D.; Balcou, Ph.; Rey, G.; Chambaret, J. P.; Teubner, U.; Klopfel, D. & Theobald, W. (2000). Generation of high-order spatially coherent harmonics from solid targets by femtosecond laser pulses, *Phys. Rev. A*, Vol. 62, 023816-1-6, ISSN 1050-2947

Ueshima, Y.; Kishimoto, Y.; Sasaki, A. & Tajima,T. (1999). Laser Larmor X-ray radiation from low-Z matter, *Laser Part. Beams*, Vol. 17, 45-48, ISSN 0263-0346

Uiberacker, M.; Uphues,Th.; Schultze, M.; Verhoef, A. J.; Yakovlev, V.; Kling, M. F.; Rauschenberger, J.; Kabachnik, N.; Schröder, M. H.; Lezius, M.; Kompa, K. L.; Muller, H.-G.; Vrakking, M. J. J.; Kleineberg, S. U.; Heinzmann, U.; Drescher , M. & Krausz, F. (2007). Attosecond real-time observation of electron tunnelling in atoms, *Nature*, Vol. 446, 627-632, ISSN 0028-0836

Vachaspati (1962). Harmonics in the Scattering of Light by Free Electrons, *Phys. Rev*, Vol. 128, 664-666, ISSN

Vos, M. H. & Martin, J.-L. (1999). Femtosecond processes in proteins, *BBA-Bioenergetics*, Vol. 1411, 1-20, ISSN 0005-2728

Wagner, R. E.; Su, Q. & Grobe, R. (1999). High-order harmonic generation in relativistic ionization of magnetically dressed atoms, *Phys. Rev. A*, Vol. 60, 3233-3243, ISSN 1050-2947

Weller, H. R. & Ahmed, M. W. (2003). The HIγS Facility: A Free-Electron Laser Generated Gamma-Ray Beam for Research in Nuclear Physics, *Mod. Phys. Lett. A*, Vol. 18, 1569-1590, ISSN 0217-7323

Yu, W.; Yu, M. Y.; Ma, J. X. & Xu, Z. (1998). Strong frequency up-conversion by nonlinear Thomson scattering from relativistic electrons, *Phys. Plasmas*, Vol. 5, 406-409, ISSN 1070-664X

Zhou, J.; Peatross, J.; Murnane, M. M. & Kapteyn, H. C. (1996). Enhanced High-Harmonic Generation Using 25 fs Laser Pulses, *Phys. Rev. Lett.*, Vol. 76, 752-755, ISSN 0031-9007

Zewail, A. (2000). femtochemistry:atomic-scale dynamics of the chemical bond, *J. Phys. Chem. A*, Vol. 104, 5660-5694, ISSN 0021-9606

23

Radiation Dynamics from the Ultra-Intense Field Ionization of Atoms

Isaac Ghebregziabher and Barry Walker
University Of Delaware
USA

1. Introduction

The interaction of high-intensity and ultrashort lasers with matter produces high energy particles and photons. Recent experiments include the generation of high quality GeV electron (Kneip et al., 2009; Leemans et al., 2006) and multi-MeV proton (Clark et al., 2000; Snavely et al., 2000) beams in laser plasma interactions. Different schemes can now be followed to produce ultra-short x-ray radiation. The production of keV x-rays from betatron oscillations of relativistic electrons in plasma channels was demonstrated 5 years ago (Rousse et al., 2004). Femtosecond x-ray beams have also been detected when high energy laser pulses cross relativistic electrons from a conventional linear electron accelerator (LINAC) (Schoenlein et al., 1996). Recently, an ultrashort x-ray pulse has been produced from the scattering of an ultrashort laser pulse off a laser generated relativistic electron beam (Schwoerer et al., 2006). The primary processes of x-ray generation are the ionization of an atom/ion and the subsequent acceleration of photoelectrons to relativistic energies in the intense laser field. A detailed description of radiation from laser-atom interaction is necessary to fully understand the mechanism behind x-ray generation from laser-plasma interactions and there is a growing interest in the physics of the radiation from laser acceleration of photoionization at relativistic intensities (Chowdhury et al., 2005).
In this chapter we focus on the generation of x-ray photons by Thomson scattering from atomic ionization in ultra-strong laser fields. When an atom is submitted to an ultra-strong laser field, the laser electric field $\mathbf{E}_{laser}$ suppresses the Coulomb barrier and the outermost electron ionizes as it tunnels through the suppressed barrier. Photoelectrons ionized in an ultra-strong field $1 \times 10^{18}\ W/cm^2$ are accelerated to a relativistic speed in a fraction of an optical cycle, the magnetic field of the ionizing laser field becomes significant and the motion of these ionized free electrons is nonlinear. Through its relativistic and non-linear motion, the electron may emit radiation at high harmonics of the drive field frequency in a process known as non-linear Thomson scattering.
During tunneling ionization, the ultra-strong laser field leaks out an electron wave packet out of an atom at a finite rate, typically every half cycle of the laser field. For each ionization event, the electron wave packet spreads to a spatial dimension from nm to μm over the interaction time of the laser field. The electron may also radiate as it is relativistically accelerated in the ultra-strong laser field. It becomes natural to ask how this spatially extended relativistically driven electron wave packet may radiate. Different parts of the

electron wave function experience different phases of the laser field and the emitted radiation from each segment of the ionization current could interfere constructively/ destructively to enhance/suppress the emitted radiation in the far field. In sections 3 and 4 results and discussion will be presented on the effects of electron wave-function spreading and interference on radiation from atomic ionization in ultrastrong laser fields. We compare emitted radiation calculated with a classical electron (point charge) and semi-classical model which allows interference. We wish to clarify the impact of an incoherent versus a coherent treatment of the radiation from photoelectrons when the ionization, i.e., tunneling, is coherently driven. Though there are some indications from plasma experiments the process may be incoherent (Chen et al., 1998; Phuoc et al., 2003), it is not known experimentally whether radiation from coherently ionized and field accelerated photoelectrons in ultrastrong fields is incoherent, coherent, or partially coherent. The results are presented here for 800 nm laser radiation, but similar results are expected for other visible and IR laser wavelengths (Ueshima et al., 1999; Tate et al., 2007).

In the semi-classical model, for each ionization event, the ionization probability is multiplied by an electron charge to approximate the extended electron wave function in the continuum. This approximate model of the quantum electron doesn't allow us to investigate the effect of phase of the electronic wave function on the emitted radiation. An exact quantum mechanical treatment of the problem is required to investigate phase effects on radiation. However, to a good approximation the effects of electron wave packet spreading and interference from an extended charge distribution could well be captured with in our semi-classical model since the quantum mechanical charge distribution is usually represented by the modulus squared of the electron wave-function which erases any phase information associated with the electron wave-function.

Previous works (Mocken & Keitel, 2005) have shown deviations in the emitted radiation spectrum from a classical and an arbitrary prepared electron wave packet in a plane wave laser field. Recently, one paper (Peatross et al., 2008) has considered emitted radiation from a single electron wave packet by establishing a relation between the exact quantum formulation and its classical counterpart via the electron's Wigner function and has asserted that the electron radiates like a point like emitter. Both of these works ignore ionization and calculate the emitted radiation from a free electron interacting with an intense laser field. Future experiments on radiation from atomic ionization would shed light on how a photoelectron radiates and would vaildate/invalidate our semi-classical approach to the calculation of the radiation (Ghebregziabher & Walker, 2008).

1.1 Exact treatment of ultra-strong laser fields

Another fundamental question arises on how to describe the ultrastrong laser field when calculating the emitted radiation from photoelectrons accelerated in ultrastrong laser focus. To achieve high intensities the laser field is usually focused to a miniature focal spot, typically several μm in size at full width half maximum (FWHM). In strong field physics, electron excursion extends out to nm length scale and the plane wave approximation to the laser field is valid. Because in strong fields $\beta \approx 0$, electron dynamics is mainly affected by the electric field component of the laser field. The magnetic field component of the laser may be neglected and photoelectron dynamics is easily obtained by solving non-relativistic equation motion.

In ultra-strong laser fields photoelectron dynamics is relativistic ($\beta \neq 0$) and non-linear. With in a plane wave approximation to the laser field an analytic solution to the relativistic photoelectron dynamics is readily available (Lau et al., 2003). Figure 1 shows the coordinates of an electron born with zero initial speed at $t_{in} = 0$ and $z_{in} = 0$ in a plane wave laser field with an 800*nm* central wavelength and field strengths $a = 0.5$, $a = 1.53$, and $a = 3$ corresponding to laser intensities of $5 \times 10^{17}\ W/cm^2$, $5 \times 10^{18}\ W/cm^2$, and $2 \times 10^{19}\ W/cm^2$, respectively. The coordinates are shown for 3 optical cycles. The figure shows for laser intensities where the field strength parameter $a > 1$, the width of the photoelectron drift along the propagation direction (z-axis) is greater than the quiver width along the polarization axis (x-axis). Moreover, electron quiver width for $a > 1$ is greater than 0.3 μm. specifically, for a field strength $a = 3$, the quiver width is 0.8 μm. Realistic laser focus has a finite spatial extent. Typically, in ultrastrong fields, the full width at half maximum (FWHM) beam diameter could be 2 to 3 microns, which is only a factor of 2 times the electron excursion width in an optical cycle. In reality the laser focus has curved wave fronts that depend on space. Therefore, adopting plane wave approximation in an ultrastrong laser field is not valid.

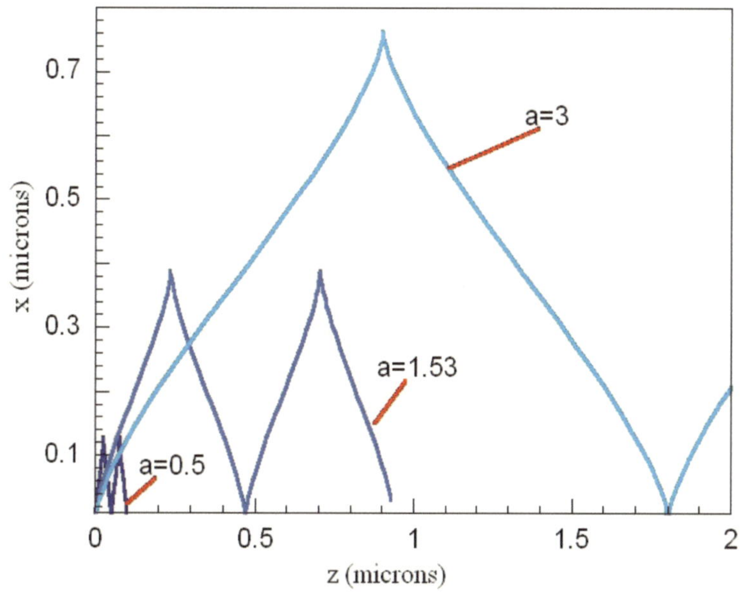

Fig. 1. Parametric plot of x versus z coordinates for an electron ionized at t=0 and z=0 for three different laser intensities corresponding to field strength of 0.5, 1.53, and 3.

To a leading order the paraxial approximation to the laser field may be adopted to solve the dynamics more accurately. In the paraxial approximation, the longitudinal field components are neglected, i.e., for a laser field polarized along the x-axis and propagating along the z-axis, the non-zero field components are E_x and B_y. These transverse field components confine the electron dynamics in the x-z plane. Such an approximation may be valid when describing photoelectron dynamics inside a laser focus with electron excursion much smaller than the waist diameter of the laser focus. However, to achieve ultrastrong laser fields, the laser field is focused to a small spot size. For instance, the diffraction limited focal spot size of a laser field with 800 nm central wavelength when focused with an f/# 1.5 parabolic mirror is about 2.42μm. In such a tight laser focus consistency with Maxwell's equations requires non-zero longitudinal fields. Previous works, see for example (Barton &

Alexander, 1989; Davis, 1979) have shown that the laser field could be described to any desired accuracy as a series expansion of the diffraction parameter $\epsilon = \lambda/(2\pi s_0)$, where λ is the carrier wave length and s_0 is waist diameter of the beam at $\exp^{-2}$. Here we give the field components accurate on the order of ϵ^3 for a laser field polarized along the x-axis and propagating in the z-axis.

$$
\begin{aligned}
E_x &= \tilde{E}\left[\cos(\Phi+\alpha)+\varepsilon^2\left\{\frac{2x^2+r^2}{s(z)^2}\cos(\Phi+3\alpha)-\frac{r^4}{s(z)^3 s_0}\cos(\Phi+4\alpha)\right\}\right]\\
E_y &= \varepsilon^2\tilde{E}\frac{2xy}{s(z)^2}\cos(\Phi+3\alpha)\\
E_z &= -2\tilde{E}\varepsilon\frac{x}{s(z)}\left[\sin(\Phi+2\alpha)+\varepsilon^2\frac{r^2}{s(z)^2}\left\{3\sin(\Phi+4\alpha)-\frac{r^2}{s(z)s_0}\cos(\Phi+5\alpha)\right\}\right]\\
B_x &= E_y\\
B_y &= \tilde{E}\left[\cos(\Phi+\alpha)+\varepsilon^2\left\{\frac{2y^2+r^2}{s(z)^2}\cos(\Phi+3\alpha)-\frac{r^4}{s(z)^3 s_0}\cos(\Phi+4\alpha)\right\}\right]\\
B_z &= E_z\frac{y}{x},
\end{aligned}
\tag{1}
$$

where,

$$
\begin{aligned}
\tilde{E} &= E_0\frac{s_0}{s(z)}\exp\left(-\frac{r^2}{s(z)^2}\right)\\
r^2 &= x^2+y^2\\
\Phi &= \omega_0 t - kz - \frac{zr^2}{z_R s(z)^2}\\
\alpha &= \arctan\left(\frac{z}{z_R}\right), z_R = \frac{ks_0^2}{2}.
\end{aligned}
$$

Where E_0 is the laser field amplitude, z_R is the rayleigh length and $s(z)$ is the beam diameter at a given propagation distance. Figure 2 (a) shows contour plot of the spatial profile of the nonparaxial laser field E_z and the largest electric field component E_x (b). The longitudinal field component E_z is a factor of 6 less than the transverse field component E_x, see Fig. 2. The figure also shows that longitudinal field components are not only non-zero but also exhibit different spatial profile compared to the transverse field components. This is evident from figure 2 where the field component E_z is asymmetric with respect to reflection about the origin while E_x preserves symmetry with reflection. Moreover, E_z vanishes on axis, i.e., $E_z = 0$ for $x = 0$ while E_x has the largest amplitude.

These complicated field structures expose photoelectrons to rather complicated dynamics. In section 5 of this chapter we compare the traditional plane wave approximation treatment of the laser field with our new results that include the 3-dimensional nature of the focused laser field with curved wave fronts to provide framework for experimental events, i.e., when should observations be compared to the plane wave results and what is the primary impact of the focus geometry used in experiments. This treatment includes the fact that electrons ionized at the same axial position but different radial positions will be accelerated by different phases of the laser field. The results presented here extend previous works of an electron interacting with an intense electromagnetic field (Lee et al., 2005; Esarey et al., 1993; Keitel & Knight, 1995; Hu & Starace, 2002) by including the ionization process itself as it affects the phase of the driving field for the photoelectrons and coherence properties of the emitted radiation.

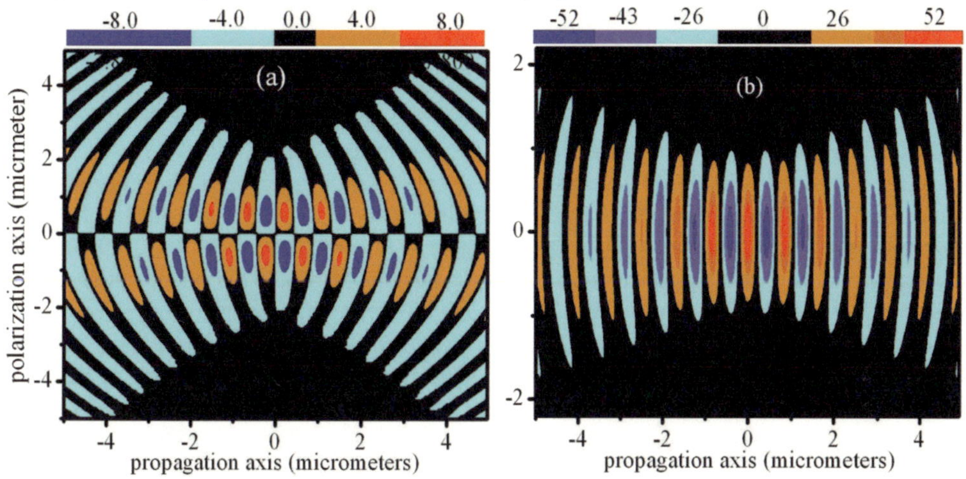

Fig. 2. Contour plot of the spatial profile of E_z (x, y = 0, z) (a) and E_x (x y = 0, z) (b) (with field strengths given in atomic units). The laser beam waist diameter is 1.52 μm and the peak laser intensity is $1 \times 10^{20} W/cm^2$.

2. The radiation model

Every aspect of the radiation would be appropriately predicted if one solves the Dirac equation in 3+1 dimensions given by

$$i\hbar\frac{\partial\psi}{\partial t} = c\vec{\alpha}\cdot\left(\frac{\hbar}{i}\vec{\nabla} - \frac{q}{c}\vec{A}\right)\psi + (V(r) + \beta mc^2)\psi \tag{2}$$

where $\vec{A}$ is the laser field vector potential, q is the charge of an electron, m is the rest mass of an electron, c is the speed of light, $V(r)$ the Coulomb potential and α^i and β are the usual Dirac matrices (i = 1, 2, 3). It requires large spatial grid sizes with exteremely small time steps to solve eqn. 2 numerically. One paper (Mocken & Keitel, 2005) based on Dirac charge current and classical electrodynamics outlines a procedure to calculate the radiation spectrum emitted by an arbitrarily prepared Dirac wave packet and has shown deviations to the purely classical calculations in the high frequency part of the radiated spectrum. To the best of our knowledge there exists no fully quantum mechanical calculation of radiated field from atomic ionization that takes into account ionization dynamics and the 3-dimensional nature of the ionizing ultrastrong laser field. We circumvent this problem by employing a semiclassical trajectory ensemble model to represent the electron wavefunction in the continuum, which has been successful in predicting correlated multielectron ionization (Palaniyappan et al., 2006) and the cutoff photon energy associated with higher-order harmonic generation (Corkum, 1993). With the simplified semiclassical trajectory ensemble representation of the electron wave function, we developed a four step ultrastrong radiation model divided as follows:

1. Atomic photoionization
2. Relativistic dynamics for the electron
3. Calculation of radiation from relativistic electrons
4. Superposition of the radiation across the laser focus

2.1 Classical and semiclassical ionization dynamics

In the first part of the radiation model, atomic photoionization, classical and semiclassical ionization models are considered. For classical ionization, the strong field tunneling ionization is merged with a Monte Carlo (MC) technique which has been described in (Ghebregziabher & Walker, 2007). In the classical model, the ionization rate during the laser pulse is first calculated according to the the ionization based on hydrogen orbitals extended to complex atoms by Ammosov, Delone, and Krainov (ADK) (Ammosov et al., 1986) (ADK) tunneling ionization model for each phase of the laser field. The ADK theory is essentially an extension of Perelomov, Popov, and Terentev (PPT) (Perelomov et al., 1966) theory where states of complex atoms are characterized by effective principal and orbital quantum numbers. The ionization rate is then normalized to the peak rate at the center of the focus, peak of the pulse. If the normalized rate exceeds 10^{-3} a comparison is made with a randomly generated number between zero and one. When the normalized ionization rate is greater than the randomly generated number, ionization is allowed 50% of the time (again determined by chance) and a single classical electron is liberated to the continuum from its bound state. Over many events this assignment of ionization provides a linear mapping of rate to ionization events. The tunneling ionization treatment used in this paper limits the maximum intensity to approximately $1 \times 10^{20}\ W/cm^2$, beyond this intensity the laser magnetic field and relativistic effects may affect the fundamental ionization mechanism (Popov, 2004). No experimental measurements have verified the atomic ionization mechanism above a few times $1 \times 10^{20}\ W/cm^2$.

For semiclassical calculations, ionization for a single atom is calculated according to ADK tunneling for each phase of the laser pulse. Rather than a discrete ionization event as in the MC ADK case, here a trajectory ensemble weighted by the fractional ionization probability for that phase is liberated to the continuum. In this semiclassical ionization model, the trajectory ensemble weighted by the ionization probability is an approximation for the tunneling probability current of the quantum electron. In Fig. 3(a) the appearance of the electron probability in the continuum is plotted as a function of time and shown to be very similar for the classical MC ADK and semiclassical ADK calculation. It is important to note the MC ADK classical ionization case (which for a single event is a step function) is shown for many MC events and so represents an average. In Fig. 3(b,c) the spatial distribution of the electron probability after a single atom interacts with the field for few cycles is shown. While classical ionization gives rise to a single point electron (b) the tunneling ionization current appears continuously (c) with maximum bursts at the peaks of the electric field.

The ionization contribution is generally considered from a single charge state; Ne^{8+} at $2\times10^{17}\ W/cm^2$, Ar^{10+} at $2 \times 10^{18}\ W/cm^2$, Ar^{16+} at $1.2 \times 10^{19}\ W/cm^2$, Na^{10+} at $5 \times 10^{19}\ W/cm^2$, and Na^{11+} at $1.2 \times 10^{20}\ W/cm^2$. Results for ionization from multiple charge states preceding the peak intensity have also been calculated and will be presented in the final sections of the chapter. These ions were chosen because they have a net ionization probability at the specified intensity of typically 10^{-3} to 10^{-2}. A cluster of 32 Opterons with 2.4 GHz processors were used for these calculations.

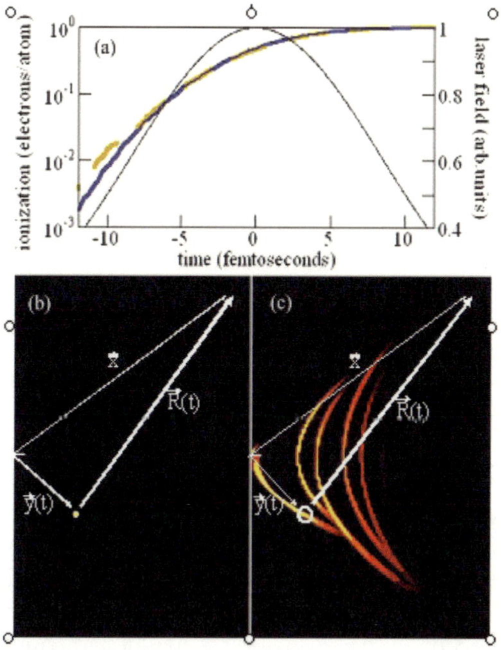

Fig. 3. Plot of ionization probability (a) as a function of time across the laser pulse (thin line) for the MC-tunneling (dashed) classical model and tunneling ionization probability current (wide line) semiclassical tunneling. Shown in (b,c) are 2D plots of the electron density for a classical electron (dot, b) and the tunneling ionization probability current (c) in a 75nm (horizontal, k) by 800nm (vertical, E) frame from an atom ionizing (located at the white tick on the left-center of each frame) a few cycles after ionization begins. Superimposed on the frames are the vector x, y, and R for the case when the atom is located at the origin. In (c) a representative charge segment $f_j q$ is circled in white.

2.2 Photoelectron continuum dynamics

For both classical and semi-classical ionization methods, the photoelectron dynamics are calculated by solving the relativistic equations of motion given as:

$$\begin{aligned}
\frac{dp_x}{dt} &= q\left(E_x + \frac{p_y}{\gamma m_0 c}B_z - \frac{p_z}{\gamma m_0 c}B_y\right)\\
\frac{dp_y}{dt} &= q\left(E_y + \frac{p_z}{\gamma m_0 c}B_x - \frac{p_x}{\gamma m_0 c}B_z\right)\\
\frac{dp_z}{dt} &= q\left(E_z + \frac{p_x}{\gamma m_0 c}B_y - \frac{p_y}{\gamma m_0 c}B_x\right)\\
\frac{dx}{dt} &= \frac{p_x}{m\gamma}\\
\frac{dy}{dt} &= \frac{p_y}{m\gamma}\\
\frac{dz}{dt} &= \frac{p_z}{m\gamma}
\end{aligned} \tag{3}$$

Where m_0 is the rest mass of an electron, q the charge of an ensemble member, **p** the particles momentum, **E** and **B** are the laser magnetic and electric field vectors, and γ is the well known relativistic lorentz factor. Implicit in the notation of the laser electric and magnetic fields is the time and space dependence of the fields, for example $E_x = E_x(\mathbf{r}, t)$. Notice that non-paraxial field components are included in equation 3.

We use a Runge-Kutta ordinary differential equation solver with relative error tolerance threshold of 10^{-6}, local error threshold of 10^{-12}, and time step typically of the order $10^{-4} fs$ to solve photoelectron dynamics (Eqn. 3). The laser pulse used for these calculations ($\mathbf{E}_{laser}$, $\mathbf{B}_{laser}$ linear polarization, 800 nm central wavelength, and 20 fs full width at half maximum Gaussian temporal profile) is comparable to current high field experiments (Albert et al., 2000). Two different spatial modes are considered: a plane wave and a f/# 1.5 TEM_{00} nonparaxial laser focus calculated to a field accuracy of 0.5%, i.e., third order non-paraxial terms. Fifth order non-paraxial field terms have also been included but the resulting dynamics didn't change from the dynamics calculated with third order accuracy. Since the scattered radiation comes from relativistic photoelectron dynamics in the laser focus, the accuracy of the calculated radiation yield will be limited by how much photoelectron dynamics is impacted when approximating the laser field with paraxial solution.

Since the motion is on the scale of nm to μm and the electron energies are of order 10 times the ionization potential energy, the Coulomb field of the core atom or ion does not affect the dynamics. Also the interaction is considered in the low-density limit so space charge effects and hard collisions (collisions with low impact parameters) are neglected. Radiation damping is not accounted for in the calculation since the ratio of the total radiated energy per cycle to the average kinetic energy of the electron is less than 10^{-6}.

In the limit of low fields ($< 1 \times 10^{16}\ W/cm^2$) or weakly focused geometries, the paraxial approximation to the laser field is valid and simplifications occur in the relativistic equations of motion given by Eqn. 3. In this case, a relativistically invariant relationship between the longitudinal and transverse momenta of the photoelectron at the end of the laser pulse exists, i.e., $p_z = p_x^2/2mc$ or equivalently $\tan(\theta) = \sqrt{2/(\gamma-1)}$ where θ is the polar angle from the propagation axis (z). Figure 4 is a snapshot of the electron probability projected on the x-z plane from ionization of Ar^{8+} at the peak of the pulse with an intensity of $2 \times 10^{18}\ W/cm^2$ and Ar^{15+} at an intensity of $1 \times 10^{19}\ W/cm^2$. As one can see from figure 4(a) photoelectron dynamics is relativistic at an intensity of $2 \times 10^{18}\ W/cm^2$ as shown by drift along z. However, as the extent of the probability distribution is near the center of the focus the continuum may be approximately described by $p_z = p_x^2/2mc$ dynamics with a spatially uniform electric and magnetic field. For the electric charge distributions in Fig. 4(a), $\mathbf{k} \cdot \mathbf{r}$ is less than 1 and γ is less than 1.1. With these conditions the electron will oscillate in phase with the laser field and the parabolic relationship between p_x and p_z will result in a similar relationship between the x and z components of the drift velocity causing the wave fronts to become parabolic.

Photoelectron dynamics for the higher laser intensity ($1 \times 10^{19}\ W/cm^2$) is clearly different, as can be seen from Fig. 4(b), with the electron probability from each tunneling ionization "burst" distributed over a $10^{-6}m$ length scale, which is comparable to the focus and wavelength dimension. Although near the center of the f/# 1.5 TEM_{00} mode laser focus the field is nearly paraxial, the large electron velocity at $1 \times 10^{19} W/cm^2$ $\gamma \simeq 2$ destroys the parabolic relationship between the x and z components of the drift velocities. As shown in

Fig. 4(b), a significant amount of electron probability approaches the Raleigh length while the field is still large, a scenario made possible by large **k** · **r**, the motion is dominated by non-paraxial fields, which are very different from the paraxial fields in phase and amplitude.

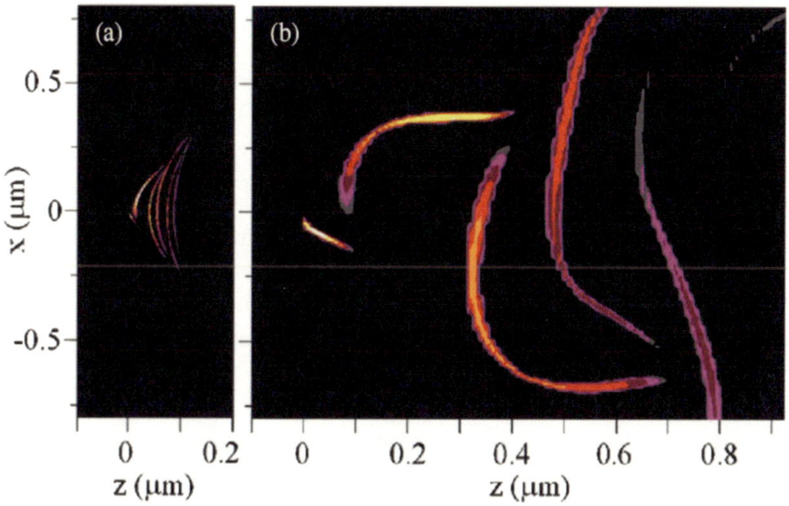

Fig. 4. Semiclassical ensemble trajectory electron probability from atomic ionization at $2 \times 10^{18} W/cm^2$ (a), and $1\times10^{19} W/cm^2$ (b). The electron probability is shown in a normalized log color scale with one color per factor exp(-1).

2.3 Radiation from relativistic charges

Once the dynamics is solved with the above equations, the radiated energy from a relativistic charge could be obtained with Lienard-Wiechert potential (Jackson, 1975). The radiated electric field is obtained with the equation given by:

$$\vec{E}(\vec{R},t) = \frac{q}{4\pi c}\left(\frac{\hat{n}\times((\hat{n}-\vec{\beta})\times\dot{\vec{\beta}})}{(1-\hat{n}\cdot\vec{\beta})|\vec{R}|}\right). \tag{4}$$

where $R(t) = |\vec{x} - \vec{y}(t)|, \hat{n} = \dfrac{\vec{x}-\vec{y}(t)}{R(t)}, \vec{\beta(t)} = \dfrac{\vec{y(t)}}{c} = \dfrac{1}{c}\dfrac{dy}{dt}(t)$ and c is the speed of light. Since $|\vec{x}| \gg |\vec{y}|, R^2 \cong |\vec{x}|^2$. For the purposes of comparison, all reported fields and radiated energies are normalized for a yield expected from a single electron. The vectors $\vec{x}, \vec{y}$, and $\vec{R(t)}$ are shown superimposed on Fig. 3 (b,c). With the temporal and spatial intensity changes in a laser focus there are corresponding changes in the photoelectron velocity and acceleration. From eqn. 4 it is easy to see that the intensity of the radiated photons from a relativistically moving charge is proportional to the following parameter:

$$f = \frac{\hat{n}\times\left[\left(\hat{n}\times\vec{\beta}\right)\times\dot{\vec{\beta}}\right]}{\left(1-\hat{n}\cdot\vec{\beta}\right)^3} = \frac{\hat{n}\cdot\dot{\vec{\beta}}(\hat{n}-\vec{\beta}) - \dot{\vec{\beta}}(1-\hat{n}\cdot\vec{\beta})}{\left(1-\hat{n}\cdot\vec{\beta}\right)^3} \tag{5}$$

where $R(t) = |\vec{x} - \vec{y}(t)|, \hat{n} = \dfrac{\vec{x} - \vec{y}(t)}{R(t)}, \vec{\beta(t)} = \dfrac{\vec{y(t)}}{c} = \dfrac{1}{c}\dfrac{dy}{dt}(t)$ and c is the speed of light. In the limit where v/c approaches unity, the parameter f given in eqn. 5 is largest for electron velocities aligned with the detector at $\vec{x}$, i.e. $\vec{\beta} \cong \hat{n}$, and accelerations perpendicular to the line of sight to the detector i.e., $\hat{n} \cdot \dot{\vec{\beta}} = 0$. For these instances, $(\vec{\beta} \cong \hat{n}, \hat{n} \cdot \dot{\vec{\beta}} = 0)$ the parameter given in eqn. 5 simplifies to the form.

$$\lim_{v/c \to 1} = \frac{\dot{\vec{\beta}}}{\left(1 - \hat{n} \cdot \vec{\beta}\right)^2} \tag{6}$$

The parameter f is used here to provide insight into the dynamics and show which regions of the laser focus are responsible for the highest radiation yields. This parameter calculated for 5000 electron trajectories is shown in Fig. 5 for laser intensities of $2 \times 10^{17}\, W/cm^2$, $2 \times 10^{18}\, W/cm^2$ and $1.2 \times 10^{19}\, W/cm^2$. At an intensity of $2 \times 10^{17}\, W/cm^2$ (Fig. 5(a)) the electron motion is largely non-relativistic with a peak $\gamma = \left(\sqrt{1 - \frac{v^2}{c^2}}\right)^{-1} = 1.05$ and the denominator of equation (6) may be neglected. The motion associated with radiation via the parameter f has a distribution that mimics the ionization distribution and spatial intensity profile of the laser focus. The radiation in this case comes from the center of the focus where the photoelectron experiences the highest intensity.

At an intensity of $2 \times 10^{18}\, W/cm^2$ (Fig. 5b) relativistic effects with a peak $\gamma = \left(\sqrt{1 - \frac{v^2}{c^2}}\right)^{-1} = 1.5$ start to kick in and the motion associated with radiation via the parameter f has a distribution that extends beyond the peak region of highest intensity.

As the intensity is increased to $1.22 \times 10^{20}\, W/cm^2$ (see Fig. 5(c)), the electron motion is highly relativistic and the dynamics change with an electron excursion comparable to the size of the laser focus. As the electron leaves the focus, the laser imparts an extra boost of speed in a process known as „surfing" coined from the wave like „winged" pattern for f in Fig. 5(c). Counter intuitively, the regions of the highest laser intensities near the center of the focus do not result in the best conditions for radiation. Rather, at ultrahigh intensities and focused geometries the radiation may be expected to come when the photoelectron is off the peak of the laser focus.

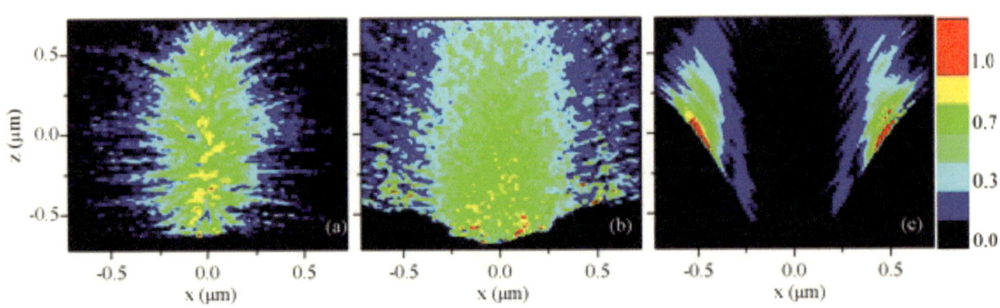

Fig. 5. Contour plot of the parameter f for laser intensities of $2 \times 10^{17}\, W/cm^2$(a), $2 \times 10^{18}\, W/cm^2$(b) and $1.22 \times 10^{20}\, W/cm^2$ (c). The magnitude scale for f is linear and normalized in each case.

2.4 Superposition of the radiation

With the emitted field obtained with Eqn. 4, the total radiated electric field is calculated by summing coherently and incoherently across the ionization current and laser focus in the focused geometry case. In the plane wave (1D) classical ionization case the yield we report in this chapter represents an average of hundreds to thousands of Monte Carlo photoelectrons for atomic ionization across a broad range of phases. Most of the radiation comes from electrons ionized at the peak of the field, however, the method accurately accounts for „off peak" ionization rates. For a single MC-ADK ionization event, the emitted electric field is given by Eq. 4. For a number of atoms ionized N, the total radiated energy is the classical total radiated energy, W_C, per unit solid angle normalized to an average yield per photoelectron given by:

$$\frac{dW_c}{d\Omega} = \left(\frac{c}{4\pi}\right) \frac{R^2}{N} \sum_{i=1}^{N} \int_{-\infty}^{\infty} \left|\vec{E}(\vec{R},t)\right|^2 dt \tag{7}$$

In the plane wave, semi-classical treatment a single atom is adequate to accurately capture the emitted field response from the ionization current since ionization occurs over all phases; however, the normalization for the fractional ionization must be correctly taken into account. For a total number of M phase steps that comprise the semi-classical ensemble, the radiated electric field from the jth phase step of the ensemble $\vec{E}_j(\vec{R},t)$ is obtained by replacing q in Eq. 4 with the weighted charge $f_j q$, where f_j is the fractional ionization probability at a given laser phase (see Fig. 3(c)). For incoherent and coherent superposition of the radiation from the tunneling probability current, the sum of the radiation and normalization must be done differently. For the incoherent case, the total radiated energy per unit solid angle from the SC tunneling probability ionization current is given by:

$$\frac{dW_{sc-incoherent}}{d\Omega} = \left(\frac{c}{4\pi}\right) R^2 \int_{-\infty}^{\infty} \left\{ \frac{\Sigma_j^M \left(\vec{E}(\vec{R},t)\right)^2}{\Sigma_j^M f_j^2(t)} \right\} dt \tag{8}$$

In the case of a coherent superposition of the radiation from the tunneling probability current, $\vec{E}_j(\vec{R},t)$ must be summed before being squared to allow for interference, i.e., $\vec{E}_{sc-coherent}(\vec{R},t) = \Sigma_{i=1}^{M} \vec{E}_j(\vec{R},t)$. The total radiated power per single electron ionization event is then calculated by normalizing to the total ionization probability squared,

$$\frac{dW_{sc-coherent}}{d\Omega} = \left(\frac{c}{4\pi}\right) R^2 \int_{-\infty}^{\infty} \left\{ \frac{\left(\Sigma_j^M \vec{E}(\vec{R},t)\right)^2}{\left(\Sigma_j^M f_j(t)\right)^2} \right\} dt \tag{9}$$

Here again M is the sum over the SC ensemble, not the sum over the ionization from different atoms.

The previous cases (Eq. 8- 9) address the radiation from the interaction of the atom with a plane wave. Calculations for atoms distributed in a three dimensional laser focus (3D) include an addition spatial sum for the radiation across all the ionizing atoms in the focus. Spatial summations for radiation between different atoms in all cases (i.e. W_C, $W_{SC\text{-}incoherent}$, $W_{SC\text{-}coherent}$) are done coherently; the total radiation field is equal to the sum of the fields from all the atoms. For the classical case, the form of Eq. 7 is the same but the index N represents a sum of all ionization events across the focus and not an average of many atoms at a single location as for the plane wave case. In the SC tunneling probability current case, an additional sum is included over the N atoms ionizing in the laser focus incorporated as shown in Eqs. 10 and 11. It is important to note that $W_{SC\text{-}incoherent}$ in the focal geometry case involves an incoherent sum of the radiation from the tunneling ionization probability for a single atom but a coherent sum of the radiation between different atoms. For the fully coherent case $W_{SC\text{-}coherent}$, the radiation sums from the tunneling probability current and between atoms are treated coherently.

$$\frac{dW_{sc-incoherent}}{d\Omega} = \left(\frac{c}{4\pi}\right) R^2 \int_{-\infty}^{\infty} \left\{ \frac{\sum_i^N \sum_j^M \left(\vec{E}(\vec{R},t)\right)^2}{\sum_i^N \sum_j^M f_j^2(t)} \right\} dt \tag{10}$$

$$\frac{dW_{sc-coherent}}{d\Omega} = \left(\frac{c}{4\pi}\right) R^2 \int_{-\infty}^{\infty} \left\{ \frac{\left(\sum_i^N \sum_j^M \vec{E}(\vec{R},t)\right)^2}{\sum_i^N \left(\sum_j^M f_j(t)\right)^2} \right\} dt \tag{11}$$

With respect to scaling, without normalization the radiated power scales as N for the incoherent W_C and $W_{SC\text{-}incoherent}$ cases and as N^2 for the $W_{SC\text{-}coherent}$ case.

To understand the difference between coherent and incoherent summation of the radiated field, it is important to make an analogy between the one-photon, two-slit experiment and summation of the radiation across the tunneling ionization current. To accurately describe the one-photon, two-slit experiment, one must have knowledge of the field amplitude on the slits and the relative phase difference for the path from the slits to the detector. The ionization of the electron over space and time creates several paths for the radiation from the electron. At the detector, when the field collapses and the radiated photon is detected, contributions from all paths for the electron probability have to be considered. In fact, as shown in figure 4, the electron probability may extend over 10^{-6} m length scales and introduces a spatial component to the phase of the radiation that could result in interferences. Figure 2.4 shows the radiated electric field for three trajectories from ionization at 1×10^{19} W/cm^2. The figure illustrates that propagation delays can lead to destructive interference of the radiated electric field (when summing coherently) just as the negative and positive fields for the three cases in the figure overlap.

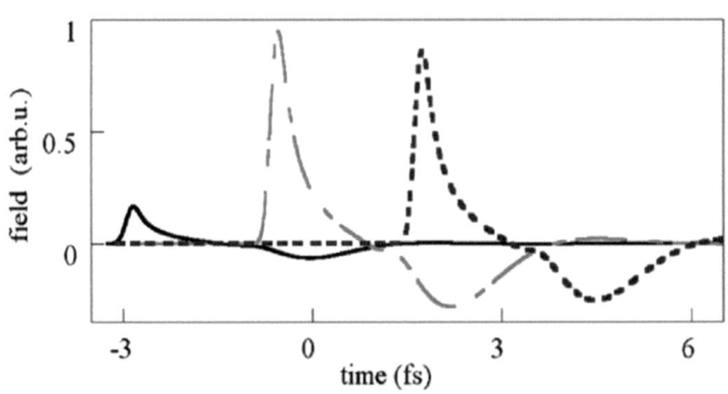

Fig. 6. The E_x field ($\theta = 50°$, $\phi = 0°$) for three trajectories from ionization at $1 \times 10^{19} W/cm^2$.

3. Radiation from ionization in a pulsed, plane wave laser field

3.1 Total energy and power spectrum

Figure 7 shows the total radiated energy from a photoelectron in a one dimensional plane wave laser pulse with the three different treatments of the radiation: classical photoionization and the tunneling probability current summed incoherently and coherently. The radiation yield in the figure is given in units of energy per photoelectron (eV/electron). The results are discussed in terms of three different intensity regimes (Ghebregziabher & Walker, 2007) based on the relative size of the electron quiver motion ($\alpha_0 = a_0/2\pi$, where $a_0 = eE_0/m\omega c$) to the wavelength of the emitted radiation: $\alpha_0 < \lambda/10$ Region I, $\lambda/10 < \alpha_0 < \lambda$ Region II, and $\lambda < \alpha_0$ Region III.

At intensities of $1 \times 10^{17}\ W/cm^2$ ($\alpha_0 = 38\ nm$) and less, highlighted as Region I in the figure, the quiver amplitude is small. In this region, the total radiated energy from a photoelectron, whether classical, semi-classical, coherent, or incoherent, is identical. The quiver amplitude is sufficiently small that interference effects in the radiation are negligible. For laser intensities in the range $1 \times 10^{17}\ W/cm^2$ to $1 \times 10^{19}\ W/cm^2$, Fig. 7 region II) the electron quiver may be up to 300 nm. The total radiated energy from a classically ionized electron is still identical to the incoherently summed SC tunneling probability current. However, interference effects in the coherent sum reduce the radiated energy from one-half to one-fifth of the incoherent treatments.

Finally, for laser intensities greater than $1\times10^{19}\ W/cm^2$ (Fig. 7 region III), the electron quiver width exceeds the fundamental wavelength. Figure 7 shows for this intensity region the radiated energy from a classically ionized electron and incoherent sum of the tunneling probability current are indistinguishable. Continuing the trend from Region II the coherently summed radiation from the tunneling probability current is fifty times smaller than incoherently summed radiation at an intensity of $1 \times 10^{19}\ W/cm^2$. This is attributed to the 1 micron electron quiver amplitude that results in significant interference effects.

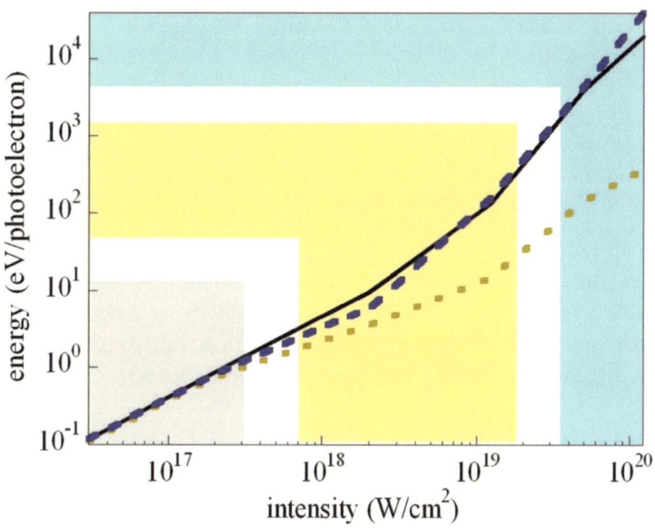

Fig. 7. Radiated energy versus laser intensity from a classically ionized electron (solid, black), incoherently summed tunneling probability current (dashed, blue) and coherently summed tunneling probability current (dotted, dark yellow).

Figure 8 shows the power spectrum of the radiated energy (Fig.7) for intensities (a) 2×10^{17} W/cm^2, (b) 1.2×10^{19} W/cm^2, and (c) 5×10^{19} W/cm^2 with the three models of ionization: MC-ADK, SC-incoherent, and SC-coherent. From the figure, one can see the power spectrum calculated with classical MC-ADK ionization and SC-incoherent model are identical. The radiated power drop at higher intensities seen in Fig. 7 for the coherent sum are manifested as a progressive decrease in the relative high frequency radiation in Fig. 8(a-c). While in Fig. 8 (a) the power spectrum for all cases is nearly identical, in Fig. 8(b,c) the higher frequency radiation is lower by an order of magnitude. This observation corroborates the earlier interpretation of destructive interference for relativistic intensities where the photoelectron excursion is equal to or exceeds the wavelength of the radiation and results in a significant phase shift. The spectral amplitude from coherent averaging over the tunneling probability current is about a factor of 10 less than that from a classically ionized electron. For higher energy photons with $\lambda < \alpha_0$, the spectral amplitude from a classically ionized electron is as large as 35 times that of a coherent average over an ionization probability current. In all cases, up to the maximum intensity of 1×10^{20} W/cm^2 studied, there is no observed difference in the integrated scattered fundamental radiation.

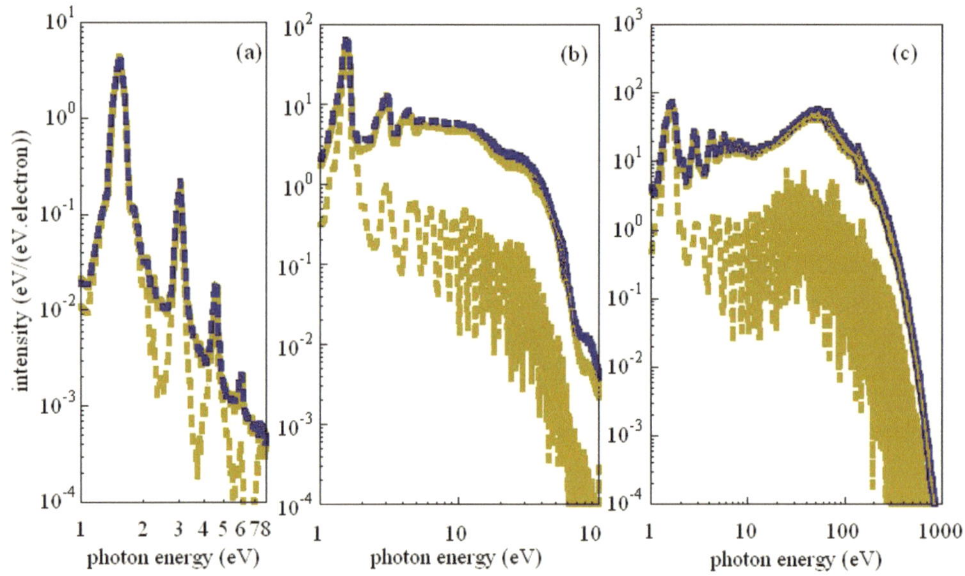

Fig. 8. Total spectral amplitude of the radiated field from atomic ionization at intensities (a) 2×10^{17} W/cm^2, (b) 1.2×10^{19} W/cm^2, and (c) 5×10^{19} W/cm^2 from a classically ionized electron (solid, dark yellow), incoherently summed tunneling probability current (dashed, blue) and coherently summed tunneling probability current (dotted, black).

3.2 Angle resolved radiated energy

The polar angle (θ measured from the $\vec{k}$ vector of the drive laser field) resolved total radiated energy is shown in Fig. 9 from a single classical electron and a tunneling probability current summed coherently in a one-dimensional plane wave laser pulse. (The incoherently summed ionization current result is indiscernible from the classical case). The radiation yield in the figure is given in units of energy per photoelectron per unit angle (eV/photoelectron-degree). At laser intensities from 3×10^{16} W/cm^2 to 2×10^{17} W/cm^2 (Fig.

9(a,b)), the electron motion is non-relativistic with $\gamma = \left(\sqrt{1-\frac{v^2}{c^2}}\right)^{-1}$ = 1.01 to 1.05 and the interaction can be treated in the dipole limit, i.e. the radiated energy shows angular symmetry with maximum radiated energy at a polar angle θ=0° and minimum at θ=90°. As the intensity is increased to 2×10^{18} W/cm^2 the interaction becomes relativistic ($\gamma \approx$ 1.5), which destroys the polar angle dipole radiation pattern symmetry. Furthermore, a clear distinction can be made between a classically ionized electron and coherent radiation treatment from a tunneling probability current. Since the radiation is identical for both methods at θ= 0° (Fig. 9(c-f)) and is increasingly different as one looks in away from k, one can infer the distinction between the two cases has an origin in interference since there is no phase difference in the propagation direction.

As the laser intensity is increased to 1.2×10^{19} W/cm^2 ($\gamma \approx$ 4) the radiation is peaked at θ=60° and resembles more closely the radiation pattern expected for relativistic accelerated charge. For a laser intensity of 5 × 10^{19} W/cm^2 ($\gamma \approx$ 12) the radiation yield from a classically ionized electron is singly peaked at a polar angle θ=30° while radiation from semi-classical ionization treatment with coherent averaging has two peaks located at θ=0° and θ=30°. As mentioned, the radiation yield from the two treatments of ionization at a polar angle θ= 0° is identical; however, as one sweeps to larger polar angles towards the secondary peak at θ= 30° the radiation yield from coherently summed tunneling probability current is drowned by a factor of approximately 50 due to interference. In Fig. 9(f) at 1.2 × 10^{20} W/cm^2 the electron motion is ultra-relativistic with a peak γ= 30, radiation emitted at θ= 20° and an incredible factor of 100 difference in the peak radiation yield from coherent tunneling ionization probability current compared to that of a classically ionized electron.

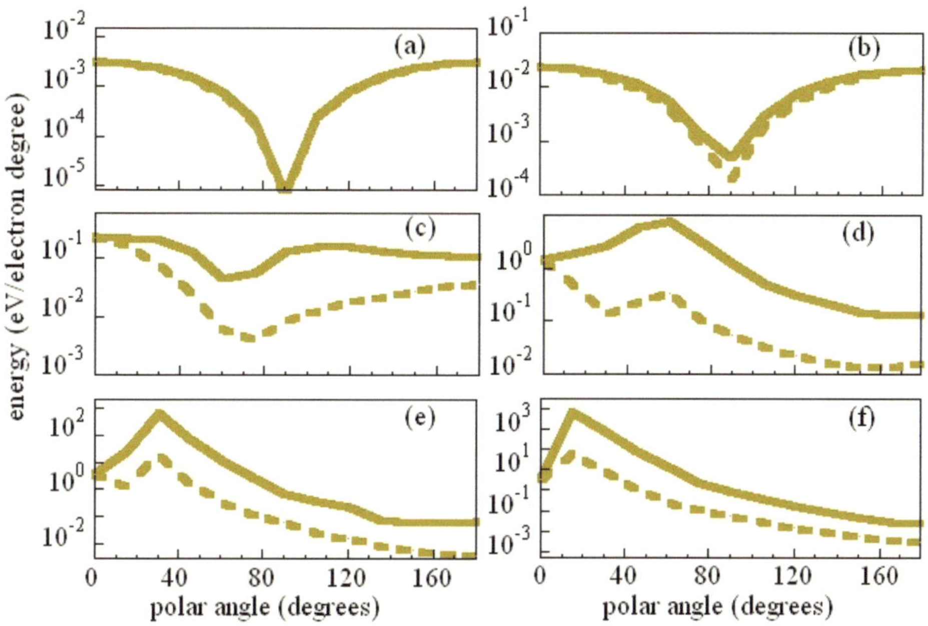

Fig. 9. Angle resolved radiated energy for ionizations in plane wave laser field with peak intensities (a) 3 ×10^{16} W/cm^2, (b) 2 ×10^{17} W/cm^2, (c) 2×10^{18} W/cm^2, (d) 1.2 ×10^{19} W/cm^2, (e) 5 × 10^{19} W/cm^2, and (f) 1.2 × 10^{20} W/cm^2 for a classical photoionization (solid) and coherently summed tunneling probability current (dotted).

4. Radiation from ionization in a pulsed, focused laser field

4.1 Total energy and power spectrum

Until now we have discussed only the results for single ionization event in a 1D plane wave laser pulse. In this section we include radiation in a three-dimensional high intensity laser focus where multiple atoms/ions are distributed randomly and uniformly in the laser focus. The density of atoms is varied to converge on a density independent result while avoiding inter-atomic spacing effects.

Figure 10 shows the total radiated energy from classical MC-ADK ionization and SC-coherent ionization in a three dimensional laser pulse. The radiation yield is normalized in the figure to the total amount of photoionization with units of energy per photoelectron (eV/electron). Figure 10 shows at intensities of $1 \times 10^{17}\ W/cm^2$ and less, highlighted as Region I in the figure, there is a slightly more of a difference between W_c and $W_{sc\text{-}coherent}$ compared to Fig. 11(a). Overall though, the results are consistent with the 1D plane wave results. Any difference from the plane wave case is expected to be rooted in the increased drift energy of the electron from the acceleration of the photoelectron out of the focus. For laser intensities in the range ($2 \times 10^{17}\ W/cm^2$ to $5 \times 10^{19}\ W/cm^2$, Figs. 10,11), the results in the focal geometry case are qualitatively identical to those for the one-dimensional analysis, i.e. a reduction in the high frequency radiation due to interference in the extended tunneling ionization probability current compared to the „perfect" coherence of a point, classical electron.

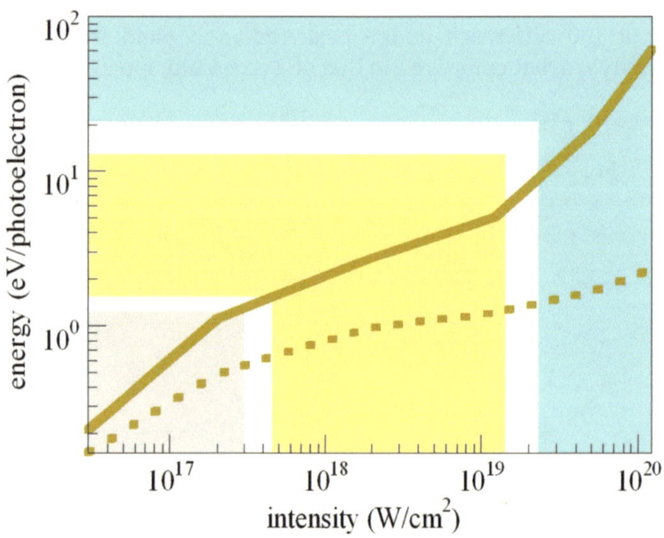

Fig. 10. Radiated energy versus laser intensity from classical photoionization (solid), and coherently summed tunneling probability current (dotted). Three regions (see text) are highlighted in the figure.

Closer examination of Fig. 7 and Fig. 8, however, reveals a significant difference between the idealized 1D case and the 3D focal geometry used in experiments. In Region II and III, the quantitative yields do not agree. Beginning at about $1 \times 10^{18}\ W/cm^2$ the yields diverge. While in the one-dimensional case the radiated energy for the incoherent (coherent) mechanism increases as $I^{1.5}$ ($I^{0.8}$) in Region II to $I^{1.8}$ ($I^{1.1}$) for Region III, for the 3D case the intensity dependence in Region II is only $I^{0.34}$ ($I^{0.11}$) and $I^{1.36}$ ($I^{0.4}$) in Region III. By a few times 1×10^{19}

W/cm^2, all radiation yields are 10 times lower for the focal geometry case than the respective yields in the plane wave case. This difference in the intensity dependence is due to electron-electron interferences between atoms ionized at different locations in a laser focus. Comparing the results at the peak intensity of $1 \times 10^{20}\ W/cm^2$ the coherent superposition of the radiation across the focal geometry has a larger effect than any difference in the way the ionization is treated; the degree of the coherence of the radiation across the focus can affect the radiation yield to a greater extent than the incoherent or coherent treatment of the radiation during the ionization process. Figure 6 is a plot of the power spectrum for the radiated harmonics at three different focal intensities. As one compares Fig. 8 to Fig. 11(c) the absolute difference in the radiation is 100 times for first few harmonics and 1000 times for the radiation near $100eV$.

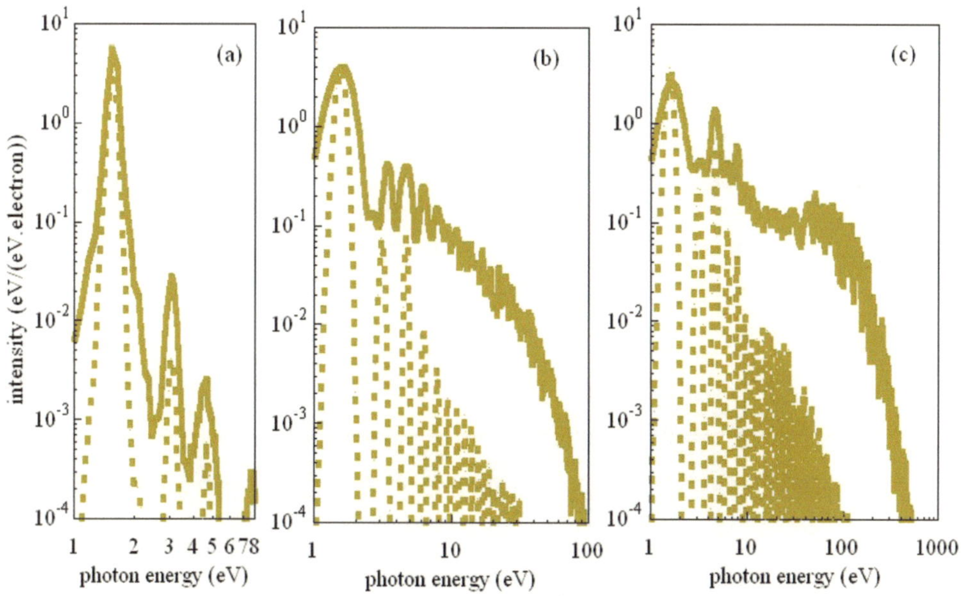

Fig. 11. Total spectral amplitude of the radiated field from atomic ionization at intensities (a) $2 \times 10^{17} W/cm^2$ (b) $1.2 \times 10^{19} W/cm^2$ and (c) $5 \times 10^{19} W/cm^2$ from classical photoionization (solid) and coherently summed tunneling probability current (dotted).

4.2 Angle resolved radiated energy

The angle and energy resolved radiation yields are presented in Fig. 12 for the focal geometry case. The plots are comparable to Fig. 8 but offer the additional derivative of the yields as a function of frequency. The radiation from classically ionized photoelectrons (essentially synonymous with the incoherently summed tunneling probability current) are shown in Fig. 12 (a, c, e, g, i) and the coherent tunneling probability current results in Fig. 12 (b, d, f, h, j). The logarithmic scale used in the plot is normalized to the peak radiation yield from classically ionized electrons in the respective laser field intensity.

An inspection of Fig. 12 reveals as one progresses beyond the dipole response (Fig. 12(a,b)) to ultrahigh fields coherently interfering radiation consistently occurs at larger angles in θ compared to incoherently summed or classical ionization. Furthermore the cutoff, or high frequency radiation, is always lower for coherently summed radiation by factors of 3 to 4

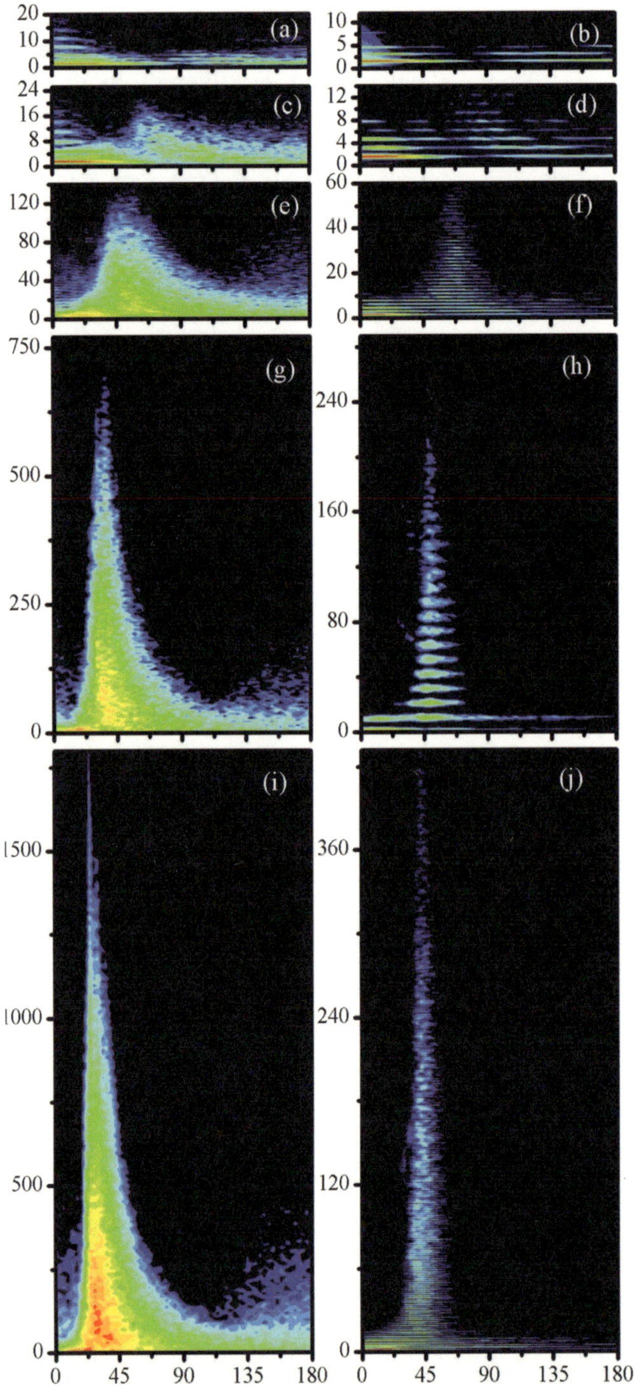

Fig. 12. Angle-and frequency resolved total radiation yield for (a,b) $2 \times 10^{17} W/cm^2$, (c,d) $2 \times 10^{18} W/cm^2$, (e,f) $1.2 \times 10^{19} W/cm^2$, (g,h) $5 \times 10^{19} W/cm^2$, and (i,j) $1.2 \times 10^{20} W/cm^2$ for classical (a,c,e,g,i) and coherently summed tunneling probability current (b,d,f,h,j).

when the intensities are between $1\times10^{19}W/cm^2$ and $1\times10^{20}W/cm^2$. This poses some difficulty for experiments since the difference between coherent and incoherent radiation is similar to the difference one would expect by a simple change in intensity. The possibility for confusion in this case of alternate causality can be seen by comparing, for example, Fig. 12(h) for coherent radiation at $5 \times 10^{19}W/cm^2$ with Fig 12(e) for incoherent radiation at $1.2 \times 10^{19}W/cm^2$. Similar comparisons can be made for Fig. 12(j) to Fig. 12(g). This observation combined with the N^2 effect of the focal geometry, the clearest identification of coherence from radiation in ultrastrong fields may come from the intensity dependence of the radiated power.

The radiation yield from the highest intensities studied involve photoionization at a peak intensity of $1.2 \times 10^{20}W/cm^2$. The radiation is peaked at a polar angle ($\theta \sim 34°$) for classically ionized electrons and ($\theta \sim 45°$) semi-classical ionization case with a cutoff photon energy for classically ionized electrons extending out to ($h\nu \sim 1.75KeV$) while for semi-classical ionization it extends out to ($h\nu \sim 420eV$).

5. Effect of focal geometry on radiation

In this section the primary effect of the 3-dimensional nature of the laser focus is quantified by comparing emitted radiation in a plane wave approximation and focused geometry cases calculated with the MC-ADK radiation model discussed above.

5.1 Laser intensity dependence of total radiated energy and spectrum

Figure 13(a) shows the total radiated energy as calculated using plane wave approximation of the laser field (dotted) and the three-dimensional laser field (solid). The radiation yield is normalized in the figure to the total amount of photoionization with units of energy per photoelectron (eV/electron). Fig. 13(a) shows at intensities of 1×10^{17} W/cm^2 and less, highlighted as region I in the figure, there is very little difference between radiation using a plane wave or focus geometry. This is associated with the small electron quiver amplitude ($\alpha_0 = a_0\lambda/2\pi$, where $a_0 = eE_0/m$) at $1 \times 10^{17}W/cm^2$, which extends out to 38 nm or less than a few percent of the focal waist radius (half-width at half-maximum $r_0 = \frac{2\sqrt{\ln 2}f/\#\lambda}{\pi}$). The limit where $\alpha_0 \leq r_0/10$, which we call region I, is also shown in Fig. 13(b) where the ratio of the electron excursion α_0 to the half width half maximum of the laser intensity at the focus r_0 is plotted as a function of intensity in the focus. In region I neglecting the spatial variation of the laser focus and adopting a one-dimensional laser field for radiation calculations is corroborated by the spectral amplitudes of the radiated field shown in Fig. 14(a). Out to the cutoff in the radiation at about 10 eV there is no difference between the spectra for photoionization in a plane wave or laser focus.

For laser intensities in the range (2×10^{17} W/cm^2 to 1.5×10^{19} W/cm^2,, Fig. 13 region II) the electron quiver width may be up to 300 nm, which is in the range of 0.1 to 0.5 times the beam waist radius of the laser focus. Since the electron travels non-negligible distances compared to the beam waist radius of the laser focus, it effectively experiences an average field that is lower than the peak field. In region II, adopting the plane wave approximation of the laser field overestimates the total radiated energy compared to the 3D focal geometry. Despite this drop in the total radiated energy from the first few harmonics, the extra velocity from surfing begins to play a role and the result is the higher photon energy persists in region II.

As one can in Fig. 14(b), for $h\nu > 7\ eV$, the radiation may be enhanced in the 3D focus case due to the acceleration of the photoelectron as it exits the focus. In addition, the laser focus and small α_0 confine the laser radiation to less than a cubic micron leading to a more coherent radiation than the plane wave approximation, where the sources of radiation span infinite space.

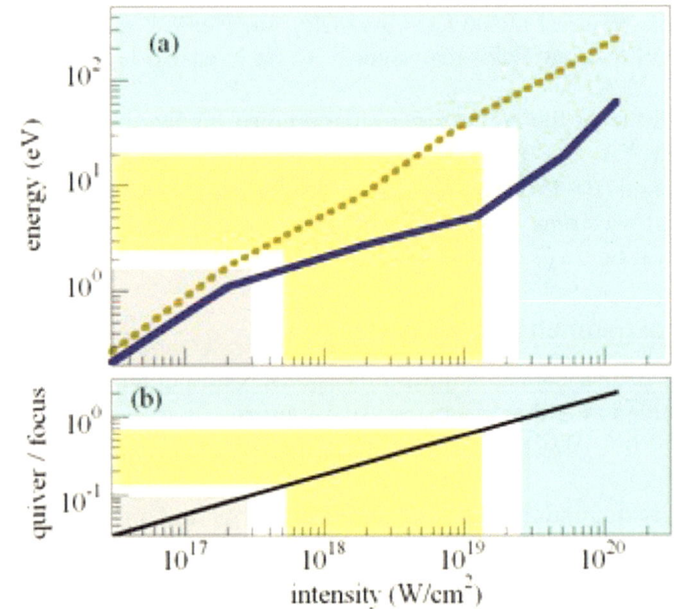

Fig. 13. Radiated energy versus laser intensity (a) with plane wave approximation (dotted, yellow-orange), and non-paraxial approximation (solid, blue). Three regions (see text) are highlighted in the figure.

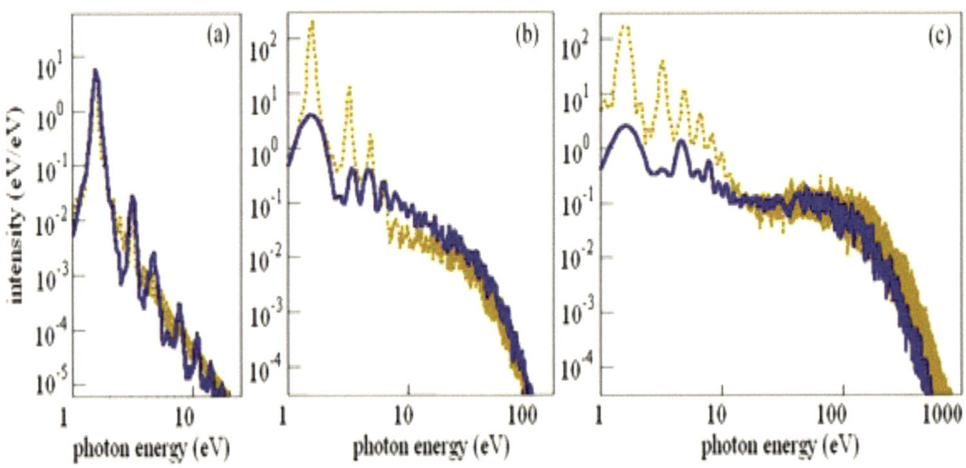

Fig. 14. Total spectral amplitude of the radiated field from atomic ionization at intensities $2\times10^{17}\ W/cm^2$ (a), $1.22\times10^{19}\ W/cm^2$(b), and $5\times10^{19}\ W/cm^2$ (c) calculated using the plane wave approximation (dotted, yellow-orange), non-paraxial approximation (solid-blue).

Figures 14 shows the spectral amplitude of the radiated field for laser light intensities of $2 \times 10^{17}\, W/cm^2$ (a), $1.22 \times 10^{19}\, W/cm^2$ (b) and $5 \times 10^{19}\, W/cm^2$ (c) calculated using the plane wave (dotted) and focal geometry (solid). The marked reduction in the intensity of the first three harmonics can be seen in Fig. 14(b) and is responsible for the lower radiated power in Fig. 13. One can also see in Fig. 14(b) the aforementioned increase in the highest photon energy radiation ($h\nu > 7\, eV$).

For laser intensities greater than $1.22 \times 10^{19}\, W/cm^2$ (Fig. 13 region III), the electron quiver width approaches 1 μm, which is larger than the beam waist radius of the laser focus. In region III, as with region II the photoelectron experiences an effectively smaller field compared to a plane wave at the same intensity. The effect is greater however with a reduced acceleration of the electron for its entire trajectory. This also means the electron will gain an extra boost of speed as it surfs out of the laser focus as shown in Fig. 5(b); but, the extra boost is not able to overcome the reduction in field from the prompt exit of the electron. Compared to region I and II, region III and radiation from ultrastrong fields comes from the largest spatial volume connected with the large α_0. Overall the total radiation from the focus geometry is significantly less than in the plane wave case.

6. Sequential ionization charge state contribution to the radiation yield

In this final section we discuss the expected results for radiation observed from a charge state distribution rather than the single, highest intensity charge state previously discussed. Here we include the impact of radiation as the intensity increases in the pulse and across the intensity distribution of the focus from the charge states generated when the laser ionizes neutral atoms up to the highest charge state. The total radiation yield is normalized by the total number of photoelectrons generated from all charge states in the respective laser foci.

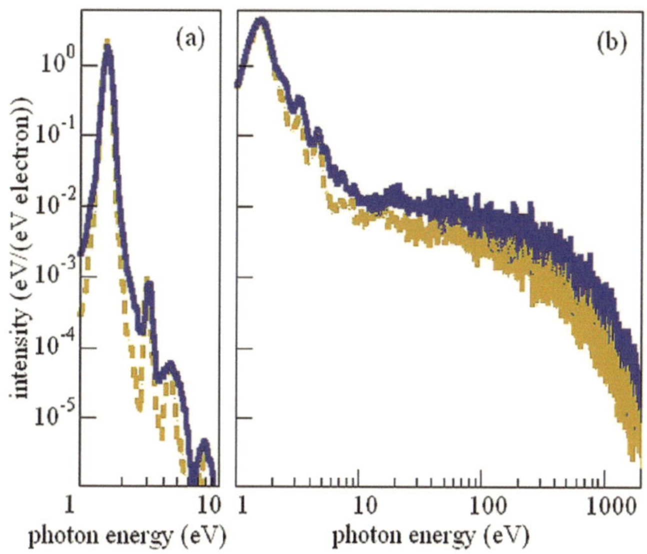

Fig. 15. Total radiated spectral intensity with full focal volume averaged (dotted, orangeyellow) and final charge state only (solid, blue) calculated with non-paraxial approximation at laser intensities of $2 \times 10^{17}\, W/cm^2$ (a) and $1.22 \times 10^{20}\, W/cm^2$(b).

The numerical values of the peak radiation from ionizing neutral atoms to highest charge state and ionizing highest charge state only, are the same. In both cases, the radiation yield is the highest for the first few lowest harmonics. The small differences, seen for example in Fig. 15(a, b), are the natural result of averaging over multiple charge states for species with varying ionization potentials. In Fig. 15(a, b) this results in narrower more resolved peaks for the harmonics. Also, due to the increase in photoelectrons from species ionized at low intensities the first harmonic radiation is enhanced, which appears as a decrease in the 3 eV to 2000 eV portion of the spectrum due to the normalization choice on the highest intensity/first harmonic.

The angle and energy resolved radiation yields from sequentially ionized neutrals (Ar and Na atoms) to their possible highest charges states Ar^{16+}, Na^{10+} and Na^{11+} in laser intensities $1.22 \times 10^{19}\ W/cm^2$, $5 \times 10^{19}\ W/cm^2$, and $1.22 \times 10^{20}\ W/cm^2$, respectively, are shown in Fig. 16 (a, c, e). Also shown in the figure is the radiation contribution from the highest charge state only (Fig. 16 (b, d, e)). The peak radiation from photoionization of these atoms/ions in a laser focus occurs at the fundamental frequency.

The logarithmic scale used in the plot is normalized to the peak radiation yield. The contribution to the radiation yield from the lower charge states is in the lowest harmonics since photoelectrons from low charge states are ionized in the beginning of the pulse, and leave the laser focus without experiencing the peak of the laser field, or are in the spatial wings of the pulse where the peak intensity is lower. To make valid comparison of the full ionization case, a logarithmic scale plot is shown similar to Fig. 12 for the polar angle dependence. Comparison of the results (Fig. 16(a, b), 16(c, d), 16(e, f)) shows focal averaging doesn't have noticeable effect on cutoff energies, polar angle peaking and angular divergence of the radiation yield.

7. Conclusion

These results on the radiation from ultrahigh field ionized atoms have characterized the role of coherence on the yield from the dipole response up to the currently known limit of tunneling ionization at $1 \times 10^{20}\ W/cm^2$. The results reported here expand our understanding in ultrastrong fields to include (1) a more detailed classical Monte Carlo ADK ionization mechanism for the photoionization, (2) the 3D focal geometry common to high field experiments, (3) electron wave packet spreading and coherence, and (4) the results when including radiation from the sequential ionization lower charge states in route to the ionization of the final charge state at the peak field. Our results are interpreted with the ratio of the photoelectron excursion to the laser focus and wavelength.

We find a classical view of ionization results in the same yield as a tunneling ionization probability current treatment when the radiation sum across the current is done incoherently. In the case that one assumes perfect coherence of the radiation across the tunneling current, the observed radiation decreases when the excursion of the photoelectron is comparable to the wavelength of the emitted radiation due to interference in the far field. When comparing the plane wave case to the focal geometry, the coherence of the radiation across the micrometer spatial extend of the focus has a larger affect on the observed yields the degree of coherence within the tunneling ionization current and interference from the photoelectron excursion for a single photoionization event. Additional difficulties may be faced by experimental studies because changes in the coherence and changes in the intensity can have a similar effect on the observed radiation.

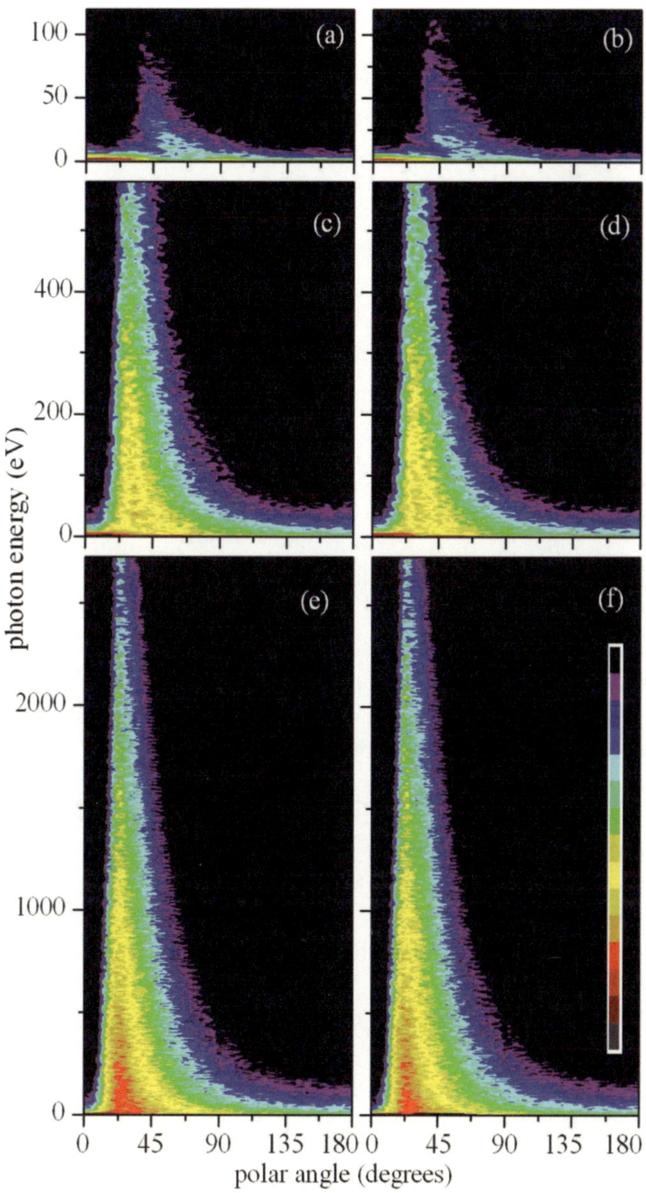

Fig. 16. Angle-and frequency resolved total radiation yield for intensities $1.22 \times 10^{19}\,W/cm^2$ (a, b), $5 \times 10^{19}\,W/cm^2$ (c, d), and $1.22 \times 10^{20}\,W/cm^2$ (e, f) from full sequential ionization of Ne-Ne^{+7} (a), Na-Na^{+10} (c), Na-Na^{+11} (e) and the highest charge state only (b, d, f). The magnitude scale inset in (f) ranges from unity to exp(-14) in steps of exp(-1) for all plots.

In fields where the excursion is less than the focus waist (Fig. 13, region I) the radiation from a focus and plane wave are indistinguishable. In fields as high as $1 \times 10^{19}\,W/cm^2$ (Fig. 13, region II), where the single cycle laser driven excursion of the photoelectron is comparable to the laser focus, the radiated power in a focal geometry drops by a factor between 2 and 10 for the lowest harmonics compared to a plane wave; but, the expected radiation near cutoff is very comparable for the two cases. In the highest fields (Fig. 13, region III), the photoelectron is unable to experience the peak laser intensity when radiation is probable and the observed cutoff radiation and angle are reduced in the focus geometry case to reflect the overall lower effective intensity. For a given peak field, the focal geometry is analogous to results one may expect at half the peak intensity for a plane wave. The results are not significantly changed if one considers only the radiation from photoelectrons near the peak of the field or when radiation from sequential ionization of the preceding charge states of the atom are included.

This material is based upon work supported by the Army Research Office under Award No. W911NF-09-1-0390 and the National Science Foundation under Award No. 0757953.

8. References

Albert, O., Wang, H., Liu, D., Chang, Z. & Mourou, G. (2000). Generation of relativistic intensity pulses at a kilohertz repetition rate, *Optics Letters* 25: 1125.

Ammosov, M. V., Delone, N. B. & Krainov, V. P. (1986). Tunnel ionization of complex atoms and of atomic ions in an alternating electromagnetic field, *Sov.Phys.JETP* 64: 1191.

Barton, J. & Alexander, D. (1989). Fifth-order corrected electromagnetic field components for a fundamental Gaussian beam, *Journal of Applied Physics* 66: 2800–2802.

Chen, S.-Y., Maksimchuk, A. & Umstadter, D. (1998). Experimental observation of relativistic nonlinear Thomson scattering, *Nature* 396: 653.

Chowdhury, E. A., Ghebregziabiher, I. & Walker, B. C. (2005). Larmor radiation from the ultra-intense field ionization of atoms, *Journal of Physics B Atomic Molecular Physics* 38: 517–524.

Clark, E. L., Krushelnick, K., Davies, J. R., M. Zepf, M. T., Beg, F. N., Machacek, A., Norreys, P. A., Santala, M. I. K., Watts, I. & Dangor, A. E. (2000). Measurements of energetic proton transport through magnetized plasma from intense laser interactions with solids, *Phys. Rev. Lett.* 84: 670.

Corkum, P. (1993). Plasma perspective on strong field multiphoton ionization, *Phys.Rev.Lett.* 71: 1994–1997.

Davis, L.W. (1979). Theory of electromagnetic beams, *Physical Review A* 19(3): 1177.

Esarey, E., Ride, S. K. & Sprangle, P. (1993). Nonlinear Thomson scattering of intense laser pulses from beams and plasmas, *Physical Review E* 48(4): 3003.

Ghebregziabher, I. & Walker, B. C. (2007). Effect of focal geometry on radiation from atomic ionization in an ultrastrong and ultrafast laser field, *Phys.Rev.A* 76(2): 023415.

Ghebregziabher, I. & Walker, B. C. (2008). Impact of coherence in radiation from ultrahigh-field atomic ionization, *Phys. Rev. A* 77: 023417.

Hu, S. X. & Starace, A. F. (2002). GeV Electrons from Ultraintense Laser Interaction with Highly Charged Ions, *Physical Review Letters* 88(24): 245003.

Jackson, J. (1975). *Classical Electrodynamics*, second edn, John wiley and Sons, New York.

Keitel, C. H. & Knight, P. L. (1995). Monte Carlo classical simulations of ionization and harmonic generation in the relativistic domain, *Phys.Rev.A* 51: 1420–1430.

Kneip, S., Nagel, S. R., Martins, S. F., Mangles, S. P. D., Bellei, C., Chekhlov, O., Clarke, R. J., Delerue, N., Divall, E. J., Doucas, G., Ertel, K., Fiuza, F., Fonseca, R., Foster, P., Hawkes, S. J., Hooker, C. J., Krushelnick, K., Mori, W. B., Palmer, C. A. J., Phuoc, K. T., Rajeev, P. P., Schreiber, J., Streeter, M. J. V., Urner, D., Vieira, J., Silva, L. O. & Najmudin, Z. (2009). Near-gev acceleration of electrons by a nonlinear plasma wave driven by a self-guided laser pulse, *Phys. Rev. Lett.* 103(13): 035002.

Lau, Y., He, F., Umstadter, D. & Kowalczyk, R. (2003). Nonlinear thomson scattering: A tutorial, *Physics of Plasmas* 10: 2155–2162.

Lee, K., Kim, B. H. & Kim, D. (2005). Coherent radiation of relativistic nonlinear Thomson scattering, *Physics of Plasmas* 12: 043107.

Leemans,W. P., Nagler, B., Gonsalves, A. J., Tth, C., Nakamura, K., Geddes, C. G. R., Esarey, E., Schroeder, C. B. & Hooker, S. M. (2006). Gev electron beams from a centimetre-scale accelerator, *Nature Physics.* 2: 696.

Mocken, G. & Keitel, C. (2005). Radiation spectra of laser-driven quantum relativistic electrons, *Computer Physics Communications* 166: 171–190.

Palaniyappan, S., DiChiara, A., Ghebregziabher, I., Huskins, E. L., Pajerowski, A. F. D. & Walker, B. C. (2006). Multielectron ultrastrong laser field ionization of Ar^{n+}, Kr^{m+} and Xe^{l+} ($n \leq 9, m \leq 9$, $l \leq 12$) at intensities from 1×10^{15} Wcm^{-2} to 1×10^{18} Wcm^{-2}, *Journal of Physics B Atomic Molecular Physics* 39: 357.

Peatross, J., Muller, C., Hatsagortsyan, K. Z. & Keitel1, C. H. (2008). Photoemission of a singleelectron wave packet in a strong laser field, *Phys. Rev. Lett.* 100: 153601.

Perelomov, A., Popov, V. & Terentev, M. (1966). Ionization of atoms in an alternating electric field, *Sov.Phys.JETP* 64: 1191.

Phuoc, K. T., Rousse, A., Pittman, M., Rousseau, J., Malka, V., Fritzler, S., Umstadter, D. & Hulin, D. (2003). X-Ray Radiation from Nonlinear Thomson Scattering of an Intense Femtosecond Laser on Relativistic Electrons in a Helium Plasma, *Physical Review Letters* 91(19): 195001.

Popov, V. S. (2004). Tunnel and multiphoton ionization of atoms and ions in a strong laser field (keldysh theory), *Physics-Uspekhi* 47(9): 855–885.

Rousse, A., Phuoc, K. T., Shah, R., Pukhov, A., Lefebvre, E., Malka, V., Kiselev, S., Burgy, F., Rousseau, J.-P., Umstadter, D. & Hulin, D. (2004). Production of a kev x-ray beam from synchrotron radiation in relativistic laser-plasma interaction, *Phys. Rev. Lett.* 93(13): 135005.

Schoenlein, R.W., Leemans,W. P., Chin, A. H., Volfbeyn, P., Glover, T. E., Balling, P., Zolotorev, M., Kim, K. J., Chattopadhyay, S. & Shank, C. V. (1996). Femtosecond x-ray pulses at 0.4 å generated by 90° thomson scattering: A tool for probing the structural dynamics of materials, *Science* 274(5285): 236.

Schwoerer, H., Liesfeld, B., Schlenvoigt, H. P., Amthor, K. & Sauerbrey, R. (2006). Thomsonbackscattered x rays from laser-accelerated electrons, *Phys. Rev. Lett.* 96: 014802.

Snavely, R. A., Key, M. H., Hatchett, S. P., Cowan, T. E., Roth, M., Phillips, T., Stoyer, M. A., Henry, E. A., Sangster, T. C., Singh, M. S., Wilks, S. C., MacKinnon, A., Offenberger, A., Pennington, D., Yasuike, K., Langdon, A. B., Lasinski, B. F., Johnson, J., Perry,

M. D. & Campbell, E. M. (2000). Intense high-energy proton beams from petawattlaser irradiation of solids, *Phys. Rev. Lett.* 85: 2945.

Tate, J., Auguste, T., Muller, H., Salières, P., Agostini, P. & Dimauro, L. (2007). Scaling of Wave-Packet Dynamics in an Intense Midinfrared Field, *Physical Review Letters* 98(1): 013901.

Ueshima, Y., Kishimoto, Y., Sasaki, A. & Tajima, T. (1999). Laser Larmor X-ray radiation from low Z matter, *Laser and Particle Beams* 17: 45–58.

24

Laser-based Particle Acceleration

Hans-Peter Schlenvoigt[1,2], Oliver Jäckel[1],
Sebastian M. Pfotenhauer[1,3], and Malte C. Kaluza[1]
[1]*IOQ, Friedrich-Schiller-Universität, Jena,*
[2]*LULI, École Polytechnique, Palaiseau,*
[3]*MIT, Cambridge, MA,*
[1]*Germany*
[2]*France*
[3]*United States*

1. Introduction

The acceleration of charged particles relies on the use of longitudinal electric fields. These fields are typically generated either via capacitors as static electric fields, or via microwaves carried in resonators. Accelerator devices of this sort are, however, constrained by their material damage thresholds: If the electric field in the cavities exceeds ~ 50 MV/m, conduction band electrons from the material are field-ionized, leading to a breakdown of the accelerator. As a consequence, the energy gain per unit length in conventional accelerators is limited, and high particle energies can only be realized through additional acceleration length, which explains the sometimes impressively large dimensions of state-of-the-art accelerators.

This chapter is dedicated to a radically new approach towards particle acceleration based on laser-produced plasma. In contrast to conventional accelerator cavities, plasmas can generate and support electric fields up to TV/m. In the present case, the plasma is created by an ultra-short, ultra-intense laser pulse impinging on a target, whereupon the free particles interact immediately with the strong electromagnetic fields of the laser. Since the energy density of such ultra-short pulses is extremely high, efficient energy coupling processes exist between the laser field and the plasma, and the particle may gain substantial kinetic energy within very short distances. As of today, laser-plasma accelerators are capable of producing beams of 1 GeV electron energy over 3 cm of acceleration length (Leemans et al., 2006; Karsch et al., 2007), as well as ion beams of several 10 MeV per nucleon over a distance of some ten microns (Snavely et al., 2000). These laser-produced particle beams possess a number of outstanding properties, such as ultra-short pulse duration of the order of the laser pulse (Pukhov & Meyer-ter-Vehn, 2002; van Tilburg et al., 2006; Fuchs et al., 2006), high peak currents and excellent emittance values (Cowan et al., 2004). Given these unique beam properties and the compactness of the acceleration scheme, the field of laser-based particle acceleration has recently attracted much attention for its great potential for exciting new applications in fundamental and applied physics.

The present chapter will be organized as follows: First, we briefly review the principle of chirped pulse amplification as the fundamental optical technology underlying the generation of ultra-intense laser pulses, as well as some fundamentals of ionization processes at high intensities. We then present the central mechanisms of laser-matter interactions at ultrahigh intensities, such as nonlinear Thomson scattering, the ponderomotive force and plasma heating processes. The main part of this chapter will be devoted to a detailed description of laser-based acceleration processes, both for electrons ("Laser Wake-Field Acceleration") and ions ("Target Normal Sheath Acceleration"). Particular attention will thereby be paid to the generation of quasi monoenergetic particle pulses with relatively moderate experimental requirements, which is an essential prerequisite for almost all applications. The final section will discuss a number of potential applications for laser particle accelerators. These include the production of short-pulsed synchrotron radiation from laser-accelerated electrons passing through an undulator as well as other small scale radiation sources for structural analysis, the possibility of laser-based radiation therapy in oncology, and applications in high energy-density physics like astrophysical studies or laser-based fusion energy sources.

2. Fundamentals of laser-plasma-physics

2.1 High-intensity laser systems

Ultra-high light intensities are produced by releasing a certain amount of laser energy W_L in a very short time span τ_L and focusing it to a very small spot size A_{foc}:

$$I_L = \frac{W_L}{\tau_L \cdot A_{foc}}. \quad (1)$$

Today's state-of-the-art laser system are capable of applying about 8 Joules of energy within a pulse duration of 30 fs and a focus area of about 5 μm, achieving intensities in excess of 10^{20}W/cm^2 while still fitting on one optical table. These intensities are equivalent to all the sun light reaching the earth surface, focused on the tip of a pencil. At present, the only technique capable of generating such intensities is the the so-called *Chirped Pulse Amplification* (CPA) (Strickland & Mourou, 1985). A mode-locked, dispersion-compensated optical oscillator generates ultra-short but spectrally broad laser pulses with pulse durations as short as $\tau_L \sim 10$ fs. These pulses contain initially only Nanojoules of energy. In order to amplify them without damaging the amplifier by too high pulse powers or intensities, the pulses are at first temporally stretched with help of dispersing elements by imprinting a positive chirp (which is equivalent to 2nd-order dispersion) onto the pulse. This easily leads to stretched pulse durations of 100's of ps. Subsequently, the long and less powerful pulses are amplified to several Joules. In order to re-compress them afterwards to ultra-short durations, all spectral components must experience the same gain during the amplification process. This condition restricts the available laser media to only a few ones like Ti:Sapphire, Nd:Glass, Yb:FP-Glass or Yb:CaF_2 and therefore the laser wavelengths λ_L to the near-infrared range. Finally, the pulses are re-compressed to ultra-short pulse durations by applying the inverse dispersion. The resulting pulses then have multi-TW to PW power and would interact with any transmissive material. Hence, only reflective elements are allowed and the beam must be guided in vacuum.

In order to obtain ultra-high intensities, these high-power laser pulses are focused with off-axis parabolic mirrors. For a given laser facility, degrees of freedom are mainly the pulse energy (since the amplification can be easily controlled) and the focal spot size (determined by the focusing optic). The pulse duration can easily be increased by detuning the pulse compressor (reducing the final pulse power), but reducing the pulse duration below a certain level, determined by the laser system or – more precisely – the spectral bandwidth of the amplified pulses, is usually not possible.

2.2 Ionization

Peak intensities necessary for laser-based particle acceleration are above $I_L \sim 10^{18}$ W/cm², which constitutes the so-called "relativistic threshold" for near-infrared wavelengths. Under such conditions, the primary mode of interaction between electromagnetic waves and matter is no longer resonant excitation of atomic states, but any material is readily ionized through *multi-photon-*, *tunnel-* and *optical-field-ionization* (Gibbon, 2005). For the laser pulses under consideration, ionization occurs already during the rising edge of the laser pulse, when the intensities exceed the range of $I_L \sim 10^{10}$ W/cm². In the multi-photon-regime, several photons are absorbed by the atom within a time period shorter than the relaxation time of the virtual excitation state. Hence, the photon energies add and appear like a single energetic photon. In the tunnel- and field-ionization regimes, the laser's electric field amplitude is comparable to the binding Coulomb field of the nucleus. Bound electrons can either tunnel through the finite potential well, formed by the superposition of the external laser field and the Coulomb field, or the potential well is completely suppressed and electrons are released into the vacuum instantaneously.

2.3 Plasma properties

Among the most important properties of plasma are its ability to screen charges, to support oscillations of the electron distribution against the positively charged ion background, and to react quickly on electromagnetic fields. These features arise mainly from the plasma's composition of quasi-free light electrons and heavy ions. Ions are usually treated as immobile, only electrons will react on internal or external electromagnetic fields.
The fundamental parameters of a plasma are its *electron density* n_e and *electron temperature* T_e. The temperature is usually defined via the mean kinetic energy of the electrons, $T_e = \langle W_{kin} \rangle / k_B$. Note that even a non-Maxwellian energy distribution yields a certain value of T_e. A first deduced quantity is the electron *plasma frequency*

$$\omega_p = \sqrt{\frac{e_0^2 n_e}{\varepsilon_0 m_e}} \qquad (2)$$

which characterizes oscillations of electrons around the immobile ions in case of a certain displacement from the equilibrium position. A second important deduced quantity is the *Debye length*

$$\lambda_D = \sqrt{\frac{\varepsilon_0 k_B T_e}{e_0^2 n_e}} \qquad (3)$$

which describes the distance after which the potential of a local charge density perturbation is reduced to $1/e$ of its initial value. Here, e_0 is the electron charge, m_e the electron rest mass, ε_0 the permittivity of free space and k_B the Boltzmann constant and e Euler's number.
The propagation of light with frequency ω_L and wave number $k_L = 2\pi/\lambda_L$ is determined by the dispersion relation

$$\omega_L^2 = \omega_p^2 + k_L^2 c^2 \,, \tag{4}$$

where c is the speed of light. Electromagnetic waves cannot propagate in plasma if $\omega_L \le \omega_p$. The limit $\omega_L = \omega_p$ corresponds to a maximum electron density called *critical density,*

$$n_c = \frac{\varepsilon_0 m_e \omega_L^2}{e_0^2} = \frac{\varepsilon_0 m_e 4\pi^2 c^2}{e_0^2 \lambda_L^2}, \tag{5}$$

which must not be exceeded to allow light propagation in the plasma. Plasma with electron densities $n_e < n_c$ are called *underdense,* otherwise *overdense* plasma.
Due to the dispersion relation (4), the plasma has an *index of refraction*

$$\eta = \sqrt{1 - \frac{\omega_p^2}{\omega^2}} = \sqrt{1 - \frac{n_e}{n_c}} < 1 \,. \tag{6}$$

For dilute plasma, $n_e \ll n_c$, Eq. (6) may be approximated by $\eta \approx 1 - 0.5\, n_e/n_c$.

2.4 Relativistic motion of single electrons

The motion of electrons in electromagnetic fields is described by the Lorentz force,

$$\frac{\mathrm{d}\mathbf{p}}{\mathrm{d}t} = -e_0(\mathbf{E}(\mathbf{r}) + \dot{\mathbf{r}} \times \mathbf{B}(\mathbf{r})), \tag{7}$$

where $\mathbf{p}(t) = \gamma m_e\, \dot{\mathbf{r}}\,(t)$ is the relativistic electron momentum, γ the relativistic Lorentz factor $\gamma = (1 - \dot{\mathbf{r}}^2/c^2)^{-1/2} = [1 + \mathbf{p}^2/(m_e c)^2]^{1/2}$, $\mathbf{E}(\mathbf{r}, t)$ the electric and $\mathbf{B}(\mathbf{r}, t)$ the magnetic field at the electron's actual position $\mathbf{r}(t)$ and $\dot{\mathbf{r}}\,(t)$ the electron velocity.
As a start, let's consider the motion in an infinite, plane electromagnetic wave, linearly polarized in the x-z-plane and propagating in z-direction, $\mathbf{E}(\mathbf{r}, t) = \hat{\mathbf{x}}\, E_0 \sin(kz - \omega t)$. We replace the electric field amplitude by the amplitude of the *normalized vector potential,*

$$a_0 = \frac{e_0 E_0}{m_e \omega c} \approx 0.85 \times \frac{\lambda_L}{\mu\mathrm{m}} \times \sqrt{\frac{I_L}{10^{18} W/cm^2}} \,. \tag{8}$$

This quantity represents the "relativisticness" of the motion. If $a_0 \ll 1$, one can neglect the influence of the $\dot{\mathbf{r}} \times \mathbf{B}$-term and finds that a_0 is the ratio of the amplitude of the quiver velocity to c. Intensities yielding $a_0 \gtrsim 1$ are called "relativistic intensities" and will be considered in the following.
To solve Eq. (7), we perform a variable transformation, $\tau = t - z(t)/c$, which describes the phase of the wave and is also the proper time t/γ of the electron. Finally, we find

$$x(\tau) = \frac{c a_0}{\omega} \sin(\omega\tau) \tag{9a}$$

$$y(\tau) = 0 \tag{9b}$$

$$z(\tau) = \frac{c a_0^2}{4}\left(\tau + \frac{1}{2\omega}\sin(2\omega\tau)\right). \tag{9c}$$

To the first order of the field amplitude a_0, the electron oscillates with ω parallel to the electric field vector, x. But to second order and thus dominant for relativistic intensities, there is an oscillation with double frequency 2ω combined with a drift motion along the propagation direction of the wave, z. This motion induces a characteristic light emission called *nonlinear Thomson scattering* (Lau et al., 2003).

Figure 1 shows trajectories according to Eq. (9) for an electron, in a frame co-moving with the electron's drift velocity $c\ a_0^2\ /(4 +\ a_0^2)$ along the laser propagation direction (a) and (b) in the lab frame, for three different intensities (or vector potential amplitudes a_0). The black curve is for $a_0 = 0.215$ which corresponds at $\lambda = 800$ nm to an intensity of $I = 10^{17}$ W/cm². This is almost a classical linear oscillation. At higher intensities, $a_0 = 0.6$ (green) and $a_0 = 0.96$ (blue), the motion in the co-moving frame becomes a figure-of-8, and the drift motion along the laser propagation direction is more pronounced in comparison to the transverse oscillation. Note that for $a_0 \approx 1$ (blue curve), the drift distance during one laser oscillation in the electron's time frame τ is already a quarter of the laser wavelength, which corresponds to a relativistic average forward velocity and again clearly shows the relativistic character of the motion.

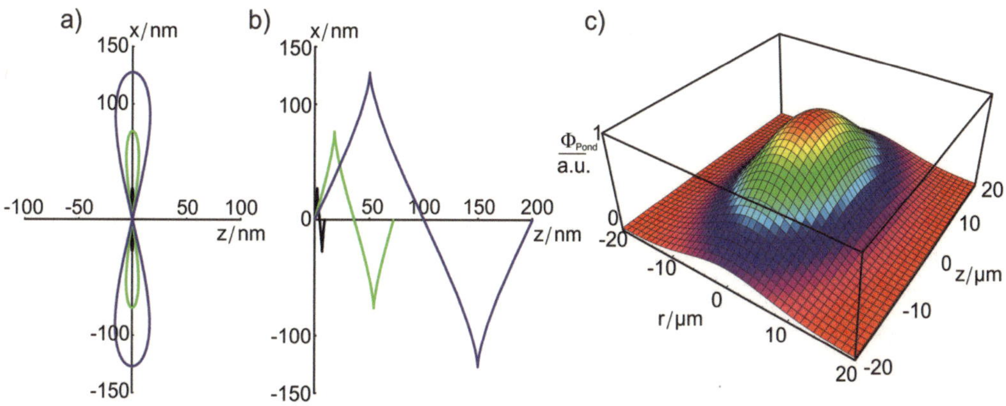

Fig. 1. Electron trajectories in the average rest frame (a) and the laboratory frame (b) of nonlinear Thomson scattering, for different laser intensities and during one laser period experienced by the electron; (c) the ponderomotive potential of a laser pulse of 15 μm full width at half maximum (FWHM) diameter and 85 fs FWHM pulse duration.

2.5 Ponderomotive force

The previously discussed case exhibits no net particle acceleration, caused by the assumption of an infinite plane wave. Even for finite pulse durations, the electron is shifted in forward direction but comes to a rest again after the laser pulse has passed. This is different for focused laser pulses with transverse dimensions and therefore steep *intensity gradients*. Averaging the Lorentz force in time over the fast oscillatory contributions, $\langle \ldots \rangle_T$, i.e. regarding just the pulse envelope, yields the *ponderomotive force* (Kruer, 1988),

$$\mathbf{F}_{\text{Pond}} = \langle \frac{\mathrm{d}\mathbf{p}}{\mathrm{d}t} \rangle_T = \frac{-e_0^2}{2m_e\omega^2} \nabla \langle E^2 \rangle_T = \frac{-e_0^2}{2\varepsilon_0 m_e \omega^2 c} \nabla I(\mathbf{r})\,. \tag{10}$$

The related ponderomotive potential reads as

$$\Phi_{\text{Pond}} = \frac{e_0^2}{2\varepsilon_0 m_e \omega^2 c} I(\mathbf{r}) = \frac{a_0^2}{4} m_e c^2 . \qquad (11)$$

Figure 1 (c) shows the ponderomotive potential of a laser with 15 μm FWHM diameter and 85 fs FWHM pulse duration. The shape of the ponderomotive potential is effectively dominated by the intensity envelope of the focused laser pulse.

We see that the ponderomotive force creates an effective net push in direction away from high intensities, being proportional to the intensity gradient. This allows electrons to gain net momentum, especially in laser propagation direction. Ponderomotive pressure is one of the key mechanisms for electrons to acquire kinetic energy at relativistic intensities and plays a central role in both electron and ion acceleration.

2.6 Effects at relativistic intensities: Relativistic optics

The previous sections have shown that electrons exposed to relativistic intensities are subject to a relativistic quiver motion and a net force. The first aspect leads – for quantities based on time-scales larger than the laser period – to an effective *relativistic mass increase* of the electrons. Therefore, in Eqs. (2), (5), (6) and (10), the electron rest mass must be replaced by the average relativistic mass, $m_e \mapsto \langle\gamma\rangle_T m_e$. The average relativistic factor is given by $\langle\gamma\rangle_T = (1 + a_0^2/2)^{1/2}$. The second effect, the ponderomotive force, leads to local *changes of the plasma electron density*. The derivation of the correct relativistic ponderomotive potential is a little more complicated (Meyer-ter-Vehn et al., 2001). Solving the equation of motion in consideration of additional plasma fields and currents (not only the laser field), one finds

$$\Phi_{\text{Pond}} = \left(\sqrt{1+a_0^2} - 1\right) m_e c^2 , \qquad (12)$$

what is just the electron's relativistic kinetic energy acquired from the laser field. Note that this corresponds to $\Phi_{\text{Pond}} = (\langle\gamma\rangle_T - 1) m_e c^2$ with $\langle\gamma\rangle_T = (1 + a_0^2/2)^{1/2}$ which is slightly different from the formula given above above.

Both effects alter the optical properties of the plasma via Eqs. (2) and (6) at regions where the laser pulse is present. By that, the laser pulse propagation and ultimately the laser pulse shape itself can be altered, too. This mutual interaction leads to a feedback loop (Mori, 1997). Most noteworthy effects in this context are self-focusing and self-modulation.

Self-focusing arises from changes of the refractive index such that at regions at high intensity the refractive index is increased. This leads to a focusing density profile which counteracts natural diffraction and may prolong the interaction region. Self-focusing occurs when the laser pulse power exceeds the following limit: $P_L > P_{SF} \approx n_c/n_e \cdot 17.4$ GW. *Self-modulation* has a similar underlying effect but leads to longitudinal bunching and compression of the laser pulse. It is caused by periodic modulations of the plasma density which arise from the ponderomotive force and the plasma's ability to perform oscillations. Further effects of these *modulation instabilities* are Raman forward scattering, photon acceleration, hosing and filamentation. The name arises from the fact that these effects inhibit stable laser propagation and modulate certain quantities of the laser pulse (Mori, 1997).

2.7 Energy absorption processes

Up to now, an underdense plasma was considered, being a rather transparent and—depending on the intensities – nonlinear medium without significant energy loss. The situation becomes different in case of reaching the critical density, $\omega_L = \omega_p$, or resonance between the laser pulse envelope and the plasma frequency (cf. Sec. 3.1).

Here we consider an interface from vacuum to overdense plasma. The electron density may be modelled by an exponential decay, $n_e(z) = n_{e,0} \cdot \exp(z/L_P)$ for $z < 0$ (in front of the target) and $n_e(z) = n_{e,0}$ for $z > 0$ (inside the target), cf. Fig. 2 (a). The quantity L_P is called *scale length* of the plasma at the target front side. This plasma originates from ionization of the target material by the leading edge of the laser pulse and is called *preplasma*. The steeper the leading edge, the later the preplasma is formed and the steeper the density gradient. The case of $L_P \to 0$ resembles a perfect step-function-like density gradient. Coming back to the laser-plasma interaction, we find that the plasma shows a high reflectivity at low intensities which is due to the large amount of free charges, similar to a metal mirror. However, at high intensities, significant absorption is observed.

At first sight, one can introduce a damping term into the Lorentz equation of the electron motion that represents *collisions* between laser-influenced electrons and ions. It can be shown that the collision frequency scales like $\nu_{ei} \sim (k_B T_e)^{-3/2}$, accordingly collisional damping becomes less effective for higher electron temperatures what starts at intensities of about 10^{15} W/cm^2 (Gibbon, 2005). Therefore, other *collisionless* mechanisms have to be taken into consideration. The most relevant collisionless absorption processes are resonance absorption and Brunel heating as well as the already introduced ponderomotive acceleration (cf. Sec. 2.5), which is closely related also to $\mathbf{v} \times \mathbf{B}$ heating.

All processes finally lead to an electron fraction with much higher mean kinetic energy than the residual, unaffected plasma electrons. Hence, these electrons give rise to a *suprathermal* component which is simply called "hot electrons", and the processes are referred to as "heating processes".

Resonance absorption and Brunel heating A *p*-polarized laser pulse with oblique incidence on the target, i.e. $\theta_L > 0$, has an electric field component which is directed parallel to the preplasma gradient. Under oblique incidence, due to the **k**-vector conservation (Kruer, 1988), reflection actually occurs before the critical density is reached. This implies, that electrons, forced by the laser to oscillate, generally do not match the local plasma frequency in the density profile, cf. Fig. 2. The electron density occurring at the turning point of the incident wave is deduced by $n_{e,\mathrm{refl}} = n_c \cos^2 \theta_L$ which is less than the critical density for incidence angles larger than zero.

At the turning point, the electric field component of the incident wave generates a standing wave pattern oscillating in target-normal direction (z). However, in case of a sufficiently small distance between the turning point and the position of the critical density, the evanescent electric field suffices to excite an electron oscillation parallel to the electron-density gradient, i.e. in longitudinal direction, within the region of the critical density. This very localized and quasi-standing plasma wave is resonantly excited by the laser field and can therefore be driven so strongly that it breaks (cf. Sec. 3.1.3). Then, a sufficient number of hot electrons is pushed into the target and transport energy obtained from the laser. This process is called *resonance absorption*. Absorption rates up to $\eta_{\mathrm{abs}}^{\mathrm{RA}} = 60$ % are reported for optimized laser and plasma parameters (Gibbon, 2005).

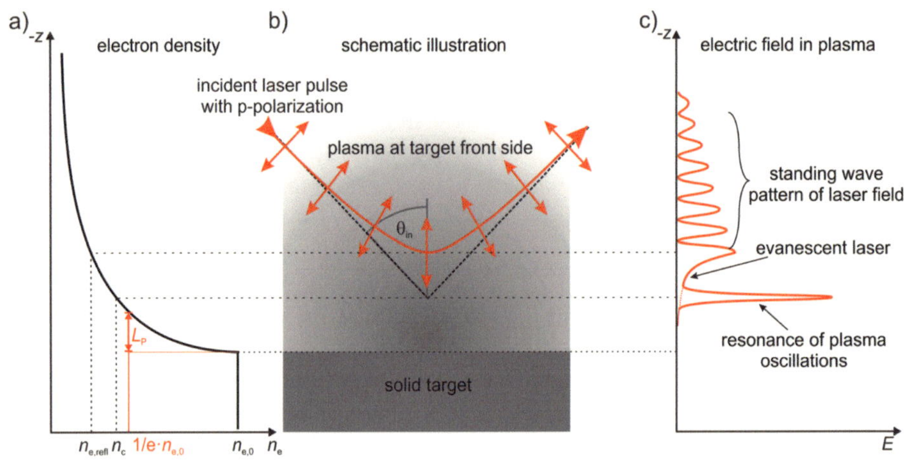

Fig. 2. Reflection at an exponential plasma profile. A *p*-polarized laser pulse obliquely incidences the preplasma profile at the target front. Reflection occurs already before the critical density has been reached. If the distance between the standing wave pattern at the turning point and the position of the critical density is short enough, the evanescent decaying electric field can excite electron oscillations at this position in longitudinal direction (into the target). Since this oscillation is excited resonantly, it can be driven so strongly that it breaks and a sufficient number of hot electrons is pushed inside the target.

Brunel heating, (Brunel, 1987), takes place at very steep plasma gradients and benefits from the fact that the laser penetration depth is smaller than the maximum displacement of the oscillating electrons. The acceleration process now takes place in the vacuum space in front of the preplasma profile or the target, respectively.

During the first half cycle of a laser period, the electrons move away from the target into the vacuum, turn around, and are accelerated back towards the target. But once they enter the target, they soon get screened from the laser fields which again penetrate the plasma only evanescently. By that, the electrons retain most of their kinetic energy and carry laser energy into the plasma. The absorption rate via Brunel absorption just depends on the incidence angle of the laser, the plasma properties are more or less fixed due to the assumption of a very steep plasma gradient. Hence, $\theta_L = 45°$ accords to an absorption rate of $\eta_{abs}^{Brunel} = 60\ \%$. This very large value has to be discussed rather critically: Firstly, the Brunel mechanism does not account for finite plasma scale lengths, but typical experimental conditions do not reveal a very steep plasma gradient. Furthermore, numerical studies, assuming very short but finite preplasma scale lengths (Gibbon, 2005), show that η_{abs}^{Brunel} saturates at about 10–15 %.

Ponderomotive force and v ×B heating The ponderomotive force allows that electrons from the preplasma can be pushed along the laser propagation direction. However, the laser pulse cannot follow the electrons into overdense regions since it is reflected at the critical density or even before, but electrons remain moving along the initial direction. The mean kinetic energy of the electrons was given by the ponderomotive potential, see Sec. 2.5. Typically, mean kinetic energies of a few MeV can be obtained (Gibbon, 2005).

The last effect to be discussed is **v** × **B** heating (Gibbon, 2005). Electrons oscillating in the electric field of a laser pulse with relativistic intensity experience the influence of the **v** × **B** term of the Lorentz force as discussed in Sec. 2.4. For steep plasma gradients, this effect can contribute to the absorption rate analogously as the Brunel mechanism does: During the first (or third) quarter of the optical cycle, electrons are accelerated into the vacuum, turn around, and are accelerated back to the target where they become screened from the electromagnetic wave and retain with their kinetic energy. In contrast to the Brunel mechanism depending only on the electric field of the laser, **v** × **B** heating acts along the laser propagation direction due to the magnetic field contribution and not along the polarisation direction. Hence, **v** × **B** heating favours small incidence angles of the laser. The kinetic energy achievable via **v** × **B** heating is deduced by the a_0^2 -dependence of the electron momentum in longitudinal direction, cf. Eq. (9c). Again, the MeV-range is reached. Due to the relation to the Brunel mechanism, similar absorption rates of 10–15 % are expected after reaching the regime of relativistic intensities (Kruer & Estabrook, 1985).

3. Laser-based electron acceleration

In this section we will point out the physical concept of how a plasma can be used to generate high-energy electron bunches. Although we already presented principles which can transfer energy from the laser to plasma electrons, the generation of high-energy, narrowband and collimated electron beams (similar to those of conventional accelerators) is different. At first, we will introduce the fundamental theoretical model, which is split into different subtypes. Secondly, we will show original experimental data: We will depict a setup, explain the course of action for the production of high-quality electron beams and present characteristics of produced electron beams. Finally, we will describe possible extensions, improvements and potential future developments.
The generation of highly energetic and collimated electron beams happens in a underdense plasma. Therefore, the highly efficient absorption mechanisms (cf. Sec. 2.7) cannot be used. Only the direct interaction of the laser's electromagnetic field with plasma electrons or the ponderomotive force are in action. This leads to 2 physically different regimes of electron acceleration: The *laser wakefield acceleration* (LWFA) (Tajima & Dawson, 1979; Modena et al., 1995) and *direct laser acceleration* (DLA). In the DLA regime (Gahn et al., 1999), a direct energy transfer from the transverse laser fields to a forward electron momentum takes place. However, beams with comparably large divergence angles and broadband spectra can be produced. For that reason we will not discuss DLA.

3.1 The model of Laser Wakefield Acceleration

The LWFA regime (Tajima & Dawson, 1979) relies on two steps:

- An intense laser pulse drives a plasma wave by the ponderomotive force, like a boat generates its wake on the sea. This plasma wave is accompanied by strong electric fields and is co-moving with the laser pulse.
- If electrons are injected at the peak of the wake, they can gain energy from the space charge electric field of the plasma wave, similar to the way a surfer does from an ocean wave.

This regime has different sub-types (Esarey et al., 1996), most important are self-modulated LWFA (SM-LWFA) and bubble acceleration (Pukhov & Meyer-ter-Vehn, 2002), which will be discussed later in Sec. 3.2. All these variants of LWFA are in principle capable of producing collimated electron pulses with narrowband spectra.
Before we go into details it must be noted that the following description of the LWFA regime considers plane plasma waves which corresponds to a 1-dimensional (1D) approach where only variations along the propagation direction are taken into account. This simplifies

the physical picture but is a bit unrealistic since in reality the plasma wave amplitude varies with the radial distance due to the transverse intensity profile of the laser pulse. Those transverse effects are difficult to model analytically but play an important role and will be discussed later in Sec. 3.2.2.

3.1.1 Plasma wave excitation

Let us start by recalling that the ponderomotive force is directed anti-parallel to the intensity gradient, see Eq. (10). Hence, plasma electrons are pushed away from intense laser pulses. Now, consider a laser pulse approaching a certain region in the plasma. Electrons are at first pushed forward. A local charge separation is formed since the ions are not affected by the laser pulse and remain at rest. The electron density changes locally by an amount of δn_e which depends mainly on the laser intensity gradient. Space charge forces start to pull the electrons back. After some time, the laser pulse overtakes the electrons since the forward and back-pointing forces compensate. The electrons will then experience a kick backwards from the ponderomotive force. The density depression becomes compensated by electrons streaming back to the depression. However, the inertia of the moving electrons hinders a stable charge density compensation. Instead, there is an overshoot, and then the electrons swing back and forth. In other words, the laser pulse has excited a plasma oscillation at the considered region, like a guitarist picking a string.

Since the laser pulse moves through the plasma, it excites oscillations all along its path. Furthermore, between the oscillations, there is a certain phase relation given by their individual time of excitation. Therefore, all oscillations together form a *travelling plasma wave* with a phase velocity equal the group velocity of the laser pulse v_{gr}. This is similar to the wave pattern induced by a speedboat, which is also co-moving with the boat. The corresponding wave length of this electron density oscillation is called plasma wavelength and is given by

$$\lambda_p = \frac{2\pi v_{gr}}{\omega_p} \approx \frac{2\pi c}{\omega_p}. \tag{13}$$

The excitation is most effective if the laser pulse fits into the first half-period of the plasma wave (Tajima & Dawson, 1979). This results in the resonance condition

$$\omega_p \cdot \tau_L = \pi. \tag{14}$$

This is a delicate condition, as it relates the laser pulse duration to the electron density via Eq. (2) and therefore puts severe constraints on the experimental conditions.

3.1.2 Nonlinear plasma waves

In the case of strong excitation, the electron density variation may become large, $\delta n_e \sim n_e$. This leads to ultra-high space charge fields in the plasma and relativistic longitudinal quiver velocities of electrons during the plasma oscillations. Hence, the plasma frequency is altered during the oscillation and the oscillation becomes anharmonic as shown in Figure 3. The longitudinal electric field in the excited plasma wave is given by (Esarey et al., 1996)

$$E_{wake} \approx \frac{a_0^2/2}{\sqrt{1+a_0^2/2}} \underbrace{c m_e \omega_p / e_0}_{E_p}. \tag{15}$$

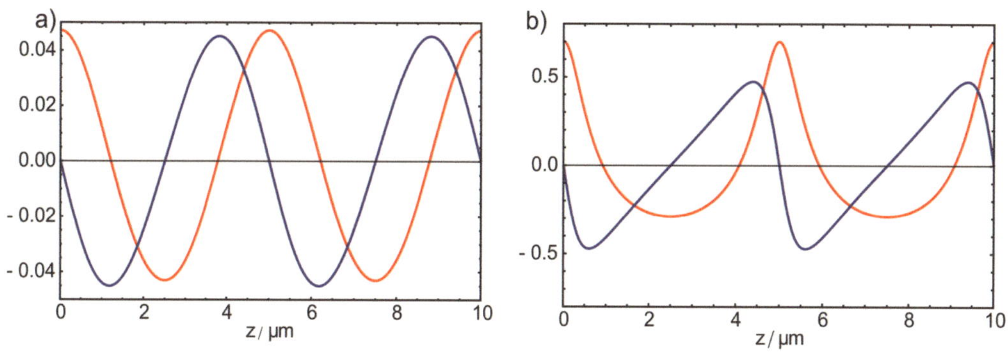

Fig. 3. Electron density perturbation $\delta n_e / n_e$ (red) and longitudinal electric field E_{wake}/E_p (blue) for (a) a weakly relativistic laser pulse ($a_0 \sim 0.3$) and and (b) a relativistically intense pulse ($a_0 \sim 1$) in a plasma with 5 μm plasma wavelength. Calculated according to Sec. 4.1 in (Gibbon, 2005).

3.1.3 Wave-breaking and injection

The plasma wave closely follows the laser pulse and travels almost with the speed of light. As a result, electrons must be injected in the travelling electric field with significant forward velocity in order to be caught by the wave and subsequently to be accelerated further by the field. However, as the velocity of the wave is smaller than c, the electrons can stay in the accelerating region of the plasma wave for a significant period of time and thus gain considerable amounts of kinetic energy. Among several possibilities (see later Sec. 3.4), a simple option for injection is *wave-breaking*.

Wave-breaking occurs when the individual particle velocities within the wave reach the phase velocity of the wave. In low-amplitude plasma waves, electrons just swing back and forth. In large-amplitude waves, however, particles can slip out of the wave, similar to whitecaps at the ocean's surge. Thereby, wave-breaking sets on the one hand the limit of the plasma wave amplitude and the electric fields inside the wave. On the other hand, it happens at the right position of the wake (its maximum) and particle velocities suffice to catch the wave. For this reasons, wave-breaking is called *self-trapping* of electrons. Furthermore, injection of electrons reduces the longitudinal electric fields and can inhibit further electron injection. This is called *beam loading*, which leads to an inherent control of the accelerated electron beam (Geddes et al., 2005).

In a nonlinear plasma wave, the wave-breaking limit limit is given by (Esarey et al., 1996)

$$E_{wb} = \sqrt{2(\gamma_p - 1)} E_p, \tag{16}$$

where γ_p is the relativistic factor related to the phase velocity of the wake and may be approximated to $\gamma_p \approx \omega_L / \omega_p = \sqrt{n_c / n_e}$. In order to reach wave-breaking, $E_{wake} \geq E_{wb}$, the laser pulse amplitude must significantly exceed the relativistic limit, $a_0^2 \gg 1$:

$$\underbrace{\frac{a_0^2 / 2}{\sqrt{1 + a_0^2 / 2}}}_{\approx a_0 / \sqrt{2}} \geq \sqrt{2(\sqrt{n_c / n_e} - 1)}\,. \tag{17}$$

3.1.4 Energy gain considerations

In general, the acceleration process can be limited by the following processes:

- Diffraction: the laser beam has a finite Rayleigh length
- Depletion: the plasma wave excitation consumes the laser pulse energy
- Dephasing: after some acceleration, electrons move faster than the plasma wave and slip into a decelerating phase.

Diffraction is usually not a problem. The depletion limit can be roughly estimated by equating the initial laser pulse energy to the energy left behind in the plasma wave and is (Esarey et al., 1996)

$$L_{\rm depl} \sim \frac{n_{\rm c}}{n_{\rm e}} \lambda_{\rm p} \begin{cases} 1/a_0^2 & a_0^2 \ll 1 \\ a_0/(3\pi) & a_0^2 \gg 1. \end{cases} \tag{18}$$

The dephasing limit is reached when electrons have travelled half a plasma wavelength (in the frame of the wake) and is given by

$$L_{\rm deph} \sim \frac{n_{\rm c}}{n_{\rm e}} \lambda_{\rm p} \begin{cases} 1 & a_0^2 \ll 1 \\ 2a_0^2/\pi & a_0^2 \gg 1. \end{cases} \tag{19}$$

Hence, the maximum energy gain is limited by the dephasing length at low intensities but by the depletion length at high intensities and may be estimated together with the maximum electric field $E_{\rm p}$ to

$$\gamma_{\max} \sim \sqrt{2} \frac{n_{\rm c}}{n_{\rm e}} \begin{cases} \pi a_0 & a_0^2 \ll 1 \\ a_0^2/3 & a_0^2 \gg 1. \end{cases} \tag{20}$$

3.1.5 Example

In order to get a feeling of the influence of the different parameters, let us consider the following example. A Ti:Sapphire laser system shall be used for electron acceleration. The wavelength is $\lambda_{\rm L} = 800$ nm and the pulse duration $\tau_{\rm L} = 100$ fs. The laser pulse energy may be chosen freely as well as the focusing optic.

1. Given by Equation (5), the critical density is $n_{\rm c} \approx 1.7 \cdot 10^{21}$ cm^{-3}, and from Eq. (14), the electron density required for resonant plasma wave excitation is $n_{\rm e} \approx 3.1 \cdot 10^{17}$ cm^{-3}. The ratio is $n_{\rm c}/n_{\rm e} \approx 5 \cdot 10^3$. The density corresponds to a plasma period of $\lambda_{\rm p} \approx 60$ μm and the characteristic field strength is $E_{\rm p} \approx 50$ GV/m.
2. The wave-breaking limit may be calculated from Eq. (16) to $E_{\rm wb} \approx 600$ GV/m. This gives via Eq. (17) a minimum intensity of $I_{\rm L} \gtrsim 6 \cdot 10^{20}$ W/cm^2 ($a_0 \gtrsim 16.7$) in order to achieve wave-breaking and self-trapping of electrons.
3. At those intensities ($a_0 \gg 1$), the acceleration length is limited by the depletion length which is of the order of 50 cm. The maximum electron energy is then 300 GeV.

The question now becomes: How can those conditions be achieved? First, the laser must retain its intensity over the whole length. Matching the Rayleigh length $z_{\rm R} = \pi w_{\rm L}^2/\lambda_{\rm L}$ to the depletion length yields a waist of $w_{\rm L} \sim 400$ μm. This determines the focal spot size $A_{\rm foc}$. Now, the pulse energy can be calculated, ending up at $W_{\rm L} > 250$ kJ – a value completely out of range for 100 fs lasers.

Assuming more realistic laser parameters, say $W_L \lesssim 10$ J, one would have to focus down to $w_L \sim 2\ \mu$m in order to reach the wave-breaking limit. In that case, the acceleration length would be limited by the Rayleigh length which is $z_R \sim 20\ \mu$m, leading to a maximum electron energy of 10 MeV.
Considering now nonlinear effects, the self-focusing limit leads for the given conditions to a minimum laser power of $P_L \approx 90$ TW which corresponds to a laser pulse energy of $W_L \gtrsim 9$ J. This is still at the upper limit of common pulse energies for 100 fs laser systems. However, if the density is increased, the self-focusing threshold can be lowered and the overall experimental constraints can be loosened significantly. This is the route pursued by the SM-LWFA regime.

3.2 Variants of LWFA

3.2.1 Self-modulated laser wakefield acceleration

In the self-modulated LWFA regime (SM-LWFA), a combination of laser pulse length and plasma density is considered where the pulse is too long to hold condition (14). In this case, the laser pulse still excites a plasma wave by its ponderomotive force, but not as strong as in the resonant case of LWFA. However, the long laser pulse overlaps with the excited wave which represents a periodic density perturbation co-moving with the laser pulse. This density modulation affects the laser pulse. Parts of the laser pulse at positions with reduced electron density propagate faster than parts at positions with increased density, cf. Eqs. (2) and (6). This leads to bunching of the laser pulse, the laser pulse envelope becomes modulated with a periodicity of the plasma wavelength λ_p. As this modulation leads to locally higher intensities, the generation of the plasma wave is enhanced. The density modulation grows, which in turn amplifies the laser pulse modulation. Thus, the pulse splits into pulselets, each separated by the plasma wavelength and now fulfilling the LWFA condition. We see that thereby the LWFA regime can be initiated at significantly less strict conditions than proposed by Eq. (14). In general it is necessary that the pulse power suffices for self-focusing, which in turn can be enabled by high plasma densities.
Let's re-consider the example for the LWFA regime with the following parameters: Again, a Ti:Sapphire laser system shall be used with the same pulse duration of $\tau_L = 100$ fs. However, we choose a density of $n_e = 6 \cdot 10^{18}$ cm^{-3}.

1. Now, the ratio is $n_c/n_e \approx 280$ so that the plasma period is reduced to $\lambda_p \approx 15\ \mu$m and the characteristic field strength is increased to $E_p \approx 240$GV/m.
2. The wave-breaking limit is also increased to $E_{wb} \gtrsim 1$ TV/m. However, from Eq. (17) follows a reduced minimum intensity of $I_L \gtrsim 1.5 \cdot 10^{20}$ W/cm^2 ($a_0 \approx 8$) to achieve wave-breaking and self-trapping of electrons.
3. Furthermore, the power threshold for self-focusing is reduced to 5 TW. Hence, laser pulse energies above 500 mJ will lead to self-focusing which overcomes the limitation of acceleration length by the Rayleigh length.
4. If the threshold for self-focusing is passed, also other nonlinearities become effective, making an analytical description difficult. Under self-focusing, the laser diameter is approximately the plasma wavelength—which can be altered further during the interaction. In addition, self-modulation and related effects can lead to effective shortening of the pulse which increases the power. Hence, a 10 J laser pulse will likely reach the wave-breaking limit and could produce 4 GeV electron beams.

As we can see, the SM-LWFA regime is easier to implement than the LWFA regime and may produce higher electron energies for a given laser system (Modena et al., 1995; Esarey et al., 1996). For a predefined pulse duration, a plasma of higher density may be used with a less energetic laser pulse focused more tightly. The interaction length is prolonged by the laser pulse itself via self-focusing, higher longitudinal electric fields are present for acceleration and the intensity threshold is reduced.

Note that if injection shall be accomplished via wave-breaking, there is no possibility to avoid powers and intensities above the threshold for the occurrence of plasma instabilities. Hence, some of the instabilities like self-focusing will always occur, even for the LWFA regime – the transition between the regimes is smooth. However, in the SM-LWFA regime the nonlinear interaction is inteded which makes the whole acceleration process difficult to control. Mostly, exponential electron spectra are obtained due to plasma heating from various instabilities and altered injection conditions for hot plasmas.

3.2.2 The bubble regime

The bubble regime was discovered via computer simulations (Pukhov & Meyer-ter-Vehn, 2002). The simulations include 3D-effects of the plasma wave formation like a tightly focused laser pulse or large transverse oscillations of the electrons but also the mutual interaction between plasma and laser. This regime uses intense ($a_0 > 1$) and very short ($\tau_L <$ 10 fs) laser pulses which propagate through relatively dense plasma ($n_e \sim 10^{19}$ cm^{-3}). The simulations show the formation of a bubble-like cavity behind the laser pulse, formed by the transverse deflection of electrons due to the ponderomotive force. This bubble exhibits strong electric fields pointing toward the bubble's centre. Electrons are injected at the rear side of the bubble (wave-breaking) and gather at a small region inside. This can lead to a partial compensation of the electric fields and stop further electron injection (beam loading). The resulting electron beam has, due to this localization, an ultra-short pulse duration, a narrowband energy spectrum and is well-collimated.

From numerous simulations, covering a wide interaction parameter range, simple analytic expressions for the electron bunch properties were derived (Gordienko & Pukhov, 2005). The maximum electron energy is determined by

$$\gamma_{\max} \approx 0.65\sqrt{\frac{P}{P_{\mathrm{rel}}}\frac{c\tau_{\mathrm{L}}}{\lambda_{\mathrm{L}}}}. \tag{21}$$

Here, $P_{\mathrm{rel}} \approx 8.7$ GW is the relativistic power unit. The number of electrons in the bunch is

$$N \approx 0.28\frac{\lambda_{\mathrm{L}}}{r_{\mathrm{e}}}\sqrt{\frac{P}{P_{\mathrm{rel}}}}, \tag{22}$$

where $r_e = e_0^2/(4\pi\varepsilon_0 m_e c^2) \approx 2.8$ fm is the classical electron radius. For laser parameters as discussed in the examples above and now with only 500 mJ of pulse energy (P = 5 TW), the maximum electron energy is 300 MeV and the bunch charge 300 pC.

3.3 Experimental studies of laser-electron acceleration

3.3.1 Setup of the laser-driven electron accelerator

A simple but robust and reliable setup for laser-electron acceleration deploys a pulsed gas nozzle in order to generate a *gas jet*, see Fig. 4. The emitted gas density profile changes on the order of microseconds and can therefore be considered constant during the interaction. The gas is ionized by the leading edge of the laser pulse. Helium is used in order to warrant inert chemical conditions and a fully ionized plasma when the pulse peak arrives. The fraction of laser energy consumed for ionization is negligible. Thus, the high-intensity part of the pulse always interacts with a fully ionized plasma.

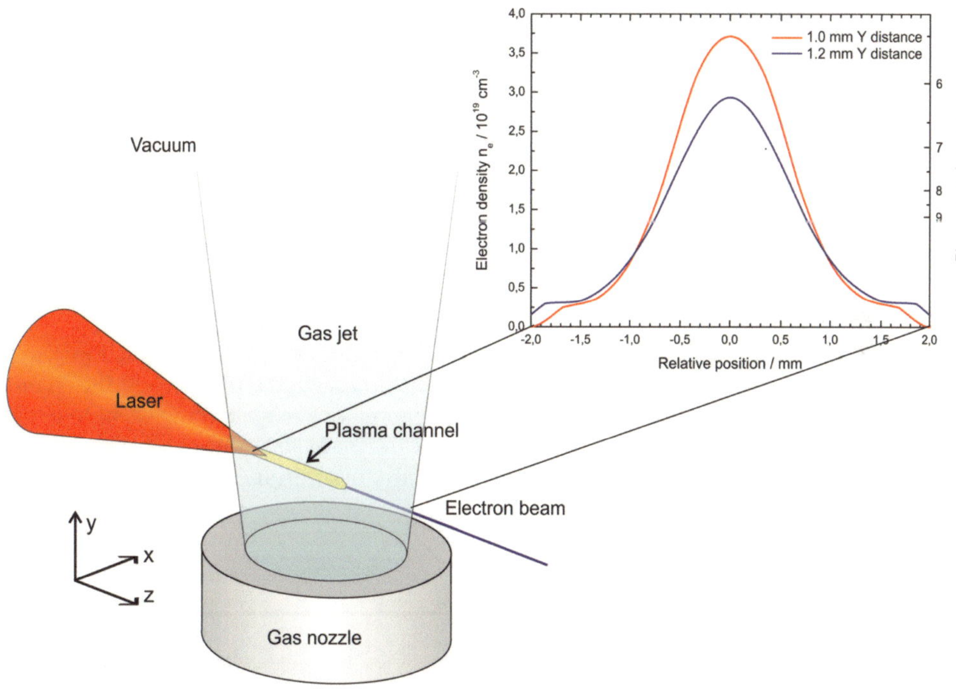

Fig. 4. Schematic of the setup for a laser-driven electron accelerator. An intense laser pulse is focused into a gas jet. Along the laser's path, the gas is ionized. Due to the interaction between laser and plasma at relativistic intensities, the laser pulse propagates in a self-formed channel. In the channel, the laser pulse generates strong plasma waves, wherein electrons are accelerated. All interactions depend strongly on the initial gas density profile, which may be controlled by different nozzle designs. The inset shows typical electron density profiles along the channel for two different vertical distances above the nozzle orifice.

The *electron density profile* along the laser's path therefore depends not on the intensity but on various other parameters: the nozzle orifice shape, the gas flow velocity, the backing pressure, the nozzle opening time and opening duration and, most importantly, the relative position of the laser focus. All these parameters allow for subtle fine tuning of the plasma parameters which turns out to be crucial for a well-controlled laser-driven electron acceleration.

In the here-described experiments, nozzles were used generating gas jets with cylindrical symmetry. The radial density profile has a Gaussian shape and the peak of the density decreases with increasing distance from the orifice. The inset of Figure 4 shows typical electron density profiles for the presented experiments, measured by interferometry of the gas distribution where full ionization of the helium gas was assumed. The nozzle was cylindrical, had 1.2mm inner diameter, 50 bar backing pressure, 700 μs opening time and a delay between opening and laser incidence of 850 μs.

For these conditions we can estimate which regime of electron acceleration can be obtained. The plasma wavelength changes along the laser propagation and is in the range of $\lambda_p = 6\text{–}20\ \mu$m (see Fig. 4). The total pulse energy on target is typically ~ 500 mJ and the pulse duration is 85 fs. The focal spot size (FWHM) is about $A_{foc} \sim 100\ \mu m^2$. The spot average laser power is $P_{FWHM} \approx 2$ TW and the intensity on the order of $I_{FWHM} \sim 10^{18}\ W/cm^2$.

We see that i) the intensity is just at the relativistic limit, ii) the pulse length is too long for resonant excitation (the densities are too high) and that iii) the power suffices for self-focusing in those quite high plasma densities. Thus, direct access to the LWFA regime is not possible, but self-focusing and self-modulation will take place. Earlier experiments with the same laser and subsequent simulations have shown (Hidding et al., 2006) that the mechanism of self-modulation can even lead to the bubble regime. Following the example in Sec. 3.2.2, we can expect collimated electron beams with energies up to hundreds of MeV.

3.3.2 Plasma imaging

The use of gas nozzles has the advantage of unobstructed access to the interaction region. This free access can be exploited by the experimentalist to image the plasma and the channel onto a camera, to image the self-emission from the channel onto a spectrometer, or to deploy short backlighting probe pulses (typically fractions of the main laser pulse) for time-resolved studies. Those *probing techniques* can be shadowgraphy, interferometry or polarimetry and can either use short pulses in a multi-shot scanning regime or chirped pulses for single-shot measurements.

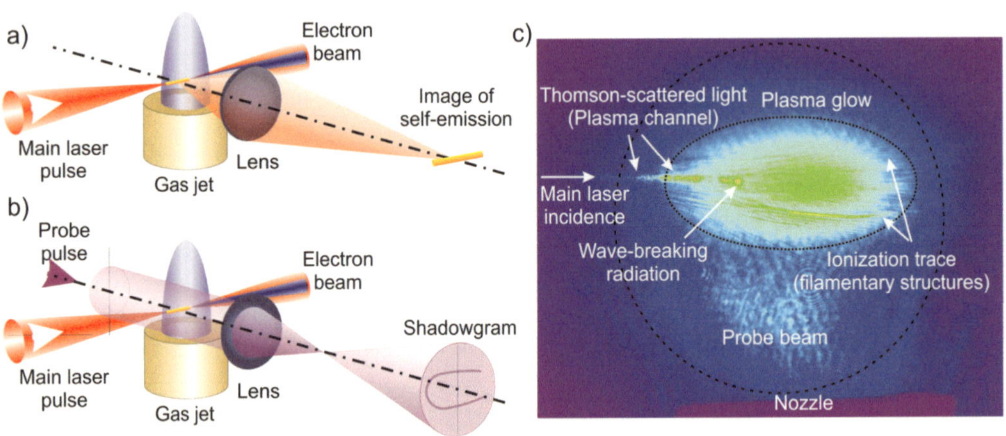

Fig. 5. Setups for imaging of the plasma. Fig. (a) shows mere imaging of the self-emitted light, Fig. (b) depicts imaging of the interaction region with a backlighting probe pulse. Fig. (c) is an image of the interaction region in pseudo-colour representation. This image was taken during setup of the gas nozzle with low plasma density (~ $10^{18}\ cm^{-3}$) and with a bandpass filter at $2\omega_L$ (400 nm ± 5 nm). The probe beam (its transverse profile is indicated by the black circle) was activated in order to see the nozzle via its shadow. The laser is incident from the left. The relativistic channel can be observed because of nonlinear Thomson-scattered light. The ionization of the gas is visible from shadowgraphy as approximately horizontal streaks. In the centre of the gas jet (above the nozzle centre), plasma glow is observed.

Figure 5 shows exemplary setups for imaging of the plasma channel (a) and (b) probing of the interaction region with a second, synchronized, much weaker ultra-short laser pulse. The main difference between both is basically whether or not the probe pulse is activated, which can be accomplished with a simple shutter. Experimentally, of course, one has to take care of attenuation with neutral density filters, selection or suppression of wavelengths with interference filters or dispersion of the imaging lens.

Figure 5 (c) shows an exemplary image of the interaction region in pseudo-colour representation. The main laser is incident from the left. The gas nozzle is visible as a shadow at the bottom. The laser pulse undergoes self-focusing and forms a relativistic channel. This plasma channel is partially visible from the nonlinear Thomson-scattered light. After a small gap where the channel is not visible, wave-breaking radiation is emitted (Thomas et al., 2007). The large intense region is the plasma glow. The probe beam was attenuated in order to record both the self-emitted light and a shadowgram. The streaky structures are likely to be ionization traces.

3.3.3 Electron beam properties

In order to characterize the produced electron beams, two diagnostics were deployed. The first one is a retractable scintillating screen which can be inserted into the electron beam path. The induced light signal is proportional to the charge density and is recorded by a camera. This screen allows for measurements of spatial properties of the electron beams. The second device is a dipole magnet spectrometer which deflects electrons depending on their energy. For fixed entrance position and angle, the output position of deflected electrons is a measure for their energy. The detection of electrons is again accomplished by a scintillating screen. These diagnostics allow for real-time acquisition of large amounts of data, which in turn allows for thorough statistical analysis.

Beam profiles and stability Figure 6 shows 4 exemplary electron beam profiles obtained with the scintillating screen for nominally identical experimental conditions. As can be seen, there can be shots with well-collimated beams (a), shots with hardly any electrons (b), fairly collimated and/or multiple beams (c) or rather uncollimated beams (d). For our experimental conditions, this is a typical behaviour for this type of laser-electron accelerator. Slight shot-to-shot fluctuations of the laser beam profile, energy and wavefront may occur and the gas jet exhibits a turbulent flow which means that for consecutive shots there are always small changes of the density profile. Furthermore, the whole acceleration relies on nonlinear interaction: self-focusing, self-modulation, nonlinear plasma wave formation and wave-breaking. The combination of all these factors leads to considerable shot-to-shot fluctuations of the electron beams.

Dependence on the gas jet position The occurrence of well-collimated beams like the one shown in Fig. 6 (a) depends strongly on the interaction conditions which can be easily varied by the gas nozzle position. Basically, the laser can hit the gas jet at any position (x, y, z), whereas typically an axis going through the centre of symmetry of the gas jet is chosen for the laser incidence, i.e. (x = 0, y, z), which corresponds to a radial crossing of the laser through the gas jet. Then, scans in the z-direction are carried out for certain heights y. The vertical distance can be measured from shadowgrams where the nozzle is visible. The condition $x = 0$ is usually set via the brightest emission of Thomson-scattered light. The distance in laser propagation direction can only be determined with help of a separate experiment via a Hartmann aperture in the main beam. The accuracy of all those methods is about 50 μm.

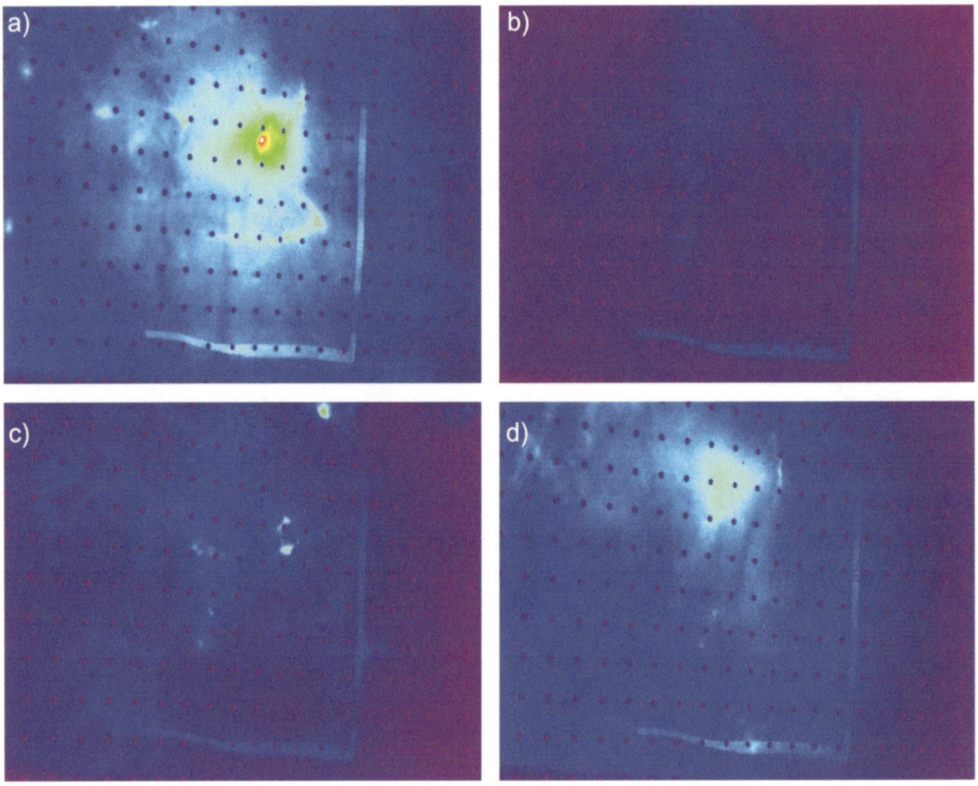

Fig. 6. Example sequence of electron beam profiles in pseudo-colour representation, recorded 30 cm behind the gas jet. The point grid has a 5 mm spacing in the vertical direction.

A typical result of such scan is shown in Figure 7. Positive values for the z-position indicate that the laser focus (in vacuum) is behind the gas jet centre. At each position, 10 to 20 shots were carried out and the aiming screen images were averaged. From such an averaged image, the sum brightness (sum over all pixel values) is a measure for the average charge per shot. From the averaged image itself, the mean direction of electrons can be inferred. The average charge as function of gas jet position is plotted in Figure 7 (a). Furthermore, from the single images of the aiming screen, the relative frequency of well-collimated beams can be counted and plotted in the same manner, see Figure 7 (b). Then, through systematic scanning in z and y, the optimum position can be found, either for high average charge or frequent occurrence of well-collimated beams or both.

Generally, a too small vertical distance y between laser incidence and gas nozzle (and therefore a high peak plasma density, cf. Fig 4) leads to a higher charge but less frequently collimated beams. Instead, electrons are distributed in a considerably large cone of about 30 mrad full opening angle. This is reasonable since at too high gas densities, the self-modulation becomes stronger and the regime of DLA is entered which is not desired for collimated and potentially monoenergetic electron bunches. In the longitudinal direction, an optimum is found slightly behind the point where the vacuum focus coincides with the gas jet centre. That means that a relatively long path in an increasing density profile is required for self-modulation and self-focusing.

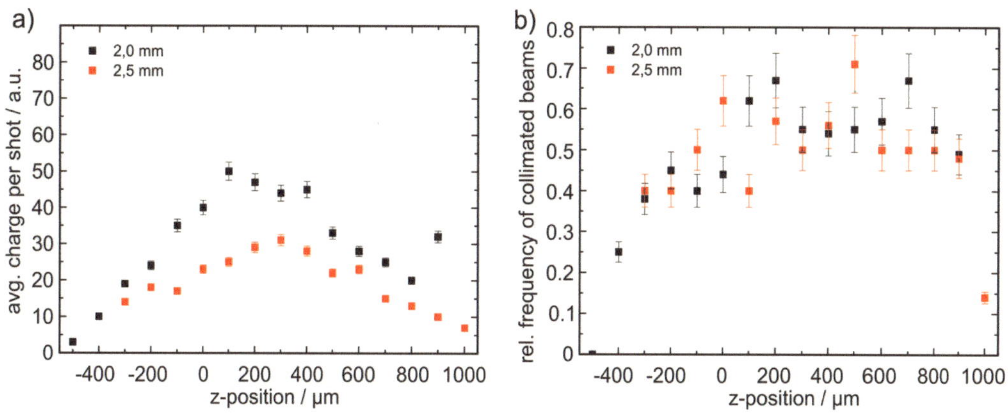

Fig. 7. Optimization of the interaction geometry. Shown are the average charge per shot (a) and the relative frequency for well-collimated beams (b) as a function of the longitudinal position z for two vertical distances y between nozzle and laser focus (black and red squares). Note that the difference of the vertical distance in contrast to Fig. 4 is due to a different nozzle.

Spatial characteristics of well-collimated beams Well-collimated beams may be also characterized in terms of size and spatial (angular) distribution. This is shown in Figure 8 for a nozzle position optimized for frequent occurrence of well-collimated beams. As can be seen, the beams propagate in average close to the on-axis direction but with a mean deviation of about 40 mrad. This is considerably high for applications but is sufficient for studies and can be reduced by other experimental approaches (see Sec. 3.4). The divergence of electron beams can be as low as 2 mrad. In comparison with results from other groups, this is remarkably low for gas jet targets and is suited for applications.

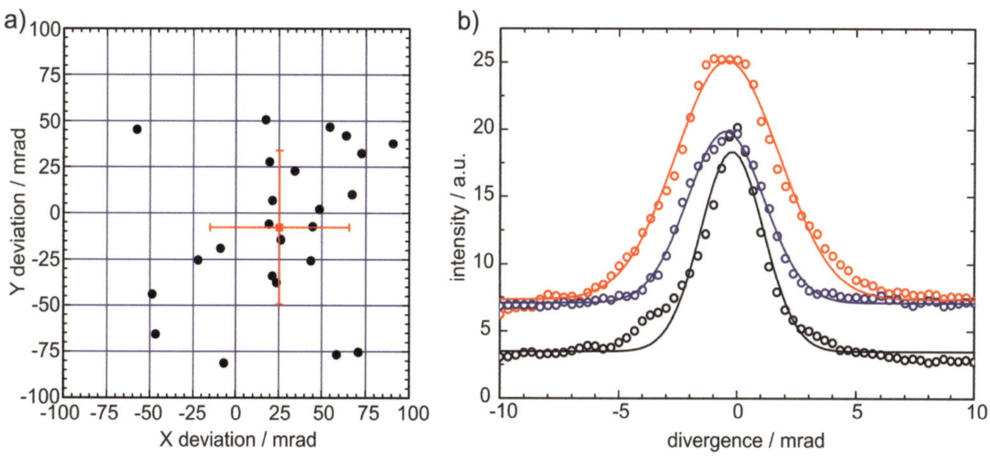

Fig. 8. Pointing (a) and divergence (b) of well-collimated electron beams under optimized nozzle conditions. Shown in (a) is the pointing direction with respect to the laser axis. The red cross represents the average position and the standard deviation. For (b), lineouts of the scintillating screen images were taken. The Gaussian fits have widths of 2–4 mrad.

Spectral characteristics Considering the spatial fluctuations of well-collimated beams and the acceptance angle of the electron spectrometer, we expected to rely on the high repetition rate and the fast online diagnostics in order to record sufficient electron spectra for a comprehensive statistical analysis. It turned out that electron spectra were measured much more frequently than expected. This was caused by the fact that also electrons contributing to the cloudy background on the scintillating screens were measured with the spectrometer and showed monoenergetic features which both was not expected.

Figure 9 shows typical electron spectra. Raw images from the scintillating screen of the electron spectrometer are displayed in Fig. (a)-(c). Figure (d) shows the corresponding calibrated spectra. Note that the shadow in the left part of each image is caused by a frame inside the spectrometer, used for calibration. The energy range shown in the spectra corresponds to the part right of the shadow.

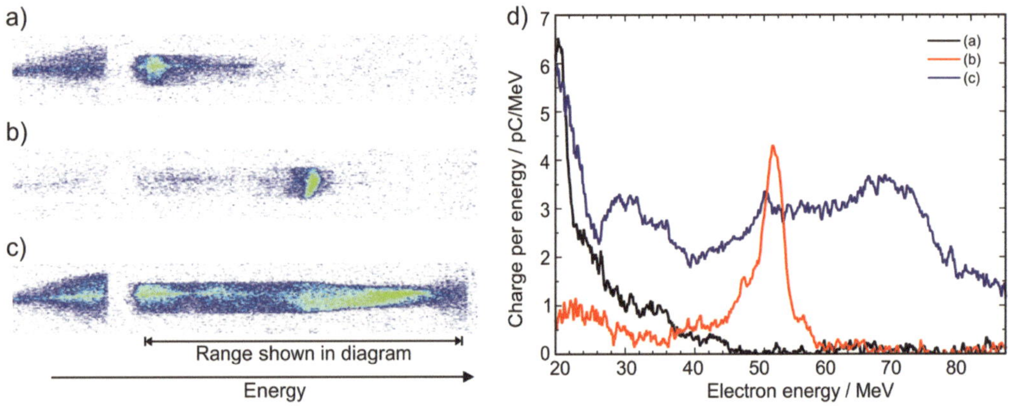

Fig. 9. Typical electron spectra, as raw data from the scintillating screen of the spectrometer (a)-(c) and as calibrated spectra (d). The gap in the left part of each image is caused by a frame inside the electron spectrometer and ranges from 16 MeV to 20 MeV.

Shot (a) shows an exponential spectrum, in this case quite intensive and therefore ranging to comparably high energies. Shot (b) shows a typical monoenergetic spectrum at 50 MeV. The bandwidth shown in the spectrum is about 4 MeV FWHM. Spectrum (c) is very broad which is a rare event. It reaches the high-energy limit of the spectrometer which is at 85 MeV.

Electron spectra can be categorized (similar to electron beam profiles) into exponential, monoenergetic, "empty" and irregular (all other) spectra. Out of a typical set of 200 laser shots at an optimized gas nozzle position, only 29 shots produced almost no electrons detectable by the spectrometer, either caused by insufficient charge at all or collimated beams propagating off axis. Other 33 shots showed exponential spectra which usually extended to 20-25 MeV. 47 shots were dominated by monoenergetic features. The remaining 91 shots could not be designated unambiguously into one of those categories. Hence, about every fourth shot is monoenergetic, however, charge and peak energy are fluctuating.

The monoenergetic spectra were analysed further. Figure 10 shows the distribution of charge (a) and energy (b) of 47 shots producing monoenergetic spectra. Figure 10 (a) shows a broad plateau up to 45 MeV for the energy distribution. The charge distribution, Fig. 10 (b), shows also a roughly constant plateau up to 40 pC. Higher charges of within a monoenergetic peak are rare. Note that a plot of peak charge versus peak energy revealed no correlation between both quantities.

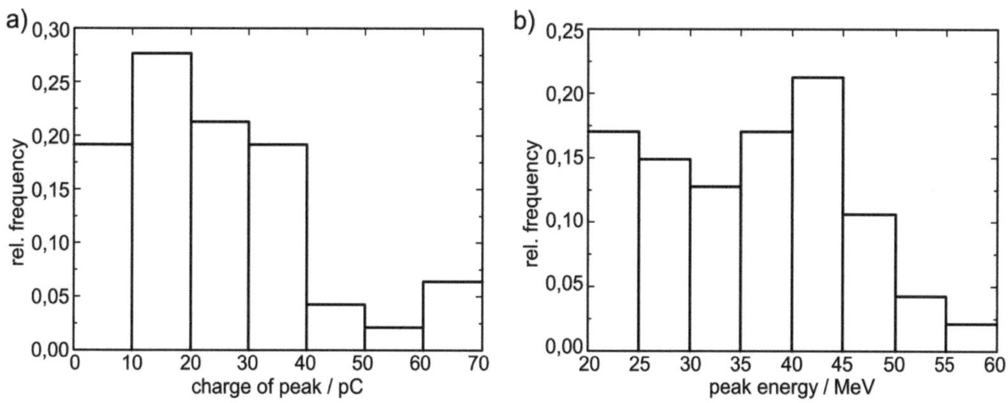

Fig. 10. Histograms for monoenergetic electron spectra, showing the frequency distributions of charge (a) and energy (b).

3.4 Ways to improved electron acceleration

The here-described laser-plasma accelerator relies on nonlinear interaction and feedback between laser pulse and plasma wave. The laser pulse a) guides itself (self-focusing), is meanwhile b) shortened by the plasma (self-modulation) and can then drive a plasma wave so strongly that c) injection via wave-breaking occurs. The setup is relatively simple and robust and allows for various investigations of the interaction region. However, the produced electron beams are subject to non-negligible fluctuations in terms of pointing and spectral parameters, which is largely due to the afore-mentioned nonlinear interactions.

Other groups in the field of laser particle acceleration proposed and demonstrated different experimental setups which produce electron beams more stably. A key feature of all these setups is that more powerful laser systems are used (more pulse energy and/or shorter pulses) and the laser pulse and its interaction with the plasma are not solely used for all the 3 steps a)–c). In particular, there is a general trend towards quasi-linear regimes (not being very reliant on nonlinearities) and to dedicated processes or means (even physically separated sections or stages) for the steps a)–c). A few but important examples shall be described in the following.

3.4.1 Shorter laser pulses

Many experiments have been carried out at laser facilities with shorter laser pulses, typically $\tau_L \sim 30$ fs, where electron energies of a few 100 MeV were obtained from gas jet targets (Mangles et al., 2004; Malka et al., 2005). Being initially shorter, the pulses have to undergo less nonlinear self-forming processes which smooths the interact. Moreover, those laser systems produce comparable pulse energies of ~ 1 J, but due to the shorter pulse the power is higher. Consequently, longer focal lengths can be used which was found to be beneficial for stable electron beam generation. Recently, electron beams were produced using a sub-10-fs laser system which directly accessed the bubble regime, i.e. without self-modulation (Schmid et al., 2009).

3.4.2 Guiding

In addition to self-focusing, external means can be used to counteract diffraction and thereby prolong the interaction length. Over the last few years, the use of capillary discharge waveguides was successfully implemented by several groups. Electron beams of up to 1 GeV kinetic energy within a beam of less than 2 mrad divergence and with negligible pointing jitter were obtained due to longer acceleration lengths (Leemans et al., 2006; Karsch et al., 2007; Rowlands-Rees et al., 2008), which is considered as a major breakthrough in laser-driven electron acceleration.

This type of plasma waveguide relies on thermal equilibrium of a gas discharge in a narrow capillary. The discharge current ionizes and heats the plasma which is cooled at the walls. Thereby, the temperature in the centre rises, the plasma density becomes thinner and a focusing density profile is formed. However, this approach increases the experimental complexity. In addition it should be noted that any external guiding structure is just an additional or assisting feature since a relativistically intense laser pulse always exhibits self-focusing.

3.4.3 Homogeneous plasma

The above-mentioned guiding experiments use a capillary with two gas injection slots close to the entrance and exit of the capillary. In recent experiments, similar capillaries were used as gas cell (Osterhoff et al., 2008), i.e. without discharge and therefore without external guiding. It was shown that electron beams can be produced very stably with 200 MeV peak energy, 2 mrad divergence and 1.4 mrad pointing jitter. This result is attributed to a very smooth plasma density profile. After filling, there is no gas flow between the gas injection slots. In contrast, gas jets always exhibit a gas flow where turbulences may be present. Those weak density fluctuations may lead, regarding the nonlinear nature of the underlying laser-plasma interaction, to less stable results and considerable shot-to-shot fluctuations. This may also explain the good results of capillary-guiding experiments.

3.4.4 Advanced injection techniques

It has been described before (cf. Sec. 3.1.3) that the electron acceleration relies on the injection of electrons into the strongly excited plasma. In our presented experiments, this was achieved by driving the wave to the nonlinear regime of wave-breaking which is accompanied by strong fluctuations. In contrast, if the injection of electrons is triggered externally, there is no need to drive the plasma wave so strongly that it reaches the nonlinear regime and the wave-breaking limit.

One way is beatwave injection (Faure et al., 2006). In addition to a strong laser pulse driving a non-breaking plasma wave, a weak counter-propagating laser pulse is used which interferes with the strong one inside the plasma. A standing wave is formed at the edge of the intense pulse. At this region, electrons oscillate relativistically in laser propagation direction. A fraction of electrons obtains sufficient energy in order to co-propagate with the plasma wave driven by the strong laser pulse. It was shown that monoenergetic electron beams can be generated very reliably. The energy can be chosen by the delay between and therefore by the overlap position of the two pulses. The longer the distance from the laser overlap to the end of the plasma, the higher the energy. Furthermore, by changing the amplitude of the standing wave, charge and energy spread of the electron bunch can be controlled.

Another possibility is ionization by the main pulse. If the gas is not fully ionized before the main pulse arrives, the residual ionization is accomplished by the pulse peak. Evidence was found that this late ionization leads to more stable electron beams. In one experiment,

capillary discharge waveguides were used. It was observed that the discharge did not fully ionize the gas and that ionization by the laser assisted the injection process (Rowlands-Rees et al., 2008). In very recent experiments, spurious amounts of gases like Nitrogen or Neon were added to the Helium of the gas jet. Hence, those atoms or partially ionized ions become ionized by the main pulse. Simulations show that tunnel ionization might be responsible. However, the observations are not yet fully explained and more work is under way.
A third route towards more controlled injection is the use of an electron density downramp (Geddes et al., 2008). If the plasma density decreases along the laser propagation direction, the plasma wavelength increases. This causes the plasma wave fronts to fall further behind as the laser propagates and effectively decreases the wake velocity. Hence, electrons require lower velocities to catch the wake and thereby the injection threshold is lowered. It was shown that with high reproducibility, low energy electrons (~ 0.8 MeV) with very low absolute energy spread (~ 0.2 MeV) can be produced, additionally with low transverse spread, good beam pointing and high charge (~ 1 nC). Very recently, this was combined with a second interaction stage for electron acceleration. The result is that the low-energy electrons are accelerated to much higher energies whereas the absolute energy spread remains small, yielding a very small relative energy spread. This represents a remarkable step towards future applications, allowing for small energy spread combined with high electron energies and stable beams with a single laser pulse focused into a single device.

4. Laser-based ion acceleration

In this section we will describe the physical concept of ion acceleration using high-intensity laser systems. In contrast to electrons, ions of mass $m_i = A \cdot 1836 \cdot m_e$ are hardly affected directly by the laser pulse. Hence, with today's laser intensities there is no effective energy transfer possible. Therefore, laser-based ion acceleration takes place as a secondary effect and is typically realized with thin foil targets of a few micrometer thickness, whereas various mechanisms have been identified as the source of the energetic ion beams (Snavely et al., 2000; Sentoku et al., 2003; Fuchs et al., 2005). The most widely investigated and used mechanism is *Target Normal Sheath Acceleration* (TNSA), which relies on three steps.

- An energetic, hot electron contribution is created by the interaction of a laser pulse with overdense plasma.
- The electrons propagate into the plasma. Usually, thin metal foils of a few microns thickness are used as target.
- At the target foil rear side, electrons leave the target and form a kind of capacitor. In the electric field arising from this charge separation, atoms from the rear side become ionized and subsequently accelerated. This is the key process of TNSA.

After presenting this standard model for laser-driven ion acceleration, we will extended it for the generation of ion beams with intrinsically narrowband spectra. After that, we will present experimental results of producing narrowband beams. We will show the special experimental setup, explain the necessary steps and discuss the obtained results. Most notably, we performed systematic studies from which we derived a scaling law of proton peak energy as function of the laser pulse energy.

4.1 TNSA – Target normal sheath acceleration of ions

The physical picture of Target Normal Sheath Acceleration of ions is displayed in Fig. 11: An intense laser pulse irradiates the front of a thin solid target. A relativistic electron population is generated that propagates through the target. If the target is chosen thin enough (i.e. typically a metal foil target of a few μm thickness), the electrons reach the back surface without considerable energy loss. The fastest electrons escape the target which thereby becomes positively charged up. The main part of the population is trapped by the electric field generated as a result of charge separation, forming a *sheath* of hot electrons at the back surface of the target, see Fig. 11 (b). As mentioned in Sec. 2.3, the electron distribution can shield the positive charge of the target only to the order of the Debye length λ_D. The available electric field strength due to this uncompensated positive charge is of the order of 10^{12} V/m. Such field strengths suffice to field-ionize atoms (cf. Sec. 2.2) present at the back surface of the target. Those atoms can be the target foil material itself, layers of contaminants (comprising hydrocarbons and water vapour), or additionally deposited material, which constitute the *ion source*. As soon as ions are generated, they are accelerated according to their charge to mass ratio q/m. Hence, protons as the lightest ion species will be most accelerated. A selection of ion species is in general only possible if the source's chemical composition is controlled, i.e. contaminants are suppressed. As shown in Fig. 11 (c), at a later time the ion front expanded into the laser-driven electron sheath at the target rear side. Since the electric field is oriented perpendicular to the conducting surface of the target, the ions are accelerated in *target normal direction* which lead to the name of Target Normal Sheath Acceleration.

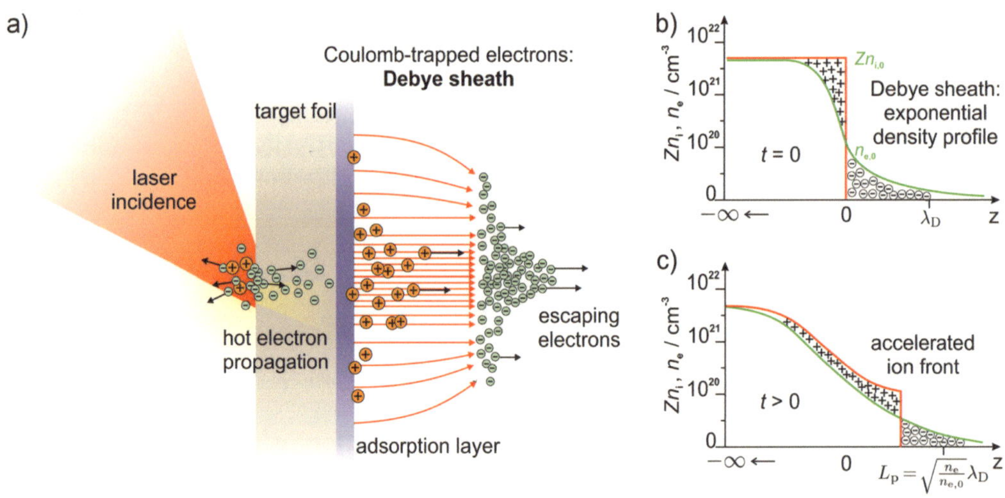

Fig. 11. The schematics of the TNSA process (a) and electron and ion density distribution at the beginning (b) and at a later time (c) of the acceleration. Here, protons are considered as the only ion species present.

4.1.1 Electron transport through the target foil

The total number of electrons produced by the absorption and electron heating processes at the target front side (cf. Sec. 2.7) can be deduced with the help of a simple estimation. As an example, consider a 500 mJ, 100 fs laser pulse, applying an intensity of $2 \cdot 10^{19}$ W/cm^2 ($a_0 \approx 3$) to the target and producing a hot electron population with an energy conversion

efficiency of about 20 % (cf. Sec. 2.7). From the laser's ponderomotive potential, Eq. (12), one can estimate the temperature of the electron population to be 1.1 MeV. Energy conservation leads to the total number of hot electrons of $N_e = 5.7 \cdot 10^{11}$.
This electron bunch, travelling through the target, forms a current of $J = e_0 N_e / \tau_L \sim 1$ MA, a value substantially exceeding the Alfvén limit $J = \beta\gamma \cdot 17$ kA, i.e. the maximum electron current that can propagate freely in vacuum (Alfvén, 1939). However, the propagation of such a hot electron current becomes possible in a conductor due to return currents that compensate the self-induced azimuthal magnetic fields around the relativistic electron forward current. In addition, the background electron density is of the order of the solid density $n_e \approx 10^{23}$ cm^{-3} and consequently much higher than the density of the relativistic electron beam, $n_e < n_c \approx 10^{21}$ cm^{-3}. Hence, the return currents are provided by a cold electron population obtaining just a slow drift. Furthermore, by using a conducting target material, the influence of magnetic field fluctuations or Weibel instabilities can be neglected since free charges are available to provide the necessary return currents. Consequently, a smooth electron beam with Gaussian temporal and spatial shape due to the laser pulse properties will propagate through the target (Honrubia et al., 2005). However, ohmic losses via collisions with the target foil atoms are still present and lead a reduction of the electron temperature at the rear side and a spread-out of the electron bunch.

4.1.2 Analytical model of the sheath field acceleration

The TNSA process can be analytically modelled as an isothermal plasma expansion into vacuum (Mora, 2003), considering only the target normal direction and disregarding transverse dependencies (1D model). This expansion is depicted in Fig. 11 (b) and (c). One can derive the maximum ion energy,

$$W_{\max} = 2Zk_B T_e \left[\ln\left(\tau + \sqrt{\tau^2 + 1}\right) \right]^2, \tag{23}$$

as well as the ion spectrum (per surface unit),

$$\frac{dN}{dW} = \frac{Zn_{e,0}t}{\sqrt{2Wm_i}} \exp\left[-\sqrt{\frac{2W}{Zk_B T_e}} \right]. \tag{24}$$

Here, Z is the charge state, $\tau = \omega_{p,i}\, t / \sqrt{2e}$ a normalized time correlated to the acceleration time $t = \tau_{acc}$, $n_{e,0}$ the initial electron density of the hot electron distribution driving the whole ion acceleration process and $\omega_{p,i}$ the plasma frequency (cf. Eq. (2) but with m_i instead of m_e) of the considered ion species.
The acceleration time τ_{acc} as one of the main free parameters has been found to be in good accordance with the following formula:

$$\tau_{acc} = \alpha(\tau_L + 60\,\text{fs}), \tag{25}$$

with $\alpha = 2.6$ decreasing linearly from 10^{19} W/cm² to the constant value of $\alpha = 1.3$ for $I \geqq 3 \times 10^{19}$ W/cm² (Fuchs et al., 2007). Following the example above, Eq. (25) gives an acceleration time of 310 fs.

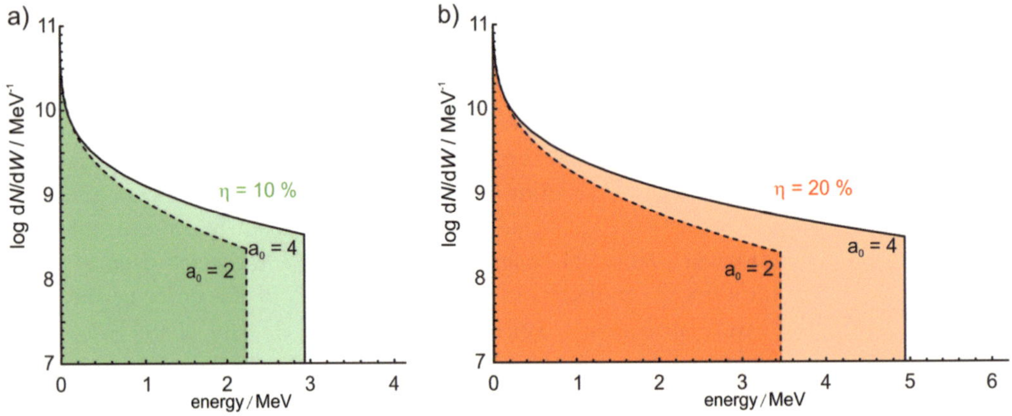

Fig. 12. Proton energy spectra calculated according to Eqs. (23) and (24) for different laser intensities a_0. The green spectra on the left hand side (a) consider a conversion efficiency η of 10 % whereas the red spectra (b) stand for η = 20 %. Taken from (Jäckel, 2009).

Regarding the second major free parameter of the model, the initial electron density, one may take the above-estimated total electron number (cf. Sec. 4.1.1), the bunch length $c\tau_L$ and needs to consider a spreading of the hot electron beam propagating through the foil, leading to an effective area covered by the electron bunch when it exits the target, $A_{Sheath} = \pi\,[r_{eff} + d_{eff}\tan(8°)]^2$ (Kaluza et al., 2004), to arrive at the following expression:

$$n_{e,0} = \frac{N_e}{A_{Sheath}\, c\tau_L}. \tag{26}$$

Here, a typical divergence of the electron beam inside the target of 8° can be assumed in good accordance with simulations and experiments. The parameters r_{eff} and d_{eff} take into account the incidence angle of the laser pulse onto the target, which is mostly 45°: $d_{eff} = d_{target}/\cos(45°)$ and $r_{eff} = w_L/\cos(45°)$ (Honrubia et al., 2005). Typical values of the initial electron density of the target rear side sheath are of the order of 10^{20} cm^{-3}.
Using τ_{acc} and $n_{e,0}$ and assuming a laser intensity equivalent to a_0 = 2–4, one obtains the proton spectra displayed in Fig. 12. The spectra show a quasi-thermal distribution with a distinct maximum energy. This maximum energy depends on both the absorption efficiency and the laser intensity.

4.1.3 Influence of the target thickness

In order to extend the discussion of the electron transport through the target, the influence of the target foil thickness on the produced ion beams will be introduced in the following. Fig. 13 shows measured proton spectra, produced from Titanium foils of different thicknesses (d_{target} = 1 - 20 μm). The number of protons is given per 0.01 MeV energy interval and 1 μsr solid angle of detection and is plotted against the proton energy. The laser parameters are identical to those of the results presented later in Sec. 4.3. We observe quasi-thermal proton spectra with distinct maximum proton energies, as discussed in the previous section. However, we find, now regarding the target thickness, a specific trend: with decreasing foil thickness, we observe an increase of both the total proton number and the maximum proton energy for each spectrum until an optimum value of d_{target} = 2 μm.

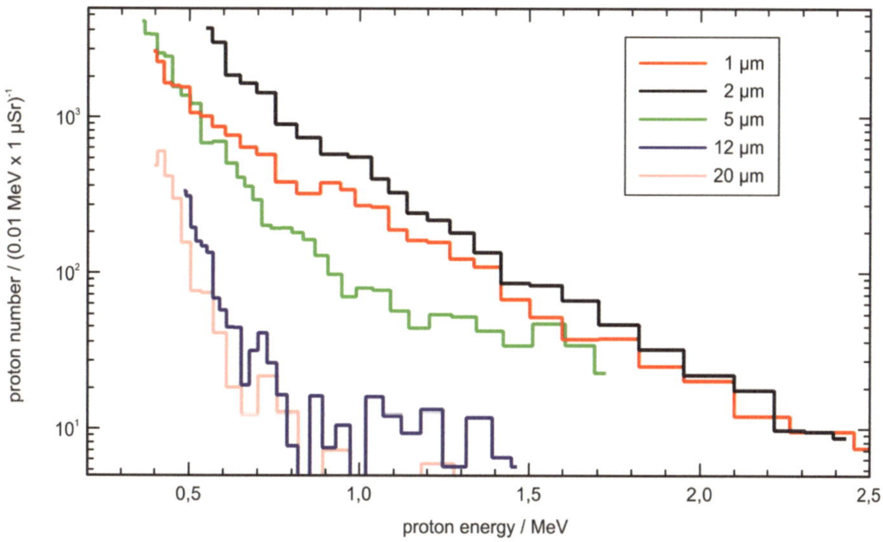

Fig. 13. Proton spectra for different target foil thicknesses.

This characteristic behaviour can be explained as follows. For decreasing foil thickness, the electron transport becomes more effective since the influence of the resistivity decreases for a shorter propagation length through the foil material, and hence the electron temperature remains higher when the electrons reach the target back surface. Furthermore, the electron sheath behind the target foil has a higher electron density since the electron beam is less transversely spread throughout the shorter propagation path, cf. (26). Thus, both parameters which characterise the initial electric field strength, $E_{\mathrm{TNSA}} = \sqrt{2/e \cdot k_B T_e \cdot n_e / \varepsilon_0}$, being responsible for laser ion acceleration, contribute positively and amplify the acceleration process. In principle, this trend can be shifted to ever thinner targets and is only limited by the temporal contrast of the laser pulse. The laser pulse contrast determines the preplasma formation at the target front side, which eventually can break through the target foil if it is very thin. In the present case, the laser pulse conditions dictate an optimum target thickness of $d_{\mathrm{target}} = 2\ \mu\mathrm{m}$. A further decrease of the thickness gives only spectra with lower proton number and maximum energy. However, this concept of decreasing target thickness and increasing laser contrast is an important route of today's research on laser ion acceleration, see (Henig et al., 2009) for example.

4.2 Generation of monoenergetic ion beams – Confined TNSA

The proton (or ion) beams produced via TNSA typically show an exponential energy spectrum with a distinct cutoff energy $W_{\max}$ of typically a few MeV, which depends strongly on the laser and target parameters, cf. Fig. 13 and (Kaluza et al., 2004). This broad distribution can be explained mainly by two contributions: First, the accelerating electric field decays when reaching into regions of higher charge density. Deeper-sited protons in the source will thus be partially screened from the electric field by their predecessors. Thus, the proton's final energy is determined by its initial longitudinal position. Secondly, the accelerating sheath field typically possesses a bell-shaped symmetry and is inhomogeneous in the transverse direction (which is actually not included in the mathematical model but does not substantially diminish its explanatory power). That means that the maximum energy a proton can gain is also determined by its radial distance from the laser spot. Protons located in the centre of the field are accelerated the most up to the maximum energy W_{max}, whereas protons outside experience a lower electric field strength and consequently are accelerated to lower energies. As a consequence of both the screening effect and the transverse inhomogeneity of the electric field, the resulting spectrum has a strong correlation to the initial distribution of the protons to be accelerated from the target.

Following this understanding, Esirkepov and Bulanov proposed an acceleration scheme for the generation of monoenergetic beams where all protons are *radially confined* to a "dot" source within the central homogeneous region of the TNSA field (Esirkepov et al., 2002). If the proton source is furthermore sufficiently thin so that screening effects are negligible, all protons experience the same potential and are accelerated to a monoenergetic distribution.

This is depicted in Figure 14 (a). Similar to standard TNSA, an intense laser pulse impinges onto a thin target foil and generates a sheath of hot electrons at the target back side. But now, there is a dot instead of transversely extended layers, constituting a thin, radially confined ion source being located in the central homogeneous field region.

Recent theoretical studies (Robinson & Gibbon, 2007) showed that the limitation of the source thickness is actually not a critical criterion, but that for "thicker" dots, the formation of monoenergetic spectra is supported by *charge separation effects* between two ion species with different charge to mass ratio q/m, e.g. protons and carbon, cf. Fig. 14 (b)–(d). Robinson & Gibbon showed in a detailed study regarding micro-dot acceleration that monoenergetic spectra can in fact be produced from dots of up to micrometer thickness and may even vanish if the source layer is chosen too thin.

So how does the charge separation mechanism work? Consider a confined ion source consisting of two ion species, say protons and C^{4+}. Owed to their different q/m-ratios, the two ion species are accelerated differentially according to $\ddot{z} = q/m \cdot E_z$, and will hence separate at an early stage of the acceleration process. Consequently, two particle distributions with distinct particle fronts will propagate away from the target surface. This situation is shown in Fig. 14 (b): The fast proton front (red) and the slower carbon front (green) each terminate at a sharp z-position, whereas the carbon front overlaps with the low energy part of the proton distribution. Fig. 14 (b) represents a snapshot taken from a 1D-PIC simulation by Alex Robinson (Robinson et al., 2009) for a 1 μm thick target located at $z = 0$, at a laser intensity of $I_L = 10^{20}$ W/cm^2. The total charge density is plotted in Fig. 14 (c). It displays two distinct discontinuities, one at the proton front ($z \approx 6$ μm), and one at the carbon front ($z \approx 4$ μm). Recall that TNSA is driven by charge separation at the solid-vacuum interface, where hot electrons are accelerated across the proton front and trapped to constitute the TNSA sheath field. Obviously, the same holds true for the charge discontinuity at the heavy ion front: The discontinuity represents a boundary where charge neutrality cannot be retained, and passing hot electrons will hence form a second sheath field across the carbon front. This is illustrated in Fig. 14 (d), where the electric field behind the target is plotted. The two strong peaks correspond to two electrostatic shocks resulting from the sheath fields at the proton and heavy ion front, respectively. Between the peaks, there exists a zero-field region ($z \approx 5.5$ μm), at which the proton front sheath field is already screened such that protons, accompanied by a colder electron population, may propagate undisturbed in quasi-neutrality. This zero-field region is where protons accumulate to form a stable monoenergetic bunch. Similar to the TNSA process at the proton front, the light protons from the low energy part of the spectrum are efficiently accelerated across the

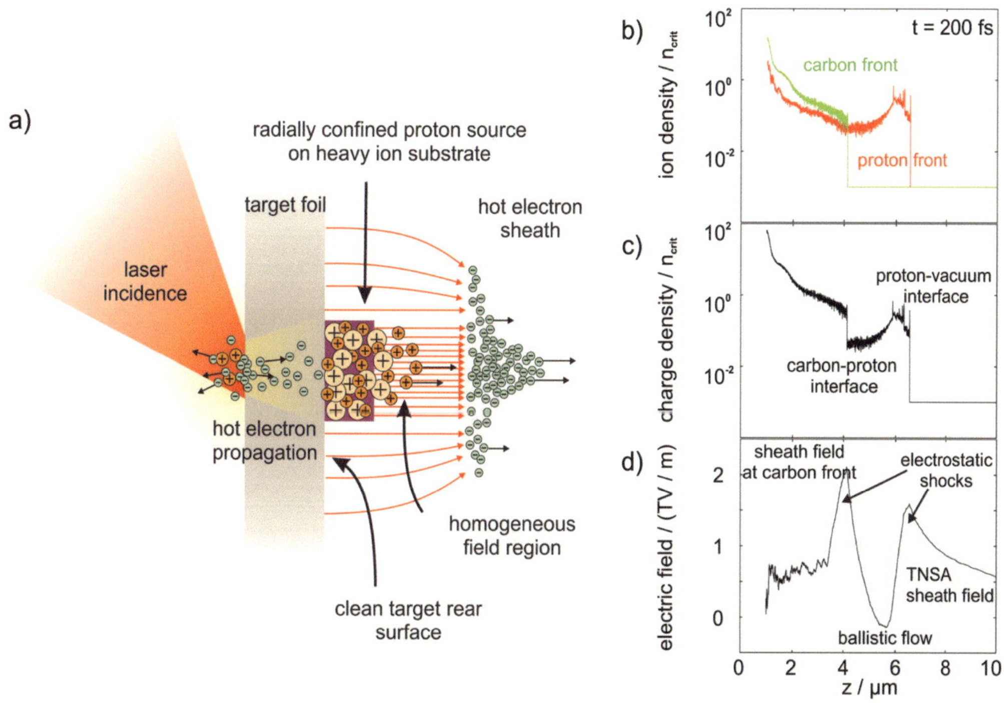

Fig. 14. The schematics of the Confined TNSA process (a) and results from a simulation of the acceleration process (b)-(d) at 200 fs. Simulation data courtesy of A.P.L. Robinson (Pfotenhauer, 2009; Robinson et al., 2009).

carbon boundary by the second sheath field, whereupon they immediately enter the zero-field region and remain in a state of motion of uniform ballistic flow. Note that the potential of charge separation for the spectral modulation of laser-accelerated ion beams had been identified in earlier works already (Tikhonchuk et al., 2005; Robinson et al., 2006).
The confined TNSA scheme has been confirmed for the first time experimentally by our group (Schwoerer et al., 2006; Pfotenhauer et al., 2008). Several other techniques for the generation of monoenergetic ion beams have been developed. These include the use of ultrathin Carbon layers for the generation of monoenergetic C^{5+} and C^{6+} beams (Hegelich et al., 2006), the generation of monoenergetic deuteron beams from droplet targets (Ter-Avetisyan et al., 2006), and active beam shaping and spectral selection of standard-TNSA generated proton beams using a fs-laser driven micro lens (Toncian et al., 2006).

4.3 Experimental studies of laser-ion acceleration
4.3.1 Setup for ion acceleration via confined TNSA

Fig. 15 (a) shows the schematics of the experimental realisation of the Confined TNSA regime. Like for standard TNSA ion acceleration, an intense laser pulse is focused onto a thin metal foil in order to generate the hot electron sheath at the target rear side. Contrary to the regular TNSA, however, a small microdot (typically made from polymer) has to be positioned exactly opposite to the laser incidence position. This is realized by manufacturing targets carrying an array of microdots on their backside and translating the target with the help of an adequate alignment procedure. Furthermore, since the ion source in standard

TNSA is constituted from contaminants being present also under vacuum, a target backside cleaning must be carried out straight before the main laser shot which has to remove the adsorbates completely but must preserve the dot. For that purpose, we use an additional, weakly focused, ns-pulsed, ablation laser.

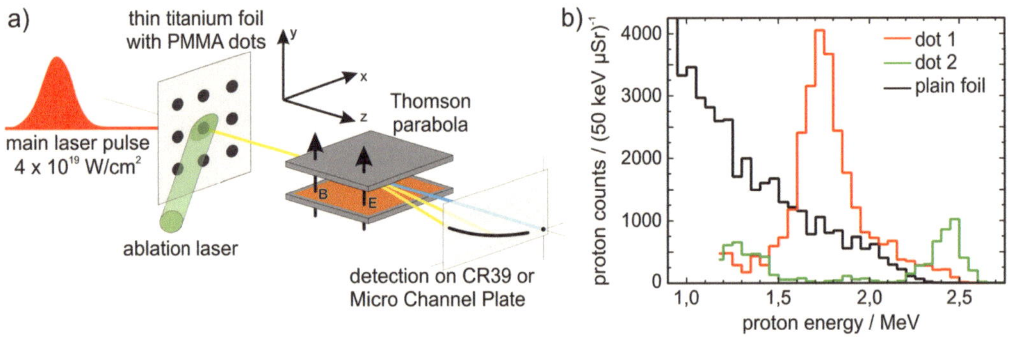

Fig. 15. Experimental setup (a) and generated proton spectra (b) of Confined TNSA. The main laser pulse hits the thin foil target at the front side exactly opposite to a micro-dot. Protons from the dot are accelerated within the central, homogeneous field region of the TNSA field and analysed with a Thomson spectrometer. The ions can be detected either with CR39 track detection plastics or an online imaging system (MCP). A second laser, which hits the target on the back side concentrically with respect to the main laser, is used for the cleaning of the target from residual contamination layer protons.

All that is highly non-trivial: The manufacturing of reproducible and resistant microstructures on a 2–5 micrometer thin metal foil represents a great engineering challenge. Secondly, the alignment of the dots with the laser incidence is equally difficult and crucial, as this means to align the laser pulse impinging on the target front side with something that is visible only from the back side. For this purpose, the back side was observed with a long distance microscope with micrometer resolution. In addition, the ablation laser must hit the target backside co-centred with the main laser incidence position. This laser was weakly focused on the surface with a diameter of typically $d_{foc} \approx 600\ \mu m$ in order to cover the whole proton source area around the dot. It was found that the careful regulation of the fluence of the ablation laser Φ_{abl}, the number of ablation shots and the time delay before the main laser pulse have a strong impact on the reliability of the generation narrowband proton beams (Pfotenhauer et al., 2008; Pfotenhauer, 2009).

The generated proton and ion beams propagated towards the spectrometer chamber, where the beam diameter was reduced by a pinhole of variable size (0.3 mm, 1 mm or 3 mm). Subsequently, the limited beam was dispersed in a Thomson spectrometer. The energy resolution of the spectrometer was defined by the pinhole diameter to $\Delta W_{res} = 50$ keV for 1 MeV protons using the 1 mm aperture. The particles were detected either with CR39 track detection plates or with an online imaging system based on micro channel plates (MCP) in a chevron setup.

4.3.2 Microstructured targets

The fabrication of adequate targets for Confined TNSA was realized in two steps. First, a layer of PMMA was applied on a thin carrier foil. The polymer was evenly spread on the back side of the foil using a spin coating technique, which yielded well-defined layer thicknesses of 50–1000 nm.

In a second step, the coated target surface was microstructured via laser ablation using a focussed Ti:Sapphire laser beam (λ_L = 800 nm, W_L = 5 μJ, τ_L = 60 fs) with 1 kHz repetition rate. Square dots with a minimum size of 10 × 10 μm^2 were carved out from the PMMA surface, see Fig. 16 (a) and (b). Alternatively, target microstructuring was also performed via photo-lithography. In this case, the PMMA surface was irradiated with a UV argon-fluoride excimer laser (λ_L = 193 nm, W_L = 3 mJ, τ_L = 15 ns) through a custom-designed chrome-layered quartz glass mask. An example target is shown in Fig. 16 (c).

The PMMA microstructures fulfil all the requirements put forth by the Confined TNSA model. That is, they provide a radially limited proton source in the form of a compound with a heavier ion species (carbon) to be subject to the charge separation effects discussed above. Together with a robust target cleaning procedure they provide the means to pursue monoenergetic proton acceleration.

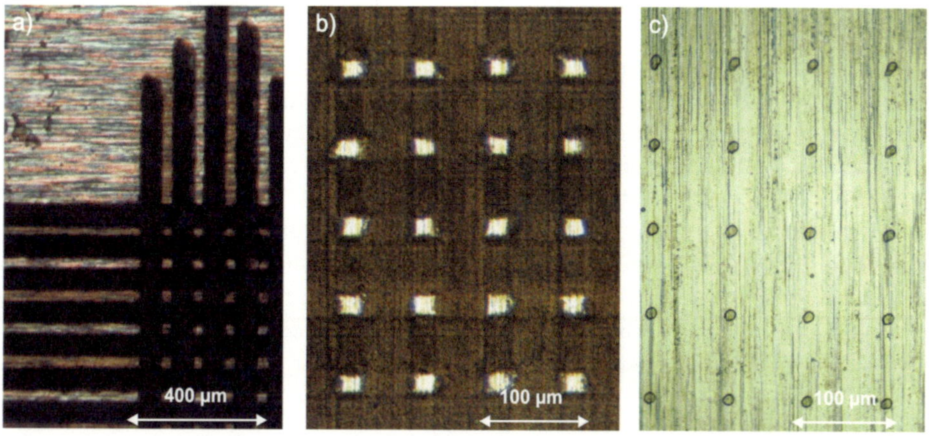

Fig. 16. (a) + (b) Fabrication of microstructured targets via fs-laser ablation. For the fabrication of microstructured targets suitable for Confined TNSA, a layer of polymer (PMMA) with 50 – 1000 nm thickness was first deposited to titanium foils via spin coating. From the polymer-covered surface, small square dots of 20×20 μm^2 extent and 80 μm separation were "carved out" with the help of a femtosecond laser system. Fig. (a) clearly shows the perpendicular carving traces ranging into the polymer. Note the well-defined carving edges typical for fs-ablation. (c) Lithography targets. Round dots of 10 μm diameter with 80 μm separation were generated via UV-lithography with a pulsed excimer laser. This alternative technique allows for a more flexible fabrication of micro-dots, which are, however, considerably more sensitive to target cleaning than those produced with the femtosecond system.

4.3.3 Properties of narrowband ion beams

Figure 14 (b) shows exemplary proton spectra obtained from microstructured targets. Spectra were obtained from the irradiation of PMMA micro-dots after 10 consecutive laser ablation shots at the threshold fluence of Φ_{abl} = 1.2 J/cm^2 (red and green lines). In comparison to plain (unstructured) 5 μm Titanium foil without laser ablation of contaminants (black line), the proton spectra from the polymer dots show distinct peaks at W_{center} = 1.7 MeV with an energy width of ΔW_{FWHM} = 0.25 MeV = 15% for dot 1 and W_{center} = 2.5 MeV with ΔW_{FWHM} = 0.2 MeV = 8 % for dot 2. The ablation has suppressed the parasitic low-energy component of the spectrum and enables the acceleration of monoenergetic protons from the confined dot source. Narrowband features appear consistently once an ablation threshold fluence of Φ_{thr} = 1.2 J/cm² at 532 nm is surpassed (Pfotenhauer et al., 2008). In contrast, the laser-irradiation of a plain Titanium foil (black line) yields a thermal distribution typical for standard TNSA. The contrast between proton peak and thermal spectrum is typically 4:1. The peaks itself stand out well from the background signal, they contain typically between 80–90 % of all protons.

The number of protons contained in the spectra of Fig. 14 (b) was limited by the solid angle of observation of the setup, 1 μsr, which in turn was determined by a 1 mm pinhole in front of the ion spectrometer. In previous experiments (Schwoerer et al., 2006), the divergence of proton beams from plain, unstructured foils was measured to be much larger, amounting to 360 msr at a proton energy of 1.6 MeV, or 60 msr at 2.5 MeV, which implies that only a small fraction of the total number of protons is actually detected.

However, simulation studies confirm that the narrow band feature in the spectrum is not limited to the small angle of observation but appears equally over the whole emission angle (Robinson et al., 2009). Another set of 2D-PIC simulations carried out by Timur Esirkepov indicates that the angle of emission for micro-dots is approximately 24 msr for a peak at 1.1 MeV and hence somewhat lower than the experimental values measured for plain targets. Using this solid angle of 24 msr as a conservative estimation for the total number of protons, one finds that the peak of dot 1 of Fig. 14 (b) contains about $N_p \approx 4.1 \cdot 10^8$ in the FWHM peak. From integration over the narrowband spectrum it follows that the *conversion efficiency* of laser energy (0.78 J on target) into kinetic energy of protons amounts to $\eta_p = 1.4 \cdot 10^{-4}$ for the FWHM peak.

4.3.4 Peak position scaling

A subject of great interest for the generation of monoenergetic proton beams is the scalability of the acceleration mechanism to higher laser powers. Having established the high degree of reproducibility discussed in the previous section, the peak parameters for many hundred monoenergetic spectra were evaluated as a function of the main laser pulse energy, keeping all other parameters (dot dimensions, focal spot size, laser pulse duration, prepulse conditions, ablation conditions etc.) fixed. It shall be emphasized that the use of an online detection system like an MCP clearly facilitates the collection of such large amounts of data. Fig. 17 shows a subset of data where the main laser pulse energy was increased from 0.5 to 0.8 J on target (black squares). The four data points comprise a total of 140 spectra, all taken from 5 μm titanium foils carrying PMMA dots of 200 nm thickness after 10 consecutive shots of ablation at the threshold fluence Φ_{thr} = (1.2 ± 0.3) J/cm². Statistical analysis yields that the average peak position increases from 1.42 to 1.63 MeV over this energy range. The error bars represent the standard deviation of the sub-sets and thus give a measure for the shot-to-shot fluctuation. Note that the energy range covered by the four data points was constrained by the available laser energy (upper boundary) and the detection range of the spectrometer (lower boundary). The four data points strongly suggest a linear dependency between the peak position and the laser energy for the observed energy range.

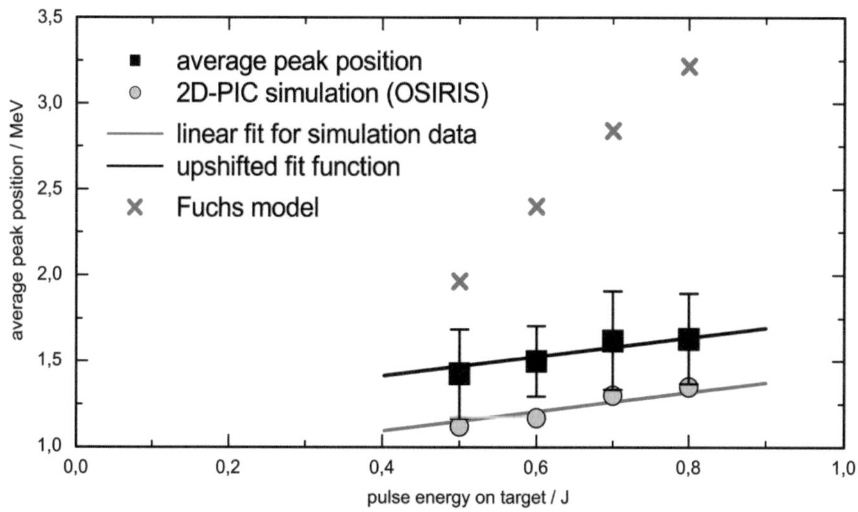

Fig. 17. Dependency of monoenergetic proton peak position on the laser pulse energy. The laser pulse energy on target was varied from 0.5 to 0.8 J, which corresponds to intensities of $2.9 - 4.7 \cdot 10^{19}$ W/cm^2. Consequently, the average peak position rose from 1.42 to 1.63 MeV (black squares). All other parameters were kept fixed. The four data points include a total of 140 monoenergetic spectra, lending sufficient statistical significance to the results. The error bars represent the standard deviation of the statistical sample and delineate the shot-to-shot fluctuation. The observed proportionality is excellently reproduced by 2D-PIC simulations carried out for our experimental parameters (grey circles). The calculated peak positions lie slightly outside the standard deviation, but are well within statistical range of the sample sets. All previous scaling laws refer to the cutoff energy of thermal spectra from plain foils and fail to account for our data. For example, the model by (Fuchs et al., 2006) predicts much higher energies and a steeper slope (grey crosses) for the current parameters.

In order to support this evidence, we carried out 2D PIC simulations for the different laser energies applied in the statistical analysis. The results are shown in Fig. 17 as grey circles: Similar to the experimental data, the calculated peak position increases in a linear manner, rising from 1.17 to 1.35 MeV over the observed energy interval. An additional simulation was carried out for a pulse energy of W_L = 15 J in order to investigate whether this behaviour holds true also at higher laser energies. This run yielded a monoenergetic peak at 9.3 MeV in very good agreement with the linear extrapolation of the four calculated spectra between 0.5 and 0.8 J. Together, the five simulation data points follow the linear function

$$W_{\text{peak}} / \text{MeV} = 0.56 \cdot W_L / \text{J} + 0.87, \tag{27}$$

represented by the grey line in Fig. 17. The numerical results appear systematically lower than the experimental values. However, it is striking how closely the slope of the simulation data matches that of the experimental data. The black line in Fig. 17 depicts the same linear fit function (27), upshifted only by 320 keV. This excellent accordance lends authority to the deduced linear scaling, which hence represents a scaling law for the generation of monoenergetic proton beams from microdot assisted laser proton acceleration.

The importance of this scaling law becomes evident when comparing it to other scalings proposed for laser proton acceleration. Various dependencies have been suggested to connect the proton energy to the laser power or intensity (Mora, 2003; Fuchs et al., 2006; Schreiber et al., 2006; Robson et al., 2007), and it is known from previous measurements that for example the model by Fuchs et al. successfully predicts the observed maximum energies for standard TNSA conditions. However, all of these scaling laws refer solely to the cutoff energy of thermal spectra from plain foils. Contrarily, it has been shown above that the peak position from Confined TNSA does *not* coincide with the cutoff energy of the corresponding thermal spectra, but is bound to the slower heavy ion front. It is therefore not ad hoc intelligible if and how the peak formation scales with laser energy. Clearly, the thermal scaling laws cannot be expected to account for the peaks, and in fact a comparison of our data with the scaling law from (Fuchs et al., 2006) shows a strong discrepancy for the observed narrow parameter interval already. This deviation is illustrated by the grey crosses in Fig. 17, showing that the model by Fuchs et al. predicts much higher energy values and an approximately 8 times steeper slope. Eq. (27) hence represents the first scaling law specifically for monoenergetic proton spectra, and verifies for the first time that techniques for the generation of intrinsically monoenergetic ion beams can be extrapolated to higher laser energies. Furthermore, it indicates that the peak energy scales slower than the thermal cutoff energies and that thermal energy scalings overestimate the potential of laser ion acceleration when it comes to monoenergetic beams, which are the prerequisite for most applications. Nonetheless, today's capability of reliably generating $\sim 10^9$ quasi-monoenergetic protons with less than 10 % bandwidth by means of a scalable technique marks an important step towards applications.

5. Summary and outlook

The main motivation of investigating laser-based particle acceleration is the possibility of using much higher accelerating fields, resulting in smaller accelerators. Thus, before giving a small outlook onto possible applications, it seems well worthwhile to summarize the achievements of laser accelerators as of today and to compare them to benchmarks of conventional accelerators. Even though this is a tempting task, it should be noted that a mere comparison of numbers might be misleading, as certain quantities have been defined with certain applications in mind. In conventional accelerator science, comparisons are often made by historically grown "figures-of-merit" (FOMs): current, transverse emittance (focusability), longitudinal emittance (compressibility) and brilliance. Without doubt, these FOMs have their qualifications but should be seen in their specific context. For instance, the quantity "spectral peak brilliance", defined as particles per pulse length, per unit area, per unit solid angle and per relative spectral bandwidth, is often used to compare synchrotron facilities (cf. Sec. 5.2). It presupposes that the electron beam can be focused to a small and collimated beam, which in turn already requires a small spectral bandwidth in order to avoid chromatic aberration. In addition, peak brilliance is rather relevant for single shot applications where a sufficiently strong signal must be provided by a single shot. This measure fails, however, if a large number of shots is required, as is the case for example in most collider experiments, a situation which would be captured better by the "average brilliance".

In view of the relatively young field of laser-particle acceleration, it seems more sincere to provide rather "primary" quantities: peak energy, spectral width, bunch charge, pulse duration, bunch repetition rate, beam pointing, beam size and divergence. Nonetheless, it is instructive to dare the comparison in the conventional accelerators' "own terms", so-to-say, in order to point out the advantages and disadvantages of the novel approach introduced in this chapter. After presenting in the following first a status of the current abilities of laser-produced particle beams, we will then point the spotlight onto a few prominent and important application-related research fields.

5.1 Comparison to conventional accelerators

Generally, laser-plasma accelerators produce significantly shorter pulses than conventional ones, which follows directly from the use of ultra-short laser pulses as drivers. This is a key feature for laser-produced particle beams and leads in general to the effect, that all "peak" beam quantities are comparably high. Thus, laser-based accelerators are predestined for time-resolved or single-shot applications. On the other hand, average beam properties are typically quite low for laser-based accelerators since today's laser repetition rates are limited to 10 Hz, whereas conventional accelerators can operate up to MHz rates. This is likely to continue in the future since current CPA laser technology cannot handle the required average optical powers and thermal loads.

A second major difference is the reduced acceleration length, which is generally seen as the most promising aspect of this new generation of particle accelerators. This may potentially reduce the dimensions of accelerator facilities significantly – which is an important economic argument. On the other hand, due to their meter-scales, conventional accelerators allow for relatively easy and independent beam control. In particular, they consist not only of accelerating units (RF resonators) but also of beam forming devices (e.g. quadrupole magnets). Currently, it is quite difficult to transfer similar schemes to laser-plasma accelerators. In the bubble regime of electron acceleration, there are transverse electric fields present which accomplish longitudinal bunching and transverse beam focusing. However, this is inherently coupled to the acceleration process – being far away from external control. The Confined TNSA with the use of two ion species and a region of zero-field strength is similar but resembles as intentional, artificial manipulation of the plasma. On top of that, very recently, we were able to realize a staged acceleration setup, where at a second target foil a second TNSA field was generated which enabled us to post-accelerate protons from a first foil target (Jäckel, 2009). Depending on the distance of the two target foils and timing of the TNSA fields, one can select the energy of the protons being post-accelerated. This temporal gating allows for spectral control, and the staging itself is an efficient method to increase the proton energy. Thus, glancing over the rapid progress in the recent years in the domain of laser-based particle acceleration, the challenges of beam control will be taken and turn into productive future developments.

In addition to such general rating, we will now present relevant beam parameters. Regarding laser-driven electron acceleration, 1 GeV electron beams can be generated within 3 cm (Leemans et al., 2006) whereas typical synchrotrons, operating at a few GeV, have diameters of up to 100 m. This milestone clearly shows the general potential of laserbased particle acceleration as more compact accelerators. The pulse lengths are also very promising. Although exact pulse length measurements are still pending, upper limits are determined to 50 fs (van Tilborg et al., 2006), and simulations promise bunch durations below 10 fs (Pukhov & Meyer-ter-Vehn, 2002). Conventional accelerators usually produce ns long pulses, sub-ps pulses are extremely difficult to generate. Bunch charges from laserplasma accelerators are typically on the order of several 10 pC. In contrast, conventional accelerators can produce several 10 nC. This large discrepancy of 3 orders of magnitude is a serious problem for many applications with laser-based accelerators, and possible solutions are currently investigated. However, the peak current as a combination of both measures is at least comparable to conventional accelerators. Considering the energy spread, typical results for laser-plasma accelerators are 5 % of the peak energy. However, this value is mostly instrument-limited, thus the actual energy spread could be smaller. A relative spread of the order of 10^{-3} would suffice for most applications. Thus, the energy spread is also a point on top of the agenda. Another rather open question is the transverse emittance, which determines the focusability of the beams. From our measurements, the smallest beam divergence is $\sigma' = 2$ mrad. Assuming a source size of $\sigma \sim 2\ \mu$m (Pukhov & Meyer-ter-Vehn, 2002), yields a transverse emittance of $\varepsilon_\perp = \sigma^2\sigma'^2 \gtrsim 10^{-5}$ mm^{2}mrad^2. This corresponds to a normalized, cold-beam transverse emittance of $\varepsilon_n = \gamma\beta\sqrt{\varepsilon_\perp} \sim 0.1$–$0.5\ \mu$m for the obtained electron energies of up to ~ 70 MeV. Very recent direct measurements show the transverse emittance to be of the order of $\varepsilon_n \sim 5\ \mu$m, which must be considered rather as an upper limit. If the normalized emittance would be below 1 μm, it would be comparable to conventional accelerators. We observed a pointing stability of the beams of 40 mrad which is much too large for applications. Other groups have shown electron beams with stable beam pointing of about 1 mrad which is, especially in combination with beam focusing elements, sufficient for further use of the beams (cf. Sec. 5.2).

Regarding ion acceleration, we will exclusively consider proton beams generated via Confined TNSA as the currently only method to reproducibly generate intrinsically narrowband ion spectra. The determined charge of $4 \cdot 10^8$ protons is of the same order as the one of electron beams, thus a similar discrepancy arises. The currently generated proton beams show peaks at about 2 MeV energy with about 10 % bandwidth. This must be seen as proof-of-principle experiment and is too early to compete with conventional ion accelerators. With today's available laser technology, the 10vMeV range can be achieved easily. Furthermore, simulations suggest that the absolute peak width saturates with increasing peak energy, which might reduce the relative width effectivly. The pulse duration may be estimated from the length of the zero-field region in Fig. 14 (d). For 2 MeV proton energy, a length of 1 micron corrensponds to a time of 300 fs. However, this pulse will disperse in free space propagation simply due to its different velocities. This problem disappears if the beams move relativistically with velocities close to c. The opening angle was also derived from simulations to 24 msr which corresponds to a divergence half angle of 87 mrad. An estimation of the transverse emittance can be provided by taking into account the angle of emission and a proton source size determined by the diameter of the microdots (20 μm), which yields a transverse emittance of $\varepsilon_\perp \gtrsim 2$ mm^{2}mrad^2. The normalized emittance amounts then to $\varepsilon_n \gtrsim 0.1\ \mu$m. For laser-plasma accelerators, transverse emittance values of $\varepsilon_n < 4 \cdot 10^{-3}\ \mu$m have been measured using mesh magnification techniques (Cowan et al., 2004). The herepresented estimate is larger but still outmatches the emittances of conventional accelerators by more than one order of magnitude (e.g. $\varepsilon_n \approx 3\ \mu$m for the CERN SPS (Arduini et al., 2003)), which clearly indicates the potential benefits of laser driven particle acceleration already at this early stage.

The longitudinal emittance is defined as the product of the pulse duration and the energy spread, $\varepsilon_{long} = \Delta\tau_{pulse} \cdot \Delta W_{kin}$, and gives a measure for the re-compressibility of the particle beam, in analogy to the time-bandwidth-product of short laser pulses (cf. Sec. 2.1). For the

experiments under discussion, the FWHM energy spread of 0.2MeV together with an initial pulse duration of 300 fs yields a longitudinal emittance of $\varepsilon_{long} \approx 6 \cdot 10^{-8}$ eVs, which is an extremely low value. Even if the pulse duration is conservatively approximated by the lifetime of the TNSA sheath field of roughly 10 ps, one still finds a theoretical emittance of $\varepsilon_{long} \approx 2 \cdot 10^{-6}$ eVs, which is many orders of magnitude below the values achieved by conventional accelerators (e.g. $\varepsilon_{long} \approx 0.35$ eV s for CERN SPS (Arduini et al., 2003)).
Finally, the beam peak brilliance, defined as particle flux dN/dt per area dA, spectral bandwidth dW and solid angle $d\Omega$, may be estimated. Gathering all numbers, we ultimately end up with $B \sim 10^{27}$ $eV^{-1}s^{-1}m^{-2}sr^{-1}$. In comparison, the peak brilliance of CERN SPS amounts to $B_{SPS} = 5.3 \cdot 10^{24}$ $eV^{-1}s^{-1}m^{-2}sr^{-1}$ according to the presented parameters.
Note, however, that a large acceleration facility like CERN SPS provides considerably more protons per pulse ($\approx 10^{11}$) at a 5 orders of magnitude higher peak energy ($\approx$ 450 GeV) with a significantly smaller spectral bandwidth! The beam properties of the laser-produced proton beams discussed in this section should hence be recognized more as an illustration of the *complementary and unique physical regime* accessed by laser accelerators, rather than a competitive comparison between different machines aiming at the same parameters.

5.2 Short-pulsed radiation sources

Due to their short wavelengths, X-rays are suited to study the structure of matter on the atomic length scale. However, for many fundamental processes like phase-transitions, it is necessary to have a time-resolving investigation method. Pump-probe experiments can do so, but require i) two short pulses with ii) a precise delay between them. Conventional means to produce X-rays are x-ray tubes and synchrotron sources. The latter have matured in the recent decades and are to date the most intense sources which attract large used communities. However, up to now they rely on conventional accelerators and are large, expensive facilities. Furthermore, to achieve sufficient time resolution, huge efforts must be undertaken. Nowadays, only two facilities can produce ultra-short X-ray pulses.
For this particular application, the use of laser-accelerated electron bunches for the production of synchrotron radiation seems very promising. Firstly, the large and expensive conventional accelerator could be replaced. Secondly and most important, the produced electron pulses are inherently ultra-short and ultimately synchronized to an ultra-short pulsed laser system. By that, the demands for pump-probe experiments are perfectly met. Furthermore, the recent realization of short-pulse capabilities with conventional accelerator techniques, facing immense problems to be overcome and going to technological limits, shows the demand of scientists for short-pulse synchrotron radiation facilities.
We have shown in a proof-of-principle experiment the production of synchrotron radiation by sending laser-accelerated electrons through a magnetic undulator for the first time (Schlenvoigt et al., 2008). This technique is now used for precise characterization of electron beams (energy, energy spread and emittance) as well as – depending on the electron beams – for the production of short VUV pulses. In the near future, the wavelength range will be extended to soft X-rays. Furthermore, an experiment to measure the pulse duration should be carried out as well as a proof-of-principle experiment for the pump-probe capabilities of this scheme. In addition, the FEL operation will be tested which requires low transverse emittance, low energy spread and high peak current. The undulator may be either operated in the SASE-FEL mode, or eventually as FEL seeded by a coherent laser-driven source.

A different laser-driven source of X-rays is Betatron radiation (Rousse et al., 2004). In this regime, electrons undergo transverse plasma oscillations in the plasma channel. Again, short radiation pulses are produced, but in a compact single-stage setup. This may have advantages, but a control of the radiation is difficult. To date, rather broadband radiation is produced, being used to diagnose the average electron motion in the plasma during the acceleration process (Phuoc et al., 2006; Kneip et al., 2008). For future applications, separated accelerator-radiator schemes are conceivable. However, in contrast to laser-driven FELs, no coherence will be possible.

5.3 Cancer therapy

Radiation therapy is a widely-used method in cancer treatment. It is commonly used for the treatment of tumours in the vicinity of risk organs, where surgery would be too difficult, and can be used as palliative means. Especially the use of high-energy ion beams has several advantageous aspects compared to Bremsstrahlung or electron beams. The absorption behaviour is such that the ions propagate with quasi-constant differential energy loss up to a point where the particle is already sufficiently slow and different interaction mechanisms lead to sudden stopping. This leads to well-defined penetration depths and the so-called Bragg peak at the end of the path: a maximum of energy deposition per unit propagation length. In addition, the beams experience only little side scattering and have thus a relative sharp contour. An implication for oncology is that the healthy tissue surrounding the tumour is exposed to much less dose if an ion beam is used. Furthermore it was found that the relative effectiveness of cell damage induced by ion beams, compared to Bremsstrahlung, is up to a factor of 6 higher in the Bragg peak.

Currently, there are only about 30 ion beam therapy (IBT) facilities worldwide in operation. This is on the one hand caused by the size of the accelerators in order to produce the required particle beams, e.g. ~ 200 MeV protons or ~ 1 GeV carbon ions. On the other hand, the additional necessary infrastructure to guide the beam from the accelerator to the patient is substantial and relies on large, heavy magnetic deflection devices, precisely movable around the patient. Both requires large buildings, shielding and high investments.

The promise of laser accelerators, then, is that the whole accelerator setup is significantly smaller than that of conventional devices, and thus the accelerator could in principle be located very close to the patient or even aligned and rotated around it. The main advantage would be that the only thing to be guided over longer distance is laser light, which means considerably less radiation hazard and hence reduces the shielding requirements. Besides that, the short and intense radiation pulses might have a different – and possibly better – radio-biological effectiveness. It is currently debated if the much higher peak dose rate induces DNA damage in a different way. Above that, one can image a kind of pump-probe scheme: A first pulse generates DNA damage in the cells. The damages are recognized by the cell and repair mechanisms are started. Now it could be possible to apply the second pulse at a time when the cell is vulnerable. But these are just ideas which have to be investigated.

Based in the reliable generation of electron beams from gas jet targets (see Sec. 3.3), we performed initial experimental investigation of biological radiation effects of laser-generated particle beams (Beyreuther et al., 2009). A first goal was to establish an experimental setup which allowed us to perform in-vitro irradiation of standard cell samples where the applied dose can be monitored simultaneously. This is a crucial feature for several reasons: The laser-plasma accelerator's performance can be measured, e.g. the amount of shot-to-shot fluctuations or transient drifts of the dose (cf. Sec. 3.3). Secondly, since online monitoring systems are implemented in all medical devices, it is reasonable to begin with the development at an early stage of innovation process. We used several dose detectors at the same time which rely on different effects, which allowed us to establish reliable dose measurements of ultra-short radiation pulses.

In a second step we conducted irradiation experiments of human cell samples. Two different standard cell lines, one for normal tissue and one for tumour tissue were irradiated. The dose was varied in order to investigate the cell response as function of dose. Two different kinds of biological response were investigated: The formation of DNA double strand breaks, induced by the irradiation (which is a primary response) and the total cell survival (which is a final goal but may be different to primary DNA damage due to DNA repair mechanisms). For methodical reasons, cell samples were also irradiated with cw Bremsstrahlung from a reference source. Future work will extend these investigation to more cell lines and to the irradiation with laser-accelerated ion beams. For a more detailed review, please see (Beyreuther et al., 2009).

5.4 Fast ignition

For future energy production, only nuclear fusion has sufficient long-term capabilities. This may be either the sun itself or man-made nuclear fusion. For the latter, two main technologies are know: magnetic confinement and inertial confinement. For inertial confinement fusion, e.g. see the textbook (Atzeni & Meyer-ter-Vehn, 2004), up to hundreds of energetic (multi-kJ) long-pulse (ns) lasers are focused onto a hollow fuel capsule of a few mm diameter. A plasma corona is created where particles are heated. The hot expanding plasma pushes (due to momentum conservation) the fuel capsule into itself, it implodes. When the shell reaches its centre, it gets compressed and heated. In the "pure" ICF regime, pressure, temperature and duration of being compressed suffice for ignition of the fusion process.

This is similar to a Diesel engine. The demand is to achieve high compression in a very symmetric manner. However, the coronal plasma shows instabilities which complicate the compression. A further option is to ignite the compressed and heated core by external means, like the fuel vapour in a petrol engine with spark-ignition. This is called fast ignition (FI).

For fast ignition, several possibilities are debated and will be investigated in the next years: Heating by laser-produced electron beams, heating by laser-produced light ions, heating by laser-driven shock waves or heating by heavy-ion beams (from conventional accelerators). Ignition by laser-accelerated electron beams demands very contrary beam characteristics as radiation generation (cf. Sec. 5.2). The goal is to generate a huge number of electrons with energies below a few MeV in a small beam. Electrons with higher energies would not deposit all their energy in the small compressed core, hence initial laser energy would be wasted. But since the total energy to be deposited is fixed, the number of electrons is the only free parameter. For this reason, LWFA-based schemes do not seem to be applicable.

One route is to accelerate electrons directly from the coronal plasma. Problems arise from the quite long distance from the acceleration zone to the core. Electron beams would have to be well-collimated in order not to miss the core, and much electron beam energy will be absorbed in the coronal plasma. Hence it is desired to bring the driving laser closer to the core. A principle option is using a very intense laser pulse of PW power which can push the coronal plasma aside (hole-boring). Another possibility is to attach a solid cone to the fuel capsule. This cone endures the compression phase due to its mass inertia. Inside the cone, there is no coronal plasma. Then, a short laser pulse is focused into the cone where an electron beam is generated. However, like for ion acceleration, a preplasma filling of the cone inhibits efficient electron acceleration (Baton et al., 2008). This demand for high-contrast laser pulses will be even more difficult to satisfy than for TNSA, since the required pulse energies are higher and the pulses longer. For ion-driven FI, there are even more open questions. Based on the results with a PW laser (Snavely et al., 2000) where ~ 10 % of laser energy were transferred into a proton beam (50 J in the particle beam, from 500 J in the laser pulse), several concepts have been developed. However, and this is a general problem also for electron-driven FI, sufficiently powerful laser systems are not yet ready or have performance problems like a poor temporal contrast which generates a preplasma and attenuates the interaction.

6. Acknowledgements

Part of this work was supported by the Deutsche Forschungsgemeinschaft under contract TR18, by the Access to Research Infrastructures activity in the Sixth Framework Programme of the EU (contract RII3-CT-2003-506350, Laserlab Europe) and by the German Ministry of Education and Research (BMBF) under contracts 03ZIK052 (ultraoptics) and 03ZIK445 (onCOOPtics). We thank B. Beleites, F. Ronneberger and W. Ziegler for their excellent technical support with the experimental studies at the JETI laser. We thank all coworkers and collaborators for their support within the experiments, especially Dr. Kay-Uwe Amthor, Dr. Enrico Brunetti, Fabian Budde, Dr. Volker Dangendorf, Alexander Debus, Dr. Hans-Jörg Fuchs, Dr. Jordan Gallacher, Waltraud Gräf, Kerstin Haupt, Ronald Lauck, Dr. Ben Liesfeld, Maria Nicolai, Jens Polz, Andreas Sachtleben, Richard Shanks, Kai Tittelmeier and Dr. Mark Wiggins. We thank Prof. Timur Esirkepov, Dr. Paul Gibbon, Prof. Dino Jaroszynski, Prof. Ken Ledingham, Prof. Gerhard G. Paulus, Dr. Alex Robinson, Prof. Erich Rohwer, Prof. Roland Sauerbrey and Prof. Heinrich Schwoerer for their support and fruitful discussions.

7. References

Alfvén, H. (1939). On the Motion of Cosmic Rays in Interstellar Space. *Physical Review*, Vol. 55, pp. 425–429

Arduini, G.; Baudrenghien, P.; Bohl, T.; Collier, P.; Cornelis, K.; Höfle, W.; Linnecar, T.; Shaposhnikova, E.; Tückmantel, J. & Wenninger, J. (2003). The LHC Proton Beam in the CERN SPS: an Update. *Proceedings of the 2003 Particle Accelerator Conference.*, pp. 1718–1720, accelconf.web.cern.ch/accelconf/p03/PAPERS/TPPB050.pdf

Atzeni, S. & Meyer-ter-Vehn, J. (2004). *The physics of inertial fusion*, Oxford University Press, ISBN 0198562640, Oxford

Baton, S.; Koenig, M.; Fuchs, J.; Benuzzi-Mounaix, A.; Guillou, P.; Loupias, B.; Vinci, T.; Gremillet, L.; Rousseaux, C.; Drouin, M.; Lefebvre, E.; Dorchies, F.; Fourment, C.; Santos, J. J.; Batani, D.; Morace, A.; Redaelli, R.; Nakatsutsumi, M.; Kodama, R.; Nishida, A.; Ozaki, N.; Norimatsu, T.; Aglitskiy, Y.; Atzeni, S. & Schiavi, A. (2008). Inhibition of fast electron energy deposition due to preplasma filling of cone-attached targets. *Physics of Plasmas*, Vol. 15, p. 042706

Beyreuther, E.; Enghardt, W.; Kaluza, M.; Karsch, L.; Laschinsky, L.; Lessmann, E.; Nicolai, M.; Pawelke, J.; Richter, C.; Sauerbrey, R.; Schlenvoigt, H.-P. & Baumann, M. (2009).

Establishment of technical prerequisites for cell irradiation experiments with laser accelerated electrons. *Medical Physics*, accepted for publication

Brunel, F. (1987). Not-so-resonant, Resonant Absorption. *Physical Review Letters*, Vol. 59, p. 52

Cowan, T.; Fuchs, J.; Ruhl, H.; Kemp, A.; Audebert, P.; Roth, M.; Stephens, R.; Barton, I.; Blazevic, A.; Brambrink, E.; Cobble, J.; Fernández, J.; Gauthier, J.-C.; Geissel, M.; Hegelich, M.; Kaae, J.; Karsch, S.; Le Sage, G. P.; Letzring, S.; Manclossi, M.; Meyroneinc, S.; Newkirk, A.; Pépin, H. & Renard-LeGalloudec, N. (2004). Ultralow Emittance, Multi-MeV Proton Beams from a Laser Virtual-Cathode Plasma Accelerator. *Physical Review Letters*, Vol. 92, p. 204801

Esarey, E.; Sprangle, P.; Krall, J. & Ting, A. (1996). Overview of plasma-based accelerator concepts. *IEEE Transactions on Plasma Science*, Vol. 24, No. 2, pp. 252–288

Esirkepov, T.; Bulanov, S.; Nishihara, K.; Tajima, T.; Pegoraro, F.; Khoroshkov, V.; Mima, K.; Daido, H.; Kato, Y.; Kitagawa, Y.; Nagai, K. & Sakabe, S. (2002). Proposed doublelayer target for the generation of high-quality laser-accelerated ion beams. *Physical Review Letters*, Vol. 89, p. 175003

Faure, J.; Rechatin, C.; Norlin, A.; Lifschitz, A.; Glinec, Y. & Malka, V. (2006). Controlled injection and acceleration of electrons in plasma wakefields by colliding laser pulses. *Nature*, Vol. 444, pp. 737–739

Fuchs, J.; Sentoku, Y.; Karsch, S.; Cobble, J.; Audebert, P.; Kemp, A.; Nikroo, A.; Antici, P.; Brambrink, E.; Blazevic, A.; Campbell, E.M.; Fernandez, J.C.; Gauthier, J.-C.; Geissel, M.; Hegelich, M.; Pepin, H.; Popescu, H.; Renard-LeGalloudec, N.; Roth, M.; Schreiber J.; Stephens, R. & Cowan, T (2005). Comparison of laser ion acceleration from the front and rear surfaces of thin foils. *Physical Review Letters*, Vol. 94, p.045004

Fuchs, J; Antici, P; d'Humieres, E; Lefebvre, E; Borghesi, M; Brambrink, E; Cecchetti, C.A.; Kaluza, M; Malka, V; Manclossi, M; Meyroneinc, S; Mora, P; Schreiber, J; Toncian, T; Pepin, H & Audebert, P. (2006). Laser-driven proton scaling laws and new paths towards energy increase. *Nature Physics*, Vol. 2, No. 1, pp. 48–54

Fuchs, J.; Cecchetti, C.A.; Borghesi, M.; Grismayer, T.; d'Humieres, E.; Antici, P.; Atzeni, S.; Mora, P.; Pipahl, A.; Romagnani, L.; Schiavi, A.; Sentoku, Y.; Toncian, T.; Audebert, P. & Willi, O. (2007). Laser-foil acceleration of high-energy protons in small-scale plasma gradients. *Physical Review Letters*, Vol. 99, No. 23, p. 015002

Gahn, C.; Tsakiris, G.; Pukhov, A.; Meyer-ter-Vehn, J.; Pretzler, G.; Thirolf, P.; Habs, D. & Witte, K. (1999). Multi-MeV electron beam generation by direct laser acceleration in high-density plasma channels. *Physical Review Letters*, Vol. 83, No. 23, pp. 4772–4775

Geddes, C.; Toth, Cs.; van Tilborg, J.; Esarey, E.; Schroeder, C.; Bruhwiler, D.; Nieter, C.; Cary, J. & Leemans, W. (2005). Production of high-quality electron bunches by dephasing and beam loading in channeled and unchanneled laser plasma accelerators. *Physics of Plasmas*, Vol. 12, p. 056709

Geddes, C.; Nakamura, K.; Plateau, G.; Toth, Cs.; Cormier-Michel, E.; Esarey, E.; Schroeder, C.; Cary, J. & Leemans,W. (2008). Plasma-Density-Gradient Injection of Low Absolute-Momentum-Spread Electron Bunches. *Physical Review Letters*, Vol. 100, p. 215004

Gibbon, P. (2005). *Short pulse laser interactions with matter*, Imperial College Press, ISBN 1860941354, London

Gordienko, S. & Pukhov, A. (2005). Scalings for ultrarelativistic laser plasmas and quasi-monoenergetic electrons. *Physics of Plasmas*, Vol. 12, p. 043109

Hegelich, M.; Karsch, S.; Pretzler, G.; Habs, D.; Witte, K.; Guenther, W.; Allen, M.; Blazevic, A.; Fuchs, J.; Gauthier, J.-C.; Geissel, M.; Audebert, P.; Cowan, T. & Roth, M. (2002).

MeV ion jets from short-pulse-laser interaction with thin foils. *Physical Review Letters*, Vol. 89, p. 085002

Hegelich, M.; Albright, B. J.; Cobble, J.; Flippo, K.; Letzring, S.; Paffett, M.; Ruhl, H.; Schreiber, J.; Schulze, R. K. & Fernandez, J. C. (2006). Laser acceleration of quasi-monoenergetic MeV ion beams. *Nature*, Vol. 439, pp. 441–444

Henig, A.; Kiefer, D.; Markey, K.; Gautier, D.C.; Flippo, K.A.; Letzring, S.; Johnson, R.P.; Shimada, T.; Yin, L.; Albright, B.J.; Bowers, K.J.; Fernandez, J.C.; Rykovanov, S.G.; Wu, H.C.; Zepf, M.; Jung, D.; Liechtenstein, V.K.; Schreiber, J.; Habs, D. & Hegelich, M. (2009). Enhanced Laser-Driven Ion Acceleration in the Relativistic Transparency Regime. *Physical Review Letters*, Vol. 103, p. 045002

Hidding, B.; Amthor, K.-U.; Liesfeld, B.; Schwoerer, H.; Karsch, S.; Geissler, M.; Veisz, L.; Schmid, K.; Gallacher, J. ; Jamison, S. ; Jaroszynski, D.; Pretzler, G. & Sauerbrey, R. (2006). Generation of quasimonoenergetic electron bunches with 80-fs laser pulses. *Physical Review Letters*, Vol. 96, p. 105004

Honrubia, J.J.; Kaluza, M.; Schreiber, J.; Tsakiris, G. D. & Meyer-ter-Vehn, J. (2005). Laserdriven fast-electron transport in preheated foil targets. *Physics of Plasmas*, Vol. 12, p. 052708

Jäckel, Oliver (2009). *Characterization of ion acceleration with relativistic laser-plasma*, PhD thesis, IOQ Jena, www.physik.uni-jena.de/ioq

Kaluza, M.; Schreiber, J.; Santala, M.I.K.; Tsakiris, G.; Eidmann, K.; Meyer-ter-Vehn, J. & Witte, K.-J. (2004). Influence of the laser prepulse on proton acceleration in thin-foil experiments. *Physical Review Letters*, Vol. 93, p. 045003

Karsch, S.; Osterhoff, J.; Popp, A.; Rowlands-Rees, T.; Major, Z.; Fuchs, M.; Marx, B.; Horlein, R.; Schmid, K.; Veisz, L.; Becker, S.; Schramm, U.; Hidding, B.; Pretzler, G.; Habs, D.; Grüner, F.; Krausz, F. % Hooker, S. (2007). GeV-scale electron acceleration in a gas-filled capillary discharge waveguide. *New Journal of Physics*, Vol. 9, p. 415

Kneip, S.; Nagel, S.; Bellei, C.; Bourgeois, N.; Dangor, A. E.; Gopal, A.; Heathcote, R.; Mangles, S.; Marques, J.-R.; Maksimchuk, A.; Nilson, P.M.; Phuoc, K.T.; Reed, S.; Tzoufras, M.; Tsung, F.; Willingale, L.; Mori, W.; Rousse, A.; Krushelnick, K. & Najmudin, Z. (2008). Observation of synchrotron radiation from electrons accelerated in a petawatt-laser generated plasma cavity. *Physical Review Letters*, Vol. 100, p. 105006

Kruer, W.L. & Estabrook, K. (1985). Jxb Heating By Very Intense Laser-Light. *Physics of Fluids*, Vol. 28, pp. 430–432

Kruer, W.L. (1988). *The physics of laser plasma interactions*, Westview Press, ISBN 0813340837, Boulder, CO, USA

Lau, Y.; He, F.; Umstadter, D. & Kowalczyk, R. (2003). Nonlinear Thomson scattering: A tutorial. *Physics of Plasmas*, Vol. 10, No. 5, pp. 2155–2162

Leemans, W.; Nagler, B.; Gonsalves, A.; Toth, Cs; Nakamura, K.; Geddes, C.; Esarey, E.; Schroeder, C. & Hooker, S. (2006). GeV electron beams from a centimetre-scale accelerator. *Nature Physics*, Vol. 2, No. 10, pp. 696–699

Malka, V.; Faure, J.; Glinec, Y.; Pukhov, A. & Rousseau, J. P. (2005). Monoenergetic electron beam optimization in the bubble regime. *Physics of Plasmas*, Vol. 12, p. 056702

Mangles, S.; Murphy, C.; Najmudin, Z.; Thomas, A.; Collier, J.; Dangor, A. E.; Divall, E. J.; Foster, P.; Gallacher, J.; Hooker, C. J.; Jaroszynski, D.; Langley, A. J.; Mori, W.; Norreys, P.; Tsung, F.; Viskup, R.; Walton, B. & Krushelnick, K. (2004). Monoenergetic beams of relativistic electrons from intense laser-plasma interactions. *Nature*, Vol. 431, pp. 535–538 (see also adjacent articles pp. 538–541 and pp. 541–544)

Meyer-ter-Vehn, J.; Pukhov, A. & Sheng, Zh.-M. (2001). Relativistic Laser Plasma Interaction, In: *Atoms, Solids, and Plasmas in Super-Intense Laser Fields*, Batani, D.; Joachain, Ch. J., Martellucci, S. & Chester, A. N. (Ed.), pp. 167–192, Kluwer Academics / Plenum Publishers, ISBN 0306466155, New York

Modena, A.; Najmudin, Z.; Dangor, A.E.; Clayton, C.; Marsh, K.A.; Joshi, C.; Malka, V.; Darrow, C. B.; Danson, C.; Neely, D. & Walsh, F.N. (1995). Electron Acceleration From The Breaking Of Relativistic Plasma-Waves. *Nature*, Vol. 377, pp. 606–608

Mora, P. (2003). Plasma expansion into a vacuum. *Physical Review Letters*, Vol. 90, p. 185002

Mori, W. (1997). The Physics of the Nonlinear Optics of Plasmas at Relativistic Intensities for Short-Pulse Lasers. *IEEE Journal of Quantum Electronics*, Vol. 33, No. 11, pp. 1942–1953

Osterhoff, J.; Popp, A.; Major, Z.; Marx, B.; Rowlands-Rees, T.; Fuchs, M.; Geissler, M.; Hoerlein, R.; Hidding, B.; Becker, S.; Peralta, E. A.; Schramm, U.; Gruener, F.; Habs, D.; Krausz, F.; Hooker, S. & Karsch, S. (2008). Generation of stable, lowdivergence electron beams by laser-wakefield acceleration in a steady-state-flow gas cell. *Physical Review Letters*, Vol. 101, p. 085002

Pfotenhauer, S.; Jäckel, O.; Sachtleben, A.; Polz, J.; Ziegler, W.; Schlenvoigt, H.-P.; Amthor, K.- U.; Kaluza, M.; Ledingham, K.; Sauerbrey, R.; Gibbon, P.; Robinson, A. & Schwoerer, H. (2008). Spectral shaping of laser generated proton beams *New Journal of Physics*, Vol. 10, p. 033034

Pfotenhauer, Sebastian (2009). *Generation of monoenergetic ion beams with a laser accelerator*, PhD thesis, IOQ Jena, www.physik.uni-jena.de/ioq

Phuoc, K.T.; Corde, S.; Shah, R.; Albert, F.; Fitour, R.; Rousseau, J.P.; Burgy, F.; Mercier, B. & Rousse, A. (2006). Imaging electron trajectories in a laser-wakefield cavity using betatron x-ray radiation. *Physical Review Letters*, Vol. 97, p. 225002

Pukhov, A. & Meyer-ter-Vehn, J. (2002). Laser wake field acceleration: the highly non-linear broken-wave regime. *Applied Physics B*, Vol. 74, No. 4-5, pp. 355–361

Robinson, A.; Bell, A. R. & Kingham, R. (2006). Effect of target composition on proton energy spectra in ultraintense laser-solid interactions. *Physical Review Letters*, Vol. 96, p. 035005

Robinson, A. & Gibbon, P. (2007). Production of proton beams with narrow-band energy spectra from laser-irradiated ultrathin foils. *Physical Review E*, Vol. 75, p. 015401

Robinson, A.; Gibbon, P.; Pfotenhauer, S.; Jäckel, O. & Polz, J. (2009). Scaling of the proton density reduction scheme for the laser acceleration of proton beams with a narrow energy spread. *Plasma Physics and Controlled Fusion*, Vol. 51, p. 024001

Robson, L.; Simpson, P. T.; Clarke, R. J.; Ledingham, K.; Lindau, F.; Lundh, O.; McCanny, T.; Mora, P.; Neely, D.; Wahlström, C.-G.; Zepf, M. & McKenna, P. (2007). Scaling of proton acceleration driven by petawatt-laser-plasma interactions. *Nature Physics*, Vol. 3, pp. 58–62

Rousse, A.; Phuoc, K.T.; Shah, R.; Pukhov, A.; Lefebvre, E.; Malka, V.; Kiselev, S.; Burgy, F.; Rousseau, J.P.; Umstadter, D. & Hulin, D. (2004). Production of a keV x-ray beam from synchrotron radiation in relativistic laser-plasma interaction. *Physical Review Letters*, Vol. 93, p. 135005

Rowlands-Rees, T.; Kamperidis, C.; Kneip, S.; Gonsalves, A.; Mangles, S.; Gallacher, J.; Brunetti, E.; Ibbotson, T.; Murphy, C.; Foster, P.; Streeter, M.J.V.; Budde, F.; Norreys, P.; Jaroszynski, D.; Krushelnick, K.; Najmudin, Z. & Hooker, S. (2008). Laser-driven acceleration of electrons in a partially ionized plasma channel. *Physical Review Letters*, Vol. 100, p. 105005

Schlenvoigt, H.-P.; Haupt, K.; Debus, A.; Budde, F.; Jäckel, O.; Pfotenhauer, S.; Schwoerer, H.; Rohwer, E.; Gallacher, J. ; Brunetti, E.; Shanks, R. ; Wiggins, M. & Jaroszynski, D. (2008). A compact synchrotron radiation source driven by a laser-plasma wakefield accelerator. *Nature Physics*, Vol. 4, No. 2, pp. 130–133

Schlenvoigt, Hans-Peter (2009). *Synchrotron Radiation Sources driven by Laser Plasma Acclerators*, PhD thesis, IOQ Jena, www.physik.uni-jena.de/ioq

Schmid, K.; Veisz, L.; Tavella, F.; Benavides, S.; Tautz, R.; Herrmann, D.; Buck, A.; Hidding, B.; Marcinkevicius, A.; Schramm, U.; Geissler, M.; Meyer-ter-Vehn, J.; Habs, D. & Krausz, F. (2009). Few-Cycle Laser-Driven Electron Acceleration. *Physical Review Letters*, Vol. 102, p. 124801

Schreiber, J.; Bell, F.; Grüner, F.; Schramm, U.; Geissler, M.; Schnürer, M.; Ter-Avetisyan, S.; Hegelich, M.; Cobble, J.; Brambrink, E.; Fuchs, J.; Audebert, P. & Habs, D. (2006). Analytical model for ion acceleration by high-intensity laser pulses. *Physical Review Letters*, Vol. 97, p. 045005

Schwoerer, H.; Pfotenhauer, S.; Jäckel, O.; Amthor, K.-U.; Liesfeld, B.; Ziegler, W.; Sauerbrey, R.; Ledingham, K. & Esirkepov, T. (2006). Laser-plasma acceleration of quasimonoenergetic protons from microstructured targets *Nature*, Vol. 439, pp. 445–448

Sentoku, Y.; Cowan, T.; Kemp, A. & Ruhl, H. (2003). High energy proton acceleration in interaction of short laser pulse with dense plasma target. *Physics of Plasmas*, Vol. 10, No. 5, pp. 2009-2015

Snavely, R.A.; Key, M.H.; Hatchett, S.P.; Cowan, T.; Roth, M.; Phillips, T.W.; Stoyer, M.A.; Henry, E.A.; Sangster, T.C.; Singh, M.S.; Wilks, S.C.; MacKinnon, A.; Offenberger, A.; Pennington, D.M.; Yasuike, K.; Langdon, A.B.; Lasinski, B.F.; Johnson, J.; Perry, M.D. & Campbell, E.M. (2000). Intense High-Energy Proton Beams from Petawatt-Laser Irradiation of Solids. *Physical Review Letters*, Vol. 85, No. 14, pp. 2945–2948

Strickland, D. & Mourou, G. (1985). Compression of amplified chirped optical pulses. *Optics Communications*, Vol. 56, No. 3, pp. 219–221 Tajima, T. & Dawson, J. (1979). Laser electron-accelerator. *Physical Review Letters*, Vol. 43, No. 4, pp. 267–270

Ter-Avetisyan, S.; Schnürer, M.; Nickles, P.; Kalashnikov, M.; Risse, E.; Sokollik, T.; Sandner, W.; Andreev, A. & Tikhonchuk, V. (2006). Quasimonoenergetic deuteron bursts produced by ultraintense laser pulses. *Physical Review Letters*, Vol. 96, No. 14, p. 145006

Thomas, A.; Mangles, S.; Najmudin, Z.; Kaluza, M.; Murphy, C. & Krushelnick, K. (2007). Measurements of wave-breaking radiation from a laser-wakefield accelerator. *Physical Review Letters*, Vol. 98, p. 054802

Tikhonchuk, V.T.; Andreev, A.A.; Bochkarev, S.G. & Bychenkov, V.Y. (2005). Ion acceleration in short-laser-pulse interaction with solid foils. *Plasma Physics and Controlled Fusion*, Vol. 47, pp. B869-B877

van Tilborg, J.; Schroeder, C.; Filip, C.; Toth, Cs.; Geddes, C.; Fubiani, G.; Huber, R.; Kaindl, R.; Esarey, E. & Leemans, W. (2006). Temporal Characterization of Femtosecond Laser-Plasma-Accelerated Electron Bunches Using Terahertz Radiation. *Physical Review Letters*, Vol. 96, p. 014801

Toncian, T.; Borghesi, M.; Fuchs, J.; d'Humieres, E.; Antici, P.; Audebert, P.; Brambrink, E.; Cecchetti, C.A.; Pipahl, A.; Romagnani, L. & Willi, O. (2006). Ultrafast laser-driven microlens to focus and energy-select mega-electron volt protons. *Science*, Vol. 312, pp. 410–413

25

Laser-Driven Proton Acceleration Research and Development

Alexander S. Pirozhkov, Hiroyuki Daido,
Mamiko Nishiuchi and Koichi Ogura
*Advanced Photon Research Center and Photo-Medical Research Center,
Japan Atomic Energy Agency
Japan*

1. Introduction

Up to now the acceleration of charged particles in most cases is based on the radio-frequency based technology. Another approach is to employ ultrahigh collective fields of plasmas produced by high-power lasers. Using various schemes, these plasma fields can be used to accelerate electrons, protons, and heavier ions. Laser-driven ion beams exhibit unique properties, among which are short duration, small source size, large number of particles, low emittance, etc. The laser-driven acceleration is inherently a very compact process, because the acceleration typically takes place within a micrometer spatial and picosecond time scales; the acceleration field can exceed TV/m, which allows achieving multi-MeV/nucleon ion energies. For the detailed reviews, see (Borghesi et al. 2006; Mourou et al. 2006) and references therein. The peculiar properties of the laser-driven proton beam, namely short duration, low emittance, and large particle number have been used in the proton imaging, which allows measurement of transient field distribution in plasma experiments (Borghesi et al. 2001); the broad proton energy spectrum allows multi-frame time-resolved studies in single shot. A compact laser-based laboratory setup has been used for the proof-of-principle radiological experiment (Yogo et al. 2009), which can be applicable in the biological, medical, and space research (Murakami et al. 2009). There are many other attractive potential applications of the laser-driven ion beams, among which the hadron therapy (Bulanov et al. 2002; Bulanov & Khoroshkov 2002), fast ignition (Roth et al. 2001), injectors into conventional accelerators (Krushelnick et al. 2000; Cowan et al. 2004), material processing, isotope production (Nemoto et al. 2001; Fritzler et al. 2003), pion production (Bychenkov et al. 2001), pump-probe experiments using simultaneous production of proton beam and THz radiation (Sagisaka et al. 2008), x-rays (Orimo et al. 2007), electrons (Li et al. 2006), and so on.

For many applications, the parameters of the presently available laser-driven ion beams need to be improved. In particular, it is necessary to increase the maximum ion energy, simultaneously achieving high conversion efficiency. At present, this is an active area of research. Up to now a substantial part of results has been obtained with relatively large-scale, single-shot picosecond and sub-picosecond laser systems (Hatchett et al. 2000; Snavely et al. 2000; Fuchs et al. 2006; Robson et al. 2007; Henig et al. 2009). However, for many applications a compact, repetitive ion source is necessary. For such applications, it is attractive to use high-power femtosecond lasers, which have smaller pulse energies at the same power levels, and therefore can provide higher repetition rates. In this Chapter we present the results of proton acceleration and plasma diagnostic experiments (Pirozhkov et al. 2009c) conducted with a repetitive femtosecond J-KAREN laser at the Advanced Photon Research Center, Japan Atomic Energy Agency. In Sections 2 and 3 we describe the laser parameters and the results of the experiments and discuss the influence of several parameters on the maximum proton energy; we show the strong influence of the laser contrast on the proton acceleration process. In Section 4 we analyze the results of the on-target laser contrast measurement, which has been performed at full laser power using the properties of radiation reflected from the target. In Section 5 we describe the laser-driven proton beam manipulation using permanent magnet quadrupoles. In Section 6 we describe the radioactivation experiments using multi-MeV laser-driven proton source.

2. Lasers for proton acceleration: present status and future prospects

2.1 High-power picosecond and femtosecond lasers

Up to now, the highest proton energies, the largest proton numbers and conversion efficiencies have been obtained using high-energy ps and sub-ps lasers. The proton energies close to 60 MeV (Hatchett et al. 2000; Snavely et al. 2000; Robson et al. 2007) have been achieved using petawatt lasers based on the Chirped Pulse Amplification (CPA) technology (Strickland & Mourou 1985). In these experiments, thin foil targets were irradiated by the focused laser pulses with intensities exceeding 10^{20} W/cm^2; regardless of the target material, the accelerated ions were predominantly protons originating from the water or hydrocarbon surface contamination. If the contamination was removed e.g. by joule or laser heating, heavier ions had been accelerated (Hegelich et al. 2002; McKenna et al. 2004). The acceleration has been attributed to the process known as Target Normal Sheath Acceleration (TNSA) (Hatchett et al. 2000; Mora 2003).

The experiments in which multi-10 MeV/nucleon protons and ions were accelerated within very short acceleration distances have caused great interest to the field of laser-driven ion acceleration. One of the attractive directions of research is miniaturization of the accelerator employing compact, relatively high-repetition rate CPA lasers with even shorter (femtosecond) pulse durations; the compactness and higher repetition rate can be achieved in this case because of the smaller pulse energy required to provide similar peak power and intensity. However, at present the achievable proton energies in this case are lower. Despite this, a number of application and proof-of-principle experiments have been performed with the ion and proton beams driven by compact femtosecond lasers. Ion energies up to 10-20 MeV/nucleon (Fukuda et al. 2009) and the laser-to-proton beam conversion efficiencies up to 3% (proton energy > 0.8 MeV) (Nishiuchi et al. 2008) have been demonstrated.

2.2 J-KAREN laser

The experiments described in this Chapter (Pirozhkov et al. 2009c) have been conducted using hybrid OPCPA/Ti:Sapphire J-KAREN laser (Kiriyama et al. 2008; Kiriyama et al. 2009); here OPCPA stands for the Optical Parametric Chirped Pulse Amplification. The laser consists of a CPA oscillator with high pulse energy, which facilitates achieving higher contrast, saturable absorber, stretcher, two- or three-stage OPCPA based on type I BBO crystals, two 4-pass Ti:Sapphire amplifiers, and vacuum compressor. The cryogenic cooling of the final amplifier (down to 100 K) removes thermal lensing, which allows 10 Hz operation with high laser beam quality. The saturable absorber efficiently reduces the Amplified Spontaneous Emission (ASE) after the CPA oscillator, which is crucial for the solid target irradiation experiments, including the ion acceleration (see Sections 3 and 4). The laser provides ~30 fs, ~ 1 J pulses at the wavelength of ≈ 820 nm with the nanosecond contrast higher than 10^{10}. Employing an f/3 Off-Axis Parabola (OAP) mirror, the pulses are focused down to a 3-4 μm spot (Full Width at Half Maximum, FWHM). The short pulse duration and small focal spot allow achieving the peak irradiance of up to 10^{20} W/cm^2.

The temporal pulse shape on the femtosecond time scale (Fig. 1) was measured with the home-built Transient Grating Frequency-Resolved Optical Gating (TG FROG) system (Pirozhkov et al. 2008). The measured FWHM and effective widths of the pulses are τ_{FWHM} = 28 fs, τ_{eff} = 35 fs without and τ_{FWHM} = 35 fs, τ_{eff} = 47 fs with the saturable absorber. (The effective width is the ratio of the pulse energy to the peak power.) The pulse duration with the saturable absorber is somewhat larger, because the saturable absorber introduces some high-order dispersion into the spectral phase, which cannot be compensated by the compressor. The power shown in Fig. 1 is calculated from the FROG data using the measured on-target pulse energies. The on-target energies without/with the saturable absorber are E_L = 880±20 mJ and 720±20 mJ, respectively, which corresponds to the peak powers of $P_0 = E_L/\tau_{eff}$ = 25 TW and 15 TW.

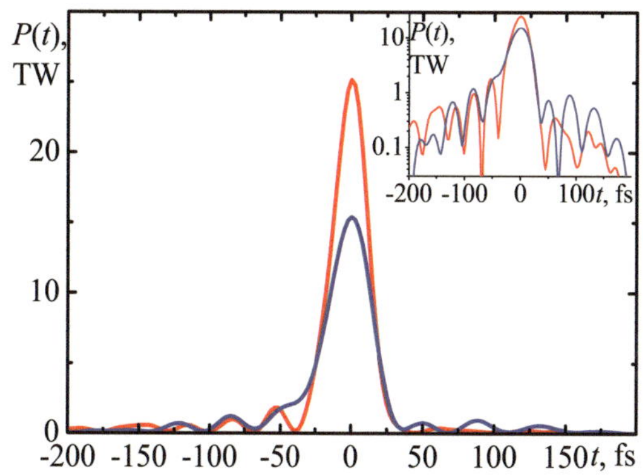

Fig. 1. Femtosecond pulse shape of J-KAREN laser measured with the Transient Grating Frequency-Resolved Optical Gating (TG FROG), the dispersion of the window and air has been subtracted. The red line corresponds to the case without the saturable absorber, the FWHM pulse duration τ_{FWHM} = 28 fs, effective pulse duration τ_{eff} = 35 fs (FROG error = 0.008, N = 128), on-target energy E_L = 880 mJ, peak power P_0 = 25 TW. The blue line corresponds to the case with the saturable absorber, τ_{FWHM} = 35 fs, τ_{eff} = 47 fs (FROG error = 0.002, N = 256), E_L = 720 mJ, P_0 = 15 TW. Inset, the same data in the log scale.

The intensity distribution of the laser pulses focused by the f/3 OAP was measured by a microscope objective at full laser power operation; the pulses were attenuated using reflection from wedges and with neutral-density filters. Without the saturable absorber, the focal spot has an elliptical shape with side lobes [Fig. 2 (a)]. The major and minor axes of the ellipse have the FWHM sizes of 5.7±0.7 μm and 2.28±0.11 μm, with the effective radius of r_{eff} = 2.82±0.10 μm (the errors are standard deviations of shot-to-shot fluctuations). With the saturable absorber, the shape of the focal spot is nearly circular [Fig. 2 (b)], with the FWHM of 3.3±0.3 μm and r_{eff} = 2.73±0.12 μm. The derived peak intensity is $I_0 = P_0/(\pi r_{eff}^2) = 1.0\times10^{20}$ W/cm^2 without and 0.7×10^{20} W/cm^2 with the saturable absorber.

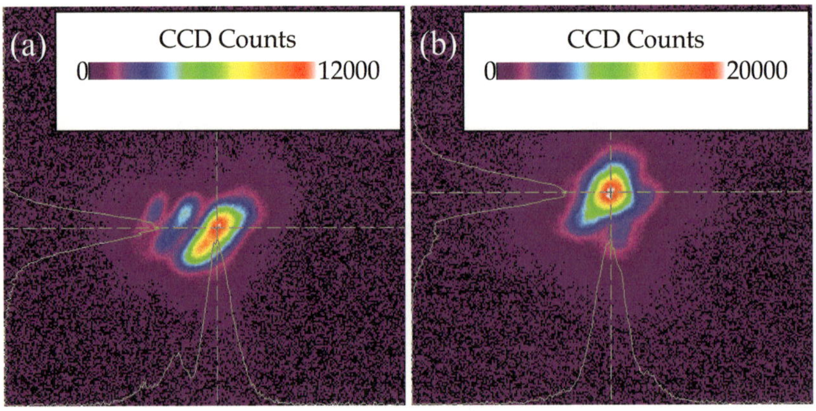

Fig. 2. Focal spot of J-KAREN laser focused with the *f*/3 off-axis parabola. The spot is measured with a microscope objective at full power mode, the laser beam is attenuated by wedges and filters. (a) The case without the saturable absorber, the spot size FWHM = 5.7 μm × 2.3 μm (major and minor axes of the fitted ellipse), effective radius r_{eff} = 2.8 μm, derived peak irradiance $I_0 = 1.0\times10^{20}$ W/cm^2. (b) The case with the saturable absorber, FWHM = 3.3 μm (nearly circular), r_{eff} = 2.7 μm, $I_0 = 0.7\times10^{20}$ W/cm^2.

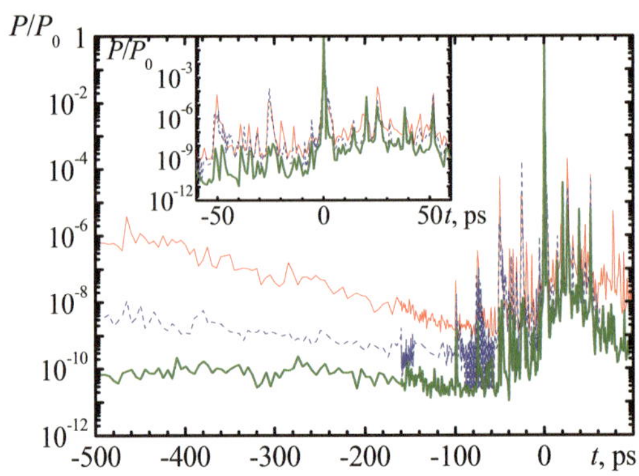

Fig. 3. Sub-nanosecond and picosecond contrast of J-KAREN laser in the three operation modes measured with the 3rd order cross-correlator: the thin red line is for the case of CPA oscillator, 3-stage OPCPA, and two multi-pass amplifiers (no saturable absorber). The blue dashed line is for the similar case but with the saturable absorber installed after the CPA oscillator. The thick green line is the same as the blue dashed line, but 2-stage (nonsaturated) OPCPA is used. The actual nanosecond contrast is ~ 2-3 times higher than shown in the figure due to the insufficient temporal resolution of the cross-correlator (~ 120 fs); the estimated actual contrasts for the three cases are ~ 5×10^6, ~ 5×10^8, and ~ 5×10^{10}. The inset shows the same data in more details within the time range of ± 60 ps.

The contrast in different operation modes has been measured with the high-dynamic range 3-rd order cross-correlator (Fig. 3). The saturable absorber improves the contrast by a factor of ~100. A similar improvement has been achieved by non-saturated OPCPA operation (two OPCPA stages instead of three) (Kiriyama et al. 2008).

3. Proton acceleration experiments with multi-10-TW J-KAREN laser

3.1 Experimental setup

A typical schematic layout of the experiments is shown in Fig. 4. The OAP focused p-polarized laser pulses onto a tape or ribbon targets at the incidence angle of 45°. The tape target (Nayuki et al. 2003) supplied fresh surface for each shot, which allowed taking advantage of the high-repetition rate laser operation. The tapes were made from polyimide with the thicknesses of 7.5 μm and 12.5 μm. The ribbon targets were made from metal (Al, Au, Pt, Pd, etc.) with the thicknesses down to 200 nm.

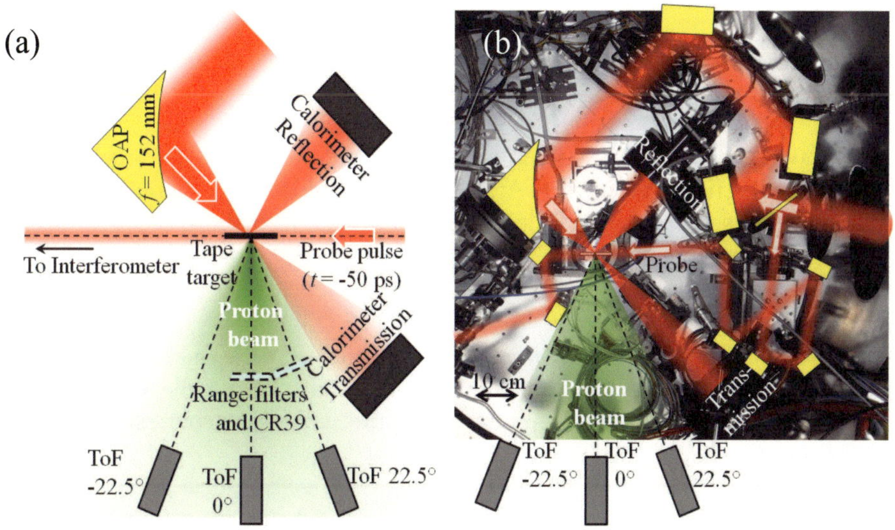

Fig. 4. Experimental setup. The p-polarized multi-10 TW laser pulses are focused by the *f*/3 off-axis parabola (OAP) to the irradiances up to 10^{20} W/cm^2, the incidence angle on the target is 45°. The target can be a polyimide tape with the thickness of 7.5 μm or 12.5 μm as well as Al, Au, Pd, or Pt ribbon with the thickness from few μm down to 200 nm. The constantly moving tape targets allow the repetitive proton acceleration at 1 Hz. The protons originating from the surface contamination are accelerated at the rear side of the target, predominantly perpendicular to the target surface. The protons are detected with several instruments, including the absolutely calibrated Time-of-Flight (ToF) spectrometers at 0° and ±22.5° from the target normal (proton energy spectrum), CR-39 nuclear track detectors with the range filters (proton beam foot-print), and Thomson parabola with CR-39 detector (energy spectra of protons and other ions) (not shown). The laser-plasma interaction is studied with the interferometry and measurements of reflectivity and transmission. (a) A schematic layout. (b) A photo of the target chamber from the top with the superposed laser beams, mirrors, calorimeters, proton beam direction, and directions of observation with the Time-of-Flight spectrometers.

The protons were detected by the Time-of-Flight (ToF) spectrometers, CR-39 nuclear track detectors, and Thomson parabola. The ToF spectrometers (Nakamura et al. 2006; Yogo et al. 2007) were installed at the target normal direction and ±22.5° from it. In order to select the proton signal and reduce noise, the ToF spectrometers contained sweeping magnets removing all but the most energetic electrons, Al filters blocking laser and other light, and relatively thin plastic scintillator and the bent light guide that prevented the transmitted x-rays to enter the photomultiplier tube. The ToF spectrometers were absolutely calibrated with the conventional accelerator. The ToF spectrometers provide the proton spectrum immediately after the shot. This together with the constantly moving tape target allowed 1 Hz shooting, which has been used for quick optimization of the proton acceleration conditions. The CR-39 detectors with the range filters were used for the angular distribution and emittance measurements. Thomson parabola ion analyzer was used to check the contribution of other ion species.

To understand the laser-plasma interaction details, we used several plasma diagnostics. The probe beam split by a pellicle from the main pulse was used for shadowgraphy at 820 nm and interferometry at 820 and 410 nm (2ω) (Sagisaka et al. 2004; Sagisaka et al. 2006). To measure the laser energy reflected from and transmitted through the target, we used in-vacuum calorimeters. The calorimeter measuring the transmitted energy was used to measure the on-target laser pulse energy, when the target situated on a motorized 3-axis stage was moved out of the beam. Simultaneously, the third calorimeter measured leakage through one of the dielectric mirrors. After such cross-calibration, the on-target energy in each shot was derived from the dielectric mirror leakage, which allowed obtaining specular reflectivity and transmission.

3.2 Results of proton acceleration

We used a broad range of laser and target parameters to study the laser-driven acceleration properties and dependences on parameters. In order to achieve the maximum proton energies, a repetitive operation was employed to optimize the laser contrast, target position, pulse duration, etc.

The dependence of the maximum proton energy on the effective pulse duration is shown in Fig. 5. The pulse duration was varied by moving a grating in the compressor. The laser pulse shape was measured using the TG FROG (Fig. 1) (Pirozhkov et al. 2008). The spectrum and spectral phase measured with the TG FROG at some grating position allow calculating the shape of the laser pulse at any other grating position with high accuracy, as it was checked by direct comparison of the measured and calculated pulses. The effective pulse duration in Fig. 5 is therefore calculated from the TG FROG data obtained with the shortest pulse. The dependence of the maximum proton energy on the pulse duration is rather smooth, with the maximum at somewhat elongated pulse. This can be understood using the model of electrostatic acceleration of protons at the rear side of the target (Mora 2003; Fuchs et al. 2006) with the additional limit that the acceleration distance equals the sheath diameter (Robson et al. 2007). The dependence of the proton energy on the pulse duration obtained from this model (the dashed curve in Fig. 5) agrees well with the measurements.

The femtosecond pulses typically have a complicated structure in time. In particular, they have several kinds of preceding light, such as the ASE, prepulses, and picosecond pedestal. The preceding light can damage the target and create the preformed plasma well before the main pulse. To understand the influence of the preformed plasma on the proton acceleration, we employed the time-resolved interferometry using the femtosecond probe pulse (Sagisaka et al. 2004; Sagisaka et al. 2006). Fig. 6 shows the results of interferometry taken at 50 ps before the main pulse arrival. The ion spectra at the target normal direction recorded simultaneously with the interferograms shown in Fig. 6 are shown in Fig. 7. In the case of a relatively low contrast ($\sim 5\times10^6$), the preplasma with the size of few hundred μm was formed [Fig. 6 (a)], and the proton energy was 2.3 MeV. On the other hand, when the high contrast mode was used (~ 5×10^{10}), the preplasma was not detectable [Fig. 6 (c)], which means it was smaller than ~20-30 μm. The maximum proton energy in this case was 3.5 MeV. The smaller proton energy in the case of lower contrast can be explained by the rear surface disruption and/or bending (Mackinnon et al. 2001; Lindau et al. 2005). Interestingly, when we adjusted the duration of the ASE in the lower-contrast case (no saturable absorber, contrast ~5×10^{6}) by shifting the gating time of the Pockels Cell (PC) by 300 ps, the proton energy increased up to 4.1 MeV in some shots; the preplasma size was in this case intermediate [Fig. 6 (b)]. The proton energy higher than in the case with the saturable absorber can be understood by larger laser power and intensity, which were achieved without the saturable absorber, by increased absorption in the preformed plasma (Gibbon & Bell 1992; Borghesi et al. 1999; Ping et al. 2008; Pirozhkov et al. 2009a), and by the laser pulse self-focusing in the preplasma with the optimum scale length (Bychenkov et al. 2001; McKenna et al. 2008).

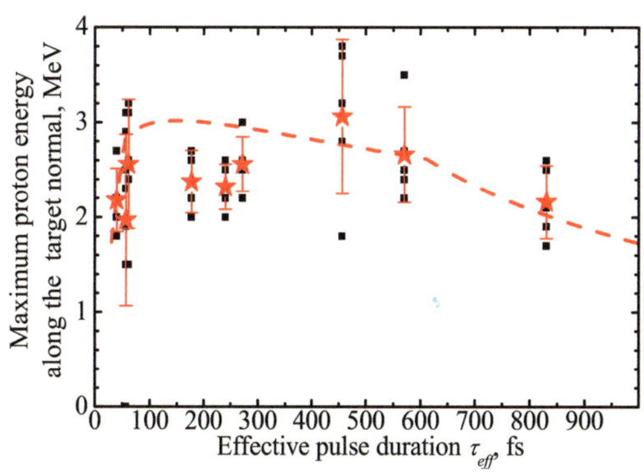

Fig. 5. Maximum proton energy at 0° (along the target normal) vs. the effective pulse duration τ_{eff}. The target is 7.5 μm polyimide tape, the saturable absorber is not used. The individual shots are shown with the squares, the several-shot average with the stars and the error bars (the standard deviation of shot-to-shot fluctuations). The dashed line shows the calculated dependence; the model used for the calculation is described in (Fuchs et al. 2006) with the additional limitation of the acceleration length equals the sheath diameter (Robson et al. 2007).

Taking advantage of the repetitive operation, we studied the stability of the proton acceleration (Fig. 8). Although the case with the optimum preplasma allowed achieving higher proton energies in some shots, the proton acceleration was less stable than in the case of high laser contrast. We measured the maximum proton energies at the target normal direction (0°) and at 22.5° from normal in 13 (40) consecutive shots without (with) the saturable absorber. In the case of higher contrast, the average proton energy was higher and fluctuations were smaller, which is important for many applications requiring stable ion beam. Further, in the high-contrast case the maximum proton energy at 0° was always larger than at 22.5°. In the case of lower contrast, the shot-to-shot fluctuations were large, and the maximum proton energies at these two angles were nearly same. Instability of the proton acceleration in the lower contrast case can be explained by the unstable nature of the preplasma itself, as well as relativistic self-focusing, which is also a kind of instability

(Litvak 1970). Angular distributions of proton beam at different contrast conditions are described in details in (Yogo et al. 2008).

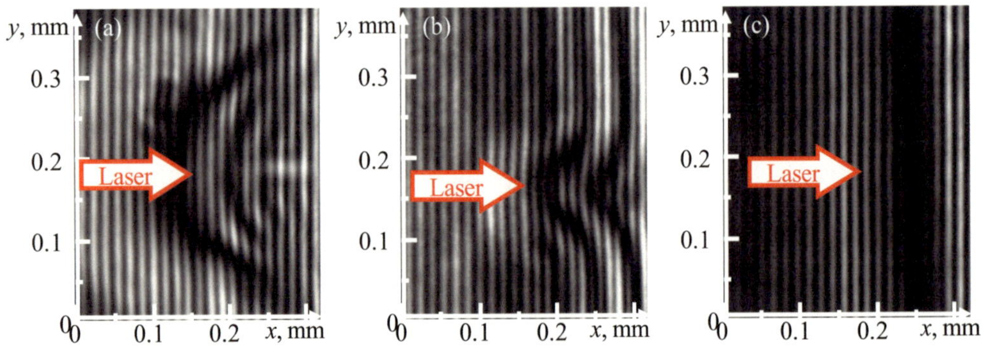

Fig. 6. Preplasma diagnostic using the interferometry with 820 nm femtosecond probe pulses 50 ps before the main laser pulse. The target is 7.5 μm thick polyimide tape. The three different laser operation modes with different contrasts are used: (a) without the saturable absorber, 3-stage OPCPA, (thin red line in Fig. 3), Pockels Cell (PC) voltage on at -1 ns; (b) same as (a), but the PC voltage on @-0.7 ns (corresponds to the ASE duration shorter by 0.3 ns); (c) with the saturable absorber and 2-stage non-saturated OPCPA (thick green line in Fig. 3), PC off (corresponds to the ASE duration of ~ 3 ns).

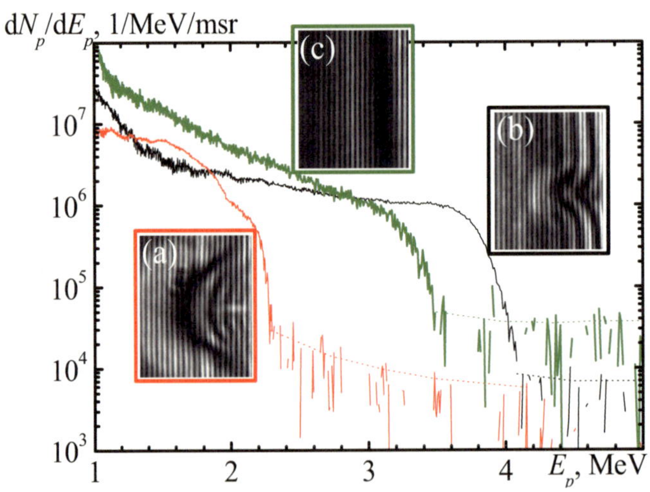

Fig. 7. Proton energy spectra at several modes of laser operation with different contrasts: The thin red line corresponds to the mode with the CPA oscillator, no saturable absorber, and 3-stage OPCPA (thin red line in Fig. 3). The thin black line corresponds to same case but with the shorter by 0.3 ns ASE duration (also thin red line in Fig. 3, because the contrast is not affected significantly by the Pockels Cell within the measurement range of -0.5 … +0.1 ns). The thick green line corresponds to the mode with the saturable absorber and 2-stage (nonsaturated) OPCPA (thick green line in Fig. 3). The dotted lines show the characteristic noise levels in the corresponding ToF spectra. The interferograms taken in the same shots [Fig. 6(a) – (c)] are shown near the spectra.

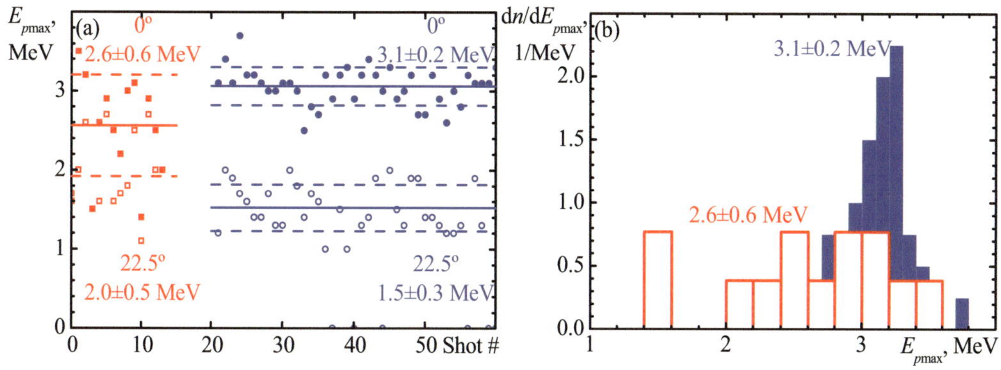

Fig. 8. Maximum proton energy in the consecutive shots with the same shooting parameters, the repetitive proton acceleration at 1 Hz employing the 7.5 μm thick polyimide tape target. (a) Maximum proton energy vs. shot number; solid symbols correspond to observation at 0° (along the target normal direction), open symbols at 22.5° off the normal in the same shots. Blue circles (red squares) correspond to the case with (without) the saturable absorber (in both cases, 3-stage OPCPA is used). (b) The histograms of maximum proton energy distributions at 0° (along the target normal) corresponding to the frame (a).

It has been shown in simulations (Esirkepov et al. 2006) and experiments (Neely et al. 2006; Antici et al. 2007; Ceccotti et al. 2007) that very thin (sub-micrometer to nanometer-scale) targets can provide higher proton energy and larger conversion efficiency, provided that the laser contrast is sufficiently high so that the target is not damaged by the preceding light. We compared the proton acceleration from ribbon targets with the thickness from 1 μm to 200 nm. The proton energies from 1 μm thick Au targets were larger than from thinner targets (Fig. 9). Although the nanosecond contrast was sufficiently high, the target was probably disturbed by the picosecond pedestal or prepulses at few tens of picoseconds before the main pulse (Fig. 3).

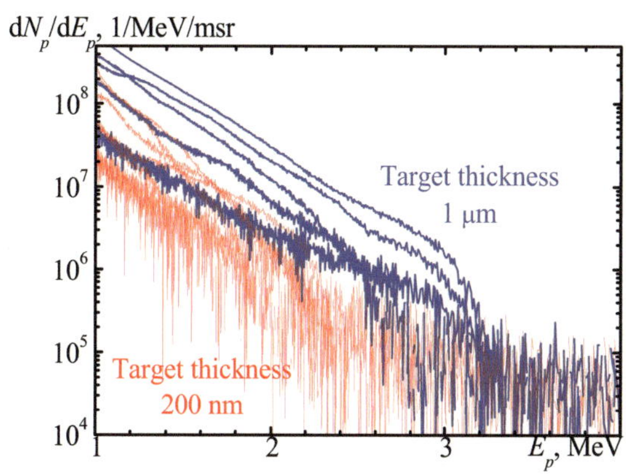

Fig. 9. Proton energy spectra at the target normal direction in several shots with different targets. The saturable absorber is used, 2-stage not-saturated OPCPA, Pockels Cell voltage off (3 ns ASE). The thick blue lines correspond to the 1 μm thick Au targets; the thin red lines correspond to the 200 nm thick Pd, Pt, and Au targets.

4. On the contrast of high-power lasers

The laser contrast strongly affects ion acceleration. In particular, in the TNSA regime higher ion energy is expected from thinner target in the ideal case, but in real experiments the rear surface of very thin targets is perturbed due to the preplasma formation, so there is an optimum target thickness (Kaluza et al. 2004). The ASE can even turn the originally solid-density target into a near-critical density plasma cloud (Matsukado et al. 2003; Yogo et al. 2008), so the acceleration mechanism changes, and the ions are accelerated by a long-living charge separation electric field, sustained by a quasistatic magnetic field of dipole vortex (Bulanov et al. 2005; Bulanov & Esirkepov 2007; Fukuda et al. 2009).
In order to understand the influence of the laser contrast in the described experiments, we performed the laser contrast diagnostics based on the target reflectivity measurement. The reflectivity was measured using calibrated calorimeters at full power operation in the real shooting conditions. The detailed description of this technique and experiments performed at three different laser systems can be found in (Pirozhkov et al. 2009a; Pirozhkov et al. 2009b). The data obtained with J-KAREN laser are shown in Fig. 10.

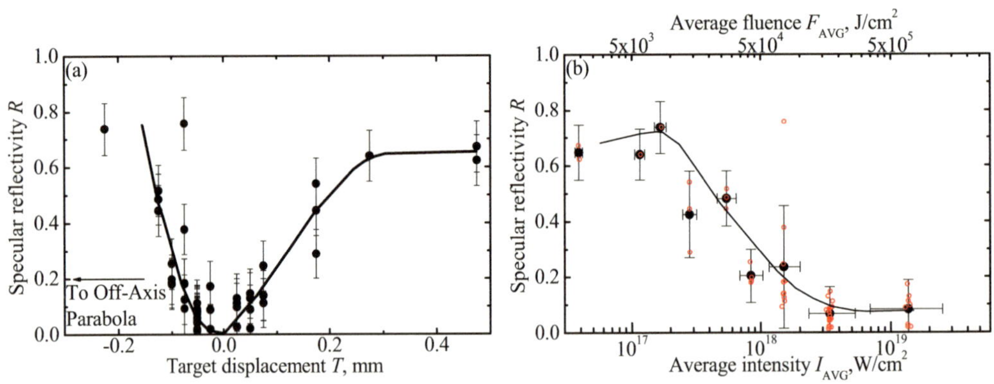

Fig. 10. On-target contrast diagnostic using the reflectivity of solid target at full laser power. The specular reflectivity R is measured with the in-vacuum calibrated calorimeter (Fig. 4). The spot size is varied by moving the target by distance T from best focus, the distance is measured along the laser beam, $T < 0$ corresponds to the target shifted toward the Off-Axis Parabola (OAP). (a) Specular reflectivity R vs. target position T. The error bars include the calibration errors and measured noise. (b) R vs. derived average intensity (bottom axis) and fluence (top axis) (the average over the FWHM spot area, accounting for the 45° incidence angle). The red open circles are the single-shot data, the black filled circles are the average values; the horizontal error bars are due to the target zero position uncertainty; the vertical error bars include shot-to-shot fluctuations, calibration errors, and measured noise.

The data were taken with the saturable absorber present and 3-stage OPCPA, the contrast was ~ 5×10^8 with the ASE duration of 3 ns. The dependence of the reflectivity on the target position [Fig. 10 (a)] exhibits a dip at best focus position $T = 0$. Such dip corresponds to the insufficient laser contrast. The specular reflectivity decreases due to both the increased absorption in the preplasma and the beam break-up. The dependence of the reflectivity on the average intensity and fluence [Fig. 10 (b)] shows that the reflectivity was still high at the average intensity of ~ 2×10^{17} W/cm^2, which is ~ 100 times smaller than the intensity at best focus. Therefore, a ~ 100-fold improvement of the contrast was necessary to avoid the preplasma formation by the ASE. This factor had been achieved by using non-saturated (2-stage) OPCPA operation, which was used to shoot sub-μm targets (Fig. 9). However, in the latter case the picosecond pedestal or prepulses might be able to produce the preplasma, which led to the decreased proton energies observed in the case of 200 nm targets.

5. Focusing of laser-driven proton beams with the permanent magnet quadrupoles

The perspective mechanisms of laser-driven ion acceleration promise to provide quasi-monoenergetic, low-divergence ion beams (Kuznetsov et al. 2001; Esirkepov et al. 2004; Esirkepov et al. 2006). However, in the present experiments the typical ion beam divergence is ~10°, and the energy spectrum is wide (energy spread ~100%) (Borghesi et al. 2006). This is disadvantageous for many applications of the laser-driven ion beams, which are possible at present. Here we can point out the radiobiological studies (Yogo et al. 2009), ion beam injector to the postaccelerator (Krushelnick et al. 2000; Cowan et al. 2004), etc. Therefore, it would be desirable to modify the ion beam divergence and spectrum. There were several proposals and demonstrations of various techniques, including the use of a curved target (Bulanov et al. 2000; Sentoku et al. 2000; Ruhl et al. 2001; Patel et al. 2003) (focusing or collimation of the proton beam), laser-driven plasma microlens (Toncian et al. 2006) (proton beam focusing and energy selection), phase rotation using a radio-frequency electric field (Nakamura et al. 2007; Ikegami et al. 2009; Wakita et al. 2009) (modification of the proton energy spectrum and beam divergence), and use of permanent magnet quadrupoles (PMQs) (Schollmeier et al. 2008; Ter-Avetisyan et al. 2008; Nishiuchi et al. 2009) (focusing or collimation of the ion beam, energy band selection). In this Section, we describe the experiment in which the efficient proton beam manipulation with the PMQs was demonstrated (Nishiuchi et al. 2009). The feature of this experiment is the 1 Hz repetition rate and the large acceptance angle of the PMQs, which allowed achieving high proton flux.
The experiment was performed with J-KAREN laser with the on-target pulse energy of 0.7 J, pulse duration of 30 fs, peak irradiance of $\sim 10^{20}$ W/cm^2, and the ASE contrast and duration of 10^7 and 1 ns, respectively. The setup was similar to the one shown in Fig. 4. However, the pair of PMQs was installed at the target rear side, as shown in Fig. 11.

Originally, the proton beam with the energy band from 1.9 to 2.8 MeV had a typical divergence of 10° (half angle), as it is demonstrated in the frame (a). The acceptance angle of the first PMQ was also ~10° (half angle). Thus, nearly all protons were accepted at this stage. The first PMQ focused the protons in the vertical and defocused in the horizontal planes. Due to the defocusing, some of the protons were not collected with the second PMQ. Those protons which were collected were further focused in the horizontal and defocused in the vertical planes with the second PMQ. As a result, the focus point was formed at the distance of 650 mm from the target. The proton beam profiles recorded with the CR-39 nuclear track detectors at the best focus as well as several other positions are shown in the frames (b) - (g) of Fig. 11 (the proton energy range is 2.2 to 3.1 MeV). At the best focus, the CR-39 image has the dimensions of ~3 mm × 8 mm. However, due to the etch pit overlapping (detector saturation), this visible spot size is overestimated, and the real spot was smaller.

Apart from the proton beam profile, we also measured the proton energy spectra with the ToF spectrometer situated at 1.93 m from the target. The original spectrum (without the PMQs) is shown in Fig. 12 (a) by the line 2. With the PMQs, the spectrum becomes quasi-monoenergetic due to the strong chromatic aberrations of the PMQs (namely, most of the

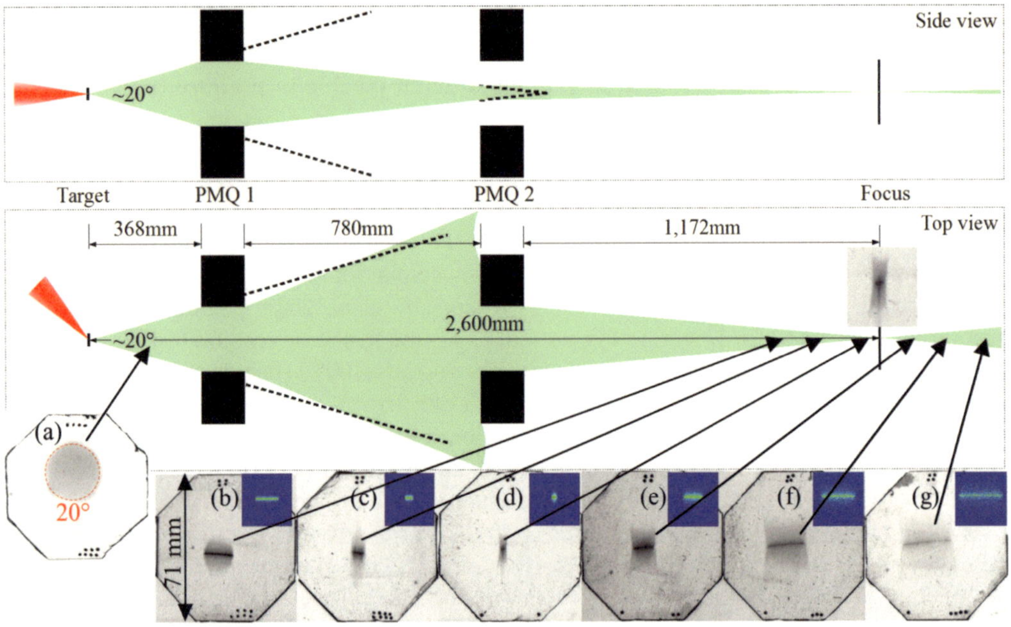

Fig. 11. Focusing of the laser-driven proton beam with two permanent magnet quadrupoles (PMQs). Schematic layout of the setup is shown in the vertical (side view) and the horizontal (top view) planes. The first PMQ has the field gradient of 55 T/m, thickness of 50 mm, and open diameter of 35 mm; the acceptance half-angle is ~10°, which is similar to the proton beam divergence. The second PMQ parameters are the field gradient of 60 T/m, thickness of 20 mm, and diameter of 23 mm. The octagonal frames (a) - (g) at the bottom of the figure show the experimental results obtained with 12.5 μm polyimide tape target; the proton beam profiles are recorded with the CR-39 nuclear track detectors covered with the Al range filter, the darker regions correspond to the higher proton density. The left CR-39 picture [frame (a)] is the original beam profile before the PMQs (Al 40 μm filter, 1.9 to 2.8 MeV). The frames (b) - (g) show the proton beam profiles at 450, 550, 650 (best focus), 750, 850, and 927 mm from the target (2.2 to 3.1 MeV). At the best focus [650 mm, frame (d)], the visible beam size is ~3 mm × 8 mm; these dimensions are overestimated due to the CR-39 saturation (etch pit overlapping) at the center of the image. The proton beam profiles in the blue insets on the frames (b) - (g) are obtained using the Monte-Carlo simulations; the spatial scales are the same as in the CR-39 images.

low-energy protons, except those near the axis, are strongly deflected by the first PMQ and do not pass through the aperture of the second PMQ).

Each frame (b) - (g) in Fig. 11 also contains the inset, which shows the result of the Monte-Carlo simulation of the proton beam profile at the corresponding position. The parameters of the original proton beam (spectrum, divergence) and the beam transport system in the simulation was set in accordance with the experimental conditions. The close similarity of the experimental beam profiles with the simulation suggests that the alignment accuracy was sufficient in the experiment; we found that the accuracy of the magnet positioning should be ~100 μm. Additional evidence of the similarity is shown in Fig. 12 (a), where the spectrum measured with the ToF is compared with the simulated one. The similarity of the experimental results and the simulation allows us to derive conclusions about the detailed parameters of the focused proton beam. In particular, we can simulate the proton energy spectrum at the focal plane (this is not same as the spectrum measured with the ToF at the distance of 1.93 m). The simulated spectrum is shown in Fig. 12 (b); here the protons passing through the 3 mm diameter aperture are included. The spectrum is quasi-monoenergetic, with the energy of 2.4±0.1 MeV. The transport efficiency of the beamline for the energy range of 2.3 - 2.5 MeV was ~0.3. The loss was due to the second PMQ, as shown in the schematic layout in Fig. 11. The efficiency of 0.3 should be compared with the transport efficiency of the divergent beam without the PMQs, which is ~ 1.7×10^{-4} (1.7×10^{3} times smaller). Simulation also allowed calculating the duration of the proton bunch, which was ~1 ns. This is much shorter than the typical achievable value, which is determined by the strong dispersion of the usual proton beam with a large energy spread.

The passive nature of the magnetic beamline led to the high stability of the spectrum after the PMQ pair. The pointing stability of the proton beam was 0.25±0.1° (standard deviation of shot-to-shot fluctuations). The instability of the proton number within the energy range of 2.4±0.1 MeV was 20%; this value included the fluctuations of the proton number, divergence, and pointing of the proton beam during the acceleration stage, originating from the laser, target, and plasma instabilities.

The demonstrated technique of the ion beam focusing can be useful in many applications of the laser-driven ion beams, including such field as high-energy density physics, radiography, nuclear physics, astrophysics, radiobiological studies, chemistry, material sciences, etc.

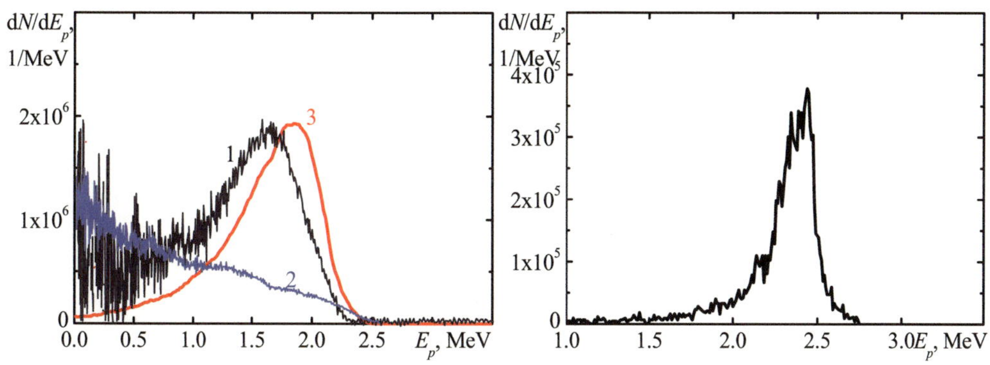

(a) (b)

Fig. 12. (a) The black line (1) is the spectrum of the proton beam focused with the PMQs; the spectrum is measured with the absolutely-calibrated Time-Of-Flight (ToF) spectrometer at the distance of 1.93 m from the target. The blue line (2) is the original spectrum (without the PMQs) at the same ToF position. The red line (3) is the simulated spectrum at the TOF position. (b) Proton energy spectrum within the 3 mm diameter circular aperture at the focal plane (simulation).

6. Radioactivation experiments

Multi-MeV laser-driven proton beams can induce nuclear reactions (Ledingham et al. 2003; McKenna et al. 2003). In fact, these nuclear reactions were used for the proton beam diagnostic in many experiments (Hatchett et al. 2000; Snavely et al. 2000; McKenna et al.

2004; Clarke et al. 2008). Another possible application of the nuclear reactions induced by laser-driven proton beams is the thin layer activation (TLA) (Racolta et al. 1995), which allows studying the wear of tools, for example cutting tools made from cubic boron nitride (BN) or artificial polycrystalline diamond (DIA) (Conlon 1985; Vasváry et al. 1994). In TLA, the thin layer of tool is activated by selecting the appropriate energy range of the proton beam. Measuring the residual activity, it is possible to determine the loss of the surface material with the high sensitivity. Various isotopes can be activated, including iron, titanium, chromium, etc., which opens wide possibilities for the industrial diagnostic. The typical reaction threshold is from few to 10 MeV, which means that compact laser systems can be used.

In this Section, we describe the results of the proof-of-principle experiment on the BN target activation (Ogura et al. 2009). The experimental setup is shown in Fig. 13 (a). The parameters of J-KAREN laser used in this experiment were as follows: the pulse energy of ~1 J, the pulse duration of 30 fs, the peak irradiance of $\sim 10^{20}$ W/cm^2, the ASE contrast of higher than 10^{10}. The target for the laser was polyimide tape. At the rear side of the tape, the BN sample was installed. The laser-driven proton beam induced the reaction ^{11}B(p,n)^{11}C, which has the threshold of 2.76 MeV. The maximum proton energy in this experiment ranged from 2.8 to 3.5 MeV, larger than the reaction threshold.

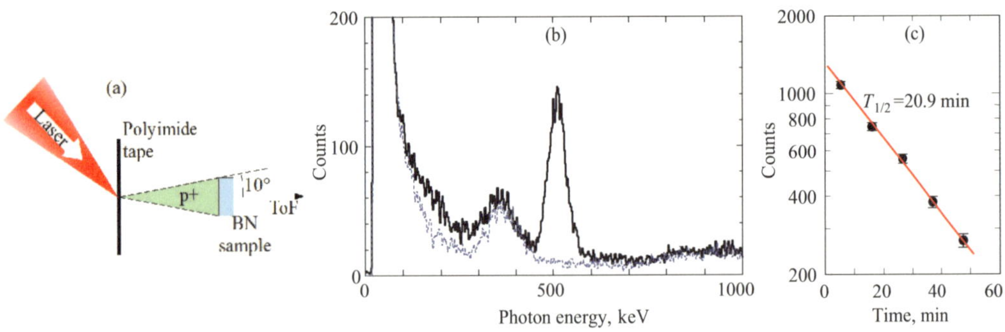

Fig. 13. (a) Experimental setup for irradiation of BN sample with the laser-driven proton beam. (b) Radiation spectrum measured using GSO [Gd_2SiO_5(Ce)] scintillation counter (solid line). The blue dotted line denotes the background. (c) The decay curve of annihilation photons with the fitted half-life time of $T_{1/2}$ = 20.9±0.7 min, which corresponds to the decay of ^{11}C ($T_{1/2}$ = 20.39±0.02 min).

The isotopes ^{11}C produced in the nuclear reaction decay through the positron emission with the half-life time $T_{1/2}$ = 20.39±0.02 min (ENSDF, NNDC Online Data Service, ENSDF database [http://www.nndc.bnl.gov/ensdf/]). The positrons annihilate with the emission of two photons with the energy of 0.511 MeV, directed at 180° to each other. The annihilation photon spectrum recorded with the GSO scintillation counter (Leo 1987) is shown in Fig. 13 (b). From the decay curve shown in the inset, the half-life time can be deduced: $T_{1/2}$ = 20.9±0.7 min, which corresponds to the decay time of ^{11}C. After 60 laser shots, the activity of the sample was measured to be 11.1±0.4 Bq. Using the 10 Hz laser operation during 10 minutes (6000 shots), the activity in the kBq range can be obtained, which is sufficient for the typical TLA diagnostic. Finally, we note that, unlike the situation with many other applications, the large divergence angle and the broad energy spectrum of the laser-induced proton beam can be convenient for the thin layer activation industrial diagnostic.

7. Conclusion

In conclusion, the laser-driven ion acceleration is the perspective new technique. The laser-driven ion beams have unique properties, such as ultrashort bunch duration, low emittance, small source size, and large particle number. The laser-driven acceleration is a very compact process, which takes place on the micrometer scale. For many applications, a repetitive ion beam is desirable, which can be achieved employing femtosecond laser systems. We described the performance of laser and properties of proton beam by the example of experiments performed in the Advanced Photon Research Center, Japan Atomic Energy Agency with J-KAREN laser. In particular, we described the dependence of the maximum proton energy on the laser pulse duration and found the optimum of few hundred femtoseconds for the given experimental conditions. We also depicted the influence of laser contrast and demonstrated the method of its diagnostic. We described two application experiments, in which the stable, repetitive (1 Hz) laser-driven proton beam was focused by the permanent magnet quadrupoles, and was used for the BN target activation, which can be applicable in the industrial testing of tools and mechanisms using the thin layer activation technique.

8. Acknowledgments

We are grateful for the expert support of J-KAREN laser team. This work was supported by the Special Coordination Funds for Promoting Science and Technology (SCF) commissioned by the Ministry of Education, Science, Sports and Culture (MEXT) of Japan and partly by Specially Promoted Education and Research by MEXT.

9. References

Antici, P., Fuchs, J., d'Humieres, E., Lefebvre, E., Borghesi, M., Brambrink, E., Cecchetti, C. A., Gaillard, S., Romagnani, L., Sentoku, Y., Toncian, T., Willi, O., Audebert, P. & Pepin, H. (2007). Energetic protons generated by ultrahigh contrast laser pulses interacting with ultrathin targets. *Phys. Plasmas*, 14, 3, (Mar) 030701, 1070-664X

Borghesi, M., Mackinnon, A. J., Gaillard, R., Willi, O. & Riley, D. (1999). Absorption of subpicosecond uv laser pulses during interaction with solid targets. *Phys. Rev. E*, 60, 6, (Dec) 7374-7381, 1063-651X

Borghesi, M., Schiavi, A., Campbell, D. H., Haines, M. G., Willi, O., MacKinnon, A. J., Gizzi, L. A., Galimberti, M., Clarke, R. J. & Ruhl, H. (2001). Proton imaging: a diagnostic for inertial confinement fusion/fast ignitor studies. *Plasma Phys. Control. Fusion*, 43, 12A A267, 0741-3335

Borghesi, M., Fuchs, J., Bulanov, S. V., Mackinnon, A. J., Patel, P. K. & Roth, M. (2006). Fast ion generation by high-intensity laser irradiation of solid targets and applications. *Fus. Sci. Tech.*, 49, 3, (Apr) 412-439, 1536-1055

Bulanov, S. V., Esirkepov, T. Zh., Califano, F., Kato, Y., Liseikina, T. V., Mima, K., Naumova, N. M., Nishihara, K., Pegoraro, F., Ruhl, H., Sentoku, Y. & Ueshima, Y. (2000).

Generation of collimated beams of relativistic ions in laser-plasma interactions. *JETP Lett.*, 71, 10 407-411, 0021-3640 (Print) 1090-6487 (Online)

Bulanov, S. V., Esirkepov, T. Zh, Khoroshkov, V., Kuznetsov, A. & Pegoraro, F. (2002). Oncological hadrontherapy with laser ion accelerators. *Phys. Lett. A*, 299, 2-3 240-247

Bulanov, S. V. & Khoroshkov, V. S. (2002). Feasibility of using laser ion accelerators in proton therapy. *Plas. Phys. Rep.*, 28, 5 453-456

Bulanov, S. V., Dylov, D., Esirkepov, T. Zh., Kamenets, F. & Sokolov, D. (2005). Ion acceleration in a dipole vortex in a laser plasma corona. *Plas. Phys. Rep.*, 31, 5, (May) 369-381, 1063-780X (Print) 1562-6938 (Online)

Bulanov, S. V. & Esirkepov, T. Zh. (2007). Comment on ``Collimated Multi-MeV Ion Beams from High-Intensity Laser Interactions with Underdense Plasma''. *Phys. Rev. Lett.*, 98, 4, (24 January) 049503-1

Bychenkov, V. Y., Sentoku, Y., Bulanov, S. V., Mima, K., Mourou, G. & Tolokonnikov, S. V. (2001). Pion production under the action of intense ultrashort laser pulse on a solid target. *JETP Lett.*, 74, 12 586-589

Ceccotti, T., Levy, A., Popescu, H., Reau, F., D'Oliveira, P., Monot, P., Geindre, J. P., Lefebvre, E. & Martin, P. (2007). Proton acceleration with high-intensity ultrahigh-contrast laser pulses. *Phys. Rev. Lett.*, 99, 18, (Nov 2) 185002, 0031-9007

Clarke, R. J., Simpson, P. T., Kar, S., Green, J. S., Bellei, C., Carroll, D. C., Dromey, B., Kneip, S., Markey, K., McKenna, P., Murphy, W., Nagel, S., Willingale, L. & Zepf, M. (2008). Nuclear activation as a high dynamic range diagnostic of laser-plasma interactions. *Nucl. Instr. Methods Phys. Res. Sect. A*, 585, 3, (Feb 1) 117-120, 0168-9002

Conlon, T. W. (1985). Nuclear physics for materials technology accelerator based methods? *Contemporary Physics*, 26, 6, (November) 521 - 558, 0010-7514

Cowan, T. E., Fuchs, J., Ruhl, H., Kemp, A., Audebert, P., Roth, M., Stephens, R., Barton, I., Blazevic, A., Brambrink, E., Cobble, J., Fernández, J., Gauthier, J. C., Geissel, M., Hegelich, M., Kaae, J., Karsch, S., Le Sage, G. P., Letzring, S., Manclossi, M., Meyroneinc, S., Newkirk, A., Pųpin, H. & Renard-LeGalloudec, N. (2004). Ultralow Emittance, Multi-MeV Proton Beams from a Laser Virtual-Cathode Plasma Accelerator. *Phys. Rev. Lett.*, 92, 20 204801

Esirkepov, T. Zh., Borghesi, M., Bulanov, S. V., Mourou, G. & Tajima, T. (2004). Highly Efficient Relativistic-Ion Generation in the Laser-Piston Regime. *Phys. Rev. Lett.*, 92, 17 175003

Esirkepov, T. Zh., Yamagiwa, M. & Tajima, T. (2006). Laser Ion-Acceleration Scaling Laws Seen in Multiparametric Particle-in-Cell Simulations. *Phys. Rev. Lett.*, 96, 10 105001-4

Fritzler, S., Malka, V., Grillon, G., Rousseau, J. P., Burgy, F., Lefebvre, E., d'Humieres, E., McKenna, P. & Ledingham, K. W. D. (2003). Proton beams generated with high-intensity lasers: Applications to medical isotope production. *Appl. Phys. Lett.*, 83, 15 3039-3041

Fuchs, J., Antici, P., D'Humieres, E., Lefebvre, E., Borghesi, M., Brambrink, E., Cecchetti, C. A., Kaluza, M., Malka, V., Manclossi, M., Meyroneinc, S., Mora, P., Schreiber, J., Toncian, T., Pepin, H. & Audebert, R. (2006). Laser-driven proton scaling laws and new paths towards energy increase. *Nature Phys.*, 2, 1, (Jan) 48-54, 1745-2473

Fukuda, Y., Faenov, A. Y., Tampo, M., Pikuz, T. A., Nakamura, T., Kando, M., Hayashi, Y., Yogo, A., Sakaki, H., Kameshima, T., Pirozhkov, A. S., Ogura, K., Mori, M., Esirkepov, T. Zh., Koga, J., Boldarev, A. S., Gasilov, V. A., Magunov, A. I., Yamauchi, T., Kodama, R., Bolton, P. R., Kato, Y., Tajima, T., Daido, H. & Bulanov, S. V. (2009). Energy Increase in Multi-MeV Ion Acceleration in the Interaction of a Short Pulse Laser with a Cluster-Gas Target. *Phys. Rev. Lett.*, 103, 16, (13 October) 165002-4

Gibbon, P. & Bell, A. R. (1992). Collisionless Absorption in Sharp-Edged Plasmas. *Phys. Rev. Lett.*, 68, 10, (Mar 9) 1535-1538, 0031-9007

Hatchett, S. P., Brown, C. G., Cowan, T. E., Henry, E. A., Johnson, J. S., Key, M. H., Koch, J. A., Langdon, A. B., Lasinski, B. F., Lee, R. W., Mackinnon, A. J., Pennington, D. M., Perry, M. D., Phillips, T. W., Roth, M., Sangster, T. C., Singh, M. S., Snavely, R. A., Stoyer, M. A., Wilks, S. C. & Yasuike, K. (2000). Electron, photon, and ion beams from the relativistic interaction of Petawatt laser pulses with solid targets. *Phys. Plasmas*, 7, (May) 2076-2082

Hegelich, M., Karsch, S., Pretzler, G., Habs, D., Witte, K., Guenther, W., Allen, M., Blazevic, A., Fuchs, J., Gauthier, J. C., Geissel, M., Audebert, P., Cowan, T. & Roth, M. (2002). MeV Ion Jets from Short-Pulse-Laser Interaction with Thin Foils. *Phys. Rev. Lett.*, 89, 8, (2 August) 085002

Henig, A., Kiefer, D., Markey, K., Gautier, D. C., Flippo, K. A., Letzring, S., Johnson, R. P., Shimada, T., Yin, L., Albright, B. J., Bowers, K. J., Fernandez, J. C., Rykovanov, S. G., Wu, H. C., Zepf, M., Jung, D., Liechtenstein, V. K., Schreiber, J., Habs, D. & Hegelich, B. M. (2009). Enhanced Laser-Driven Ion Acceleration in the Relativistic Transparency Regime. *Phys. Rev. Lett.*, 103, 4, (21 July) 045002-4

Ikegami, M., Nakamura, S., Iwashita, Y., Shirai, T., Souda, H., Tajima, Y., Tanabe, M., Tongu, H., Itoh, H., Shintaku, H., Yamazaki, A., Daido, H., Yogo, A., Orimo, S., Mori, M., Nishiuchi, M., Ogura, K., Sagisaka, A., Pirozhkov, A. S., Kiriyama, H., Kanazawa, S., Kondo, S., Yamamoto, Y., Shimomura, T., Tanoue, M., Nakai, Y. & Akutsu, A. (2009). Radial focusing and energy compression of a laser-produced proton beam by a synchronous rf field. *Phys. Rev. STAB*, 12, 6 063501-6

Kaluza, M., Schreiber, J., Santala, M. I. K., Tsakiris, G. D., Eidmann, K., Meyer-ter-Vehn, J. & Witte, K. J. (2004). Influence of the Laser Prepulse on Proton Acceleration in Thin-Foil Experiments. *Phys. Rev. Lett.*, 93, 4, (20 July) 045003

Kiriyama, H., Mori, M., Nakai, Y., Shimomura, T., Tanoue, M., Akutsu, A., Kondo, S., Kanazawa, S., Okada, H., Motomura, T., Daido, H., Kimura, T. & Tajima, T. (2008). High-contrast, high-intensity laser pulse generation using a nonlinear preamplifier in a Ti : sapphire laser system. *Opt. Lett.*, 33, 7, (Apr 1) 645-647, 0146-9592

Kiriyama, H., Mori, M., Nakai, Y., Shimomura, T., Tanoue, M., Akutsu, A., Okada, H., Motomura, T., Kondo, S., Kanazawa, S., Sagisaka, A., Ma, J., Daito, I., Kotaki, H., Daido, H., Bulanov, S. V., Kimura, T. & Tajima, T. (2009). Generation of high-contrast and high-intensity laser pulses using an OPCPA preamplifier in a double CPA, Ti:sapphire laser system. *Opt. Comm.*, 282, 4, (Feb 15) 625-628, 0030-4018

Krushelnick, K., Clark, E. L., Allott, R., Beg, F. N., Danson, C. N., Machacek, A., Malka, V., Najmudin, Z., Neely, D., Norreys, P. A., Salvati, M. R., Santala, M. I. K., Tatarakis, M., Watts, I., Zepf, M. & Dangor, A. E. (2000). Ultrahigh-intensity laser-produced plasmas as a compact heavy ion injection source. *IEEE Trans. on Plasma Sci.*, 28, 4 1184-1189, 0093-3813

Kuznetsov, A., Esirkepov, T. Zh., Kamenets, F. & Bulanov, S. (2001). Efficiency of ion acceleration by a relativistically strong laser pulse in an underdense plasma. *Plas. Phys. Rep.*, 27, 3 211-220, 1063-780X (Print) 1562-6938 (Online)

Ledingham, K. W. D., McKenna, P. & Singhal, R. P. (2003). Applications for Nuclear Phenomena Generated by Ultra-Intense Lasers. *Science*, 300, 5622, (May 16, 2003) 1107-1111

Leo, W. R. (1987). *Techniques for nuclear and particle physics experiments*, Springer-Verlag New York Inc., New York

Li, Z., Daido, H., Fukumi, A., Sagisaka, A., Ogura, K., Nishiuchi, M., Orimo, S., Hayashi, Y., Mori, M., Kado, M., Bulanov, S. V., Esirkepov, T. Zh., Oishi, Y., Nayuki, T., Fujii, T., Nemoto, K., Nakamura, S. & Noda, A. (2006). Measurements of energy and angular distribution of hot electrons and protons emitted from a p- and s-polarized intense femtosecond laser pulse driven thin foil target. *Phys. Plasmas*, 13, 4, (18 April) 043104-6

Lindau, F., Lundh, O., Persson, A., McKenna, P., Osvay, K., Batani, D. & Wahlstrom, C. G. (2005). Laser-accelerated protons with energy-dependent beam direction. *Phys. Rev. Lett.*, 95, 17, (Oct 21) 175002, 0031-9007

Litvak, A. (1970). Finite-amplitude wave beams in a magnetoactive plasma. *Sov. Phys. JETP.*, 30 344

Mackinnon, A. J., Borghesi, M., Hatchett, S., Key, M. H., Patel, P. K., Campbell, H., Schiavi, A., Snavely, R., Wilks, S. C. & Willi, O. (2001). Effect of plasma scale length on multi-MeV proton production by intense laser pulses. *Phys. Rev. Lett.*, 86, 9, (Feb 26) 1769-1772, 0031-9007

Matsukado, K., Esirkepov, T. Zh., Kinoshita, K., Daido, H., Utsumi, T., Li, Z., Fukumi, A., Hayashi, Y., Orimo, S., Nishiuchi, M., Bulanov, S. V., Tajima, T., Noda, A., Iwashita, Y., Shirai, T., Takeuchi, T., Nakamura, S., Yamazaki, A., Ikegami, M., Mihara, T., Morita, A., Uesaka, M., Yoshii, K., Watanabe, T., Hosokai, T., Zhidkov, A., Ogata, A., Wada, Y. & Kubota, T. (2003). Energetic Protons from a Few-Micron Metallic Foil Evaporated by an Intense Laser Pulse. *Phys. Rev. Lett.*, 91, 21, (18 November) 215001

McKenna, P., Ledingham, K. W. D., McCanny, T., Singhal, R. P., Spencer, I., Santala, M. I. K., Beg, F. N., Krushelnick, K., Tatarakis, M., Wei, M. S., Clark, E. L., Clarke, R. J., Lancaster, K. L., Norreys, P. A., Spohr, K., Chapman, R. & Zepf, M. (2003). Demonstration of Fusion-Evaporation and Direct-Interaction Nuclear Reactions using High-Intensity Laser-Plasma-Accelerated Ion Beams. *Phys. Rev. Lett.*, 91, 7 075006

McKenna, P., Ledingham, K. W. D., Yang, J. M., Robson, L., McCanny, T., Shimizu, S., Clarke, R. J., Neely, D., Spohr, K., Chapman, R., Singhal, R. P., Krushelnick, K., Wei, M. S. & Norreys, P. A. (2004). Characterization of proton and heavier ion acceleration in ultrahigh-intensity laser interactions with heated target foils. *Phys. Rev. E*, 70, 3, (21 September) 036405

McKenna, P., Carroll, D. C., Lundh, O., Nurnberg, F., Markey, K., Bandyopadhyay, S., Batani, D., Evans, R. G., Jafer, R., Kar, S., Neely, D., Pepler, D., Quinn, M. N., Redaelli, R., Roth, M., Wahlstrom, C. G., Yuan, X. H. & Zepf, M. (2008). Effects of front surface plasma expansion on proton acceleration in ultraintense laser irradiation of foil targets. *Laser and Particle Beams*, 26, 4, (Dec) 591-596, 0263-0346

Mora, P. (2003). Plasma expansion into a vacuum. *Phys. Rev. Lett.*, 90, 18, (May 9) 185002, 0031-9007

Mourou, G. A., Tajima, T. & Bulanov, S. V. (2006). Optics in the relativistic regime. *Rev. Mod. Phys.*, 78, 2, (Apr-Jun) 309-371, 0034-6861

Murakami, M., Demizu, Y., Niwa, Y., Miyawaki, D., Terashima, K., Arimura, T., Mima, M., Nagayama, S., Maeda, T., Baba, M., Akagi, T., Hishikawa, Y. & Abe, M. (2009). Current status of the HIBMC and results of representative diseases, *Proceedings of Laser-Driven Relativistic Plasmas Applied To Science, Industry And Medicine: 2nd International Symposium*, pp. 400-415, Kyoto (Japan), 19-23 January 2009, AIP

Nakamura, S., Iwashita, Y., Noda, A., Shirai, T., Tongu, H., Fukumi, A., Kado, M., Yogo, A., Mori, M., Orimo, S., Ogura, K., Sagisaka, A., Nishiuchi, M., Hayashi, Y., Li, Z., Daido, H. & Wada, Y. (2006). Real-time optimization of proton production by intense short-pulse laser with time-of-flight measurement. *Jpn. J. Appl. Phys.*, 45, 33-36, (Sep) L913-L916, 0021-4922

Nakamura, S., Ikegami, M., Iwashita, Y., Shirai, T., Tongui, H., Souda, H., Daido, H., Mori, M., Kado, M., Sagisaka, A., Ogura, K., Nishiuchi, M., Orimo, S., Hayashi, Y., Yogo, A., Pirozhkov, A. S., Bulanov, S. V., Esirkepov, T. Zh., Nagashima, A., Kimura, T., Tajima, T., Takeuchi, T., Fukumi, A., Li, Z. & Noda, A. (2007). High-quality laser-produced proton beam realized by the application of a synchronous RF electric field. *Jpn. J. Appl. Phys.*, 46, 29-32, (Aug) L717-L720, 0021-4922

Nayuki, T., Oishi, Y., Fujii, T., Nemoto, K., Kayoiji, T., Okano, Y., Hironaka, Y., Nakamura, K. G., Kondo, K. & Ueda, K. (2003). Thin tape target driver for laser ion accelerator. *Rev. Sci. Instr.*, 74, 7, (Jul) 3293-3296, 0034-6748

Neely, D., Foster, P., Robinson, A., Lindau, F., Lundh, O., Persson, A., Wahlstrom, C. G. & McKenna, P. (2006). Enhanced proton beams from ultrathin targets driven by high contrast laser pulses. *Appl. Phys. Lett.*, 89, 2, (Jul 10) 021502, 0003-6951

Nemoto, K., Maksimchuk, A., Banerjee, S., Flippo, K., Mourou, G., Umstadter, D. & Bychenkov, V. Y. (2001). Laser-triggered ion acceleration and table top isotope production. *Appl. Phys. Lett.*, 78, 5 595-597

Nishiuchi, M., Daido, H., Yogo, A., Orimo, S., Ogura, K., Ma, J., Sagisaka, A., Mori, M., Pirozhkov, A. S., Kiriyama, H., Bulanov, S. V., Esirkepov, T. Zh., Choi, I. W., Kim, C. M., Jeong, T. M., Yu, T. J., Sung, J. H., Lee, S. K., Hafz, N., Pae, K. H., Noh, Y. C., Ko, D. K., Lee, J., Oishi, Y., Nemoto, K., Nagatomo, H., Nagai, K. & Azuma, H. (2008). Efficient production of a collimated MeV proton beam from a polyimide target driven by an intense femtosecond laser pulse. *Phys. Plasmas*, 15, 5, (22 May) 053104-10

Nishiuchi, M., Daito, I., Ikegami, M., Daido, H., Mori, M., Orimo, S., Ogura, K., Sagisaka, A., Yogo, A., Pirozhkov, A. S., Sugiyama, H., Kiriyama, H., Okada, H., Kanazawa, S., Kondo, S., Shimomura, T., Tanoue, M., Nakai, Y., Sasao, H., Wakai, D., Sakaki, H., Bolton, P., Choi, I. W., Sung, J. H., Lee, J., Oishi, Y., Fujii, T., Nemoto, K., Souda, H., Noda, A., Iseki, Y. & Yoshiyuki, T. (2009). Focusing and spectral enhancement of a repetition-rated, laser-driven, divergent multi-MeV proton beam using permanent quadrupole magnets. *Appl. Phys. Lett.*, 94, 6, (Feb 9) 061107, 0003-6951

Ogura, K., Shizuma, T., Hayakawa, T., Yogo, A., Nishiuchi, M., Orimo, S., Sagisaka, A., Pirozhkov, A. S., Mori, M., Kiriyama, H., Kanazawa, S., Kondo, S., Nakai, Y., Shimoura, T., Tanoue, M., Akutsu, A., Motomura, T., Okada, H., Kimura, T., Oishi, Y., Nayuki, T., Fujii, T., Nemoto, K. & Daido, H. (2009). Proton-induced Nuclear Reactions Using Compact High-Contrast High-Intensity Laser. *Applied Physics Express*, 2, 6 066001

Orimo, S., Nishiuchi, M., Daido, H., Yogo, A., Ogura, K., Sagisaka, A., Li, Z., Pirozhkov, A. S., Mori, M., Kiriyama, H., Kanazawa, S., Kondo, S., Yamamoto, Y., Shimomura, T., Tanoue, M., Nakai, Y., Akutsu, A., Nakamura, S., Shirai, T., Iwashita, Y., Noda, A., Oishi, Y., Nemoto, K., Choi, I. W., Yu, T. J., Sung, J. H., Jeong, T. M., Kim, H. T., Hong, K.-H., Noh, Y.-C., Ko, D.-K. & Lee, J. (2007). Simultaneous Proton and X-ray Imaging with Femtosecond Intense Laser Driven Plasma Source. *Jpn. J. Appl. Phys.*, 46, 9A, (7 September) 5853

Patel, P. K., Mackinnon, A. J., Key, M. H., Cowan, T. E., Foord, M. E., Allen, M., Price, D. F., Ruhl, H., Springer, P. T. & Stephens, R. (2003). Isochoric Heating of Solid-Density Matter with an Ultrafast Proton Beam. *Phys. Rev. Lett.*, 91, 12 125004

Ping, Y., Shepherd, R., Lasinski, B. F., Tabak, M., Chen, H., Chung, H. K., Fournier, K. B., Hansen, S. B., Kemp, A., Liedahl, D. A., Widmann, K., Wilks, S. C., Rozmus, W. & Sherlock, M. (2008). Absorption of short laser pulses on solid targets in the ultrarelativistic regime. *Phys. Rev. Lett.*, 100, 8, (Feb 29) 085004, 0031-9007

Pirozhkov, A. S., Mori, M., Ogura, K., Nishimura, A., Murakami, H., Shimada, Y., Sagisaka, A., Orimo, S., Kimura, T. & Daido, H. (2008). Transient-Grating FROG for Measurement of Sub-10-fs to Few-ps Amplified Pulses, *Proceedings of Advanced Solid-State Photonics* (OSA Technical Digest Series), p. MC8, Nara, Japan, 27 January 2008, Optical Society of America

Pirozhkov, A. S., Choi, I. W., Sung, J. H., Lee, S. K., Yu, T. J., Jeong, T. M., Kim, I. J., Hafz, N., Kim, C. M., Pae, K. H., Noh, Y. C., Ko, D. K., Lee, J., Robinson, A. P. L., Foster, P., Hawkes, S., Streeter, M., Spindloe, C., McKenna, P., Carroll, D. C., Wahlstrom, C. G., Zepf, M., Adams, D., Dromey, B., Markey, K., Kar, S., Li, Y. T., Xu, M. H., Nagatomo, H., Mori, M., Yogo, A., Kiriyama, H., Ogura, K., Sagisaka, A., Orimo, S., Nishiuchi, M., Sugiyama, H., Esirkepov, T. Zh., Okada, H., Kondo, S., Kanazawa, S., Nakai, Y., Akutsu, A., Motomura, T., Tanoue, M., Shimomura, T., Ikegami, M., Daito, I., Kando, M., Kameshima, T., Bolton, P., Bulanov, S. V., Daido, H. & Neely, D. (2009a). Diagnostic of laser contrast using target reflectivity. *Appl. Phys. Lett.*, 94, 24, (15 June) 241102-3

Pirozhkov, A. S., Kando, M., Esirkepov, T. Zh., Fukuda, Y., Chen, L.-M., Daito, I., Ogura, K., Homma, T., Hayashi, Y., Kotaki, H., Sagisaka, A., Mori, M., Koga, J. K., Kawachi, T., Kiriyama, H., Okada, H., Kawase, K., Kameshima, T., Nishimori, N., Ragozin, E. N., Faenov, A. Y., Pikuz, T. A., Kimura, T., Tajima, T., Daido, H., Kato, Y. & Bulanov, S. V. (2009b). Demonstration of Flying Mirror with Improved Efficiency, *Proceedings of 2nd Int. Symp. Laser-Driven Relativistic Plasmas Applied to Science, Industry and Medicine* (AIP Conf. Proc.), pp. 274-284, Kyoto (Japan), 20 January 2009b, AIP

Pirozhkov, A. S., Mori, M., Yogo, A., Kiriyama, H., Ogura, K., Sagisaka, A., Ma, J. L., Orimo, S., Nishiuchi, M., Sugiyama, H., Esirkepov, T. Zh., Bulanov, S. V., Okada, H., Kondo, S., Kanazawa, S., Nakai, Y., Akutsu, A., Motomura, T., Tanoue, M., Shimomura, T., Ikegami, M., Shirai, T., Iwashita, Y., Noda, A., Choi, I. W., Lee, S.

K., Lee, J., Oishi, Y., Kimura, T., Tajima, T. & Daido, H. (2009c). Laser-driven proton acceleration and plasma diagnostics with J-KAREN laser, *Proceedings of Nonlinear Optics and Applications III* (Proc. SPIE), pp. 735414-10, Prague, Czech Republic, 20 April 2009c, SPIE

Racolta, P. M., Popa-Simil, L., Pascovici, G., Ivanov, E. A. & Alexandreanu, B. (1995). In: *Application of particle and laser beams in materials technology*, Misaelides, P. (Ed.), 415, Kluwer Academic Publishers, 978-0-7923-3324-1, London

Robson, L., Simpson, P. T., Clarke, R. J., Ledingham, K. W. D., Lindau, F., Lundh, O., McCanny, T., Mora, P., Neely, D., Wahlstrom, C. G., Zepf, M. & McKenna, P. (2007). Scaling of proton acceleration driven by petawatt-laser-plasma interactions. *Nature Phys.*, 3, 1, (Jan) 58-62, 1745-2473

Roth, M., Cowan, T. E., Key, M. H., Hatchett, S. P., Brown, C., Fountain, W., Johnson, J., Pennington, D. M., Snavely, R. A., Wilks, S. C., Yasuike, K., Ruhl, H., Pegoraro, F., Bulanov, S. V., Campbell, E. M., Perry, M. D. & Powell, H. (2001). Fast Ignition by Intense Laser-Accelerated Proton Beams. *Phys. Rev. Lett.*, 86, 3, (Jan 15) 436

Ruhl, H., Bulanov, S. V., Cowan, T. E., Liseikina, T. V., Nickles, P., Pegoraro, F., Roth, M. & Sandner, W. (2001). Computer simulation of the three-dimensional regime of proton acceleration in the interaction of laser radiation with a thin spherical target. *Plas. Phys. Rep.*, 27, 5, (May) 363-371, 1063-780X

Sagisaka, A., Daido, H., Ogura, K., Orimo, S., Hayashi, Y., Nishiuchi, M., Mori, M., Matsukado, K., Fukumi, A., Li, Z., Nakamura, S., Takagaki, K., Hazama, H., Suzuki, M., Utsumi, T., Bulanov, S. V. & Esirkepov, T. Zh. (2004). Characterization of preformed plasmas with an interferometer for ultra-short high-intensity laser-plasma interactions. *Appl. Phys. B*, 78, 7-8, (May) 919-922, 0946-2171

Sagisaka, A., Pirozhkov, A. S., Daido, H., Fukumi, A., Li, Z., Ogura, K., Yogo, A., Oishi, Y., Nayuki, T., Fujii, T., Nemoto, K., Orimo, S., Nishiuchi, M., Hayashi, Y., Mori, M., Kado, M., Nakamura, S., Noda, A., Choi, I. W., Sung, J. H., Ko, D. K. & Lee, J. (2006). Development of a two-color interferometer for observing wide range electron density profiles with a femtosecond time resolution. *Appl. Phys. B*, 84, 3, (Sep) 415-419, 0946-2171

Sagisaka, A., Daido, H., Nashima, S., Orimo, S., Ogura, K., Mori, M., Yogo, A., Ma, J., Daito, I., Pirozhkov, A. S., Bulanov, S. V., Esirkepov, T. Zh., Shimizu, K. & Hosoda, M. (2008). Simultaneous generation of a proton beam and terahertz radiation in high-intensity laser and thin-foil interaction. *Appl. Phys. B*, 90, 3, (March) 373-377, 0946-2171 (Print) 1432-0649 (Online)

Schollmeier, M., Becker, S., Geissel, M., Flippo, K. A., Blazevic, A., Gaillard, S. A., Gautier, D. C., Gruner, F., Harres, K., Kimmel, M., Nurnberg, F., Rambo, P., Schramm, U., Schreiber, J., Schutrumpf, J., Schwarz, J., Tahir, N. A., Atherton, B., Habs, D., Hegelich, B. M. & Roth, M. (2008). Controlled Transport and Focusing of Laser-Accelerated Protons with Miniature Magnetic Devices. *Phys. Rev. Lett.*, 101, 5, (Aug 1) 055004-4

Sentoku, Y., Liseikina, T. V., Esirkepov, T. Zh., Califano, F., Naumova, N. M., Ueshima, Y., Vshivkov, V. A., Kato, Y., Mima, K., Nishihara, K., Pegoraro, F. & Bulanov, S. V. (2000). High density collimated beams of relativistic ions produced by petawatt laser pulses in plasmas. *Phys. Rev. E*, 62, 5 7271

Snavely, R. A., Key, M. H., Hatchett, S. P., Cowan, T. E., Roth, M., Phillips, T. W., Stoyer, M. A., Henry, E. A., Sangster, T. C., Singh, M. S., Wilks, S. C., MacKinnon, A., Offenberger, A., Pennington, D. M., Yasuike, K., Langdon, A. B., Lasinski, B. F., Johnson, J., Perry, M. D. & Campbell, E. M. (2000). Intense High-Energy Proton Beams from Petawatt-Laser Irradiation of Solids. *Phys. Rev. Lett.*, 85, 14 2945

Strickland, D. & Mourou, G. (1985). Compression of amplified chirped optical pulses. *Opt. Comm.*, 55 447-449

Ter-Avetisyan, S., Schnurer, M., Polster, R., Nickles, P. V. & Sandner, W. (2008). First demonstration of collimation and monochromatisation of a laser accelerated proton burst. *Laser and Particle Beams*, 26, 04 637-642, 0263-0346

Toncian, T., Borghesi, M., Fuchs, J., d'Humieres, E., Antici, P., Audebert, P., Brambrink, E., Cecchetti, C. A., Pipahl, A., Romagnani, L. & Willi, O. (2006). Ultrafast Laser-Driven Microlens to Focus and Energy-Select Mega-Electron Volt Protons. *Science*, 312, 5772, (April 21, 2006) 410-413

Vasváry, L., Ditrói, F., Takács, S., Szabó, Z., Szűcs, J., Kundrák, J. & Mahunka, I. (1994). Wear measurement of the cutting edge of superhard turning tools using TLA technique. *Nucl. Instr. Methods Phys. Res. Sect. B*, 85, 1-4 255-259, 0168-583X

Wakita, A., Iwashita, Y., Shirai, T., Ikegami, M., Tongu, H., Souda, H., Mori, M., Yogo, A., Orimo, S. & Nishiuchi, M. (2009). Characteristics of a laser-produced proton beam improved by a synchronous RF field. *Nuclear Inst. and Methods in Physics Research, A*, 599, 1 15-19

Yogo, A., Daido, H., Fukumi, A., Li, Z., Ogura, K., Sagisaka, A., Pirozhkov, A. S., Nakamura, S., Iwashita, Y., Shirai, T., Noda, A., Oishi, Y., Nayuki, T., Fujii, T., Nemoto, K., Choi, I. W., Sung, J. H., Ko, D. K., Lee, J., Kaneda, M. & Itoh, A. (2007). Laser prepulse dependency of proton-energy distributions in ultraintense laser-foil interactions with an online time-of-flight technique. *Phys. Plasmas*, 14, 4, (Apr) 043104, 1070-664X

Yogo, A., Daido, H., Bulanov, S. V., Nemoto, K., Oishi, Y., Nayuki, T., Fujii, T., Ogura, K., Orimo, S., Sagisaka, A., Ma, J. L., Esirkepov, T. Zh., Mori, M., Nishiuchi, M., Pirozhkov, A. S., Nakamura, S., Noda, A., Nagatomo, H., Kimura, T. & Tajima, T. (2008). Laser ion acceleration via control of the near-critical density target. *Phys. Rev. E*, 77, 1 Pt 2, (Jan) 016401, 1539-3755 (Print)

Yogo, A., Sato, K., Nishikino, M., Mori, M., Teshima, T., Numasaki, H., Murakami, M., Demizu, Y., Akagi, S., Nagayama, S., Ogura, K., Sagisaka, A., Orimo, S., Nishiuchi, M., Pirozhkov, A. S., Ikegami, M., Tampo, M., Sakaki, H., Suzuki, M., Daito, I., Oishi, Y., Sugiyama, H., Kiriyama, H., Okada, H., Kanazawa, S., Kondo, S., Shimomura, T., Nakai, Y., Tanoue, M., Sasao, H., Wakai, D., Bolton, P. R. & Daido, H. (2009). Application of laser-accelerated protons to the demonstration of DNA double-strand breaks in human cancer cells. *Appl. Phys. Lett.*, 94, 18 181502-3